SALEM HEALTH
GENETICS
& INHERITED CONDITIONS

SALEM HEALTH
GENETICS
& INHERITED CONDITIONS

Volume I
Aarskog syndrome — Galactosemia

Editor

Jeffrey A. Knight, Ph.D.
Mount Holyoke College

SALEM PRESS
Pasadena, California   Hackensack, New Jersey

Editor in Chief: Dawn P. Dawson

Editorial Director: Christina J. Moose *Editorial Assistant:* Brett Steven Weisberg
Project Editor: Tracy Irons-Georges *Photo Editor:* Cynthia Breslin Beres
Manuscript Editor: Rebecca Kuzins *Production Editor:* Joyce I. Buchea
Acquisitions Editor: Mark Rehn *Layout:* Mary Overell

Note to Readers

The material presented in *Salem Health: Genetics and Inherited Conditions* is intended for broad informational and educational purposes. Readers who suspect that they or someone whom they know or provide caregiving for suffers from any disorder, disease, or condition described in this set should contact a physician without delay; this work should not be used as a substitute for professional medical diagnosis. Readers who are undergoing or about to undergo any treatment or procedure described in this set should refer to their physicians and other health care team members for guidance concerning preparation and possible effects. This set is not to be considered definitive on the covered topics, and readers should remember that the field of health care is characterized by a diversity of medical opinions and constant expansion in knowledge and understanding.

Library of Congress Cataloging-in-Publication Data

Genetics and inherited conditions / editor, Jeffrey A. Knight.
     p. cm. — (Salem health)
   Includes bibliographical references and index.
   ISBN 978-1-58765-650-7 (set : alk. paper) — ISBN 978-1-58765-651-4 (v. 1 : alk. paper) — ISBN 978-1-58765-652-1 (v. 2 : alk. paper) — ISBN 978-1-58765-653-8 (v. 3 : alk. paper)
 1. Genetic disorders. 2. Genetics. I. Knight, Jeffrey A., 1948-
 RB155.5.G4616 2010
 616'.042—dc22

                                                                            2010005289

First Printing

PRINTED IN THE UNITED STATES OF AMERICA

Contents

Contents

Publisher's Note

In 2003, the early completion of the Human Genome Project, an unprecedented global effort, marked a new beginning for genetics, an exciting and often controversial branch of science that is increasingly shaping our world. In the years that have followed, genetic research has grown tremendously. Many believe that this vital field will provide the keys for unlocking numerous biological secrets about life on Earth. *Salem Health: Genetics and Inherited Conditions* offers nonspecialist information about a variety of topics in genetics—from the science behind the field to diseases that can be passed down among generations. Designed for the general reader, it provides authoritative, essential information in easy-to-understand language on an often intimidating subject.

Salem Health: Genetics and Inherited Conditions is a revised and expanded version of Salem's award-winning *Encyclopedia of Genetics, Revised Edition*, published in 2004. This set is an addition to the Salem Health family of titles, which also includes both print and electronic versions of *Salem Health: Cancer* (2009), *Salem Health: Psychology and Mental Health* (2009), and the core set *Magill's Medical Guide* (revised every three years). All of them come with free online access with the purchase of the print set.

Scope and Coverage

This A-Z encyclopedia arranges 455 essays on all aspects of genetics—diseases, biology, techniques and methodologies, genetic engineering, biotechnology, ethics, and social issues. Written by professors and professional medical writers for nonspecialists, this comprehensive reference publication will interest biology and premedical students, public library patrons, and librarians building scientific collections.

Salem Health: Genetics and Inherited Conditions surveys this continually evolving discipline from a variety of perspectives, offering historical and technical background along with a balanced discussion of recent discoveries and developments. Basics of biology—from the molecular and cellular levels through the organismal level, from Mendelian principles to the latest on DNA sequencing technology—constitute the core coverage. Medical topics comprise a significant number of essays, as the genetic predisposition for many illnesses and syndromes has increasingly come to light. Genetic technologies that promise a world without hunger, disease, and disability—and promise to rewrite human values—are addressed as well. The encyclopedia's scope embraces the key social and ethical questions raised by these new genetic frontiers: from cloning to stem cells to genetically modified foods and organisms.

Revision Details

This edition, expanded to three volumes, adds 236 topics new to the encyclopedia, many of them diseases and conditions that are known to be genetic in origin or suspected of involving genetic factors. In addition, two entries from the last edition were replaced and 79 were heavily revised by experts to bring them up to date with the latest information and discoveries (the previous edition appeared just as the human genome was being decoded).

All other previously published entries were re-edited and their bibliographies updated with the latest sources, and every essay now includes the section "Web Sites of Interest." All six appendixes were updated by science experts as well.

The result of these additions is a 50 percent increase in the number of essays over the previous edition and an increase of more than 33 percent in overall new material.

Organization and Format

Essays vary in length from two to six pages. Each essay follows a standard format, including ready-reference top matter and the following standard features:

- CATEGORY lists one or more of a dozen sub-disciplines of genetics or biology under which the topic falls:
  - Bacterial genetics
  - Bioethics
  - Bioinformatics
  - Cellular biology

- Classical transmission genetics
- Diseases and syndromes
- Developmental genetics
- Evolutionary biology
- Genetic engineering and biotechnology
- History of genetics
- Immunogenetics
- Molecular genetics
- Population genetics
- Social issues
- Techniques and methodologies
- Viral genetics

Disease and syndrome essays then provide standard information in these subsections:

- DEFINITION introduces, defines, and describes the disease.
- RISK FACTORS identifies the major factors involved, both environmental and genetic, and the population affected.
- ETIOLOGY AND GENETICS identifies the genes known or suspected to be involved and their pattern of inheritance.
- SYMPTOMS states the main symptoms associated with the disease or syndrome.
- SCREENING AND DIAGNOSIS identifies the procedures used to screen for and diagnose the condition, such as physical examination, family history, and various types of testing.
- TREATMENT AND THERAPY identifies the treatment and therapy regimens, if any.
- PREVENTION AND OUTCOMES identifies any behaviors that can catch the condition early, mitigate its effect, or prevent its occurrence (including genetic counseling), as well as typical short-term and long-range outcomes.

Other types of essays are divided into these sections:

- SIGNIFICANCE provides a definition and summary of the topic's importance.
- KEY TERMS identifies and defines concepts central to the topic.
- Topical subheads, chosen by the author, divide the main text and guide readers through the essay.

All essays conclude with the following material:

- The contributor's byline lists the biologist or other area expert who wrote the essay, including advanced degrees and other credentials.
- The SEE ALSO section lists cross-references to other essays of interest within the set.
- FURTHER READING lists sources for further study, often with annotations; all bibliographical sections have been fully updated and reformatted to include the latest relevant works and full citation data for easy library access.
- WEB SITES OF INTEREST provides authoritative free sites on the Internet, including the sponsor and URL. This section, which appears in every essay and is often annotated, lists government agencies, professional or academic societies, and support organizations as well as useful, reliable Web pages with the latest information from such sources as MedlinePlus from the National Library of Medicine and the National Institutes of Health or the Online Mendelian Inheritance in Man (OMIM) database.

Some essays in the encyclopedia include sidebars, which appear in shaded boxes, that offer coverage of significant subtopics within overview essays.

SPECIAL FEATURES

The articles in *Salem Health: Genetics and Inherited Conditions* are arranged alphabetically by title; a "Complete List of Contents" appears at the beginning of each volume, while a "Category Index" arranged by area of study appears at the end. Also of help to readers trying to locate topics of interest are a "Personages Index" of important figures in genetics and a comprehensive subject index, both of which appear at the end of volume 3. Diagrams, charts, graphs, drawings, and tables elucidate complex concepts, and more than one hundred photographs illustrate the text.

In addition, a series of appendixes appear at the end of volume 3, all of which were updated for this edition: The "Biographical Dictionary of Important Geneticists" features almost two hundred scientists who had an impact on genetics. "Nobel Prizes for Discoveries in Genetics" lists Nobel Prize winners whose contributions altered the history of genetics. The "Time Line of Major Developments in Genetics" offers a chronological overview of the field's development from prehistoric times to the present. The "Glossary" provides definitions of almost 700 commonly used scientific terms and concepts. The general "Bibliography," arranged by category, offers citations for both classic and recently published sources for additional research. The importance of the Internet to bioinformatics and to general education in genetics is reflected in the "Web Sites" appendix.

ACKNOWLEDGMENTS

We wish to thank the many biologists and other scholars who contributed to previous editions and to this one; their names and academic affiliations appear in the list of Contributors that follows. Special mention must be made of consulting editor Jeffrey A. Knight, who applied his broad knowledge of genetics to shaping the book's contents and provided many of the "Etiology and Genetics" sections for the disease and syndrome essays.

***

ABOUT THE EDITOR

Jeffrey A. Knight, Ph.D., is Chair of the Department of Biological Sciences at Mount Holyoke College, where he teaches courses in genetics, molecular biology, and microbiology. He has held Visiting Professor appointments at the medical schools of the Universities of Massachusetts, Florida, and Vermont. Author of several research papers on the mitochondrial genetics of baker's yeast, he has also been a frequent contributor to Salem Press publications. He edited the first edition of the *Encyclopedia of Genetics* (Salem Press, 1999), winner of the American Library Association, Reference and User Services Association's "Outstanding Reference Source," 2000. He was also the primary editor for *Salem Health: Cancer* (2009).

Preface

The science of genetics, once the purview only of serious students and professionals, has in recent decades come of age and entered the mainstream of modern life. An unparalleled explosion of new discoveries, powerful new molecular techniques, and practical applications of theories and research findings has brought genetics and its related disciplines to the forefront of public consciousness. Animal cloning, genetically modified crop plants, gene therapy, embryonic stem cell research—these are all hot-button topics which inspire emotional responses ranging from excitement and great hope to fear, distrust, and misunderstanding. All these feelings are valid, and it is the task of scientists, ethicists, and policy makers to work together to address new questions and challenges that once were conceivable only in the realm of science fiction. Advances in our understanding of human genetics and the development of new reproductive technologies have brought an increasing demand for the newest health professionals, genetic counselors, at hospitals and medical centers around the world. As new research results reach the popular press almost daily, it is perhaps instructive to reflect a bit on the historical development of the young science of genetics.

Among many other events of historical importance, the year 1900 marked the rediscovery of the Austrian monk Gregor Mendel's experimental work on the inheritance of traits in the garden pea. Mendel had published his results thirty-four years earlier, but his work attracted little attention and soon faded into obscurity. By the close of the nineteenth century, however, much had happened on the scientific front. Chromosomes had been discovered, and the cellular processes of mitosis and meiosis had been observed under the microscope. The physical bases for understanding Mendel's principles of inheritance had been established, and the great significance of his pioneering work could finally be appreciated. The so-called chromosome theory of heredity was born, and the age of transmission genetics had arrived.

The first great geneticist to emerge (and some would still call him the greatest of the twentieth century) was Thomas Hunt Morgan, who established his "fly laboratory" at Columbia University and began studying the principles of transmission genetics, using the fruit fly as a model organism. All the major principles of transmission genetics, including single and multifactorial inheritance, chromosome mapping, linkage and recombination, sex linkage, mutagenesis, and chromosomal aberrations, were first investigated by Morgan and his students.

The subdisciplines of bacterial and molecular genetics had their beginnings in the 1940's, when bacteria and their viruses became favored genetic systems for research because of their relative simplicity and the ease with which they could be grown and manipulated in the laboratory. In particular, the common intestinal bacterium *Escherichia coli* was studied intensely, and today far more is known about the biology of this single-celled organism than about any other living system. In 1952, James Watson and Francis Crick provided the molecular model for the chemical structure of DNA, the genetic material, and the next twenty years saw great progress in the understanding of the molecular nature of essential cellular processes such as DNA replication, protein synthesis, and the control of bacterial gene expression.

The 1970's witnessed the discovery of a unique class of enzymes known as restriction endonucleases, which set the stage for the development of the exciting new technology known by various names as cloning, genetic engineering, or recombinant DNA technology. Since that time, research has progressed rapidly on several fronts, with the development of genetic solutions to many practical problems in the fields of medicine, agriculture, plant and animal breeding, and environmental biology. With the help of the new technology, many of the essential questions in cell and molecular biology that were first addressed in bacteria and viruses in the 1950's and 1960's can now be effectively studied in practically any organism.

In the eleven years since the publication of the first edition of Salem's *Encyclopedia of Genetics*, both technological advances and basic genetic research have proceeded at unprecedented levels. For example, fully automated high-speed DNA sequencers have been perfected that can accurately sequence tens of thousands of bases along a DNA molecule in a single day. Huge new computers and new software programs have been developed to sort out and interpret the wealth of data generated by these sequencers, and as a result the new field of bioinformatics has been spawned. With great international publicity, the Human Genome Project was completed in 2003—a massive exercise in bioinformatics that included the sequence of over three billion base pairs of DNA among the 24 different human chromosomes (chromosomes 1-22, X, and Y). The genomes of literally hundreds of other organisms—from bacteria and viruses to tomatoes, frogs, mice, cows, and chimpanzees—have been similarly sequenced, and the new subdiscipline known as comparative genomics has shed enormous light on suspected and previously unsuspected evolutionary relationships between species. For example, mice and humans diverged from their common ancestor about 75 million years ago, yet 99 percent of mouse protein coding genes have a homolog in humans. In fact, homologs (two similar genes in two organisms that diverged from a common ancestor) for many human genes can be identified in fruit flies, yeast cells, and even *E. coli*! Humans and chimpanzees diverged from their common ancestor about 6 million years ago, and comparative genomics tells us that there are about 35 million single nucleotide differences between the genomes. Most of these differences are in noncoding regions, and the proteins encoded by the two species are very similar. In fact, 29 percent of orthologous proteins in humans and chimpanzees are identical in sequence.

Basic genetic research continues to flourish in laboratories throughout the world, and many new "model organisms" have been identified in which particular aspects of genetics, cell biology, or biochemistry can be effectively studied. The fruit fly, baker's yeast, *E. coli*, and the laboratory mouse have lost none of their charm or popularity, but they now share the research stage with zebra fish, leeches, roundworms, hundreds of different bacterial species, and many other exotic model systems too numerous to mention. New tissue culture cell lines, along with protocols using both embryonic and adult stem cells, have moved the study of human genetic disease along at a record pace. While the early years of gene therapy were marked by few successes, widespread disappointment, and the occasional disaster, there is great new hope for many avenues of gene therapy in the near future.

And what are the major problems remaining to be solved? No doubt there are many, some of which cannot even be articulated given the present state of scientific understanding. Two important questions, however, are drawing disproportionate shares of attention in the current sphere of basic research. One of these is the problem variously referred to as "the second genetic code" or "protein folding." Scientists know how a particular molecule of DNA, with a known sequence of nucleotide subunits, can cause the production of a particular unique protein composed of a known sequence of amino acid subunits. What is not understood, however, is the process by which that protein will spontaneously fold into a characteristic three-dimensional shape in which each amino acid interacts with other amino acids to produce a functional protein that has the proper pockets, ridges, holes, protuberances, and other features that it needs in order to be biologically active. If all the rules for protein folding were known, it would be possible to program a computer to create an instant three-dimensional picture of the protein resulting from any given sequence of amino acids. Such knowledge would have great applications, both for understanding the mechanisms of action of known proteins and for designing new drugs for therapeutic or industrial use.

The second "big question" at the forefront of experimental genetic inquiry relates to the control of gene expression in humans and other higher organisms. In other words, what factors come into play in turning on or turning off genes at the proper times, either during an individual cell cycle or during the developmental cycle of an organism? How is gene expression controlled differentially—that is, how are different sets of genes turned on or off in different tissues in the same organism at the same time? Many human genetic diseases are now known or suspected to be caused by errors in gene expression—that is, too much or too little of a particular protein is made in the critical tissues at the critical developmental times—so the answers to these and related questions are sure to suggest new possibilities for gene therapy or other treatments.

The purpose of these reference volumes is two-fold. First, the editors seek to highlight some of the most exciting new advances and applications of genetic research, particularly in the field of human medical genetics. Second, we hope to provide a solid basis for understanding the fundamental principles of genetics as they have been developed over the last one hundred and ten years, along with an appreciation of the historical context in which the most important discoveries were made. It is our hope that such an understanding and appreciation might help to inspire a new generation of geneticists who will continue to expand the boundaries of scientific knowledge well into the new millennium.

Jeffrey A. Knight, Ph.D.

Contributors

Mohei Abouzied, M.D.
Millard Fillmore Hospital and Dent Neurological Institute
State University of New York, Buffalo

Barbara J. Abraham, Ph.D.
Hampton University

Linda R. Adkison, Ph.D.
Mercer University School of Medicine

Richard Adler, Ph.D.
University of Michigan-Dearborn

Jane Adrian, M.P.H., Ed.M., M.T.
(ASCP)
Scottsdale, Arizona

Oluwatoyin O. Akinwunmi, Ph.D.
Muskingum College

Rick Alan
Medical writer and editor

Cathy Anderson, R.N.
American Medical Writers Association

Jeff Andrews, M.D., FRCSC,
FACOG
Vanderbilt University Medical Center

Brent M. Ardaugh
Boston University School of Public Health

Michele Arduengo, Ph.D., ELS
Promega Corporation

Steven Matthew Atchison
Auburn University

Mihaela Avramut, M.D., Ph.D.
Verlan Medical Communications

Michelle Badash, M.S.
Wakefield, Massachusetts

J. Craig Bailey, Ph.D.
University of North Carolina, Wilmington

Carl L. Bankston III, Ph.D.
Tulane University

Carolyn K. Beam
Emory University
American Medical Writers Association

Kenneth D. Belanger, Ph.D.
Colgate University

D. B. Benner, Ph.D.
East Tennessee State University

Alvin K. Benson, Ph.D.
Utah Valley University

Gerald K. Bergtrom, Ph.D.
University of Wisconsin, Milwaukee

Janet Ober Berman, M.S., C.G.C.
Temple University School of Medicine

R. L. Bernstein, Ph.D.
New Mexico State University

Leah M. Betman, M.S.
Centreville, Virginia

Massimo D. Bezoari, M.D.
Huntingdon College

Jigna Bhalla, Pharm.D.
American Medical Writers Association

Poonam Bhandari, Ph.D.
Virginia Commonwealth University

Anna Binda, Ph.D.
American Medical Writers Association

Jennifer Birkhauser, M.D.
University of California at Irvine

Jane Blood-Siegfried, D.N.Sc.
Duke University School of Nursing

Carrie Lynn Blout, M.S., C.G.C.
University of Maryland

Sabina Maria Borza, M.A.
American Medical Writers Association

Wanda Todd Bradshaw, R.N.,
M.S.N.
Duke University

Barbara Brennessel, Ph.D.
Wheaton College

Victoria M. Breting-Garcia, M.A.
Houston, Texas

Dominique Walton Brooks, M.D.,
M.B.A.
Sugar Land, Texas

Douglas H. Brown, Ph.D.
Wellesley College

Kecia Brown, M.P.H.
Washington, D.C.

Stuart M. Brown, Ph.D.
New York University School of Medicine

Thomas L. Brown, Ph.D.
*Wright State University Boonshoft
School of Medicine*

Faith Brynie, Ph.D.
Bigfork, Montana

Alia Bucciarelli, M.S.
Arlington, Massachusetts

Jill Buchanan
Gloucester, Massachusetts

Fred Buchstein, M.A.
John Carroll University

Michael A. Buratovich, Ph.D.
Spring Arbor University

Paul R. Cabe, Ph.D.
Washington and Lee University

James J. Campanella, Ph.D.
Montclair State University

Rebecca Cann, Ph.D.
University of Hawaii at Manoa

Richard P. Capriccioso, M.D.
University of Phoenix

Adrienne Carmack, M.D.
Brenham, Texas

Pauline M. Carrico, Ph.D.
*State University of New York,
Empire State College*

Christine M. Carroll, R.N.,
B.S.N., M.B.A.
*American Medical Writers
Association*

Rosalyn Carson-DeWitt, M.D.
Durham, North Carolina

Donatella M. Casirola, Ph.D.
*University of Medicine and Dentistry-
New Jersey Medical School*

Laurie F. Caslake, M.S., Ph.D.
Lafayette College

J. Aaron Cassill, Ph.D.
University of Texas at San Antonio

Tahnee N. Causey, M.S.
*Virginia Commonwealth University
Medical Center*

Stephen Cessna, Ph.D.
Eastern Mennonite University

Robert Chandler, Ph.D.
Union College

Judy Chang, M.D., FAASM
University of Pittsburgh Physicians

Paul J. Chara, Jr., Ph.D.
Northwestern College

Kerry L. Cheesman, Ph.D.
Capital University

Richard W. Cheney, Jr., Ph.D.
Christopher Newport University

Christopher Cheyer, M.D.
Wayne State School of Medicine

Stacie R. Chismark, M.S.
Heartland Community College

Marcin Chwistek, M.D.
Fox Chase Cancer Center

Jaime S. Colomé, Ph.D.
*California Polytechnic State
University, San Luis Obispo*

Jessie Conta, M.S.
Seattle Children's Hospital

Joyce A. Corban
Wright State University

Angela Costello
*American Medical Writers
Association*

Suzanne Cote, M.S.
Costa Mesa, California

Stephen S. Daggett, Ph.D.
Avila University

Mercy M. Davidson, Ph.D.
Columbia University Medical Center

Jennifer Spies Davis, Ph.D.
Shorter College

Patrick J. DeLuca, Ph.D.
Mt. Saint Mary College

Cynthia L. De Vine
*American Medical Writers
Association*

Cherie Dewar
*American Medical Writers
Association*

Jackie Dial, Ph.D.
MedicaLink, LLC

Sandra Ripley Distelhorst
Vashon, Washington

Patricia Stanfill Edens, Ph.D.,
R.N., FACHE
The Oncology Group, LLC

David K. Elliott, Ph.D.
Northern Arizona University

Elicia Estrella, M.S., C.G.C.,
L.G.C.
Children's Hospital, Boston

Renée Euchner, R.N.
*American Medical Writers
Association*

Elisabeth Faase, M.D.
Athens Regional Medical Center

Daniel J. Fairbanks, Ph.D.
Utah Valley University

Phillip A. Farber, Ph.D.
*Bloomsburg University of
Pennsylvania*

James L. Farmer, Ph.D.
Brigham Young University

Jill Ferguson, Ph.D.
National Writers Union

Linda E. Fisher, Ph.D.
University of Michigan, Dearborn

Jesse Fishman, Pharm.D.
Children's Healthcare of Atlanta

Chet S. Fornari, Ph.D.
DePauw University

Kimberly Y. Z. Forrest, Ph.D.
*Slippery Rock University of
Pennsylvania*

Joy Frestedt, Ph.D., RAC, CCTI
Frestedt, Inc.

Michael J. Fucci, D.O.
University of Connecticut

Daniel R. Gallie, Ph.D.
University of California, Riverside

Laura Garasimowicz, M.S.
*Prenatal Diagnosis of Northern
California*

W. W. Gearheart, Ph.D.
Piedmont Technical College

John R. Geiser, Ph.D.
Western Michigan University

Valerie L. Gerlach, Ph.D.
Schering-Plough Research Institute

Soraya Ghayourmanesh, Ph.D.
Bayside, New York

Sibdas Ghosh, Ph.D.
University of Wisconsin, Whitewater

Jennifer L. Gibson, Pharm.D.
Marietta, Georgia

Sander Gliboff, Ph.D.
Indiana University

James S. Godde, Ph.D.
Monmouth College

Jessica M. Goehringer, M.S.
*Dartmouth-Hitchcock Medical Center,
Nashua Clinic*

D. R. Gossett, Ph.D.
*Louisiana State University,
Shreveport*

Daniel G. Graetzer, Ph.D.
*University of Washington Medical
Center*

Dennis W. Grogan, Ph.D.
University of Cinncinati

Anne Grove, Ph.D.
Louisiana State University

Patrick G. Guilfoile, Ph.D.
Bemidji State University

Susan Estabrooks Hahn, M.S.,
C.G.C.
*Miami Institute for Human
Genomics*

Beth M. Hannan, M.S.
Albany Medical Center

Randall K. Harris, Ph.D.
William Carey College

H. Bradford Hawley, M.D.
Wright State University

Collette Bishop Hendler, R.N.,
M.S.
Abington Memorial Hospital

Michelle L. Herdman, Ph.D.
*University of Charleston School of
Pharmacy*

Margaret Trexler Hessen, M.D.
Drexel University College of Medicine

Jane F. Hill, Ph.D.
Bethesda, Maryland

Carl W. Hoagstrom, Ph.D.
Ohio Northern University

Jenna Hollenstein, M.S., RD
Genzyme

Katherine L. Howard, M.S.
University of Colorado at Denver

Carina Endres Howell, Ph.D.
*Lock Haven University of
Pennsylvania*

Brian D. Hoyle, Ph.D.
*Square Rainbow Ltd. Science
Wordsmithing*

Lynne A. Ierardi-Curto, M.D.,
Ph.D.
Laboratory Corporation of America

Chris Iliades, M.D.
Centerville, Massachusetts

April D. Ingram
Kelowna, British Columbia, Canada

Vicki J. Isola, Ph.D.
Hope College

Domingo M. Jariel, Ph.D.
Louisiana State University

Jennifer Johnson
Oregon Health Science University

Cheryl Pokalo Jones
Townsend, Delaware

Karen E. Kalumuck, Ph.D.
The Exploratorium, San Francisco

Manjit S. Kang, Ph.D.
Louisiana State University

Susan J. Karcher, Ph.D.
Purdue University

Armand M. Karow, Ph.D.
Xytex Corporation

Kari Kassir, M.D.
Orange, California

Patricia Griffin Kellicker, B.S.N.
Upton, Massachusetts

Roger H. Kennett, Ph.D.
Wheaton College

Ing-Wei Khor, Ph.D.
Oceanside, California

Stephen T. Kilpatrick, Ph.D.
University of Pittsburgh at Johnstown

Lisa Sniderman King, M.Sc., CGC, CCGC
University of Washington

Samuel V. A. Kisseadoo, Ph.D.
Hampton University

Jeffrey A. Knight, Ph.D.
Mount Holyoke College

Marylane Wade Koch, M.S.N., R.N.
University of Memphis, Loewenberg School of Nursing

Diana Kohnle
Platte Valley Medical Center

Anna Kole, M.P.H.
European Organisation for Rare Diseases

Nicole Kosarek Stancel, Ph.D.
University of Texas Southwestern Medical Center at Dallas

Audrey Krumbach
Huntingdon College

Anita P. Kuan, Ph.D.
Woonsocket, Rhode Island

Steven A. Kuhl, Ph.D.
V & R Consulting

Jeanne L. Kuhler, Ph.D.
Auburn University

Rebecca Kuzins
Pasadena, California

Sally K. Laden, M.S.
MSE Communications LLC

William R. Lamberson, Ph.D.
University of Missouri

Jill D. Landis, M.D.
Rye, New York

Dawn A. Laney, M.S.
Emory University

Kate Lapczynski, M.S.
Motlow State Community College

Jeffrey P. Larson, P.T., A.T.C.
Northern Medical Informatics

Laurie LaRusso, M.S., ELS
University School of Nutrition Science and Policy

Craig S. Laufer, Ph.D.
Hood College

David M. Lawrence
J. Sargeant Reynolds Community College

Diana R. Lazzell
Fishers, Indiana

Michael R. Lentz, Ph.D.
University of North Florida

Lorraine Lica, Ph.D.
La Jolla, California

Lauren Lichten, M.S., C.G.C.
Tufts Medical Center

Rimas Lukas, M.D.
University of Chicago

Kimberly Lynch
University of Pennsylvania

Daniel E. McCallus, Ph.D.
Nucleonics, Inc.

Kelly L. McCoy
Green Bean Medical Writing, LLC

Krisha McCoy, M.S.
American Medical Writer's Association

Doug McElroy, Ph.D.
Western Kentucky University

Sarah Lea McGuire, Ph.D.
Millsaps College

Trudy McKanna, M.S.
Spectrum Health Genetic Services

Michael J. Mclachlan
University of South Carolina

Julie D. K. McNairn, M.D.
Children's Mercy Hospital, Kansas City

Kyle J. McQuade, Ph.D.
Mesa State College

Marianne M. Madsen, M.S.
University of Utah

Lois N. Magner, Ph.D.
Purdue University

Daus Mahnke, M.D.
Gastroenterology of the Rockies

Judy Majewski
Geneva, Illinois

Sarah Malone
Springfield, Missouri

Nancy Farm Männikkö, Ph.D.
National Park Service

Elizabeth A. Manning, Ph.D.
Akros Pharma, Inc.

Katia Marazova, M.D., Ph.D.
Paris, France

Sarah Crawford Martinelli, Ph.D.
Southern Connecticut State University

Lee Anne Martínez, Ph.D.
University of Southern Colorado

Amber M. Mathiesen, M.S.
Saint Luke's Regional Medical Center

Grace D. Matzen, M.A.
Molloy College

Maria Mavris, Ph.D.
European Organisation for Rare Diseases

J. Thomas Megerian, M.D., Ph.D., F.A.A.P.
Neurometrix, Inc.
North Shore Children's Hospital
Children's Hospital, Boston

Ulrich Melcher, Ph.D.
Oklahoma State University

Dervla Mellerick, Ph.D.
Science Word Doctor, LLC

Ralph R. Meyer, Ph.D.
University of Cincinnati

Heather F. Mikesell, M.S.
University Hospitals Case Medical Center

Randall L. Milstein, Ph.D.
Oregon State University

Eli C. Minkoff, Ph.D.
Bates College

Allison G. Mitchell, M.S.
University of Maryland School of Medicine

Beatriz Manzor Mitrzyk, Pharm.D.
Mitrzyk Medical Communications, LLC

Paul Moglia, Ph.D.
South Nassau Communities Hospital

Thomas J. Montagno, Ph.D.
Simmons College

Beth A. Montelone, Ph.D.
Kansas State University

Robin Kamienny Montvilo, Ph.D.
Rhode Island College

Randy Moore
Wright State University

Christina J. Moose
Monrovia, California

Marvin L. Morris, LAc, M.P.A.
American Medical Writers Association

Nancy Morvillo, Ph.D.
Florida Southern College

Judy Mouchawar, M.D.
University of Colorado Health Sciences Center

Crystal L. Murcia, Ph.D.
Inkwell Medical Communications LLC

Donald J. Nash, Ph.D.
Colorado State University

Mary A. Nastuk, Ph.D.
Wellesley College

Deanna M. Neff, M.P.H.
Stow, Massachusetts

Leah C. Nesbitt
Huntingdon College

Bryan Ness, Ph.D.
Pacific Union College

Diane Voyatzis Norwood, M.S., RD, CDE
McKenzie-Willamette Medical Center

Heather S. Oliff, Ph.D.
American Medical Writers Association

David A. Olle, M.S.
Eastshire Communications

Henry R. Owen, Ph.D.
Eastern Illinois University

Oladayo Oyelola, Ph.D., S.C. (ASCP)
American Medical Writers Association

Robert J. Paradowski, Ph.D.
Rochester Institute of Technology

Ellen Anderson Penno, M.D., M.S., FRCSC
Western Laser Eye Associates

Massimo Pigliucci, Ph.D.
University of Tennessee

Nancy A. Piotrowski, Ph.D.
Capella University
University of California, Berkeley

Jevon Plunkett
Washington University

Anuradha Pradhan, Ph.D.
Moffitt Cancer Center

Toni R. Prezant, Ph.D.
Esoterix Laboratory Services, Inc.

Frank E. Price, Ph.D.
Hamilton College

Nancy E. Price, Ph.D.
American Medical Writers
Association

Igor Puzanov, M.D.
Vanderbilt University

Cynthia F. Racer, M.A., M.P.H.
New York Academy of Sciences

Theodor B. Rais, M.D.
University of Toledo, Ohio

Elie Edmond Rebeiz, M.D., FACS
Tufts-New England Medical Center
Tufts University School of Medicine

Diane C. Rein, Ph.D., M.L.S.
Purdue University

Andrew J. Reinhart, M.S.
Washington University School of
Medicine

Andrew Ren, M.D.
Kaiser Permanente Los Angeles
Medical Center

Alice C. Richer, RD
Norwood, Massachusetts

Mary Beth Ridenhour, Ph.D.
State University of New York,
Potsdam

Erin Rooney Riggs, M.S.
Emory University

Brad A. Rikke, Ph.D.
University of Colorado at Boulder

Julie Riley, M.S., RD
Tufts University School of Medicine

Connie Rizzo, M.D., Ph.D.
Pace University

James L. Robinson, Ph.D.
University of Illinois at Urbana-
Champaign

James N. Robinson
Huntingdon College

Ana Maria Rodriguez-Rojas, M.S.
GXP Medical Writing, LLC

Charles W. Rogers, Ph.D.
Southwestern Oklahoma State
University

Laurie Rosenblum, M.P.H.
Education Development Center,
Massachusetts

Nadja Rozovsky, Ph.D.
Somerville, Massachusetts

Paul C. St. Amand, Ph.D.
Kansas State University

Virginia L. Salmon
Northeast State Technical
Community College

Scott J. Salsman, Ph.D.
The Selva Group

Mary K. Sandford, Ph.D.
University of North Carolina at
Greensboro

Lisa M. Sardinia, Ph.D.
Pacific University

Cathy Schaeff, Ph.D.
American University

Elizabeth D. Schafer, Ph.D.
Loachapoka, Alabama

Dianne Scheinberg, M.S., RD,
LDN
Newton, Massachusetts

Matthew M. Schmidt, Ph.D.
State University of New York,
Empire State College

Amy Scholten, M.P.H.
Inner Medicine Publishing

Tom E. Scola
University of Wisconsin, Whitewater

Rebecca Lovell Scott, Ph.D., PA-C
Massachusetts College of Pharmacy
& Health Sciences

Rose Secrest
Chattanooga, Tennessee

Bonnie L. Seidel-Rogol, Ph.D.
Plattsburgh State University

Sibani Sengupta, Ph.D.
American Medical Writers
Association

Kayla Mandel Sheets, M.S.
University of Utah

Martha Sherwood, Ph.D.
Kent Anderson Law Associates

Nancy N. Shontz, Ph.D.
Grand Valley State University

R. Baird Shuman, Ph.D.
University of Illinois,
Urbana-Champaign

Sanford S. Singer, Ph.D.
University of Dayton

Robert A. Sinnott, Ph.D.
Larreacorp, Ltd.

David A. Smith, Ph.D.
Lock Haven University

Dwight G. Smith, Ph.D.
*Southern Connecticut State
University*

Nathalie Smith, M.S.N, R.N.
Lincoln, Nebraska

Roger Smith, Ph.D.
Portland, Oregon

Lisa Levin Sobczak, R.N.C.
Santa Barbara, California

F. Christopher Sowers, M.S.
Wilkes Community College

Claire L. Standen, Ph.D.
*University of Massachusetts Medical
School*

Sharon Wallace Stark, R.N.,
A.P.R.N., D.N.Sc.
Monmouth University

Joan C. Stevenson, Ph.D.
Western Washington University

Craig E. Stone, Ph.D.
University of Pennsylvania

Diane Stresing
Kent, Ohio

Annie Stuart
Pacifica, California

Jamalynne Stuck, M.S.
Western Kentucky University

Bethany Thivierge, M.P.H.
Technicality Resources

James N. Thompson, Jr., Ph.D.
University of Oklahoma

Leslie V. Tischauser, Ph.D.
Prairie State College

Sean A. Valles
Indiana University-Bloomington

Susan A. Veals, Ph.D.
*American Medical Writers
Association*

Charles L. Vigue, Ph.D.
University of New Haven

Peter J. Waddell, Ph.D.
*University of Tokyo
University of South Carolina*

C. J. Walsh, Ph.D.
Mote Marine Laboratory

Matthew J. F. Waterman, Ph.D.
Eastern Nazarene College

Judith Weinblatt, M.A., M.S.
New York City, New York

Marcia J. Weiss, M.S., J.D.
Point Park College

David C. Weksberg, M.D., Ph.D.
Baylor College of Medicine

Patricia G. Wheeler, M.D.
*Indiana University School of
Medicine*

Kayla Williams
Huntingdon College

Steven D. Wilt, Ph.D.
Kentucky Wesleyan College

Michael Windelspecht, Ph.D.
Appalachian State University

James A. Wise, Ph.D.
Hampton University

Nicola E. Wittekindt, Ph.D.
Pennsylvania State University

Barbara Woldin
*American Medical Writers
Association*

Debra Wood, R.N.
Brewster, Massachusetts

R. C. Woodruff, Ph.D.
Bowling Green State University

Robin L. Wulffson, M.D., FACOG
Tustin, California

Geetha Yadav, Ph.D.
Bio-Rad Laboratories Inc.

Carin Lea Yates, M.S., C.G.C.
Wayne State University

Rachel Zahn, M.D.
Solana Beach, California

Ross Zeltser, M.D., FAAD
*Westchester Dermatology and Mohs
Surgery Center*

George D. Zgourides, M.D.,
Psy.D.
John Peter Smith Hospital

Ming Y. Zheng, Ph.D.
Gordon College

Susan M. Zneimer, Ph.D.
US Labs

Complete List of Contents

Volume 1

Volume 2

Volume 3

SALEM HEALTH
GENETICS
& INHERITED CONDITIONS

A

Aarskog syndrome

CATEGORY: Diseases and syndromes
ALSO KNOWN AS: Aarskog-Scott syndrome; faciodigitogenital dysplasia or syndrome; faciogenital dysplasia; shawl scrotum syndrome

DEFINITION

Aarskog syndrome is an extremely rare genetic disorder. This syndrome causes changes in the size and shape of certain bones and cartilage in the body. The face, fingers, and toes are most often affected.

RISK FACTORS

Aarskog syndrome mainly affects males. Those at risk of inheriting Aarskog syndrome are male children of mothers who do not have the disorder, but who carry the gene for it.

ETIOLOGY AND GENETICS

Aarskog syndrome results from a mutation in the *FGD1* gene, which is located on the short arm of the X chromosome at position p11.21. The inheritance pattern of this disease is typical of all sex-linked recessive mutations (those found on the X chromosome). Mothers who carry the mutated gene on one of their two X chromosomes face a 50 percent chance of transmitting this disorder to each of their male children. Female children have a 50 percent chance of inheriting the gene and becoming carriers like their mothers. Although females rarely express the syndrome fully, female carriers may occasionally show minor manifestations. Affected males will pass the mutation on to all of their daughters but to none of their sons.

The *FGD1* gene specifies a guanine exchange factor, which is one of a class of proteins that acts via an intracellular signaling network to turn on other genes at appropriate times during embryonic and fetal development. The specific genes and pathways known to be activated by the FGD1 protein are involved with the cell growth and differentiation of cells destined to become components of skin, bone, and cartilage. The mouse has proven to be a useful animal model system for the study of this disorder, and the mouse data suggest that FGD1 signaling specifically affects the development of several different skeletal cell types, including chondrocytes, osteoblasts, and mesenchymal prechondrocytes.

SYMPTOMS

The main symptoms of Aarskog syndrome are disproportionately short stature, abnormalities of the head and face (including rounded face, wide-set eyes, slightly slanted eyes, drooping eyelids, a small nose, front-facing nostrils, an underdeveloped midportion of the face, a wide groove above the upper lip, a crease below the lower lip, folding of the top portion of the ear, and/or delayed teeth growth), and in some cases cleft lip or palate. Other symptoms may include a malformed scrotum; undescended testicles; small, wide hands and feet; short fingers and toes; mild webbing of fingers and toes or a simian crease in the palm of the hand; abnormalities of the sternum (mildly sunken chest); a protruding navel; inguinal hernias; ligament problems, resulting in hyperextension of the knees; and/or mild mental deficiencies (in about one-third of those affected).

SCREENING AND DIAGNOSIS

The doctor will ask about symptoms and medical history and will perform a physical exam. The diagnosis of Aarskog syndrome is usually based on facial characteristics. The diagnosis can be confirmed by X rays of the face and skull.

TREATMENT AND THERAPY

There is no known cure for Aarskog syndrome. Treatment is limited to surgical procedures to treat conditions caused by the disorder and supportive treatment. Orthodontic treatment is often needed as well. Because researchers have located abnormalities in the *FGD1* gene in people with this syndrome, genetic testing for mutations in this gene may be available.

Treatment may include surgery to correct inguinal hernia, cleft lip or palate, or undescended testicles. In some cases, orthodontic treatment may help certain facial and dental abnormalities caused by the disorder. Supportive treatment generally includes educational assistance to those afflicted with mental deficiencies, including advice and supportive treatment for parents.

PREVENTION AND OUTCOMES

There is no known way to prevent Aarskog syndrome. Those with a family history of the disorder may be referred to a genetic counselor when deciding whether to have children.

Rick Alan; reviewed by Rosalyn Carson-DeWitt, M.D.
"Etiology and Genetics" by Jeffrey A. Knight, Ph.D.

FURTHER READING

Assumpcao, F., et al. "Brief Report: Autism and Aarskog Syndrome." *Journal of Autism and Developmental Disorders* 29, no. 2 (April, 1999): 179-181.

DiLuna, Michael, et al. "Cerebrovascular Disease Associated with Aarskog-Scott Syndrome." *Neuroradiology* 49, no. 5 (May, 2007): 457-461.

EBSCO Publishing. *Health Library: Aarskog-Scott Syndrome.* Ipswich, Mass.: Author, 2009. Available through http://www.ebscohost.com.

Pasteris, N. G., et al. "Isolation, Characterization, and Mapping of Mouse *Fgd3* Gene: A New Faciogenital Dysplasia (*FGD1*; Aarskog Syndrome) Gene Homologue." *Gene* 242, nos. 1/2 (January, 25, 2000): 237-247.

Rimoin, David L., et al. *Emery and Rimoin's Principles and Practice of Medical Genetics.* 5th ed. Philadelphia: Churchill Livingstone Elsevier, 2007.

Schwartz, C. E., et al. "Two Novel Mutations Confirm *FGD1* Is Responsible for the Aarskog Syndrome." *European Journal of Human Genetics* 8, no. 11 (November, 2000): 869.

Stevenson, Roger E. "Aarskog Syndrome." In *Management of Genetic Syndromes*, edited by Suzanne B. Cassidy and Judith E. Allanson. 2d ed. Hoboken, N.J.: Wiley-Liss, 2005.

WEB SITES OF INTEREST

About Kids Health
http://www.aboutkidshealth.ca

Genetics Home Reference
http://ghr.nlm.nih.gov

Health Canada
http://www.hc-sc.gc.ca/index-eng.php

International Birth Defects Information Systems
http://www.ibis-birthdefects.org

National Organization for Rare Disorders
http://www.rarediseases.org

Ontario March of Dimes
http://www.marchofdimes.ca/dimes

United States National Library of Medicine
http://www.nlm.nih.gov

See also: Autism; Congenital defects; Hereditary diseases.

ABO blood group system

CATEGORY: Classical transmission genetics; Immunogenetics

SIGNIFICANCE: ABO blood typing has long been known to be essential for use in blood banking and for emergency transfusions. The inheritance patterns of the various ABO blood types are well understood, and the system provides a model application of various principles of classical genetics (multiple alleles, complete dominance, codominance) as applied to an important human trait.

KEY TERMS

allele: one of two or more alternative forms of a gene

antibody: a protein produced by the immune system that recognizes a foreign substance (antigen) and binds to it, targeting it for destruction

antigen: a foreign molecule that is recognized by a particular antibody

codominance: when the two alleles are both expressed in a heterozygote; neither is dominant over the other

complete dominance: when a single allele determines

Achondroplasia

CATEGORY: Diseases and syndromes
ALSO KNOWN AS: Achondroplastic dwarfism

DEFINITION

Achondroplasia is a genetic disorder that causes dwarfism (short stature). It is a disorder in which bone and cartilage do not grow normally. It is the most common cause of dwarfism.

This condition leads to patients attaining a full-grown height of less than four feet. The greatest shortening occurs in the humerus (the bone between the shoulder and the elbow) and the femur (the bone between the hip and the knee). There may also be underdevelopment of the face.

Achondroplasia is the most common form of inherited disproportionate short stature. It occurs in 1 in 26,000 to 1 in 40,000 live births.

RISK FACTORS

Those at risk of inheriting achondroplasia are children of a parent with achondroplasia and children of normal-sized parents who carry a mutated *FGFR3* gene. Advanced paternal age can cause spontaneous mutations.

ETIOLOGY AND GENETICS

Achondroplasia is inherited as an autosomal dominant disorder, meaning that a single copy of the mutation is sufficient to cause full expression of the syndrome. An affected individual has a 50 percent chance of transmitting the mutation to each of his or her children. Most cases of achondroplasia, however, result from spontaneous new mutation, so in these instances affected individuals will have unaffected parents. Advanced paternal age has been identified as a contributing factor in many spontaneous cases, and researchers using mouse models are endeavoring to identify aspects of deoxyribonucleic acid (DNA) replication or repair during spermatogenesis that could result in a predisposition to this mutation.

Homozygous achondroplasia, in which both copies of the gene carry the mutation, is a severe disorder that is invariably fatal either before or shortly after birth. Rare reports of marriages in which both partners have achondroplasia confirm the prediction that in such families 50 percent of the children will have achondroplasia, 25 percent will be unaffected, and 25 percent will die from the severe homozygous form of the condition.

Either of two mutations at nucleotide 1138 of the *FGFR3* gene, found on the short arm of chromosome 4 at position 4p16.3, will result in achondroplasia. The normal product of this gene, fibroblast growth factor receptor 3, is a protein that exerts negative regulatory control on bone growth during development. The mutant protein, which has the amino acid glycine substituted for an arginine residue at position 380, appears to be overly active, thus leading to the defects in skeletal development and decreased bone growth that are characteristic of this disorder. In fact, since the FGFR3 protein is concentrated in the cartilage and connective tissue as well as the bone, the ligaments, tendons, and muscles of patients with achondroplasia are also affected.

SYMPTOMS

Patients with achondroplasia have short stature, a long trunk, and shortened limbs, which are noticeable at birth. Adults usually reach a height of between 42 and 56 inches. An individual's head is large and his or her forehead is prominent, and portions of the face can be underdeveloped. At birth, the legs appear straight, but as a child begins to walk, he or she develops a knock-knee or bowed-leg deformity. The hands and the feet appear large, but the fingers and toes are short and stubby; straightening of the arm at the elbow may be restricted but usually does not keep a patient with achondroplasia from doing any specific activities. Children may develop an excessive curve of the lower back and a waddling walking pattern.

Other common symptoms include weight control problems; bowed legs; middle ear infections, especially in children, which, if not treated properly, can result in hearing loss; dental problems (from overcrowding of teeth); hydrocephalus (water on the brain); and neurologic and respiratory problems. Individuals also experience fatigue, pain, and numbness in the lower back and spine. Spinal compression may occur in the upper back or where the spinal cord exits from the skull in the back of the neck; compression at this latter site may cause sleep apnea or even death if not recognized and treated early. A magnetic resonance imaging (MRI) or a computed tomography (CT) scan evaluation can help detect these complications

British actor Warwick Davis was born with achondroplasia. (Time & Life Pictures/Getty Images)

SCREENING AND DIAGNOSIS

The diagnosis for achondroplasia includes clinical evaluation and radiographs. Molecular genetic testing can be used to detect a mutation in the *FGFR3* gene; such testing is 99 percent sensitive and is performed in clinical laboratories. A doctor can usually diagnose the disorder in a newborn by observing physical symptoms. To confirm that dwarfism is caused by achondroplasia, X rays are taken.

It is important that patients follow their doctors' advice to make sure that spinal stenosis does not develop. The physician can evaluate the strength of a patient's extremities and bladder control. Weakness and loss of bladder control are both signs of developing spinal stenosis.

TREATMENT AND THERAPY

Unfortunately there is currently no treatment that can cure this condition. Because it is now known that achondroplasia is caused by an absence of growth factor receptor, scientists are exploring ways to create alternate growth factors that can bypass the missing receptor and lead to normal bone growth. Such treatments are still well in the future but may offer the possibility of enhanced stature to future families who have children with achondroplasia.

Treatment with human growth hormone has been used for more than a decade and effectively increases bone growth rate, at least in the first year of life.

There have been few studies looking at whether children treated with growth hormone achieve greater (or normal) adult heights.

Surgery is sometimes needed to correct specific skeletal deformities. Spinal fusion is a surgery to permanently connect otherwise separate vertebrae. This surgery is performed for patients with significant spinal kyphosis.

Laminectomy is a surgical procedure to open the spinal canal to relieve pressure on the compressed spinal cord from spinal stenosis. Spinal stenosis, a narrowing of the spinal canal, is the most serious complication of achondroplasia.

In an osteotomy, the bones of the leg are cut and allowed to heal in the correct anatomical position. This procedure is used for patients with severe knock-knees or bowed legs.

While osteotomy has primarily been used to correct deformities, in recent years bone-lengthening procedures have been used for many short children, including those with achondroplasia. The procedures are lengthy, traumatic, and very demanding for both children and their families. Complications, sometimes serious, are common. One center has reported an average leg length (height) gain of about seven inches and an average increase in arm length of about four inches for achondroplastic individuals who undergo surgery. The combination of growth hormone therapy followed by lengthening surgery may provide benefit in achieving near-normal stature and proportions.

PREVENTION AND OUTCOMES

Because achondroplasia is an inherited disorder, there are no preventive measures. The prognosis depends on the severity of the condition. Patients who have two copies of the deficient gene (one from each parent, also known as homozygous) generally die a few weeks to months after birth. Those with one copy (from only one parent, also known as heterozygous) have a normal life span and intelligence, although children often take longer to develop normal motor skills and there is an increased risk of death in the first year of life due to respiratory problems. Patients are usually independent in their daily life activities. Many of these patients, in fact, have gone on to do great things in life.

Rick Alan; reviewed by Rosalyn Carson-DeWitt, M.D.
"Etiology and Genetics" by Jeffrey A. Knight, Ph.D.

FURTHER READING

Aldegheri, R., and C. Dall'Oca. "Limb Lengthening in Short Stature Patients." *Journal of Pediatric Orthopaedics, Part B* 10, no. 3 (July, 2001): 238-247.

Aviezer D., M. Golembo, and A. Yayon. "Fibroblast Growth Factor Receptor-3 as a Therapeutic Target for Achondroplasia—Genetic Short Limbed Dwarfism." *Current Drug Targets* 4, no. 5 (July, 2003): 353-365.

Beers, Mark H., ed. *The Merck Manual of Medical Information.* 2d home ed., new and rev. Whitehouse Station, N.J.: Merck Research Laboratories, 2003.

Boulet, S., et al. "Prenatal Diagnosis of Achondroplasia: New Specific Signs." *Prenatal Diagnosis* 29, no. 7 (July, 2009): 697-702.

EBSCO Publishing. *Health Library: Achondroplasia.* Ipswich, Mass.: Author, 2009. Available through http://www.ebscohost.com.

Kumar, Vinay, Abul K. Abbas, and Nelson Fausto, eds. *Robbins and Cotran Pathologic Basis of Disease.* 7th ed. St. Louis: MD Consult, 2004.

Parens, Erik, ed. *Surgically Shaping Children: Technology, Ethics, and the Pursuit of Normality.* Baltimore: Johns Hopkins University Press, 2006.

Pauli, Richard M. "Achondroplasia." In *Management of Genetic Syndromes,* edited by Suzanne B. Cassidy and Judith E. Allanson. 2d ed. Hoboken, N.J.: Wiley-Liss, 2005.

Shirley, E. D., and M. C. Ain. "Achondroplasia: Manifestations and Treatment." *Journal of the American Academy of Orthopaedic Surgeons* 17, no. 4 (April, 2009): 231-241.

WEB SITES OF INTEREST

American Academy of Pediatrics
http://www.aap.org

Genetics Home Reference
http://ghr.nlm.nih.gov

Little People of America
http://www.lpaonline.org

Little People of Ontario
http://www.lpo.on.ca

March of Dimes
http://www.modimes.org

United States National Library of Medicine
http://www.nlm.nih.gov

See also: Congenital defects; Dwarfism; Hereditary diseases.

Adrenoleukodystrophy

CATEGORY: Diseases and syndromes
ALSO KNOWN AS: ALD; sudanophilic leukodystrophy; Schilder's disease; X-linked adrenoleukodystrophy

DEFINITION

X-linked adrenoleukodystrophy (ALD) is a rare inherited genetic disorder. There have been thirty-four types of ALD described; X-linked ALD is the most common category.

ALD results in degeneration of the myelin sheath, the fatty insulation covering on nerve fibers in the brain, and of the adrenal gland and surrounding adrenal cortex, which produces vital hormones.

There are six types of X-linked ALD: childhood cerebral ALD, adolescent cerebral ALD, adrenomyeloneuropathy (AMN), adult cerebral ALD, adrenal insufficiency-only, and symptomatic heterozygotes.

RISK FACTORS

Individuals whose mothers carry the defective X-linked ALD gene are at risk for the disorder. Individuals in childhood to young adulthood and males are also at risk, although females may be affected.

ETIOLOGY AND GENETICS

The most common form of ALD results from a mutation in the *ABCD1* gene (ATP-binding cassette, subfamily D, member 1), which is located on the long arm of the X chromosome at position Xq28. The inheritance pattern of this disease is typical of all sex-linked recessive mutations or those found on the X chromosome. Mothers who carry the mutated gene on one of their two X chromosomes face a 50 percent chance of transmitting this disorder to each of their male children. Female children have a 50 percent chance of inheriting the gene and becoming carriers like their mothers. Although females rarely express the syndrome fully, female carriers may occasionally show minor manifestations. Affected males will pass the mutation on to all of their daughters but to none of their sons.

The *ABCD1* gene codes for a protein that is one of a family of similar proteins called transporter proteins. It is not an enzyme but instead serves to carry an enzyme that is used to break down very-long-chain fatty acids (VLCFA) in cellular peroxisomes. A single defective gene in males or two copies of the mutated gene in females blocks this process and results in the accumulation of high levels of VLCFA, which can damage the adrenal gland and nerve cells. Loss of the myelin sheath around nerve fibers results from a mechanism that is not clearly understood. Research prospects are promising, however, since very similar transporter proteins have been found in baker's yeast, and a mouse model for the study of ALD has been recently developed.

SYMPTOMS

Symptoms can vary within the types of ALD. Childhood cerebral ALD is the most severe form. It affects only boys. Symptoms usually begin between two and ten years of age; about 35 percent of patients can have severe symptoms during the early phase. On average, death results in two years; some patients may live a couple of decades. Initial symptoms of this form of ALD include behavioral changes and poor memory. As the disease progresses, more serious symptoms develop, including vision loss, seizures, hearing loss, difficulty swallowing and speaking, and difficulty with walking and coordination. Other symptoms include vomiting, fatigue, increased pigmentation ("bronzing") of the skin due to adrenal hormone deficiency (Addison's disease), progressive dementia, and vegetative state or death.

Adolescent cerebral ALD is similar to the childhood type. This type begins around eleven to twenty-one years of age, and the progression is usually slower.

Adrenomyeloneuropathy (AMN) is the most common form. Symptoms of AMN can present in the twenties, and the disease progresses slowly. Symptoms can include weakness, clumsiness, weight loss and/or nausea, emotional disturbances or depression, muscle problems (walking problems), urinary problems or impotence, and adrenal gland dysfunction.

Symptoms of adult cerebral ALD usually do not appear until young adulthood (twenties) or middle age (fifties). This type of ALD causes symptoms similar to schizophrenia and dementia. It usually progresses quickly, and death or a vegetative state can occur in three to four years.

Symptomatic heterozygotes is a type seen only in women. Symptoms may be mild or severe, and this type of ALD usually does not affect the adrenal gland function.

SCREENING AND DIAGNOSIS

The doctor will ask about symptoms and medical history, and a physical exam will be done. The doctor may suspect ALD from its symptoms. To confirm the diagnosis, the doctor may order blood tests to look for increased amounts of VLCFA. In addition, a magnetic resonance imaging (MRI) scan of the brain may be done to look for brain involvement.

TREATMENT AND THERAPY

There is no known cure for the brain damage of ALD. However, the adrenal deficiency can be treated with cortisone replacement. ALD often causes death within ten years of the onset of symptoms.

Some therapies can help to manage the symptoms of ALD. There are also some experimental treatments. Therapies to help manage the symptoms of ALD include physical therapy, psychological therapy, and special education (for children).

Some treatments patients may want to discuss with their doctors include bone marrow transplant, a procedure that may be most helpful when given early to boys with X-linked child-onset ALD. Diet therapy includes consumption of a very low-fat diet, Lorenzo's oil—dietary supplements of glycerol trioleate and glycerol trierucate (oleic and euric acid), and lovastatin—an anticholesterol medication.

PREVENTION AND OUTCOMES

There is no way to prevent ALD. Individuals who have ALD or have a family history of the disorder can talk to a genetic counselor when deciding to have children.

Early recognition and treatment may prevent the development of clinical symptoms. This is especially true in young boys who are treated with Lorenzo's oil. New technologies may soon allow early identification through newborn screening.

Rick Alan; reviewed by Rosalyn Carson-DeWitt, M.D.
"Etiology and Genetics" by Jeffrey A. Knight, Ph.D.

FURTHER READING

Beers, Mark H., ed. *The Merck Manual of Medical Information.* 2d home ed., new and rev. Whitehouse Station, N.J.: Merck Research Laboratories, 2003.

EBSCO Publishing. *Health Library: Adrenoleukodystrophy.* Ipswich, Mass.: Author, 2009. Available through http://www.ebscohost.com.

Moser, H. W. "Therapy of X-linked Adrenoleukodystrophy." *Neurorx: The Journal of the American Society for Experimental Neurotherapeutics* 3, no. 2 (April, 2006): 246-253.

Moser, H. W., and Nga Hang Brereton. "Adrenoleukodystrophy and Other Peroxisomal Disorders." In *Pediatric Nutrition in Chronic Diseases and Developmental Disorders: Prevention, Assessment, and Treatment,* edited by Shirley W. Ekvall and Valli K. Ekvall. 2d ed. New York: Oxford University Press, 2005.

Moser, H. W., G. V. Raymond, and P. Dubey. "Adrenoleukodystrophy: New Approaches to a Neurodegenerative Disease." *Journal of the American Medical Association* 294, no. 24 (December 28, 2005): 3131-3134.

Moser, H. W., et al. "Follow-up of Eighty-nine Asymptomatic Patients with Adrenoleukodystrophy Treated with Lorenzo's Oil." *Archives of Neurology* 62, no. 7 (July, 2005): 1073-1080.

WEB SITES OF INTEREST

Canadian Directory of Genetic Support Groups
http://www.lhsc.on.ca/programs/medgenet/adrenole.htm

Genetics Home Reference
http://ghr.nlm.nih.gov

The Myelin Project of Canada
http://www.myelinprojectcanada.ca

National Center for Biotechnology Information
http://www.ncbi.nlm.nih.gov

National Institute of Neurological Disorders and Stroke
http://www.ninds.nih.gov

National Organization for Rare Disorders
http://www.rarediseases.org

United Leukodystrophy Foundation
http://www.ulf.org

See also: Adrenomyelopathy; Alexander disease; Canavan disease; Cerebrotendinous xanthomatosis; Hereditary diseases; Krabbé disease; Leukodystrophy; Metachromatic leukodystrophy; Pelizaeus-Merzbacher disease; Refsum disease; Vanishing white matter disease; X chromosome inactivation.

Adrenomyelopathy

CATEGORY: Diseases and syndromes
ALSO KNOWN AS: Adrenoleukodystrophy; adrenomyeloneuropathy; Addison disease; childhood cerebral adrenoleukodystrophy; ALD; Schilder-Addison complex

DEFINITION

Adrenomyelopathy is a category and adult form of the disease known as adrenoleukodystrophy. Adrenoleukodystrophy describes adrenomyelopathy and several other closely related inherited disorders that interrupt the metabolism of very-long-chain fatty acids (VLCFA).

RISK FACTORS

A family history of adrenoleukodystrophy that is expressed as adrenomyelopathy or related inherited disorders is the primary risk factor for the disease.

ETIOLOGY AND GENETICS

The accumulation of long-chain fatty acids in the nervous system, adrenal gland, and testes results in the disruption of normal activity. The condition is genetically passed down from parents to their children as an X-linked genetic trait. Although mostly males in their twenties or later are affected by adrenomyelopathy, some female carriers of the gene can have milder forms of the disease. Approximately 1 in 20,000 people from all races are afflicted with adrenoleukodystrophy, which includes two other major categories of disease. The childhood cerebral form is characterized by a more severe onset of neurological symptoms appearing between five and twelve years of age. The other common category is called impaired adrenal gland function, also referred to as Addison disease or Addison-like phenotype, in which the adrenal gland does not produce enough steroid hormones.

Genetically speaking, the *ALD* gene is located on the Xq28 region of the X chromosome, contains ten exons, and spans 20 kilobase pairs (kb) of genomic DNA. The gene codes for the ALD protein have been localized to the peroxisomal membrane. The gene is subject to X inactivation, placing implications for female (XX) members of a family with an X-linked inherited disorder. This X inactivation process, in which one of the two X chromosomes becomes condensed and inactive, randomly and permanently occurs at the embryonic stage. If the defective allele is on the chromosome that has been inactivated, then there will be no phenotypic manifestation of the disease. However, if the defective allele is on the active X chromosome, the other having been inactivated, then there will be a clinical expression of the disorder. Mutation analysis of the *ALD* gene in thirty-five unrelated individuals with adrenoleukodystrophy revealed that all had Xq28 *ALD* gene mutations, 6 percent had large deletions, and 17 percent had an AG deletion in axon 5. The remainder had "private" mutations that were specific for each kindred, of which 55 percent had missense mutations and 30 percent had frame-shift mutations; nonsense mutations occurred in 8 percent and splice defects in 4 percent. No correlation between the nature of the mutation and the phenotype has been detected.

SYMPTOMS

Symptoms of adrenomyelopathy include adrenal dysfunction, trouble controlling urination, muscle weakness or leg stiffness that may worsen over time, and difficulty with visual memory and rapidity of thinking.

SCREENING AND DIAGNOSIS

Laboratory investigation of blood samples may show levels of elevated very-long-chain fatty acids. Peripheral nerve biopsy has revealed characteristic inclusion bodies in the Schwann cells. Chromosome studies can be useful in demonstrating specific gene mutations. Magnetic resonance imaging (MRI) will reveal images of damaged white matter in the brain.

TREATMENT AND THERAPY

A specific treatment for X-linked adrenoleukodystrophy is not available, but adrenal dysfunction is often treated with steroids such as cortisol. A diet low in very-long-chain fatty acids is thought to lower the blood levels of very-long-chain fatty acids. Ingesting oils, particularly a substance called Lorenzo's oil, has been used as a treatment in lowering the blood levels of very-long-chain fatty acids. Bone marrow transplant is also considered as a potential treatment.

PREVENTION AND OUTCOMES

Genetic counseling is recommended for potential parents who have a family history of X-linked adrenoleukodystrophy. A test sensitive in denoting very-long-chain fatty acids and a DNA probe study by specialized laboratories can diagnose the carrier state in 85 percent of female cases. Prenatal diagnosis of X-linked adrenoleukodystrophy is also available by an amniocentesis. Outcomes of adrenomyelopathy are milder than the childhood form of X-linked adrenoleukodystrophy, which is described as progressive and leads to a long-term coma two years after neurological symptoms develop.

Jeffrey P. Larson, P.T., A.T.C.

FURTHER READING

Kolata, Gina. "Experts Join in Studying Lorenzo's Oil." *The New York Times*, September 11, 1994.

Korenke, G. C., and C. Roth. "Variability of Endocrinological Dysfunction in Fifty-five Patients with X-Linked Adrenoleukodystrophy: Clinical, Laboratory and Genetic Findings." *European Journal of Endocrinology* 137 (1997): 40-47.

Moloney, J. B. M., and J. G. Masterson. "Detection of Adrenoleukodystrophy Carriers by Means of Evoked Potentials." *Lancet* 32 (October 16, 1982): 852-853.

Moser, H. W. "Adrenoleukodystrophy: Phenotype, Genetics, Pathogenesis, and Therapy." *Brain* 120 (1997): 1485-1508.

Moser, H. W., G. V. Raymond, and P. Dubey. "Adrenoleukodystrophy: New Approaches to a Neurodegenerative Disease." *Journal of the American Medical Association* 294, no. 24 (December 28, 2005): 3131-3134.

O'Neill, B. P., L. C. Marmion, and E. R. Feringa. "The Adrenoleukomyeloneuropathy Complex: Expression in Four Generations." *Neurology* 31 (1981): 151-156.

Rosebush, Patricia I., Sarah Garside, Anthony J. Levinson, and Michael F. Mazurek. "The Neuropsychiatry of Adult-Onset Adrenoleukodystrophy." *The Journal of Neuropsychiatry and Clinical Neurosciences* 11 (August, 1999): 315-327.

Web Sites of Interest

Adrenal Gland Disorders
www.nichd.nih.gov/health/topics/Adrenal_Gland_Disorders.cfm

Endocrine Abstracts
http://www.endocrine-abstracts.org/index.aspx

Endocrine Associates
www.endocrine-associates.com

Endocrine Society
http://www.endo-society.org

Endocrineweb
http://www.endocrineweb.com

The New York Times. "Adrenoleukodystrophy"
http://health.nytimes.com/health/guides/disease/adrenoleukodystrophy/overview.html

See also: Adrenoleukodystrophy; Alexander disease; Canavan disease; Cerebrotendinous xanthomatosis; Hereditary diseases; Krabbé disease; Leukodystrophy; Metachromatic leukodystrophy; Pelizaeus-Merzbacher disease; Refsum disease; Vanishing white matter disease; X chromosome inactivation.

Agammaglobulinemia

Category: Diseases and syndromes
Also known as: Bruton's agammaglobulinemia; X-linked agammaglobulinemia (XLA); hypogammaglobulinemia

Definition

Agammaglobulinemia is a disorder of the immune system resulting from a failure of white blood cells, called B lymphocytes, to develop. These B cells are the source of the antibodies or immunoglobulins, which defend the body against infections.

Risk Factors

The disease is inherited as X-linked recessive. The defective gene is located on the X chromosome, which is one of two sex chromosomes (the other is the Y chromosome). In males, who have only one X chromosome, a defective gene causes agammaglobulinemia. In females, who have two X chromosomes, a defective gene on one chromosome is insufficient to cause disease but makes the woman a carrier capable of passing the abnormal gene to her children. Males cannot pass the disease to their sons, but they can pass the defective gene to their daughters, who will then be carriers. Rarely, spontaneous gene mutations cause the disease to appear without the mother being a carrier. These spontaneous mutations occur more often in the male gamete, and it has been observed that while the mother of a boy with XLA has an 80 percent chance of being a carrier, the maternal grandmother is a carrier in only 25 percent of cases. In the United States, the incidence at birth is 1 in 379,000.

Etiology and Genetics

The defective gene responsible for this disease is Bruton's tyrosine kinase (*Btk*) gene, which is named in honor of the physician who first described the illness in 1952. The *Btk* gene is quite large, with nineteen exons encoding the 659 amino acids of the Btk enzyme and spanning 37.5 kilobase pairs (kb) on the long arm of the X chromosome (Xq21.33-q22). The *Btk* molecular location is from base pair 100,491,097 to base pair 100,527,837 on the X chromosome. More than eight hundred different mutations have been reported on the international mutation database. The Btk enzyme belongs to the Tec family of cytoplasmic tyrosine kinases and is expressed in hematopoietic cells, predominantly B cells. Btk is necessary for the development, differentiation, and functioning (signaling) of B cells. Btk deficiency blocks B cell development from the pro-B cell to pre-B cell transition, leading to a severe reduction in the number of circulating B lymphocytes and failure of the humoral response associated with an inability to produce immunoglobulins. The specific *Btk* gene mutation may influence the severity of the illness, but environmental factors and functional aspects of other components of the immune system are also important influences.

Gross gene deletions of varying lengths may produce contiguous deletion syndrome affecting the

X22q region. The defects in contiguous genes (*TIMM8A*, *TAF7L*, and *DRP2*) can complicate the problems of XLA by adding neurological impairment, sensorineural deafness, and dystonia. There are numerous other genetic causes for hypogammaglobulinemia or agammaglobulinemia, such as autosomal recessive agammaglobulinemia, hyper-IgM syndromes, and common variable immunodeficiency.

SYMPTOMS

Patients with XLA are healthy at birth but start to have problems with infections after about three months, when the antibodies passed from the mother begin to dwindle. Patients have problems with common viral infections, but particularly with encapsulated bacteria (*Streptococcus pneumoniae* and *Haemophilus influenzae*) and a parasite (*Giardia lamblia*). Children are usually diagnosed during a hospitalization for a severe infection between the ages of two and five.

SCREENING AND DIAGNOSIS

The concentration of serum immunoglobulins can be measured, and the serum IgG level is typically less than 200 milligrams per deciliter (mg/dL) in affected individuals. IgM and IgA are often low as well. The number of B lymphocytes (CD19+ cells) in the peripheral blood is markedly reduced. Finally, molecular genetic testing for mutations in the *Btk* gene can be employed for diagnosis, carrier detection, and prenatal diagnosis.

TREATMENT AND THERAPY

Since the original patient was described by Colonel Ogden Bruton in 1952, the primary treatment has been immunoglobulin replacement. The immunoglobulin may be administered intravenously once a month or subcutaneously each week. The dosages are adjusted to maintain a trough serum IgG level of 500 to 800 mg/dL as well as a satisfactory clinical response. Antibodies and other measures are employed when needed to manage infections. Live virus vaccines, such as oral polio vaccine, should be avoided. More recently, cures have been reported using stem cell transplants from the cord blood or bone marrow of histocompatible siblings.

PREVENTION AND OUTCOMES

Genetic counseling should always be provided for the parents of an affected child. Fortunately,

early diagnosis and aggressive treatment now enable most patients to lead moderately healthy and productive lives.

H. Bradford Hawley, M.D.

FURTHER READING

Broides, Amon, Wenjian Yang, and Mary Ellen Conley. "Genotype/Phenotype Correlations in X-Linked Agammaglobulinemia." *Clinical Immunology* 118 (2006): 195-200. A study examining specific *Btk* gene mutations and severity of disease.

Conley, Mary Ellen, et al. "Primary B Cell Immunodeficiencies: Comparisons and Contrasts." *Annual Review of Immunology* 27 (2009): 199-227. Broad review that includes clinical and laboratory information.

Howard, Vanessa, et al. "The Health Status and Quality of Life of Adults with X-Linked Agammaglobulinemia." *Clinical Immunology* 118 (2006): 201-208. A survey of forty-one adults with XLA.

Mohamed, Abdalla J., et al. "Bruton's Tyrosine Kinase (Btk): Function, Regulation, and Transformation with Special Emphasis on the pH Domain." *Immunological Reviews* 228 (2009): 58-73. A state-of-the-art review of the *Btk* gene.

WEB SITES OF INTEREST

Immune Deficiency Foundation
http://www.primaryimmune.org

Jeffrey Modell Foundation/National Primary Immunodeficiency Resource Center
http://www.info4pi.org

National Library of Medicine and the National Institutes of Health. Genetics Home Reference
http://ghr.n.m.nih.gov/gene=btk

See also: Autoimmune disorders; Hereditary diseases; Immunogenetics; Waldenström macroglobulinemia (WM).

Aggression

CATEGORY: Human genetics and social issues
SIGNIFICANCE: Aggression refers to behavior directed toward causing harm to others. Aggressive antisocial behavior is highly heritable, and antiso-

cial behavior (ASB) during childhood is a good predictor of ASB in adulthood and subsequent crime. Physical acts of aggression are sometimes distinguished from the more context-sensitive "covert" ASBs, including theft, truancy, and negative peer interactions.

KEY TERMS

antisocial behavior (ASB): behavior that violates rules or conventions of society and/or personal rights

impulsivity: a tendency to act quickly without planning or a clear goal in mind

irritability: a tendency to overreact to minor stimuli; short-temperedness or volatility

liability: the risk of exhibiting a behavior; the higher one's score for a measure of liability, the greater one's risk of exhibiting the behavior

serotonin: a neurotransmitter, 5-hydroxytryptamine (5-HT), present in blood platelets, the gastrointestinal tract, and certain regions of the brain, that plays roles in initiating sleep, blood clotting, and stimulating the heartbeat, and levels of which have been correlated with aggressive behavior as well as depression and panic disorder

AGGRESSION AND RELATED BEHAVIORS

Aggression or agonistic behavior in animals is usually an adaptive response to specific environmental situations during competition for resources, as in establishing dominance and a territory or in sexual competition. Rat and mice studies indicate it is partly genetic, because selective breeding produces strains that differ in levels of aggression. Human aggression can also represent a variety of natural responses to challenging situations. Measures of aggression vary, but of greatest concern are antisocial behaviors (ASBs), such as crime and delinquency, and whether some individuals are more likely to engage in these behaviors than are others.

The earliest evidence for a genetic contribution to these complex behaviors comes from twin and adoptee studies. Genes also increase the liability for many clinical conditions that include aggressive behaviors, such as conduct disorder (physically aggressive acts such as bullying or forced sexual activity) and antisocial personality disorder (persistent violation of social norms, including criminal behavior) and for personality traits that often accompany aggression, such as impulsivity and irritability. Differences in measuring ASBs partly account for the vari-

ability in heritability estimates, which range from 7 to 81 percent, but many studies indicate a heritability for genetic influences of 0.40-0.50, a minor influence of shared environment, and a much more significant influence of nonshared environment (environment unique to the individual).

From the more than one hundred studies that have sought to determine the extent of genetic influence on behavior, meta-analyses have shown a 40 to 50 percent genetic contribution, with the remaining 50 to 60 percent of the influence coming from environmental factors under typical conditions without societal interventions. However, these environmental influences may have a greater effect on individuals with a genetically determined vulnerability to ASB. Behavioral-genetic research is attempting to clarify this issue.

AGGRESSION AND HUMAN DEVELOPMENT

Aggressive behavior develops in children through a complex interaction of many environmental and biological factors. Also increasing liability for aggression and perhaps criminality are such factors as low socioeconomic status and parental psychopathology. A consistent finding is that the measure of serotonin activity of the central nervous system correlates inversely with levels of lifetime aggression, tendency to physically assault, irritability, and impulsivity. Some of the implicated genes regulate serotonin synthesis, release, and reuptake as well as metabolism and receptor activation and vary from individual to individual. Serotonergic dysfunction is also noted in alcoholism with aggression and in those who attempt and complete suicide. Brain injuries can also exacerbate tendencies to exhibit ASBs.

Some aggression, however, is a normal part of development. Thus, researcher Terrie Moffitt and colleagues distinguish between "adolescent-limited aggression"—times when most adolescents are rebelling against adult authority—and "life-course persistent" ASB, which likely reflects neuropsychological deficits and specific temperaments that are often exacerbated in unsupportive family settings. Genetic factors play a smaller role in adolescent delinquency and are consistent with aggression at this age as a developmental response to social context. The Centers for Disease Control and Prevention coined the term "electronic aggression" to describe the use of electronic media such as cell phones, personal digital assistants, and the Internet to "embarrass, harass, or threaten" another person. This type of ASB is becoming increasingly prevalent among teens and preteens.

SEX DIFFERENCES

A significant feature of ASB is a marked difference between the sexes. Males exhibit higher levels of physical aggression and violence at every age in all situations except in the context of partner violence (where females exceed males). More males than females are diagnosed with conduct disorder at every age. More males than females begin acts of theft and violence at every age. Males also exhibit higher rates of risk factors, such as impaired neurocognitive status, increased hyperactivity, and difficulties with peers. Females are rarely identified with the life-course persistent form of ASB; the male-to-female sex ratio is 10:1. Antisocial male and female adolescents tend to associate and often marry and reproduce at younger ages.

The role that hormones, particularly testosterone, may play in these differences is not clear in humans. Animal studies have shown that testosterone is significantly correlated with certain forms of aggression, such as intermale challenge in resident-intruder tests, by modulating levels of various neurotransmitters, especially serotonin, to elicit arousal and response. However, because the same experiments performed on animals cannot be performed on humans and because human behavior patterns differ from those of animals, extrapolation of the results of animal studies to humans cannot be reliably made.

IMPACT

There is much controversy surrounding the efforts to identify genes associated with aggression or crime, especially now that genome sequencing is easier than ever. The single *D4R4* gene has been related to the personality trait of novelty seeking, which in turn has been related to criminality. However, there is no one gene responsible for aggression, and simplistic answers are unlikely to be found; complex interactions among genes and between genes and environmental stimuli remain to be studied and clarified. The question has subtly shifted from "Do genes contribute to aggression?" to "Who is genetically vulnerable to environmental factors eliciting aggression?" Many people demand that the privacy of individuals be protected because the presence of specific genes does not dictate behavioral outcomes:

Genes do not determine socially defined behaviors but only act on physiological systems. Genetic testing must carry the same legal protections as other sensitive medical information, especially in cases where other genetic diseases or disorders may inadvertently be uncovered.

In addition, what constitutes acceptable or unacceptable behavior for individuals is culturally defined. Biological and environmental risk factors may increase an individual's liability to commit an act of aggression or crime, but the behavior must be interpreted within its specific context. Criminal law presumes that behavior is a function of free will, and most attempts to use genes as a mitigating factor in the courtroom have been unsuccessful. Efforts to prevent crime and violence must include consideration of all factors. Family milieu and parental competence are just as important as impaired cognitive mechanisms such as reduced serotonin activity. An imbalance in brain chemistry leading to impulsivity or aggression may be ameliorated by a supportive home setting, by medication, or by adequate nutrition and sleep.

Joan C. Stevenson, Ph.D.;
updated by Bethany Thivierge, M.P.H.

FURTHER READING

Fishbein, Diana H., ed. *The Science, Treatment, and Prevention of Antisocial Behaviors: Application to the Criminal Justice System.* Kingston, N.J.: Civic Research Institute, 2000. An excellent set of reviews on aggression and the many associated behaviors and mental disorders.

Lesch, Klaus Peter, and Ursula Merschdorf. "Impulsivity, Aggression, and Serotonin: A Molecular Psychobiological Perspective." *Behavioral Sciences and the Law* 18, no. 5 (2000): 581-604. A wonderful review of the interacting factors, including all the elements of the serotonin system.

Moffitt, Terrie E. "The New Look of Behavioral Genetics in Developmental Psychopathology: Gene-Environment Interplay in Antisocial Behaviors." *Psychological Bulletin* 105, no. 4 (2005): 533-554. This monograph moves beyond the question of whether antisocial behavior is an inherited trait to look at how genetic factors create a predisposition for adverse reactions to environmental influences.

Moffitt, Terrie E., Avshalom Caspi, Michael Rutter, and Phil A. Silva. *Sex Differences in Antisocial Behaviour: Conduct Disorder, Delinquency, and Violence in* the *Dunedin Longitudinal Study.* New York: Cambridge University Press, 2001. Sex differences are documented as children grow up.

Nelson, Randy J., ed. *Biology of Aggression.* New York: Oxford University Press, 2005. Current and future directions in the study of aggression in humans and other animals in light of the advances in pharmacology and gene-targeting techniques.

Siegel, Allan. *The Neurobiology of Aggression and Rage.* London: Informa Healthcare, 2004. The author examines the anatomical, physiological, neurochemical, and genetic mechanisms underlying the expression and control of violent behavior.

WEB SITES OF INTEREST

Centers for Disease Control and Prevention
http://www.cdc.gov/features/dsElectronic
Aggression
Offers statistics and additional information for greater understanding of this new and growing type of antisocial behavior known as electronic aggression.

Human Genome Project. Behavioral Genetics
http://www.ornl.gov/sci/techresources/
Human_Genome/elsi/behavior.shtml
Presents an easy-to-understood overview of the science of behavioral genetics and its application in topics such as intelligence, aggression, and homosexuality.

National Institutes of Health, National Institute of Mental Health
http://www.nimh.nih.gov/publicat/
violenceresfact.cfm
Provides information on child and adolescent violence and antisocial behavior, including research into the possible genetic factors of aggression.

University of Delaware. "Genetic Predisposition to Criminality: Should It Be Monitored?"
http://www.udel.edu/chem/C465/senior/fall00/
GeneticTesting/intro1.htm
Provides information on various aspects of this issue and argues against genetic screening for aggression, including the legal ramifications.

See also: Aging; Behavior; Biological determinism; Criminality; DNA fingerprinting; Forensic genetics; Sociobiology; Steroid hormones; XYY syndrome.

Aging

CATEGORY: Human genetics and social issues

SIGNIFICANCE: In the light of modern science and medicine, it has become apparent that the roots of aging lie in genes; therefore, the genetic changes that take place during aging are the source of the major theories of aging currently being proposed.

KEY TERMS

antioxidant: a molecule that preferentially reacts with free radicals, thus keeping them from reacting with other molecules that might cause cellular damage

free radical: a highly reactive form of oxygen in which a single oxygen atom has a free, unpaired electron; free radicals are common by-products of chemical reactions

mitochondrial DNA (mtDNA): the genome of the mitochondria, which contain many of the genes required for mitochondrial function

pleiotropy: a form of genetic expression in which a gene has multiple effects; for example, the mutant gene responsible for cystic fibrosis causes clogging of the lungs, sterility, and excessive salt in perspiration, among other symptoms

WHY STUDY AGING?

Biologists have long suspected that the mechanisms of aging would never be understood fully until a better understanding of genetics was obtained. As genetic information has exploded, a number of theories of aging have emerged. Each of these theories has focused on a different aspect of the genetic changes observed in aging cells and organisms. Animal models, from simple organisms such as *Tetrahymena* (a single-celled, ciliated protozoan) and *Caenorhabditis* (a nematode worm) to more complex organisms like *Drosophila* (fruit fly) and mice, have been used extensively in efforts to understand the genetics of aging. The study of mammalian cells in culture and the genetic analysis of human progeroid syndromes (that is, premature aging syndromes) such as Werner's syndrome and diseases of old age such as Alzheimer's disease have also improved the understanding of aging. From these data, several theories of aging have been proposed.

GENETIC CHANGES OBSERVED IN AGING CELLS

Most of the changes thus far observed represent some kind of degeneration or loss of function. Many comparisons between cells from younger and older individuals have shown that more mutations are consistently present in older cells. In fact, older cells seem to show greater genetic instability in general, leading to chromosome deletions, inversions, and other defects. As these errors accumulate, the cell cycle slows down, decreasing the ability of cells to proliferate rapidly. These genetic problems are partly a result of a gradual accumulation of mutations, but the appearance of new mutations seems to accelerate with age due to an apparent reduced effectiveness of DNA repair mechanisms.

Cells that are artificially cultured have been shown to undergo a predictable number of cell divisions before finally becoming senescent, a state where the cells simply persist and cease dividing. This phenomenon was first established by Leonard Hayflick in the early 1960's when he found that human fibroblast cells would divide up to about fifty times and no more. This phenomenon is now called the Hayflick limit. The number of divisions possible varies depending on the type of cell, the original age of the cell, and the species of organism from which the original cell was derived. It is particularly relevant that a fibroblast cell from a fetus will easily approach the fifty-division limit, whereas a fibroblast cell from an adult over age fifty may be capable of only a few divisions before reaching senescence.

The underlying genetic explanation for the Hayflick limit appears to involve regions near the ends of chromosomes called telomeres. Telomeres are composed of thousands of copies of a repetitive DNA sequence and are a required part of the ends of chromosomes due to certain limitations in the process of DNA replication. Each time a cell divides, it must replicate all of the chromosomes. The process of replication inevitably leads to loss of a portion of each telomere, so that with each new cell division the telomeres get shorter. When the telomeres get to a certain critical length, DNA replication seems to no longer be possible, and the cell enters senescence. Although the process discussed above is fairly consistent with most studies, the mechanism whereby a cell knows it has reached the limit is unknown.

A result of these genetic changes in aging humans is that illnesses of all kinds are more common, partly because the immune system seems to function more slowly and less efficiently with age. Other diseases, like cancer, are a direct result of the relentless accumulation of mutations. Cancers generally develop after a series of mutations or chromosomal rearrangements have occurred that cause the mutation of or inappropriate expression of proto-oncogenes. Proto-oncogenes are normal genes that are involved in regulating the cell cycle and often are responsible for moving the cell forward toward mitosis (cell division). Mutations in proto-oncogenes transform them into oncogenes (cancer genes), which results in uncontrolled cell division, along with the other traits displayed by cancer cells.

PROGEROID SYNDROMES AS MODELS OF AGING

Several progeroid syndromes have been studied closely in the hope of finding clues to the underlying genetic mechanisms of aging. Although such studies are useful, they are limited in the sense that they display only some of the characteristics of aging. Also, because they are typically due to a single mutant gene, they represent a gross simplification of the aging process. Several genetic analyses have identified the specific genetic defects for some of the progeroid syndromes, but often this has only led to more questions.

Down syndrome is the most common progeroid syndrome and is usually caused by possession of an extra copy of chromosome 21 (also called trisomy 21). Affected individuals display rapid aging for a number of traits such as atherosclerosis and cataracts, although the severity of the effects varies greatly. The most notable progeroid symptom is the development of Alzheimer's disease-like changes in the brain such as senile plaques and neurofibrillary tangles. One of the genes sometimes involved in Alzheimer's disease is located on chromosome 21, possibly accounting for the common symptoms.

Werner syndrome is a very rare autosomal recessive disease. The primary symptoms are severe atherosclerosis and a high incidence of cancer, including some unusual sarcomas and connective tissue cancers. Other degenerative changes include premature graying, muscle atrophy, osteoporosis, cataracts, and calcification of heart valves and soft tissues. Death, usually by atherosclerosis, often occurs by fifty or sixty years of age. The gene responsible for Werner syndrome has been isolated and encodes a DNA helicase (called WRN DNA helicase), an enzyme that is involved in helping DNA strands to separate during the process of replication. The faulty enzyme is believed to cause the process of replication to stall at the replication fork, the place where DNA replication is actively taking place, which leads to a higher-than-normal mutation rate in the DNA, although more work is needed to be sure of its mechanism.

Hutchinson-Gilford progeria shows even more rapid and pronounced premature aging. Effects begin even in early childhood with balding, loss of subcutaneous fat, and skin wrinkling, especially noticeable in the facial features. Later, bone loss and atherosclerosis appear, and most affected individuals die before the age of twenty-five. The genetic inheritance pattern for Hutchinson-Gilford progeria is still debated, but evidence suggests it may be due to a very rare autosomal dominant gene, which may represent a defect in a DNA repair system.

Cockayne syndrome, another very rare auto-somal recessive defect, displays loss of subcutaneous fat, skin photosensitivity (especially to ultraviolet, or UV, light), and neurodegeneration. Age of death can vary but seems to center around forty years of age. The specific genetic defect is known and involves the action of a few different proteins. At the molecular level, the major problems all relate to some aspect of transcription, the making of messenger RNA (mRNA) from the DNA template, which can also affect some aspects of DNA repair.

Another, somewhat less rare, autosomal recessive defect is ataxia telangiectasia. It displays a whole suite of premature aging symptoms, including neurodegeneration, immunodeficiency, graying, skin wrinkling, and cancers, especially leukemias and lymphomas. Death usually occurs between forty and fifty years of age. The specific defect is known to be loss of a protein kinase, an enzyme that normally adds phosphate groups to other proteins. In this case, the kinase appears to be involved in regu-

In April, 2003, fifteen-year-old John Tacket announced the discovery of a gene that causes the disease he suffers from, progeria, a syndrome that accelerates aging. (AP/Wide World Photos)

lating the cell cycle, and its loss causes shortening of telomeres and defects in the repair of double-stranded breaks in DNA. One of the proteins it appears to normally phosphorylate is *p53*, a tumor-suppressor gene whose loss is often associated with various forms of cancer.

Although the genes involved in the various progeroid syndromes are varied, they do seem to fall into some common functional types. Most have something to do with DNA replication, transcription, or repair. Other genes are involved in control of some part of the cell cycle. Although many other genes remain to be discovered, they will likely also be involved with DNA or the cell cycle in some way. Based on many of the common symptoms of aging, these findings are not too surprising.

GENETIC MODELS OF AGING

The increasing understanding of molecular genetics has prompted biologists to propose a number of models of aging. Each of the models is consistent with some aspect of cellular genetics, but none of the models, as yet, is consistent with all evidence. Some biologists have suggested that a combination of several models may be required to adequately explain the process of aging. In many ways, understanding of the genetic causes of aging is in its infancy, and geneticists are still unable to agree on even the probable number of genes involved in aging. Even the extent to which genes control aging at all has been debated. Early studies based on correlations between time of death of parents and offspring or on the age of death of twins suggested that genes accounted for 40 to 70 percent of the heritability of longevity. Later research on twins has suggested that genes may only account for 35 percent or less of the observed variability in longevity, and for twins reared apart the genetic effects appear to be even less.

Genetic theories of aging can be classified as either genome-based or mutation-based. Genome-based theories include the classic idea that longevity is programmed, as well as some evolution-based theories such as antagonistic pleiotropy, first proposed by George C. Williams, and the disposable soma theory. Mutation-based theories are based on the simple concept that genetic systems gradually fall apart from "wear and tear." The differences among mutation-based theories generally involve the causes of the mutations and the particular genetic systems involved. Even though genome-based and mutation-

based theories seem to be distinct, there is actually some overlap. For example, the antagonistic pleiotropy theory (a genome-based theory) predicts that selection will "weed out" lethal mutations whose effects are felt during the reproductive years, but that later in life lethal mutations will accumulate (a mutation-based theory) because selection has no effect after the reproductive years.

GENOME-BASED THEORIES OF AGING

The oldest genome-based theory of aging, sometimes called programmed senescence, suggested that life span is genetically determined. In other words, cells (and by extrapolation, the entire organism) live for a genetically predetermined length of time. The passing of time is measured by some kind of cellular clock and when the predetermined time is reached, cells go into a self-destruct sequence that eventually causes the death of the organism. Evidence for this model comes from the discovery that animal cells, when grown in culture, are only able to divide a limited number of times, the so-called Hayflick limit discussed above, and then they senesce and eventually die. Further evidence comes from developmental studies where it has been discovered that some cells die spontaneously in a process called apoptosis. A process similar to apoptosis could be responsible for cell death at old age. The existence of a cellular clock is consistent with the discovery that telomeres shorten as cells age.

In spite of the consistency of the experimental evidence, this model fails on theoretical grounds. Programmed senescence, like any complex biological process, would be required to have evolved by natural selection, but natural selection can only act on traits that are expressed during the reproductive years. Because senescence happens after the reproductive years, it cannot have developed by natural selection. In addition, even if natural selection could have been involved, what advantage would programmed senescence have for a species?

Because of the hurdles presented by natural selection, the preferred alternative genome-based theory is called antagonistic pleiotropy. Genes that increase the chances of survival before and during the reproductive years are detrimental in the postreproductive years. Because natural selection has no effect on genes after reproduction, these detrimental effects are not "weeded" out of the population. There is some physiological support for this in that sex hormones, which are required for reproduction earlier in life, cause negative effects later in life, such as osteoporosis in women and increased cancer risks in both sexes.

The disposable soma theory is similar but is based on a broader physiological base. It has been noted that there is a strong negative correlation among a broad range of species between metabolic rate and longevity. In general, the higher the average metabolic rate, the shorter lived the species. In addition, the need to reproduce usually results in a higher metabolic rate during the reproductive years than in later years. The price for this high early metabolic rate is that systems burn out sooner. This theory is not entirely genome-based, but also has a mutation-based component. Data on mutation rates seem to show a high correlation between high metabolic rate and high mutation rates.

One of the by-products of metabolism is the production of free oxygen radicals, single oxygen atoms with an unpaired electron. These free radicals are highly reactive and not only cause destruction of proteins and other molecules, but also cause mutations in DNA. The high metabolic rate during the reproductive years causes a high incidence of damaging DNA mutations that lead to many of the diseases of old age. After reproduction, natural selection no longer has use for the body, so it gradually falls apart as the mutations build up. Unfortunately, all attempts so far to assay the extent of the mutations produced have led to the conclusion that not enough mutations exist to be the sole cause of the changes observed in aging.

MUTATION-BASED THEORIES OF AGING

The basic premise of all the mutation-based theories of aging is that the buildup of mutations eventually leads to senescence and death, the ultimate cause being cancer or the breakdown of a critical system. The major support for these kinds of theories comes from a number of studies that have found a larger number of genetic mutations in elderly individuals than in younger individuals, the same pattern being observed even when the same individual is assayed at different ages. The differences among the various mutation-based theories have to do with what causes the mutations and what kinds of DNA are primarily affected. As mentioned above, the disposable soma theory also relies, in part, on mutation-based theories.

The most general mutation-based theory is the somatic mutation/DNA damage theory, which relies on background radiation and other mutagens in the environment as the cause of mutations. Over time, the buildup of these mutations begins to cause failure of critical biochemical pathways and eventually causes death. This theory is consistent with experimental evidence from the irradiation of laboratory animals. Irradiation causes DNA damage, which, if not repaired, leads to mutations. The higher the dose of radiation, the more mutations result. It has also been noted that there is some correlation between the efficiency of DNA repair and life span. Further support comes from observations of individuals with more serious DNA repair deficiencies, such as those affected by xeroderma pigmentosum. Individuals with xeroderma pigmentosum have almost no ability to repair the type of DNA damage caused by exposure to UV light, and as a result they develop skin cancer very easily, which typically leads to death.

The major flaw in this theory is that it predicts that senescence should be a random process, which it is not. A related theory called error catastrophe also predicts that mutations will build up over time, eventually leading to death, but it suffers from the same flaw. Elderly individuals do seem to possess greater amounts of abnormal proteins, but that does not mean that these must be the ultimate cause of death.

The free radical theory of aging is more promising and is probably one of the most familiar theories to the general public. This theory has also received much more attention from researchers. The primary culprit in this theory is free oxygen radicals, which are highly reactive and cause damage to proteins, DNA, and RNA. Free radicals are a natural by-product of many cellular reactions and most specifically of the reactions involved in respiration. In fact, the higher the metabolic rate, the more free radicals will likely be produced. Although this theory also involves a random process, it is a more consistent and predictable process, and through time it can potentially build on itself, causing accelerated DNA damage with greater age.

Significant attention has focused on mitochondrial DNA (mtDNA). Because free radicals are produced in greater abundance in respiration, which takes place primarily in the mitochondria, mtDNA should show more mutations than nuclear DNA. In addition, as DNA damage occurs, the biochemical pathways involved in respiration should become less efficient, which would theoretically lead to even greater numbers of free radicals being produced, which would, in turn, cause more damage. This kind of positive feedback cycle would eventually reach a point where the cells could not produce enough energy to meet their needs and they would senesce. Assays of mtDNA have shown a greater number of mutations in the elderly, and it is a well-known phenomenon that mitochondria are less efficient in the elderly. Muscle weakness is one of the symptoms of these changes.

The free radical theory has some appeal, in the sense that ingestion of increased amounts of antioxidants in the diet would be expected to reduce the number of free radicals and thus potentially delay aging. Although antioxidants have been used in this way for some time, no significant increase in life span has been observed, although it does appear that cancer incidence may be reduced.

FROM THEORY TO PRACTICE

Many of the genetic theories of aging are intriguing and even seem to be consistent with experimental evidence from many sources, but none of them adequately addresses longevity at the organismal level. Although telomeres shorten with age in individual cells, cells continue to divide into old age, and humans do not seem to die because all, or most, of their cells are no longer able to divide. Cells from older individuals do have more mutations than cells from younger individuals, but the number of mutations observed does not seem adequate to account for the large suite of problems present in old age. Mitochondria, on average, do function more poorly in older individuals and their mtDNA does display a larger number of mutations, but many mitochondria remain high functioning and appear to be adequate to sustain life.

Essentially, geneticists have opened a crack in the door to a better understanding of the causes of aging, and the theories presented here are probably correct in part, but much more research is needed to sharpen the understanding of this process. The hope of geneticists, and of society in general, is to learn how to increase longevity. Presently, it seems all that is possible is to help a larger number of people approach the practical limit of 120 years through lifestyle modification and medical inter-

vention. Going significantly beyond 120 years is probably a genetic problem that will not be solved for some time.

Bryan Ness, Ph.D.

FURTHER READING

Arking, Robert, ed. *Biology of Aging: Observations and Principles*. 3d ed. New York: Oxford University Press, 2001. An updated edition of a 1990 text that examines such topics as defining and measuring aging, changes in populations, genetic determinants of longevity, and aging as an intracellular process.

Macieira-Coelho, Alvaro. *Biology of Aging*. New York: Springer, 2002. A solid text that includes many figures, tables, charts, and illustrations.

Manuck, Stephen B., et al., eds. *Behavior, Health, and Aging*. Mahwah, N.J.: Lawrence Erlbaum, 2000. Examines a host of health care dilemmas associated with the elderly. One section considers the basic tenets of genetic and molecular biology, including some of the methods of looking at heritable differences in health and well-being. Illustrated.

Medina, John J. *The Clock of Ages: Why We Age, How We Age—Winding Back the Clock*. New York: Cambridge University Press, 1996. Designed for the general reader. Covers aging on a system-by-system basis and includes a large section on the genetics of aging.

Read, Catherine Y., Robert C. Green, and Michael A. Smyer, eds. *Aging, Biotechnology, and the Future*. Baltimore: Johns Hopkins University Press, 2008. Collection of essays describing how advances in medicine and technology are affecting the aging process and the lives of elderly persons.

Ricklefs, Robert E., and Caleb E. Finch. *Aging: A Natural History*. New York: W. H. Freeman, 1995. A good general introduction to the biology of aging by two biologists who specialize in aging research.

Timiras, Paola S. *Physiological Basis of Aging and Geriatrics*. 3d ed. Boca Raton, Fla.: CRC Press, 2003. Divided into three main sections, this text addresses the basic processes of biogerontology, surveys the aging of body systems, and provides a synopsis of pharmacologic, nutritional, and physical exercise guidelines for preserving physical and mental health into senescence. Illustrated with numerous tables and graphs.

Toussaint, Olivier, et al., eds. *Molecular and Cellular Gerontology.* New York: New York Academy of Sciences, 2000. Elucidates the molecular mechanisms of aging.

Vijg, Jan. *Aging of the Genome: The Dual Role of the DNA in Life and Death.* New York: Oxford University Press, 2007. Critically reviews the concept of genomic instability as a possible cause of aging, placing the concept within the context of a holistic understanding of genome functioning in complex organisms.

Yu, Byung Pal, ed. *Free Radicals in Aging.* Boca Raton, Fla.: CRC Press, 1993. An in-depth discussion of the importance of free radicals in aging.

WEB SITES OF INTEREST

Alliance for Aging Research
http://www.agingresearch.org
Provides information on genetics and the aging process, including how the Human Genome Project will affect the future of health and health care.

American Federation of Aging Research
http://www.afar.org
Includes a section called Infoaging.org that provides information on research about the biology and diseases of old age and other aspects of aging.

American Geriatrics Society
http://www.americangeriatrics.org
This national society of health care providers for older persons posts information on genetic screening for such disorders as Alzheimer's disease.

Gerontological Society of America
http://www.geron.org
The society is devoted to research, education, and practice in the field of aging.

National Institute on Aging
http://www.nia.nih.gov
The institute supports research programs on the biology and genetics of aging and provides the public with information on aging.

See also: Alzheimer's disease; Autoimmune disorders; Biochemical mutations; Biological clocks; Biological determinism; Cancer; Chemical mutagens; Developmental genetics; Diabetes; DNA repair; Genetic engineering: Medical applications; Heart disease; Human genetics; Human growth hormone; Immunogenetics; Insurance; Mitochondrial genes; Mutation and mutagenesis; Oncogenes; Stem cells; Telomeres; Tumor-suppressor genes.

Alagille syndrome

CATEGORY: Diseases and syndromes
ALSO KNOWN AS: AG; arteriohepatic dysplasia; syndromic bile duct paucity

DEFINITION
Alagille syndrome (AGS) is a multisystem disorder that includes involvement of the liver, heart, eyes, skeleton, and unusual facies (facial expressions). These features are highly variable within and between affected families.

RISK FACTORS
AGS is an autosomal dominantly inherited syndrome. Offspring of an affected parent are at 50 percent risk of having AGS. Inherited mutations are found in 30 to 50 percent of affected individuals. De novo mutations are seen in 50 to 70 percent of individuals with AGS. Parents of a child with a de novo mutation have a small risk for a second affected child because of germ-line mosaicism. The prevalence of AGS is 1 in 70,000; however, this may be an underestimate as a result of the variable expressivity and reduced penetrance seen in this disorder.

ETIOLOGY AND GENETICS
AGS was shown to be autosomal dominant using family history. The syndrome was mapped to chromosome 20 by cytogenetic deletions found in individuals with AGS. Subsequently, the gene *JAG1* (20p12) has been shown to be associated with AGS. More than two hundred mutations in *JAG1* have been identified as causing AGS in about 88 percent of affected individuals. About 7 percent of affected individuals have a cytogenetically detectable microdeletion on chromosome 20p12 that includes the *JAG1* gene. A second gene, *NOTCH 2*, which is found in the same pathway as *JAG1*, has recently been shown to be associated with less than 1 percent of AGS families.

SYMPTOMS

AGS is characterized by reduction in the number of bile ducts, leading to cholestasis (bile blockage). Since neonatal cholestasis is common, additional findings (unusual facies, eye, heart, and skeletal malformations) must be seen to make the diagnosis of AGS. Liver symptoms, present in the newborn, range from mild (jaundice) to severe (liver failure). About 15 percent of affected individuals will need liver transplantation. Conversely, a small number of individuals have no detectable liver disease. Cardiac defects are seen in more than 90 percent of AGS patients, the most common being pulmonary stenosis (67 percent).

Posterior embryotoxin (a thickened ring around the cornea) is the most common eye defect (80 percent) seen in AGS; it is difficult to use this as a diagnostic tool, however, as 15 percent of the general population has posterior embryotoxin. Other common eye findings include both Axenfeld and Rieger anomalies. Although a large portion of patients with AGS have eye defects, most do not have vision problems.

The most common skeletal malformation is butterfly vertebrae (clefting of the vertebral body). Butterfly vertebrae are found in 50 percent of AGS patients and, although they cause no symptoms, provide a good diagnostic tool for affected patients and families. The unusual facies seen in AGS are characterized by a prominent forehead, deep-set eyes, moderate hypertelorism, a saddle-shaped or straight nose with bulbous tip, and a pointed chin. These features give the face an inverted triangular appearance seen in 95 percent of affected individuals.

Early reports claimed that 30 percent of affected individuals had mental retardation or developmental delay. Mental retardation is now reported in 2 percent and developmental delay in 16 percent of affected individuals. This difference is believed to be the result of more aggressive nutritional management and intervention.

Less common findings associated with Alagille syndrome include renal abnormalities, pancreatic insufficiency, growth failure, neurovascular accidents, delayed puberty, high-pitched voice, and craniosynostosis.

SCREENING AND DIAGNOSIS

The most important feature of AGS is bile duct paucity. A clinical diagnosis can be made if an individual has three of five major clinical features (cholestasis, cardiac defect, butterfly vertebrae, eye abnormalities, unusual facies) and bile duct paucity. An individual with an affected first-degree relative and one or more of the five major clinical features also receives a diagnosis of AGS. To confirm this diagnosis, sequence analysis of the *JAG1* gene should be considered first. If a *JAG1* mutation is not identified, then fluorescence in situ hybridization (FISH) can be used to look for intragenic deletions. If deletions including *JAG1* are identified, then a full cytogenetic study is warranted to rule out additional chromosomal translocations. This is especially important if developmental delay or mental retardation are identified in an individual with AGS. Lastly, *NOTCH2* genetic testing should be considered if all other molecular testing is negative but clinical suspicion remains high.

TREATMENT AND THERAPY

A multidisciplinary team is the best approach to management of individuals with AGS, including specialists in genetics, gastroenterology, nutrition, cardiology, and ophthalmology. Mortality is 10 percent, with cardiac defects the cause of most neonate deaths. Liver failure and vascular accidents account for most later-onset morbidity and mortality. Liver transplantation has an 80 percent five-year survival rate. Catchup growth and improved liver functions are seen in 90 percent of transplanted patients. AGS patients with liver disease should avoid alcohol and contact sports. Cardiac disease can vary from asymptomatic, nonprogressive murmur to complex structural defects (such as tetralogy of Fallot) requiring surgical intervention. Growth in AGS individuals should be closely monitored. Nutritional optimization should be used to maximize growth potential and prevent developmental delay. Head injuries and/or neurologic symptoms in individuals with AGS should be treated aggressively. Magnetic resonance imaging and magnetic resonance angiography can identify aneurysms, dissections, or bleeds in symptomatic individuals. These therapies are being debated for use in presymptomatic individuals with AGS.

PREVENTION AND OUTCOMES

Genetic counseling, prenatal diagnosis, and preimplantation genetic diagnosis (PGD) are available to affected or at-risk family members for the prevention of AGS. Prenatal diagnosis and PGD are options in families and individuals with a molecular diagnosis.

Elicia Estrella, M.S., C.G.C., L.G.C.

FURTHER READING

Kim, B. J., and A. B. Fulton. "The Genetics and Ocular Findings of Alagille Syndrome." *Seminars in Ophthalmology* 22 (2007): 205-210.

Oda, T., A. G. Elkahloun, and B. L. Pike et al. "Mutations in the Human Jagged1 Gene Are Responsible for Alagille Syndrome." *Nature Genetics* 16, no. 3 (1997): 235-242.

WEB SITES OF INTEREST

Alagille Syndrome Alliance
http://www.alagille.org

GeneTests at NCBI
www.genetests.org

See also: Fluorescence in situ hybridization (FISH); Hereditary diseases.

Albinism

CATEGORY: Diseases and syndromes

DEFINITION

Albinism is the absence of pigment such as melanin in eyes, skin, hair, scales, or feathers. It is a direct result of decreased or nonexistent pigmentation of the skin, hair, and eyes.

RISK FACTORS

Tyrosine, an amino acid, is normally converted by the body to a variety of pigments called melanins, which give an organism its characteristic colors in areas such as the skin, hair, and eyes. Albinism results when the body is unable to produce melanin because of defects in the metabolism of tyrosine. Those with albinism can be divided into two subgroups: tyrosinase-negative (those who lack the enzyme tyrosinase) and tyrosinase-positive (those in whom tyrosinase is present but inactive). The most serious case is that of complete albinism or tyrosinase-negative oculocutaneous albinism, in which there is a total absence of pigment.

ETIOLOGY AND GENETICS

Albinism appears in various forms and may be passed to offspring through autosomal recessive, autosomal dominant, or X-linked modes of inheritance. In the autosomal recessive case, both parents of a child with autosomal recessive albinism are carriers; that is, they each have one copy of the recessive form of the gene and are therefore not albino themselves. When both parents are carriers, there is a one-in-four chance that the child will inherit the condition. On the other hand, X-linked albinism occurs almost exclusively in males, and mothers who carry the gene will pass it on 50 percent of the time.

SYMPTOMS

People with this condition have white hair, colorless skin, red irises, and serious vision defects. The red irises are caused by the lack of pigmentation in the retina and subsequent light reflection from the blood present in the retina. These people also display rapid eye movements (nystagmus) and suffer from photophobia, decreased visual acuity, and, in the long run, functional blindness. People with this disorder sunburn easily, since their skin does not tan. Partial albinos have a condition known as piebaldism, characterized by the patchy absence of skin pigment in places such as the hair, the forehead, the elbows, and the knees.

Ocular albinism is inherited and involves the lack of melanin only in the eye, while the rest of the body shows normal or near-normal coloration. This condition reduces visual acuity from 20/60 to 20/400, with African Americans occasionally showing acuity as good as 20/25. Other problems include strabismus (crossed eyes or "lazy eye"), sensitivity to brightness, and nystagmus.

Several complex diseases are associated with albinism. Waardenburg syndrome is identified by the presence of a white forelock (a lock of hair that grows on the forehead) or the absence of pigment in one or both irises, Chediak-Higashi syndrome is characterized by a partial lack of pigmentation of the skin, and tuberous sclerosis patients have only small, localized depigmented areas. A more serious case is the Hermansky-Pudlak syndrome, a disorder that includes bleeding.

SCREENING AND DIAGNOSIS

A physical examination will reveal the nature and extent of albinism. In ocular albinism, the color of

At the Santa Lucia school in Guatemala City, an albino girl, Maria del Carmen Quel, eats a snack as she plays on a swing. Albinism is frequently associated with blindness. (AP/Wide World Photos)

the iris may be any of the normal colors, but an optician can easily detect the condition by shining a light from the side of the eye.

TREATMENT AND THERAPY

Albinism can affect an individual's lifestyle. Treatment of the disease involves reduction of the discomfort the sun creates. Thus photophobia may be relieved by sunglasses that filter ultraviolet light, while sunburn may be reduced by the use of sun protection factor (SPF) sunscreens and by covering the skin with clothing.

In ocular albinos, the light shines through the iris because of the absence of the light-absorbing pigment. Children with this condition have difficulty reading what is on a blackboard unless they are very close to it. Surgery and the application of optical aids appear to have had positive results in correcting such problems.

PREVENTION AND OUTCOMES

Albinism has long been studied in humans and captive animals. Since albinism is basically an inherited condition, genetic counseling is of great value to individuals with a family history of albinism. Albinism has not been found to affect expected life span among humans. However, albino humans are susceptible to sunburns and skin cancer.

Albinism has also been detected in wild animals, but such animals often have little chance of survival because they cannot develop normal camouflage colors, important for protection from predators. Animals in which albinism has been recorded include deer, giraffes, squirrels, frogs, parrots, robins, turtles, trout, and lobsters. Partial albinism has also been reported in wildlife. In other cases, such as the black panther of Asia, too much melanin is formed and the disorder is called melanism.

Albinism has also been observed in plants, but their life span rarely goes beyond seedline state, because without the green pigment chlorophyll, they cannot obtain energy using photosynthesis. A few species of plants, such as Indian pipes (*Monotropa*), are normally albino and obtain their energy and nutrition from decaying material in the soil.

Soraya Ghayourmanesh, Ph.D.

FURTHER READING

Gahl, William A., et al. "Genetic Defects and Clinical Characteristics of Patients with a Form of Oculocutaneous Albinism (Hermansky-Pudlak Syndrome)." *New England Journal of Medicine* 338, no. 18 (April 30, 1998): 125. Discusses several aspects of Hermansky-Pudlak syndrome. Details the diagnosis of this syndrome in forty-nine patients of Puerto Rican descent and patients from the mainland United States. Two charts.

Gershoni-Baruch, R., et al. "Dopa Reaction Test in Hair Bulbs of Fetuses and Its Application to the Prenatal Diagnosis of Albinism." *Journal of the American Academy of Dermatology* 24, no. 2 (February, 1991): 220-222. Describes the hair bulb tyrosinase test.

King, Richard A., and C. Gail Summers. "Albinism and Hermansky-Pudlak Syndrome." In *Management of Genetic Syndromes*, edited by Suzanne B. Cassidy and Judith E. Allanson. Hoboken, N.J.: Wiley-Liss, 2005. Discusses incidence, diagnostic criteria, etiology, and diagnostic testing for albinism.

King, Richard A., et al. "Albinism." In *The Metabolic and Molecular Bases of Inherited Disease*, edited by C. R. Scriver et al. 7th ed. New York: McGraw-Hill, 1995. A solid introduction and overview of albinism.

Pollier, Pascale. *Journal of Audiovisual Media in Medicine* 24, no. 3 (September, 2001): 127. Examines the medical, biological, and genetic causes of albinism and provides notes from the author's attendance at a conference of the Albinism Fellowship.

Salway, Jack G. "Amino Acid Disorders: Maple Syrup Urine Disease, Homocystinuria, Alkaptonuria, and Albinism." In *Medical Biochemistry at a Glance*. 2d ed. Malden, Mass.: Blackwell, 2006. Provides a concise description of albinism and other amino acid disorders. Illustrated with tables and diagrams.

Scriver, Charles, et al., eds. *The Metabolic and Molecular Bases of Inherited Disease*. 8th ed. 4 vols. New York: McGraw-Hill, 2001. These authoritative volumes on genetic inheritance survey all aspects of genetic disease. The eighth edition has been thoroughly updated; more than half of its contents are new.

Tomita, Yasuchi. "Molecular Bases of Congenital Hypopigmentary Disorders in Humans and Oculocutaneous Albinism 1 in Japan." *Pigment Cell Research* 13, no. 5 (October, 2000): 130. Presents a study that identified the molecular bases of congenital hypopigmentary disorders in humans and oculocutaneous albinism (OCA)-1 in Japan. Piebaldism, Waardenburg syndrome, Hermansky-Pudlak syndrome, and tyrosinase gene-related OCA-1 are closely examined.

Web Sites of Interest
Albinism Database
http://albinismdb.med.umn.edu
Lists mutations associated with all major known forms of oculocutaneous and ocular albinism.

Genetics Home Reference, Oculocutaneous Albinism
http://ghr.nlm.nih.gov/condition=oculocutaneousalbinism
Describes the genetic basis of oculocutaneous albinism and provides other information about the condition.

Mayo Clinic.com
http://www.mayoclinic.com/health/albinism/DS00941
This section of the Mayo Clinic's Web site focuses on albinism, providing information on symptoms, causes, tests and diagnosis, treatments, and self-care.

National Organization for Albinism and Hypopigmentation
http://www.albinism.org
A volunteer organization for albinos and for those who care for people with albinism, providing resources for self-help and promoting research and education.

See also: Biochemical mutations; Chediak-Higashi syndrome; Complete dominance; Dihybrid inheritance; Inborn errors of metabolism; Monohybrid inheritance.

Alcoholism

CATEGORY: Diseases and syndromes
ALSO KNOWN AS: Alcohol dependence, alcohol abuse, persistent drunkenness, frequent intoxication

DEFINITION
Alcoholism is a disease or addiction that occurs when a person is experiencing serious health and/or social problems related to the use of alcohol. The physiological and psychological dependence on alcohol that is alcoholism is formally described in the *Diagnostic and Statistical Manual of Mental Disorders: DSM-IV-TR*, or *DSM* (rev. 4th ed., 2000), issued by the American Psychiatric Association.

RISK FACTORS
Certain characteristics or factors make persons more likely to develop alcoholism; they include family history of the disorder, negative environment, emotional stress, access to alcohol, young age at first use, age range of eighteen to twenty-nine, male gender, and low level of education. Persons of certain races or ethnic origins, the unmarried, and children of alcoholics are more likely to become dependent

on alcohol. Mental disorders such as major depression, anxiety, bipolar disorder, and antisocial personality disorder are associated with the development of alcoholism and may also be hereditary. Antisocial personality disorder has been referred to as the most important risk factor for alcoholism.

ETIOLOGY AND GENETICS

Alcoholism is a relatively common chronic and relapsing disorder that results in significant health and social consequences. Alcohol has a relatively high addictive potential in the general population and is even higher in susceptible individuals. Several epidemiology studies have been conducted to attempt to categorize genes and characteristics related to alcohol dependence. However, much is still not known about this disease and the role of genetics in the development, course, and outcome of alcoholism.

Genes under investigation for their potential role in this disorder are typically grouped by involvement in the metabolism of alcohol, rewarding circuits, and response to alcohol dependence treatment. The enzymes responsible for hepatic alcohol metabolism are alcohol dehydrogenase (ADH) and aldehyde dehydrogenase (ALDH); the corresponding candidate genes are *ADH2*2*, *ADH3*1*, and *ALDH2*2*. Protective genes (associated with reduced alcohol consumption) are *ADHB*2* and are found in Asians and Israeli Jews, while *ADH1C*2* appears to protect against complications related to alcoholism, such as cirrhosis and pancreatitis. Gamma-aminobutyric acid (GABA) receptors are the most important inhibitory receptors and are involved in the rewarding circuit; alcohol acts as an agonist or activator of the $GABA_A$ receptor. Scots, Germans, Native Americans, and Finns have variants of *GABRA6* and *GAD* that are associated with alcoholism.

Alcohol increases the concentration of dopamine in the brain and is important in reinforcing its effects. A variant in the *DRD2* gene of the dopamine D2 receptor may be a vulnerability gene for alcohol dependence; however, data are conflicting. Presence of the *DAT1* gene (which codes for the dopamine transporter and is responsible for dopamine reuptake) is associated with worse outcomes during alcohol withdrawal, such as seizures. Many other genes are linked to physiologic markers and to the diagnosis of alcohol dependence, including *HTT*, *CRF*, *CRF1*, *CYP2EI*, *GABRA1*, *COMT*, and *DRD1*. Psychiatric illness and high-risk behaviors may also be genetically linked and direct persons to select environments associated with dependence.

SYMPTOMS

Early behavioral symptoms of alcoholism include frequent intoxication, a pattern of heavy drinking, drinking alone or in secret, or drinking alcohol in high-risk situations (such as drinking and driving). Erratic or dramatic changes in behavior with alcohol consumption, "blacking out," or not remembering events that occurred while drinking, may also be signs of the disorder. Symptoms of alcohol dependence may become worse over time. The physical symptoms of alcoholism are many and can include jaundice (yellowing of skin or eyes), hepatitis (enlarged liver), abdominal pain, nausea and vomiting, infections, malaise (not feeling well), weight loss, fluid retention, problems with memory, and anorexia (decreased eating). Laboratory analyses may reveal increased liver enzymes, low potassium levels, low hemoglobin and hematocrit (indicating anemia), and vitamin deficiencies.

Symptom expression of alcohol dependence may differ by culture and ethnicity, because people of different cultures may express physical and mental ailments differently. Ethnic and racial groups may respond differently to alcohol and medications used to treat alcohol dependence. Some groups may even enjoy greater protection against alcoholism as a result of their genetics. Certain ethnic groups may be more susceptible to alcoholism or related complications. For instance, vulnerability to cardiomyopathy and Wernicke-Korsakoff's syndrome may be heritable and may vary by ethnicity. Latino men, for example, tend to show greater susceptibility to alcohol-related liver damage than do white men.

Early alcohol exposure and its interaction with genetics may lead to problems in fetal and child development. Fetal alcohol syndrome may result when a pregnant woman drinks even a moderate amount of alcohol. Exposure to alcohol in the womb can cause mild to severe facial and dental abnormalities, mental impairment, and bone and heart problems that become more obvious and problematic as the child grows. Vision, hearing, and attention problems are also common. Children of alcoholic fathers also can have difficulties in learning, language, and temperament because of hereditary and environmental effects of growing up in an unstable home with an alcoholic parent. In sum, parents who drink may increase the likelihood that their children will develop alcoholism through both genetic and environmental factors.

SCREENING AND DIAGNOSIS

Various questionnaires are commonly used to screen a person for alcoholism. Questions typically ask about the amount of alcohol consumed, how often drinking occurs, how much time is spent thinking about drinking, if withdrawal occurs after stopping drinking, and effects of drinking on personal life and health. Most persons with alcoholism will deny having the disorder, and family and friends may be questioned to support the diagnosis. Alcohol dependence is rarely diagnosed in a routine office visit. Diagnosis typically follows after a major negative health or social event occurs, such as liver disease or a motor vehicle accident. After diagnosis, patients may be subgrouped into type-I or type-II alcoholism; type-II is highly heritable (88 percent) while type-I has a relatively low heritability (21 percent).

TREATMENT AND THERAPY

Three candidate genes for alcohol treatment response are *OPRM1*, *HTT*, and *COMT*. Currently, treatment for alcoholism includes the use of medications such as naltrexone (opioid antagonist), acamprosate (taurine analog), or disulfuram (alcohol deterrent). Whether persons will respond to a certain medication and how long they will abstain from alcohol use may vary based on genetic makeup. Psychoanalysis and behavior modification are important parts of alcohol dependence treatment. Alcoholics Anonymous has a twelve-step support program for persons with alcohol dependence.

Continued study of the genes associated with different patterns of alcohol problems, protective genetic effects in populations with exceptionally low rates of alcoholism, and genetically based interventions (such as matching pharmacotherapies to different populations of individuals to forestall the development of the problem) are assured. The study of genetics and alcoholism is also likely to encourage growth in the field of ethnopharmacology, the study of how different therapeutic drugs differentially affect members of specific ethnic groups.

PREVENTION AND OUTCOMES

The best way to prevent alcoholism is to avoid the use of alcohol. Alcoholism is associated with an increased risk of hepatitis, liver cancer, abuse of other substances (such as marijuana, cocaine, sedatives, and stimulants), sexually transmitted diseases and other infections, malnutrition, psychiatric illness, and premature death. Persons with alcohol dependence are more likely to gamble, smoke cigarettes, or engage in other risky behaviors.

The presence of alcohol in modern life may have genetic roots. Historically, it helped those who could tolerate its taste and effects to survive and be selected for when others who could not do so perished as a result of consuming contaminated water. Alcohol has a complex relationship to human life, and alcoholism will be studied for some time.

Nancy A. Piotrowski, Ph.D.;
updated by Beatriz Manzor Mitrzyk, Pharm.D.

FURTHER READING

American Psychiatric Association. *Diagnostic and Statistical Manual of Mental Disorders: DSM-IV-TR.* Rev. 4th ed. Washington, D.C.: Author, 2000. This American professional manual describes all major psychiatric disorders. There is a chapter devoted to substance use disorders.

Goldman, David, Gabor Oroszi, and Francesca Ducci. "The Genetics of Addictions." *Nature Reviews Genetics* 6 (2005): 521-531. An article that argues that addictions are moderately to highly heritable.

Gorwood, Philip, Mathias Wohl, Yann L. Strat, and Frederic Rouillon. "Gene-Environment Interactions in Addictive Disorders: Epidemiological and Methodological Aspects." *Comptes rendus Biologies* 330 (2007): 329-338. Describes how "the gene-environment interactions approach could explain some epidemiological and clinical factors associated with addictive behaviours."

Plomin, Robert, and Gerald E. McClearn, eds. *Nature, Nurture, and Psychology.* Washington, D.C.: American Psychological Association, 1993. The topic of alcoholism is discussed, among other topics, with an emphasis on comparing the roles of genetics versus social processes and the environment.

Strat, Yann L., Nicolas Ramoz, Gunter Schumann, and Philip Gorwood. "Molecular Genetics of Alcohol Dependence and Related Endophenotypes." *Current Genomics* 9 (2008): 444-451. The authors claim that "Predisposition to alcohol dependence is affected by multiple environmental and genetic factors in a complicated way."

WEB SITES OF INTEREST

American Psychological Association (APA)
http://www.apa.org
This site provides access to PsycNET and Psyc-ARTICLES, which can be searched for published information on the genetics of alcoholism.

*National Center for Biotechnology Information (NCBI)
and Online Mendelian Inheritance in Man (OMIM).
Alcohol Dependence*
http://www.ncbi.nlm.nih.gov/entrez/
dispomim.cgi?id=103780
Provides detailed information about the genetics of alcoholism.

*National Institute on Alcohol Abuse and Alcoholism,
ETOH*
http://etoh.niaaa.nih.gov
ETOH is the chemical abbreviation for ethyl alcohol. This site includes reports related to alcohol dependence, including epidemiology, etiology, prevention, policy, and treatment.

The Pharmagenomics Knowledge Database (PharmGKB)
http://www.pharmgkb.org/index.jsp
PharmGKB's mission is "to collect, encode, and disseminate knowledge about the impact of human genetic variations on drug response. We curate primary genotype and phenotype data, annotate gene variants and gene-drug-disease relationships via literature review, and summarize important PGx genes and drug pathways." There is a link to access Collaborative Study on the Genetics of Alcoholism (COGA) data.

See also: Aggression; Behavior; Congenital defects; Criminality; Eugenics; Genetic testing: Ethical and economic issues; Hereditary diseases; Sociobiology; Thalidomide and other teratogens.

Alexander disease

CATEGORY: Diseases and syndromes
ALSO KNOWN AS: Leukodystrophy with Rosenthal fibers

DEFINITION

Alexander disease is a type of leukodystrophy, or disorder of the white matter of the brain, caused by a genetic mutation. This rare progressive neurological condition may develop in infancy, childhood, or adulthood. It is variable in severity yet typically fatal and was first described by W. Stewart Alexander in 1949.

RISK FACTORS

The only risk factor is a known familial mutation. Because variability in symptoms and age of onset exists even within the same family, other unknown genes and environmental factors must play a role in how the disease is expressed. Alexander disease has been reported across many different ethnicities and does not appear to occur more commonly in any one ethnic group. Men and women are equally likely to be affected. Fewer than five hundred cases of Alexander disease have been reported, although not all have had confirmatory genetic testing.

ETIOLOGY AND GENETICS

Alexander disease typically occurs as a result of a sporadic new mutation, meaning one that is not inherited from a parent, in the glial fibrillary acidic protein (*GFAP*) gene on chromosome 17q21. The mutation most likely occurs in the father's sperm but may occur in the mother's egg. Rarely, individuals are suspected to have Alexander disease but do not have a detectable *GFAP* mutation. Mutations are typically missense, where one amino acid is changed to another, and may cause abnormal formation of the glial acidic fibrillary protein, which is necessary for proper myelination of nerve cells. Myelin acts as insulation for these cells. If myelination is disrupted, then the nervous system may not work properly. Abnormal protein may accumulate as Rosenthal fibers, which may be detected in certain brain cells.

Parents of one affected infant or child are not typically at risk of having another child with Alexander disease, since neither parent commonly has the mutation. Alexander disease is inherited in an autosomal dominant manner, meaning offspring of an individual with a mutation are at a 50 percent risk to inherit the same mutation. Individuals with a *GFAP* mutation known to cause earlier onset forms of the disease will likely develop the condition. There are reports of adults with a *GFAP* mutation that have yet to show signs of the condition, meaning that onset may occur relatively late in life or not at all. Current research focuses on understanding the effects of the *GFAP* mutations and developing animal models for treatment studies.

SYMPTOMS

The infantile form is the most common and most severe, while the adult-onset form is least common and most variable. In infancy (the first two years of life), symptoms progress rapidly and can include delay in developmental milestones, loss of developmental skills, seizures, extra fluid in the brain, increase in head size, and coordination difficulties. Children with juvenile onset (two to thirteen years of age) may have problems with coordination, swallowing, and speech; frequent vomiting; seizures; intellectual decline; and loss of motor skills. Symptoms of the adult form (late teens through adulthood) may include problems with speech, swallowing, and walking; eye movement abnormalities; abnormal movement of the palate; incontinence; constipation; sweating and blood pressure abnormalities; sleep apnea; and seizures. Abnormal spinal curvature, diabetes, and problems with growth may occur. Brain biopsy or autopsy may show Rosenthal fibers, which are common with Alexander disease but are also known to occur with other conditions. Routine medical examinations are recommended to monitor for disease progression and medical complications.

SCREENING AND DIAGNOSIS

Diagnostic criteria for brain MRI have been developed and may be helpful in reaching a diagnosis. Brain biopsy and autopsy are available to identify Rosenthal fibers. Genetic testing has replaced the need for biopsy for diagnosis, however. Other clinical studies may be beneficial to detect medical problems and complications.

TREATMENT AND THERAPY

There is no cure for this disease, and treatment is supportive. Medications may be available to alleviate some neurologic symptoms, but individual response varies. Assistive devices and therapy may be beneficial.

PREVENTION AND OUTCOMES

Diagnostic genetic testing can provide an accurate diagnosis in affected individuals. With a known familial mutation, predictive genetic testing is available to determine whether an unaffected adult has inherited the gene mutation and may be at risk to develop the adult-onset form of Alexander disease.

There is no known prevention for developing Alexander disease once an individual is known to have a *GFAP* mutation. Routine medical care and therapies may help to prevent other complications. Prenatal diagnosis and preimplantation genetic diagnosis may be possible with identified familial mutations.

Alexander disease is typically fatal for infantile and juvenile-onset forms. Life expectancy is usually up to about ten years, although those with juvenile or adult onset may survive longer.

Katherine L. Howard, M.S.

FURTHER READING

Nussbaum, Robert L., Roderick R. McInnes, and Huntington F. Willard. *Thompson and Thompson Genetics in Medicine.* 7th ed. New York: Saunders, 2007. Thorough review of genetics for professionals and the layperson.

Parker, James N., and Philip M. Parker. *The Official Parent's Sourcebook on Alexander Disease: Updated Directory for the Internet Age.* Rev. ed. San Diego: Icon Group International, 2003. Provides trusted sources for research and review of Alexander disease for the layperson.

Watts, Ray, and William C. Koller. *Movement Disorders: Neurologic Principles and Practices.* 2d ed. New York: McGraw-Hill, 2004. Reviews movement disorders for the professional.

WEB SITES OF INTEREST

GeneTests at NCBI
http://www.genereviews.org

The Myelin Project: Alexander's Disease
http://www.myelin.org/en/cms/405

Online Mendelian Inheritance in Man: Alexander Disease
http://www.ncbi.nlm.nih.gov/entrez/dispomim.cgi?id=203450

United Leukodystrophy Foundation: Alexander Disease
www.ulf.org/types/Alexander.html

The Waisman Center: Alexander Disease
www.waisman.wisc.edu/alexander/index.html

See also: Adrenoleukodystrophy; Canavan disease; Cerebrotendinous xanthomatosis; Hereditary diseases; Krabbé disease; Leukodystrophy; Metachromatic leukodystrophy; Pelizaeus-Merzbacher disease; Refsum disease; Vanishing white matter disease.

Alkaptonuria

CATEGORY: Diseases and syndromes
ALSO KNOWN AS: Black urine disease; AKU; alcaptonuria; homogentisate oxidase deficiency; homogentisic aciduria

DEFINITION

Alkaptonuria is an inherited autosomal recessive disorder of the degradative metabolism of the amino acid tyrosine. Alkaptonuria describes the excretion in urine of homogentisate, a metabolic product of tyrosine. Urine turns black in air from the oxidation of homogentisate.

RISK FACTORS

To have the disorder, a person must have received one defective copy of the gene for homogentisate 1,2-dioxygenase from each parent. The disorder is present from birth and is largely unaffected by treatment or lifestyle. The disease incidence is very rare, except in people whose ancestry is derived from Slovakia or the Dominican Republic.

ETIOLOGY AND GENETICS

In 1902, the physician Archibald E. Garrod called alkaptonuria an "inborn error of metabolism," meaning an inherited disorder of normal metabolism. This statement for the first time linked human disease with the recently rediscovered rules for the inheritance of traits in pea plants, originally discovered by Gregor Mendel in 1865. Among the hundreds of human genetic diseases that also fit this description are phenylketonuria, galactosemia, and

Tay-Sachs disease. Though the terms "gene," "DNA," and "enzyme" were unknown in 1902, Garrod correctly implied that a single human gene determined the enzymatic ability to metabolize "alkapton" (now called homogentisate) and that the disease was caused by a defect in the gene. By 1908, Garrod had further identified the human disorders albinism, cystinuria, and pentosuria as inborn errors of metabolism analogous to alkaptonuria.

Alkaptonuria is completely caused by the inherited inability to convert homogentisate, a degradative metabolic intermediate of tyrosine, to its product 4-maleylacteoacetate. The conversion normally occurs mostly in the liver. In persons with alkaptonuria, accumulation of homogentisate in the liver leads to excretion into the blood and ultimately into the urine. Homogentisate deposits may occur in joints and other tissues over time. The enzyme homogentisate 1,2-dioxygenase is responsible for the metabolic conversion. It is encoded by the *HGD* gene on the long arm of chromosome 3, band 3q13.33. The enzyme is inactive in persons with alkaptonuria as a consequence of defective alleles (gene copies) of the *HGD* gene, one from each parent. (Parents do not have the disorder, because they each have one normal functioning *HGD* allele.)

The *HGD* gene has been cloned and sequenced. Mutations in the form of altered DNA sequence of the coding portion of the *HGD* gene, leading to amino acid substitutions that alter and inactivate the homogentisate 1,2-dioxygenase protein, have been identified in persons with alkaptonuria. Mutations at different locations in the *HGD* gene have been found in alcaptonurics living in the United States, Slovakia, Spain, Finland, Iraq, and Turkey, clearly indicating that the mutations and the disease arose multiple times in human history independently.

In most of the world, the incidence of alkaptonuria is between 1 in 200,000 and 1 in 1 million births. The defective gene frequency is highest in populations whose ancestry is derived from Slovakia and the Dominican Republic. In Slovakia, the disease occurs about once in 20,000 births. It occurs somewhat more often in the Dominican Republic, though the frequency is not precisely known.

SYMPTOMS

The outstanding sign of the disease is urine that turns black shortly after excretion. Discolored urine may be the only sign in infants and young children.

Homogentisate is colorless, but air oxidation leads to a melanin-like dark pigment. Other signs and symptoms, especially in older children and adults, derive from homogentisate deposits in cartilage (ochronosis), and include dark earlobes, dark sclera of the eyes, joint and lower back pain, and kidney stones. Homogentisate deposits damage cartilage, and joint deposits can cause osteoarthritis. Deposits may also damage heart valves.

SCREENING AND DIAGNOSIS

Infants may be diagnosed because of a black or brown urine-stained diaper. Confirmation requires measuring significant homogentisate levels in blood plasma or urine. As for many genetic disorders, one aid in diagnosis is another person in the same family or a close relative who has the disease.

TREATMENT AND THERAPY

No effective treatment has been demonstrated for alkaptonuria. Dietary restriction of phenylalanine (as for phenylketonuria) and tyrosine has been tried. Large doses of ascorbic acid (vitamin C) have been tried, on the idea that the vitamin can help prevent homogentisate deposits. Treatment with the herbicide nitisinone, an inhibitor of 4-hydroxyphenylpyruvate dioxygenase, an enzyme that also generates homogentisate, has also been tried, but the drug may be toxic over the long term. Joint stiffness can be delayed by exercise and physical training. Surgery can replace knee, hip, and shoulder joints damaged by arthritis in older adults.

PREVENTION AND OUTCOMES

A family that already has a child with alkaptonuria identifies the parents as a "couple at risk," who have a 25 percent chance with each pregnancy of having a child with alkaptonuria. Prenatal diagnosis and the option of abortion are feasible for these couples. Further pregnancies have a 75 percent chance of a child without alkaptonuria, which genetic testing can confirm. Prevention by routine genetic testing to identify couples at risk is not feasible in most populations, because the disease is so rare.

R. L. Bernstein, Ph.D.

FURTHER READING

Bearn, A. G. "Inborn Errors of Metabolism: Garrod's Legacy." *Molecular Medicine* 2 (1996): 271-273.

Brief overview of alkaptonuria, including treatment.

Garrod, Archibald E. "The Incidence of Alkaptonuria: A Study in Chemical Individuality." *Yale Journal of Biology and Medicine* 75 (2002): 221-231. Reprint of the classic Garrod article published in 1902.

Korf, Bruce R. *Human Genetics and Genomics.* 3d ed. Oxford, England: Wiley-Blackwell, 2006. Discussion of human genetics and diseases.

WEB SITES OF INTEREST

AKU Society
http://alkaptonuria.org/en/home.php
A support group for people with alkaptonuria.

Genetics Home Reference
http://ghr.nlm.nih.gov/condition=alkaptonuria
An essay on alkaptonuria.

OMIM
http://www.ncbi.nlm.nih.gov/entrez/dispomim.cgi?id=203500
A database of human genetic disorders.

See also: Hereditary diseases; Hereditary xanthinuria; Homocystinuria; Inborn errors of metabolism; Orotic aciduria; Paroxysmal nocturnal hemoglobinuria.

Allergies

CATEGORY: Diseases and syndromes
ALSO KNOWN AS: Atopy; allergic rhinitis; hay fever; atopic dermatitis; anaphylaxis

DEFINITION

Allergies are a disorder of the immune system. Allergic reactions occur when the immune system responds strongly to normally harmless substances, such as pet dander or pollen, which are referred to as allergens. Individuals can develop an allergy because of a genetic susceptibility inherited from their parents and subsequent exposure to that allergen; however, they also can develop allergies without any genetic risk factors.

RISK FACTORS

Individuals have a higher risk of developing allergies if they have family members with allergies or asthma. They also have a higher risk of developing an allergy if they have asthma or one or more allergies already. Children are more likely to develop allergies than are adults, and allergies are more common in firstborn children and among children in smaller families. Allergies are also more common in urban than in rural environments and more common in developed than in developing countries. Other environmental factors (such as exposure to cigarette smoke and pollution) as well as medical factors (such as infections, autoimmune disease, diet, and stress) can also affect allergy risk.

ETIOLOGY AND GENETICS

Multiple factors modulate risk for allergic diseases, without a single causal agent; however, the most important component influencing whether a person will develop allergies is genetic predisposition. Atopy, characterized by high levels of immunoglobulin E (IgE), is the condition that underlies allergic diseases and is highly influenced by genetics. People who have a genetic predisposition toward developing allergic conditions have ten to twenty times the risk of developing allergies than those who do not. If one or more of a person's parents or siblings have allergies, the risk for developing allergies is 30 to 50 percent or 25 to 35 percent, respectively. Monozygotic twins, who share 100 percent of their DNA, are more likely to have the same type of allergy than are dizygotic twins, who share 50 percent of their DNA, suggesting that genetic factors are important in allergy risk. Even in monozygotic twins, however, only about 50 to 60 percent of twins share the same allergic condition, demonstrating that nongenetic factors also influence allergies. As a result, allergy is considered a complex genetic disease because it does not follow the laws of Mendelian inheritance.

Because multiple allergic conditions exist and allergies are also influenced by exposure to allergens, determining specific genetic risk factors for allergies is challenging. Nevertheless, several candidate susceptibility genes for allergic diseases have been identified. These genes include human leukocyte antigen DRB1 (*HLA-DRB1*), high-affinity IgE receptor (*FCER1B*), interleukins 4 and 13 (*IL4*, *IL13*), and the alpha chain of the IL-4 receptor (*IL4RA*).

Several linkage studies suggest that the major histocompatibility complex class II region (MHC II) influences allergy. This genomic region contains human leukocyte antigen (HLA) genes, which encode

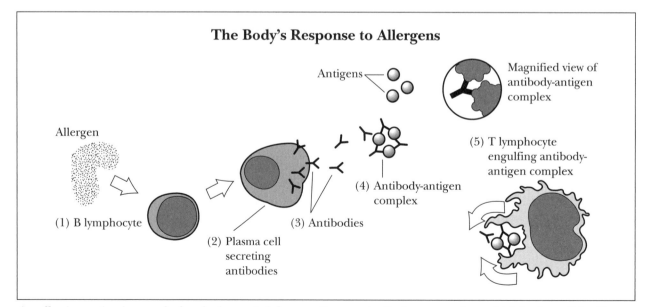

The Body's Response to Allergens

Antigens

Magnified view of antibody-antigen complex

Allergen

(5) T lymphocyte engulfing antibody-antigen complex

(4) Antibody-antigen complex

(1) B lymphocyte

(3) Antibodies

(2) Plasma cell secreting antibodies

An allergic reaction is caused when foreign materials, or antigens, enter the immune system, which produces B lymphocytes (1) that cause blood plasma cells to secrete antibodies (2). The antibodies (3) link with antigens to form antibody-antigen complexes (4), which then are engulfed and destroyed by a T lymphocyte (5). (Hans & Cassidy, Inc.)

antigen-presenting proteins on the cell surface. Genetic variation in HLA genes determines the specificities of HLA proteins and whether the immune system will respond to a particular allergen. Several HLA haplotypes have been associated with specific allergies, such as the reported association between the HLA class *II DRB1*1501* allele and ragweed pollen allergies. Other HLA haplotypes are associated more generally with allergies, such as the association between particular *HLADQB1*03* alleles and higher levels of IgE.

Other candidate genes for allergies include those related to immunoglobulins. Polymorphisms in the *FCER1B* gene that encodes for the beta chain of the high-affinity receptor for IgE affect the extent to which the immune system responds against allergens and have been associated with allergy. Additionally, polymorphisms in genes encoding the *IL13* and *IL4* receptor alpha chain are associated with increased serum IgE levels as well as allergy risk. Allergy risk has also been associated with another interleukin, the *p40* gene, which encodes for one of the two subunits of interleukin 12. Another group of immunoglobulin-related genes, the T cell immunoglobulin and mucin domain (TIM) family genes, have been associated with protection from developing allergies. The *PHF11* gene, which could be involved in immunoglobulin synthesis, is another immunoglobulin-related gene consistently linked to allergy risk.

Other genes associated with allergy in multiple studies include various components of immune response. *CD14* encodes a cell-surface receptor intended to detect bacterial proteins, but variation in this gene is also associated with allergic responses to harmless allergens. Additionally, genes encoding transcription factors involved in the development of development T regulatory cells, such as *GATA3*, which regulates Th2 cytokine responses, and *T-bet*, which regulates Th2 cytokine responses, have also been associated with allergy.

Because both genes and environment factors in combination influence allergy risk, some researchers have investigated gene-gene and gene-environment interactions. For example, individuals who had certain polymorphisms in *CD14* had high or low allergy risk depending on whether they had pets or whether they lived on a farm in childhood. Additionally, interactions between polymorphisms in different genes, such as an interaction between *GATA3* and *IL13*, can affect allergy risk.

SYMPTOMS

Allergy symptoms vary widely. Sneezing, runny nose, and sore throat are common with seasonal allergies, sometimes called hay fever or allergic rhinitis. Allergic reactions can also affect the eyes, leading to redness, watery or itchy eyes, and swelling. Sometimes, allergies may affect the skin, leading to rashes or hives. More severe allergic reactions can lead to anaphylaxis, which may include the symptoms listed above in addition to low blood pressure or shock.

SCREENING AND DIAGNOSIS

A doctor may perform a skin test or blood test to test for allergies. In a skin test, a small drop of the possible allergen is either placed onto skin followed by scratching with a needle over the drop or injected into the skin. With skin tests, if the individual is allergic to a substance, the test site will become red, swollen, and itchy within twenty minutes. Another way to test for allergies involves taking a blood sample. The medical laboratory adds the allergen to the blood and then measures the immune response to the allergen. If the body produces many antibodies to attack the allergen, then the individual is allergic to the tested substance.

TREATMENT AND THERAPY

Several medications are available to relieve allergies. Oral and nasal antihistamines, such as Benadryl and Claritin, help with allergy symptoms by blocking the action of histamine, a substance the body releases during an allergic reaction. Nasal sprays containing corticosteroids, such as Nasonex and Flonase, or nonsteroidal anti-inflammatory drugs (NSAIDs), such as NasalCrom, are sprayed into the nose to reduce inflammation. Decongestants can also be used to alleviate allergy symptoms, sometimes in combination with antihistamines, as in Allegra D. Leukotriene receptor antagonists, such as Singulair, are another treatment that may be used to reduce inflammation-related allergy symptoms.

Immunotherapy, or allergy shots, is another treatment for allergies. People who receive immunotherapy have small amounts of allergens injected into their bodies. The doses of these allergens are increased over at least three to five years in order to develop the body's immunity to them. When the patient experiences minimal symptoms for two seasons or more, the treatment is stopped.

PREVENTION AND OUTCOMES

The simplest way to prevent allergic conditions or to reduce symptoms is to minimize exposure to the problematic allergen. For example, delaying the time at which infants are first exposed to highly allergenic foods such as cow's milk and peanuts may help prevent allergy development. Eating a healthy diet and managing stress effectively can also help prevent and alleviate allergy symptoms.

Jevon Plunkett

FURTHER READING

Contopoulos-Ioannidis, D. G., I. N. Kouri, and J. P. Ioannidis. "Genetic Predisposition to Asthma and Atopy." *Respiration* 74, no. 1 (2007): 8-12.

Grammatikos, A. P. "The Genetic and Environmental Basis of Atopic Diseases." *Annals of Medicine* 40, no. 7 (2008): 482-495.

Thomsen, S. F., K. O. Kyvik, and V. Backer. "Etiological Relationships in Atopy: A Review of Twin Studies." *Twin Research and Human Genetics* 11, no. 2 (2008): 112-120.

Torres-Borrego, J., A. B. Molina-Terán, and C. Montes-Mendoza. "Prevalence and Associated Factors of Allergic Rhinitis and Atopic Dermatitis in Children." *Allergologia et Immunopathologia* 36, no. 2 (2008): 90-100.

WEB SITES OF INTEREST

Allergy and Asthma Foundation of America
http://www.aafa.org

American Academy of Allergy, Asthma, and Immunology
http://www.aaaai.org

National Institute of Allergy and Infectious Diseases (NIAID)
http://www3.niaid.nih.gov

See also: Autoimmune disorders; Immunogenetics.

Alpha-1-antitrypsin deficiency

CATEGORY: Diseases and syndromes

ALSO KNOWN AS: AAT deficiency; AATD; alpha-1; a1-antitrypsin deficiency; A1AD; A-1ATD; anti-elastase; alpha-1 proteinase inhibitor; genetic emphysema

DEFINITION

Alpha-1-antitrypsin deficiency (AATD) is caused when the liver cannot produce sufficient amounts of alpha-1-antitrypsin protein, which inactivates other proteins, including elastase. Elastase protects the lungs, but if it becomes overactive, it destroys lung tissue. People who inherit two copies of the Z form of the *SERPINA1* gene are most affected by this disease.

RISK FACTORS

AATD affects mostly Caucasians of Northern European descent. Men and women are affected in equal numbers; however, males with this disease are more likely to develop liver problems than are females with this disease. Nongenetic factors such as smoking and exposure to lung-harming chemicals or fumes affect the severity of this disease.

ETIOLOGY AND GENETICS

The *SERPINA1* (formerly *PI*) gene that causes this disease is on chromosome 14. This gene has three common forms—M, S, and Z—with M being the normal form. The severity of the disease hinges on which forms of these genes are inherited. These gene forms all have subforms (M1, M2, S1, S2, Z1, Z2, etc.) that also affect the severity of the disease expression. More than seventy forms of this gene have been identified. A common genetic mutation that creates the Z form happens when two amino acids switch places on a chromosome (lysine replaces glutamic acid at position 342 on the *SERPINA1* gene on chromosome 14).

AATD is carried in an autosomal codominant pattern, where a person must inherit an abnormal gene from each parent, each of whom has the disease or is a carrier. The most severe gene form is Z, and if a person inherits two Z forms of this gene, his or her disease may be very severe, where the liver produces only about 15 percent of the normal levels of alpha-1-antitrypsin protein. The gene form S is the next most severe form, and a person with one Z and one S form generally produces about 38 percent of the normal levels of this protein. M is the normal form of the gene, and a person with one Z and one M may produce about 60 percent of normal protein levels, depending on the subform of the gene inherited. These genes are codominant, meaning that they each affect the level of protein; for example, someone with an MZ gene type has a protein level somewhere between a person with an MM gene type and a person with a ZZ gene type.

People who have at least one M (normal) gene type generally produce enough alpha-1-antitrypsin to somewhat protect their lungs and may never exhibit any disease symptoms. These people are at greatest risk from outside factors; for example, someone with an MZ gene type may develop lung disease only if he or she smokes. Even some people who have two abnormal genes do not exhibit symptoms of this disease, depending on the subforms of the genes inherited. The subforms of these genes and how they affect the manner in which disease is expressed is a focus of research in this field.

Symptoms

In adults, this disease is characterized by breathlessness, wheezing, and early and rapid progressive lung disease, particularly in a person who does not smoke. Liver disease, such as jaundice, in children or adults may also be a symptom of AATD.

Screening and Diagnosis

The lung component of AATD is screened and diagnosed with the same tools, such as pulmonary function or lung capacity tests, as are used for other lung diseases. The liver component is also screened and diagnosed with standard liver function tests. The actual diagnosis of AATD is based on a blood test that measures the blood levels of this protein. Genetic testing can determine which gene types one carries, if any, and thus determine carrier status.

Treatment and Therapy

Slowing the progress of lung disease is the first-line treatment. Quitting smoking immediately is essential, and limiting exposure to secondhand tobacco smoke and other lung irritants can also improve outcomes.

Treatment may involve infusions of alpha-1-antitrypsin directly into the bloodstream. This treatment is not effective if lung tissue is extremely damaged, nor does it alleviate any liver problems. Treatment for AATD liver disease is a liver transplant. Lung transplant may also be a treatment option.

Treating respiratory infections quickly and receiving influenza and pneumococcal vaccinations may be helpful in managing lung disease. Avoiding alcohol, minimizing exposure to hepatitis C, and receiving hepatitis A and B vaccinations may help minimize liver problems.

This disease is often misdiagnosed and treated as another lung disease such as asthma or chronic obstructive pulmonary disease (COPD); minimal response to the standard therapies for these diseases may be an indication of AATD.

Prevention and Outcomes

There is no way to prevent this disease, though its severity can be limited by quitting smoking or avoiding irritants. Life expectancy depends on the severity of symptoms. Those who smoke or are exposed to lung irritants have much less successful outcomes than those who do not. The outlook for patients who progress to emphysema or cirrhosis is grim.

Marianne M. Madsen, M.S.

Further Reading

Köhnlein, Thomas, and Tobias Welte. *Alpha-1 Antitrypsin Deficiency: Clinical Aspects and Management.* Bremen, Germany: Uni-Med Verlag, 2007. This title is written to provide clinicians a basic understanding of how to manage patients with this disease.

_____. "Alpha-1 Antitrypsin Deficiency: Pathogenesis, Clinical Presentation, Diagnosis, and Treatment." *JAMA* 121, no. 1 (2008): 3-9. This article defines and describes recent genetic and clinical findings.

Parker, Phillip M. *Alpha-1 Antitrypsin Deficiency: A Bibliography and Dictionary for Physicians, Patients, and Genome Researchers.* San Diego: ICON Group International, 2007. Though geared toward researchers, this book is easy enough to read that patients can also use it. It defines many terms related to alpha-1-antitrypsin deficiency.

Web Sites of Interest

AlphaNet
http://www.alphanet.org

Alpha-1 Association
http://www.alpha1.org

Alpha-1 Foundation
http://www.alphaone.org

Alpha1Health
http://www.alpha1health.com

See also: Asthma; Hereditary diseases.

Alport syndrome

CATEGORY: Diseases and syndromes

ALSO KNOWN AS: AS; hereditary deafness and progressive kidney disease

DEFINITION

Alport syndrome is a genetic condition affecting one of several subunits of type IV collagen proteins. These proteins form a major part of the basement membrane in the kidney (glomeruli), inner ear (cochlea), and eye.

RISK FACTORS

This disease is most common in males and has more severe consequences in men than in women. The biggest risk for developing Alport syndrome is a family history of the disease. Incidence is 1 in 5,000 people in the United States.

ETIOLOGY AND GENETICS

Basement membranes are found in many parts of the body. In Alport syndrome, the effect is most noticeable in the glomeruli that filter blood in the kidney. There is a defect in the production of the alpha chains of type IV collagen that promotes increased splitting and destruction of the glomerular basement membrane. This process produces scarring in the kidney and eventual kidney failure. A normal basement membrane in the cochlea of the ear and the eye are also important for normal function. In this disease, progressive sensorineural hearing loss is usually present by late childhood or early adolescence.

Alport syndrome is associated with three different genetic presentations that control the production of type IV collagen proteins. More than 80 percent of cases are the X-linked form of the disease, which increases the incidence in males. The mutation is in the gene *COL4A5* and results in damage to the alpha-5 chain of type IV collagen.

In the other 20 percent of cases the defect is on chromosome 2 and is either transmitted in a recessive form or a dominant form, with males and females equally affected. The autosomal recessive form is caused by mutations in *COL4A3* or *COL4A4* genes that encode the alpha-3 or alpha-4 subunits of type IV collagen. The autosomal dominant form (less than 5 percent of cases) is caused by mutations

in these same genes but is transmitted in a dominant fashion. In both autosomal forms, kidney disease develops later in adolescence or adulthood. There is some indication that the autosomal dominant form of this disease may not be a true manifestation of Alport syndrome but another condition associated with kidney failure, deafness, and other blood abnormalities.

SYMPTOMS

The primary symptoms in Alport syndrome are associated with kidney disease. Blood in the urine (hematuria) is usually discovered during the first years of life in the X-linked form. Protein in the urine (proteinuria) is usually absent in childhood but eventually develops in the X-linked and autosomal recessive form by the teenage years or early adulthood. In the autosomal dominant form, kidney disease develops later in adulthood. If individuals do not have hematuria during the first decade of life, then they are unlikely to have Alport syndrome.

Hearing loss is not usually present at birth and does not become apparent until late childhood or early adolescence, generally before the onset of kidney failure. There is a wide range of expression with this syndrome. Some families do not have hearing loss. Eye findings include lenticonus, which are abnormal spherical or conelike protrusion on the lens of the eye, yellowish or whitish flecks in the macula of the inner eye, corneal changes, and recurrent corneal erosion.

SCREENING AND DIAGNOSIS

Alport syndrome is one of the diseases suspected when a child or young adult has recurrent microscopic or gross blood in the urine. Unique changes in the walls of the blood vessels of the glomeruli can be detected on a kidney biopsy. In the X-linked form, a skin biopsy will show abnormalities in the alpha-5 chain of type IV collagen that is normally present in the skin. Genetic testing is helpful when the diagnosis is not clear.

TREATMENT AND THERAPY

No specific treatment therapies are available. Early diagnosis can help postpone the complications of kidney disease by promoting good health care and prevention of other causes of kidney damage. Strict control of blood pressure is important. Patients who develop end-stage kidney disease usually require dialysis and/or kidney transplantation. The success rate of kidney transplantation is good, and patients often have excellent allograft survival rates.

PREVENTION AND OUTCOMES

Alport syndrome is a rare disease, and morbidity and mortality are most often related to progressive kidney disease. The prognosis varies depending on the type of inheritance, the sex of the patient, and the type of mutations in type IV collagen genes. Most patients with Alport syndrome develop end-stage kidney disease by forty years of age.

Prenatal testing is available for the X-linked and autosomal recessive forms of the syndrome. Testing for the autosomal dominant form may be available from specialized testing laboratories.

Jane Blood-Siegfried, D.N.Sc.

FURTHER READING

Kliegman, Robert M., Richard E. Behrman, Hal B. Jenson, and Bonita F. Stanton. *Nelson Textbook of Pediatrics.* 18th ed. New York: Saunders, 2007. This is a fairly easy-to-understand textbook on pediatrics with a section on Alport syndrome.

Nussbaum, Robert L., Roderick R. McInnes, and Huntington F. Willard. *Thompson and Thompson Genetics in Medicine.* 7th ed. New York: Saunders, 2007. This is a classic textbook that is very easy to read and understand.

Reilly, Robert, and Mark Perazella. *Nephrology in Thirty Days.* New York: McGraw-Hill, 2005. This book offers a fairly easy-to-read text on kidney disease with a section on Alport syndrome.

WEB SITES OF INTEREST

Alport Syndrome Organization
https://www.alportsyndrome.org

National Kidney Foundation
http://www.kidney.org

OMIM: Alport Syndrome, Autosomal Recessive
http://www.ncbi.nlm.nih.gov/entrez/
dispomim.cgi?id=203780

OMIM: Alport Syndrome, X-Linked
http://www.ncbi.nlm.nih.gov/entrez/
dispomim.cgi?id=301050

See also: Bartter syndrome; Hereditary diseases; Polycystic kidney disease.

Altruism

CATEGORY: Population genetics

SIGNIFICANCE: In a strictly Darwinian system, actions that reduce the success of individual reproduction should be selected against; however, altruism, which occurs at a cost to the altruist, is observed regularly in natural populations. This paradox may be resolved if the cost of altruism is offset by the reproductive success of relatives with which altruists share genes. Kin selection results in selection for altruistic behaviors which, if directed at relatives, preserve inclusive reproductive success, and thus Darwinian fitness.

KEY TERMS

altruism: behavior that benefits others at the evolutionary (reproductive) cost of the altruist

evolution: a change in the frequency of alleles resulting from the differential reproduction of individuals

haplodiploidy: a system of sex determination in which males are haploid (developing from unfertilized eggs) and females are diploid

inclusive fitness: an individual's total genetic contribution to future generations, comprising both direct fitness, which results from individual reproduction, and indirect fitness, which results from the reproduction of close relatives

kin selection: an evolutionary mechanism manifest in selection for behaviors that increase the inclusive fitness of altruists

maternal altruism: altruism on the part of mothers toward offspring as well as between and among members of groups comprising closely related females

natural selection: a process whereby environmental factors influence the survival and reproductive success of individuals; natural selection leads to genetic changes in populations over time

reciprocal altruism: mutual exchange of altruistic acts typically associated with highly cohesive social groups

REPRODUCTIVE SUCCESS = SURVIVAL

If evolutionary outcomes in a Darwinian world are described as natural economies, then individual reproduction is the currency of these economies and of natural selection. Given both naturally occurring genetic variation among individuals and a certain environmental dynamic, it follows that some individuals will be better adapted to locally changing environments than others. Such differential adaptation is expressed as a difference in the frequency with which individual genes pass into future generations. This simple scenario fulfills the genetic definition of evolution—change in allele frequencies in natural populations—by explaining environmental influences on these changes. Note that this argument emphasizes, as its central postulate, the importance of individual reproduction rather than simple survival. Survival of the fittest is therefore more properly viewed as the differential propagation of genes.

A challenge to such a scenario is the paradox of altruism. Altruism is defined as any behavior that benefits another at a cost to the altruist. Charles Darwin himself suggested that this problem was a "special difficulty . . . which at first appeared . . . insuperable, and actually fatal to [the] whole theory" of natural selection. The individual who pushes siblings from the track as he himself is killed by the rushing locomotive is an altruist; the colony sentinel that issues an alarm call to her cohort to take cover, despite the risk of drawing the attention of an approaching predator, is also acting altruistically. These behaviors make no sense in Darwin's econ-

omy, since they appear to decrease the likelihood of individual reproduction—unless, as W. D. Hamilton suggested, Darwinian success is not limited to the success of individual bodies harboring particular genes but may be extended to include the reproductive success of relatives who share genes with the altruist. Hamilton defined inclusive fitness as the sum of an individual's own fitness plus the influence that individual has on the fitness of relatives. Kin selection is the evolutionary mechanism that selects for behaviors that increase the inclusive fitness of altruists. Even though there are potential costs to altruistic behavior, the evolutionary economy of an altruist operates in the black because actors profit (beyond associated costs) by helping others who share their genes. The bottom line is that altruists increase their inclusive fitness through the reproduction of others.

EVIDENCE OF KIN SELECTION

One of the best evidences for kin selection is the social structure of certain groups of insects, including the *Hymenoptera* (ants, bees, and wasps). A unique system of sex determination (haplodiploidy) in which females are diploid and males are haploid predisposes some group members to behave altruistically. In certain bees, for example, the queen is diploid and fertile. Worker bees are female, diploid, and sterile. Drones are male, developed from unfertilized eggs, and haploid. Such a situation makes for unusual patterns of genetic relationship among hive members. In diploid systems the genetic relation between parents and offspring and among offspring is symmetrical. Offspring receive half their genetic complement from their mother and half from the father; sons and daughters are related to each parent by ½ and sibs (siblings) are related to each other by ½. In the haplodiploid system such genetic relationships are asymmetric. Drones are haploid and receive half of the queen's genome. Workers are diploid and share 100 percent of their paternal genes and, on average, half of their maternal genes with their sisters. Sisters are therefore related

The altruistic behaviors of honeybees and some other animal species may be a result of selection for behaviors that place the group, rather than the individual, at a reproductive advantage. (© Lukas Pobuda / Dreamstime.com)

to each other by ¾. Because sisters and their brothers share no paternal genes, and on average half of their maternal genes, they are related to drones by only ¼. In this economy it makes sense that workers should act altruistically to assist the queen in the production of sisters. What would appear to be purely altruistic acts, on the part of workers, result in greater inclusive success than if the workers had reproduced themselves. In contrast, drones contribute little to community welfare and serve only to fertilize the queen. Note that in this system there is no conscious decision on the part of workers not to reproduce; their sterility is an inherent part of this unusual system of sex determination.

A Test of Predictions

One prediction made by the kind of kin selection described above is that, assuming the queen produces male and female offspring in equal proportion, female workers should invest three times the energy in caring for sisters than they do for brothers. Because queens are related to both male and female offspring equally, one would predict that eggs are equally divided between the sexes. Because workers are related to their sisters by ¾ and to their brothers by ¼, one would predict that they should invest three times the energy in care of eggs eventually yielding sisters than they do in the care of eggs eventually yielding brothers. Remarkably, it has been shown that certain worker ants are able to identify and then selectively care for eggs containing sisters. Kin recognition has also been studied in the house mouse, *Mus musculus domesticus*, and in some cases individuals can distinguish full sibs from half sibs on the basis of their major histocompatibility complexes (glycoproteins important in immune system function). The specific MHC type is fairly unique for each mouse, but related individuals will have similar patterns and share some specific MHC glycoproteins. MHC glycoproteins are found in mouse urine, and individuals can distinguish these molecules by smell. Consistent with the foregoing hypothesis, the degree of female altruism toward the offspring of close relatives was predicted by the degree of relation based on MHC type and type recognition.

Maternal Altruism

Altruism may be observed in a variety of natural systems in which groups comprise individuals who share a high degree of genetic relatedness. A classic example of this sort occurs with Belding's ground squirrels. Males tend to disperse from colonies, while females remain to create highly related maternal groups. Members of such maternal groups demonstrate altruistic behaviors such as alarm calling to warn relatives of danger. Although truly altruistic in the sense that alarm callers may incur risk of personal injury or death, they can be reasonably assured of breaking even in this economy as long as their genes live on in the bodies of those they have saved by their actions.

Reciprocal Altruism

It would seem that altruism based on Hamilton's argument of inclusive fitness would be precluded by human social organization. Scientists have predicted, however, that reciprocal altruism should exist in systems characterized by a high frequency of interaction among member individuals and life spans long enough to allow the recipients of altruistic acts to repay altruists. Note that the theoretical basis for the existence of reciprocal altruism differs from that for kin selection, and that any system in which evidence for reciprocity is found must necessarily include the development of a complex web of sophisticated social interaction. Such systems would be expected to foster traits expressing the panoply of human emotion and the development of certain moral architectures and group cohesion.

David A. Smith, Ph.D.

Further Reading

Dugatkin, Lee Alan. *The Altruism Equation: Seven Scientists Search for the Origins of Goodness.* Princeton, N.J.: Princeton University Press, 2006. Traces the paradox of altruism's role in a world supposedly ruled by survival of the fittest from the initial theories of Darwin to those of Hamilton and subsequent thinkers.

Freeman, Scott, and Jon C. Herron. "Kin Selection and Social Behavior." In *Evolutionary Analysis.* Upper Saddle River, N.J.: Prentice Hall, 2001. A well-written and logical analysis of altruistic behavior. Arguments are supported with data and analysis from the primary literature.

Gould, Stephen Jay. "So Cleverly Kind an Animal." In *Ever Since Darwin.* New York: W. W. Norton, 1973. An elegantly expressed description of altruism and haplodiploidy in social insects.

Hrdy, Sarah Blaffer. *Mothers and Others: The Evolutionary Origins of Mutual Understanding.* Cambridge, Mass.: Belknap Press of Harvard University Press, 2009. Hrdy, a sociobiologist, argues that human cooperation is not rooted in making war but in making babies and caring for children.

Keltner, Dacher. *Born to Be Good: The Science of a Meaningful Life.* New York: W. W. Norton, 2009. Examines human emotions within the context of evolution, arguing that positive emotions, such as compassion and gratitude, are at the core of human nature.

Keltner, Dacher, Jason Marsh, and Jeremy Adam Smith, eds. *The Compassionate Instinct: The Science of Human Goodness.* New York: W. W. Norton, 2010. Collection of essays by scientists and science writers who examine human goodness. These essays originally were published in *Greater Good*, the journal of the Great Good Science Center at the University of California at Berkeley.

Volpe, E. Peter, and Peter A. Rosenbaum. "Natural Selection and Social Behavior." In *Understanding Evolution.* 6th ed. Boston: McGraw-Hill, 2000. Solid analysis of the theoretical basis for kin selection, including consideration of genetic asymmetries associated with haplodiploidy.

WEB SITES OF INTEREST

Greater Good Science Center
http://peacecenter.berkeley.edu
In the words of its Web site, the research center is dedicated to the "scientific understanding of happy and compassionate individuals, strong social bonds, and altruistic behavior."

Stanford Encyclopedia of Philosophy
http://plato.stanford.edu
This site's article "Biological Altruism" provides a solid overview of the topic, including discussion of altruism and level of selection, kin selection, and reciprocal altruism.

See also: Behavior; Evolutionary biology; Homosexuality; Natural selection; Population genetics; Sociobiology.

Alzheimer's disease

CATEGORY: Diseases and syndromes

DEFINITION

Alzheimer's disease is a progressive neurodegenerative disorder that causes a gradual, irreversible loss of memory, language, visual-spatial perceptions, and judgment. Approximately 5.3 million Americans have the disease, a number that is expected to increase to between 11 and 16 million by 2050 if means of preventing or effectively treating it are not discovered. The annual costs of caring for persons with Alzheimer's disease and costs to businesses for lost productivity from caregivers totaled $148 billion in 2005. The number of persons with Alzheimer's disease and the associated economic burden is expected to rise dramatically as the baby boomer generation ages.

RISK FACTORS

Individual features and environmental influences may either cause a disease or increase the risk of developing that disease. Aging is a well-established risk factor for Alzheimer's disease, and the rate of Alzheimer's disease doubles every five years after age sixty-five. According to data collected in 2008, 2.4 million American women over the age of seventy-one had Alzheimer's disease or another form of dementia, compared with 1 million American men in the same age group.

Alzheimer's disease is a complex disorder, and many scientists believe that a combination of variations in some genes, possibly acting in conjunction with external factors, may increase the risk. Several genetic factors are known to cause Alzheimer's disease, but they are extremely rare and account for a very small minority of cases. Persons with early-onset Alzheimer's disease (which develops before the age of sixty) who also have multiple family members with Alzheimer's disease from at least three generations are considered to have familial early-onset Alzheimer's disease, which is very rare (less than 2 percent of all persons with the disease). Familial early-onset Alzheimer's disease is caused by mutations in the gene for amyloid precursor protein (APP) gene and the presenilin 1 (*PSEN1*) and presenilin 2 (*PSEN2*) genes. Down syndrome (Trisomy 21) is another genetically determined cause of

Alzheimer's disease. Persons with Down syndrome have evidence of amyloid pathology at an early age, and those who live to their forties will develop Alzheimer's disease.

Risk factors for the much more common late-onset Alzheimer's disease (occurring after age sixty-five) are less clear-cut. Some scientists hypothesize that late-onset Alzheimer's disease is caused by amyloid plaque accumulation in the brain or by enhanced degradation of the tau protein leading to development of neurofibrillary tangles. To date, no single factor has been identified that definitely causes late-onset Alzheimer's disease. The risk of this form is increased (but not caused) by the presence of susceptibility genes. Persons without susceptibility genes can develop Alzheimer's disease, however, just as individuals who carry a susceptibility gene may never develop the disease. The most thoroughly studied susceptibility gene for Alzheimer's disease is apolipioprotein E (*APOE*) ε4. Individuals who carry one or more *APOE ε4* alleles are at increased risk compared with noncarriers. Environmental factors that may increase the chance of Alzheimer's disease include diabetes mellitus, hypercholesterolemia, hypertension, depression, traumatic brain injury, and a lower level of education.

Etiology and Genetics

The hallmark lesions in the brains of persons with Alzheimer's disease are extracellular amyloid plaques and intraneuronal neurofibrillary tangles. Accumulation of the amyloid-beta (A-beta) peptide in the brain triggers a series of events that culminate in the development of Alzheimer's disease. The A-beta peptide is a sticky substance that forms clumps (or aggregates) called amyloid plaques that surround nerve cells. Amyloid plaques concentrate in the hippocampus and other brain regions that control memory and cognition. A-beta is produced by a series of steps that convert APP to neurotoxic forms of A-beta. APP is broken down by the beta-secretase and gamma-secretase enzymes, which results in the formation of toxic A-beta peptides that aggregate and form plaque. Another enzyme, alpha-secretase, is believed to protect against A-beta production. Accumulation of A-beta, either by overproduction or

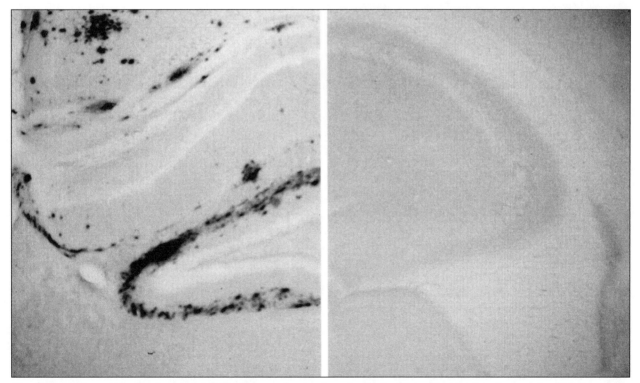

Two images of mouse-brain tissue, one (left) engineered to produce the dark protein deposits that characterize Alzheimer's disease, and the other normal. (AP/Wide World Photos)

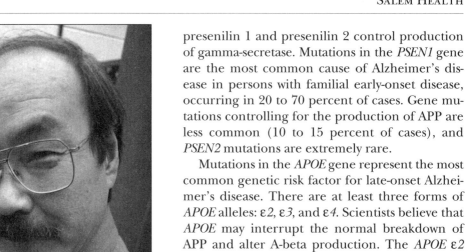

Etsuro Uemura, a professor of veterinary science who has been working on Alzheimer's disease since the early 1970's, before the disease was named, displays tissue cultures from rat brains that he has been using in his research. (AP/Wide World Photos)

reduced clearance from the brain or both, leads to inflammation, poorly functioning nerve cells, hyperphosphorylation of tau proteins (causing formation of neurofibrillary tangles and destruction of nerve cell structure). Together, these pathological events result in death of nerve cells in the brain, decreased function of the chemical messengers in the brain (neurotransmitters), loss of gray matter, dementia, and death.

The rare, familial early-onset Alzheimer's disease is caused by mutations in the *APP*, *PSEN1*, and *PSEN2* genes. Genes that encode for APP regulate the production of beta-secretase, and genes encoding for presenilin 1 and presenilin 2 control production of gamma-secretase. Mutations in the *PSEN1* gene are the most common cause of Alzheimer's disease in persons with familial early-onset disease, occurring in 20 to 70 percent of cases. Gene mutations controlling for the production of APP are less common (10 to 15 percent of cases), and *PSEN2* mutations are extremely rare.

Mutations in the *APOE* gene represent the most common genetic risk factor for late-onset Alzheimer's disease. There are at least three forms of *APOE* alleles: ε2, ε3, and ε4. Scientists believe that *APOE* may interrupt the normal breakdown of APP and alter A-beta production. The *APOE* ε2 allele, which is rare and develops later in life, may protect individuals against Alzheimer's disease. The *APOE* ε3 allele is believed to play a neutral role in Alzheimer's disease risk. The *APOE* ε4 allele occurs in approximately 25 percent of all individuals. Persons who carry the *APOE* ε4 allele have a two- to fourfold increased risk of developing Alzheimer's disease compared with noncarriers. Individuals with two copies of the *APOE* ε4 allele are more than ten times more likely to develop Alzheimer's disease than noncarriers. *SORL1* and GRB-associated binding protein 2 (*GAB2*) are other susceptibility genes that may increase the risk of developing Alzheimer's disease. The *SORL1* gene found on chromosome 11 encodes for normal APP breakdown and generation of A-beta. Single nucleotide polymorphisms (SNPs) that have been identified on the *SORL1* gene are related to Alzheimer's disease, but the risk relationship is not yet completely understood. *GAB2* is a susceptibility gene that increases the risk of Alzheimer's disease for individuals who also carry the *APOE* gene.

The National Institute on Aging (NIA) and other branches of the National Institutes of Health (NIH) sponsor studies designed to identify genes associated with late-onset Alzheimer's disease.

SYMPTOMS

Early symptoms of Alzheimer's disease are difficult to discern. Mild cognitive impairment, which is characterized by subtle memory loss, impaired language skills, or other minor deficits in mental function that are noticed by friends and family but are not so severe as to impair the individual's ability to perform basic activities of daily living, is believed to precede Alzheimer's disease. In persons who will eventually develop the disease, the symptoms of mild cognitive impairment become progressively more severe and are eventually manifest as loss of higher brain activities, inability to speak, failure to understand the spoken and written word, and inability to perform even the simplest activities of daily living. Agitation, aggression, nighttime wandering (sundowning), and other behavioral disturbances often accompany advanced Alzheimer's disease and may result in institutionalization. While survival following a diagnosis of Alzheimer's disease has been reported to range from four to six years, patients may live for as long as twenty years after being diagnosed.

SCREENING AND DIAGNOSIS

In 1906, a German physician, Dr. Alois Alzheimer, reported on the rapid mental decline and death of his patient, Frau August D, and at autopsy described plaques and neurofibrillary tangles in her brain. Today, observation of plaques and tangles at autopsy remains the only way to definitively diagnose Alzheimer's disease. A presumptive diagnosis of Alzheimer's disease is based on clinical observation of symptoms and progressive deterioration.

Early diagnosis of Alzheimer's disease is essential to ensure that treatable causes of memory loss and cognitive impairment and other diseases, such as depression, drug interactions, nutritional deficiencies, or endocrine disorders are ruled out. The Risk Evaluation and Education for Alzheimer's Disease (REVEAL) multiphase study sponsored by the National Human Genome Research Institute (NHGRI) is investigating the impact of genetic testing and disclosure of *APOE* gene status to the adult children and siblings of persons with Alzheimer's disease. Findings from the REVEAL study will help inform patients' and clinicians' decisions about genetic counseling and guide actions taken after learning about this status.

TREATMENT AND THERAPY

Despite a huge research effort, disease-modifying therapies that prevent, slow, or halt disease progression have yet to be identified. Currently, cholinesterase inhibitors and an N-methyl-D-aspartate (NMDA) receptor antagonist are in widespread use. These drugs slow the rate of symptom development for some patients, but they do not significantly impact disease progression. Other drugs such as nonsteroidal anti-inflammatory drugs (NSAIDs), cholesterol-lowering statins, ginko biloba, estrogen, and vitamin E have not been shown to be effective. Treatments that interfere with tau pathology, prevent clumping of A-beta, or boost the immune response to A-beta are being actively studied in clinical trials.

The NIA Alzheimer's Disease Prevention Initiative seeks to accelerate the rate of new drug discovery and development. The Alzheimer's Association is a national organization that provides patient advocacy and funding for research of potential new therapies.

PREVENTION AND OUTCOMES

Genetic counseling is useful for those rare persons with familial early-onset Alzheimer's disease, but there is no consensus among experts about the benefits of testing other individuals. Issues of patient confidentiality remain germane to genetic testing for Alzheimer's disease. Confidentiality may be compromised and patients may face employment, insurance, or healthcare discrimination if genetic testing becomes part of the medical record.

Sharon Wallace Stark, R.N., A.P.R.N., D.N.Sc.;
updated by Sally K. Laden, M.S.

FURTHER READING

Bird, T. "Genetic Aspects of Alzheimer Disease." *Journal of Medical Genetics* 10 (2008): 231-239. This state-of-the-art paper reviews the genetic factors associated with Alzheimer's disease.

Farrer, L., L. Cupples, J. Haines, et al. "Effects of Age, Gender, and Ethnicity on the Association Between Apolipoprotein E Genotype and Alzheimer Disease: A Meta-analysis." *JAMA* 278 (1997): 1349-1356. Designed for medical professionals, this paper discusses the most likely cause of the most common form of the disease, late-onset Alzheimer's disease.

Food and Drug Administration. "Head Injury Linked to Increased Risk of Alzheimer's Disease." *FDA Consumer,* January/February, 2001, 8. Discusses research that focuses on the link between head injuries and dementias, including Alzheimer's disease.

Gauthier, S., ed. *Clinical Diagnosis and Management of Alzheimer's Disease.* 2d ed. London: Martin Dunitz, 2001. A collection of discussions concerning symptoms, genetics, diagnosis, and treatment of Alzheimer's disease.

Green, R. "Implications of Amyloid Precursor Protein and Subsequent Beta-Amyloid Production to the Pharmacotherapy of Alzheimer's Disease." *Pharmacotherapy* 22 (2002): 1547-1563. Identifies causes, genetic risks, diagnosis, and treatment of Alzheimer's disease.

Hamdy, Ronald, James Turnball, and Joellyn Edwards. *Alzheimer's Disease: A Handbook for Caregivers.* New York: Mosby, 1998. Causes, symptoms, stages, and treatment options for Alzheimer's disease are discussed.

Leon, J., C. Cheng, and P. Neumann. "Alzheimer's Disease Care: Costs and Potential Savings." *Health Affiliates,* November/December, 1998, 206-216. Identifies the economic impact for caring for and treating those with Alzheimer's disease and reasons for identifying a cure.

Mace, M., and P. Rabins. *The Thirty-six-Hour Day: A Family Guide to Caring for Persons with Alzheimer Disease, Related Dementing Illnesses, and Memory Loss in Later Life.* Baltimore: Johns Hopkins University Press, 1999. Discusses what dementia is, physical and psychological effects on caregivers, financial and legal issues, and long-range care planning for Alzheimer's patients.

Powell, L., and K. Courtice. *Alzheimer's Disease: A Guide for Families and Caregivers.* Cambridge, Mass.: Perseus, 2001. Provides information about early signs, tests, diagnosis, and treatment for Alzheimer's disease.

St. George-Hyslop, Peter H. "Piecing Together Alzheimer's." *Scientific American,* December, 2000, 76-83. Good description of Alzheimer's disease, including symptoms, support, and ongoing research in the quest for a cure.

Terry, R., R. Katzman, K. Bick, and S. Sisodia. *Alzheimer Disease.* 2d ed. Philadelphia: Lippincott Williams & Wilkins, 1999. An in-depth review of hereditary links, signs and symptoms, diagnosis, and treatment for Alzheimer's disease.

Waring, S., and R. Rosenberg. "Genome-Wide Association Studies in Alzheimer Disease." *Archives of Neurology* 65 (2008): 329-334. This technical review article covers the various susceptibility genes being studied for late-onset Alzheimer's disease and discusses the use of genome-wide association studies in the field of Alzheimer's disease research.

WEB SITES OF INTEREST

Alzheimer's Association
http://www.alz.org
This site provides a wealth of information for patients and their families as well as for healthcare professionals. Useful publications are available at no charge, including the "2009 Alzheimer's Disease Facts and Figures," other informative patient publications, and information about the Alzheimer's Disease Genetics Initiative, a study conducted by the Alzheimer's Association and the National Institute of Aging.

Alzheimer's Disease Education and Referral Center, National Institutes of Health
http://www.alzheimers.org
A good general starting place for information and links to standard resources. Includes the "Alzheimer's Disease Genetics" fact sheet that covers the genetic influences in Alzheimer's disease in a user-friendly format.

Alzheimer Research Forum
http://www.alzforum.org
This site is a comprehensive source of research updates and clinical trials in Alzheimer's disease. The AlzGene section of this website provides an updated, searchable database for genetic studies.

ClinicalTrials.gov
http://clinicaltrials.gov
A registry of clinical studies on the genetics of Alzheimer's disease as well as studies of new treatments. This site is easily searched for specific topics related to the disease.

Dolan DNA Learning Center, Your Genes Your Health
http://www.ygyh.org
Sponsored by the Cold Spring Harbor Laboratory, this site, a component of the DNA Interactive Web site, offers information on more than a dozen inherited diseases and syndromes, including Alzheimer's disease.

See also: Aging; Alcoholism; Behavior; Biological clocks; Cancer; Diabetes; Down syndrome; Genetic testing: Ethical and economic issues; Heart disease; Heriditary diseases; Hypercholesterolemia; Insurance; Prion diseases: Kuru and Creutzfeldt-Jakob syndrome; Proteomics; Stem cells; Telomeres.

Amniocentesis

CATEGORY: Techniques and methodologies

SIGNIFICANCE: Amniocentesis is the needle aspiration (withdrawal) of fluid from the amniotic sac (the fluid-filled sac surrounding a fetus developing in the uterus). Fetal cells in the fluid are then analyzed for chromosomal abnormalities such as Down syndrome (extra chromosome 21) or trisomy 18 (extra chromosome 18). Other analyses of the amniotic sac can be performed, depending on the clinical situation.

KEY TERMS

chromosome: an organized structure of DNA that contains genetic coding

deoxyribonucleic acid (DNA): a nucleic acid that contains genetic instructions

Down syndrome: a genetic disorder characterized by mental retardation and physical abnormalities

genetic counseling: parental counseling to explain amniocentesis results

prenatal diagnosis: the diagnosis of a genetic abnormality before birth

triple test: a blood test that screens for genetic defects

trisomy: an extra chromosome (such as trisomy 21)

TECHNIQUE

Amniocentesis is most commonly done between the fourteenth and sixteenth week of pregnancy. Ultrasound is used to determine a safe location for insertion of a needle through the mother's abdomen and into the amniotic sac. Before insertion of the needle, a disinfectant such as Betadine solution is applied to the skin and a local anesthetic is injected. Approximately 20 milliliters of amniotic fluid are aspirated for testing. The fluid is sent to a genetics laboratory, where the fetal cells are separated from it. After growing the cells in a culture medium, the chromosomes are extracted and microscopically examined for abnormalities. The process usually takes from six to ten days; if faster results are needed, a fluorescence in situ hybridization (FISH) analysis, which takes one to two days, can be done. The most common abnormalities detected involve extra or missing chromosomal material: trisomy 21 (Down syndrome), trisomy 18 (Edwards syndrome), and single X chromosome (Turner syndrome). If an abnormal gene is suspected, then gene sequencing can be done. The levels of substances in the amniotic fluid are often measured. Some genetic disorders can be detected by abnormal levels of certain substances in the amniotic fluid. Commonly measured substances include alpha-fetoprotein, unconjugated estriol, and human chorionic gonadotrophin. Abnormal levels of these substances can suggest the possibility of a genetic abnormality. Other substances can be measured in cases where a specific type of abnormality is suggested. The ultrasound examination conducted during the amniocentesis also checks for fetal anomalies.

A common indication (medical reason) for amniocentesis is advanced maternal age (thirty-five or older). Other indications include a fetal abnormality found on ultrasound, polyhydramnios (excess amniotic fluid), intrauterine growth retardation, an abnormal triple screen test, a previous child with a chromosome abnormality, and family history of a genetic disorder.

RELATED PROCEDURES

Chorionic villus sampling involves removing a small sample of the chorionic villus (placental tissue) and analyzing the sample for genetic abnormalities. It can be performed earlier in pregnancy than amniocentesis; however, it carries a slightly higher risk of miscarriage.

A common screening test is the triple test, which involves drawing a small sample of the mother's blood; it is offered to all women less than thirty-five years of age. The levels of three different substances are measured: alpha-fetoprotein, unconjugated estriol, and human chorionic gonadotrophin. Alpha-fetoprotein is a protein produced by the developing fetus; it is increased in defects in the spinal cord (such as spina bifida) and abdominal wall (omphalocele); it is decreased in Down syndrome. Unconjugated estriol is produced by the placenta; its level is decreased in Down syndrome and trisomy 18. Human chorionic gonadoptrophin is produced by the placenta; its level is low in trisomy 18 and high in trisomy 21. Another screening test is the quad test, which includes measurement of the substances included in the triple test as well as a fourth hormone produced by the placenta, inhibin-A. Measurement of this hormone increases the detection rate of Down syndrome.

A survey conducted by the Society for Maternal-Fetal Medicine compared Down syndrome screening between 2001 and 2007. The study found that Down syndrome screening evolved during those years with an increasing emphasis on the first trimester (first three months of pregnancy). The authors noted that with the increase in earlier procedures, the number of invasive procedures (amniocentesis and chorionic villus sampling) has declined.

RISKS AND COMPLICATIONS

Possible complications include infection of the amniotic sac from the needle and failure of the puncture to heal properly, which can result in leakage or infection. Serious complications can result in miscarriage. Other possible complications include preterm labor and delivery, respiratory distress, fetal injury, and Rh disease. The Department of Fetal Medicine, Copenhagen University Hospital, conducted a national registry study to determine the fetal loss rate after chorionic villus sampling and amniocentesis. The study group was comprised of 32,852 women who underwent amniocentesis and 31,355 women who underwent chorionic villus sampling between 1996 and 2006. The miscarriage rates were 1.4 percent after amniocentesis and 1.9 percent after chorionic villus sampling. The miscar-

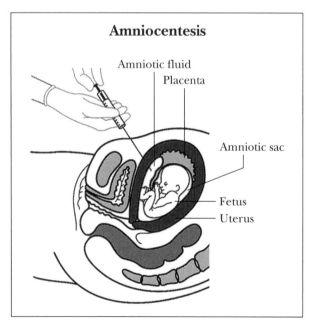

Amniocentesis

Amniotic fluid
Placenta
Amniotic sac
Fetus
Uterus

Removal and analysis of fluid from the amniotic sac that surrounds a fetus during gestation can be used to rule out or confirm the presence of serious birth defects or genetic diseases. (Hans & Cassidy, Inc.)

riage rate was significantly higher for both procedures in departments that performed less than 500 procedures during the study period than it was in those that performed more than 1,500 during that period.

OTHER REASONS FOR AMNIOCENTESIS

Preterm delivery is delivery before thirty-seven weeks of pregnancy. The risks for respiratory problems, neurologic problems, and fetal death increase in relation to the degree of prematurity. Furthermore, the cost of neonatal (newborn) care can be high because of the special medical care required. Currently, researchers are evaluating the measurement of substances in amniotic fluid that can signal an increased risk of preterm delivery. These substances include low urocortin levels, high human beta defensins 2 (HBD2) and 3 (HBD3) levels, and increased thrombin generation. Urocortin, HBD2, and HBD3 are peptides (short amino acid chains). Thrombins are substances involved in blood clotting.

Stem cells are undifferentiated cells that have the potential of developing into a wide variety of structures and organs. Embryonic stem cells have received much scientific and media attention recently. Critics of the use of embryonic cells claim that obtaining them results in the destruction of an embryo. Another source of stem cells is amniotic fluid; this source does not have the disadvantage associated with embryonic stem cells.

IMPACT

Amniocentesis can detect a wide range of genetic disorders, ranging from mild to severe. For example, Down syndrome is characterized by mental retardation and physical deformities such as an enlarged tongue, poor muscle tone, and cardiac abnormalities. Trisomy 18 is characterized by profound physical deformities and mental retardation; about 95 percent of affected fetuses die in the uterus before birth and those that live rarely survive beyond infancy. Amniocentesis can provide invaluable information for both parents and health care professionals. Genetic counseling can give the parents a clear understanding of the situation. In some cases, particularly with severely affected fetuses, an abortion can be performed. In cases in which the parents deem an abortion to be unacceptable, counseling can assist them in preparation for dealing with a child with special needs. Preparation for a problem is always preferable to the sudden impact on parents of being told that their newborn child has a major genetic disorder. Unfortunately, the information can be misused, as some parents have opted for abortion of a normal child if it is not of the desired gender.

Robin L. Wulffson, M.D., FACOG

FURTHER READING

Cunningham, F. *Williams Obstetrics.* 22d ed. New York: McGraw-Hill Professional, 2005. A comprehensive textbook for medical professionals.

Lewis, Ricki. *Human Genetics.* 8th ed. New York: McGraw-Hill, 2007. A basic human genetics reference text written by a practicing genetic counselor.

Rapp, Rayna. *Testing Women, Testing the Fetus: The Social Impact of Amniocentesis in America.* New York: Routledge, 2000. Examines the social impact and cultural meaning of currently available prenatal tests.

Scriver, Charles. *The Metabolic and Molecular Bases of Inherited Disease.* 4 vols. 8th ed. New York: McGraw-Hill Professional, 2007. A comprehensive reference indispensible to those in the field, as well as to a much broader audience.

WEB SITES OF INTEREST

American Pregnancy Association
www.americanpregnancy.org

Amniocentesis Report
www.amniocentesis.org

March of Dimes
www.marchof dimes.com

See also: Bioethics; Chorionic villus sampling; Down syndrome; Fluorescence in situ hybridization (FISH); Genetic screening; Genetic testing; Genetic testing: Ethical and economic issues; Prenatal diagnosis.

Amyotrophic lateral sclerosis

CATEGORY: Diseases and syndromes
ALSO KNOWN AS: ALS; Lou Gehrig's disease; motor neuron disease

DEFINITION

Amyotrophic lateral sclerosis (ALS) is a progressive nervous system disorder that gradually destroys the nerves responsible for muscle movement. Over time, ALS leads to total paralysis of muscle movement, including respiration.

Prognosis is poor in most cases because of the progressive nature of the condition due to eventual respiratory failure. After patients are diagnosed with the disease, their life span ranges from two to five years. The five-year survival rate is 25 percent, while 10 percent of patients will survive more than ten years. In general, the younger the age of onset, the slower the disease progresses.

RISK FACTORS

Individuals who have a family member with ALS are at risk for the disease. Military personnel may have an increased risk, and persons with certain genetic mutations are also at risk.

ETIOLOGY AND GENETICS

While 90-95 percent of new cases of ALS are sporadic, with no known genetic basis, the remaining 5 to 10 percent of cases are termed "familial" because their inheritance pattern has been identified. Among the several types of familial ALS, the most common type results from a mutation in the *SOD1* gene, found on the long arm of chromosome 21 at position 21q22.1. This mutation appears to be inherited in an autosomal dominant fashion, meaning that a single copy of the mutation is sufficient to cause full expression of the disease. An affected individual has a 50 percent chance of transmitting the mutation to each of his or her children.

Mutations in at least nine other genes on several different chromosomes have been associated with rare forms of familial autosomal dominant ALS in other families. In addition, mutations in three other genes appear to cause an autosomal recessive form of familial ALS, and at least one gene on the X chromosome is also a candidate for involvement.

The *SOD1* gene encodes the protein Cu/Zn superoxide dismutase, which is an enzyme responsible for breaking down superoxides and other toxic free radicals. These highly reactive molecules accumulate in cells as a result of normal cellular metabolism, and if they are not destroyed they can lead to permanent damage of the deoxyribonucleic acid (DNA). Interestingly, research suggests that motor neuron death does not result from dysfunctional dismutase activity in ALS patients, which suggests that the dominant mutation results in a toxic gain of function for the enzyme. An excellent mouse model system for the study of *SOD1* gene activity has been developed, so prospects for a greater understanding of the underlying molecular mechanisms remain high.

SYMPTOMS

Symptoms of ALS include progressive weakness in arms and legs (at first often on only one side) over weeks to months without changes in sensory abilities. The initial presentation of the disease may be a wrist or foot drop. Other symptoms include trouble holding things without dropping them, frequent tripping while walking, shrunken muscles, twitchy muscles, unpredictably changing emotions, clumsiness, overactive reflexes, slurred speech, and hoarseness. Trouble chewing and swallowing, resulting in frequent choking and gagging; weight loss due to trouble eating; trouble breathing; and excess salivation and drooling are also symptoms. An individual's cognition and sensation are intact. Some individuals also have trouble coughing, resulting in the development of pneumonia.

SCREENING AND DIAGNOSIS

The doctor will ask about symptoms and medical history and will perform a physical exam. There are no tests that definitively diagnose ALS, but tests may be used to rule out other medical conditions.

Tests may include an electromyogram (EMG) to look for progressive muscle weakness and twitching; a computed tomography (CT) scan, a type of X ray that uses a computer to make pictures of the structures inside the head; and a magnetic resonance imaging (MRI) scan, a test that uses magnetic waves to make pictures of the structures inside the head. Blood tests can rule out metabolic, heavy metal exposure, or rarely infections such as Lyme disease or human immunodeficiency virus (HIV). A lumbar puncture—a procedure to collect cerebrospinal fluid (CSF)—and urine tests may also be conducted.

TREATMENT AND THERAPY

There is currently no cure for ALS. A multidisciplinary approach may work best for patients and their families. This approach may include taking medications, working with therapists and joining a support group, and participating in religious and social activities.

Treatment options include medications. The drug riluzole has been approved for ALS, with a clinical trial revealing a modest lengthening of survival. The drug may slightly improve functioning, but it does not stop the disease from progressing. A study, however, showed that the addition of lithium carbonate (a medication used to treat mood disorders) to riluzole may slow the progression of ALS and prolong survival. Other drugs are also being studied.

A doctor may prescribe medication for symptoms. Diazepam (Valium), baclofen (Lioresal), or dantrolene are used to reduce spasticity; nonsteroidal anti-inflammatory drugs (NSAIDs) and other pain medications may also be prescribed. Atropine (AtroPen), scopolamine (Isopto), or antihistamine are used to reduce heavy drooling. Antidepressants and antianxiety medications may also be used. Physical therapy is used to reduce pain associated with muscle cramping and spasticity.

In some cases, patients may need to receive a mixture of air and oxygen from a machine. If the patients cannot move enough air in and out of their lungs, they may need surgery to have a tube inserted into their airways.

The doctor may make changes in a patient's diet. In some cases, getting nutrition through tube-feeding is needed.

Speech therapy may be used to optimize communication. Therapy can include exploring alternative methods of communication.

PREVENTION AND OUTCOMES

There are no guidelines for preventing ALS because the cause is unknown.

Rosalyn Carson-DeWitt, M.D.;
reviewed by J. Thomas Megerian, M.D., Ph.D., F.A.A.P.
"Etiology and Genetics" by Jeffrey A. Knight, Ph.D.

FURTHER READING

Bradley, Walter G., et al., eds. *Neurology in Clinical Practice.* 5th ed. 2 vols. Philadelphia: Butterworth-Heinemann/Elsevier, 2008.

EBSCO Publishing. *DynaMed: Amyotrophic Lateral Sclerosis.* Ipswich, Mass.: Author, 2009. Available through http://www.ebscohost.com/dynamed.

_____. *Health Library: Amyotrophic Lateral Sclerosis.* Ipswich, Mass.: Author, 2009. Available through http://www.ebscohost.com.

Fauci, Anthony S., et al., eds. *Harrison's Principles of Internal Medicine.* 17th ed. New York: McGraw-Hill Medical, 2008.

Fornai, F., et al. "Lithium Delays Progression of Amyotrophic Lateral Sclerosis." *Proceedings of the National Academy of Sciences of the United States of America* 105, no. 6 (February 12, 2008): 2052-2057. Available through *EBSCO DynaMed Systematic Literature Surveillance* at http://www.ebscohost.com/dynamed.

Goetz, Christopher G., ed. *Textbook of Clinical Neurology.* 3d ed. Philadelphia: Saunders Elsevier, 2007.

Miller, R. G., et al. "Riluzole for Amyotrophic Lateral Sclerosis (ALS)/Motor Neuron Disease (MND)." *Amyotrophic Lateral Sclerosis and Other Motor Neuron Disorders* 4, no. 3 (September, 2003): 191-206.

Samuels, Martin A., and Steven K. Feske, eds. *Office Practice of Neurology.* 2d ed. Philadelphia: Churchill Livingstone, 2003.

Walling, A. D. "Amyotrophic Lateral Sclerosis: Lou Gehrig's Disease." *American Family Physician* 59, no. 6 (March 15, 1999): 1489-1496.

WEB SITES OF INTEREST

ALS Association
http://www.alsa.org

ALS Society of British Columbia
http://www.alsbc.ca

ALS Society of Canada
http://www.als.ca

Genetics Home Reference
http://ghr.nlm.nih.gov

National Institute of Neurological Disorders and Stroke
http://www.ninds.nih.gov

See also: Arnold-Chiari syndrome; Ataxia telangiectasia; Parkinson disease; Vanishing white matter disease.

Ancient DNA

CATEGORY: Evolutionary biology; Molecular genetics
SIGNIFICANCE: In 2009, scientists from the Max Planck Institute for Evolutionary Anthropology announced during the 2009 Annual Meeting of the American Association for the Advancement of Science (AAAS) that they had generated a first draft sequence equivalent to more than 60 percent of the complete Neanderthal genome and were comparing it to that of their closest relatives, human beings, which might shed light on the origin of humankind as well as the evolutionary process. What makes this composite sequence of 3 billion bases remarkable is that it is derived from 30,000-year-old samples of ancient DNA. There is no doubt that the sequencing technology developed for the Human Genome Project in 2000, which provided the biochemical code for all the genes in the human genome, and subsequent refinements in sequencing technology, has played a crucial role in helping to decipher the Neanderthal genome. Moreover, advances in the polymerase chain reaction (PCR) have made possible amplification of the small, fragile samples of ancient Neanderthal DNA so they could be sequenced and analyzed; in 1983, R. K. Saiki and Kary B. Mullis were the first researchers to amplify a gene using PCR, which revolutionized molecular genetics and captured the imagination of researchers and students of ancient DNA. Along with advances in phylogenetics, PCR and the sequencing technology, which allow for detailed genetic analyses of ancient DNA, have helped researchers determine the origins of both novel and ancient strains of microorganisms; establish evolutionary paths of plant, fish, and animal species including human beings; document new species; and classify those that are endangered.

KEY TERMS
DNA: a long linear polymer found in the nucleus of a cell, formed from nucleotides and shaped like a double helix.
DNA polymerase: the enzyme that produces a complementary strand of DNA using a DNA template
mitochondrial DNA (mtDNA): believed to be of bacterial origin before it was transferred to the eukaryotic nucleus; in the most multicellular organisms, maternally inherited, most likely present in ancient specimens, and useful in tracing human origin; used to compare closely related species such as Neanderthals and human beings
primer: an oligonucleotide (short strand of nucleotides typically eighteen to thirty bases long) used

as a starting point for Taq polymerase to produce a complementary copy of a DNA template strand; two primers are needed that flank the DNA sequence being amplified

ribosomal RNA (rRNA): the central component of the ribosome, the protein manufacturing machinery of all living cells, thereby providing a mechanism for decoding messenger RNA (mRNA) into amino acids; in contrast to mtDNA, rRNA genes are used to compare distantly related species

Taq polymerase: the enzyme that was first isolated from the hot spring bacterium *Thermus aquaticus*, which is stable at temperatures close to the boiling point of water (100 degrees Celsius, or 212 degrees Fahrenheit); it is used as the DNA polymerase in PCR cycling, which must reach high temperatures to separate the strands of DNA

thermal cycler: a machine that raises and lowers the temperature of the DNA sample in preprogrammed steps; high temperatures separate the double strands of double-stranded DNA (dsDNA), and low temperatures are used to anneal the strands

INTRODUCTION: DINOSAURS AND *JURASSIC PARK*

Ancient DNA research is characterized as the retrieval of DNA sequences from museum specimens, archaeological finds, and fossil and mummified remains, as well as ancient microorganisms that were embedded in ice, rock, or amber, where only a miniscule amount of the original matter is present. For years, researchers struggled with their tiny and often degraded samples of ancient DNA, and the fruits of their labor were often equivocal.

In 1993, filmmaker Stephen Spielberg produced the highly popular motion picture *Jurassic Park*, based on Michael Crichton's 1990 novel wherein scientists were able to bring dinosaurs to life using PCR and other biotechnologies. The film captured the imagination of the public as well as the scientific community, sparking renewed interest in ancient DNA. In a communication in 2009, Dr. Alan Cooper, at the University of Adelaide, Australia, noted that while is not possible to retrieve DNA from dinosaurs that became extinct more than 65 million years ago, per the unrealistic science of *Jurassic Park* (the upper limit of ancient DNA survival even under the most amenable, deep frozen conditions is approximately 500,000 years), genetic data gleaned from ancient DNA taken together with fossil records

make it possible to study much older events by extrapolating backward.

As such, ancient DNA may be employed to study how species and populations evolved when impacted by climate change and mass extinctions. For example, while much data are available from ice cores to gather information about past environmental conditions, paleontologists are now more likely to use ancient DNA to assess the effects of climate change on various species; research on permafrost-preserved megafauna in Alaska and Canada showed that the last glacial ice age, which occurred more than 22,000 years ago, ended with the extinction of the giant bison and other species.

ISOLATION AND ANALYSIS OF ANCIENT DNA

The degree of DNA degradation of the ancient specimen is a function of age and the environmental conditions under which it was preserved; samples a few thousand years old will typically yield very viable DNA, while a woolly mammoth frozen shortly after death will yield more and better-preserved DNA than do the bones of a turtle that had weathered years before its demise. Of greater concern is that the ancient DNA may become contaminated with modern DNA; fossils and ancient remains (teeth and bones) are potentially at risk of contamination from pollen, bacteria, fungi, or the skin cells of the person extracting the DNA, and even minute quantities of modern DNA may contaminate the sample, resulting in faulty data. If the sample had been encased in ice or rock, contamination is unlikely; on the other hand, if the sample was exposed to air, contamination is highly possible.

Since ancient DNA is usually found in fossils in very small amounts, amplification with PCR is a necessity. However, contamination of ancient DNA is possible during the PCR process itself; erroneous nucleotides may be introduced and mistaken for mutations in the ancient DNA when compared to a sequence of modern DNA that is extant. Thus, the researcher must take special precautions to protect the specimen from contamination from the time it is extracted and isolated, through analysis including the PCR process, and during sequencing; upon retrieval, specimens should be placed into airtight, sterile containers by researchers wearing sterile gloves and face masks and using sterilized laboratory equipment and reagents in a "clean room," which has a low level of environmental contaminants. Detailed guidelines exist that scientists may follow to prevent contamination of ancient DNA during PCR and sequencing.

In 2008, Dr. Melanie Pruvost at the University of Paris and coauthors outlined protocols to be used with PCR that cover the prevention of mutations, the removal of contamination from earlier PCR and cloning procedure, and the prevention of contamination of fossil teeth and bones with modern DNA. In 2009, Dr. Svante Pääbo and his group at the Max Planck Institute made significant technological advances in preserving ancient specimens and protecting them from contamination during sequencing. The group created genome sequencing "libraries" (a set of DNA of fragments containing the whole genome of an organism) under "clean room" conditions; DNA "sequence tags" with unique markers attached to molecules of ancient DNA in "clean rooms"; and radioactively labeled DNA to identify and alter steps in the sequencing procedure where losses of ancient DNA had been shown to occur. Taken together, these steps dramatically reduced contamination as well as the amount of fossil material required for sequencing, so that less than half a gram of bone was required to produce the draft sequence of three billion bases in the Neanderthal genome.

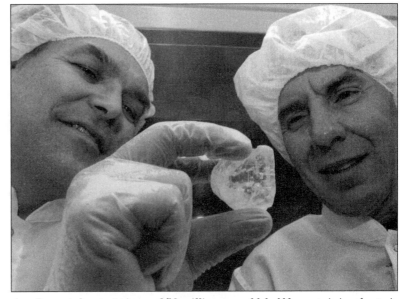

A salt crystal entrapping a 250-million-year-old bubble containing bacteria was excavated from 1,850 feet below ground near Carlsbad, New Mexico; it offers scientists Russell Vreeland (left) and William Rosenzweig the opportunity to study ancient DNA. (AP/Wide World Photos)

ANCIENT MICROBIAL DNA

One of the most fertile areas of ancient DNA analysis has been in the study of the origins of human diseases. In the past, archaeologists relied on physical evidence such as bone scars, deformities, and dental remains to determine whether an ancient human had suffered from a particular disease. More recently, the ability to recover ancient bacterial DNA from Egyptian mummies helped establish the presence of skeletal tuberculosis (TB) in finds dated as early as 3000 B.C.E. The bubonic plague, also known as the Black Death, which is caused by the bacillus *Yersinia pestis*, was thought to be responsible for a series of epidemics that occurred during the sixteenth century but could not be confirmed

without adequate medical records. In 1998, however, French researchers unearthed the skeletal remains of persons who presumably died from the plague in the sixteenth century; using PCR to amplify a gene from *Y. pestis* extracted from dental pulp, along with sequencing, the scientists obtained proof that the plague did indeed exist at the end of the sixteenth century in France.

In 1999, Charles L. Greenblatt, at Hebrew University in Jerusalem, Israel, reported the isolation of DNA in several types of bacteria from 120-million-year-old amber. Comparisons of the DNA sequences of ribosomal RNA (rRNA) lent credence to the claim that he and his colleagues had actually isolated ancient DNA and not contaminants. While studies from 1994 to 2006 had shown that amber is a useful medium for conservation of soft-bodied microorganisms, viable specimens older than 135 million years were very rare and had not included microbes. In 2006, however, a group of German and Italian researchers discovered a "micro world" in Triassic amber as old as the first dinosaurs displaying the diversity of 220-million-year-old microbial life; droplets of amber were found to contain protozoa, fungi, bacteria, and algae comparable to extant genera, thus providing insight into the evolution and paleoecology of the Lower Mesozoic (the Triassic, Jurassic and Cretaceous periods from 251 to 200 million years ago) microorganisms, which have apparently undergone little or no change from the Triassic to the modern era. The largest known deposit of Triassic amber was discovered near the Italian Dolomites, with bacteria the most prevalent microorganism. Examples of all trophic (nutritional) levels were found; bacteria and algae as producers and food sources, protozoa as consumers, and fungi as decomposers, for example. The researchers' discovery of association among various protozoa or with other one-cell organisms with shells indicates that they had settled outside huge bodies of water. Unchanged since the lower Mesozoic period, these protozoa had obviously survived the age of the dinosaurs as well as the diversification of mammals and birds.

EVOLUTION AND ANCIENT DNA

Neanderthals, the closest relatives of modern humans dwelled in Europe and Asia until they became extinct approximately 30,000 years ago; for more than a hundred years, anthropologists and paleontologists have attempted to demonstrate their evolutionary relationship to modern humans, who emerged about 400,000 years ago. Dr. Pääbo made the first major contributions to the understanding the genetic relationship of modern humans to Neanderthals when he sequenced Neanderthal mtDNA in 1997. In 2009, Dr. Pääbo, who heads an international consortium of researchers called the Neanderthal Genome Project, announced that the group had completed a first draft of the complete Neanderthal genome, which can now be compared to earlier sequences of human and chimpanzee genomes in order to obtain initial insights as to how the Neanderthal genome differs from that of modern humans. The genome, generated from DNA extracted from the bones of three Croatian Neanderthal fossils, may help to elucidate the evolutionary relationship between humans and Neanderthals, as well as identify the genetic changes that enabled humans to migrate from Africa across the globe about 100,000 years ago.

Also in 2009, evolutionary anthropologists lent credence to the theory that a Neanderthal known as Shanidar 3 whose skeletal remains were unearthed in the late 1950's and early 1960's was killed by a human being who was capable of using a projectile weapon, not another Neanderthal whose weapon of choice would have been a thrusting spear. This finding contributed to a body of evidence that contact between the two species was most often violent, eventually resulting in the extinction of the Neanderthal; Fernando Rozzi of the Centre Nationale de Recherche in Paris found evidence that humans were both violent and cannibalistic with their Neanderthal neighbors. Many conflicting theories explaining Neanderthal extinction exist, however, such as that climate change adversely affected their hunting grounds, causing the species to starve. Another theory posits that Neanderthals became extinct because they bred with human beings, a premise based on a 2006 discovery of 30,000-year-old skeletal remains in Romania that had both Neanderthal and human characteristics as determined by genetic analysis. However, it may be that the extinction of the Neanderthals was a foregone conclusion and the species was apparently doomed by its genome; in July, 2009, scientists from the Max Planck Institute for Evolutionary Anthropology found scant genetic diversity among DNA samples gleaned from six Neanderthal fossils and concluded that the species "teetered on the brink of extinction" with a population that never exceeded ten thousand.

Studying rates of evolution in Adélie penguins has been accomplished by studying ancient DNA. Adélie penguins, a species common to the Antarctic coast and neighboring islands, dwelled in the same areas of Antarctica for many thousands of years. Through excavation of various colony sites, researchers retrieved partially fossilized bones covering a range of up to almost seven thousand years. By comparing the sequences of a portion of mtDNA among samples of various ages with modern samples, researchers were able to estimate rates of evolutionary change in Adélie penguins. Because the samples were only thousands of years old, the results were deemed more reliable than those from older fossils. Where did these penguins come from, and why were these mobile birds bound to the Southern Hemisphere? Competing hypotheses posit that the penguins either originated in tropical, warm temperate waters or species-diverse cool temperate regions, or even in Gondwanaland about 100 million years ago when it was located farther north. (Gondwanaland or Gondwana was a composite continent, comprised of Antarctica, South America, Madagascar, Africa, India, parts of south Asia, and Australia and at one time even encompassed Florida and most of southern Europe).

To test their hypotheses, researchers in 2005 constructed a phylogeny of extant penguins from 5,851 base pairs of both mitochondrial and nuclear DNA and came to the conclusion that an Antarctic origin of extant taxa was highly likely; using molecular dating techniques, it was estimated that penguins originated about 71 million years ago in Gondwanaland when it was farther south and cooler. The researchers hypothesized that, as Antarctica became covered with ice, the modern penguin migrated via the circumpolar current to oceanic islands that are close by Antarctica and later to the southern continents. Thus, global cooling played a major role in penguin evolution, as it had in vertebrates in general; penguins reached cooler tropical waters in the Galápagos only about 4 million years ago and have not crossed the equatorial thermal barrier.

FUTURE RESEARCH: NEW AND ENDANGERED SPECIES, CLIMATE CHANGE

In 2009, the International Union for Conservation of Nature estimated there are 8 to 14 million plants and animals in the world, of which only 1.8 million had been documented; every year, scientists across the globe discover new animal and plant species. Recently documented were an Ecuadorian salamander that resembles the film character E.T., a jumping spider, and the fossil of the oldest gecko species trapped in amber, dating back 100 million years ago. A flying frog, the world's smallest deer, and an emerald green viper are among the more than 350 new species found in a decade of research in the eastern Himalayas; the discoveries from 1998 to 2008 put the region on par with the island of Borneo in Indonesia as a "biological hotspot." The findings point to the importance of protecting the area, which covers northern Myanmar and India, Bhutan, Nepal, and Tibet. One can ask, however, how species are defined and why is it important to identify them.

In the nineteenth century, Charles Darwin studied species by observation using taxonomic systems: Did the animal have fur, fins, or feet? It was not until the early twentieth century that scientists began to compare the genetic differences among species. This led to the notion that a species could be defined by the barrier to reproducing with other species. However, since some species do not reproduce,

one of the strongest notions that rivaled that of biological "species" was the phylogenetic concept, which replaced sexual reproduction with origin from a common ancestor; today, biological diversity can be ascertained by obtaining both modern and ancient DNA samples and tracking how a species descended from such an ancestor.

In a 2008 review article, Dr. Carl Zimmer at the University of California at Berkeley illustrated the vagaries of species classification using wolves versus coyotes as an example, allowing as how difficult it is to determine where one species ends and another begins. However, while the debate is ongoing, the question of "What is a species?" must be answered in the near future in order to determine which species are considered "endangered"; in the face of the current global climate change, which has already led to the extinction of many plant and animal species, the current goal is to preserve the biological diversity of planet Earth and to prevent existing species from being lost forever.

Bryan Ness, Ph.D.;
updated by Cynthia F. Racer, M.A., M.P.H.

FURTHER READING

Desalle, R., and D. Lindley. *The Science of "Jurassic Park" and "The Lost World."* New York: Harper-Collins, 1998. A critical look at the "science" used in the first two films based on Michael Crichton's novels; for the nonscientist.

Jones, M. *The Molecule Hunt: Archaeology and the Search for Ancient DNA.* New York: Arcade, 2002. An overview of the history of ancient DNA analysis and other molecules.

Lane, N. *Power, Sex, Suicide.* Oxford, England: Oxford University Press, 2005. Follows the role of mitochondria throughout the life cycle.

Mithen, S. *Singing Neanderthals.* London: Weidenfeld & Nicholson, 2005. Presents theories on the origins of language and music.

Pennisi, E. "A Shaggy Dog History." *Science* 298, no. 5598 (2003): 1540-1542. DNA extracted from ancient dog remains point to an Old World origin for New World dogs.

Teletchea, F. "Molecular Identification Methods of Fish Species: Reassessment and Applications." *Reviews in Fish Biology and Fisheries* 19, no. 3 (2009): 265-293. Provides an overview of PCR methods for fish species classification.

Wayne, R. K., J. A. Jennifer, and A. Cooper. "Full of

Sound and Fury: The Recent History of Ancient DNA." *Annual Review of Ecology and Systematics* 30 (1999): 457-477. Includes excellent examples in which DNA originally thought to be "ancient" was found to be "recent."

Zimmer, C. *Evolution: The Triumph of an Idea.* New York: HarperCollins, 2006. Helpful for understanding the various theories of evolution.

WEB SITES OF INTEREST

Australian Centre for Ancient DNA (ACAD)
www.adelaide.edu.au/acad

Understanding Evolution
http://evolution.berkeley.edu

See also: DNA structure and function; Evolutionary biology; Genetic code; Molecular clock hypothesis; Mutation and mutagenesis; Polymerase chain reaction; Punctuated equilibrium; RNA structure and function; RNA world.

Andersen's disease

CATEGORY: Diseases and syndromes

ALSO KNOWN AS: Glycogen storage disease type IV; GSD IV; brancher enzyme deficiency, glycogen branching enzyme deficiency; GBE1 deficiency; glycogenosis IV; amylopectinosis; adult polyglucosan body disease; APBD

DEFINITION

Andersen's disease, also known as glycogen storage disease type IV (GSD (GSD IV), is a clinically heterogeneous disorder resulting from the accumulation of structurally abnormal glycogen in the body. Hepatic, neuromuscular, and multisystem subtypes are characterized by the tissue(s) affected, clinical presentation, and age of onset. Hypoglycemia is not a common feature of GSD IV, as in other GSDs.

RISK FACTORS

Since GSD IV is inherited as an autosomal recessive trait, the parents of a child with GSD IV have a 25 percent recurrence risk for an affected child with each pregnancy. The adult-onset neuromus-cular form, called adult polyglucosan body disease (APBD), is is more common in individuals of Ashkenazi Jewish ancestry.

ETIOLOGY AND GENETICS

Glycogen storage diseases are a group of disorders characterized by the deficiency of specific enzymes involved in the formation or breakdown of glycogen. Glycogen is a complex carbohydrate that is stored in various tissues in the body and converted into glucose for energy. GSD IV is an uncommon form of GSD, accounting for only 3 percent of all cases. Patients with GSD IV form an abnormal glycogen molecule with fewer branch points and longer outer branches resembling amylopectin, the major storage polysaccharide in legumes (beans, peas).

An essential step in the synthesis of glycogen is the formation of branch points at regular intervals along the glycogen molecule. Normally, glycogen branching enzyme (GBE) catalyzes the transfer of at least six glucose units from the outer end of a glycogen chain to produce a branch point on the same or a neighboring chain. This branched structure of glycogen allows it to form a compact, soluble molecule in the cytoplasm.

Patients with GSD IV have decreased or absent GBE activity in one or more tissues. The particular subtype of GSD IV reflects the distribution of GBE activity and the accumulation of abnormal glycogen. Although the glycogen concentration in affected tissues is not increased, the reduced number of branch points leads to decreased solubility and aggregation of this abnormal glycogen in the cytoplasm of affected cells. Thus, decreased GBE activity and abnormal glycogen accumulation occur primarily in liver cells of patients with classic GSD IV and in nerve cells of patients with APBD. The presence of this insoluble glycogen induces a foreign-body reaction that triggers the immune system and leads to cellular injury and organ dysfunction. This immune-mediated reaction is responsible for the severe scarring (cirrhosis) of the liver seen in patients with classic GSD IV.

GSD IV results from mutations in the *GBE1* gene, which encodes for the glycogen branching enzyme. The *GBE1* gene is located on chromosome band 3p12. GSD IV is different from other GSDs in the spectrum of subtypes and clinical presentations (hepatic, neuromuscular, multisystem) that result from different mutations in the same gene. Correlations between phenotype (clinical presentation) and genotype (mutation type) are not well defined. In general, subtypes with severe symptoms and earlier onset have been associated with *GBE1* mutations resulting in absent or low (0 to 10 percent) levels of GBE activity. A homozygous *GBE1* mutation (Tyr329Ser) is found in all APBD patients of Ashkenazi Jewish descent.

SYMPTOMS

Hepatic subtypes of GSD IV include the classic form, characterized by progressive liver cirrhosis during early childhood, and a milder, nonprogressive hepatic form. Neuromuscular subtypes of GSD IV range from a perinatal form with severe hypotonia and cardiomyopathy to an adult-onset form with dementia, urinary incontinence, and walking difficulties.

SCREENING AND DIAGNOSIS

The diagnosis of a hepatic or neuromuscular form of GSD IV is initially based upon the constellation of abnormal clinical findings and laboratory evidence of organ dysfunction. Characteristic microscopic findings of affected tissues include PAS-positive, diastase-resistant cytoplasmic material and polyglucosan bodies by electron microscopy. Definitive diagnosis of GSD IV relies on the demonstration of decreased or absent GBE activity in affected tissues. Sequence analysis of the *GBE1* gene has identified different mutations associated with the subtypes of GSD IV.

TREATMENT AND THERAPY

Patients with classic GSD IV initially require medical treatment for progressive liver dysfunction and associated complications, such as portal hypertension. Liver transplantation is currently the definitive treatment for these patients. Long-term success, however, is limited by complications from transplantation and the possibility of disease progression in other organs. Treatment for APBD patients includes medications and catheterization for spastic bladder, walking devices, and cognitive aids. Supportive care is necessary for patients with GSD IV subtypes with severe multisystem involvement.

PREVENTION AND OUTCOMES

Classic GSD IV leads to end-stage liver failure and death in affected children by age five, unless liver transplantation is performed. The perinatal and in-fantile subtypes are fatal. Patients with cardiomyopathy develop progressive heart failure. Patients with later-onset neuromuscular subtypes have a better long-term prognosis, although they may have progressive disability.

Lynne A. Ierardi-Curto, M.D., Ph.D.

FURTHER READING

Chen, Y. T. "Glycogen Storage Diseases." In *The Metabolic and Molecular Bases of Inherited Disease*, edited by C. R. Scriver, A. L. Beaudet, W. S. Sly, and D. Valle. 8th ed. New York: McGraw-Hill, 2001.

Moses, S. W., and R. Parvari. "The Variable Presentations of Glycogen Storage Disease Type IV: A Review of Clinical, Enzymatic, and Molecular Studies." *Current Molecular Medicine* 2, no. 2 (2002): 177-188.

Özen H. "Glycogen Storage Diseases: New Perspectives." *World Journal of Gastroenterology* 13, no. 18 (2007): 2541-2553.

WEB SITES OF INTEREST

Adult Polyglucosan Body Disease (APBD) Research Foundation
http://www.apbdrf.org

Association for Glycogen Storage Disease
http://www.agsdus.org

See also: Forbes disease; Glycogen storage diseases; Hereditary diseases; Hers' disease; Inborn errors of metabolism; McArdle's disease; Pompe disease; Tarui's disease; Von Gierke disease.

Androgen insensitivity syndrome

CATEGORY: Diseases and syndromes
ALSO KNOWN AS: Testicular feminization syndrome

DEFINITION

The sex of a baby is usually determined at conception by the sex chromosomes, but other genetic events can alter the outcome. One such condition is androgen insensitivity syndrome, which causes a child with male chromosomes to be born with feminized genitals.

RISK FACTORS

Individuals whose mothers carry a mutated copy of the *AR* gene on one of their two X chromosomes are at risk for androgen insensitivity syndrome. About two-thirds of all cases of this disorder are inherited from mothers with this altered gene. The remaining cases result from a new mutation that can occur in the mother's egg cell before the child's conception or during early fetal development.

ETIOLOGY AND GENETICS

Introductory biology courses teach that a fertilized egg that receives two X chromosomes at conception will be a girl, whereas a fertilized egg that receives an X and a Y chromosome will become a boy. However, other factors can also affect the development of a person's gender. Gender development in mammals begins at conception with the establishment of chromosomal sex (the presence of XX or XY chromosomes). Even twelve weeks into development, male and female embryos have the same external appearance. Internal structures for both sexes are also similar. However, the machinery has been set in motion to cause the external genitals to become male or female, with corresponding internal structures of the appropriate sex. The baby is usually born with the proper phenotype to match its chromosomal sex. However, development of the sex organs is controlled by several genes. This leaves a great deal of room for developmental errors to occur.

The *AR* gene, the primary gene involved in sex determination, is carried on the Y chromosome. This gene is responsible for converting the early unisex gonads into testes. Once formed, the testes then produce the balance of androgen and estrogen that pushes development in the direction of the male phenotype. In the absence of this gene, the undetermined gonads become ovaries, and the female phenotype emerges. Therefore, the main cause of sex determination is not XX or XY chromosomes, but rather the presence or absence of the gene that promotes testis differentiation.

In order for the male hormones to have an influence on the development of the internal and external reproductive structures, the cells of those structures must receive a signal that they are part of a male animal. The androgens produced by the testes are capable of entering a cell through the cell membrane. Inside the cell, the androgens attach to specific protein receptor molecules (androgen receptors). Attachment causes the receptors to move from the cytoplasm into the nucleus of the cell. Once in the nucleus, the receptor-steroid complexes bind to DNA near genes that are designed to respond to the presence of these hormones. The binding event is part of the process that turns on specific genes—in this case, the genes that direct the process of building male genitals from the unisex embryonic structures, as well as those that suppress the embryonic female uterus and tubes present in the embryo's abdomen.

In cases of complete androgen insensitivity syndrome, androgen receptors are missing from male cells. This is the result of a recessive allele located on the X chromosome. Because normal males have only one X, the presence of a recessive allele on that X will result in no production of the androgen receptor in that individual. The developing embryo is producing androgen in the testes; without the receptor molecules, however, the cells of the genitals are unable to sense the androgen and respond to it.

SYMPTOMS

In individuals with this disorder, the cells of the genitals are still capable of responding to estrogen from the testes. As a result, the genitals become feminized: labia and clitoris instead of a scrotum and penis, and a short, blind vagina. To the obstetrician and parents, the baby appears to be a healthy girl. An internal examination would show the presence of testes rather than ovaries and the lack of a uterus and Fallopian tubes, but there would normally be no reason for such an examination.

SCREENING AND DIAGNOSIS

Several events may lead to the diagnosis of this condition. The attempted descent of the testes into a nonexistent scrotum will cause pain that may be mistaken for the pain of a hernia; the presence of testes in the child will be discovered when the child undergoes repair surgery. In other cases, the child may seek medical help in the midteen years because she does not menstruate. Exploratory surgery would then reveal the presence of testes and the absence of a uterus. Androgen insensitivity syndrome can be detected by blood tests, which check levels of testosterone and other hormones, pelvic ultrasound, and genetic testing.

TREATMENT AND THERAPY

As a general rule, the testes are left in the abdomen until after puberty because they are needed as a source of estrogen to promote the secondary sex characteristics, such as breast development. Without this estrogen, the girl would remain childlike in body form. After puberty, the testes are usually removed because they have a tendency to become cancerous.

As a result of its phenotypic sex, an infant with androgen insensitivity syndrome is normally raised as a girl whose only problem is an inability to bear children. If the girl has athletic ability, however, other problems may arise. Since 1966, female Olympic athletes have had to submit to a test for the presence of the correct chromosomal sex. In the past, this has meant microscopic examination of cheek cells to count X chromosomes. In 1992, this technique was replaced by a test for the Y chromosome. Individuals who fail the "sex test," including those with androgen insensitivity syndrome, cannot compete against other women. Proponents argue that androgens aid muscle development, and the extra testosterone produced by the testes of a normal male would provide an unfair physical advantage. However, because people with androgen insensitivity syndrome are lacking androgen receptors, their muscle development would be unaffected by the extra androgen produced by the testes, and thus they would not be any stronger than well-conditioned women.

PREVENTION AND OUTCOMES

There currently is no cure for or way in which to prevent this condition. However, the prognosis is good if the testes are removed before they become cancerous.

Nancy N. Shontz, Ph.D.

FURTHER READING

Callahan, Gerald N. *Between XX and XY: Intersexuality and the Myth of Two Sexes.* Chicago: Chicago Review Press, 2009. Discusses numerous intersex conditions, including androgen insensitivity syndrome.

Chen, Harold. "Androgen Insensitivity Syndrome." In *Atlas of Genetic Diagnosis and Counseling.* Totowa, N.J.: Humana Press, 2006. Provides an overview of the syndrome, including its genetics, clinical features, diagnostic tests, and genetic counseling issues.

Goodall, J. "Helping a Child to Understand Her Own Testicular Feminisation." *Lancet* 337, no. 8732 (January 5, 1991): 33. Discusses how communicating with children in stages about their androgen insensitivity syndrome helps them cope emotionally.

Lemonick, Michael. "Genetic Tests Under Fire." *Time* 139, no. 8 (February 24, 1992): 65. Discusses the syndrome and its relationship to athletes and athletic performance.

Mange, Elaine Johansen, and Arthur P. Mange. *Basic Human Genetics.* 2d ed. Sunderland, Mass.: Sinauer Associates, 1999. Provides a detailed discussion of androgen insensitivity syndrome. Illustrations (some color), maps, laser optical disc, bibliography, index.

Meschede, D., H. M. Behre, and E. Nieschlag. "Disorders of the Androgen Target Organs." In *Andrology: Male Reproductive Health and Dysfunction,* edited by E. Nieschlag et al. 2d ed. New York: Springer, 2001. Includes information about androgen insensitivity syndrome.

WEB SITES OF INTEREST

Androgen Insensitivity Syndrome Support Group
http://www.aissg.org
The group, based in the United Kingdom, provides information about the syndrome on its Web site.

Genetics Home Reference, Androgen Insensitivity Syndrome
http://ghr.nlm.nih.gov/condition
=androgeninsensitivitysyndrome
An overview of the condition, focusing on the genes related to it and its inheritance patterns.

Intersex Society of North America
http://www.isna.org
The society is "devoted to systemic change to end shame, secrecy, and unwanted genital surgeries for people born with an anatomy that someone decided is not standard for male or female." Its Web site includes links to information on androgen insensitivity syndrome and other conditions.

Johns Hopkins University, Division of Pediatric Endocrinology, Syndromes of Abnormal Sex Differentiation
http://www.hopkinschildrens.org/intersex
Site provides a guide to the science and genetics

of sex differentiation, with information about complete androgen syndrome and other syndromes of sex differentiation.

Medline Plus, Androgen Insensitivity Syndrome
https://www.nlm.nih.gov/medlineplus/ency/article/001180.htm
Provides an overview of the condition, with information on causes, symptoms, diagnosis, and treatment.

See also: Fragile X syndrome; Gender identity; Hereditary diseases; Hermaphrodites; Klinefelter syndrome; Metafemales; Pseudohermaphrodites; Steroid hormones; XYY syndrome.

Animal cloning

CATEGORY: Genetic engineering and biotechnology
SIGNIFICANCE: Animal cloning is the process of generating a genetic duplicate of an animal starting with one of its differentiated cells. Sheep, mice, cattle, goats, pigs, cats, and dogs are among the animals that have been cloned. While currently an inefficient process that may pose risks to the clone, animal cloning offers the benefits of replicating valuable animals.

KEY TERMS

asexual reproduction: reproduction not requiring fusion of haploid gametes as a first step
clone: a genetic replica of a biological organism
differentiated cell: a somatic cell with a specialized function
mitochondrial genome: DNA found in mitochondria, coding for forty genes, involved in energy metabolism, and maternally inherited
nuclear genome: DNA found in the nucleus, coding for 30,000 genes in higher organisms, half inherited from each parent
telomere: a specialized structure at the chromosome end, which shortens in somatic cells with age

CLONES AND CLONING

Asexual reproduction occurs in numerous bacteria, fungi, and plants, as well as some animals, leading to genetically identical offspring or clones. In addition, humans can assist in such reproduction. For instance, cuttings from plants generate thousands of replicates. Dividing some animals, such as earthworms or flatworms, allows them to regenerate. However, most vertebrates, including all mammals, reproduce sexually, requiring fertilization of an ovum by sperm. In such species, clones occur, as in the case of identical twins, when an embryo splits into two early in development. This process can be instigated artificially using microsurgical techniques to divide a harvested early-stage embryo and reimplanting the halves into surrogate dams (mothers). While this can be considered animal cloning, the term should be reserved for cloning from nonembryonic cells.

CLONING PROCEDURE

Animal cloning typically refers to mammals or other higher vertebrates and involves creating a duplicate animal starting from a differentiated cell. Although such a cell only has the ability to perform its specialized function, its nucleus retains all genetic information for the organism's development. Animal cloning requires that such information be reprogrammed into an undifferentiated cell that can reinitiate the developmental process from embryo to birth and beyond.

In theory, the process is straightforward. It consists of taking a differentiated cell from an adult animal, inserting its diploid nucleus into a donor ovum whose own haploid nucleus has been removed, initiating embryonic development of this ovum, inserting the resultant embryonic mass into a receptive surrogate dam and allowing it to proceed to term. In practice, the technique is difficult and was thought to be impossible until 1997. It also appears fraught with species specificity. Various differentiated cells have been used as the starting source; mammary cells were used in the first case, while skin fibroblasts and cumulus cells are now often used. The preparation of the anucleate ovum is an important step. A limitation to cloning dogs appears to be the difficulty in obtaining ova suitable for nuclear transfer. The technique for inserting the nucleus is crucial, as is the conversion to the undifferentiated embryonic state. Transfer of the embryonic cells to a receptive surrogate dam is generally a well-developed technology, although more than four viable embryos are necessary to maintain pregnancy in pigs.

Dolly the Sheep

In 1997, the world was taken aback when a group of scientists headed by embryologist Ian Wilmut at the Roslin Institute in Scotland announced the successful cloning of a sheep named Dolly. Scientists had already cloned cows and sheep, but they had used embryo cells. Dolly was the first vertebrate cloned from the cell of an adult vertebrate.

The feat was accomplished by removing cells from the udder of a six-year-old ewe and placing them in a laboratory dish filled with nutrients, where they were left to grow for five days. Then the nutrients were reduced to 5 percent of what the cells needed to continue growing, which caused the cells to enter a state resembling suspended animation, making them more receptive to becoming dedifferentiated. When the nuclei of these cells were placed in the ova of host sheep, the cytoplasm of each ovum directed the nucleus it received to enter an undifferentiated state, thus causing the cell to develop into an embryo.

Of an initial 277 adult cells introduced into sheep ova, thirteen resulted in pregnancy, and only one, Dolly, was carried to full term. Dolly was a genetic replica of the sheep from whose udder the original cells were extracted. Environmental factors would make Dolly, like any other clone, individual, but genetically she would never have the individuality that an organism produced by usual reproductive means would possess. Over the next six years, she gave birth to several, apparently healthy, offspring. In 2002, at the age of six, Dolly became lame in her left hind leg, a victim of arthritis. Although sheep commonly suffer arthritis, a veterinarian noted that both the location and the age of onset were uncommon. Then, in February, 2003, she was euthanized after the discovery of a progressive lung disease.

Dolly's health problems led to speculations about premature aging in clones but are complicated by her unique experiences as well. As Wilmut noted, in the early years following the announcement of her cloning, she became something of a celebrity, which led to overfeeding by visitors and in turn a period of obesity, later corrected. More significant were the discovery of her arthritis and then her lung disease—conditions not uncommon in sheep but that tend to emerge later (sheep typically live to be eleven or twelve years old). Theories of premature aging are supported by the fact that Dolly's telomeres were shorter than normal. These cell structures function as "caps" that prevent "fraying" at the ends of DNA cells. As a cell ages, its telomeres become progressively shorter, until finally they disappear altogether and are no longer able to protect the cell, which then dies.

Was Dolly older genetically than she was chronologically? The answer to the question of whether Dolly was completely "normal" or aged prematurely as a result of being a clone must await tracking of her offspring's lives and monitoring of other vertebrate clones through their life spans.

Dolly, the first animal cloned from an adult vertebrate cell, in 1997. (AP/Wide World Photos)

R. Baird Shuman, Ph.D.;
updated by Christina J. Moose

Furthermore, the genetic makeup of a putative clone must be verified, to ensure that it is indeed a replica of its progenitor and not an unintended offspring of either the donor of the ovum or the surrogate dam. DNA fingerprinting via microsatellite analysis at a number of polymorphic sites is an unambiguous way to establish its genetic identity.

The first cloned mouse and its "parent" are displayed at a news conference in July, 1998. (AP/Wide World Photos)

IDENTICALNESS

Such a clone is not absolutely identical, because of mitochondrial differences and environmental effects. While the nuclear genome must be identical to its progenitor, the mitochondrial genome of the clone will invariably be different, because it comes from the ovum used. While mitochondria make a minor contribution to the total genetic makeup, they can influence phenotypic expression. In addition, the prenatal environment can affect some traits. Coat color and color pattern are characteristics that can be developmentally influenced; the first cloned cat was not an exact duplicate of its progenitor in coloration. Some behavioral features are also impacted during intrauterine development.

CLONED ANIMALS

The first cloned animal was a sheep named Dolly. While she was the only live offspring generated from 277 attempts, her birth showed that animal cloning was possible. Shortly thereafter, mice and cattle were cloned. Reproducible cloning of mice is more difficult than imagined, whereas more cattle were cloned in the first five years after Dolly's birth than any other species. Goats, pigs, and a cat were among the animals that were subsequently cloned.

PROBLEMS AND POTENTIAL BENEFITS

Prominent among the problems with animal cloning is its inefficiency. Although this may not be surprising as the technology is still under development, no more than 2 percent of embryos generated lead to viable offspring. Additionally, most cloned animals are larger than normal at birth, often requiring cesarian delivery, and some have increased morbidity and mortality. Some have had smaller telomeres and shorter lives. Dolly exhibited this trait and lived for only six years (although she was euthanized, she clearly would not have lived much longer)—half of the average life span. Conversely, some cloned mice do not exhibit shortened telomeres or premature aging, even through six consecutive cloned generations. Further research will establish whether these problems are inherent to cloning, are consequences of some aspect of the current procedure, or are attributable to the small numbers of cloned animals studied.

The benefits of animal cloning would involve duplicating particularly valuable animals. Livestock with highly valued production characteristics could be targets for cloning. However, the technique is likely to be most beneficial in connection with transgenesis, to replicate animals that yield a therapeutic agent in high quantities or organs suitable for transplantation into humans. If animal cloning can be made efficient and trouble-free, its potential benefits could be fully developed.

James L. Robinson, Ph.D.

FURTHER READING

Houdebine, Louis-Marie. *Animal Transgenesis and Cloning.* Translated by Louis-Marie Houdebine et al. Hoboken, N.J.: John Wiley & Sons, 2003. Describes the molecular biological techniques used to clone animals and create transgenic animals and the limits and risks of cloning, gene therapy, and transgenesis.

Panno, Joseph. *Animal Cloning: The Science of Nuclear Transfer.* New York: Facts On File, 2005. An overview designed for the general reader. Provides the history and basic facts of animal cloning, describes the cloning of Dolly the sheep, and examines the ethical and legal issues surrounding the creation of cloned animals.

Patterson, Lesley, William Richie, and Ian Wilmut. "Nuclear Transfer Technology in Cattle, Sheep and Swine." In *Transgenic Animal Technology, A Laboratory Handbook*, edited by Carl A. Pinkert. 2d ed. London: Academic Press, 2002. Describes the detailed protocol needed to clone three livestock species, as well as the limitations to increased efficiency.

Pennisi, Elizabeth, Gretchen Vogel, and Dennis Normile. "Clones: A Hard Act to Follow." *Science* 288, no. 5472 (2000): 1722-1727. Reviews the status of animal cloning, three years after the announcement of Dolly. The problems, questions, and concerns are presented in a highly readable text.

Wilmut, Ian, Keith Campbell, and Colin Tudge. *The Second Creation: The Age of Biological Control by the Scientists That Cloned Dolly.* London: Headline, 2000. The story of the scientific collaboration between an agricultural scientist and a cell biologist, describing the perseverance and the serendipity that led to the creation of Dolly, the first cloned sheep.

WEB SITES OF INTEREST

ActionBioScience.org
http://www.actionbioscience.org/biotech/pecorino.html

Features the article "Animal Cloning: Old MacDonald's Farm Is Not What It Used To Be" and several useful links to the animal cloning debate.

Human Genome Project, Cloning Fact Sheet
http://www.ornl.gov/sci/techresources/Human_Genome/elsi/cloning.shtml

A basic overview of the subject, including a description of the technologies of DNA, reproductive, and therapeutic cloning.

Roslin Institute
http://www.roslin.ac.uk

The site of the oldest cloning group in the world, founded in 1919, which cloned Dolly the sheep. Includes information on genomics and animal breeding.

U.S. Food and Drug Administration (FDA), Animal Cloning
http://www.fda.gov/AnimalVeterinary/SafetyHealth/AnimalCloning/default.htm

In 2001, the FDA began examining the safety of food from cloned animals and their offspring. This page provides access to the agency's findings, released in 2008, that these foods were safe for human consumption.

See also: Biopharmaceuticals; cDNA libraries; Cloning; Cloning: Ethical issues; Cloning vectors; DNA replication; DNA sequencing technology; Genetic engineering; Genetic engineering: Agricultural applications; Genetic engineering: Historical development; Genetic engineering: Industrial applications; Genetic engineering: Medical applications; Genetic engineering: Risks; Genetic engineering: Social and ethical issues; Knockout genetics and knockout mice; Mitochondrial genes; Parthenogenesis; Polymerase chain reaction; Restriction enzymes; Reverse transcriptase; Shotgun cloning; Telomeres; Transgenic organisms; Xenotransplants.

Aniridia

CATEGORY: Diseases and syndromes
ALSO KNOWN AS: Irideremia; WAGR syndrome; Gillespie syndrome

DEFINITION

Aniridia, meaning "without iris," is a rare congenital disorder characterized by the partial or complete absence of the iris (the colored circular part of the eye that surrounds the pupil). The disorder is frequently associated with secondary symptoms contributing to severe visual impairment.

RISK FACTORS

No known risk factors exist. The incidence of aniridia is between 1 in 50,000 and 1 in 100,000 and the prevalence between 1 in 100,000 and 9 in 100,000. The disease affects males and females equally.

ETIOLOGY AND GENETICS

Aniridia is most often caused by a mutation of the *PAX6* gene located on chromosome 11 on band

11p13. The *PAX6* gene and its product, PAX6 protein, play significant roles in the development of the eye. In aniridics, only one functional copy of the *PAX6* gene exists, and as such an insufficient amount of PAX6 protein is made (known as haploinsufficiency), stopping eye development too early and ultimately leading to the incomplete development of the iris.

In approximately 70 percent of cases, the disease is familial, or passed down from parent to offspring genetically. In 30 percent of cases, it is sporadic, or noninherited and occurring as a single random case. Some sporadic cases of aniridia are caused by larger chromosome deletions that affect not only the *PAX6* gene but also adjacent regions on 11p13. In these cases, aniridia is either an isolated symptom or occurs as part of Wilms' tumor-aniridia-genitourinary anomalies-mental retardation (WAGR) syndrome, in which more extensive deletions affecting the nearby Wilms tumor locus (WT1) occur.

Familial aniridia is usually transmitted as an autosomal dominant trait, meaning that only one defective *PAX6* gene on an autosomal chromosome (non-sex chromosome) needs to be passed down for aniridia to be expressed. Very rarely, aniridia is inherited as an autosomal recessive trait, in which two copies of another defective gene are passed down. In these cases, the disorder is inherited as part of an unusual association of aniridia, cerebellar ataxia, and mental deficiency (Gillespie syndrome), and the underlying genetic defect is not linked to a *PAX6* gene mutation.

SYMPTOMS

In unaffected individuals, the iris controls the amount of light let into the eye by opening and closing like the lens of a camera. The inability to control light entry because of partial or complete lack of an iris makes aniridics very light-sensitive. Secondary conditions associated with aniridia include low vision (visual acuity of 20/200 considered legally as blindness), lens dislocation, nystagmus (constant involuntary movement of the eyeball), and glaucoma (elevated pressure in the eyeball) in more than half of cases, usually occurring during the teenage years. Additional secondary conditions include cataracts (clouding in the lens of the eye), corneal disease (a variety of conditions that affect the cornea, the small transparent part of the eye that covers the pupil and iris) and optic nerve disease (conditions that affect the nerve connecting the eye to the brain, making vision possible).

SCREENING AND DIAGNOSIS

Aniridia can be diagnosed by clinical examination techniques including slit-lamp examination, iris fluorescein angiography, optical coherence tomography (OCT), and high-frequency ultrasound biomicroscopy (UBM). Examination of a patient's chromosomes (known as karyotyping) and investigation into other family cases of aniridia can determine whether the disorder is familial or sporadic. If inherited, isolated aniridia (cases not presenting as a part of another syndrome) can also be diagnosed with molecular genetic testing techniques. Deletion testing (testing the absence of a segment in the *PAX6* gene) and sequence analysis (a process by which the nucleotide sequence is determined for a segment of the *PAX6* gene) can be used to identify mutations in or near this gene associated with isolated aniridia and help diagnose patients.

TREATMENT AND THERAPY

No specific treatments for aniridia exist. There are, however, treatments available for many of the associated complications. They include but are not limited to low-vision aids such as glasses and magnifiers; computer software elements such as text enhancers and text-to-speech programs; devices that block light such as sunglasses and contact lenses with an artificial iris; and medical and surgical treatments such as cataract extraction, transplants of corneal tissue created from stem cells, antiglaucoma medication, or drainage tube surgery.

PREVENTION AND OUTCOMES

When available, both genetic counseling and preimplantation genetic diagnosis, a procedure in which embryos created with in vitro fertilization techniques are screened for genetic disease before being implanted into the woman's uterus, may prevent the inheritance of genetic diseases by offspring. Ocular testing for offspring at risk of inheriting aniridia and surveillance of secondary conditions (such as glaucoma screening, visual assessment, monitoring for aniridic fibrosis, and renal examination for patients with WAGR syndrome) cannot prevent the disease itself but can help to avoid further complications. While secondary symptoms of aniridia such as cataract, glaucoma, and corneal and optic nerve disease develop in early adulthood, abnormalities such as malformed iris and nystagmus are typically apparent earlier, usually by six weeks of age.

Anna Kole, M.P.H.

FURTHER READING

Lee, H., R. Khan, and M. O'Keefe. "Aniridia: Current Pathology and Management." *Acta Ophthalmologica* 86, no. 7 (November, 2008): 708-715.

Scriver, C. R., et al., eds. *The Metabolic and Molecular Bases of Inherited Disease.* 8th ed. New York: McGraw-Hill, 2007.

WEB SITES OF INTEREST

Aniridia Foundation International
www.aniridia.net

Aniridia Network International
www.aniridia.org

European Organisation for Rare Diseases (EURORDIS)
www.eurordis.org

GeneTests at NCBI
http://www.genetests.org

National Organization for Rare Disorders
www.rarediseases.org

Orphanet Database of Rare Diseases and Orphan Drugs
www.orpha.net

See also: Best disease; Choroideremia; Corneal dystrophies; Hereditary diseases.

Ankylosing spondylitis

CATEGORY: Diseases and syndromes
ALSO KNOWN AS: Marie-Strumpell disease

DEFINITION

Ankylosing spondylitis is a chronic inflammatory disease that causes arthritis of the spine and hips. It can also affect other joints, such as the knees, and can cause inflammation of the eyes, lungs, or heart valves.

RISK FACTORS

Males and individuals between the ages of fifteen and thirty-five are at risk for ankylosing spondylitis, as are individuals whose family members have the disease. Other risk factors include having the *HLA-B27* gene and having inflammatory bowel disease, ulcerative colitis, or Crohn disease.

ETIOLOGY AND GENETICS

The causes of ankylosing spondylitis are not well understood, but it seems clear that both genetic and environmental factors play contributing roles. Approximately 90 percent of affected individuals carry the *HLA-B27* gene, but not all individuals who express this gene will develop the disease. For example, while 50 percent of the children of an affected parent will inherit the *HLA-B27* gene, only about 25 percent of them will develop spondyloarthritis.

The *HLA-B27* gene is one of a family of genes located at the major histocompatability locus on the short arm of chromosome 6 at position 6p21.3. It encodes a protein that is present on the surface of almost all cells and functions to display protein fragments (peptides) that have been exported from the cell to components of the immune system. If the antigens are recognized as foreign, an inflammatory response is triggered. The conditions under which the HLA-B27 protein initiates an inflammatory response resulting in disease are not clear, and theories range from the improper presentation of peptides to the misfolding of the protein itself.

Two additional genes, *IL23R* and *ARTS1*, have been shown to have an association with ankylosing spondylitis, but the molecular nature of the association is unknown. Both of these genes also play a role in immune function, yet neither has been shown to be involved with other autoimmune diseases, such as rheumatoid arthritis or lupus. This supports the intriguing possibility that ankylosing spondylitis may not be an autoimmune disease after all but rather may result from an altered response to infection.

SYMPTOMS

The severity of an individual's symptoms can vary from mild to very severe. Common symptoms may include stiffening and pain (arthritis) of the lower back and the sacroiliac joint, where the back and hip meet, possibly radiating down the legs. The pain is often worse at night; stiffness is worse in the morning. Symptoms may improve with exercise or activity.

Individuals may occasionally experience pain and stiffness in other joints, including the knees, upper back, rib cage, neck, shoulders, and feet. Another symptom is chest pain, which may suggest heart, heart valve (aortic insufficiency), or lung involvement; eye pain, visual changes, and increased tearing may suggest eye involvement (uveitis). Less common symptoms may include fatigue, loss of appetite or weight loss, fever, and numbness (if arthritic spurs compress the spinal nerves).

SCREENING AND DIAGNOSIS

The doctor will ask about a patient's symptoms and medical history and will perform a physical exam. Diagnosis is based on common symptoms of ankylosing spondylitis, such as dramatic loss of flexibility of the lower back and spine (limitation of motion of the low back), pain in the lower back, and limited chest expansion when taking deep breaths.

Diagnostic tests may include X rays of the lower back and hips to check for characteristic changes and occasionally a magnetic resonance imaging (MRI) scan or a computed tomography (CT) scan of the involved joints. Blood tests can check for the *HLA-B27* gene marker, anemia, an elevated sedimentation rate, and the presence of other autoimmune markers.

TREATMENT AND THERAPY

There is no cure for ankylosing spondylitis. Treatment is aimed at providing education and relieving the symptoms and may include medication, including nonsteroidal anti-inflammatory drugs (NSAIDs) to control pain and inflammation. In recent years, a number of newer anti-inflammatory medications have been discovered.

Physical therapy techniques can prevent progression and worsening of symptoms and may include learning proper posture and the best positions for sleeping. Daily exercise is another treatment and can include abdominal and back exercises (to decrease back stiffness and maintain good posture), stretching exercises, swimming exercises, and breathing exercises (in cases where the rib cage is affected). In severe cases, hip or joint replacement surgery may be needed to relieve pain and restore mobility. In some instances spinal surgery is needed to allow the person to maintain an upright posture.

PREVENTION AND OUTCOMES

There are no guidelines for preventing ankylosing spondylitis because the cause is unknown.

Rick Alan; reviewed by Julie D. K. McNairn, M.D.
"Etiology and Genetics" by Jeffrey A. Knight, Ph.D.

FURTHER READING

Beers, Mark H., ed. *The Merck Manual of Medical Information.* 2d home ed., new and rev. Whitehouse Station, N.J.: Merck Research Laboratories, 2003.

Braun, J., and X. Baraliakos. "Treatment of Ankylosing Spondylitis and Other Spondyloarthritides." *Current Opinion in Rheumatology* 21, no. 4 (July, 2009): 324-334.

Brown, M. A. "Genetics and the Pathogenesis of Ankylosing Spondylitis." *Current Opinion in Rheumatology* 21, no. 4 (July, 2009): 318-323.

EBSCO Publishing. *Health Library: Ankylosing Spondylitis.* Ipswich, Mass.: Author, 2009. Available through http://www.ebscohost.com.

Firestein, Gary S., ed. *Kelley's Textbook of Rheumatology.* 8th ed. Philadelphia: Saunders/Elsevier, 2009.

Royen, Barend J. van, and Ben A. C. Dijkmans, eds. *Ankylosing Spondylitis: Diagnosis and Management.* New York: Taylor & Francis, 2006.

Toussirot, Eric A. "Management of Ankylosing Spondylitis and Related Spondylarthritis: Established Treatments, New Pharmacological Options, and Anti-TNF Therapies." In *Arthritis Research*, edited by Frank Columbus. New York: Nova Biomedical Books, 2005.

WEB SITES OF INTEREST

American College of Rheumatology
http://www.rheumatology.org

The Arthritis Society
http://www.arthritis.ca/custom%20home/default.asp?s=1

Canadian Spondyloarthritis Association
http://www.spondylitis.ca/en

Genetics Home Reference
http://ghr.nlm.nih.gov

National Ankylosing Spondylitis Society
http://www.nass.co.uk

Peh, Wilfred C. G. "Ankylosing Spondylitis." *Emedicine*
http://emedicine.medscape.com/article/386639
-overview

Schaffert, Alan. "Ankylosing Spondylitis." *Emedicine*
http://emedicine.medscape.com/article/1145824
-overview

Spondylitis Association of America
http://www.spondylitis.org

*University of Washington Orthopedics and Sports
    Medicine Department*
http://www.orthop.washington.edu

See also: Autoimmune disorders; Hereditary diseases.

Anthrax

CATEGORY: Bacterial genetics

SIGNIFICANCE: Anthrax has plagued humankind for thousands of years. Naturally occurring anthrax spores have caused disease in livestock and wildlife more often than in humans, but with the rise of genetic technologies anthrax has become amenable to manipulation as an agent of bioterrorism and biowarfare.

KEY TERMS

plasmids: extrachromosomal DNA, found most commonly in bacteria, which can be transferred between bacterial cells

polymerase chain reaction (PCR): a process in which a portion of DNA is selected and repeatedly replicated

single nucleotide polymorphism (SNP): the difference in a single nucleotide between the DNA of individual organisms

variable number tandem repeat (VNTR): the difference in the number of tandem repeats (short sequences of DNA repeated over and over) between the DNA of individual organisms

HISTORY

A disease killing cattle in 1491 B.C.E., likely to have been anthrax, is recounted in the Book of Genesis. In Exodus 9, the Lord instructs Moses to take "handfuls of ashes of the furnace" and "sprinkle it toward the heaven in the sight of the Pharaoh." Moses performed the deed and "it became a boil breaking forth with blains upon man and upon beast." This may represent the first use of anthrax as a biological weapon. Greek peasants tending goats suffered from anthrax; the Greek word from which "anthrax" derives means coal, referring to the coal-black center of the skin lesion.

Anthrax became the first pathogenic bacillus to be seen microscopically when described in infected animal tissue by Aloys-Antoine Pollender in 1849. Studies by Robert Koch in 1876 resulted in the four postulates that form the basis for the study of infectious disease causation. In 1881, Louis Pasteur demonstrated the protective efficacy of a vaccine for sheep made with his attenuated vaccine strain.

THE DISEASE

Anthrax is primarily a disease of herbivorous animals that can spread to humans through association with domesticated animals and their products. Herbivorous animals grazing in pastures with soil contaminated with anthrax endospores become infected when the spores gain entry through abrasions around the mouth and germinate in the surrounding tissues. Omnivores and carnivores can become infected by ingesting contaminated meat. Human infection is often a result of a close association with herbivores, particularly goats, sheep, or cattle (including their products of hair, wool, and hides).

The most common clinical illness in humans is skin infection (cutaneous anthrax), acquired when spores penetrate through cuts or abrasions. After an incubation period of three to five days, a papule develops, evolves into a vesicle, and ruptures, leaving an ulcer that dries to form the characteristic black scab. Inhaled spores reach the alveoli of the lung, where they are engulfed by macrophages and germinate into bacilli. Bacilli are carried to lymph nodes, where release and multiplication are followed by bloodstream invasion and the infection's spread to other parts of the body, including the brain, where it causes meningitis. The symptoms of the illness, which begin a few days after inhalation, resemble those of the flu and may be associated with substernal discomfort. Cough, fever, chills, and respiratory distress with raspy, labored breathing ensue. The least common type of infection is that of the gastrointestinal tract.

An effective vaccine is available for prevention, and antibiotics have been used when immediate protection is needed. Antibiotics can also successfully treat the infection. Inhalational anthrax is nearly always fatal if untreated, and even with treatment the mortality ranges from 40 to 80 percent. Mortality from treated cutaneous anthrax is less than 1 percent.

THE ANTHRAX BACTERIUM

The *Bacillus anthracis* bacterium is large (1-1.2 × 3-10 microns), encapsulated, gram-positive, and rod-shaped. It produces spores and exotoxins (toxins that are released from the cells). Spores are ellipsoidal or oval (1-2 microns) and located within the bacilli. The endospores have no reproductive significance, as only one spore is formed by each bacillus and a germinated spore yields a single bacillus. Spores form in soil and dead tissue and with no measurable metabolism may remain dormant for years. They are resistant to drying, heat, and many disinfectants.

The genetic composition of *B. anthracis* differs little from the other *Bacillus* species, and studies have demonstrated remarkable similarity within *B. anthracis* strains. The resting stage of sporulation may

have contributed to the extremely similar DNA of all strains of *B. anthracis*. The circular chromosomal DNA is composed of 5.2 million base pairs and codes for metabolic function, cell repair, and the sequential process of sporulation. Comparative genome sequencing has uncovered only four differences between the single-copy chromosomal DNA of two strains. In addition to the single-copy DNA, comprising the majority of the genome, a remaining portion consists of repetitive DNA sequences that are either dispersed or clustered into satellites. The satellite repeats occur in tandem. The number of tandem repeats varies among different strains; six chromosomal marker loci have been identified by multiple-locus variable number tandem repeat (VNTR) analysis.

In addition to its chromosome, *B. anthracis* has two large plasmids that carry genes necessary for pathogenesis. The pXO1 plasmid has 181,654 base pairs and contains the structural genes for the anthrax toxins *cya* (edema factor), *lef* (lethal factor), and *pagA* (protective antigen). The pXO2 plasmid consists of 96,231 base pairs and carries three genes required for synthesis of the capsule. These plasmids contain a much greater number of single nucleotide polymorphisms (SNPs) and VNTRs among strains than the chromosomal genome. There are a variety of reference strains, such as Pasteur (which lacks the XO1 plasmid), Sterne (which lacks the XO2 plasmid), and Ames (which has both plasmids and is fully virulent).

BIOTERRORISM

Anthrax spores can be easily packaged to act as aerosoled (airborne) agents of war, and the genome may be bioengineered to alter the virulence or effectiveness of current vaccines. Knowledge of the genetic composition of *B. anthracis* has facilitated the investigation of anthrax attacks. In 1993, the Aum Shinrikyo cult aerosoled a suspension of anthrax near Tokyo, Japan. Molecular studies of the genome from this strain revealed it to be devoid of the pXO2 plasmid (Sterne strain), explaining why only a bad

Anthrax cells invading the spleen of a monkey, in an undated electronmicrograph from the U.S. Department of Defense Web site. (AP/Wide World Photos)

odor rather than illness was the fortunate conse- quence. In 2001, analysis of material from letter-based attacks with anthrax in the United States demon- strated the source to be the Ames strain. Further- more, as a result of the extensive laboratory stud- ies associated with these attacks, a sensitive and specific three-target (two-plasmid and one-chromo- some) assay has been developed for rapid detection and identification of *B. anthracis*, including bioengi- neered strains, from both patients and the environ- ment.

H. Bradford Hawley, M.D.

FURTHER READING

Dixon, Terry C., et al. "Anthrax." *New England Jour- nal of Medicine* 341, no. 11 (September 9, 1999): 815-826. Details the disease and its pathogenesis.

Emerging Infectious Diseases 8, no. 10 (October, 2002). This issue is devoted to an examination of bioter- rorism-related anthrax, and it summarizes the in- vestigation following the 2001 bioterrorism at- tacks in the United States.

Holmes, Chris. *Spores, Plagues, and History: The Story of Anthrax.* Dallas: Durban House, 2003. Holmes, a medical epidemiologist, recounts the history of anthrax and its effects on human beings from the time of Moses through its use of the disease as a weapon of bioterrorism in the twenty-first cen- tury.

Miller, Judith, Stephen Engelberg, and William Broad. *Germs: Biological Weapons and America's Se- cret War.* New York: Simon & Schuster, 2001. This book, written by three *New York Times* reporters, explores the ideas and actions of scientists and politicians involved in the past, present, and fu- ture of germ warfare. Includes forty-two pages of notes and a select bibliography.

Read, Timothy D., et al. "Comparative Genome Se- quencing for Discovery of Novel Polymorphisms in *Bacillus anthracis*." *Science* 296, no. 5575 (June 14, 2002): 2028-2033. Describes the complete se- quencing of the anthrax genome.

Tucker, Amy E., and Jimmy D. Ballard. "Anthrax Toxin and Genetic Aspects Regulating Its Expres- sion." In *Microbial Protein Toxins*, edited by Manfred J. Schmitt and Raffael Schaffrath. New York: Springer, 2005. Explains how protein tox- ins, such as anthrax, have developed strategies that enable them to enter, permeate, and kill tar- geted cells.

WEB SITES OF INTEREST

Center for Biosecurity
http://www.upmc-biosecurity.org/website/ index.html
The Web site of the center, which is located at the University of Pittsburgh Medical Center, contains a fact sheet on *Bacillus anthracis* and other informa- tion about anthrax.

Centers for Disease Control, Public Health Emergency Preparedness and Response
http://www.bt.cdc.gov
This comprehensive site offers information on how to recognize illness caused by anthrax exposure and more. Available in Spanish.

Nature
http://www.nature.com
The online version of the premier science jour- nal *Nature* includes links to research articles on the genetics of anthrax.

World Health Organization
http://whqlibdoc.who.int/publications/2008/ 9789241547536_eng.pdf
This is the online version of the fourth edition of the organization's book *Anthrax in Humans and Ani- mals* (2008).

See also: Bacterial genetics and cell structure; Bac- terial resistance and super bacteria; Biological weap- ons; Plasmids; Smallpox.

Antibodies

CATEGORY: Immunogenetics

SIGNIFICANCE: Antibodies provide the main line of defense (immunity) in all vertebrates against in- fections caused by bacteria, fungi, viruses, or other foreign agents. Antibodies are used as ther- apeutic agents to prevent specific diseases and to identify the presence of antigens in a wide range of diagnostic procedures. Large quantities of antibodies have also been produced in plants for use in human and plant immunotherapy. Be- cause of their importance to human and animal health, antibodies are widely studied by geneticists seeking improved methods of antibody production.

KEY TERMS

B cells: a class of white blood cells (lymphocytes) derived from bone marrow responsible for antibody-directed immunity

B memory cells: descendants of activated B cells that are long-lived and that synthesize large amounts of antibodies in response to a subsequent exposure to the antigen, thus playing an important role in secondary immunity

helper T cells: a class of white blood cells (lymphocytes) derived from bone marrow that prompts the production of antibodies by B cells in the presence of an antigen

lymphocytes: types of white blood cells (including B cells and T cells) that provide immunity

plasma cells: descendants of activated B cells that synthesize and secrete a single antibody type in large quantities and also play an important role in primary immunity

ANTIBODY STRUCTURE

Antibodies are made up of a class of proteins called immunoglobulins (Ig's) produced by plasma cells (descendants of activated B cells) in response to a specific foreign molecule known as an antigen. Most antigens are also proteins or proteins combined with sugars. Antibodies recognize, bind to, and inactivate antigens that have been introduced into an organism by various pathogens such as bacteria, fungi, and viruses.

The simplest form of antibody molecule is a Y-shaped structure with two identical, long polypeptides (substances made up of many amino acids joined by chemical bonds) referred to as "heavy chains" and two identical, short polypeptides referred to as "light chains." These chains are held together by chemical bonds. The lower portion of each chain has a constant region made up of similar amino acids in all antibody molecules, even among different species. The remaining upper portion of each chain, known as the "variable region," differs in its amino acid sequence from other antibodies. The three-dimensional shape of the tips of the variable region (antigen-binding site) allows for the recognition and binding of target molecules (antigens). The high-affinity binding between antibody and antigen results from a combination of hydrophobic, ionic, and van der Waals forces. Antigen-binding sites have specific points of attachment on the antigen that are referred as "epitopes" or "antigenic determinants."

ANTIBODY DIVERSITY

There are five classes of antibodies (IgG, IgM, IgD, IgA, and IgE), each having a distinct structure, size, and function (see the table headed "Classes, Locations, and Functions of Antibodies"). IgG is the principal immunoglobulin and constitutes up to 80 percent of all antibodies in the serum.

The human body can manufacture a limitless number of antibodies, each of which can bind to a different antigen; however, human genomes have a limited number of genes that code for antibodies. It has been proposed that random recombination of DNA segments is responsible for antibody variability. For example, one class of genes (encoding light chain) contains three regions: the L-V (leader-variable) region (in which each variable region is separated by a leader sequence), the J (joining) region, and the C (constant) region. In the embryonic B cells, each gene consists of from one hundred to three hundred L-V regions, approximately six J regions, and one C region. These segments are widely separated on the chromosome. As the B cells mature, one of the L-V regions is randomly joined to one of the J regions and the adjacent C region by a recombination event. The remaining segments are cut from the chromosome and subsequently destroyed, resulting in a fusion gene encoding a specific light chain of an antibody. In mature B cells, this gene is then transcribed and translated into polypeptides that form a light chain of an antibody molecule. Genes for the other class of light chains as well as heavy chains are also made up of regions that undergo recombination during B-cell maturation. These random recombination events in each B cell during maturation lead to the production of billions of different antibody molecules. Each B cell has, however, been genetically programmed to produce only one of the many possible variants of the same antibody.

PRODUCTION OF ANTIBODIES: IMMUNE RESPONSE

Immunity is a state of bodily resistance brought about by the production of antibodies against an invasion by an antigen. The immune response is mediated by white blood cells known as lymphocytes that

Classes, Locations, and Functions of Antibodies

Class	Location	Functions
IgG	Blood plasma, tissue fluid, fetuses	Produces primary and secondary immune responses; protects against bacteria, viruses, and toxins; passes through the placenta and enters fetal bloodstream, thus providing protection to fetuses.
IgM	Blood plasma	Acts as a B-cell surface receptor for antigens; fights bacteria in primary immune response; powerful agglutinating agent; includes anti-A and anti-B antibodies.
IgD	Surface of B cells	Prompts B cells to make antibodies (especially in infants).
IgA	Saliva, milk, urine, tears, respiratory and digestive systems	Protects surface linings of epithelial cells, digestive, respiratory, and urinary systems.
IgE	In secretion with IgA, skin, tonsils, respiratory and digestive systems	Acts as receptor for antigens causing mast cells (often found in connective tissues surrounding blood vessels), to secrete allergy mediators; excessive production causes allergic reactions (including hay fever and asthma).

are made in the bone marrow. There are two types of lymphocytes: T cells, which are formed when lymphocytes migrate to the thymus gland, circulate in the blood, and become associated with lymph nodes and the spleen; and B cells, which are formed in bone marrow and move directly to the circulatory and the lymph systems. B cells are genetically programmed to produce antibodies. Each B cell synthesizes and secretes only one type of antibody, which has the ability to recognize with high affinity a discrete region (epitope or antigenic determinant) of an antigen. Generally, an antigen has several different epitopes, and each B cell produces a set of different antibodies corresponding to one of the many epitopes of the same antigen. All of the antibodies in this set, referred to as "polyclonal" antibodies, react with the same antigen.

The immune system is more effective at controlling infections than the nonspecific defense response (bodily defenses against infection—such as skin, fever, inflammation, phagocytes, natural killer cells, and some other antimicrobial substances—that are not part of the immune system proper). The immune system has three characteristic responses to antigens: diverse, which effectively neutralizes or destroys various foreign invaders, whether they are microbes, chemicals, dust, or pollen; specific, which effectively differentiates between harmful and harmless antigens; and anamnestic, which has a memory

component that remembers and responds faster to a subsequent encounter with an antigen. The primary immune response involves the first combat with antigens, while the secondary immune response includes the memory component of a first assault. As a result, humans typically get some diseases (such as chicken pox) only once; other infections (such as cold and influenza) often recur because the causative viruses mutate, thus presenting a different antigenic face to the immune system each season.

An antibody-mediated immune response involves several stages: detection of antigens, activation of helper T cells, and antibody production by B cells. White blood cells known as macrophages continuously wander through the circulatory system and the interstitial spaces between cells searching for antigen molecules. Once an antigen is encountered, the invading molecule is engulfed and ingested by a macrophage. Helper T cells become activated by coming in contact with the antigen on the macrophage. In turn, an activated helper T cell identifies and activates a B cell. The activated T cells release cytokines (a class of biochemical signal molecules) that prompt the activated B cell to divide. Immediately, the activated B cell generates two types of daughter cells: plasma cells (each of which synthesizes and releases approximately two thousand to twenty thousand antibody molecules per second into the bloodstream during its life span of four to five days) and B memory cells (which have a life span of a few months to a year). The B memory cells are the component of the immune memory system that, in response to a second exposure to the same type of antigen, produces antibodies in larger quantities and at faster rates over a longer time frame than the primary immune response. A similar cascade of events occurs when a macrophage presents an antigen directly to a B cell.

POLYCLONAL AND MONOCLONAL ANTIBODIES

Plasma cells originating from different B cells manufacture distinct antibody molecules because each B cell was presented with a specific portion of the same antigen by a helper T cell or macrophage. Thus a set of polyclonal antibodies is released in response to an invasion by a foreign agent. Each member of this group of polyclonal antibodies will launch the assault against the foreign agent by recognizing different epitopes of the same antigen. The polyclonal nature of antibodies has been well recognized in the medical field.

In the case of multiple myeloma (a type of cancer), one B cell out of billions in the body proliferates in an uncontrolled manner. Eventually, this event compromises the total population of B cells of the body. The immune system will produce huge amounts of IgG originating from the same B cell, which recognizes only one specific epitope of an antigen; therefore, this person's immune system produces a set of antibodies referred to as "monoclonal" antibodies. Monoclonal antibodies form a population of identical antibodies that all recognize and are specific for one epitope on an antigen. Thus, someone with this condition may suffer frequent bacterial infections because of a lack of antibody diversity. Indeed, a bacterium whose antigens do not match the antibodies manufactured by the overabundant monoclonal B cells has a selective advantage.

The high-affinity binding capacity of antibodies with antigens has been employed in both therapeutic and diagnostic procedures. It is, however, unfortunate that the effectiveness of commercial preparations of polyclonal antibodies varies widely from batch to batch. In some instances of immunization, certain epitopes of a particular antigen are strong stimulators of antibody-producing cells, whereas at other times, the immune system responds more vigorously to different epitopes of the same antigen. Thus one batch of polyclonal antibodies may have a low level of antibody molecules directed against a major epitope and not be as effective as the previous batch. Consequently, it is desirable to produce a cell line that will produce monoclonal antibodies with a high affinity for a specific epitope on the antigen for commercial use. Such a cell line would provide a consistent and continual supply of identical (monoclonal) antibodies. Monoclonal antibodies can be produced by hybridoma cells, which are generated by the fusion of cancerous B cells and normal spleen cells obtained from mice immunized with a specific antigen. After initial selection of hybridoma clones, monoclonal antibody production is maintained in culture. In addition, the hybridoma cells can be injected into mice to induce tumors that, in turn, will release large quantities of fluid containing the antibody. This fluid containing monoclonal antibodies can be collected periodically and may be used immediately or stored for future use. Various systems used to produce monoclonal antibodies include cultured lymphoid cell lines, yeast cells, *Trichoderma reese* (ascomycetes), insect cells, *Escherichia coli*, and monkey and Chinese hamster ovary cells. Transgenic plants and plant cell cultures have been explored as potential systems for antibody expression.

IMPACT AND APPLICATIONS

The high-affinity binding capacity of antibodies may be used to inactivate antigens in vivo (within a living organism). The binding property of antibodies may also be employed in many therapeutic and diagnostic applications. In addition, it is a very effective tool in both immunological isolation and detection methods.

Monoclonal antibodies may outnumber all other products being explored by various biotechnology-oriented companies for the treatment and prevention of disease. For example, many strategies for the treatment of cancerous tumors as well as for the inhibition of human immunodeficiency virus (HIV) replication are based on the use of monoclonal antibodies. HIV is a retrovirus (a virus whose genetic material is ribonucleic acid, or RNA) that causes acquired immunodeficiency syndrome (AIDS). Advances in plant biotechnology have made it possible to use transgenic plants to produce monoclonal antibodies on a large scale for therapeutic or diagnostic use. Indeed, one of the most promising applications of plant-produced antibodies in immunotherapy is in passive immunization (for example, against *Streptococcus mutans*, the most common cause of tooth decay). Large doses of the antibody are required in multiple applications for passive immunotherapy to be effective. Transgenic antibody-producing plants may be one source that can supply huge quantities of antibodies in a safe and cost-effective manner. It has been demonstrated that a hybrid IgA-IgG molecule produced by transgenic plants prevented colonization of *S. mutans* in culture, which appears to be how the antibody prevents colonization of this bacterium in vivo.

It has been estimated that antibodies expressed in soybeans at a level of 1 percent of total protein may cost approximately one hundred dollars per kilogram of antibody, which is relatively inexpensive in comparison with the cost of traditional antibiotics. Transgenic plants have also been used as bioreactors for the large-scale production of antibodies with no extensive purification schemes. In fact, antibodies have been expressed in transgenic tobacco roots and then accumulated in tobacco seeds. If this technology could be employed to obtain stable accumulation of antibodies in more edible plant organs such as potato tubers, it could potentially allow for long-term storage as well as a safe and easy delivery of specific antibodies for immunotherapeutic applications. In addition, plant-produced antibodies may be more desirable for human use than microbial-produced antibodies, because plant-produced antibodies undergo eukaryotic rather than the prokaryotic (bacterial) post-translational modifications. Human glycosylation (a biochemical process whereby sugars are attached onto the protein) is more closely related to that of plants than that of bacteria.

The potential use of antibody expression in plants for altering existing biochemical pathways has also been demonstrated. For example, germination mediated by a phytochrome (a biochemical produced by plants) has been altered by utilizing plant-produced antibodies. In addition, antibodies expressed in plants have been successfully used to immunize host plants against pathogenic infection; for example, tobacco plants have already been immunized with antibodies against viral attack. This approach has great potential to replace the traditional methods (use of chemicals) in controlling pathogens.

Sibdas Ghosh, Ph.D., and Tom E. Scola

FURTHER READING

Coico, Richard, and Geoffrey Sunshine. "Antibody Structure and Function." In *Immunology: A Short Course*. 6th ed. Hoboken, N.J.: Wiley-Blackwell, 2009. Describes the role of antibodies within the broader context of immunology.

Dübel, Stefan, ed. *Handbook of Therapeutic Antibodies*. 3 vols. Weinheim, Germany: Wiley-VCH, 2007. Comprehensive information about the development, production, and therapeutic application of antibodies. Volume 1 focuses on established techniques and clinical applications; volume 2 covers emerging technologies; volume 3 provides specific details about each currently approved type of antibody.

Glick, Bernard R., and Jack J. Pasternak, eds. *Molecular Biotechnology: Principles and Applications of Recombinant DNA*. Washington, D.C.: ASM Press, 1998. Discusses the structure and function of antibodies, as well as the role of biotechnology in the use of antibodies. Covers both the underlying scientific principles and the wide-ranging industrial, agricultural, pharmaceutical, and biomedical applications of recombinant DNA technology. Numerous illustrations and figures in both color and black and white.

Harlow, Ed, and David Lane, eds. *Using Antibodies: A Laboratory Manual*. Rev. ed. Cold Spring Harbor, N.Y.: Cold Spring Harbor Laboratory Press, 1999. A standard manual, providing a detailed account of different methods involved in the production and application of antibodies.

Kontermann, Roland, and Stefan Dübel, eds. *Antibody Engineering*. New York: Springer, 2001. Serves as a lab manual for antibody engineers, demonstrating the state of the art and covering all essential technologies in the field. Designed both to lead beginners in this technology and to keep experienced engineers current with the most detailed protocols. Includes color and halftone illustrations.

Mayforth, Ruth D. *Designing Antibodies*. San Diego: Academic Press, 1993. A practical introduction to designing antibodies for use in medicine or science. Explains such aspects as making monoclonal antibodies, designing them for human therapy, targeting, idiotypes, and catalytic antibodies.

Raz, E. *Immunostimulatory DNA Sequences*. New York: Springer, 2001. Includes chapters on the introduction and discovery of immunostimulatory DNA sequences, mechanisms of immune stimulation by bacterial DNA, and multiple effects of immunostimulatory DNA on T cells and the role of type I interferons.

Smith, Mathew D. "Antibody Production in Plants." *Biotechnology Advances* 14, no. 3 (1996): 267-281. Summarizes production and applications of plant-produced antibodies.

Wang, Henry Y., and Tadayuki Imanaka, eds. *Antibody Expression and Engineering.* Washington, D.C.: American Chemical Society, 1995. Among other topics, examines antibody production and expression in insect cells, plants, myeloma and hybridoma cells, and proteins.

WEB SITES OF INTEREST

Microbiology and Immunology On-Line
http://pathmicro.med.sc.edu/book/welcome.htm

An online textbook prepared by the University of South Carolina School of Medicine. "Immunology—Chapter 4: Immunoglobulins—Structure and Function" focuses on antibodies.

Mike's Immunoglobulin Structure/Function Home Page
http://www.path.cam.ac.uk/~mrc7/mikeimages
.html

Mike Clark, a lecturer in the department of pathology at Cambridge University, prepared this collection of articles and images about immunoglobulin structure. The colored drawings and animations of the Ig antibodies in humans and mice are particularly useful.

See also: Allergies; Autoimmune disorders; Biopharmaceuticals; Blotting: Southern, Northern, and Western; Diabetes; Hybridomas and monoclonal antibodies; Immunogenetics; Molecular genetics; Multiple alleles; Oncogenes; Organ transplants and HLA genes; Synthetic antibodies.

Antisense RNA

CATEGORY: Molecular genetics

SIGNIFICANCE: Antisense RNA and RNA interference are powerful modifiers of gene expression that act via RNA-RNA binding through comple-

mentary base pairing. This provides a flexible mechanism for specific gene regulation and has great potential for experimental studies and therapeutic action. RNA interference, a specialized form of antisense RNA, even mimics the immune system, for example, targeting RNA viruses within a cell. Processes involving antisense RNA appear in eukaryotes, eubacteria, and archaea.

KEY TERMS

antisense: a term referring to any strand of DNA or RNA that is complementary to a coding or regulatory sequence; for example, the strand opposite the coding strand (the sense strand) in DNA is called the antisense strand

down-regulation: a process of gene expression in which the amount that a gene is transcribed and/or translated is reduced

gene silencing: any form of genetic regulation in which the expression of a gene is completely repressed, either by preventing transcription (pre-transcriptional gene silencing) or after a messenger RNA (mRNA) has been transcribed (post-transcriptional gene silencing

RNA interference (RNAi): sequence-specific degradation of messenger RNA (mRNA) caused by complementary double-stranded RNA

up-regulation: a process of gene expression in which the amount that a gene is transcribed and/or translated is increased

DISCOVERY

There are many kinds of RNA molecules in addition to the three main types of messenger RNA (mRNA), transfer RNA (tRNA), and ribosomal RNA (rRNA). Some have an effect on mRNA molecules through complementary binding. When this type of RNA binds to an mRNA, it effectively blocks translation of the mRNA and can therefore be described as having an antisense action; that is, it blocks the expression of the message in the mRNA. Antisense RNA was discovered in 1981 as a mechanism regulating copy number of bacterial plasmids. Some RNAs, such as small nuclear and small nucleolar RNAs (snRNA and snoRNA), splice and edit other RNAs, guided by complementary base pairing.

Various forms of gene down-regulation were discovered throughout the 1990's, including post-transcriptional gene silencing (prevention of mRNA translation) in plants, gene silencing (pre-

Dr. Phillip A. Sharp, who with Richard J. Roberts won the 1993 Nobel Prize in Physiology or Medicine. Sharp acknowledges that the discovery of antisense RNA and RNA interference has changed his cancer research. The process could theoretically offer ways of "silencing" the genes associated with cancer. (AP/Wide World Photos)

vention of gene transcription) in fungi, and RNA interference in the nematode *Caenorhabditis elegans.* The importance of noncoding RNA molecules, including antisense RNA, is becoming clear. They add a previously unknown level of genetic complexity, and the extent of their influence is yet to be determined fully.

NATURAL FUNCTION

Antisense RNA is utilized in a number of ways by bacterial plasmids. Replication of ColE1 plasmids requires an RNA preprimer, called RNA II, that interacts with the origin of replication and forms a particular secondary structure. This allows an enzyme to cut and form the mature primer needed for DNA replication. Antisense RNA I can bind to RNA II, preventing the formation of the necessary struc-

ture. In the R1 plasmid, the CopA antisense RNA can bind and prevent the translation of the RNA transcript for replication initiation protein RepA. Thus, change in plasmid number is controlled by changing levels of antisense RNA, modifying the ability of plasmids to replicate.

Many plasmids use antisense RNA to ensure their maintenance within bacteria. The R1 plasmid transcribes Hok toxin mRNA, but interaction with antisense Sok RNA prevents its translation. Sok RNA is less stable than Hok RNA, so plasmid loss leads to Sok degradation but leaves some Hok transcripts. These transcripts are translated into a toxin that kills the cell in an elegant mechanism of selection for plasmid propagation. Antisense regulation has also been found in association with transposons and bacteriophages.

Bacteria use antisense RNA to regulate particular genes. Such RNA is often encoded in a region different from that of the target and may affect multiple genes. For example, the OxyS RNA, induced by oxidative stress, inhibits translation of fhlA mRNA, involved in formate metabolism. In conjunction with the protein Hfq, OxyS RNA binds near the ribosome-binding site in fhlA mRNA, preventing translation. MicF RNA is induced under cellular stress and binds to the mRNA of membrane pore protein ompF to prevent its translation.

One of the first examples of antisense regulatory mechanisms in eukaryotes came from the nematode *C. elegans*. Small antisense RNA molecules lin-4 and let-7 show imperfect base-pairing to the 3′ untranslated region of their target gene mRNAs. This results in translational inhibition and is important for normal development. These small antisense RNAs are members of the microRNA (miRNA) class of small, single-stranded molecules found throughout eukaryotes. They are produced by cleavage of longer molecules containing partial self-complementarity that produces a hairpin structure.

Antisense RNA has been implicated in other processes. Imprinted genes are often associated with antisense transcripts from the same locus, and double-stranded RNA may be capable of affecting DNA chromatin structure through methylation of homologous sequences.

RNA INTERFERENCE

RNA interference (RNAi) causes sequence-specific gene silencing in response to the presence of double-stranded RNA. The pioneers of RNAi research, Andrew Fire and Craig Mello, were awarded the 2006 Nobel Prize in Physiology or Medicine for their work involving the injection of sense/antisense RNA pairs into *C. elegans* and observation of the resulting phenotypes. They found that injection of double-stranded RNA led to efficient loss of targeted homologous mRNA by a post-transcriptional mechanism. The process of gene silencing by RNAi is proposed to have evolved as a mechanism of avoiding viral infection and limiting replication of transposable elements and repeat sequences, as all of these can involve double-stranded RNA and are recognized by the RNAi system as foreign nucleic acid molecules. RNA silencing by mechanisms involving RNAi is therefore part of the immune defense of many organisms, including plants, worms

and flies. The relevance of RNAi in vertebrate defense against viruses and transposable elements is less clear.

The mechanism of RNA silencing by RNAi is present in a wide variety of eukaryotes (including mice and probably humans), and the steps involved are likely to be similar. The process begins with the recognition of a long double-stranded RNA molecule by the conserved RNase III-type endonuclease enzymes Drosha and Dicer, which cut the long RNA to produce a double-stranded RNA about 20 to 24 nucleotides long with overhanging 3′ ends). This molecule is unwound by the helicase activity of Argonaute proteins recruited by Dicer. The mature, single-stranded small interfering RNA (siRNA), produced by these actions acts as a guide molecule for the RNA-induced silencing complex (RISC), which contains multiple nucleases and uses the antisense strand of the siRNA to recognize complementary RNA sequences. These sequences are then either translationally repressed or cleaved and degraded. Some organisms, including plants, fungi and *C. elegans*, use an RNA-dependent RNA polymerase (RdRP) to amplify the siRNA signal by producing secondary siRNAs to be processed by Dicer. *C. elegans* and plants show evidence of a systemic response, whereby initial silencing in one cell spreads to other cells by transport of siRNA (occurring via phloem in plants).

Several different types of RNAi, including siRNA and miRNA, have been discovered, and all take advantage of the properties of complementary antisense RNA by binding target RNA, particularly mRNA, via noncovalent base pairing. For siRNA, binding of target sequences is governed by a critical region in the siRNA sequence called the "seed region," the ribonucleotides encompassing positions 2 through 7. It is this region that gives siRNA its target specificity and allows RISC to bind, leading to target cleavage or repression. While the seed region is important in target recognition, complementarity in other regions is critical for target cleavage.

IMPACT

Many disease states are a result of abnormal gene expression and are therefore potential targets for gene therapy. One therapeutic approach involves the use of antisense RNA or RNAi. Antisense oligonucleotides containing CpG sequences, for example, have been shown to be immunostimulatory and are being studied in clinical trials for cancer, asthma, and allergies, and as vaccine adjuvants.

Cancer cells often show overexpression of genes involved in growth and proliferation. These genes, as well as mutated genes, can be targeted using antisense RNA to decrease expression of proteins encoded by these genes to prevent tumor growth. mRNA from a mutant allele may be targeted for degradation in a heterozygous patient, allowing expression of the correct protein from only the wild-type allele. Several RNAi molecules have been evaluated in clinical trials for cancer and other diseases. Antisense techniques are being studied for the targeting of viruses and prevention of their replication and could eventually be used to correct aberrant splicing of gene transcripts. Many of these potential uses have been successfully demonstrated in experimental systems, including cell culture and mouse models.

Many issues remain to be addressed before antisense RNA therapeutics are truly feasible. Effective delivery systems are needed to produce sustained effects in appropriate cell types. The systemic transport of RNAi in some organisms may facilitate therapeutic applications. The safety of such approaches remains to be established, but, overall, therapeutic uses of antisense RNA are promising.

Peter J. Waddell, Ph.D.;
updated by Scott J. Salsman, Ph.D.

FURTHER READING

Brantl, S. "Antisense-RNA Regulation and RNA Interference." *Biochimica et Biophysica Acta* 1575 (2002): 15-25. A detailed survey of the wide variety of ways in which antisense RNAs operate.

Naqvi, A. R., M. N. Islam, N. R. Choudhury, and Q. M. R. Haq. "The Fascinating World of RNA Interference." *International Journal of Biological Sciences* 5 (2009): 97-117. A thorough review of the known types of small interfering RNA and how they originate and function.

Rayburn, E. R., and R. Zhang. "Antisense, RNAi, and Gene Silencing Strategies for Therapy: Mission Possible or Impossible?" *Drug Discovery Today* 13 (2008): 513-521. An objective report on the attempts at producing therapies based on antisense and RNAi mechanisms, both those that have failed and those that still show promise.

Shuey, D. J., D. E. McCallus, and T. Giordano. "RNAi: Gene Silencing in Therapeutic Interven-

tion." *Drug Discovery Today* 7 (2002): 1040-1046. A look at challenges ahead for using RNAi in a medical setting.

WEB SITES OF INTEREST

Broad Institute of MIT and Harvard. The RNAi Consortium
http://www.broad.mit.edu/rnai/trc

National Center for Biotechnology Information. "RNA Interference: A Guide to Computational Resources"
http://www.ncbi.nlm.nih.gov/genome/RNAi

See also: Gene regulation: Eukaryotes; Human genetics; Model organism: *Caenorhabditis elegans*; Noncoding RNA molecules; RNA structure and function; RNA transcription and mRNA processing; RNA world; Viral genetics.

APC gene testing

CATEGORY: Molecular genetics; Techniques and methodologies

SIGNIFICANCE: *APC* gene testing is used to identify individuals at extreme risk for colon cancer caused by a germline mutation in the adenomatous polyposis coli gene. Introduced in 1994, *APC* gene testing was the first service commercially developed for presymptomatic diagnosis of an adult-onset disease; it illustrates well the utility and limitations of such testing.

KEY TERMS

adenoma: benign tumor originating from secretory (glandular) epithelial cells

adenomatous polyposis coli (APC): a familial adenomatous polyposis caused by mutations in the *APC* gene; APC can also refer to the protein encoded by the *APC* gene; when italicized, APC refers to the *APC* gene

APC *gene:* a tumor-suppressor gene; mutations of this gene cause several closely related colorectal polyposis syndromes including FAP, Gardner syndrome, and Turcot syndrome

autosomal dominant: mutation on a nonsex chromosome in which only one mutant allele is sufficient to produce the phenotype (disease)

familial adenomatous polyposis (FAP): a precancerous, genetically inherited polyposis of the epithelial lining of the large intestine (colon) or rectum; subclassified as classic, attenuated, or autosomal recessive, with classic and attenuated FAP caused by mutations of the *APC* gene and autosomal recessive FAP caused by mutations in the DNA repair gene *MUTYH*

missense mutation: a single base substitution resulting in a codon specifying an incorrect amino acid

nonsense mutation: a single base substitution producing a premature stop codon

polyp: abnormal growth or tumor projecting from a mucous membrane

polyposis: the presence of many adenoma polyps, generally more than one hundred

tumor-suppressor gene: a gene in which functional loss of both alleles in a somatic cell causes tumorogenesis

APC STRUCTURE AND FUNCTION

First identified (cloned) in 1991, the *APC* gene is located on the long arm of human chromosome 5 (megabase position 112.1). The gene encompasses 108 kilobases of DNA, making it a relatively large target for mutations. Its fifteen exons produce an mRNA transcript of about 10 kilobases that encodes a protein of 2,843 amino acids (312 kilodaltons). This huge protein, nine times larger than average, consists of binding sites for a variety of proteins including APC itself, axin, beta-catenin, conductin, EB1, tubulin, and protein kinases.

APC functions in a variety of tissues as part of a large protein complex that targets beta-catenin for degradation. In the absence of this degradation, beta-catenin translocates to the nucleus to promote cell division. APC also interacts with proteins necessary for chromosome segregation, cell adhesion, and cytoskeletal structure; the absence of these interactions could potentially promote tumor growth as well.

APC MUTATIONS

Roughly 70 percent of colorectal tumors have mutations in the *APC* gene. Less than 1 percent of cases are caused by germline *APC* mutations; the remainder are somatic mutations present only in the tumor. A mutation in the germline, because it is present in all cells, leads uniquely to hundreds or thousands of polyps. This occurs because individu-als with only one functional copy of *APC* are highly susceptible to losing all normal APC activity in a cell by chance mutation of the second copy, especially in tissues with a high turnover rate such as the colorectal epithelium—loss of APC function in a single cell is sufficient to initiate polyp development. Consequently, inheriting a nonfunctional or dysfunctional *APC* allele results inevitably in polyps. Approximately 25 percent of FAP cases are not inherited; they are caused by a new (de novo) germline mutation of the *APC* gene.

More than nine hundred *APC* mutations causing FAP have been identified, with almost 90 percent producing a truncated protein. Almost 50 percent are small deletions, about 25 percent are nonsense mutations, about 15 percent are large deletions or rearrangements, about 10 percent are small insertions, and about 3 percent are missense mutations. Two of the most common mutations occur at codons 1061 (about 10 percent frequency) and 1309 (about 15 percent frequency). All these mutations lead to polyposis, and several additional somatic mutations of other genes have to occur for the polyps to become cancerous. When there are many polyps, however, the lifetime risk of malignancy is 100 percent, with cancer typically occurring between thirty-five and forty-five years of age.

MUTATION TESTING

Individuals with a high familial risk for FAP typically undergo *APC* gene testing between ten and sixteen years of age. In cases where the *APC* mutation of relatives has not already been identified, *APC* testing has a 70 to 90 percent chance of discovering a causal mutation. The location of the mutation also provides important diagnostic information: Classic FAP is typically associated with mutations between codons (amino acid positions) 168 and 1580, whereas mutations outside this region are often associated with attenuated FAP (fewer than one hundred polyps and older age of onset); severe polyposis is more likely when mutations occur between codons 1250 and 1464; and a number of mutations tend to be associated with additional pathologies outside the colon and rectum.

A variety of methods are used for *APC* mutation testing, usually conducted on DNA from a peripheral blood sample. The most thorough method, used in high-throughput laboratories, is to sequence the entire coding region and splice junctions. Another method used by low- to medium-throughput laboratories is denaturing high-performance liquid chromatography (DHPLC). This method uses DNA fragments generated by the polymerase chain reaction (PCR) and hybridizes them to complementary fragments with no mutations. Sequence mismatches alter the migration rate of the hybrid DNA through an HPLC column. A limitation of this method is that large deletions or rearrangements may prevent PCR amplification. A third procedure is real-time reverse transcription polymerase chain reaction (RT-PCR), which measures the relative RNA copy number of each exon, with a reduced copy number indicating a deletion or truncation. A similar approach is the RNA-based protein truncation test (also called an in vitro synthesized-protein assay).

BENEFITS AND CAVEATS

Colorectal cancer is highly preventable when polyps are detected and removed as soon as possible. In the case of FAP, treatment ordinarily means removing the entire colon and perhaps rectum. APC gene testing helps physicians decide how soon and frequently patients should be monitored for polyps and how aggressively to intervene surgically. Testing also provides the important benefit of identifying family members who are not carrying the APC mutation.

Like all genetic tests, APC testing has limitations. One is that a negative result does not rule out a mutation in families where the mutation has not previously been identified. A second, is that even with surgery and subsequent monitoring, patients still have an above-normal risk of cancer. Testing for an APC mutation can also cause anxiety and depression regardless of test outcome: those testing positive because of the emotional burden and those testing negative because of possible survivor's guilt.

IMPACT

In Western societies, colorectal cancer is the leading cause of cancer among nonsmokers; the death rate in the United States averages 140 people per day. APC was the first colorectal cancer gene identified, and APC testing has contributed to a better understanding of colorectal cancer and saved many lives.

Brad A. Rikke, Ph.D.

FURTHER READING

Chapman, P. D., and J. Burn. "Colorectal Cancer." In *Molecular Genetics of Cancer*, edited by J. K. Cowell. 2d ed. Oxford, England: BIOS Scientific, 2001. This book is a comprehensive survey of genetic factors predisposing to many different kinds of cancers.

Harrington, Susan M., and Malcolm G. Dunlop. "Familial Colon Cancer Syndromes and Their Genetics." In *Genetic Predisposition to Cancer*, edited by R. A. Eeles, D. F. Easton, B. A. J. Ponder, and C. Eng. 2d ed. New York: Oxford University Press, 2004. This book is a compendium of articles by experts covering biological, social, and ethical aspects of cancer genetics.

Hegde, Madhuri R., and Benjamin B. Roa. "Detecting Mutations in the *APC* Gene in Familial Adenomatous Polyposis (FAP)." In *Current Protocols in Human Genetics*. New York: John Wiley & Sons, 2006. A laboratory manual detailing several methods for detecting *APC* gene mutations.

WEB SITES OF INTEREST

GeneTests at NCBI
http://www.genetests.com

Human Genome Mutation Database
http://www.hgmd.cf.ac.uk/ac/index.php

Online Mendelian Inheritance in Man #175100:
Adenomatous Polyposis of the Colon
http://www.ncbi.nlm.nih.gov/entrez/dispomim.cgi?id=175100

U.S. National Library of Medicine, Gene Reviews
http://www.ncbi.nlm.nih.gov/bookshelf/br.fcgi?book=gene&part=fap

U.S. National Library of Medicine, Genetics Home Reference
http://ghr.nlm.nih.gov/gene=apc

See also: Cancer; Colon cancer; Genetic screening; Genetic testing; Mutagenesis and cancer; Mutation and mutagenesis; Oncogenes; Tumor-suppressor genes.

Apert syndrome

CATEGORY: Diseases and syndromes
ALSO KNOWN AS: Acrocephalosyndactyly type 1 (ACS1)

DEFINITION

Apert syndrome is a rare genetic condition causing abnormal growth and development of bone and cutaneous tissue, resulting in a characteristic appearance of the skull, face, hands, and feet. It is one of a group of disorders known as craniofacial/limb anomalies. Apert syndrome is inherited as an autosomal dominant trait. A significant number of cases occur with no prior family history of the disease. The syndrome is named for French physician Eugene Apert, who first described it in 1906.

RISK FACTORS

Apert syndrome may be inherited from a parent who has the disorder, but it most frequently occurs as a sporadic (new) mutation. Because the trait is autosomal dominant, a parent with Apert syndrome has a 50 percent risk of transmitting the disease. An unaffected child of a parent with the condition is no more likely to have a child with Apert syndrome than the general population. Research has established an association between advanced paternal age (greater than fifty years) and Apert syndrome. Prevalence is approximately 1 in 65,000 births.

ETIOLOGY AND GENETICS

The genetic mutation responsible for Apert syndrome affects a single gene on chromosome 10, band q26. It is known as fibroblast growth factor receptor 2 (*FGFR2*). All individuals have two copies of this gene, which consists of a sequence of approximately two thousand DNA building blocks, or base pairs. Apert syndrome results when one of these is replaced by an incorrect pair. Two different mutations are known to cause the disorder. Both lead to abnormal development of the skull, face, and limbs. The presence of one mutation appears to increase the severity of face and skull abnormalities, while the other may lead to a more severe form of fusion of the hands and feet.

As an autosomal dominant trait, a single copy of the abnormal gene is all that is required for expression of the disease. The other copy of the gene is entirely normal. However, familial cases are relatively rare, likely because of decreased fertility as a result of the significant physical and mental disabilities suffered by those with the disorder. More than 98 percent of cases arise from new mutations.

Mutations in the *FGFR2* gene lead to increased bone formation, particularly of the skull, during fetal life. This causes premature closure of the cranial sutures and abnormal growth of the skull and face. Development of the hands and feet are also affected. The result is the characteristic appearance of a flattened skull, sunken midface, and protruding eyes. Partial to complete fusion of the digits of the hands and feet (syndactyly) may also occur as a result of abnormal bone and soft tissue growth. Patients with one of the known mutations, *Pro253Arg*, have a more severe form of syndactyly, while patients with the *Ser252Trp* mutation have a higher rate of cleft palate and visual impairment.

SYMPTOMS

The major features of Apert syndrome are craniofacial malformation and syndactyly of fingers and toes. Proptosis (bulging eyes), a flattened head shape, and a small and concave middle third of the face result, Apert syndrome is distinguished from similar craniofacial conditions by coexisting syndactyly, often affecting both fingers and toes.

Additional associated symptoms include mental retardation (variable), brain malformation, hearing loss and ear abnormalities, vision loss, heart defects, airway narrowing, hydrocephalus, sleep apnea, and severe acne. Affected infants are usually identified at birth because of the characteristic physical features.

SCREENING AND DIAGNOSIS

Since most cases arise from new mutations, prenatal screening for Apert syndrome is largely based on ultrasound imaging to identify typical cranial and limb malformations. If Apert syndrome is suspected, then pregnant patients should be referred to a specialized center for further evaluation and screening for related abnormalities. If the gene defect has been identified in an affected parent, then DNA testing is available by amniocentesis or chorionic villus sampling. Postnatal molecular analysis is available to confirm mutations in the *FGFR2* gene.

Newborn diagnosis is confirmed by physical examination and skull X ray to evaluate craniosynostosis (premature fusion of the cranial sutures). Other imaging studies can identify additional problems for future intervention. Hearing assessment, genetic counseling, and parent support is recommended.

TREATMENT AND THERAPY

Treatment begins immediately after birth. A multidisciplinary team is best equipped to address the wide variety of issues. Early surgery is often necessary to release the cranial sutures and allow normal brain growth. Several procedures are available to improve the structure and shape of the face. Fingers and toes may also require surgical separation. Other associated symptoms may be treated as required.

PREVENTION AND OUTCOMES

There is currently no known strategy to prevent the sporadic occurrence of Apert syndrome. Prenatal diagnosis and counseling may allow parents and their health care team to prepare an early treatment plan.

Prognosis depends on the severity of symptoms and early surgical treatment. Advances in craniofacial management allow children with Apert syndrome to maximize social and intellectual functioning. Approximately 50 percent will have intelligence close to the normal range.

Rachel Zahn, M.D.

FURTHER READING

Children's Craniofacial Association. *A Guide to Understanding Apert Syndrome.* Dallas: Author, 2008.

Liptak, Gregory S., and Joseph M. Serletti. "Consultations with the Specialist: Pediatric Approach to Craniosynostosis." *Pediatrics in Review* 19 (1998): 352-359.

Robin, Nathaniel H. *Genetics: Its Application to Speech, Hearing, and Craniofacial Disorders.* San Diego: Plural, 2008.

WEB SITES OF INTEREST

eMedicine: Apert Syndrome
http://emedicine.medscape.com/article/941723

Faces: The National Craniofacial Association
http://faces-cranio.org

Teeter's Page
http://apert.org

See also: Brachydactyly; Congenital defects; Hereditary diseases; Polydactyly.

Archaea

CATEGORY: Cellular biology

SIGNIFICANCE: Archaea are diverse prokaryotic organisms distinct from the historically familiar bacteria. Archaea have certain molecular properties previously thought to occur only in eukaryotes, and others commonly associated with bacteria. Many archaea require severe conditions for growth, and their genetic processes have adapted to these extreme conditions in ways that are not fully understood.

KEY TERMS

conjugation: the process by which one bacterial cell transfers DNA directly to another

domain: the highest-level division of life, sometimes called a superkingdom

extreme halophiles: microorganisms that require extremely high salt concentrations for optimal growth

insertion sequence: a small, independently transposable genetic element

methanogens: microorganisms that derive energy from the production of methane

prokaryotes: unicellular organisms with simple ultra-structures lacking nuclei and other intracellular organelles

small subunit ribosomal RNA (ssu rRNA): the RNA molecule found in the small subunit of the ribosome; also called 16S rRNA (in prokaryotes) or 18S rRNA (in eukaryotes)

GENE SEQUENCES MEASURE THE DIVERSITY OF PROKARYOTES

Prokaryotic microorganisms have been on earth for as many as 3.5 billion years and have diverged tremendously in genetic and metabolic terms. Unfortunately, the magnitude of this divergence has made it difficult to measure the relatedness of prokaryotes to one another. In the 1970's, Carl R. Woese addressed this problem using a method of reading short sequences of ribonucleotides from a highly conserved RNA molecule, the small subunit ribosomal RNA (ssu rRNA). Because this RNA is present in all organisms and has evolved very slowly, any two organisms have at least a few of these short nucleotide sequences in common. The proportion of shared sequences thus provided a quantitative index of similarity by which all cellular organisms could, in principle, be compared.

When the nucleotide sequence data were used to construct an evolutionary tree, eukaryotes (plants, animals, fungi, and protozoa) formed a cluster clearly separated from the common bacteria. Unexpectedly, however, a third cluster emerged that was equally distinct from both eukaryotes and common bacteria. This cluster consisted of prokaryotes that lacked biochemical features of most bacteria (such as a cell wall composed of peptidoglycan); possessed other features not found in any other organisms (such as membranes composed of isoprenoid ether lipids); and occurred in unusual, typically harsh, environments. Woese and his coworkers eventually designated the three divisions of life represented by these clusters, or "domains," naming the nonbacterial prokaryotes the domain *Archaea*.

BIOLOGY OF THE DOMAIN *ARCHAEA*

Archaea tend to require unusual conditions for growth, which has made it challenging to determine their genetic properties. The methanogens, for example, live by converting hydrogen (H_2) and carbon dioxide (CO_2) or other simple carbon compounds into methane and are killed by even trace amounts of oxygen. The extreme halophiles, in contrast, normally live in brine lakes and utilize oxygen for growth. However, they require extremely high concentrations of salt to maintain their cellular structure. A third class of archaea, the extreme thermophiles, occur naturally in geothermal springs and grow best at temperatures ranging from 60-105 degrees Celsius (140-221 degrees Fahrenheit). Many derive energy from the oxidation or reduction of sulfur compounds. Sequencing of DNA fragments recovered from "moderate" environments, such as ocean water or soil, has revealed many additional archaeal species that presumably do not require unusual environmental conditions but have never been cultured in the laboratory.

THE GENETIC MACHINERY OF ARCHAEA

Because bacteria and eukaryotes differ greatly with respect to gene and chromosome structure and the details of gene expression, molecular biologists have examined the same properties in archaea and have found a mixture of "bacterial" and "eukaryotic" features. The organization of DNA within archaeal cells is bacterial, in the sense that archaeal

chromosomes are circular DNAs of between 2 million and 4 million base pairs having single origins of replication, normally replicated bidirectionally. As in bacteria, the genes are densely packed and often grouped into clusters of related genes transcribed from a common promoter. The promoters themselves, however, resemble the TATA box/BRE element combination of eukaryotic DNA polymerase II (Pol II) promoters, and the RNA polymerases have the complex subunit composition of eukaryotes rather than the simple composition found in bacteria. Furthermore, archaea initiate transcription by a simplified version of the process seen in eukaryotic cells. Transcription factors (TATA-binding protein and a TFIIB) first bind to regions ahead of the promoter, then recruit RNA polymerase to attach and begin transcription. Introns are rare in archaea, however, and do not interrupt protein-encoding genes, but have been found to interrupt RNA-coding genes. Also, the regulation of transcription in archaea seems to depend heavily on the types of repressor and activator proteins found in bacteria; however, regulatory proteins of the eukaryotic type, and those totally unique to archaea, have also been found.

GENOMES OF ARCHAEA

The availability of complete DNA sequences now enables archaeal genomes to be compared to the genomes of bacteria and eukaryotes. One pattern that emerges from these comparisons is that most of the archaeal genes responsible for the processing of information (synthesis of DNA, RNA, and proteins) resemble their eukaryotic counterparts, whereas most of the archaeal genes for metabolic functions (biosynthetic pathways, for example) resemble their bacterial counterparts. The genomes of archaea also reveal probable cases of gene acquisition from distant relatives, a process called lateral gene transfer.

A third pattern to emerge from genome comparisons is that some archaea are missing genes thought to be important or essential. For example, the genomes of at least two methanogenic archaea do not encode an enzyme that charges transfer RNA (tRNA) with cysteine. These archaea instead use a novel strategy for making cysteinyl tRNA. Some of the seryl tRNA made by these cells is converted to cysteinyl tRNA by a specialized enzyme. A more severe example of gene deficiency is provided by *Nanoarchaeum equitans*, the first reported parasitic or symbiotic archaea that grows attached to an *Igni-*

coccus, another hyperthermophile. *N. equitans* has been reported to have a volume approximating 1 percent of an *Escherichia coli* cell, the smallest non-viral cellular genome (0.49 Mbp), and numerous 16S rRNA nucleotide base substitutions even in regions normally conserved in other archaeal species. This extremely small genome lacks genes necessary for numerous metabolic functions including genes coding for lipid, amino acid, nucleotide, and enzyme cofactor biosynthesis. This includes genes coding for vital catabolic pathways including glycolysis. It has been suggested but not proven that these functions are supplied by its *Ignicoccus* host. Even more intriguing is the much longer list of archaea—all of which happen to be hyperthermophiles, which grow optimally at 80 degrees Celsius (176 degrees Fahrenheit) or above—that lack genes for the DNA mismatch repair proteins found in all other organisms.

UNIQUE GENETIC PROPERTIES?

This last observation raises an important question: Has an evolutionary history distinct from all conventional genetic systems, combined with the special demands of life in unusual environments, resulted in unique genetic properties in archaea? Although basic genetic assays can be performed in only a few species, the results help identify which genetic properties of cellular organisms are truly universal and which ones may have unusual features in archaea.

The methanogen *Methanococcus voltae* transfers short pieces of chromosome from one cell to another, using particles that resemble bacterial viruses (bacteriophages). This means of gene transfer has been seen in only a few bacteria. In other methanogens, researchers have used more conventional genetic phenomena, such as antibiotic-resistance genes, plasmids, and transposable elements, to develop tools for cloning or inactivating genes. As a result, new details about the regulation of gene expression in archaea and the genetics of methane formation are now coming to light.

The extreme halophile *Halobacterium salinarum* exhibits extremely high rates of spontaneous mutation of the genes for its photosynthetic pigments and gas vacuoles. This genetic instability reflects the fact that insertion sequences transpose very frequently into these and other genes. A distantly related species, *Haloferax volcanii*, has the ability to transfer chromosomal genes by means of conjugation. Although many bacteria engage in conjugation, the mechanism used by *H. volcanii* does not resemble the typical bacterial system, since no plasmid seems to be involved, and there is no apparent distinction between donor strain and recipient strain in the transfer of DNA.

Genetic tools for the archaea from geothermal environments are less well developed, but certain selections have made it possible to study spontaneous mutation and homologous recombination in some species of *Sulfolobus*. At the normal growth temperatures of these aerobic archaea, 75-80 de-

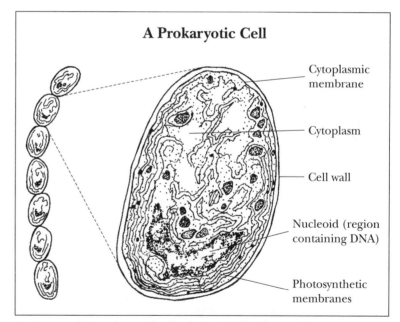

A Prokaryotic Cell

Cytoplasmic membrane

Cytoplasm

Cell wall

Nucleoid (region containing DNA)

Photosynthetic membranes

Archaea and bacteria are the simplest and oldest forms of life, consisting of prokaryotic cells, which differ from the cells that form higher organisms (fungi, algae, protozoa, plants, and animals), called eukaryotic cells. Based on an electron microscope image of one cell in a string forming a cyanobacterium, this depiction shows the basic features of a prokaryote. Note the lack of a defined nucleus and organelles (there are no plastids or mitochondria)—the components that house genetic information in eukaryotes. Instead, genetic material in prokaryotes is located in an unbound region called the nucleoid. (Kimberly L. Dawson Kurnizki)

grees Celsius (167-176 degrees Fahrenheit), spontaneous chemical decomposition of DNA is calculated to be about one thousand times more frequent than in the organisms previously studied by geneticists. In spite of this, *Sulfolobus acidocaldarius* exhibits the same frequency of spontaneous mutation as *E. coli* and significantly lower proportions of base-pair substitutions and deletions. This indicates especially effective mechanisms for avoiding or accurately repairing DNA damage, including mismatched bases, despite the fact that no mismatch repair genes have been found in *Sulfolobus* species. Also, *S. acidocaldarius*, like *H. volcanii*, has a mechanism of conjugation that does not require a plasmid or distinct donor and recipient genotypes. The transferred DNA recombines efficiently into the resident chromosome, as indicated by frequent recombination between mutations spaced only a few base pairs apart.

Finding two similar and unusual mechanisms of conjugation in two dissimilar and distantly related archaea (*H. volcanii* and *S. acidocaldarius*) raises questions regarding the possible advantages of this capability. Population genetic theory predicts that organisms that reproduce clonally (as bacteria and archaea do) would benefit from occasional exchange and recombination of genes, because this accelerates the production of beneficial combinations of alleles. Such recombination may be particularly important for archaea such as *Haloferax* and *Sulfolobus* species, whose extreme environments are like islands separated by vast areas that cannot support growth. For these organisms, frequent DNA transfer between cells of the same species may provide an efficient way to enhance genetic diversity within small, isolated populations.

IMPACT

Woese's monumental discovery that two very different prokaryotic groups (bacteria and archaea) exist based on DNA sequencing of a conserved macromolecule (rRNA) led to a complete reevaluation of the evolution of not only bacteria (previously including archaea) but also eukaryotes. Before, it was thought that eukaryotes evolved from prokaryotes. His data definitively showed that all three were derived from a common ancestor. Even more surprising was the fact the all eukaryotes were found to be more closely related to each other and distantly related to both bacteria and archaea. Finally, his data

showed that archaea were more closely related to eukaryotes than bacteria and were the most ancient organisms derived first from the common ancestor.

Since then, the DNA sequences of numerous archaeal isolates of different groups have been compared to each other and also to those of both bacteria and eukaryotes, further delineating the evolution of different diverse archaeal groups. These data suggest that a hyperthermophile was probably the common ancestor of *Archaea*. Also, the properties of numerous types of archaea have been studied. Many are found in extremely harsh environments that normal bacteria and eukaryotes cannot tolerate. This has led to molecular and genetic studies of the macromolecules including proteins and lipids that allow survival of these organisms, so that their mechanisms can be elucidated. Finally, various DNA polymerases naturally found in these organisms have been employed in modern DNA analysis techniques that require enzymes to be heat-stable, thus providing a practical result of the study of these important organisms.

Dennis W. Grogan, Ph.D.;
updated by Steven A. Kuhl, Ph.D.

FURTHER READING

Cavicchioii, Richard, ed. *Archaea: Molecular and Cellular Biology*. New York: ASM Press, 2007. This book contains a series of review articles written by experts in the field highlighting all aspects of molecular and cellular biology of this group of organisms that allow them to survive under extreme conditions.

Garrett, Roger A., and Hans-Peter Klenk, eds. *Archaea: Evolution, Physiology, and Molecular Biology*. New York: Wiley-Blackwell, 2007. This book contains a series of broad review articles exploring all aspects of archaea and specialist articles concentrating on the molecular biological aspects of the organism.

Madigan, Michael T., and John M. Martinko. *Brock Biology of Micro-organisms*. 11th ed. Upper Saddle River, N.J.: Prentice Hall, 2006. Chapter 13 of this popular microbiology text provides an accurate and well-illustrated overview of the biological diversity of the archaea.

Olsen, Gary, and Carl R. Woese. "Archaeal Genomics: An Overview." *Cell* 89 (1997): 991-994. This mini-review article, along with several accompanying articles, summarizes for specialists numerous molecular differences and similarities be-tween archaea and bacteria or eukaryotes, based on the first archaeal genomes to be sequenced.

Woese, Carl R. "Archaebacteria." *Scientific American* 244 (1981): 98-122. A clear, though somewhat dated, description of the archaea and the various lines of evidence for their status as a "third form of life."

WEB SITES OF INTEREST

GNN Genome News Network
http://www.genomenewsnetwork.org/articles/
05_02/amino_acid.shtml

GOLD Genomes Online Database
http://www.genomesonline.org

See also: Antisense RNA; Bacterial genetics and cell structure; Gene regulation: Bacteria; Lateral gene transfer; Noncoding RNA molecules.

Arnold-Chiari syndrome

CATEGORY: Diseases and syndromes
ALSO KNOWN AS: Arnold-Chiari malformation; type II Chiari malformation; cerebellomedullary malformation syndrome

DEFINITION

Arnold-Chiari syndrome is a defect in the formation of the cerebellum (the small, bottom portion of the brain) and brainstem. This defect can prevent the passage of blood from the brain into the spinal canal. Arnold-Chiari syndrome is usually accompanied by a myelomeningocele, which is a form of spina bifida. There are four types of Arnold-Chiari syndrome, with different degrees of severity. Type 2 is the one that is associated with spina bifida.

RISK FACTORS

Spina bifada and hydrocephalus are commonly associated with Arnold-Chiari syndrome but are not thought to be a cause of the disorder.

ETIOLOGY AND GENETICS

Primary Arnold-Chiari malformations typically result from structural defects in the spinal cord and brain which occur during fetal development. There has been suspicion for many years that this malformation results at least in part from a genetic defect because medical literature has reported about several families in which more than one family member is affected. Despite these observations, no specific candidate gene or genes have been identified and there is no clear basis for categorizing this condition as a genetic disorder (as opposed to a congenital birth defect).

Some researchers have suggested that the developmental trigger is a lack of particular vitamins or specific nutrients in the maternal diet. Others suggest that since the base of a patient's skull is often quite small, resulting in the cerebellum being forced downward, the genes regulating this aspect of skull development must be involved. A third group suggests that the primary cause is the overgrowth of the cerebellum and that genes involved with cerebellar development are responsible. It is likely that many genes play small contributing roles and that these, along with other significant environmental or developmental factors, all combine to induce the development of the malformation.

SYMPTOMS

Symptoms exhibited in infants may include vomiting, mental impairment, weakness, and paralysis of the limbs. Symptoms in adolescents are usually milder and may include dizziness, fainting, weakness of the legs, headaches, double vision, and deafness. Other adolescent symptoms may include swelling of the optic nerve region, rapid eye movement, lack of muscular coordination, uncontrolled shaking or trembling, walking problems, and numbness or tingling in the extremities.

SCREENING AND DIAGNOSIS

The doctor will ask about a patient's symptoms and medical history and will perform a physical exam. The doctor may also perform a magnetic resonance imaging (MRI) scan and/or a computed tomography (CT) scan to view the inside of the brain.

Cerebrospinal fluid (CSF) is a vital fluid that surrounds the brain and spine. Special studies to evaluate the flow of CSF may be performed.

TREATMENT AND THERAPY

Patients should discuss the best plans with their doctors. Among the treatment options, surgery is usually required to correct any obstruction in the brain.

Depending on the symptoms associated with Arnold-Chiari syndrome, other treatments may be beneficial. For example, physical or occupational therapy can help improve muscular coordination and trembling. Braces or a wheelchair may be needed. Speech therapy may also be beneficial.

PREVENTION AND OUTCOMES

There is no known way to prevent Arnold-Chiari syndrome. Parents of a child with this condition may benefit from genetic counseling.

Krisha McCoy, M.S.; reviewed by Rimas Lukas, M.D.
"Etiology and Genetics" by Jeffrey A. Knight, Ph.D.

FURTHER READING

Chen, Harold. "Chiari Malformation." In *Atlas of Genetic Diagnosis and Counseling*. Totowa, N.J.: Humana Press, 2006.

EBSCO Publishing. *Health Library: Arnold-Chiari Syndrome*. Ipswich, Mass.: Author, 2009. Available through http://www.ebscohost.com.

Kumar, Praveen, and Barbara K. Burton, eds. "Chiari Malformations." In *Congenital Malformations: Evidence-Based Evaluation and Management*. New York: McGraw-Hill Medical, 2008.

Rubin, Jonathan M., and William F. Chandler. *Ultrasound in Neurosurgery*. New York: Raven Press, 1990.

Viana, G. M., et al. "Association of HTLV-I with Arnold Chiari Syndrome and Syringomyelia." *Brazilian Journal of Infectious Diseases* 12, no. 6 (December, 2008): 536-537.

WEB SITES OF INTEREST

Canadian Neurological Sciences Federation
http://www.ccns.org

Health Canada
http://www.hc-sc.gc.ca/index-eng.php

March of Dimes Birth Defects Foundation
http://www.marchofdimes.com

National Institute of Neurological Disorders and Stroke
http://www.ninds.nih.gov

See also: Amyotrophic lateral sclerosis; Congenital defects; Vanishing white matter disease.

Artificial selection

CATEGORY: Evolutionary biology; Population genetics

SIGNIFICANCE: Artificial selection is the process through which humans have domesticated and improved plants and animals. It continues to be the primary means whereby agriculturally important plants and animals are modified to improve their desirability. However, artificial selection is also a threat to the genetic diversity of agricultural organisms, as uniform and productive strains replace the many diverse, locally produced varieties that once existed around the globe.

KEY TERMS

genetic merit: a measure of the ability of a parent to contribute favorable characteristics to its progeny

genetic variation: a measure of the availability of genetic differences within a population upon which artificial selection has potential to act

heritability: a proportional measure of the extent to which differences among organisms within a population for a particular character result from genetic rather than environmental causes (a measure of nature versus nurture)

NATURAL VS. ARTIFICIAL SELECTION

Selection is a process through which organisms with particular genetic characteristics leave more offspring than do organisms with alternative genetic forms. This may occur because the genetic characteristics confer upon the organism a better ability to survive and ultimately produce more offspring than individuals with other characteristics (natural selection), or it may be caused by selective breeding of individuals with characteristics valuable to humans (artificial selection). Natural and artificial selection may act in concert, as when a genetic characteristic confers a disadvantage directly to the organism. Dwarfism in cattle, for example, not only directly reduces the survival of the affected individuals but also reduces the value of the animal to the breeder. Conversely, natural selection may act in opposition to artificial selection. For example, a genetic characteristic that results in the seed being held tightly in the head of wheat grass is an advantage to the farmer, as it makes harvesting easier, but it would be a disadvantage to wild wheat because it would limit seed dispersal.

EARLY APPLICATIONS

Artificial selection was probably conducted first by early farmers who identified forms of crop plants that had characteristics that favored cultivation. Seeds from favored plants were preferentially kept for replanting. Characteristics that were to some degree heritable would have had the tendency to be passed on to the progeny through the selected seeds. Some favored characteristics may have been controlled by a single gene and were therefore quickly established, whereas others may have been controlled by a large number of genes with individually small effects, making them more difficult to establish. Nevertheless, seeds selected from the best plants would tend to produce offspring that were better than average, resulting in gradual improvement in the population. It would not have been necessary to have knowledge of the mechanisms of genetics to realize the favorable effects of selection.

The beefalo is created by breeding a cow and a bison, and then breeding the offspring again to a cow. Such hybridization, in both plants and animals, is a form of artificial selection that has been practiced by humans for thousands of years to meet agricultural needs. (AP/Wide World Photos)

Likewise, individuals who domesticated the first animals for their own use would have made use of selection to capture desirable characteristics within their herds and flocks. The first of those characteristics was probably docile behavior, a trait known to be heritable in contemporary livestock populations.

FROM PEDIGREES TO GENOME MAPS

Technology to improve organisms through selective breeding preceded an understanding of its genetic basis. Recording of pedigrees and performance records began with the formal development of livestock breeds in the 1700's. Some breeders, notably Robert Bakewell, began recording pedigrees and using progeny testing to determine which sires had superior genetic merit. Understanding of the principles of genetics through the work of Gregor Mendel enhanced but did not revolutionize applications to agricultural plant and animal improvement.

Development of reliable methods for testing the efficiency of artificial selection dominated advances in the fields of plant and animal genetics during the first two-thirds of the twentieth century. Genetic merit of progeny was expected to be equal to the average genetic merit of the parents. More effective breeding programs are dependent on identifying potential parents with superior genetic merit. Computers and large-scale databases have greatly improved selection programs for crops and livestock. However, selection to improve horticultural species and companion animals continued to rely largely on the subjective judgment of the breeder to identify superior stock.

Plant and animal genome-mapping programs have facilitated the next leap forward in genetic improvement of agricultural organisms. Selection among organisms based directly on their gene sequences promised to allow researchers to bypass the time-consuming data-recording programs upon which genetic progress of the 1990's relied. Much effort has gone into identifying quantitative trait loci (QTLs), which are regions of chromosomes or genes that may play a role in the diversity of a trait. QTL analysis is often used along with marker-assisted selection (MAS), where markers associated with the gene of interest are used as surrogates for the actual gene. An example of a biochemical marker is a protein that is encoded by a specific gene. An example of a gene marker is a single nucleotide polymorphism (SNP). Regarding SNPs, there are some cases in which a trait may be controlled by a gene that has a different allele (also known as an alternative DNA sequence) depending on a change in one single nucleotide, and this change may or may not result in a different characteristic. A benefit to using QTL and MAS is that plants and animals can undergo genetic screening to determine whether a desired trait has been artificially selected into the new progeny, rather than waiting until the plant or animal has matured to see if the trait has been passed on.

DIVERSITY VS. UNIFORMITY

In addition to identifying different alleles, SNP analysis can be used to learn more about the genetic history of plants and animals and to determine which genes have remained consistent over time and which have varied due to artificial selection. For example, an evolutionary study found that approximately 1,200 genes in the modern maize genome were affected by artificial selection during the domestication from wild grass teosinte.

The ultimate limit to what can be achieved by selection is the exhaustion of genetic variants. One example of the extremes that can be accomplished by selection is evident in dog breeding: The heaviest breeds weigh nearly one hundred times as much as the lightest breeds. Experimental selection for body weight in insects and for oil content in corn has resulted in variations of similar magnitudes.

However, most modern breeding programs for agricultural crops and livestock seek to decrease variability while increasing productivity. Uniformity of the products enhances the efficiency with which they can be handled mechanically for commercial purposes. As indigenous crop and livestock varieties are replaced by high-producing varieties, the genetic variation that provides the source of potential future improvements is lost. Widespread use of uniform varieties may also increase the susceptibility to catastrophic losses or even extinction from an outbreak of disease or environmental condition. The lack of biodiversity in the wake of such species loss could threaten entire ecosystems and human beings themselves.

IMPACT

As the genomes of different species are sequenced and analyzed, databases of gene mapping are becoming available. These gene maps help to link the QTLs to specific genes by sequencing and functional analysis, and help to connect different alleles to different SNPs. It is predicted that this enhanced genetic knowledge will improve artificial selection by facilitating the selective breeding of animals and plants with greater valued commercial traits. However, many traits are controlled by several genes, making QTL analysis quite complex.

William R. Lamberson, Ph.D.;
updated by Elizabeth A. Manning, Ph.D.

FURTHER READING

Dekkers, J. C. "Commercial Application of Marker- and Gene-Assisted Selection in Livestock: Strategies and Lessons." *Journal of Animal Science* 82 E-Suppl (2004): E313-328. A scientific review discussing the pros and cons of the impact of MAS.

Lurquin, Paul F. *The Green Phoenix: A History of Genetically Modified Plants.* New York: Columbia University Press, 2001. Gives equal weight to the science behind developing improved crop strains and the multinational corporations marketing the results.

Rissler, Jane, and Margaret Mellon. *The Ecological Risks of Engineered Crops.* Cambridge, Mass.: MIT Press, 1996. A scientific and policy assessment of the dangers. Outlines the risks of artificial selection and suggests ways to minimize them.

Tudge, Colin. *The Engineer in the Garden: Genes and Genetics, from the Idea of Heredity to the Creation of Life.* New York: Hill and Wang, 1995. A British science journalist follows the history of human manipulation of genetics to explore the social ramifications of artificial selection, which has resulted in genetic "advances" that have not been adequately exposed to ethical and policy considerations. *Library Journal* praised the book as "a good balance between overall breadth of coverage and the intelligent, readable synthesis of the myriad issues" of genetic research.

Williams, J. L. "The Use of Marker-Assisted Selection in Animal Breeding and Biotechnology." *Revue Scientifique et Technique* 24, no. 1 (April, 2005): 379-391. A review of QTL identification and applications.

Wright, S. I., et al. "The Effects of Artificial Selection on the Maize Genome." *Science* 308, no. 5726 (May 27, 2005): 1310-1314. An investigation into the genetics of maize domestication.

Zohary, Daniel, and Maria Hopf. *Domestication of Plants in the Old World: The Origin and Spread of*

Cultivated Plants in West Asia, Europe, and the Nile Valley. 3d ed. New York: Oxford University Press, 2001. Reviews information on the beginnings of agriculture, particularly utilizing new molecular biology findings on the genetic relations between wild and domesticated plant species.

WEB SITES OF INTEREST

North Dakota State University. "Mapping Quantitative Trait Loci"
http://www.ndsu.nodak.edu/instruct/mcclean/plsc731/quant/quant1.htm

University of Illinois at Urbana-Champaign. Without Miracles: "14. The Artificial Selection of Organisms and Molecules."
http://faculty.ed.uiuc.edu/g-cziko/wm/14.html

See also: Eugenics; Eugenics: Nazi Germany; Evolutionary biology; Gene therapy; Genetic engineering: Agricultural applications; Genetic engineering: Historical development; Genomes; Genome libraries; Hardy-Weinberg law; High-yield crops; Inbreeding and assortative mating; Natural selection; Pedigree analysis; Polyploidy; Population genetics; Punctuated equilibrium; Quantitative inheritance; Sociobiology; Speciation.

Asthma

CATEGORY: Diseases and syndromes

DEFINITION

Asthma is an inflammation and narrowing of the bronchial tubes. Air travels in and out of the lungs through these tubes.

RISK FACTORS

Patients should tell their doctors if they have any of the risk factors that increase their chance of developing asthma. These factors include living in a large urban area, regularly breathing in cigarette smoke (including secondhand smoke), and regularly breathing in industrial or agricultural chemicals. Other risk factors include having a parent who has asthma, having a history of multiple respiratory infections during childhood, low birth weight, being overweight, and having gastroesophageal reflux disease (GERD).

ETIOLOGY AND GENETICS

The causes of asthma are complex and poorly understood, but it is clear that both genetic and environmental factors are involved. The prevailing notion seems to be that an individual inherits the tendency to develop asthma, but the disease itself is triggered by the exposure to environmental stimuli. More than a dozen genes have been implicated in one way or another in the etiology of asthma, so it is not surprising that there is no clear pattern or predictability for the inheritance of the disease.

The first gene to which a known link to asthma was established is *ADAM33*, found on the short arm of chromosome 20 at position 20p13. The protein encoded by this gene helps regulate airway hyperresponsiveness by a mechanism that is not well understood. It is expressed in multiple tissues within the lung but most strongly in lung fibroblasts and bronchial smooth muscle, two cell types that are most critical to airway hyperresponsiveness.

A second gene with a clear association with asthma is *CHI3L1*, found on the long arm of chromosome 1 at position 1q32.1. Its gene product, YKL-40, is a protein which attaches to and destroys chitin, a structural component found in fungi, crustaceans, and insects. Studies show that patients with severe asthma often have highly elevated levels of YKL-40 in their blood relative to nonasthmatic individuals.

Other genes that may contribute to a person's susceptibility to asthma have been identified on chromosomes 5, 6, 11, 12, 13, and 14. Of these, research is concentrated most heavily on a gene cluster on chromosome 5 that encodes key molecules involved in the inflammatory response (cytokines and growth factor receptors) and on the major histocompatability cluster of related genes on chromosome 6.

SYMPTOMS

Symptoms include wheezing, tightness in the chest, trouble breathing, shortness of breath, a cough, chest pain, self-limited exercise, and difficulty keeping up with one's peers.

SCREENING AND DIAGNOSIS

The doctor will ask about a patient's symptoms and medical history and will perform a physical exam. Tests may include a peak flow examination, in which a patient blows quickly and forcefully into a special instrument that measures the output of air; pulmonary function tests (PFTs), in which a patient breathes into a machine that records information about the function of the lungs; and bronchoprovocation tests—lung function tests performed after exposure to methacholine, histamine, or cold or dry air to stimulate asthma. These latter tests can help to confirm asthma in unclear cases. Other tests include exhaled nitric oxide (a marker of airway inflammation) to suggest the diagnosis and manage medications, and allergy tests, usually skin or sometimes blood tests, to find out if allergies may be contributing to the symptoms.

TREATMENT AND THERAPY

The treatment approach to asthma is fourfold: regular assessment and monitoring; control of con-

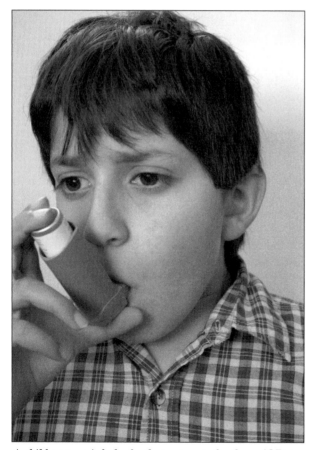

A child uses an inhaler for the treatment of asthma. (©Peter Elvidge/Dreamstime.com)

tributing factors, such as gastroesophageal reflux and sinusitis, and avoidance of allergens or irritants; patient education; and medications.

Some medications are used to control the condition and avoid asthma attacks but are not used to treat an acute attack. These medications include inhaled corticosteroid, used daily to reduce inflammation in the airways. Long-acting beta agonists, such as inhaled salmeterol, can be used daily to prevent asthma attacks, but they should not be taken without an inhaled corticosteroid. Long-acting beta agonists may increase the risk of asthma-related death, intubation (putting a tube in the windpipe to breathe), and hospitalization. Patients who have any concerns about this medication should talk to their doctors.

Cromolyn sodium or a nedocromil sodium inhaler can be used daily to prevent asthma flare-ups or to prevent exercise-induced symptoms; zafirlukast, zileuton, and montelukast may be taken daily to help prevent asthma attacks; and omalizumab (Xolair), a monoclonal antibody against immunoglobulin E (IgE), may be given as an injection under the skin, used along with other medications. Theophylline may be taken daily to help prevent asthma attacks, but this drug is not as commonly used because of interactions with other drugs.

Some medications can be used to treat an asthma attack. Quick-acting beta agonists, such as inhaled albuterol or Xopenex (levalbuterol), relax a patient's airways so that they become wider again and may be used to avoid exercise-induced asthma attacks. Anticholinergic agents, or inhaled medications, such as ipratropium, that function as a bronchodilator typically are used only in an emergency setting.

Corticosteroids in the form of pills, injections, or intravenous (IV) medications are given to treat an acute flare-up of symptoms. Pills may be taken for a longer period of time if the patient has severe asthma that is not responding to other treatments. A shot of epinephrine can be given to stop an asthma attack.

PREVENTION AND OUTCOMES

There are no guidelines for preventing asthma because the cause is unknown. However, patients can help prevent asthma attacks by avoiding things that trigger the attacks. Some general guidelines include keeping windows closed; considering the use of high efficiency particulate air (HEPA) filters for heating/cooling systems and vacuum cleaners;

keeping humidity in the house down; and avoiding strenuous outdoor exercise during days with high air pollution, a high pollen count, or a high ozone level. Patients should get a yearly flu shot and should treat allergies and sinusitis; they should not smoke, should avoid breathing in chemicals or secondhand smoke, and should not regularly use a wood-burning stove. Researchers have reported that heating systems that are more efficient and nonpolluting can help to reduce asthma symptoms in children. If allergies trigger a patient's asthma attacks, he or she should ask the doctor about allergy shots.

Patients should also talk to their doctors about an appropriate level of exercise and about ways to track asthma; tracking will help patients to identify and treat flare-ups right away.

Rosalyn Carson-DeWitt, M.D.;
reviewed by Julie D. K. McNairn, M.D.
"Etiology and Genetics" by Jeffrey A. Knight, Ph.D.

FURTHER READING

Bailey, E. J., et al. "Culture-Specific Programs for Children and Adults from Minority Groups Who Have Asthma." *Cochrane Database of Systematic Reviews* (2009): CD006580. Available through *EBSCO DynaMed Systematic Literature Surveillance* at http://www.ebscohost.com/dynamed.

Campbell, G. Douglas, Jr., and Keith Payne, eds. *Bone's Atlas of Pulmonary and Critical Care Medicine.* 2d ed. Philadelphia: Lippincott Williams & Wilkins, 2001.

Conn, H. F., and R. E. Rakel. *Conn's Current Therapy.* 53d ed. Philadelphia: W. B. Saunders, 2001.

EBSCO Publishing. *Health Library: Asthma.* Ipswich, Mass.: Author, 2009. Available through http://www.ebscohost.com.

Goldman, Lee, and Dennis Ausiello, eds. *Cecil Medicine.* 23d ed. Philadelphia: Saunders Elsevier, 2008.

Howden-Chapman, P., et al. "Effects of Improved Home Heating on Asthma in Community Dwelling Children: Randomised Controlled Trial." *British Medical Journal, Clinical Research Edition* 337 (2008): 1411. Available through *EBSCO DynaMed Systematic Literature Surveillance* at http://www.ebscohost.com/dynamed.

Johnston, Sebastian L., ed. *Asthma: An Atlas of Investigation and Management.* Ashland, Ohio: Clinical, 2007.

Kleigman, Robert M., et al., eds. *Nelson Textbook of Pediatrics.* 18th ed. Philadelphia: Saunders, 2007.

Levenson, M. "Long-Acting Beta-Agonists and Adverse Asthma Events Meta-Analysis." Joint Meeting of the Pulmonary-Allergy Drugs Advisory Committee, Drug Safety and Risk Management Advisory Committee and Pediatric Advisory Committee. Available through *EBSCO DynaMed Systematic Literature Surveillance* at http://www.ebscohost.com/dynamed.

Patterson, Alina V., and Pauline N. Yeager, eds. *Asthma: Etiology, Pathogenesis, and Treatment.* New York: Nova Biomedical Books, 2008.

Rees, John, and Dipak Kanabar. *ABC of Asthma.* 5th ed. Oxford, England: BMJ Books, 2006.

WEB SITES OF INTEREST

Allergy Asthma Information Association
http://aaia.ca

American Academy of Allergy, Asthma, and Immunology
http://www.aaaai.org

American Lung Association
http://www.lungusa.org

Asthma and Allergy Foundation of America
http://www.aafa.org

The Lung Association
http://www.lung.ca

See also: Alpha-1-antitrypsin deficiency; Autoimmune disorders; Immunogenetics.

Ataxia telangiectasia

CATEGORY: Diseases and syndromes
ALSO KNOWN AS: A-T; ATM (ataxia telangiectasia mutated); Louis-Bar syndrome; Boder-Sedgwick syndrome

DEFINITION

Ataxia telangiectasia is a rare, autosomal recessive neurodegenerative and immunodeficiency disease caused by mutations in the *AT* gene on chromosome 11q22-23. It is characterized by lack of motor coordination (ataxia), dilated small blood vessels in the eyes and skin (telangiectasia), hypersensitivity to ionizing radiation, respiratory infections, and high incidence of cancer.

RISK FACTORS

The prevalence of ataxia telangiectasia is estimated to be 1 in 40,000-100,000. About 1 to 2 percent of the population is heterozygous for an ataxia telangiectasia mutation (carriers). These individuals have an increased risk of cancer.

ETIOLOGY AND GENETICS

The *AT* gene is 160 kilo base pairs (kbp) long, containing sixty-six exons, coding for a 13 kbp mature transcript with a 9,168 nucleotide-long open reading frame that translates into a protein of approximately 370 kilodaltons (kDa). The AT protein is a serine/threonine protein kinase (enzyme that adds a phosphate group to other proteins) with multiple functions and protein targets. Many different proteins are phosphorylated and thereby regulated by the AT kinase. Lack of functioning AT protein leads to defects in DNA repair, cancer, and neurodegeneration. AT kinase activates repair proteins in response to double-stranded breaks in DNA. If there is too much DNA damage for the repair system, then AT activates p53 and Chk1 to cause cell-cycle arrest or programmed cell death (apoptosis).

Ataxia telangiectasia patients have lack of function mutations because of truncated AT proteins or splice-site mutations that result in short, unstable AT proteins. Because patients have a decreased ability to repair double-strand DNA breaks, they are very sensitive to ionizing radiation (X and gamma rays). In the normal immune system, rearrangements of DNA occur to create immunoglobins (for example, VDJ recombination in B and T cells). AT kinase plays a role in the breaks that occur in this rearrangement process. Individuals who lack functioning AT will have immunodeficiencies. Deficiencies in DNA repair and loss of regulation of the cell cycle can result in cancer. These deficiencies also lead to degeneration of postmitotic neurons of the cerebellum, the part of the brain that controls voluntary body movements.

Individuals who are carriers for the ataxia telangiectasia mutated (*ATM*) gene—have a single defective copy of the *AT* gene—have an increased risk of developing breast, lung, and blood cancers.

SYMPTOMS

Since ataxia telangiectasia may show incomplete penetrance; severity of symptoms or age of occurrence of symptoms varies. Symptoms include developmental delay of motor skills in the young child, difficulty in coordinating movements, and poor balance. As patients age, the problems with motor control progressively worsen and also include lack of control of limb movements. Patients may have slurred speech and difficulty swallowing. Telangiesctasias (visible blood vessels) in the eyes typically occur by age five, though not all patients develop them. Neck and extremities may also develop telangiesctasias. From a young age, patients have frequent infections, especially of the sinuses and lungs as a result of immune system defects. Patients have an increased incidence of developing cancers, especially lymphomas and leukemias.

SCREENING AND DIAGNOSIS

Patients with ataxia telangiectasia are generally diagnosed between the ages of two and seven. A clinical diagnosis is based on the observation of ataxia and telangiectasia of the eyes. Patients have elevated serum levels of alpha-fetoprotein. Tests include sensitivity of cells to X-ray damage, chromosome instability in the patient's lymphocytes, an antibody for the protein, and sequencing of the *AT* gene. Cerebellar atrophy may be seen in MRI or CT scans. General indicators of ataxia telangiectasia are increased ionizing radiation sensitivity and lack of the AT protein.

TREATMENT AND THERAPY

There is no cure for ataxia telangiectasia, nor are there treatments that are able to slow the progression of the disease. Symptoms are treated. Antibiotics and gammaglobulins are given to fight recurrent respiratory infections. Physical therapy helps the patient maintain flexibility. Speech therapy may be indicated for individuals who develop slurred speech. Psychological counseling may help individuals with ataxia telangiectasia. Patients generally have normal intelligence but may not perform well on tests that require visual motor coordination. Because of the hypersensitivity to ionizing radiation, patients should limit exposure to X rays. Diagnostic X rays should be used only when there is no alternative to obtain a diagnosis.

PREVENTION AND OUTCOMES

Patients with ataxia telangiectasia are often confined to wheelchairs by their teens and generally die in their teens to early twenties, though some live into their forties. They die from recurring respiratory infections (more than 50 percent) and cancer (between 30 and 50 percent), especially leukemia and lymphomas. Couples with the *ATM* gene in their families can receive genetic counseling. Prenatal testing can be done to determine if a fetus has a mutated *AT* gene. Following a family with *ATM*, linkage analysis and microsatellite markers are used to screen the fetus. Direct testing for the mutated gene (from known ataxia telangiectasia patients in the family) is used to determine whether the fetus has *ATM*.

Susan J. Karcher, Ph.D.

FURTHER READING

Gorospe, Myriam, and Rafael de Cabo. "AsSIRTing the DNA Damage Response." *Trends in Cell Biology* 18, no. 2 (2008): 77-83. Review of *ATM* in cell stress response.

Lavin, Martin F. "Ataxia-telangiectasia: From a Rare Disorder to a Paradigm for Cell Signalling and Cancer." *Nature Reviews: Molecular Cell Biology* 9 (2008): 759-769. Includes a time line of ataxia telangiectasia discoveries.

Staropoli, John F. "Tumorigenesis and Neurodegeneration: Two Sides of the Same Coin?" *BioEssays* 30 (2008): 719-727. Discusses DNA repair mutations role in neural degeneration.

Turnpenny, Peter, and Sian Ellard. *Emery's Elements of Medical Genetics.* 12th ed. Philadelphia: Elsevier, 2005. Gives a brief summary of ataxia telangiectasia.

WEB SITES OF INTEREST

A-T Children's Project
http://www.communityatcp.org/Page.aspx?pid=1200

National Institute of Neurological Disorders and Stroke: Ataxia Telangiectasia Information Page
http://www.ninds.nih.gov/disorders/a_t/a-t.htm

OMIM: Online Mendelian Inheritance in Man: Ataxia-telangiectasia
http://www.ncbi.nlm.nih.gov/entrez/dispomim.cgi?id=208900

See also: Amyotrophic lateral sclerosis; Autoimmune disorders; Friedreich ataxia; Hereditary diseases; Immunogenetics.

Atherosclerosis

CATEGORY: Diseases and syndromes

DEFINITION

Atherosclerosis is hardening of a blood vessel from a buildup of plaque. Plaque is made of fatty deposits, cholesterol, and calcium. It builds on the inside lining of arteries. This causes the artery to narrow and harden. It affects large and medium-sized arteries.

As plaque builds up it can slow and even stop blood flow. This means the tissue supplied by the affected artery is cut off from its blood supply. This often leads to pain or decreased function. This condition can cause a number of serious health problems. Depending on the location of the blockage, it can cause coronary heart disease—a loss of blood to areas of the heart; stroke—a loss of blood to areas of the brain; and/or peripheral vascular disease, characterized by leg pain with walking.

In addition, a hardened artery is more likely to be damaged. Repeated damage to the inner wall of an artery causes blood clots to form. The clots are called thrombi. They can lead to a further decrease in blood flow. In some cases, a thrombus may become so large that it completely closes off the artery. It could also break into clumps, called emboli. These clumps travel through the bloodstream and lodge in smaller arteries, blocking them off. In these cases, the tissue supplied by the artery receives no oxygen and quickly dies. When this occurs in the heart, it is called a heart attack. In the brain, it is called a stroke.

Long-term atherosclerosis can also cause arteries to weaken. In response to pressure, they may bulge. This bulge is called an aneurysm. If untreated, aneurysms can rupture and bleed.

RISK FACTORS

There are two types of factors that increase an individual's chance of atherosclerosis: risk factors the individual cannot control and risk factors the individual can control.

Some of the risk factors that cannot be controlled are having a father or brother who developed complications of atherosclerosis before age fifty-five, or having a mother or sister who developed complications of atherosclerosis before age sixty-five. Men forty-five years of age or older and women fifty-five or older are also at risk. Men have a greater risk of heart attack than women.

Risk factors that can be controlled include having high cholesterol, especially low-density lipoprotein (LDL), or "bad" cholesterol, and low high-density lipoprotein (HDL), or "good" cholesterol; having high blood pressure; cigarette smoking; diabetes Type I and Type II; being overweight or obese; and a lack of physical activity.

Metabolic syndrome is a combination of three out of the following five findings: low HDL cholesterol (also called "good" cholesterol), high triglycerides, elevated blood sugar, elevated blood pressure, and an increased waist circumference (greater than 40 inches in men and 35 inches in women).

Etiology and Genetics

Multiple environmental and genetic factors play a contributing part in atherosclerosis. Some individuals are genetically predisposed to developing the condition, yet a detailed genetic analysis and prediction of inheritance patterns are not possible, since so many different genes seem to be implicated. One estimate suggests that more than one hundred different genes may be involved. There are at least twenty-one identifiable diseases or syndromes, including atherosclerosis, among the symptoms that are known to result from mutations in single genes.

One gene with a clear association with atherosclerosis is *APOE*, found on the long arm of chromosome 19 at position 19q13.2. *APOE* encodes the protein apolipoprotein E, which functions to carry excess cholesterol from the blood to receptors on the surface of cells in the liver. Some mutations in the gene lead to altered protein products that lack the ability to bind to the receptors, resulting in a marked increase in an individual's blood cholesterol.

Studies using deoxyribonucleic acid (DNA) microarray analyses have implicated another gene, *EGR1* (early growth response gene 1), as a contributor to some cases of atherosclerosis. Found on the long arm of chromosome 5 at position 5q31.1, this gene encodes a protein that is an important part of the body's vascular repair system. When inappropriately active in coronary arteries, the effect is the slow closure of the arteries, leading to angina and possible starvation of heart muscle. This important discovery has opened new avenues of research designed to develop drugs targeted to inhibit *EGR1* gene expression.

Symptoms

There are no symptoms in early atherosclerosis. As the arteries become harder and narrower, symptoms may begin to appear. If a clot blocks a blood vessel or a large embolus breaks free, symptoms can occur suddenly.

Symptoms depend on which arteries are affected. For example, coronary (heart) arteries may cause symptoms of heart disease, such as chest pain; arteries in the brain may cause symptoms of a stroke, such as weakness or dizziness; and arteries in the lower extremities may cause pain in the legs or feet and trouble walking.

Screening and Diagnosis

Most patients are diagnosed after they develop symptoms. However, patients can be screened and treated for risk factors.

A patient who has symptoms will be asked questions by his or her doctor; these questions will help to determine which arteries might be affected. The doctor will also need to know a patient's full medical history, and a physical exam will be conducted. Tests will depend on which arteries may be involved; these tests will be decided based on the patient's symptoms, physical exam, and/or risk factors.

Many of these tests detect problems with the tissue that is not getting enough blood. Two common tests that directly evaluate the atherosclerotic arteries are angiography and ultrasound. In angiography, a tube-like instrument is inserted into an artery. Dye is injected into the vessel to help determine the degree of blood flow. When done in the heart, this test is called cardiac catheterization. Ultrasound is a test that uses sound waves to examine the inside of the body. In this case, the test examines the size and shape of arteries.

Treatment and Therapy

An important part of treatment for atherosclerosis is reducing risk factors. Beyond that, treatment depends on the area of the body most affected.

Treatment may include medications, such as drugs to interfere with the formation of blood clots, like aspirin or clopidogrel (Plavix); drugs to control blood pressure, if elevated; drugs to lower cholesterol, if elevated; and drugs that improve the flow of blood through narrowed arteries, such as cilostazol (Pletal) or pentoxifylline (Trental).

Procedures involving a thin tube, called a catheter, can also be used. The catheter is inserted into an artery. Catheter-based procedures are most often done for arteries in the heart; they may be used to treat atherosclerosis elsewhere in the body. These procedures include balloon angioplasty, in which a balloon-tipped catheter is used to press plaque against the walls of the arteries, increasing the amount of space for the blood to flow.

Stenting is usually done after angioplasty. In this procedure, a wire mesh tube is placed in a damaged artery; it will support the arterial walls and keep them open.

In an atherectomy, instruments are inserted via a catheter. They are used to cut away and remove plaque so that blood can flow more easily. This procedure is not often performed.

Surgical options include endarterectomy—removal of the lining of an artery obstructed with large plaques. This procedure is often done in the carotid arteries of the neck; these arteries bring blood to the brain.

Arterioplasty can repair an aneurysm; it is usually done with synthetic tissue. Bypass is the creation of an alternate route for blood flow using a separate vessel.

PREVENTION AND OUTCOMES

There are a number of ways to prevent, as well as reverse, atherosclerosis. They include eating a healthful diet that should be low in saturated fat and cholesterol and rich in whole grains, fruits, and vegetables. Patients should exercise regularly, maintain a healthy weight, and lose weight if they are overweight. They should not smoke; if they smoke, they should quit. Patients should also control their diabetes.

If a doctor recommends it, a patient should take medication to reduce his or her risk factors. This may include medicine for high blood pressure or high cholesterol. Patients should also talk to their doctors about screening tests for atherosclerotic disease of the heart (coronary artery disease) if they have risk factors.

Laurie Rosenblum, M.P.H.;
reviewed by Igor Puzanov, M.D.
"Etiology and Genetics" by Jeffrey A. Knight, Ph.D.

FURTHER READING

Ballantyne, Christie M., James H. O'Keefe, and Antonio M. Gotto. *Dyslipidemia and Atherosclerosis Essentials.* 4th ed. Sudbury, Mass.: Jones and Bartlett, 2009.

Beers, Mark H., ed. *The Merck Manual of Medical Information.* 2d home ed., new and rev. Whitehouse Station, N.J.: Merck Research Laboratories, 2003.

EBSCO Publishing. *Health Library: Atherosclerosis.* Ipswich, Mass.: Author, 2009. Available through http://www.ebscohost.com.

Roberts, Robert, Ruth McPherson, and Alexandre F. R. Stewart. "Genetics of Atherosclerosis." In *Cardiovascular Genetics and Genomics*, edited by Dan Roden. Hoboken, N.J.: Wiley-Blackwell, 2009.

Stephenson, Frank H. "Atherosclerosis." In *DNA: How the Biotech Revolution Is Changing the Way We Fight Disease.* Amherst, N.Y.: Prometheus Books, 2007.

Triffon, Douglas W., and Erminia M. Guarneri. "Dyslipidemia and Atherosclerosis." In *Food and Nutrients in Disease Management*, edited by Ingrid Kohlstadt. Boca Raton, Fla.: CRC Press, 2009.

WEB SITES OF INTEREST

American College of Radiology
http://www.acr.org

American Heart Association
http://www.americanheart.org

Canadian Cardiovascular Society
http://www.ccs.ca/home/index_e.aspx

Heart and Stroke Foundation of Canada
http://ww2.heartandstroke.ca

National Heart, Lung, and Blood Institute
http://www.nhlbi.nih.gov

See also: Barlow's syndrome; Cardiomyopathy; Heart disease.

Attention deficit hyperactivity disorder (ADHD)

CATEGORY: Diseases and syndromes

ALSO KNOWN AS: ADHD; attention deficit disorder (ADD); hyperkinetic syndrome; hyperkinetic impulse disorder

DEFINITION

Attention deficit hyperactivity disorder (ADHD) is a chronic behavioral disorder. It is behavior that is hyperactive, impulsive, and/or inattentive. This behavior must persist for at least six months and be present in two environments (home, work, or school). ADHD affects children, adolescents, and adults.

RISK FACTORS

Males and individuals having a parent or sibling with ADHD are at risk for this disorder.

ETIOLOGY AND GENETICS

Like other neurobehavioral developmental disorders, such as autism or bipolar affective disorder, ADHD is a complex condition whose expression depends on both genetic and environmental determinants. Genetics appears to be the major contributing factor, with reports of heritability ranging from 0.75 to 0.92. In most cases, development of ADHD involves many genes, with each gene providing a small to moderate contribution to the overall phenotype. Twin studies have suggested concordance rates as high as 92 percent in monozygotic (identical) twins and 33 percent in dizygotic (fraternal) twins. While there is no consistent pattern of inheritance, 10-35 percent of children with ADHD have a first-degree relative who is also affected, and approximately half of the parents with ADHD will have one or more children with the disorder.

Recent molecular genetics studies based in part on deoxyribonucleic acid (DNA) sequence data from the Human Genome Project have identified a number of genes that appear to play contributing roles in the development of ADHD. Many of these are associated with neurotransmitters or other proteins that serve as message carriers in the brain. For example, three different ADHD susceptibility genes are known to specify dopamine receptor proteins (*DRD2* at position 11q23.1, *DRD4* at position 11p15, and *DRD5* at position 4p16), another encodes the dopamine transporter protein (*DAT1* at position 5p15), and yet another specifies the dopamine beta-hydroxylase enzyme (*DBH* at position 9q34). Serotonin is another neurotransmitter protein, and associated ADHD susceptibility genes include the serotonin 1B receptor gene (*HTR1B*, at position 6q13) and the serotonin transporter genes *SLC6A3* and *SLC6A4* (at chromosomal locations 5p15 and 17q11.1-q12, respectively).

Five additional ADHD susceptibility genes have been identified: *SNAP25* (at position 20p11.2), which encodes the synaptosomal associated protein; *ADRA2A* (at position 10q24), which specifies the adrenergic alpha-2A acceptor; *SCN8A* (at position 12q13), which encodes the sodium channel alpha polypeptide; *TPH2* (at position 12q21), the gene for tryptophan hydroxylase; and *COMT* (at position 22q11), which specifies catechol-O-methyltransferase.

SYMPTOMS

All children display some of the symptoms of ADHD. Children with ADHD have symptoms that are more severe and consistent. They often have difficulty in school and with their family and peers. ADHD can last into adulthood. It can cause problems with relationships, job performance, and job retention.

There are three types of ADHD: inattentive (classic ADD), hyperactive-impulsive, and combined. Individuals with inattentive (classic ADD) are easily distracted by sights and sounds, do not pay attention to detail, do not seem to listen when spoken to, make careless mistakes, and do not follow through on instructions or tasks. These individuals also avoid or dislike activities that require longer periods of mental effort, lose or forget items necessary for tasks, and are forgetful in day-to-day activities.

Individuals with the hyperactive-impulsive type of the disorder are restless, fidget, and squirm; run and climb and are unable to stay seated; blurt out answers before hearing the entire question; have difficulty playing quietly; talk excessively; interrupt others; and have difficulty waiting in line or waiting for a turn.

Combined ADHD is the most common type of the disorder; individuals with this type have a combination of the symptoms in the inattentive and hyperactive-impulsive types. In addition, many people with ADHD often have depression, anxiety, conduct disorder, oppositional defiant disorder, learning disorders, and substance abuse.

SCREENING AND DIAGNOSIS

There is no standard test to diagnose ADHD. Diagnosis is done by a trained mental health professional; family and teachers are also involved.

The American Academy of Pediatrics recommends that the following guidelines be used for diagnosis in children six to twelve years of age: Diagnosis should be initiated if a child shows signs of difficulty in school, academic achievement, and relationships with peers and family. During diagnosis, the following information should be gathered directly from parents, caregivers, teachers, or other school professionals: an assessment of symptoms of ADHD in different settings (home and school), the age at which symptoms started, and how much the behavior affects the child's ability to function. The professional should examine the child for other conditions that might be causing or aggravating symptoms, learning and language problems, aggression, disruptive behavior, depression or anxiety, psychotic symptoms, and personality disorder. In order for a diagnosis of ADHD to be made, symptoms must be present in two or more of the child's settings, must interfere with the child's ability to function for at least six months, and must fit a list of symptoms detailed in the most recent version of the *Diagnostic and Statistical Manual of Mental Disorders* (DSM) of the American Psychiatric Association.

TREATMENT AND THERAPY

The goal of treatment is to improve the child's ability to function. Doctors should work with parents and school staff; by working together, they can set realistic goals and evaluate the child's response.

Treatments include medications that can help control behavior and increase attention span. Stimulants are the most common choice for ADHD; they increase activity in parts of the brain that appear to be less active in children with ADHD. Stimulant medications include methylphenidate (Ritalin, Concerta, Metadate, and Daytrana), dextroamphetamine (Dexedrine), amphetamine (Adderall), and atomoxetine (Strattera). Lisdexamfetamine (Vyvanse) has been approved to treat adults with ADHD; it can also be used to treat children aged six to twelve years.

Parents should talk to their child's doctor if they have any questions about ADHD medication. There are possible risks with these medications, including cardiovascular events (stroke and heart attack) and psychiatric problems (hearing voices and becoming manic). Because of the rare risk of serious heart problems, the American Heart Association suggests that children have an electrocardiogram (ECG) before starting stimulant medication for ADHD.

Other drugs used to treat ADHD include antide-

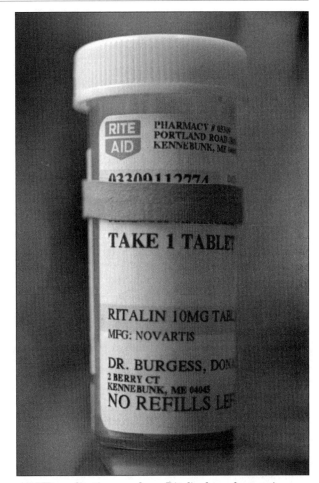

ADHD medications such as Ritalin have become increasingly common in schools. (AP/Wide World Photos)

pressants, such as imipramine (Janimine and Tofranil), venlafaxine (Effexor), and bupropion (Wellbutrin). Clonidine (used for Tourette's syndrome) can treat impulsivity.

Children who take medication and go to behavior therapy do better than those who only use medication. Therapy sessions focus on practicing social and problem-solving skills. Counselors will also teach parents and teachers to help the child through positive reinforcement, which could involve changes in the classroom and in parenting style. Often, daily report cards are exchanged between parents and teachers.

Other tools, like the Disc'O'Sit cushion, may be helpful in improving a child's attention in class. The Disc'O'Sit is a dome-shaped cushion filled with air on which the child balances.

PREVENTION AND OUTCOMES

There are no guidelines for preventing ADHD because the cause is unknown. Proper treatment can prevent problems later in life.

Julie Riley, M.S., RD;
reviewed by Rosalyn Carson-DeWitt, M.D.
"Etiology and Genetics" by Jeffrey A. Knight, Ph.D.

FURTHER READING

American Psychiatric Association. *Diagnostic and Statistical Manual of Mental Disorders: IV-TR.* Rev. 4th ed. Washington, D.C.: Author, 2000.

EBSCO Publishing. *Health Library: Attention Deficit Hyperactivity Disorder and Attention Deficit Disorder.* Ipswich, Mass.: Author, 2009. Available through http://www.ebscohost.com.

Hughes, Lesley, and Paul Cooper. *Understanding and Supporting Children with ADHD: Strategies for Teachers, Parents, and Other Professionals.* Thousand Oaks, Calif.: Paul Chapman, 2007.

McBurnett, Keith, and Linda Pfiffner, eds. *Attention Deficit Hyperactivity Disorder: Concepts, Controversies, New Directions.* New York: Informa Healthcare, 2008.

Pfeiffer, B., et al. "Effectiveness of Disc 'O' Sit Cushions on Attention to Task in Second-Grade Students with Attention Difficulties." *American Journal of Occupational Therapy* 62, no. 3 (May/June, 2008): 274-281.

Rappley, M. D. "Child Practice: Attention Deficit-Hyperactivity Disorder." *New England Journal of Medicine* 352, no. 2 (January 13, 2005): 165-173.

Timimi, Sami, and Jonathan Leo, eds. *Rethinking ADHD: From Brain to Culture.* New York: Palgrave Macmillan, 2009.

WEB SITES OF INTEREST

About Kids Health
http://www.aboutkidshealth.ca

American Academy of Child and Adolescent Psychiatry
http://www.aacap.org

American Academy of Pediatrics
http://www.aap.org

American Psychiatric Association
http://www.psych.org

Attention Deficit Disorder Association
http://www.add.org

Attention Deficit Disorder Resources
http://www.addresources.org

Canadian Psychiatric Association
http://www.cpa-apc.org

Children and Adults with Attention-Deficit/Hyperactivity Disorder (CHADD)
http://www.chadd.org

Mental Health America
http://www.nmha.org

National Institute of Mental Health
http://www.nimh.nih.gov

See also: Aggression; Autism; Behavior; Bipolar affective disorder; Developmental genetics; Dyslexia.

Autism

CATEGORY: Diseases and syndromes
ALSO KNOWN AS: Pervasive development disorder; PDD

DEFINITION

Autism represents a spectrum of complex brain disorders resulting in social, behavioral, and language problems. People with autism have difficulty communicating and forming relationships. They may be preoccupied, engage in repetitive behaviors, and exhibit marked inflexibility. Autism occurs in about one in one thousand children.

RISK FACTORS

Boys are four times more likely than girls to have autism. Siblings of a child with autism have a 3-7 percent chance of being autistic.

A number of other conditions are associated with autism, although the relationships among them are not clear. These conditions include neurofibromatosis, tuberous sclerosis, fragile X syndrome, phenylketonuria (PKU), Möbius syndrome, epilepsy, herpes encephalitis, and cytomegalovirus.

Children born to women who have problems during pregnancy or delivery, or who contract rubella during pregnancy, are also at risk. These conditions primarily affect the central nervous system.

ETIOLOGY AND GENETICS

Autism is a complex condition whose expression is determined by a host of genetic, developmental, and environmental factors. While the genetic determinants appear to play the predominant role, they are difficult to quantify since so many different genes are apparently involved.

The first genetic region shown to have an association with autism is a deoxyribonucleic acid (DNA) sequence on chromosome 5 that is between two genes that specify cell-adhesion molecules. These proteins are important components of nerve synapses, the junctions between nerve cells. In subsequent years, many additional genes or genetic regions have been implicated in autism occurring in some families. A 2009 report identifies no fewer than twenty-seven different genetic regions with rare copy number variations (duplications or deletions), located on twenty different chromosomes, that were found in children with autism but not in the control group. Specific genes that are thought to play a contributing role in some cases include *Shank3* (at position 22q13.3), *BZRAP1* (at position 17q23.2), *MDGA2* (at position 14q21), *MECP* (at position Xq28), and *PTEN* (at position 10q23.31).

One classical estimate of the extent to which genetics plays a role in the development of a complex trait or disease is the measurement of concordance rates in monozygotic (identical) twins as compared with those for dizygotic (fraternal) twins. Such measurements for autism have been reported as 70 percent in monozygotic twins and 5 percent in dizygotic twins, confirming the importance of genetic factors in the etiology of the disease.

SYMPTOMS

Autism first appears in children aged three and younger. The severity of symptoms varies over a wide spectrum. Behaviors and abilities may differ from day to day; symptoms may decrease as the child grows older. Children with autism may exhibit a combination of abnormal behaviors.

Symptoms include avoiding social contact, loss of language, using words incorrectly, changing the meaning of a common word, gesturing frequently, avoiding eye contact, and having trouble with nonverbal communication. Other symptoms include lack of interest in normal activities for their age; spending a lot of time alone; not playing imaginatively; not starting pretend games; not imitating others; sensitivity to sound, smell, taste, sights, and touch; responding to stimulation in an abnormal

A child with autism is taught how to play by pulling another child in a wagon. Some children can benefit from early intervention. (Time & Life Pictures/Getty Images)

way; and not reacting to smiles in the manner expected. Hyperactivity; passiveness; tantrums; single-mindedness; aggression; hurting themselves or self-mutilation; repetitive movements, such as rocking or flapping a hand; resisting change; forming odd attachments to objects; sniffing or licking toys; and not understanding other people's feelings and needs are additional symptoms.

Some people with autism suffer from other disorders as well, including seizures, mental retardation, and genetic disorders, such as fragile X syndrome.

Some people with autism have unusual abilities. For example, they may memorize things or be able to play a musical instrument without lessons. Children with autism may show varying signs of cognitive impairment but have normal intelligence. According to the Autism Information Center, children with autism may be very good at putting puzzles together or solving problems, but instead have trouble in other areas, like talking or making friends. Autism, a group of developmental disabilities caused by abnormality in the brain, is a highly individualized disorder.

SCREENING AND DIAGNOSIS

Doctors who specialize in autism will observe a child's behavior, social contacts, and communication abilities. They will assess mental and social development and ask parents about the child's behavior. Some doctors ask parents to bring in videotapes of the child at home.

Tests may include neuropsychological tests, questionnaires and observation schedules, and intelligence quotient (IQ) tests. Medical tests can rule out other conditions that cause similar symptoms. Blood tests; urine tests; deoxyribonucleic acid (DNA) testing; and an electroencephalogram (EEG), a test that records the brain's activity by measuring electrical currents through the brain, are among these medical tests.

TREATMENT AND THERAPY

There is no cure for autism. The severity of symptoms may decrease over the years, but the condition lasts for life. Children with autism and their families may benefit from early intervention. Children with autism respond well to a structured, predictable schedule. With help, many children with autism learn to cope with their disabilities. Most need assistance and support throughout their lives. Others

are able to work and live independently when they grow up.

Interventions to help children with autism include special education programs; these programs are designed to meet the child's special needs and improve the odds of learning. Children with autism may have trouble with assignments, concentration, and anxiety. Teachers who understand the condition can build on the child's unique abilities. Programs should incorporate the child's interests. Some children do better in a small-group setting; others do well in regular classrooms with special support. Vocational training can help prepare young adults for jobs.

Therapy services, such as speech, physical, and occupational therapies, may improve speech and activities. Children with autism need help developing social skills.

Professional support helps a family cope with caring for a child with autism. Counselors help parents learn how to manage behaviors. Caring for a child with autism can be exhausting and frustrating; arranging occasional respite care is essential, so that the main caregiver can have some breaks.

Although there are no drugs to treat autism, several drugs are used to help manage symptoms. For example, drugs prescribed for anxiety and depression can help tone down obsessive and aggressive behaviors.

PREVENTION AND OUTCOMES

There are no guidelines for preventing autism because the cause is unknown. Scientists are actively searching for its underlying causes.

Debra Wood, R.N.; reviewed by Rimas Lukas, M.D.
"Etiology and Genetics" by Jeffrey A. Knight, Ph.D.

FURTHER READING

Boucher, Jill. *The Autistic Spectrum: Characteristics, Causes, and Practical Issues.* Los Angeles: Sage, 2009.

Dodd, Susan. *Understanding Autism.* New York: Elsevier, 2005.

EBSCO Publishing. *Health Library: Autism.* Ipswich, Mass.: Author, 2009. Available through http://www.ebscohost.com.

Moldin, Steven O., and John L. R. Rubenstein, eds. *Understanding Autism: From Basic Neuroscience to Treatment.* Boca Raton, Fla.: CRC/Taylor & Frances, 2006.

Rapin, I. "An Eight-Year-Old Boy with Autism." *Journal of the American Medical Association* 285, no. 13 (April 4, 2001): 1749-1757.

Sykes, N. H., and J. A. Lamb. "Autism: The Quest for the Genes." *Expert Reviews in Molecular Medicine* 9, no. 24 (2007): 1-15.

Thompson, Travis. *Making Sense of Autism.* Baltimore: Paul H. Brookes, 2007.

WEB SITES OF INTEREST

Autism Canada Foundation
http://www.autismcanada.org/home.htm

Autism Information Center, Centers for Disease Control and Prevention
http://www.cdc.gov/ncbddd/autism/index.htm

Autism Society of America
http://www.autism-society.org

Autism Society Canada
http://www.autismsocietycanada.ca

Genetics Home Reference
http://ghr.nlm.nih.gov

National Institute of Mental Health
http://www.nimh.nih.gov

National Institute of Neurological Disorders and Stroke
http://www.ninds.nih.gov

See also: Aarskog syndrome; Aggression; Behavior; Bipolar affective disorder; Fragile X syndrome; Sociobiology.

Autoimmune disorders

CATEGORY: Diseases and syndromes
ALSO KNOWN AS: Autoimmune diseases

DEFINITION

Autoimmune disorders are chronic diseases that arise from a breakdown of the immune system's ability to distinguish between the body's own cells and foreign substances. This causes an immune attack against the organs or tissues of the individual's own body. Autoimmune disorders can be caused by both genetic and environmental factors or, more likely, a combination of these factors.

RISK FACTORS

The precise origin of most autoimmune disorders is not clearly understood. Researchers have shown that most autoimmune disorders occur more frequently in females than in males and that the development of autoimmune disorders often requires both a genetic susceptibility and additional stimuli such as exposure to a toxin. For example, cigarette smoking has been associated with rheumatoid arthritis, but an underlying genetic predisposition is likely to cause disease development. Psoriasis flares are often associated with stress, trauma, or infection. Geographical location can also influence the incidence of some conditions: Multiple sclerosis is much more common in Western and Northern Europe and North America than in Asia, Africa, or South America. Family history of any autoimmune disease increases the risk for the same or other autoimmune disorders in relatives. This clustering of diseases in some families may be attributable to shared genes, similar environmental exposures, or other combined factors.

ETIOLOGY AND GENETICS

Autoimmune disorders involve a large group of chronic and potentially life-threatening diseases that are initiated by an individual's own immune system attacking the organs or tissues of his or her own body. The main function of the immune system is to defend against invading microorganisms such as bacteria, fungi, viruses, protozoa, and parasites by producing antibodies or lymphocytes that recognize and destroy the harmful agent. The ability to distinguish normal body constituents (self) from foreign substances (nonself) is crucial to appropriate immune functioning. Loss of this ability to distinguish between self and nonself can lead to serious damage to the affected organs and tissues.

Autoimmune disorders are generally categorized as organ-specific diseases and non-organ-specific (also called systemic) diseases. Organ-specific autoimmune diseases involve an attack directed against one main organ and have been documented for essentially every organ in the body. Common examples include multiple sclerosis (brain), insulin-dependent diabetes mellitus (pancreas), Addison's disease (adrenal glands), Graves' disease (thyroid), pernicious anemia (stomach), myasthenia gravis (muscles), autoimmune hemolytic anemia (blood), primary biliary cirrhosis (liver), pemphigus vulgaris (skin), and glomerulonephritis (kidneys). Non-organ-specific autoimmune diseases involve an attack by the immune system on several body areas, potentially causing diseases such as systemic lupus erythematosus, rheumatoid arthritis, polyarteritis nodosa, scleroderma, ankylosing spondylitis, and rheumatic fever.

Some evidence suggests that other conditions (such as certain types of eye inflammation and male and female infertility) may be autoimmune related. Allergies involve hypersensitivity reactions that result in immune reactions that can lead to inflammation and tissue damage. Environmental antigens such as pollen, dust mites, food proteins, and bee venom may cause allergic reactions such as hay fever, asthma, and food intolerance in sensitive individuals via the antibody class known as immunoglobulin E (IgE). Medications such as antibiotics may also be recognized as chemical antigens, causing adverse allergic reactions.

Of the numerous theories proposed for the cause of autoimmunity development, three models have received the most consideration by clinical researchers. The first theory, that of clonal deletion, suggests that autoimmunity develops if autoreactive T or B cell clones are not eliminated during the fetal period or very soon after birth. The body normally does not react to its own fetal or neonatal antigens, which are recognized because the corresponding T and B cell clones are eliminated from the immune system. In the unfortunate event that "forbidden clones" of autoreactive cells remain active, antibodies are produced that are directed against its own antigens, and autoimmunity develops, frequently involving the loss of the helper T cells' ability to regulate B-cell function. A second theory suggests that some antigens that are normally nonimmunogenic (hidden antigens) somehow become autoimmunogenic and stimulate the immune system to react against itself. A third theory suggests that autoimmunity can be initiated by an exogenous antigen, assuming that the antibodies produced to fight it cross-react with a similar determinant on the body's own cells.

The observation that autoimmune disorders cluster in some families suggests that genetic predisposition is an important factor in protection from, or susceptibility to, autoimmune diseases. It is likely that a number of genes are involved in determining an individual's risk for autoimmune disorders. A few common human leukocyte antigen (HLA) class II genes have been linked to autoimmune disorders, but research on non-HLA related genes has been limited in the past. However, with recent advancing technologies and methodologies, significant progress has been made in the search for autoimmune predisposition genes. A large family study undertaken by the Multiple Autoimmune Disease Genetics Consortium (MADGC) discovered a change in the *PTPN22* gene that corresponded to risks for Type I diabetes, rheumatoid arthritis, systemic lupus erythematosus, and Hashimoto thyroiditis, suggesting a potential common genetic basis for these diverse conditions. This gene change was not significant for multiple sclerosis, suggesting that there may be a different underlying genetic mechanism for this condition.

Genome-wide association studies have more recently found susceptibility loci for ulcerative colitis, systemic lupus erythematosus, and rheumatoid arthritis, among other autoimmune disorders. While these genetic loci give researchers an area to focus on to locate specific predisposition genes, it is still clear that the genetic factors of autoimmune disorder susceptibility are complex, and not limited to single gene determinants. Isolation of these susceptibility genes, however, is likely to open the doors to unlocking some of this complex etiology which ideally would result in more targeted treatments and therapies.

SYMPTOMS

Given the variability, diversity, and range of autoimmune disorders, no consistent underlying symptoms are suggestive of all autoimmune disorders. General symptoms that may represent an autoimmune disorder can include dizziness, general malaise, muscle weakness, muscle or joint pain, fatigue, vision changes, skin changes (hyperpigmentation, rashes, dryness), temperature irregularities (fevers, low body temperature), and neurological symptoms (tremors, numbness, balance or coordination problems).

SCREENING AND DIAGNOSIS

Diagnosis of autoimmune disorders generally begins with the often difficult task of documenting autoantibodies and autoreactive T cells. A condition suspected to be caused by autoimmunity can also be confirmed by a number of other direct and indirect methods, such as a favorable response to immunosuppressive, corticosteroid, or anti-inflammatory drug treatment along with several other immunologic techniques.

The relatively limited information on predisposition genes involved in autoimmune disorders means that presymptomatic or diagnostic genetic testing is not available for most of these conditions. However, screening for some genes involved in specific autoimmune disorders is clinically available. Ankylosing spondylitis has been strongly associated with the HLA-B27 allele. However, the presence of this allele does not guarantee the development of this disease, nor can it predict disease onset or severity. Clinical genetic testing is also available for *CARD15/NOD2*, a gene linked to Crohn disease. 60 percent of individuals with Crohn disease have at least one of the four mutations in this gene that have been associated with disease development. Since roughly 40 percent of those with Crohn disease do not have any mutations in this gene, however, genetic testing for this condition is by no means conclusive. Numerous research studies for conditions such as multiple sclerosis, lupus, scleroderma, and psoriasis are underway in an attempt to begin providing similar genetic screening and testing options.

Treatment and Therapy

Treatment strategies lag behind the ability to diagnose autoimmune disorders. Initial management involves the control and reduction of both pain and loss of function. Correction of deficiencies in hormones such as insulin or thyroxin caused by autoimmune damage to glands is often performed first. Replacing blood components by transfusion is also considered, but treatment effectiveness is often limited by the lack of knowledge of the precise disease mechanisms. Suppression of the immune system is also often attempted, but achieving a delicate balance between controlling the autoimmune disorder and maintaining the body's ability to fight disease in general is critical.

Medication therapy commonly includes corticosteroid drugs, with more powerful immunosuppressant drugs such as cyclophosphamide, methotrexate, azathioprine, chloroquine derivatives, and small doses of antimetabolic or anticancer drugs often required. A majority of these medications can rapidly damage dividing tissues such as the bone marrow and thus must be used with caution. Plasmapheresis (removal of toxic antibodies) is often helpful in diseases such as myasthenia gravis, while other treatments involve drugs that target immune system cells such as the cyclosporines. Fish oil and antioxidant supplementation have been shown to be an effective anti-inflammatory intervention and may help suppress autoimmune disorders such as rheumatoid arthritis and systemic lupus erythematosus. Early clinical trials on a medication called fampridine have demonstrated improved mobility in some people with multiple sclerosis. This medication is believed to work by blocking channels for potassium ions on the surface of nerve cells.

Prevention and Outcomes

There is no known prevention for most autoimmune disorders. Most autoimmune disorders are chronic, but many can be controlled with treatment. Many of these conditions exhibit flareups, in which symptoms reappear and eventually abate or disappear for a period of time. Some medications can lessen the severity of flareups or increase time intervals between flares. With a better understanding of the environmental and genetic triggers for autoimmune disorders, long-term prognosis and management for these conditions in general would be expected to improve.

Daniel G. Graetzer, Ph.D., and Bryan Ness, Ph.D.;
updated by Trudy McKanna, M.S.

Further Reading

Abbas, Abul K., and Richard A. Flavell, eds. *Genetic Models of Immune and Inflammatory Diseases.* New York: Springer, 1996. Individual chapters provide descriptions of research results or minireviews of transgenic and targeted gene disruption models used to study autoimmune and inflammatory diseases. Illustrations.

Bona, Constantin A., et al., eds. *The Molecular Pathology of Autoimmune Diseases.* 2d ed. New York: Taylor and Francis, 2002. A review of the latest research carried out in immunology, particularly at the molecular level. Examines developments in the diagnosis and treatment of such conditions as hemolytic anemia, diabetes, Graves' disease, Addison's disease, multiple sclerosis, inflammatory bowel disease, and autoimmune hepatitis.

Criswell, Lindsey A., et al. "Analysis of Families in the Multiple Autoimmune Disease Genetics Consortium (MADGC) Collection: The PTPN22 620W Allele Associates with Multiple Autoimmune Phenotypes." *American Journal of Human Genetics* 76, no. 4 (2005): 561-571. Findings that suggest a common underlying pathway for some autoimmune disorders.

Fernandes, Gabriel, and Christopher A. Jolly. "Nutrition and Autoimmune Diseases." *Nutrition Reviews* 56 (January, 1998). Summarizes several topics from a conference on nutrition and immunity and relates striking benefits of fish oil and antioxidant supplementation on gene and T-cell subsets.

Gutierrez-Roelens, I., and B. R. Lauwerys. "Genetic Susceptibility to Autoimmune Disorders: Clues from Gene Association and Gene Expression Studies." *Current Molecular Medicine* 8, no. 6 (2008): 551-561. Describes advances in research into several autoimmune disorders, such as arthritis and lupus.

Lachmann, P. J., et al., eds. *Clinical Aspects of Immunology.* Malden, Mass.: Blackwell, 1993. Provides an excellent overview of immunology for the clinician, covering new areas of immunologic diagnosis, mechanisms, and techniques.

Rioux, J. D., and A. K. Abbas. "Paths to Understanding the Genetic Basis of Autoimmune Disease." *Nature* 435 (2005): 584-589. Discusses genetic sequence variants that may contribute to autoimmune diseases.

Theofilopoulos, A. N., ed. *Genes and Genetics of Autoimmunity.* New York: Karger, 1999. Examines a broad spectrum of topics related to autoimmunity, including the chapters "Immunoglobulin Transgenes in B Lymphocyte Development, Tolerance, and Autoimmunity," "The Role of Cytokines in Autoimmunity," "Genetics of Human Lupus," and "Genetics of Multiple Sclerosis." Illustrated.

Vyse, Timothy J., and John A. Todd. "Genetic Analysis of Autoimmune Disease." *Cell* 85 (May, 1996). Describes how ongoing study of the entire human genetic code will assist in the isolation and correction of aberrant genes that cause immunological disease and presents evidence that environmental factors have only minor effects on immune system abnormalities.

WEB SITES OF INTEREST

American Autoimmune Related Diseases Association
http://www.aarda.org

Site of the only national organization devoted specifically to autoimmune disorders and chronic illness. Includes a medical glossary, research articles, information on the science of autoimmune-related diseases, and links to related resources.

Multiple Autoimmune Disease Genetics Consortium (MADGC)
www.madgc.org

This group is involved in genetic research into nine common autoimmune diseases.

National Institute of Allergy and Infectious Diseases
www3.niaid.nih.gov/topics/autoimmune

Contains information on research, clinical trials, and general information on autoimmune diseases.

See also: Aging; Allergies; Ankylosing spondylitis; Antibodies; Asthma; Autoimmune polyglandular syndrome; Celiac disease; Developmental genetics; Diabetes; Hybridomas and monoclonal antibodies; Immunogenetics; Organ transplants and HLA genes; Stem cells; Steroid hormones.

Autoimmune polyglandular syndrome

CATEGORY: Diseases and syndromes
ALSO KNOWN AS: APS; polyglandular autoimmune syndrome (PGA); polyglandular failure syndrome

DEFINITION

The term "autoimmune polyglandular syndrome," often called polyglandular autoimmune syndrome (PGA), is best described as a set of multiple endocrine system failures or insufficiencies. It also is known as polyglandular failure syndrome, and there are two further classifications, denoted as type I and type II. PGA is an autoimmune disease that destroys endocrine gland tissues, shows multiple ectodermal disorders, and is responsible for a chronic case of mucocutaneous candidiasis, which is the medical term for a yeast infection. The genetic mode of transmission is autosomal recessive inheritance.

RISK FACTORS

The incidence of PGA depends greatly on location and ethnicity. In the United States, the disease is rare, with an occurrence ratio of 1 in 100,000. There is an increased occurrence overseas, with Iranian Jews being the highest risk group at 1 in 6,500-9,000. Sardinians have a ratio of 1 in 14,000, and the Finns have a ratio of 1 in 25,000. Research has shown that it does affect both sexes almost equally, but with a slight female preponderance.

ETIOLOGY AND GENETICS

Research on PGA began as early as the mid-nineteenth century, when Thomas Addison first started classifying the pathology behind adrenocortical failure with pernicious anemia. Since that time, the combined research of endocrinologists and immunologists has helped further explain both the pathophysiology and pathogenesis of PGA. Polyglandular autoimmune syndrome is caused by an autosomal recessive gene inheritance, with the short arm of chromosome 21 near markers D21s49 and D21s171 on band 21p22.3 being the genetic locus. Mutations in the autoimmune regulator (*AIRE*) gene are to blame, due to the mutated encoding of the AIRE protein, which then acts as a transcription factor. The mutation *R257X* has been shown to be responsible for 82 percent of all cases. Although the exact mechanisms are still rather poorly understood at the moment, there is a pathway that has been postulated. First, a patient must have a predisposed genetic susceptibility and then must be exposed to some type of autoimmune trigger. This trigger could be an environmental one or even an intrinsic factor. Next this trigger imitates the structure of a body's self-antigen. At this point, the self-antigen reproduces in an organ, and organ-specific antibodies are replicated. Autoimmune activity increases in that organ to the extent that there is notable glandular destruction. This infection, if left untreated, can continue to spread until excessive organ damage has occurred because of autoimmune activity and the organ is filled with chronic inflammatory infiltrate, which is composed primarily of lymphocytes.

SYMPTOMS

In order for a diagnosis of PGA type I to be given, at least two out of the following three symptoms must be present: chronic mucocutaneous candidiasis (CMC), autoimmune adrenal gland insufficiency, or chronic hypoparathyroidism. Other diseases that have been observed when PGA type I is present are vitiligo, alopecia, hypogonadism, chronic hepatitis, malabsorption, keratoconjunctivitis, autoimmune Addison's disease (AAD), and chronic atrophic gastritis. CMC is the first symptom to become visible and usually attacks the skin, but it has also been known to spread to the mouth, esophagus, vagina, nails, and intestines. The second overall symptom to appear is the endocrine disease hypoparathyreosis.

SCREENING AND DIAGNOSIS

Most of the symptoms for PGA type I become evident in the first twenty years of life, and the other diseases mentioned tend to occur by the age range of forty to fifty. Most cases of PGA type II involve the combination of Addison's disease with Hashimoto's thyroiditis, (a thyroid autoimmune disease) while the least amount of diagnosed PGA type II cases are a combination of Addison's disease, Type I diabetes mellitus, and Graves' disease (a type of autoimmune hyperthyroidism). In order to classify a patient with PGA type II, there must be an occurrence of both AAD in combination with Type I diabetes mellitus and/or thyroid autoimmune diseases. Other diseases that occur along with those three are celiac disease, pernicious anemia, and myasthenia gravis. There is also a PGA type III, which occurs when autoimmune thyroiditis is present without any form of autoimmune adrenalitis. The other diseases that can display symptoms are alopecia, Sjogren's syndrome, pernicious anemia, and Type I diabetes mellitus.

TREATMENT AND THERAPY

Treatment options vary depending on type, but since all cases are chronic, lifelong diseases, there is no cure. PGA type I requires hormone replacement therapy, as a result of the use of ketoconazol (an antifungal medication) to combat the CMC. Ketoconazol inhibits the production of testosterone and cortisol, which can have a major impact on the function of the adrenal gland of patients who already exhibit a lower-than-average pituitary-adrenal reserve. Chronic hepatitis can be treated with immunosuppressive therapy by using medications such as prednisone or azathioprine. Other treatment options are available, but all depend on the combination of ailments and their symptoms.

PREVENTION AND OUTCOMES

Although PGA cannot be cured, patients can continue to live a normal life if the infections caused by PGA can be controlled. This control can best be achieved if the hormone deficiencies are corrected, or by treating the yeast infections, or treating diabetes through the use of insulin.

Jeanne L. Kuhler, Ph.D., and Steven Matthew Atchison

FURTHER READING

Gibson, Toby, Chenna Ramu, Christina Gemund, and Rein Aasland. "The APECED Polyglandular Autoimmune Syndrome Protein, AIRE-1, Contains the SAND Domain and Is Probably a Transcription Factor." *Trends in Biochemical Sciences* 23, no. 7 (July 1, 1998): 242-244. An overview of polyglandular autoimmune syndrome.

Husebye, E. S., J. Perheentupa, R. Rautemaa, and O. Kämpe. "Clinical Manifestations and Management of Patients with Autoimmune Polyendoceine Syndrome Type I." *Journal of Internal Medicine* 265 (2009): 514-529. Recent research results pertaining to the mechanism of genetic transmission.

Zlotogora, J., and M. S. Shapiro. "Polyglandular Autoimmune Syndrome Type I Among Iranian Jews." *Journal of Medical Genetics* 29 (1992): 824-826. A detailed description of the genetic transmission of polyglandular autoimmune syndrome.

WEB SITES OF INTEREST

Medicinenet
http://www.medterms.com

Medscape
http://emedicine.medscape.com

See also: Autoimmune disorders; Hereditary diseases; Immunogenetics.

B

Bacterial genetics and cell structure

CATEGORY: Cellular biology; Bacterial genetics

SIGNIFICANCE: The study of bacterial structure and genetics has made tremendous contributions to the fields of genetics and medicine, leading to the development of drugs for the treatment of disease, the discovery of DNA as the master chemical of heredity, and knowledge about the regulation of gene expression in other organisms, including humans.

KEY TERMS

cloning: the generation of many copies of DNA by replication in a suitable host

eukaryote: an organism made up of cells having a membrane-bound nucleus that contains chromosomes

mutation: the process by which a DNA base-pair change or a change in a chromosome is produced; the term is also used to describe the change itself

prokaryote: an organism lacking a membrane-bound nucleus

recombinant DNA: a DNA sequence that has been constructed or engineered from two or more distinct DNA sequences

BACTERIA AND THEIR STRUCTURE

The old kingdom Monera contained what has now been classified into the domains *Bacteria* and *Archaea.* Organisms in these domains are unicellular (one-celled) and prokaryotic (lacking a membrane-bound nucleus). Bacteria are among the simplest, smallest, and most ancient of organisms, found in nearly every environment on earth. While some bacteria are autotrophic (capable of making their own food), most are heterotrophic (forced to draw nutrients from their environment or from other organisms). For most of human history, the existence of bacteria was unknown. It was not until the late 1800's that bacteria were first identified. Their role in nature is that of decomposers: They break down organic molecules into their component parts. Along with fungi, they are the major recyclers in nature. They are also capable of changing atmospheric nitrogen to a form that is usable by plants and animals.

It has long been known that some bacteria are pathogens, or causers of disease. Scientists have expended tremendous effort in describing the role bacteria play in disease and in creating agents that could kill them. Other bacteria, such as *Escherichia coli,* may be part of a mutualistic relationship with another organism, such as humans. Bacteria have been used extensively in genetics research because of their small size and because they reproduce rapidly; some bacteria produce a new generation every twenty minutes. Since they have been so thoroughly studied, a great deal is known about their structure and genetics.

Most bacteria are less than one micron (one-millionth of a meter) in length. They do not contain mitochondria (organelles that produce the energy molecule adenosine triphosphate, or ATP), chloroplasts (plant organelles in which the reactions of photosynthesis take place), lysosomes (organelles that contain digestive enzymes), or interior membrane systems such as the endoplasmic reticulum or Golgi bodies. They do, however, contain RNA, ribosomes (organelles that serve as the sites of protein synthesis), and DNA, which is organized as part of a single, circular chromosome. The circular chromosome is centrally located within the cell in a region called the nucleoid region and is capable of supercoiling. Bacteria often have additional genes carried on small circular DNA molecules called plasmids, which have been used extensively in genetic research. Some plasmids carry genes that impart antibiotic resistance to the cells that contain them.

Bacteria have three basic morphologies, or cell shapes. Bacteria that are spherical are called cocci. Some coccus bacteria form clusters (staphylococcus), while others may form chains (streptococcus). Bacteria that have a rodlike appearance are called bacilli. Spiral or helical bacteria are called spirilla (sometimes called spirochetes).

CLASSIFICATION OF BACTERIA

Bacteria fall into three basic types: those that lack cell walls, those with thin cell walls, and those with thick cell walls. Mycoplasmas lack cell walls entirely. The bacteria that cause tuberculosis, *Mycobacterium tuberculosis*, do have cell walls and, unlike *Archaea*, their cell walls are composed of peptidoglycan, a complex organic molecule made of two unusual sugars held together by short polypeptides (short chains of amino acids). In 1884, Hans Christian Gram, a Danish physician, found that certain bacterial cells absorbed a stain called crystal violet, while others did not. Those cells that absorb the stain are called gram-positive, and those that do not are called gram-negative. It has since been found that gram-positive bacteria have thick walls of peptidoglycan, while gram-negative bacteria have thin peptidoglycan walls covered by a thick outer membrane. It is this thick outer membrane that prevents crystal violet from entering the bacterial cell. Distinguishing between gram-positive and gram-negative bacteria is an important step in the treatment of disease since some antibiotics are more effective against one class than the other.

By contrast with bacteria, members of *Archaea* have cell walls that do not contain peptidoglycan. Members of *Archaea* are usually found in extreme environments, such as hot springs, extremely saline environments, and hydrothermal vents. Methanogens are the most common and are strict anaerobes, which means that they are killed by oxygen. They live in oxygen-free environments, such as sewers and swamps, and produce methane gas as a waste product of their metabolism. Halobacteria live in only those environments that have a high concentration of salt, such as salt ponds. Thermoacidophiles grow in very hot or very acidic environments.

Bacteria can be further differentiated by the presence or absence of certain surface structures. Some strains produce an outer slime layer called a "capsule." The capsule permits the bacterium to adhere to surfaces (such as human teeth, for example, where the build-up of such bacteria causes dental plaque) and provides some protection against other microorganisms. Some strains display pili, which are fine, hairlike appendages that also allow the bacterium to adhere to surfaces. Some pili, such as F pili in *E. coli*, are involved in the exchange of genetic material from one bacterium to another in a process called conjugation. Some bacterial strains have one or more flagella, which allow them to be motile (capable of movement). Any bacterium may have one or more of these surface structures.

Research in molecular genetics is continuing to expand insight into bacterial classification and gene function. Many researchers have been actively sequencing the genomes of bacteria from a broad spectrum. The number of species that have been sequenced is now in the hundreds and includes many human pathogens, such as those that cause tuberculosis, bacterial pneumonia, ulcers, bacterial influenza, leprosy, and Lyme disease. The genomes of a wide range of nonpathogenic bacteria have also been sequenced. Comparisons among the genomes that have been sequenced are beginning to show extensive evidence that bacteria of different species have transferred genes back and forth many times in the past, thus making it difficult to trace their evolutionary lineages.

BACTERIAL REPRODUCTION

Bacteria reproduce in nature by means of binary fission, wherein one cell divides to produce two daughter cells that are genetically identical. As bacteria reproduce, they form clustered associations of cells called colonies. All members of a colony are genetically identical to one another, unless a mutagen (any substance that can cause a mutation) has changed the DNA sequence in one of the bacteria. Changes in the DNA sequence of the chromosome often lead to changes in the physical appearance or nutritional requirements of the colony. While a bacterium is microscopic, bacterial colonies can be seen with the naked eye; changes in the colonies are relatively easy to perceive. This is one of the reasons bacteria have been favored organisms for genetic research.

For the most part, there is very little genetic variation between one bacterial generation and the next. Unlike higher organisms, bacteria do not engage in sexual reproduction, which is the major source of genetic variation within a population. In laboratory settings, however, bacteria can be induced to engage in a unidirectional (one-way) exchange of genetic material via conjugation, first observed in 1946 by biochemists Joshua Lederberg and Edward Tatum. The unidirectional nature of the gene transfer was discovered by William Hayes in 1953. He found that one bacterial cell was a donor cell while the other was the recipient. In the 1950's, molecular biologists François Jacob and Elie Wollman used conjugation and a technique called "interrupted mating" to map genes onto the bacterial chromosome. By breaking apart the conjugation pairs at intervals and analyzing the times at which donor genes entered the recipient cells, they were able to determine a correlation between time and the distance between genes on a chromosome. The use of this technique led to a complete map of the sequence of genes contained in the chromosome. It also led to a surprise: It was use of interrupted mating with *E. coli* that first demonstrated

the circularity of the bacterial chromosome. The circular structure of the chromosome was in striking contrast to eukaryotic chromosomes, which are linear.

TRANSFORMATION AND TRANSDUCTION

The bacterium *Streptococcus pneumoniae* was used in one of the early studies that eventually led to the identification of DNA as the master chemical of heredity. Two strains of *S. pneumoniae* were used in a study conducted by microbiologist Frederick Griffith in 1928. One strain (S) produces a smooth colony that is virulent (infectious) and causes pneumonia. The other strain (R) produces a rough colony that is avirulent (noninfectious). When Griffith injected mice with living type R bacteria, the mice survived and no bacteria were recovered from their blood. When he injected mice with living type S, the mice died, and type S bacteria were recovered from their blood. However, if type S was heat-killed before the mice were injected, the mice did not die, and no bacteria were recovered from their blood.

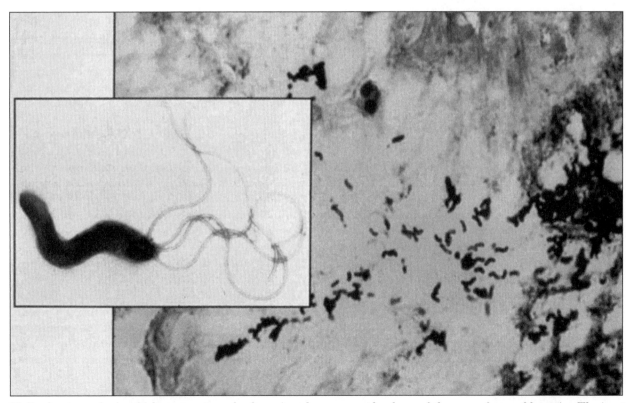

Helicobacter pylori, which causes stomach ulcers, is only one example of one of the many forms of bacteria. The inset shows a single bacterium. (AP/Wide World Photos)

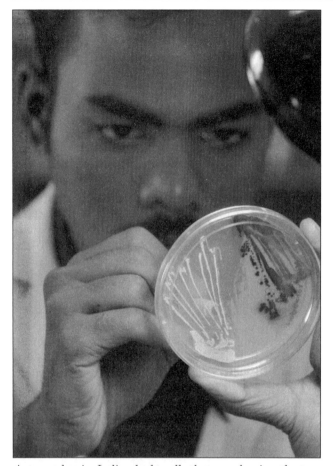

A researcher in India checks cells that are showing the presence of enzymes in bacteria. (AP/Wide World Photos)

mained. They placed some of the mixture onto agar plates (glass dishes containing a gelatin growth medium). At this point, transformation still occurred; therefore, it was clear that one of these three molecules was the transforming agent. They treated their extract with protein-degrading enzymes, which denatured (destroyed) all the proteins in the extract. Despite the denaturing of the proteins, transformation still occurred when some of the extract was plated; had protein been the transforming agent, no transformation could have occurred. Protein was eliminated as the transforming agent. The next step was to determine which of two nucleic acids was responsible for the transformation of the R strain into the S strain. They introduced RNase, an enzyme that degrades RNA, to the extract. The RNA was destroyed, yet transformation took place. RNA was thus eliminated. At this point, it was fairly obvious that DNA was the transforming agent. To conclusively confirm this, they introduced DNase to the extract. When the DNA was degraded by the enzyme, transformation did not take place, showing that DNA was the transforming agent.

Another way that genetic material can be exchanged between bacteria is by transduction. Transduction requires the presence of a bacteriophage (a virus that infects bacteria). A virus is a simple structure consisting of a protein coat called a capsid that contains either RNA or DNA. Viruses are acellular, nonliving, and extremely small. To reproduce, they must infect living cells and use the host cell's internal structures to replicate their genetic material and manufacture viral proteins. Bacteriophages, or phages, infect bacteria by attaching themselves to a bacterium and injecting their genetic material into the cell. Sometimes, during the assembly of new viral particles, a piece of the host cell's DNA may be enclosed in the viral capsid. When the virus leaves the host cell and infects a second cell, that piece of bacterial DNA enters the second cell, thus changing its genetic makeup. Generalized transduction (the transfer of a gene from one bacterium to another) was discovered by Joshua and Esther Lederberg and Norton Zinder in 1952. Using *E. coli* and a bacteriophage called *P1*, the Lederbergs and Zinder were able to show that transduction could be used to map genes to the bacterial chromosome.

This confirmed what Griffith already knew: Only living type S *S. pneumoniae* caused lethal infections. Something interesting happened when Griffith mixed living type R with heat-killed type S, however: Mice injected with this mixture died, and virulent type S bacteria were recovered from their blood. An unknown agent apparently transformed avirulent type R into virulent type S. Griffith called the agent the "transforming principle." It was his belief that the transforming principle was a protein.

Sixteen years later, in 1944, bacteriologists Oswald Avery, Colin MacLeod, and Maclyn McCarty designed an experiment that showed conclusively that the transforming principle was DNA rather than protein. They showed that R bacteria could be transformed to S bacteria in a test tube. They then progressively purified their extract until only proteins and the two nucleic acids, RNA and DNA, re-

Hershey-Chase Bacteriophage Experiments

The use of bacteriophages has been instrumental in confirming DNA as the genetic material of living cells. Alfred Hershey and Martha Chase devised a series of experiments using *E. coli* and the bacteriophage *T2* that conclusively established DNA as genetic material in 1953. Bacteria are capable of manufacturing all essential macromolecules by utilizing material from their environment. Hershey and Chase grew cultures of *E. coli* in a growth medium enriched with a radioactive isotope of phosphorus, phosphorus 32. DNA contains phosphorus; as the succeeding generations of bacteria pulled phosphorus from the growth medium to manufacture DNA, each DNA strand also carried a radioactive label. *T2* phages were used to infect the cultures of *E. coli*. When the new *T2* viruses were assembled in the bacterial cells, they too carried the radioactive label phosphorus 32 on their DNA. A second culture of *E. coli* was grown in a medium enriched with radioactive sulfur 35. Proteins contain sulfur (but no phosphorus). *T2* viruses were used to infect this culture. New viruses contained the sulfur 35 label on their protein coats.

Since the *T2* phage consists of only protein and DNA, one of these two molecules had to be the genetic material. Hershey and Chase infected unlabeled *E. coli* with both types of radioactive *T2* phages. Analysis has shown that the phosphorus 32 label passed into the bacterial cells, while the sulfur 35 label was found only in the protein coats that did not enter the cells. Since the protein coat did not enter the bacterial cell, it could not influence protein synthesis. Therefore, protein could not be the genetic material. The Hershey-Chase experiment confirmed DNA as the genetic material.

Restriction Enzymes and Gene Expression

Using the aforementioned methods, it has been possible to construct a complete genetic map showing the order in which genes occur on the chromosome of *E. coli* and other bacteria. Certain genes are common to all bacteria. There are also several genes that are shared by bacteria and higher life-forms, including humans. Further research showed that genes can be either inserted into or deleted from bacterial DNA. In nature, only bacteria contain specialized enzymes called restriction enzymes. Restriction enzymes are capable of cutting DNA at specific sites called restriction sites. The function of restriction enzymes in bacteria is to protect against invading viruses. Bacterial restriction enzymes are designed to destroy viral DNA without harming the host DNA. Hundreds of different restriction enzymes have been isolated from bacteria, and each is named for the bacterium from which it comes. The discovery and isolation of restriction enzymes led to a new field of biological endeavor: genetic engineering.

Use of these enzymes has made gene cloning possible. Cloning is important to researchers because it permits the detailed study of individual genes. Restriction enzymes have also been used in the formation of genomic libraries (a collection of clones that contains at least one copy of every DNA sequence in the genome). Genomic libraries are valuable because they can be searched to identify a single DNA recombinant molecule that contains a particular gene or DNA sequence.

Bacterial studies have been instrumental in understanding the regulation of gene expression, or the translation of a DNA sequence first to a molecule of messenger RNA (mRNA) and then to a protein. Bacteria live in environments that change rapidly. To survive, they have evolved systems of gene regulation that can either "turn on" or "turn off" a gene in response to environmental conditions. François Jacob and Jacques Monod discovered the *lac* operon, a regulatory system that permits *E. coli* to respond rapidly to changes in the availability of lactose, a simple sugar. Other operons, such as the tryptophan operon, were soon discovered as well. An operon is a cluster of genes whose expression is regulated together and involves the interaction of regions of DNA with regulatory proteins. The discovery of operons in bacteria led to searches for them in eukaryotic cells. While none has been found, several other methods of regulating the expression of genes in eukaryotes have been described.

Impact and Applications

Diabetes mellitus is a disease caused by the inability of the pancreas to produce insulin, a protein hormone that is part of the critical system that controls the body's metabolism of sugar. Prior to 1982, people who suffered from diabetes controlled their disease with injections of insulin that had been isolated from other animals, such as cows. In 1982, human insulin became the first human gene product to be manufactured using recombinant DNA. The technique is based on the knowledge that genes can be inserted into the bacterial chromosome; that once inserted, the gene product, or protein, will be produced; and that once produced, the protein can be purified from bacterial extracts. Human proteins are usually produced by inserting a human gene into a plasmid vector, which is then inserted into a bacterial cell. The bacterial cell is cloned until large quantities of transformed bacteria are produced. From these populations, human proteins, such as insulin, can be recovered.

Many proteins used against disease are manufactured in this manner. Some examples of recombinant DNA pharmaceutical products that are already available or in clinical testing include atrial natriuretic factor, which is used to combat heart failure and high blood pressure; epidermal growth factor, which is used in burns and skin transplantation; factor VIII, which is used to treat hemophilia; human growth hormone, which is used to treat dwarfism; and several types of interferons and interleukins, which are proteins that have anticancer properties.

Bacterial hosts produce what are called the "first generation" of recombinant DNA products. There are limits to what can be produced in and recovered from bacterial cells. Since bacterial cells are different from eukaryotic cells in a number of ways, they cannot process or modify most eukaryotic proteins, nor can they add sugar groups or phosphate groups, additions that are often required if the protein is to be biologically active. In some cases, human proteins produced in prokaryotic cells do not fold into the proper three-dimensional shape; since shape determines function in proteins, these proteins are nonfunctional. For this reason, it may never be possible to use bacteria to manufacture all human proteins. Other organisms are used to produce what are called the second generation of recombinant DNA products.

The impact of the study of bacterial structures and genetics and the use of bacteria in biotechnology, cannot be underestimated. Bacterial research has led to the development of an entirely new branch of science, that of molecular biology. Much of what is currently known about molecular genetics, the expression of genes, and recombination comes from research involving the use of bacteria. Moreover, bacteria have had and will continue to have applications in the production of pharmaceuticals and the treatment of disease. The recombinant DNA technologies developed with bacteria are now being used with other organisms to produce medicines and vaccines.

Kate Lapczynski, M.S.; updated by Bryan Ness, Ph.D.

FURTHER READING

Birge, Edward A. *Bacterial and Bacteriophage Genetics.* 5th ed. New York: Springer, 2006. Examines how genetic investigations and manipulations of bacteria and bacteriophages have made vital contributions to the basic understanding of living cells and to the development of genetic engineering and biotechnology.

Drlica, Karl. *Understanding DNA and Gene Cloning: A Guide for the Curious.* 4th ed. Hoboken, N.J.: Wiley, 2004. A basic overview designed to help general readers understand molecular biology and recombinant DNA technology. Good illustrations and graphics.

Goldberg, Joanna B., ed. *Genetics of Bacterial Polysaccharides.* Boca Raton, Fla.: CRC Press, 1999. Gives background on the field's history, polysaccharide diversity, research gaps, and nomenclature issues. Nine chapters by international researchers present the genetic analysis of polysaccharides from various bacteria pathogens to humans and one symbiotic with legumes.

Hacker, J., and J. B. Kaper, eds. *Pathogenicity Islands and the Evolution of Pathogenic Microbes.* 2 vols. New York: Springer, 2002. Explores pathogenicity islands, plasmids, and bacteriophages, which are able to carry genes whose products are involved in pathogenic processes. Shows how such elements and their products play an important role in pathogenesis due to the intestinal *E. coli,* as well to *Shigellae.*

Hatfull, Graham F., and William R. Jacobs, Jr., eds. *Molecular Genetics of Mycobacteria.* Washington, D.C.: ASM Press, 2000. Surveys all aspects of mycobacterial genetics, starting with the development of mycobacterial genetics and then presenting the molecular genetics of mycobacteria in sections on genomes and genetic exchange, gene expression, metabolism, and genetic strategies.

Russell, Peter J. *Fundamentals of Genetics.* 2d ed. San Francisco: Benjamin Cummings, 2000. Introduces the three main areas of genetics: transmission genetics, molecular genetics, and population and quantitative genetics. Reflects advances in the field, such as the structure of eukaryotic chromosomes, alternative splicing in the production of mRNAs, and molecular screens for the isolation of mutants.

Schumann, Wolfgang. *Dynamics of the Bacterial Chromosome: Structure and Function.* Weinheim, Germany: Wiley-VCH, 2006. Provides an overview of bacterial genetics, bacterial genome projects, and gene technology. Includes discussion of bacterial cell structure, organization of the bacterial chromosome, replication, and recombination.

Schumann, Wolfgang, S. Dusko Ehrlich, and Naotake Ogasawara, eds. *Functional Analysis of Bacterial Genes: A Practical Manual.* New York: Wiley, 2001. Follows two teams of laboratories that analyze thousands of newly discovered bacterial genes to try to discover their functions. Addresses the biology of *Bacillus subtilis.*

Snyder, Larry, and Wendy Champness. *Molecular Genetics of Bacteria.* 3d ed. Washington, D.C.: ASM Press, 2007. Comprehensive textbook on bacterial genetics. Focuses on *E. coli* and *Bacillus subtilis* but provides information about other bacteria that have a medical, ecological, or biotechnological significance.

Thomas, Christopher M., ed. *The Horizontal Gene Pool: Bacterial Plasmids and Gene Spread.* Amsterdam: Harwood Academic, 2000. International geneticists, biologists, and biochemists discuss the various contributions plasmids make to horizontal gene pools: replication, stable inheritance, and transfer modules; the phototypic markers they carry; how they evolve; how they contribute to their host population; and approaches for studying and classifying them.

Watson, James D., et al. *Recombinant DNA: A Short Course.* 3d ed. New York: W. H. Freedman, 2007. A classic account by one of three men who shared a Nobel Prize in Physiology or Medicine for describing the molecular structure of DNA.

WEB SITES OF INTEREST

E. coli Genome Project, University of Wisconsin
http://www.genome.wisc.edu

The genome research center that sequenced the organism's complete K-12 genome now maintains and updates that sequence, as well as those of other strains and other pathogenic *Enterobacteriaceae.*

Pseudomonas Genome Database
http://www.pseudomonas.com

Describes the work of British Columbia-based researchers who are studying the genomics of the *Pseudomonas aeruginosa* bacterium. People with cystic fibrosis, burn victims, and other patients in intensive care units are at risk of disease from aeruginosa infection; the research project aims to accelerate the discovery of *Pseudomonas aeruginosa* drug targets and vaccines.

See also: Archaea; Bacterial resistance and super bacteria; Biopharmaceuticals; Cholera; Chromosome walking and jumping; Cloning; Diabetes; Gene regulation: Bacteria; Gene regulation: *Lac* operon; Gene regulation: Viruses; Genetic code, cracking of; Lateral gene transfer; Model organism: *Escherichia coli*; Molecular genetics; Plasmids; Restriction enzymes; Transposable elements.

Bacterial resistance and super bacteria

CATEGORY: Bacterial genetics
SIGNIFICANCE: Antibiotic-resistant bacteria have become a significant worldwide health concern. Some strains of bacteria (called super bacteria) are now resistant to most, if not all, of the available antibiotics and threaten to return health care to a preantibiotic era. Understanding how and why bacteria become resistant to antibiotics may aid treatment, the design of future drugs, and efforts to prevent other bacterial strains from becoming resistant to antibiotics.

KEY TERMS

antimicrobial drugs: chemicals that destroy disease-causing organisms without damaging body tissues; chemicals made naturally by bacteria and fungi are also known as antibiotics
plasmids: small, circular pieces of DNA that can exist separately from the bacterial chromosome; plasmids can be transferred among bacteria, and they may carry more than one R factor
resistance factor (R factor): a piece of DNA that carries a gene encoding for resistance to an antibiotic

transposons: also known as jumping genes, transposons are pieces of DNA that carry R factors and can integrate into a bacterial chromosome; they are also responsible for the spread of drug resistance in bacteria and fungi, and, like plasmids, each transposon may carry more than one R factor

HISTORY OF ANTIBIOTICS

Throughout history, illnesses such as cholera, pneumonia, and sexually transmitted diseases have plagued humans. However, it was not until the early twentieth century that antibiotics were discovered. Until then, diseases such as diphtheria, cholera, and influenza were serious and sometimes deadly. With the advent of the antibiotic era, it appeared that common infectious diseases would no longer be a serious health concern. A laboratory accident led to the discovery of the first mass-produced antibiotic. In 1928, Scottish bacteriologist Alexander Fleming grew *Staphylococcus aureus* in petri dishes, and the plates became contaminated with a mold. Before Fleming threw out the plates, he noticed that there was no bacterial growth around the mold. The mold, *Penicillium notatum,* produced a substance that was later called penicillin, which was instrumental in saving the lives of countless soldiers during World War II. From the 1950's until the 1980's, antibiotics were dispensed with great regularity for most bacterial infections, for earaches, for colds, and as a preventive measure.

As the twentieth century progressed, however, it became apparent that the initial promise of antibiotics was mitigated by the ability of microorganisms to evolve quickly, given their relatively short life spans. Emerging infectious diseases such as multidrug-resistant tuberculosis, vancomycin-resistant enterococci, and penicillin-resistant gonorrhea became serious global health care concerns. The problem was exacerbated by the seemingly haphazard dispensing of antibiotics for viral infections (against which antibiotics are ineffective, although often prescribed as a hedge against secondary infections or simply to palliate patients).

THE RISE OF BACTERIAL RESISTANCE

On average, bacteria can replicate every twenty minutes. Several generations of bacteria can reproduce in a twenty-four-hour period. This quick generation time leads to a rapid adaptation to changes in the environment. English naturalist Charles Darwin's *On the Origin of Species by Means of Natural Selection* (1859) first explained the theory of natural selection, the process whereby this adaptation occurs. If an organism has an advantage over other organisms (such as the ability to grow in the presence of a potentially harmful substance), that organism will survive to pass that characteristic on to its offspring while the other organisms die. The emergence of antibiotic-resistant bacteria is an excellent example of Darwin's theory of natural selection at work.

In the early twentieth century, German microbiologist Paul Ehrlich coined the term "magic bullet" in reference to chemotherapy (the treatment of disease with chemical compounds). For a drug such as an antibiotic to be a "magic bullet," it must have a specific target that is unique to the disease-causing agent and cannot harm the host in the process of curing the disease. In 1910, Ehrlich discovered that arsphenamine (Salvarsan), a derivative of arsenic, could be used to treat syphilis, a common sexually transmitted disease in the early twentieth century. Until that time syphilis had no known cure. The use of Salvarsan did cure some patients of syphilis, but, since it was a rat poison, it killed other patients. Generally speaking, antimicrobials have specific targets (or modes of action) within bacteria. They target the following structures or processes: synthesis of the bacterial cell wall, injury to the plasma membrane, and inhibition of synthesis of proteins, DNA, RNA, and other essential metabolites (all of these substances are building blocks for the bacteria). A good antibiotic will have a target that is unique to the bacteria so the host (the patient) will not be harmed by the drug.

Bacteria and fungi are, of course, resistant to the antibiotics they naturally produce. Other bacteria have the ability to acquire resistances to antimicrobials, and this drug resistance occurs either through a mutation in the DNA or resistance genes on plasmids or transposons. Plasmids are small, circular pieces of DNA that can exist within or independently of the bacterial chromosome. Transposons, or "jumping genes," are pieces of DNA that can jump from one bacterial species to another and be integrated into the bacterial chromosome. The spread of plasmids and transposons that carry antibiotic resistance genes has led bacteria to become resistant to many, if not all, currently available antibiotics.

Several antimicrobial resistance mechanisms allow bacteria to become drug resistant. The first mechanism does not allow the drug to enter the bacterial cell. A decrease in the permeability of the cell wall will inhibit the antimicrobial drug from reaching its target. An alteration in a penicillin-binding protein (pbp), a protein found in the bacterial cell wall, will allow the cell to "tie up" the penicillin. Also, the pores in the cell wall can be altered so the drug cannot pass through. A second strategy is to pump the drug out of the cell after it has entered. Such systems are found in pathogenic *Escherichia coli, Pseudomonas aeruginosa,* and *Staphylococcus aureus.* These pumps are usually nonspecific and can cause bacteria to become resistant to more than one antibiotic at a time. Another method of resistance is through chemical modification of the drug. Penicillin is inactivated by breaking a chemical bond found in its ring structure. Other drugs are inactivated by the addition of other chemical groups. Finally, the target of the drug can be altered in such a manner that it is no longer affected by the drug. For example, *Mycobacterium tuberculosis,* which causes tuberculosis, became resistant to the drug rifampin by altering the three-dimensional structure of a specific protein.

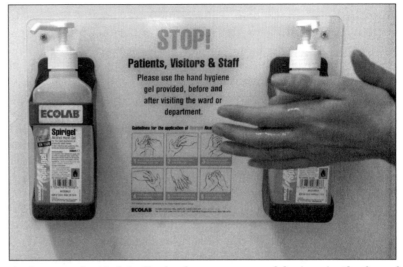

A sign at a hospital in England encourages good hygiene in the face of "superbugs" such as MRSA bacteria. (AP/Wide World Photos)

ANTIBIOTIC MISUSE AND DRUG RESISTANCE

The misuse of antibiotics over several decades has caused many strains of bacteria to become resistant. For some bacterial infections, only one or no effective drugs are available for treatment. Many different factors of misuse, overuse, and abuse of antibiotics have led to drug-resistant diseases. Perhaps one of the most important factors in the emergence of drug-resistant bacteria is the overprescription or inappropriate use of antibiotics. Another major factor is misuse by the patient. After several days of taking an antibiotic, a patient may begin to feel better and decide not to finish all of the prescription. By not completing the full course of treatment, the patient merely kills the bacteria that are sensitive to the antibiotic, leaving the resistant bacteria to grow, multiply, pass on their resistant genes, and cause the same infection. This time, another antibiotic (if there is one available that is effective) must be used.

Another contributing factor is the ease with which the newest and best antibiotics may be obtained in many countries. In several countries in Central America, for example, one can walk into the local pharmacy and receive any antibiotic without a prescription. Another factor in the worldwide spread of drug-resistant infectious diseases is the ease of travel. Infected people can carry bacteria from one continent to another in a matter of hours and infect anyone with whom they come in contact.

The use of antibiotics is not limited to humans. They also play an important role in agriculture. Antibiotics are added to animal feed on farms to help keep herds healthy, and they are also used on fish farms for the production of fish for market. Antibiotics are used to treat domestic animals such as cats, dogs, birds, and fish and are readily available in pet stores to clear up fish aquariums. This widespread use of antibiotics allows bacteria in all environmental niches the possibility of becoming resistant to potentially useful drugs.

EMERGING RESISTANT INFECTIONS AND SUPER BACTERIA

The misuse of antibiotics over the decades has led to more infectious diseases becoming resistant to the current arsenal of drugs. Some diseases that could be treated effectively in the 1970's and 1980's can no longer be controlled with the same drugs. Two very serious problems have emerged: vancomycin-resistant enterococci and multidrug-resistant tuberculosis. The enterococcus is naturally resistant to many types of antibiotics, and the only effective treatment has been vancomycin. With the appearance of vancomycin-resistant enterococci, however, there are no reliable alternative treatments. The fear that vancomycin resistance will spread to other bacteria such as *staphylococci* seems well founded: A report from Japan in 1997 indicated the existence of a strain of staphylococcus that had become partially resistant to vancomycin. If a strain of methicillin-resistant *Staphylococcus aureus* (MRSA) also becomes resistant to vancomycin, there will be no effective treatment available against this super bacterium.

A second problem is the appearance of multidrug-resistant tuberculosis. *Mycobacterium tuberculosis* is a slow-growing bacterium that requires a relatively long course of antibiotic therapy. Tuberculosis (TB) is spread easily, and it is a deadly disease. In the United States in 1900, tuberculosis was the number-one cause of death. In the 1990's, it was still a leading cause of death worldwide. Treatment of multidrug-resistant tuberculosis requires several antibiotics taken over a period of at least six months, with a success rate of approximately 50 percent; on the other

Multiple-Resistant Bacteria

They lurk in schools, nursing homes, and hospitals—perhaps even in your home. Often, you cannot see them to avoid them. Increasingly, they are a global health problem. What are these unseen purveyors of disease? Antibacterial soaps.

Antibacterial soaps contain antibacterials, a subclass of antimicrobials, which kill or inhibit the growth of bacteria and other microorganisms. Antiseptics are antimicrobial agents that are sufficiently nontoxic to be applied to human tissue. Antibiotics are chemicals that inhibit a specific pathway or enzyme in a bacterium and are critical to the treatment of a bacterial infection. When bacteria are exposed to sublethal concentrations of an antibiotic, resistance can develop through the elimination of normal bacteria, allowing the resistant ones to survive and reproduce. The question has been whether exposure to antibacterial products can promote antibiotic resistance. The answer is that the use of antibacterial products may actually *increase* the prevalence of antibiotic-resistant bacteria.

Antibiotic resistance is irreversible and unavoidable, due to the selective pressure on bacteria to become resistant. This selection is in large part a result of the widespread use of antibiotics to increase growth rates in livestock, as well as unnecessary and improper use of antibiotics to restore and maintain human health. The indiscriminate use or overuse of antibiotics has been widely blamed for the appearance of so-called super bacteria—bacteria that are unaffected by more than one antibiotic. In addition, a widely used antibacterial agent used in toothpaste, kitchen utensils and appliances, clothing, cat litter, and toys could cause resistant strains of bacteria to develop.

Triclosan is a good example of the potent antibacterial and antifungal agents that are increasingly used to produce "germ-free" consumer products. Until recently, triclosan was considered a broad-spectrum antiseptic rather than a true antibiotic. As a general biocide, triclosan was not expected to have a specific target in the bacterial cell. However, Stuart Levy and his colleagues at Tufts University School of Medicine determined that triclosan specifically interferes with an enzyme important in the synthesis of plasma membrane lipids. As triclosan kills off normal bacteria, it could make way for the growth of strains with triclosan-insensitive enzymes. More troubling, one of the front-line antibiotics commonly used to treat tuberculosis, isoniazid, targets the same enzyme, raising the possibility that the use of triclosan will lead to new drug-resistant strains of *Mycobacterium tuberculosis*.

Consumers are convinced that use of products with antimicrobial chemicals will lower their risk of infection. While this has not been demonstrated scientifically, effective handwashing has been demonstrated to prevent illness. However, the key to effective handwashing is the length of time (15-30 seconds) spent scrubbing, not the inclusion of antibacterials in the soap. Regular soap, combined with scrubbing action, physically dislodges and removes microorganisms. The constant exposure of bacteria to sublethal concentrations of triclosan promotes development of resistance; the substitution of antibacterial soap for proper handwashing techniques will eventually render triclosan ineffective. In the battle of the soaps, "plain" wins.

Laurie F. Caslake, M.S., Ph.D.

hand, susceptible strains of TB have a cure rate of nearly 100 percent.

Another contributing factor to the emergence of drug-resistant infectious diseases is the lack of basic knowledge about some bacteria. Funding for basic genetic research on tuberculosis was reduced dramatically in the mid-twentieth century when it appeared that TB would be eradicated just as smallpox had been. The appearance of multidrug-resistant tuberculosis caught scientists and physicians unprepared. Little was known about the genetics of tuberculosis or how drug resistance occurred.

Another concern about drug-resistant infections is how to control them. Hospitals are vigilant, and, in some cases, very proactive in screening for drug-resistant infections. People can be asymptomatic carriers (that is, they carry the disease-causing organism but are still healthy) of a disease such as methicillin-resistant *Staphylococcus aureus* and could infect other people without knowing it. The role of the infection-control personnel is to find the source of the infection and remove it.

IMPACT AND APPLICATIONS

There is little encouraging news about the availability of new antibiotics. The crisis of super bacteria has altered the view that few new antibiotics would be needed. Pharmaceutical companies are scrambling to discover new antimicrobial compounds and modify existing antibiotics. Policy decisions of the 1970's and 1980's requiring more and larger clinical trials for antibiotics before they are approved by the Food and Drug Administration for use have increased the prices of antibiotics and the amounts of time it takes to market them. It may take up to ten years from the time of "discovery" for an antibiotic to be approved for use. The scientific community has therefore had to meet the increase of drug-resistant bacterial strains with fewer and fewer new antibiotics.

The emergence of antibiotic-resistant bacteria and super bacteria is a serious global health concern that will lead to a more prudent use of available antibiotics. It has also prompted pharmaceutical companies to search for potentially new and novel antibiotics in the ocean depths, outer space, and other niches. "Rational" drug design—or RDD, drug design based on knowledge of how bacteria become drug resistant—will also be important. Exactly how scientists and physicians will be able to

combat super bacteria is a question that remains to be answered. Until a more viable solution is found, prudent use of antibiotics, surveillance of drug-resistant infections, and well-orchestrated worldwide monitoring and containment of emerging diseases appear to be the answers.

Mary Beth Ridenhour, Ph.D.

FURTHER READING

Fond, I. W., and Karl Drlica, eds. *Antimicrobial Resistance and Implications for the Twenty-first Century.* New York: Springer, 2008. Collection of essays analyzing antimicrobial resistance in specific types of bacteria, viruses, and fungi.

Franklin, T. J., and G. A. Snow. *Biochemistry and Molecular Biology of Antimicrobial Drug Action.* 6th ed. New York: Springer, 2005. Provides an explanation of the chemistry of antimicrobials and how bacteria may become resistant to their effects.

Gould, Ian M., and Jos W. M. van der Meer, eds. *Antibiotic Policies: Fighting Resistance.* New York: Springer, 2007. Collection of essays examining the key issues that society and hospitals are facing as a result of epidemic drug resistance and the shortage of new antibiotics.

Levy, Stuart. *The Antibiotic Paradox: How the Misuse of Antibiotics Destroys Their Curative Powers.* 2d ed. Cambridge, Mass.: Perseus, 2002. Provides an overview of antibiotic resistance in bacteria. Discusses mechanisms of resistance, reasons for the spread of antimicrobial resistance, and ways to combat this spread.

Murray, Patrick, ed. *Manual of Clinical Microbiology.* 9th ed. Washington, D.C.: ASM Press, 2007. Presents a direct approach to organizing information with thorough but concise treatments of all the major areas of microbiology, including new microbial discoveries, changing diagnostic methods, and emerging therapeutic challenges facing clinicians.

Salyers, Abigail A., and Dixie D. Whitt. *Revenge of the Microbes: How Bacterial Resistance Is Undermining the Antibiotic Miracle.* Washington, D.C.: ASM Press, 2005. Traces the history of antibiotics and of bacteria's eventual resistance to these drugs. Designed for the general reader.

Tortora, Gerard J., Berdell R. Funke, and Christine L. Case. *Microbiology: An Introduction.* 10th ed. San Francisco: Pearson Benjamin Cummings, 2010. An accessible introduction to the basic principles of microbiology, the interaction between microbe and host, and human diseases caused by microorganisms. Provides a general overview of antibiotics and how bacteria become resistant to them.

Wax, Richard G., et al., eds. *Bacterial Resistance to Antimicrobials.* 2d ed. Boca Raton, Fla.: CRC Press, 2008. Provides a history of antimicrobial agents and bacterial resistance. Examines the social, economic, and medical issues related to bacterial resistance.

WEB SITES OF INTEREST

Centers for Disease Control and Prevention, Antimicrobial Resistance
http://www.cdc.gov/drugresistance

The CDC's Web site contains numerous pages about antimicrobial resistance, including a question and answer page about the subject, antibiotic fact sheets, and information to help patients know when antibiotics will—and will not—work.

Evolution of Antibiotic Resistance
http://www.pbs.org/wgbh/evolution/library/10/4/l_104_03.html

One of the programs in *Evolution: The Evolutionary Arms Race*, a television series aired on the Public Broadcasting System (PBS), focused on drug-resistance. The Web site for WGBH, the Boston-based PBS affiliate, contains a page with background information on the evolution of antibiotic resistance and a brief animated video that follows the progression of this resistance.

National Consumers League
http://www.nclnet.org/Resisttext.html

The league's brochure, "Bacterial Resistance: When Antibiotics Don't Work," provides background information about bacterial resistance and advice to help consumers protect their health.

See also: Archaea; Bacterial genetics and cell structure; Chromosome walking and jumping; DNA replication; Emerging and reemerging infectious diseases; Gene regulation: bacteria; Gene regulation: *Lac* operon; Lateral gene transfer; Mutation and mutagenesis; Model organism: *escherichia coli*; Natural selection; Plasmids; Transposable elements.

Barlow's syndrome

CATEGORY: Diseases and syndromes
ALSO KNOWN AS: Mitral valve prolapse; floppy valve syndrome; click-murmur syndrome

DEFINITION

Barlow's syndrome, also known as mitral valve prolapse (MVP), is a common, usually benign heart disorder. The mitral valve controls blood flow between the upper (atrium) and lower (ventricle) chambers on the left side of the heart. Normally, blood should flow only in one direction, from the upper chamber into the lower chamber. In MVP, the valve flaps do not work properly; part of the valve balloons into the atrium, which may be associated with blood flowing in the wrong direction, or leaking back into the atrium.

RISK FACTORS

Individuals with a family history of MVP, individuals between the ages of fourteen and thirty, and females are at risk for the disorder. Other risk factors include having scoliosis, a thin chest diameter, low body weight, low blood pressure, chest wall deformities, Marfan syndrome, or Graves' disease.

ETIOLOGY AND GENETICS

While the majority of cases of Barlow's syndrome appear sporadically with no demonstrable genetic origin, it has been recognized since the first description of this disorder in the 1960's that in some families there is a distinct hereditary component. With the completion of the Human Genome Project and other molecular genetic studies in the last decade, at least three autosomal loci and one X-linked gene have been identified in which mutations can lead to development of the disease. The autosomal genes are found on the short arm of chromosome 11 (at position 11p15.4), the long arm of chromosome 13 (at position 13q31-32), and the short arm of chromosome 16 (at position 16p11.2-12.1). Although the genes have been localized to these specific chromosomal regions, little is currently known about the gene products or their specific association with disease symptoms. In all three cases, inheritance is transmitted as an autosomal dominant, meaning that a single copy of the mutated gene is sufficient to cause full expression of the syndrome. An affected individual has a 50 percent chance of transmitting the mutation to each of his or her children.

The rare X-linked form of Barlow's syndrome is inherited as a sex-linked recessive trait. Mothers who carry the mutated gene on one of their two X chromosomes face a 50 percent chance of transmitting this disorder to each of their male children. Female children have a 50 percent chance of inheriting the gene and becoming carriers like their mothers. Although females rarely express the syndrome fully, female carriers may occasionally show minor manifestations. Affected males will pass the mutation on to all of their daughters but to none of their sons.

The *filamin A* gene, located at position Xq28, has been identified as the gene responsible for the X-linked form of Barlow's syndrome, which is also known as familial cardiac valvular dystrophy. Filamin A is a phosphoprotein that acts in cells to help bind the actin cytoskeleton to the plasma membrane. It is unknown at present how mutations in this gene affect MVP, but research suggests the possibility that filamin A may affect heart valve development by regulating transforming growth factors or affecting some aspect of the signaling pathways for these growth factors.

SYMPTOMS

People with MVP often have no symptoms at all. If symptoms do occur, they may include irregular heartbeat, fatigue, chest pain, panic attacks or anxiety, rapid heartbeat (palpitations), a sensation of missed heartbeats, shortness of breath, dizziness, and intestinal problems (such as irritable bowel syndrome).

SCREENING AND DIAGNOSIS

MVP can be heard through a stethoscope. A small blood leakage will sound like a murmur. When the mitral valve balloons backward, it may produce a clicking sound. Both murmurs and clicks are telltale signs of MVP. An echocardiogram can confirm the diagnosis. A patient may be asked to wear a Holter monitor for a day or two to continuously record the electrical activity of his or her heart (electrocardiograph).

TREATMENT AND THERAPY

In most cases, no treatment is necessary. Patients should ask their doctors whether they should take antibiotics prior to dental work or surgery; antibiotics may help to prevent endocarditis, an infection of the membrane that covers the inside of the heart.

If symptoms include chest pain, anxiety, or panic attacks, a beta-blocker medication can be prescribed. Patients should also ask their doctors if they may continue to participate in their usual athletic activities. In very rare cases, the blood leakage may become severe, requiring surgery to repair and replace the mitral valve.

PREVENTION AND OUTCOMES

There are no guidelines for preventing MVP of unknown or genetic origin. Patients may be able to prevent symptoms, however, through certain lifestyle changes. They can limit their intake of caffeine; avoid medications, such as decongestants, that speed up their heart rates; and exercise regularly, following their healthcare providers' recommendations for intensity.

Michelle Badash, M.S.;
reviewed by Michael J. Fucci, D.O.
"Etiology and Genetics" by Jeffrey A. Knight, Ph.D.

FURTHER READING

Durante, James F., Cheryl L. Durante, and John G. Furiasse. *The Mitral Valve Prolapse Syndrome, Dysautonomia Survival Guide.* Oakland, Calif.: New Harbinger, 2002.

EBSCO Publishing. *Health Library: Barlow's Syndrome.* Ipswich, Mass.: Author, 2009. Available through http://www.ebscohost.com.

Frederickson, Lyn. *Confronting Mitral Valve Prolapse Syndrome.* New York: Warner Books, 1992.

WEB SITES OF INTEREST

American Heart Association
http://www.americanheart.org

Canadian Cardiovascular Society
http://www.ccs.ca/home/index_e.aspx

Canadian Family Physician
http://www.cfpc.ca/cfp

MyHeartCentral.com
http://www.healthcentral.com/heart-disease

National Heart, Lung, Blood Institute
http://www.nhlbi.nih.gov

See also: Atherosclerosis; Heart disease; Hereditary diseases.

Bartter syndrome

CATEGORY: Diseases and syndromes
ALSO KNOWN AS: Potassium wasting disease; Gitelman syndrome

DEFINITION

Bartter syndrome describes a group of rare, inherited disorders that share a defect in the kidney's reabsorption system known as the thick ascending loop of Henle. The abnormality causes excessive loss of fluid, potassium, sodium, and chloride in the urine, resulting in electrolyte imbalance, muscle weakness, and growth retardation. Three clinical types of Bartter syndrome have been identified; the neonatal type, the classic type, and Gitelman syndrome, a milder variant. The syndrome is named for Frederic Bartter, who described the combination of fluid loss, salt-wasting, and growth and muscle abnormalities in two patients in 1962.

RISK FACTORS

Bartter syndrome may be inherited by autosomal recessive transmission in which both parents carry a defective copy of the gene responsible for proteins that transport electrolytes across cell membranes in the nephron. Most cases appear to occur sporadically as new mutations, however, and are not familial.

Prevalence varies and may be related to the incidence of consanguineous marriage in the countries studied. In Costa Rica, the incidence is approximately 1.2 cases per 100,000 live births. Incidence is higher in Kuwait at 1.7 per 100,000. In Sweden, the prevalence is approximately 1.2 cases per 1,000,000. It is quite rare in the United States, and the precise incidence is unknown. There is no racial or gender preference.

ETIOLOGY AND GENETICS

In all three major Bartter types, defects in proteins responsible for the transport of fluid and electrolytes across cell membranes cause large volumes of urine and salts to pass through without being reabsorbed.

Neonatal Bartter syndrome is the most severe type and may be caused by defects in the *NKCC2* (neonatal Type I) and *ROMK* (neonatal Type II) genes. Cases are often diagnosed prenatally as a result of the presence of excess amniotic fluid (polyhydramnios). After delivery, infants urinate excessively and may become critically dehydrated. They are not able to sustain normal serum electrolyte levels because of sodium, chloride, and potassium loss through the kidneys. A specific neonatal clinical syndrome (Types IV and V) is associated with sensorineural deafness and results from mutations in the *BSND* and *CLCNKA-B* genes.

Classic Bartter syndrome may present by age two but is often discovered later. The defect results from a mutation in the chloride-channel (*CLCNKB*) gene. This leads to the loss of sodium chloride and large volumes of fluid in the urine, as well as increased levels of the hormones angiotensin and prostaglandin E2.

Gitelman syndrome is a milder variant of Bartter syndrome and may appear in the teen or adult years. While it shares many of the characteristics of the classic type, it is differentiated by a consistent loss of serum magnesium caused by a defect in the *NCCT* gene.

SYMPTOMS

Neonatal Bartter syndrome typically appears in the last trimester of pregnancy with maternal polyhydramnios (excess amniotic fluid) due to increased volume of fetal urine. If left untreated, infants will urinate excessively and become seriously dehydrated. Thirst increases, resulting in increased fluid intake. Vomiting is common, causing further dehydration and electrolyte imbalance. Short stature and growth delay are accompanied by typical triangular facies (facial expressions) with protruding ears. Mental retardation may occur in severe cases.

Patients with classic Bartter syndrome have excess urine production and increased fluid intake, along with low serum electrolytes, particularly potassium, but symptoms are milder than in the neonatal type. Blood pressure is typically low to normal and kidney function remains normal if the disease is treated. However, there have been cases proceeding to end-stage renal failure. Metabolic alkalosis, weakness, and vomiting may occur.

Gitelman syndrome is commonly asymptomatic and may be found incidentally when routine blood studies demonstrate low serum magnesium.

SCREENING AND DIAGNOSIS

There is currently no prenatal genetic screening for Bartter syndrome. If it is suspected prenatally, amniotic fluid can be assessed for elevated electrolyte levels. Renal ultrasound may show renal calcium deposits, as well as enlarged kidneys as a result of polyuria.

Evaluation of potassium, sodium, calcium, and magnesium levels in blood and urine is essential. Patients are likely to have low serum levels (most notably of potassium) and elevated urine levels. Low serum magnesium levels are typical of Gitelman syndrome. A complete blood count (CBC) may show hemoconcentration as a result of fluid loss. Renal function may be normal or may decrease over time as a result of chronic potassium wasting. Genetic analysis may pinpoint the specific gene defect on a case-by-case basis.

TREATMENT AND THERAPY

There is no cure for Bartter syndrome. Medical treatment focuses on electrolyte and fluid replacement. Electrolyte supplements are given, as well as specific medications to counteract increases in aldosterone and prostaglandins. Short stature and growth retardation have been successfully treated with growth hormone supplementation.

Patients who have received kidney transplants for end-stage renal failure (usually for reasons other than Bartter syndrome) have found their abnormalities corrected following the transplant. This may be because the genetic defect in Bartter syndrome is present only in the kidneys.

PREVENTION AND OUTCOMES

There is no known prevention for Bartter syndrome. Early diagnosis and treatment of children with Bartter syndrome may prevent short stature and normalize electrolyte imbalance. The severity of disease depends on the degree of dysfunction in the loop of Henle. With proper management, prognosis is good and patients may lead normal lives.

Rachel Zahn, M.D.

FURTHER READING

Bartter, F. C., P. Pronove, and J. R. Gill. "Hyperplasia of the Juxtaglomerular Complex with Hyperaldosteronism and Hypokalemic Alkalosis." *American Journal of Medicine* 33 (1962): 811-828.

Kleta, R., and D. Bockenhauer. "Bartter Syndromes and Other Salt-Losing Tubulopathies." *Nephron Physiology* 104, no. 2 (2006): 73-80.

Rudin, A. "Barrter's Syndrome: A Review of Twenty-eight Patients Followed for Ten Years." *Acta Medica Scandinavica* 224, no. 2 (1988): 165-171.

WEB SITES OF INTEREST

AllRefer Health: Barrters Syndrome
http://health.allrefer.com/health/bartters-syndrome-info.html

BarrterSite.org
http://barttersite.org

See also: Alport syndrome; Hereditary diseases; Polycystic kidney disease.

Batten disease

CATEGORY: Diseases and syndromes
ALSO KNOWN AS: Neuronal ceroid lipofuscinosis

DEFINITION

Batten disease is the most common form of a group of rare disorders known as neuronal ceroid lipofuscinoses (NCLs). Batten disease is an inherited genetic disorder that causes a buildup of lipopigments in the body's tissue. "Batten disease" refers to the juvenile form of NCL, but the other forms of NCL can also be referred to as "Batten disease." About 2 to 4 of every 100,000 births are affected. The forms of NCL include infantile NCL, late infantile NCL, juvenile NCL, and adult NCL.

RISK FACTORS

Since Batten disease is an inherited condition, people at risk include children of parents with Batten disease and children of parents not afflicted with Batten disease, but who carry the abnormal genes that cause the disease.

ETIOLOGY AND GENETICS

The infantile, late infantile, and juvenile forms of Batten disease, while caused by mutations in three different genes, are all inherited in an autosomal recessive manner. Both copies of the particular gene must be deficient in order for the individual to be afflicted. Typically, an affected child is born to two unaffected parents, both of whom are carriers of the recessive mutant allele. The probable outcomes for children whose parents are both carriers are 75 percent unaffected and 25 percent affected, but two-thirds of the unaffected children will be carriers like their parents. A simple blood test is available to screen for and identify the most common carrier phenotype for juvenile Batten disease.

Adult NCL, which can be caused by mutations in at least two different genes, is inherited as either an autosomal recessive or in an autosomal dominant fashion. In autosomal dominant inheritance, a single copy of the mutated gene is sufficient to cause full expression of the disease. An affected individual has a 50 percent chance of transmitting the mutation to each of his or her children.

The juvenile form of Batten disease is caused by mutations in the *CLN3* gene, found on the long arm of chromosome 16 at position 16p12.1. The CLN3 protein, also known as palmitoyl-protein delta-9 desaturase, has been shown to participate in the membrane-associated modification of proteolipids. Lack of this enzyme activity may result in the accumulation of proteolipids in cells, leading to the neurodegeneration that is characteristic of the disorder.

Candidate genes associated with the infantile and late infantile forms of the disease have also been identified, and these have been named *CLN1* and *CLN2*, respectively. *CLN1*, which is found on the short arm of chromosome 1 at position 1p32, encodes a protein called palmitoyl-protein thioesterase. *CLN2*, located on the short arm of chromosome 11 at position 11p15.5, encodes an acid protease, an enzyme that hydrolyzes specific proteins.

SYMPTOMS

Symptoms of Batten disease include vision loss (an early sign) and blindness, muscle incoordination, mental retardation or decreasing mental function, emotional disturbances or difficulties, seizures, muscle spasms, deterioration of muscle tone, and movement problems.

These symptoms are similar in each type of the disease. However, the time of appearance, severity, and rate of progression of symptoms can vary depending on the type of the disease.

In infantile NCL (Santavuori-Haltia disease), symptoms begin to appear between the ages of six months and two years and progress rapidly. Children with this type generally live until midchild-

hood (about age five), though some survive in a vegetative state a few years longer.

In late infantile NCL (Jansky-Bielschowsky disease), symptoms begin to appear between ages two and four and progress rapidly. Children with this type usually live until ages eight through twelve. Symptoms of juvenile NCL (Spielmeyer-Vogt-Sjogren-Batten disease) begin to appear between ages five and eight and progress less rapidly. Those afflicted usually live until their late teens or early twenties and, in some cases, into their thirties.

Symptoms of adult NCL (Kufs disease or Party's disease) usually begin to appear before age forty. Symptoms progress slowly and are usually milder. However, this form of the disease usually does shorten a person's life span.

SCREENING AND DIAGNOSIS

Batten disease is often difficult to diagnose because it is so rare. Vision problems are often the first symptoms. Therefore, an initial diagnosis may result from an eye exam. To confirm the diagnosis, tests are taken. These include testing to look for evidence of a buildup of lipopigments, such as blood tests, urine tests, and tissue biopsies examined with an electron microscope.

Imaging tests can look for specific brain abnormalities. These include a magnetic resonance imaging (MRI) scan, which uses magnetic waves to make pictures of the inside the body; a computed tomography (CT) scan, a type of X ray that uses a computer to take pictures inside the body; and an electroencephalogram (EEG), a test that records the brain's activity by measuring electrical currents through the brain. Electrical studies of the eyes can look for vision problems associated with the disease, while deoxyribonucleic acid (DNA) analysis can look for the abnormalities that may cause this disease.

TREATMENT AND THERAPY

There is no known treatment that will stop the progression or effects of Batten disease. Therefore, treatment aims to reduce symptoms.

Patients who have seizures can be given anticonvulsant medications to help control seizures. In addition, physical and/or occupational therapy can help people continue functioning for a longer period of time.

One experimental therapy is supplementation

with vitamins C and E combined with a diet low in vitamin A. This may slow the progression of the disease in children; however, there is no evidence that it will halt the ultimate progression of the disease. Parents should talk to their child's doctor before trying this therapy.

Very early trials of stem cell treatment for infantile and late infantile disease are now underway. There is hope that these or other forms of gene therapies may have an effect on the progression of Batten and Jansky-Bielschowsky diseases.

PREVENTION AND OUTCOMES

There is no known way to prevent Batten disease. Individuals who have Batten disease or have a family history of the disorder can talk to a genetic counselor when deciding to have children.

Rick Alan; reviewed by Rosalyn Carson-DeWitt, M.D.
"Etiology and Genetics" by Jeffrey A. Knight, Ph.D.

FURTHER READING

Dawson, G., and S. Cho. "Batten's Disease: Clues to Neuronal Protein Catabolism in Lysosomes." *Journal of Neuroscience Research* 60, no. 2 (April 15, 2000): 133-140.

EBSCO Publishing. *Health Library: Batten Disease.* Ipswich, Mass.: Author, 2009. Available through http://www.ebscohost.com.

Mole, S. E. "Batten's Disease: Eight Genes and Still Counting?" *Lancet* 354, no. 9177 (August 7, 1999): 443-445.

Sondhi, D., et al. "Feasibility of Gene Therapy for Late Neuronal Ceroid Lipofuscinosis." *Archives of Neurology* 58, no. 11 (November, 2001): 1793-1798.

Taupin, P. "HuCNS-SC (StemCells)." *Current Opinion in Molecular Therapeutics* 8, no. 2 (April, 2006): 156-163.

Wisniewski, Krystyna E., and Nanbert Zhong, eds. *Batten Disease: Diagnosis, Treatment, and Research.* San Diego: Academic Press, 2001.

WEB SITES OF INTEREST

Batten Disease Support and Research Association
http://www.bdsra.org

Genetics Home Reference
http://ghr.nlm.nih.gov

MedHelp: Batten Disease
http://www.medhelp.org/lib/batten.htm

National Institute of Neurological Disorders and Stroke
http://www.ninds.nih.gov

Office of Rare Diseases Research
http://rarediseases.info.nih.gov

United States National Library of Medicine
http://www.nlm.nih.gov

See also: Amyotrophic lateral sclerosis; Ataxia telangiectasia; Epilepsy; Essential tremor; Hereditary diseases; Parkinson disease.

Beckwith-Wiedemann syndrome

CATEGORY: Diseases and syndromes
ALSO KNOWN AS: Exomphalos-macroglossia-gigantism syndrome; Wiedemann-Beckwith syndrome; BWS

DEFINITION

Beckwith-Wiedemann syndrome (BWS) is a complex genetic disorder affecting growth. Abnormalities associated with the disorder include enlarged body size, enlarged organs, and presence of specific types of tumors. The etiology and inheritance of BWS are not well understood. The condition is thought to be sporadic in 85 percent of cases and inherited in an autosomal dominant manner in 15 percent of cases.

RISK FACTORS

BWS is panethnic and usually sporadic. However, in familial cases, family history may be a risk factor, especially those demonstrating autosomal dominant inheritance patterns. Chromosome abnormalities involving chromosome 11, such as translocations, inversions, or duplications, may increase the risk. Paternal uniparental disomy (UPD) for chromosome 11 may also increase the risk.

ETIOLOGY AND GENETICS

Most cases of BWS are thought to involve abnormal expression or transcription of genes in the 11p15 region. The region is part of an imprinted domain; some genes are normally expressed only from the paternally derived alleles, while other genes in the region are expressed only from the maternally derived alleles. When there is a disruption, deletion, or mutation in one copy of the imprinted alleles, BWS occurs.

About 50 to 60 percent of cases of BWS are caused by methylation defects in the KCNQ1OT1 (*DMR2*) gene in the 11p15 region. Methylation is a process in which a methyl group is added to a piece of DNA to inactivate the segment. When *DMR2* is not properly methylated, the genes cannot properly be regulated, which leads to the overgrowth and other features associated with BWS. Similarly, another 6 to 7 percent of cases of the disorder are caused by methylation defects in the H19 (*DMR1*) gene. The cases of BWS involving methylation defects are almost exclusively sporadic cases with a low recurrence risk.

Uniparental disomy (UPD) accounts for 10 to 20 percent of cases of BWS. This results in loss of maternal alleles in the region and increased expression of paternal alleles in the region. The resulting underexpression and overexpression of involved genes results in the phenotype associated with BWS. Cases of UPD are usually a result of recombination events during mitosis. Since these events occur after conception, patients with UPD may be mosaic for the changes and are considered sporadic cases with low recurrence risks.

Mutations in the maternal allele of the *CDKN1C* gene account for 5 to 10 percent of sporadic cases and 40 percent of familial (autosomal dominant) cases of BWS. Chromosomal abnormalities, including maternal translocations involving chromosome 11, inversions of chromosome 11, or duplication of paternally derived 11p15, account for about 1 to 2 percent of BWS. Autosomal dominant cases have up to a 50 percent risk of recurrence, while chromosomal abnormalities vary in their recurrence risks. The remaining causes of BWS are not yet known.

SYMPTOMS

The majority of symptoms related to BWS are related to overgrowth of various tissues. Prenatal overgrowth occurs in about 50 percent of cases, which can be associated with prematurity. Also frequently seen is macrosomia (large body size), macroglossia (large tongue), and visceromegaly (large internal organs). Cardiomyopathy is also common. Children with BWS are at an increased risk for developing embryonal tumors, including Wilms' tumor and hepatoblastoma. Renal abnormalities are found in about 50 percent of cases. Omphalocele, an abdominal

wall defect, can also be seen in association with BWS. There are significant risks of mortality, mostly as a result of complications of prematurity. Hemihyperplasia may be seen, resulting in unbalanced growth of body segments or specific tissues; this may resolve over time. The growth rate usually slows in late childhood. Developmental delay is not common unless there is a chromosome abnormality present.

SCREENING AND DIAGNOSIS

BWS occurs in every 1 in 12,000-13,700 live births. The diagnosis is typically acheived through clinical symptoms. Genetic studies for suspected diagnoses should be offered for confirmation. Molecular studies for diagnosis include methylation studies, karyotyping with fluorescence in situ hybridization (FISH), microsatellite analysis, and sequencing of the *CDKN1C* gene. Prenatal testing is available by chorionic villus sampling or amniocentesis when the molecular abnormality has been previously identified in the family.

TREATMENT AND THERAPY

Treatment of BWS involves management of presenting symptoms, which may include surgery to correct the physical consequences of omphalocele, hemihyperplasia, macroglossia, and renal malformations. Periodic screening for embryonal tumors is essential in detection because of their fast growing and potentially dangerous nature. Speech therapy, physical therapy, and occupational therapy may be helpful in overcoming obstacles associated with macroglossia and hemihyperplasia.

PREVENTION AND OUTCOMES

Except for the familial cases where prenatal diagnosis can be performed, prevention of BWS is not possible. Genetic counseling should be offered to all families with a person confirmed or suspected to be affected with BWS. About 20 percent of infants with BWS die as a result of prematurity and cardiac complications. Beyond infancy, mortality rates depend on the presence of symptoms and appropriate management.

Leah M. Betman, M.S.

FURTHER READING

Jones, Kenneth Lyons, and David W. Smith. *Smith's Recognizable Patterns of Human Malformation.* 5th ed. New York: Saunders, 1997.

Jorde, Lynn B., et al. *Medical Genetics.* 3d ed. Philadelphia: Mosby, 2006.

Nussbaum, Robert L., Roderick R. McInnes, and Huntington F. Willard. *Thompson and Thompson Genetics in Medicine.* 7th ed. New York: Saunders, 2007.

WEB SITES OF INTEREST

Beckwith-Wiedemann Children's Foundation
http://www.beckwith-wiedemannsyndrome.org

Gene Reviews: Beckwith-Wiedemann Syndrome
http://www.ncbi.nlm.nih.gov/bookshelf/
br.fcgi?book=gene&part=bws

Genetics Home Reference: Beckwith-Wiedemann Syndrome
http://ghr.nlm.nih.gov/condition=
beckwithwiedemannsyndrome

See also: Fluorescence in situ hybridization (FISH); Hereditary diseases; Wilms' tumor; Wilms' tumor aniridia-genitourinary anomalies-mental retardation (WAGR) syndrome.

Behavior

CATEGORY: Population genetics

SIGNIFICANCE: Of the many long-standing questions pondered by biologists is to what extent genes control the way people behave. To date, researchers have identified human genes that have been linked to such behavioral characteristics as depression, anxiety, psychosis, and alcoholism; however, these and other genetic findings are complicated by methodological questions and by the problem of distinguishing between the effects of genetic and environmental factors. This "nature versus nurture" debate continues along with genetic research, including the Human Genome Project.

KEY TERMS

eugenics: a process in which negative genetic traits are removed from the population and positive genetic traits are encouraged, by controlling, in some manner, who is allowed to reproduce

genome: the entire set of genes required by an organism; a set of chromosomes

heritability: the probability that a specific gene or trait will be passed from parent to offspring, rendered as a number between 0 and 100 percent, with 0 percent being not heritable and 100 percent being completely heritable

Human Genome Project (HGP): an international genetics project developed to identify and map the human genome with its approximate 24,000 genes, the first assembly of which was completed by the UCSC Genome Bioinformatics Group in 2003

linkage: a relation of gene loci on the same chromosome; the more closely linked two loci are, the more often the specific traits controlled by these loci are expressed together

neurotransmitter: a chemical messenger that transmits a neural impulse between neurons

population genetics: the discipline within the field of evolutionary biology concerned with the study of changes in gene frequency, including how this relates to human groups

BRAIN BIOLOGY

As the first organ system to begin development and the last to be completed, the vertebrate nervous system—brain, spinal cord, and nerves—with the brain at the control, remains something of an enigma to biologists and other scientists. The vertebrate brain comprises, among other structures, neurons, which are special cells that generate and transmit bioelectrical impulses via a number of different neurotransmitters. The brain consists of three major neural structures: the brain stem, the cerebellum, and the cerebrum. A reptilian brain consists of only the brain stem, while the mammalian brain has all three, including a well-developed cerebrum (the two large hemispheres on top). The brain stem controls basic body functions such as breathing and heart rate, while the cerebrum is the ultimate control center. Consisting of billions of neurons (commonly called brain cells), the cerebrum controls such higher-level functions as memory, speech, hearing, vision, and analytical skills.

Scientists have long sought to understand the complex relationship between the brain, behavior, and genetics. Decades of research have led to a general consensus that fundamental to human behavior, cognition, and emotions is the functioning of the cerebral cortex (that is, for higher-brain functions) and the limbic system, which includes the amygdala, septum, cingulate, hippocampus, anterior thalamic nuclei, fornix, and mammillary bodies. In particular, the human brain is governed by the frontal lobes of the cerebral cortex, which controls cognitive processes. And connecting the brain stem and the cortex is the limbic system, the center that mediates motivated behaviors, emotional states, and memory, as well as regulates temperature, blood-sugar levels, and blood pressure.

Clearly, neurotransmitters influence behavior and cognition by modulating the activity of neurons. More specifically, the brain—at all levels—is an exceedingly complex network of billions of neurons. As messages enter the brain stem from the spinal cord, groups of neurons either respond directly or transfer information to higher levels. In order to communicate with other neurons, each individual neuron generates impulses much like the impulse that carries a digital signal over a fiber-optic cable, and this message travels from the beginning to the end of each neuron. At the end of one neuron and the beginning of the next in line, a small open space exists. Here the message is carried across to the next neuron by a chemical known as a neurotransmitter.

Neurotransmitters are of several biochemical classifications, including acetylcholines, amines, amino acids, and peptides. An individual neuron and an entire neuronal circuit may fire or not fire an impulse based on the messages carried by these neurotransmitters. For example, the signal for pain is transmitted from neuron to neuron by a peptide-based neurotransmitter known as substance P, while another peptide transmitter (endorphin) acts as a natural painkiller. Thought, memory, and behavior, then, are produced by the activity along neuronal circuits. A genetic link occurs here, since neurotransmitters are expressed either directly or indirectly based on information in genes.

By birth, the collection of approximately 24,000 genes in humans has guided the development of the nervous system. At birth, the brain consists of approximately 100 billion neurons and trillions of supporting glial cells to protect and nourish neurons. However, the intricate wiring between these neurons, including exactly how a virtual multitude of neural signals eventually translates into thinking and behavior, remains to be determined.

While many scientists have held to a strictly genetic model in order to explain human behavior and cognitive functioning, others have suggested that the critical networking and circuit formation between these billions of neurons that control later brain function are determined not from genes but from environmental input and experiences from birth until the brain is fully developed around age seven. In other words, controversy has arisen as scientists have attempted to explain behavior from a genetic, inherited perspective versus from a social and environmental one. Typically the nature-versus-nurture debate, which has been ongoing for centuries, has been a dichotomy: either nature or nurture. What has developed is a sense that both play a role, with the controversy centering around to what degree either nature or nurture predominates as the primary causal factor in any given trait outcome.

Just how many human traits and abilities are innate or acquired through interactions with the environment is unknown, however. On one end of the continuum is John Locke's concept of the *tabula rasa*, in which the human brain of the newborn is thought to be a "blank slate" that will be differentiated only by sensory experiences. On the other end is today's biological determinism, in which behavior is thought to be strictly innate and heritable. A majority of experts subscribe to a middle-of-the-road view that rests between these two extremes of this nature-versus-nurture argument. For example, a spider phobia might be considered to result from a combination of an innate evolution-driven fear of potentially dangerous spiders, a heritable tendency toward anxiety, and conditioning through prior "bad" experiences with insects.

GENES AND BEHAVIOR

Traditionally, the field of behavioral genetics has emphasized evaluating how much population variance is determined by environmental or hereditary factors. From the perspective of human development and behavior, the issue also becomes one of how this process is expressed within cultural constraints—by what means genetic and social surroundings reciprocate to yield obvious outcomes.

Genes make proteins, and proteins cause biochemical responses in cells. The behavior of an animal takes place under the combined influences of its genes, expressed through the actions of proteins, and its environment. A good example is the phenomenon of mating seasons in many animals. As day length gradually increases toward spring and summer, a critical length is reached that signals the release of hormones that result in increased sexual activity, with the ultimate goal of seasonal mating. The production and activity of hormones involve genes or gene products. If the critical number of daylight hours is not reached, the genes will not be activated, and sexual behavior will not increase.

Each neuron making up the intricate networks and circuits throughout the cerebrum (about 80 percent of the human brain) has protein receptors (chemoreceptors) that respond to specific signaling molecules. The production of the receptors and signaling molecules used for any type of brain activity is directly tied to genes. A slightly different gene may lead to a slightly different signaling molecule or receptor and thus a slightly different cell (neuron) response. A larger difference among genes may lead to a larger difference among signaling molecules or receptors and thus a larger variation in cell response. Since human behavior involves the response of neurons and neuron networks in the brain to specific signals, and because the response of neurons occurs from the interaction between a signaler and a receptor built by specific genes, the genetic link seems straightforward: input, signal, response, behavior. However, when the slight variations between genes are added to the considerable variation among noncoding or regulatory sequences of DNA, the genetic connection to behavior becomes much less direct. Because a gene is under the control of one or several regulatory sequences that in turn may be under the control of various environmental inputs, the amount of genetic variation among individuals is compounded by two other critical factors: the environmental variations under which the brain develops and the daily environmental variations to which the individual is exposed. A convenient way to think of genetics and behavior is to consider that genes allow humans to respond to a specific stimulus by building the pathway required for a response, while behavior is defined by the degree and the manner of human response.

Eugenics

Eugenics is the categorization of a specific human behavior to an underlying genetic cause. People inherit specific genes to build specific pathways that allow them to respond in certain ways to environmental input. With variations possible—from the gene-to-gene regulators to the final cellular response—it is virtually impossible to disconnect the nature-versus-nurture tie that ultimately controls human behavior. Genes are simply the tools by which the environment shapes and reshapes human behavior. There is a direct correlation between gene and protein: Change the gene, change the protein. However, there is no direct correlation between gene and behavior: Changing the gene does not necessarily change the behavior. Behavior is a multifaceted, complex response to environmental influences that is only partially related to genetic makeup. Most studies conducted on humans based on twin and other relative data suggest that most behavioral characteristics have between a 30 and 70 percent genetic basis, leaving considerable room for environmental influence.

Another important fact is that almost no behaviors are controlled by a single gene locus, and the more complex the behavior, the more likely that it is controlled by several to many genes. Hence, not only do environmental effects cloud the picture, but each gene involved in more complex behavioral traits represents just a small part of the genetic basis for the trait as well. The study of the genetic basis for complex traits, therefore, involves the search for quantitative trait loci (QTLs), rather than for single genes.

Searching for QTLs requires that a large number of genetic markers be identified in the human genome, and the Human Genome Project has provided numerous such markers. A QTL is identified by looking for "linkage" between a specific genetic marker and the trait being studied. Linkage occurs when a marker is close to one of the genes that control the trait. Practically speaking, this means that individuals with the behavioral trait have the marker, and those who do not have the trait lack the marker. Thus, geneticists are not directly identifying the genes involved, but are identifying the approximate locations of the genes. Unfortunately, the more genes that control a trait, the harder it is to identify QTLs. Environmental effects can also mask the existence of QTLs, causing some people to have the trait that lack a QTL and others to lack the trait but have a QTL. In spite of these difficulties, QTLs have been identified for a number of behavioral traits, such as aggression, depression, and a number of other mental disorders.

Single-Gene Behavioral Traits

Although behavioral traits controlled by a single gene have been identified, they probably require interaction with other genes in order to produce the specific characteristics of the behavior. On top of this are laid environmental effects. The most dramatic case of a single gene that controls a complex behavior was the discovery in the early twenty-first century of a gene that controls honeybee social status. This same gene is found in fruit flies and affects how actively fruit flies seek food. Bees with a more actively expressed form of the gene (called the *for* gene) were much more likely to forage than bees with a less active *for* gene. Not surprisingly, the *for* gene produces a protein that acts as a cell-signaling molecule.

In humans, only a few behavioral traits are clearly controlled by a single gene. The best examples are Huntington's disease (a rare, autosomal dominant gene), early-onset Alzheimer's disease (also a rare, autosomal dominant gene), and fragile X syndrome (actually involves two genes). The remaining traits, as far as has been determined to date, probably represent multigene traits where one primary QTL has been identified as primarily responsible.

Several genes were identified, beginning in the late 1980's, with possible direct behavioral links. Genes have been implicated in such behaviors as anxiety, depression, hostility, and impulsiveness. One such gene produces a protein that transports a chemical called serotonin, across neuronal membranes. Serotonin is a neurotransmitter and is the chemical that is affected by the antidepressant drug Prozac and other selective serotonin reuptake inhibitors (SSRIs). Scientists have also identified a gene that may be related to schizophrenia and a gene that may determine how well alcohol is cleared from the brain after overindulgence.

One of the more recent, and in some ways controversial, discoveries involved a gene for antisocial behavior (ASB). The study followed the lives of more than one thousand boys from birth. Children who grew up in abusive environments were more likely to display antisocial behavior later, which is not a surprise. However, about half of the boys were found to have lower levels of an enzyme called monoamine oxidase A (MAOA), which is involved in the metabolism of several neurotransmitters. The boys with the lower MAOA activity were twice as likely to have been diagnosed with conduct disorder and were three times more likely to have been convicted of a violent crime by age twenty-six. It should be noted that lower MAOA activity alone was not

enough; the boys also had to be exposed to abusive upbringings. Although the link seems strong, it has not been proved, with continued study needed.

In short, a more thorough understanding of single-gene behavioral traits could open the way to more accurate diagnoses and better treatments.

MULTIPLE-GENE BEHAVIORAL TRAITS

Geneticists concede that for many behavioral traits it may never be possible to sort out the details of the underlying genetic causes. Still, theories abound and researchers continue to speculate. Some genes may play such a minor role that the search for some QTLs will be fruitless. Nevertheless, geneticists have been able to discover QTLs for some important behavioral traits, and the heritability of a number of traits has been determined. The better data available from the Human Genome Project have spawned the rapidly growing field of behavioral genomics, with its emphasis on identifying the specific genetic mechanisms involved in the determination of behavior.

Nonetheless, the quality of the environment matters in most cases. A practical example of this is intelligence or IQ, which is thought by many experts to involve both environmental and genetic influences, given individual abilities to adapt to social stressors. Successful adaptation requires personal coping but may also require either altering the quality of the present surroundings or locating another environment altogether. Such intentional coping also requires a number of mental processes, including sensation, perception, memory, reasoning, learning, and problem solving. The primary thrust is to avoid labeling human mental functioning as strictly nature or nurture, but rather as a selective combination of multiple adaptive processes employed for successful coping in the environment. In short, certain traits may never be fully understood from a strictly genetic perspective. Even when heritability is high, the environment also plays an important role, and numerous genes are likely involved.

More success has come from focusing on specific disorders. For example, a series of genes have been identified that may be involved in attention deficit hyperactivity disorder. Other QTLs have been identified in some studies but have not been found in others. This shows one of the frustrating aspects of studying the genetics of behavior. QTLs identified using one set of data will not be supported by another set of data. This may be true because such QTLs play such a small part in developing the trait that they are undetectable under certain environmental conditions. Genes and QTLs for dyslexia and schizophrenia have also been discovered.

For the most complex human traits, QTLs still await discovery, but twin studies have perhaps yielded the most valuable data regarding the relationship between genes and behavior. Twin studies involve comparing the traits of identical twins that were separated from birth. The assumption is that, because they have been raised in different environments, any traits they share will be primarily attributable to genetics rather than to environment. An early study of Swedish men showed that heritability of cognitive (thinking) ability was 62 percent, while spatial ability was 32 percent. Heritability of other personality traits fell somewhere between these values. Although these kinds of studies are interesting, they can be misleading unless considered in proper context. Consequently, a number of geneticists criticize such research, especially twin studies, as having some inherent statistical problems. These studies can also lead to misunderstandings, especially by nonscientists, who often interpret the numbers incorrectly. For example, saying that cognitive ability has a 62 percent heritability does not mean that a child has a 62 percent chance of being as intelligent as his or her parents but rather that, of the factors involved in determining a person's intelligence, genetics accounts for approximately 62 percent of the observed variation in the population.

THE FUTURE OF BEHAVIORAL GENETICS

Researchers continue to actively investigate the potential links between behavior and genetics in human functioning. Even when such links are found, however, the degree to which a particular gene is involved and the amount of variation among humans will likely be hard to uncover. The Human Genome Project has greatly accelerated interest in and the search for the genetic bases of behavior, yet with these new data have come an even clearer realization of the complexities of the interplay between genes and behavior. If nothing else, the future should hold more precise answers to the long-standing questions about what makes human beings who they are. The consensus among geneticists today is that behavior is determined neither solely by genes

nor solely by the environment. To this end, further research should attempt to make the relative contributions of genes and environment more understandable.

W. W. Gearheart, Ph.D., and Bryan Ness, Ph.D.;
updated by George D. Zgourides, M.D., Psy.D.

FURTHER READING

Avital, Eytan, and Eva Jablonka. *Animal Traditions: Behavioural Inheritance in Evolution.* New York: Cambridge University Press, 2000. Broadens the evolutionary approach to behavior by arguing that the transfer of learned information across generations is indispensable.

Badcock, C. R. *Evolutionary Psychology: A Critical Introduction.* Malden, Mass.: Polity Press in association with Blackwell, 2000. An introductory text that addresses such topics as selection and adaptation, survival of the fittest, the benefits and costs of brain evolution, psychological conflict between parent and child, language, and development and conflict.

Benjamin, Jonathan, Richard P. Ebstein, and Robert H. Belmaker, eds. *Molecular Genetics and the Human Personality.* Washington, D.C.: American Psychiatric Association, 2002. Provides a comprehensive overview of the genetic basis for human personality. Eighteen chapters, each of which ends with a reference section. Index.

Briley, Mike, and Fridolin Sulser, eds. *Molecular Genetics of Mental Disorders: The Place of Molecular Genetics in Basic Mechanisms and Clinical Applications in Mental Disorders.* Malden, Mass.: Blackwell, 2001. Explores the role of molecular genetics in the understanding of mental disorders and how molecular genetics might help in the development of new drugs for mental illness. Illustrations.

Burnham, Terry, and Jay Phelan. *Mean Genes: From Sex to Money to Food—Taming Our Primal Instincts.* Cambridge, Mass.: Perseus, 2000. In examining the issues that most affect people's lives—body image, money, addiction, violence, and relationships, friendship, love, and fidelity—Burnham argues that struggles for self-improvement are, in fact, battles against one's own genes.

Carson, Ronald A., and Mark A. Rothstein. *Behavioral Genetics: The Clash of Culture and Biology.* Baltimore: Johns Hopkins University Press, 1999. Experts from a range of disciplines—genetics, ethics, neurosciences, psychiatry, sociology, and law—address the cultural, legal, and biological underpinnings of behavioral genetics.

Cartwright, John. *Evolution and Human Behavior: Darwinian Perspectives on Human Nature.* Cambridge, Mass.: MIT Press, 2000. Offers an overview of the key theoretical principles of human sociobiology and evolutionary psychology and shows how they illuminate the ways humans think and behave. Argues that humans think, feel, and act in ways that once enhanced the reproductive success of our ancestors.

Clark, William R., and Michael Grunstein. *Are We Hardwired? The Role of Genes in Human Behavior.* New York: Oxford University Press, 2000. Explores the nexus of modern genetics and behavioral science, revealing that few elements of behavior depend upon a single gene; instead, complexes of genes, often across chromosomes, drive most of human heredity-based actions. Asserts that genes and environment are not opposing forces but work in conjunction.

DeMoss, Robert T. *Brain Waves Through Time: Twelve Principles for Understanding the Evolution of the Human Brain and Man's Behavior.* New York: Plenum Trade, 1999. Provides an accessible examination on what makes humans unique and delineates twelve principles that can explain the rise of humankind and the evolution of human behavior.

Kim, Yong-Kyu. *Handbook of Behavior Genetics.* New York: Springer, 2009. A thorough and fascinating look at how the many aspects of behavioral sciences and genetics and interrelate.

Plomin, Robert, et al. *Behavioral Genetics.* 4th ed. New York: Worth, 2001. Introductory text that explores the basic rules of heredity, its DNA basis, and the methods used to find genetic influence and to identify specific genes.

Rosen, David H., and Michael C. Luebbert, eds. *Evolution of the Psyche.* Westport, Conn.: Praeger, 1999. Surveys a range of scientific and theoretical approaches to understanding some of the most important markers connected with the evolution of the psyche, including sex and mating, evolution and creativity and humor, the survival value of forgiveness, and the evolutionary significance of archetypal dreams.

Rutter, Michael. *Genes and Behavior: Nature-Nurture Interplay Explained.* New York: Wiley-Blackwell, 2006. A brilliant commentary, from a biopsycho-social perspective, on the co-action of genetic and environmental factors at all stages of human development.

Wright, William. *Born That Way: Genes, Behavior, Personality.* New York: Knopf, 1998. Uses twin and adoption studies to trace the evolution of behavioral genetics and discusses the corroborating research in molecular biology that underlines the links between genes and personality.

WEB SITES OF INTEREST

Behavior Genetics Association (BGA)
http:www.bga.org
Devoted to the scientific study of the interrelationship of genetic mechanisms and behavior.

Human Genome Project Information, Behavioral Genetics
http://www.ornl.gov/sci/techresources/Human
_Genome/elsi/behavior.shtml
This site includes information on the basics of behavioral genetics and links to related resources.

Institute for Behavioral Genetics (IBG)
http://ibgwww.colorado.edu
Founded in 1967, the IBG is a research unit of the University of Colorado dedicated to researching the genetic and environmental bases of human behavior.

National Institute of Mental Health, Center for Genetic Studies
http://zork.wustl.edu/nimh
A technical site on the collecting of clinical data to help determine the possible genetic bases of certain mental disorders.

See also: Aggression; Alcoholism; Altruism; Attention deficit hyperactivity disorder (ADHD); Autism; Biological clocks; Biological determinism; Bipolar affective disorder; Criminality; Developmental genetics; Eugenics; Gender identity; Genetic engineering: Medical applications; Genetic engineering: Social and ethical issues; Genetic screening; Genetic testing; Genetic testing: Ethical and economic issues; Heredity and environment; Homosexuality; Human genetics; Inbreeding and assortative mating; Intelligence; Klinefelter syndrome; Knockout genetics and knockout mice; Miscegenation and antimiscegenation laws; Natural selection; Sociobiology; Steroid hormones; Twin studies; XYY syndrome.

Best disease

CATEGORY: Diseases and syndromes

ALSO KNOWN AS: Best's vitelliform macular dystrophy (early-, juvenile-, and adult-onset); BVMD; vitelline dystrophy; vitelliruptive degeneration; vitelliform macular degeneration; macular degeneration, polymorphic vitelline; early-onset macular degeneration

DEFINITION

Best disease is a rare hereditary disorder that causes macular dystrophy. The disease destroys the macular, a very small area in the center of the retina. Best disease occurs most often in children and usually results in partial or complete blindness. The disorder is caused by the *BEST1* gene on chromosome 11. The disease may also be affected by another gene mutation, *VMD2*, which causes age-related vitelliform macular dystrophy and macular degeneration in adults.

RISK FACTORS

The only known risk factor of Best disease is a family history of the *BEST1/VMD2* gene. Best disease affects children of European, African, and Hispanic backgrounds. Male and female children are equally affected. Diet and activity do not appear to influence disease progression.

ETIOLOGY AND GENETICS

Best disease is an autosomal dominant disorder. It has been mapped to the long arm of chromosome 11 (11q12-q13). The *BEST1/VMD2* gene encodes the protein bestrophin-1, an integral membrane protein located in the retinal pigment epithelium (RPE). The protein plays a role in regulating voltage-dependent calcium channels and ocular development. The *VMD1* gene, mapped to chromosome 8 by earlier genetic markers, no longer exists as a genetic locus or most likely as a disease. The related disorder was known as atypical vitelliform macular dystrophy.

Pathology of the *BEST1/VMD2* gene remains unknown. However, one study demonstrated a change on the signal transduction pathway that affects the light peak on the electro-oculgram, a diagnostic test. Despite the disease effect on the RPE, blindness most likely comes from scarring of the macula, which contains millions of cone cells that provide central vision, sharp visual acuity, and color vision.

There are more than 120 known mutations of the *BEST1/VMD2* gene, resulting in many ocular phenotypes. Although symptoms of Best disease are erratic, the disorder appears to have complete genetic penetrance, and most affected children have some visual disturbances. However, there is also evidence for genetic nonpenetrance: In some cases, vision remains unaffected. Symptoms of Best disease neither progress nor decrease from one generation to the next, indicating that the disorder has no genetic anticipation.

Best disease passes from one generation to the next via just one parent who carries the *BEST1/VMD2* gene. Each child of an affected parent has a 50 percent chance of also carrying the mutation. There does not appear to be a standard correlation between the gene and resultant symptoms (genotype-phenotype correlation). Usually, one or more family members may be affected in each generation, although it is not uncommon for Best disease to skip a generation.

SYMPTOMS

The disease has six stages of progression. Sight is usually not affected in the first three stages, and the child may have 20/20 to 20/50 visual acuity for several years. During the previtelliform phase (stage 1), there is abnormal electrical potential during eye positioning and movements. In stage 2, yellow round spots ("egg-yolk" lesions) are evident on eye examination. Absorption of the lesions is identified in stage 3. Surprisingly, sight is not affected as the lesions continue to grow, but rather when the lesions break up (stage 4) and cause a "scrambled egg" appearance or scarring of the macula. During stage 5, vision is significantly affected, and subretinal scarring occurs in stage 6.

Best disease takes on different characteristics depending on the child. The disease may not progress, and vision loss may be barely noticeable. The disease may also progress rapidly, and vision may deteriorate markedly. Vision loss first occurs as blurring (decreased acuity) and object distortion (metamorphopsia). Children often have an easier time seeing objects that are close rather than far away (hyperopia). Peripheral vision and dark adaptation are usually not affected. Best disease is often bilateral, although it may occur in just one eye.

SCREENING AND DIAGNOSIS

Several tests may be used to identify Best disease. The electroretinogram measures the retina electrical response when stimulated by light. The electrooculogram evaluates poorly defined macular lesions. The diagnosis of Best disease is unusual in that results of the electroretinogram are nearly normal, while the electro-oculargraphic findings are abnormal.

The disease is usually diagnosed between three and fifteen years of age, with many children being diagnosed around six years of age. Best disease has also been known to occur first in adulthood.

TREATMENT AND THERAPY

Currently, there is no treatment for this condition. Direct laser photocoagulation may be used for revascularization and bleeding. Genetic counseling is available at clinics that specialize in macular diseases or low vision. Various services, including lighting, visual aids, and visual assistance, are also available to help people with low vision or sight loss.

PREVENTION AND OUTCOMES

Cessation of cigarette smoking may help the retina revascularize.

Renée Euchner, R.N.

FURTHER READING

Boon, C. J., et al. "Clinical and Genetic Heterogeneity in Multifocal Vitelliform Dystrophy." *Archives of Ophthalmology* 125 (2007): 1100-1106. Clinical assessment of this genetic condition.

Kliegman, Robert M., Richard E. Behrman, Hal B. Jenson, and Bonita F. Stanton. *Nelson Textbook of Pediatrics.* 18th ed. Philadelphia: Saunders/Elsevier, 2007. Includes a table with a dozen genetic reference Web sites.

Yanoff, Myron, and Jay S. Duker. *Ophthalmology.* 3d ed. St. Louis: Mosby/Elsevier, 2008. Includes latest advances in genetics and its relationship to ophthalmology.

WEB SITES OF INTEREST

Association for Macular Disease, Inc.
www.macula.org

Foundation Fighting Blindness
www.blindness.org

International Association for Fighting Best Disease
www.best.org.il

Macular Disease Society
www.maculardisease.org

Partially Sighted Society
www.patient.co.uk/showdoc/26739030

Royal National Institute of Blind People (RNIB)
www.rnib.org.uk

See also: Aniridia; Choroideremia; Corneal dystrophies; Hereditary diseases.

Biochemical mutations

CATEGORY: Human genetics and social issues; Molecular genetics

SIGNIFICANCE: The study of the biochemistry behind a particular phenotype is often necessary to understand the modes of inheritance of mutant genes. Knowledge of the biochemistry of mutant individuals is especially useful in determining treatments for genetic diseases.

KEY TERMS

allele: a form of a gene at a specific gene locus; a locus in an individual organism typically has two alleles

biochemical pathway: the steps in the production or breakdown of biological chemicals in cells; each step usually requires a particular enzyme

genotype: the genetic characteristics of a cell or organism, expressed as a set of symbols representing the alleles present

heterozygous: a genotype in which a locus has two alleles that are different

homozygous: a genotype in which a locus has two alleles that are the same

phenotype: expressed or visible characteristics of a genotype; different genotypes often are expressed as different phenotypes but may have the same phenotype

PROTEINS AND SIMPLE DOMINANT AND RECESSIVE ALLELES

In order to understand how certain genotypes are expressed as phenotypes, knowledge of the biochemistry behind gene expression is essential. It is known that the various sequences of nitrogenous bases in the DNA of genes code for the amino acid sequences of proteins. How the proteins act and interact in an organism determines that organism's phenotype.

Simple dominant and recessive alleles are the easiest to understand. For example, in the genetic disease phenylketonuria (PKU), two alleles of the PKU locus exist: p^+, which codes for phenylalanine hydroxylase, an enzyme that converts phenylalanine (a common amino acid in proteins) to tyrosine (another common amino acid); and p, which is unable to code for the functional form of the enzyme. Individuals with two normal alleles, p^+p^+, have the enzyme and are able to perform this conversion. However, individuals with two abnormal alleles, pp, do not have any of this enzyme and are unable to make this conversion. Since phenylalanine is not converted to tyrosine, the phenylalanine accumulates in the organism and eventually forms phenylketones, which are toxic to the nervous system and lead to mental retardation. The heterozygote, p^+p, has one normal and one abnormal allele. These individuals have phenylalanine and tyrosine levels within the normal range, since the enzyme can be used over and over again in the conversion. In other words, even when there is only one normal allele present, there is enough enzyme produced for the conversion to proceed at the maximum rate.

Many other inborn errors of metabolism follow this same pattern. In the case of albinism, for example, afflicted individuals are missing the enzyme necessary to produce the brown-black melanin pigments. Galactosemics are missing an essential enzyme for the breakdown of galactose.

OTHER SINGLE-GENE PHENOMENA

Many other genetic phenomena can be explained by looking at the biochemistry behind them. For example, the "chinchilla coat" mutation in rabbits causes a gray appearance in the homozygous state, $c^{ch}c^{ch}$. This occurs because the c^{ch} allele codes for a pigment enzyme that is partially defective. The partially defective enzyme works much more slowly than the normal enzyme, and the smaller amount of pigment produced leads to the gray phenotype. When this allele is heterozygous with the fully defective c allele, $c^{ch}c$, there is only half as much of an enzyme that works very slowly. As one might expect, there is less pigment produced, and the phenotype is an even lighter shade of gray called light chinchilla. The enzyme concentration does affect the rate of the reaction and, ultimately, the amount of product made. This phenomenon is known as incomplete, or partial, dominance. Genes for the red pigments in such

Many genetic phenomena can be explained by looking at the biochemistry behind them. For example, the chinchilla coat in rabbits such as this one at the Dallas Zoo is caused by a mutant allele that, in the homozygous state, codes for a pigment enzyme that is partially defective. This partially defective enzyme works much more slowly than the normal enzyme, and the smaller amount of pigment produced leads to the gray phenotype. (AP/Wide World Photos)

flowers as four-o'clocks and snapdragons show incomplete dominance, as do the hair, skin, and eye pigment genes of humans and the purple pigment genes of corn kernels.

Sometimes a mutation occurs that creates an enzyme with a different function instead of creating a defective enzyme. The B allele in the ABO blood-group gene codes for an enzyme that adds galactose to a short sugar chain that exists on the blood cell's surface, forming the B antigen. The A allele codes for an enzyme that adds N-acetylgalactosamine to the same previously existing sugar chain, forming the A antigen. Anyone with two B alleles, $I^B I^B$, makes only the B antigen and is type B. Those with two A alleles, $I^A I^A$, make only the A antigen and are type A. Heterozygotes, $I^A I^B$, have the enzymes to make both antigens, and they do. Since they have both antigens on their blood cells, they are classified as type AB. This phenomenon is known as codominance and is also seen in other blood-type genes.

Biochemistry can also explain other single-gene phenomena such as the pigmentation pattern seen in Siamese cats and Himalayan rabbits. The Siamese-Himalayan allele codes for an enzyme that is so unstable that it falls apart and is completely nonfunctional at the normal body temperature of most mammals. Only at cooler temperatures can the enzyme retain its stability and function. Since mammals have lower temperatures at their extremities, it is there that the enzyme produces pigment; at more centrally located body areas, it cannot function. This leaves a pattern of dark pigmentation on the tail, ears, nose, feet, and scrotum, with no pigmentation at other areas.

MULTIPLE-GENE PHENOMENA

Few genes act completely independently, and biochemistry can be used to explain gene interactions. One simple interaction can be seen in fruit-fly eye pigmentation. There are two separate biochemical pathways to make pigment. One produces the red pteridines, and the other produces the brown omochromes. If *b* is an allele that cannot code for an enzyme necessary to make red pigments, a bbr^+r^+ fly would have brown eyes. If *r* is an allele that cannot code for an enzyme necessary to make brown pigment, a b^+b^+rr fly would have red eyes. When mated, the resulting progeny would be b^+br^+r. They would make both brown and red pigments and have

the normal brick-colored eyes. Interbreeding these flies would produce some offspring that were *bbrr.* Since these offspring make neither brown nor red pigments, they would be white-eyed.

Another multigene phenomenon that is seen when looking at the genes of enzymes that are in the same biochemical pathway is epistasis. Consider the following pathway in dogs:

colorless → brown → black

The a^+ allele codes for the enzyme that converts colorless to brown, but the *a* allele cannot, and the b^+ allele codes for the enzyme that converts brown to black, but the *b* allele cannot. The phenotype of an organism that is aab^+b^+ depends only on the *aa* genotype, since an *aa* individual produces no brown and the b^+b^+ enzyme can make black only by converting brown to black. The cross $a^+ab^+b \times a^+ab^+b$ would be expected to produce the normal $9a^+\_b^+\_$ (black) : $3a^+bb\_$ (brown) : $3aab^+\_$ (white) : $1aabb$ (white) phenotypic ratio of the classic dihybrid cross, but this is more appropriately expressed as 9 black : 3 brown : 4 white ratio. (The symbol "_" is used to indicate that the second gene can be either dominant or recessive; for example, $A\_$ means that both AA and Aa will result in the same phenotype.) Other pathways give different epistatic ratios such as the following pathway in peas:

white → white → purple

If *A* codes for the first enzyme, *B* codes for the second enzyme, and *a* and *b* are the nonfunctional alleles, both *AAbb* and *aabb* are white. Their progeny when they are crossed, *AaBb*, is purple because it has both of the enzymes in the pathway. Interbreeding the *AaBb* progeny gives a ratio of 9 purple to 7 white.

Human pigmentation is another case in which many genes are involved. In this case, the various genes determine how much pigment is produced by nonalbino individuals. Several gene loci are involved, and the contributions of each allele of these loci is additive. In other words, the more functional alleles one has, the darker the pigmentation; the fewer one has, the lighter. Since many of the genes involved for skin, eye, and hair color are independent, ranges of color in all three areas are seen that may or may not be the same. In addition, there are genes that code for enzymes that produce chemicals that modify the expression of the pigment genes (for example, to change blue eyes to gray, convert hazel eyes to green, or change brown hair to auburn). This gives rise to the great diversity of pigmentation in humans. Add to these many possible expression patterns at the biochemical level the effect of the environment, and it is clear why such great variation in phenotypic expression is possible.

Richard W. Cheney, Jr., Ph.D.

Further Reading

Clark, David P. "Mutations." In *Molecular Biology.* Boston: Elsevier Academic Press, 2005. Places genetic mutation in the overall context of molecular biology and the genetic revolution. Colored illustrations.

Neumann, David, et al. *Human Variability in Response to Chemical Exposures: Measures, Modeling, and Risk Assessment.* Boca Raton, Fla.: CRC Press, 1998. Addresses genetic evidence for variability in the human response to chemicals associated with reproductive and developmental effects, the nervous system and lungs, and cancer.

Strachan, Tom, and Andrew Read. *Human Molecular Genetics 3.* 3d ed. New York: Garland Press, 2004. Provides introductory material on DNA and chromosomes and describes principles and applications of cloning and molecular hybridization. Surveys the structure, evolution, and mutational instability of the human genome and human genes. Examines mapping of the human genome, the study of genetic diseases, and the dissection and manipulation of genes.

Watson, James D., et al. *Molecular Biology of the Gene.* 6th ed. San Francisco: Pearson/Benjamin Cummings, 2008. Watson, a Nobel Prize winner for his discovery of DNA's structure, is one of the authors of this textbook providing an overview of molecular biology.

Web Sites of Interest

Genetics Home Reference
http://ghr.nlm.nih.gov
Sponsored by the National Institutes of Health, this site describes its contents as "consumer-friendly information about the effects of genetic variations on human health." A search for "mutation" will retrieve numerous pages discussing various aspects of the subject.

The Human Genome
http://genome.wellcome.ac.uk/doc_WTD020780 .html
The Welcome Trust, the largest charity in the United Kingdom, sponsors this Web site about genetics that includes this page, which provides a brief description of mutation and polymorphism.

See also: Chemical mutagens; Chromosome mutation; Classical transmission genetics; Complete dominance; Dihybrid inheritance; Epistasis; Inborn errors of metabolism; Incomplete dominance; Monohybrid inheritance; Mutagenesis and cancer; Mutation and mutagenesis; Oncogenes; Phenylketonuria (PKU); Tumor-suppressor genes.

Bioethics

Category: Bioethics; Human genetics and social issues

Significance: Bioethics is the practice of helping society and, more specifically, families, patients, and medical teams, make tough health care decisions. This branch of philosophy focuses on helping individuals decide what is right for them while addressing the needs of families, health care providers, and society.

Key terms

genetic testing: the use of the techniques of genetics research to determine a person's risk of developing, or status as a carrier of, a disease or other disorder

informed consent: the right of patients to know the risks of medical treatment and to determine what is done to their bodies

The Emergence of Bioethics

As early as the mid-1960's, advances in genetics and reproduction, life support, and transplantation technologies spurred an increased focus on ethical issues in medicine and scientific research. From the late 1960's through the mid-1970's, bioethicists were preoccupied with the moral difficulties of obtaining voluntary, informed consent from human subjects in scientific research. They concentrated on the development of ethical guidelines in research that

President's Council on Bioethics

President George W. Bush established the President's Council on Bioethics by executive order on November 28, 2001. Its mission was to advise the chief executive on bioethical issues emerging from advances in biomedical science and technology. Specifically mentioned in the council's mission were embryo and stem cell research, assisted reproduction, cloning, and end-of-life issues. Other ethical and social issues identified for discussion included the protection of human research subjects and the appropriate use of biomedical technologies. The council consisted of eighteen members appointed by the president, including scientists, physicians, ethicists, social scientists, lawyers, and theologians. The council was renewed in 2003, 2005, and 2007.

Deeply controversial issues constituted the subject matter of the inquiries undertaken by the council. Debate among its members as well as discussions on the floors of the Senate and House of Representatives were strongly divisive, producing heated argument and disagreement. The council's members were particularly divided on the issue of human cloning, producing two recommendations for national policy. Both recommendations would ban cloning to produce children, and ten of the eighteen council members recommended a four-year moratorium on human cloning for biomedical research while the issue continued to be studied. Declining to call for an outright ban on cloning, the divided council stated that "prudent and sensible" regulation was the best way to advance research while guarding against abuse. The minority favored regulating cloned embryos used in biomedical research, including federal licensing, oversight, and time limits on the length of time for development of cloned embryos.

President Bush stated his strong opposition to human cloning in a speech in August, 2001. The Human Cloning Prohibition Act of 2003, which banned all forms of human cloning, including cloning to create a pregnancy and cloning for medical research, passed the House of Representatives in February of 2003 by a vote of 241 to 155. It also made it a crime to "receive or import a cloned human embryo or any product derived from a cloned human embryo," punishable by $1 million in fines and ten years of imprisonment. This part of the law essentially made it illegal to harvest embryonic stem cells for medical research. On March 9, 2009, newly elected President Barack Obama fulfilled a campaign promise by reversing Bush's stem cell restrictions. On November 24, 2009, President Obama replaced the council with the Presidential Commission for the Study of Bioethical Issues.

Stem cells—undifferentiated cells that have the potential to grow into any type of tissue—are created in the first days of pregnancy. Scientists hope to direct stem cells to grow a variety of tissues for use in transplantation to treat serious illnesses such as cancer, heart disease, and diabetes. Embryos have been valued in research for their ability to produce these stem cells, but the harvesting process requires the destruction of days-old embryos (a procedure condemned by the Catholic Church, antiabortion activists, and women's rights organizations). Other research, however, points to similar promise using stem cells harvested from adults, so that no embryos are destroyed.

Marcia J. Weiss, M.S., J.D.

---

would ensure the protection of individuals vulnerable to exploitation, including mentally or physically handicapped individuals, prisoners, and children. Beginning in the mid-1970's and continuing through the mid-1980's, bioethicists became increasingly involved in discussions of the definitions of life, death, and what it means to be human. In the mid-1980's, practitioners began to focus on cost containment in health care and the allocation of scarce medical resources.

Beginning in 1992, the Joint Commission on Accreditation of Health Care Organizations, the U.S. agency that accredits hospitals and health care institutions, required these organizations to establish committees to formulate ethics policies and address ethical issues. Ethics teams within hospitals and professional organizations exist to provide consultation regarding ethical dilemmas in clinical practice and research. Such resources are critical as technological advances, particularly related to genetics and genomics, proceed more rapidly than policy. Centers for the study of biomedical ethics such as the Society for Health and Human Values and the Park Ridge Center for the Study of Health, Faith, and Ethics are important forums for public debate and research. Since completion of the Human Genome Project, an increasing number of organizations are committed to ethical research and policy making related to the use of genomic information.

The overriding principle of bioethics and U.S. law is to respect each person's right to make decisions, free of coercion, about treatments or procedures he or she will undergo. This principle is complicated when the person making the decision is considered incompetent because of youth, mental retardation, or medical deterioration. Other important principles include a patient's right to know that medical practitioners are telling the truth, the right to know the risks and benefits of proposed medical treatment, and the right to privacy of health information.

IMPACT AND APPLICATIONS

Advances in genomics and genetic testing have presented numerous dilemmas for bioethicists, patients, and health care providers. For example, as the ability to forecast and understand the genetic

code progresses, people will have to decide whether knowing the future, even if it cannot be altered or changed, is beneficial to them or their children. Knowledge of the genomic basis of common diseases has lead to the birth of direct-to-consumer marketing of testing that provides individuals with often complicated risk profiles for conditions such as diabetes and heart disease. Bioethicists are critical players in policy-making regarding this new form of personalized medicine.

Bioethicists help people determine the value of genetic testing, including the risks and benefits of genetic testing in particular situations. Factors typically considered before a person undergoes genetic testing include the nature of the test, the timing of the test, and the impact that the results will have on health and medical management. Testing can be done prenatally to detect disorders in fetuses; it can also be done before conception to determine whether a prospective parent is a carrier for a par-

Leon Kass of the University of Chicago was appointed head of the President's Council on Bioethics in November, 2001. Professor Kass headed a panel of scientists, doctors, lawyers, and ethicists who advised the George W. Bush administration on policy issues surrounding stem cell and other research in biology, medicine, and genetics. (AP/Wide World Photos)

ticular disorder or disease that could be passed to a child. Technology even allows for testing of embryos created by in vitro fertilization, thereby preventing the transmission of a genetic condition by transferring only unaffected embryos to the mother's womb. Predictive and presymptomatic genetic tests can provide information about whether an adult has an increased susceptibility to, or will ultimately manifest symptoms of, a genetic disorder. Information gained from genetic testing could help predict the nature and severity of a particular disorder as well as potential options for screening or intervention. Knowing one's genetic fate may be more of a burden than a person wants, however, particularly if nothing can be done to change or alter the risks that the person faces. Bioethicists act as guides through the complicated and often wrenching decision process.

Consumers of genetic testing must also decide whether the knowledge gained from the test is worth potential legal and social implications. On May 21, 2008, the Genetic Information Nondiscrimination Act (GINA) was signed into law. GINA provides protection against genetic discrimination in health insurance and employment, but it does not protect other insurance arenas, such as life insurance and disability insurance. Fear of discrimination may prevent some individuals from pursuing genetic testing that could provide beneficial guidance for preventive care. For example, a woman with a strong family history of breast cancer could have genetic testing to determine if she has inherited a hereditary cancer predisposition syndrome, which in turn would lead to increased vigilance with breast screening. Many women in this situation defer testing because of discrimination fears and risk detection of cancer at a much later stage, with potentially devastating consequences. Bioethicists can help guide policymakers in creating stricter protections against potential discrimination.

The Human Genome Project has provided researchers with a wealth of information, but this comes with a paucity of knowledge about the specific effects of the genetic sequence related to health and disease. Genome-wide association studies are ongoing to better understand the complicated nature of gene-gene and gene-environment interactions. Technology may very soon allow individuals to sequence their entire genome. However, the challenge to bioethicists, researchers, and the general public is how to interpret the information in a meaningful way.

Fred Buchstein, M.A.; updated by Jessie Conta, M.S.

FURTHER READING

Beauchamp, Tom, LeRoy Walters, Jeffrey Kahn, and Anna Mastroianni. *Contemporary Issues in Bioethics.* Belmont, Calif.: Thomson/Wadsworth, 2007. This updated anthology includes essays about important topics in bioethics, such as genetics, human reproduction, and human and animal research. Written by scholars in bioethics and judges in landmark legal cases.

Bulger, Ruth Ellen, Elizabeth Heitman, and Stanley Joel Reiser, eds. *The Ethical Dimensions of the Biological and Health Sciences.* 2d ed. New York: Cambridge University Press, 2002. Designed for graduate students who will be conducting research in the medical and biological sciences. Provides essays, readings, and questions to stimulate thinking about ethical issues and implications.

Caplan, Arthur. *Due Consideration: Controversy in the Age of Medical Miracles.* New York: Wiley, 1997. A leading bioethicist analyzes the moral questions regarding scientific advancements, among them cloning, assisted suicide, genetic engineering, and treating illnesses during fetal development.

Charon, Rita, and Martha Montello, eds. *Stories Matter: The Role of Narrative in Medical Ethics.* New York: Routledge, 2002. Explores the narrative interaction of the medical field—the written and verbal communication involved in doctors' notes, patients' stories, the recommendations of ethics committees, and insurance justifications—and the way in which this interaction profoundly affects decision making, patient health, and treatment.

Comstock, Gary L., ed. *Life Science Ethics.* Ames: Iowa State Press, 2002. Introduces ethical reasoning in the area of humankind's relationship with nature and presents twelve fictional case studies as a means to show the application of ethical reasoning.

Danis, Marion, Carolyn Clancy, and Larry R. Churchill, eds. *Ethical Dimensions of Health Policy.* New York: Oxford University Press, 2002. The three authors, from varied professions within the medical field, attempt to identify the goals of health care, examine how to connect ethical considerations with the making of health policy, and discuss specific areas of ethical controversy such as resource allocation, accountability, the needs of vulnerable populations, and the conduct of health services research.

Evans, John Hyde. *Playing God? Human Genetic Engineering and the Rationalization of Public Bioethical Debate.* Chicago: University of Chicago Press, 2002. Provides a framework for understanding the public debate. Details the various positions of the debate's players, including eugenicists, theologians, and bioethicists.

Kass, Leon R. *Life, Liberty, and the Defense of Dignity: The Challenge for Bioethics.* San Francisco: Encounter Books, 2002. Examines genetic research, cloning, and active euthanasia, and argues that biotechnology has left humanity out of its equation, often debasing human dignity rather than celebrating it.

Kristol, William, and Eric Cohen, eds. *The Future Is Now: America Confronts the New Genetics.* Lanham, Md.: Rowman & Littlefield, 2002. Brings together classic writings (George Orwell, Aldous Huxley) as well as more recent essays and congressional testimony about human cloning, genetic engineering, stem cell research, biotechnology, human nature, and American democracy.

May, Thomas. *Bioethics in a Liberal Society: The Political Framework of Bioethics Decision Making.* Baltimore: Johns Hopkins University Press, 2002. Takes the debate about biomedical ethics into the political realm, analyzing how the political context of liberal constitutional democracy shapes the rights and obligations of both patients and health care professionals.

Mepham, Ben. *Bioethics: An Introduction for the Biosciences.* New York: Oxford University Press, 2008. Written for students who are new to the principles of bioethics, this book provides a foundation for students to foster an objective exploration of current bioethical issues, such as genetic modification.

O'Neill, Onora. *Autonomy and Trust in Bioethics.* New York: Cambridge University Press, 2002. Examines issues surrounding reproductive and principled autonomy, trust, consent, and the media and bioethics.

Singer, Peter. *Unsanctifying Human Life: Essays on Ethics.* Edited by Helga Kuhse. Malden, Mass.: Blackwell, 2002. Singer is one of today's major bioethicists. Here he examines the role of philos-

ophers and philosophy in such questions as the moral status of the embryo, animal rights, and how people should live.

Veatch, Robert M. *The Basics of Bioethics.* 2d ed. Upper Saddle River, N.J.: Prentice Hall, 2003. In a textbook designed for students, Veatch presents an overview of the main theories and policy questions in biomedical ethics. Includes diagrams, case studies, and definitions of key concepts.

WEB SITES OF INTEREST

American Journal of Bioethics Online
http://www.bioethics.net
Provides sections on cloning basics, animal cloning, stem cells, U.S. federal and state laws, the cloning debate, news, and more.

Center for Bioethics and Human Dignity
http://www.cbhd.org/resources/index.html
Includes essays and other resources exploring bioethical issues in a variety of topics, including cloning, stem cell research, and genetics.

The Hastings Center
http://www.thehastingscenter.org
This independent nonprofit organization specializes in bioethics, and its site contains news postings, articles on bioethics and different aspects of genetic science, and announcements of events and publications.

Kennedy Institute of Ethics, Georgetown University
http://kennedyinstitute.georgetown.edu
Links to many resources on bioethics as well as a "bioethics library" that in turn leads to resources on human genetics and ethics.

National Human Genome Research Institute
http://www.genome.gov
Provides information about ongoing research, as well as a section about policy and ethics, including details about genetic antidiscrimination legislation.

National Information Resource on Ethics and Human Genetics
http://genethx.georgetown.edu
Supports links to databases, annotated bibliographies, and articles about the ethics of genetic testing and human genetics.

National Institutes of Health Bioethics Resources on the Web

http://bioethics.od.nih.gov

Resource that includes links to background information and positions on bioethics, including specific links related to genetics and genomics.

President's Council on Bioethics

http://bioethics.gov

Government arm that advises on ethical issues surrounding biomedical science and technology. Includes links to bioethics literature and other resources on ethics and human genetics.

Secretary's Advisory Committee on Genetics, Health, and Society

http://oba.od.nih.gov/SACGHS/sacghs_home.html

Advises the Secretary of Health and Human Services on ethical, legal, and social issues surrounding the development and use of genetic technologies, such as direct-to-consumer genetic testing and pharmacogenomics.

See also: Amniocentesis; Bioinformatics; Biological determinism; Chorionic villus sampling; Cloning: Ethical issues; Criminality; DNA fingerprinting; Eugenics; Eugenics: Nazi Germany; Forensic genetics; Gene therapy; Gene therapy: Ethical and economic issues; Genetic counseling; Genetic engineering: Social and ethical issues; Genetic screening; Genetic testing; Genetic testing: Ethical and economic issues; Human genetics; In vitro fertilization and embryo transfer; Insurance; Miscegenation and anti-miscegenation laws; Patents on life-forms; Paternity tests; Prenatal diagnosis; Race; Stem cells; Sterilization laws.

Biofertilizers

CATEGORY: Genetic engineering and biotechnology

SIGNIFICANCE: Growth in the global population has resulted in the need for more food. Synthetic fertilizers and pesticides have increased crop yield. Resulting long-term harm to the environment, however, makes this approach unsustainable. As the world's population grows to some nine billion over the next forty years, genetic applications of ecofriendly biofertilizers offer promising results without devastating environmental damage.

KEY TERMS

agricultural biotechnology: technology that employs scientific tools to modify the genetics of an organism for a practical agricultural purpose

biofertilizers: fertilizers that contain living organisms (bacteria or fungi) used to enhance availability and uptake of minerals in plants and improve fertility

microbial inoculants: microbes introduced into the soil or plant to build symbiotic relationships for mutual benefit between microbes and plants; used in organic farming

mycorrhiza: symbiosis that occurs between fungi and plants; fungi colonize the cortical tissue of roots in active plant growth and receive carbon from the plant while providing the plant with needed nutrients for growth

nitrogen fixers: biofertilizers that use microbes to take nitrogen from the atmosphere and turn it into usable material to promote plant growth

phosphate solubilizer: biofertilizers that use microbes to dissolve inorganic phosphorus from insoluble materials to produce increased crop yield

symbiosis: a living together, close union, or cooperative relationship of dissimilar organisms to create a state of mutualism, where each party benefits from the relationship

BIOFERTILIZERS FOR SUSTAINABLE AGRICULTURE

The increase in worldwide population has resulted in the global need for more food. The use of artificial fertilizers and synthetic pesticides has produced extensive short-term growth in crop yield and food production in developed counties. However, the direct and indirect environmental impact of these chemicals include poor quality, mineral-depleted soil; toxic chemicals and metals in the soil; air pollution; poisoning of lakes and rivers through run-off and chemical leaching; premature births; and general disruption of the ecosystem. Despite the increase in food production, the environmental damage by commercial fertilizers makes this option unsustainable for future farming. As the world's population increases, especially in underdeveloped countries, this challenge demands healthier alternatives to meet the agricultural needs of the world.

Just as the human genome has come under investigation, scientists today use agricultural biotechnology to capitalize on the genetic properties of both plants and microorganisms. Ecofriendly biofertilizers have been developed to augment or replace commercial artificial fertilizers. They consist of natural living ("bio" means "life") bacteria added to the soil, seed, or plant surface to enhance plant fertility. These beneficial microbes are often incorporated into various materials such as peat moss and applied to the plant's soil to promote the health of the microflora and produce more and better crops. They enrich soil quality, prevent infections from phytopathogens, and lessen the stress of heavy metals left in soil by commercial fertilizers. Biofertilizers offer positive production outcomes in both developed and developing countries without irreparable damage to the environment.

HOW BIOFERTILIZERS WORK

Biofertilizers work in diverse ways to support plant growth. The goal is to provide a higher level of ecofriendly bacteria than are normally found in the soil so that plants can thrive when necessary nutrients are unavailable. Biofertilizers improve the growing environment of plants by adding organic material to enrich the physical condition and texture of the soil and minimize erosion. Some use the biological process to mobilize or "fix" the nutrients needed for plants to flourish. They can also have an impact on the microbial actions in the rhizosphere (around the roots) of the plant.

There are several types of biofertilizers available today. Nitrogen-fixing biofertilizers contain bacteria that extract nitrogen (N_2) from the air and convert it to a form (N_3) that can be used by the plant. Common nitrogen fixers include _Rhizobium_, used primarily for legume inoculation where the bacteria invades the root, multiplies in the plant cortex cells, and produces nodules. This is a symbiotic relationship whereby the plant provides the _Rhizobium_ with food and energy while the resulting nodules provide nitrogen fixing. Another microbe, _Azotobacter_, uses nitrogen in cell protein synthesis that frees nitrogen for the soil at cell death. When _Azotobacter_ is applied to seeds, it improves germination and helps control plant disease. _Azospirillum_ and _Cyanobacteria_, such as blue-green algae, are also nitrogen fixers.

Phosphate solubilizers dissolve inorganic phosphorus from insoluble materials to assist with plant

growth and produce increased crop yield. Examples of phosphate solubilizers include the microbes _Pseudomonas_, _Xanthomonas_, and _Bacillus megaterium_. Mycorrhiza occurs when the fungi _Acaulospora_, _Endogone_, _Gigaspora_, and _Glomus_ receive carbon from the plant and in turn provide nutrients for the plant, again increasing crop yield.

BENEFITS OF BIOFERTILIZERS

Biofertilizers offer many benefits in agricultural biotechnology. First, they are cost-effective as compared with petroleum-based fertilizers. While synthetic fertilizers require repeated use of large quantities and produce the adverse effect of depleting the soil of nutrients, biofertilizers enrich the soil with naturally occurring microbes and stay in the soil for a longer time. A major benefit is accelerated plant growth with an increase in crop yield, usually between 5 and 30 percent. Plants supported with biofertilizers demonstrate greater pest and disease resistance, requiring less costly applications of pesticides. The soil itself shows better water-holding capacity and growing space while minimizing detrimental changes in pH levels. A major benefit of biofertilizers is that they are naturally occurring and contain no substances that can harm food and water.

IMPACT

In this world of limited resources and growing population demands, scientists are looking for ways to use genome biology and analysis to improve various aspects of human life. Microbial Genomics, a branch of the U.S. Department of Energy Office of Science, is charged with defining new ways to use microbes to uncover alternative energy sources, to define the process of biological carbon cycling, and to clean up toxic environmental wastes. So far, this group has completed genomic sequencing of more than 485 microbes. They believe that microbes offer untold benefits for applications to the environment, health, industry, and energy. One specific application of microbes is in the development and use of biofertilizers.

Biofertilizers employ microbes to achieve greater positive, ecofriendly results when compared with costly inorganic, petroleum-based fertilizers. With demands to "go green" and minimize adverse environmental damage, agricultural biotechnology provides reasonable, cost-effective alternatives for crop yield maximization. Applying genome principles and study to biofertilizers looks promising for the future of food production for developed and developing countries of the world.

Marylane Wade Koch, M.S.N., R.N.

FURTHER READING

Kannaiyan, Sadasivam. *Biotechnology of Biofertilizers.* New Delhi, India: Springer, 2002. The author explains the basic concept of biological nitrogen fixing (BNF) in symbiotic relationships provided by biofertilizers.

Palacios, Rafael, and William E. Newton, eds. *Genomes and Genomics of Nitrogen-Fixing Organisms.* Dordrecht, Netherlands: Springer, 2005. This book is the third volume of the series Nitrogen Fixation: Origins, Applications, and Research Progress. Discusses genomic science applied to nitrogen fixation.

Rai, Mahendra K. *Handbook of Microbial Biofertilizers.* New York: Food Products Press/Haworth Press, 2006. Rai provides in-depth information about biofertilizers and microbes with specific applications for sustainable agricultural achievements.

Vessey, J. Kevin. "Plant Growth Promoting Rhizobacteria as Biofertilizers." *Plant and Soil* 255 (2003): 571-586. Vessey discusses the possibilities of using plant growth-promoting rhizobacteria (PGPR) as biofertilizers to meet the agricultural challenges of the current world.

Wang, Vi-Ping, et al. *Biological Nitrogen Fixation, Sustainable Agriculture, and the Environment.* Proceedings of the 14th International Nitrogen Fixation Congress. Dordrecht, Netherlands: Springer, 2005. This book records the 14th Congress conference proceedings of Biological Nitrogen Fixation (BNF) research on genomics and plant and microbial science.

WEB SITES OF INTEREST

Biofertilizer.com
http://biofertilizer.com/biofertilizer/biofertilizer.htm

ESA Report: Ecology Society of America
http://www.biosci.ohio-state.edu/~asnowlab/Snowetal05.pdf

Food and Technology Center
http://www.agnet.org/library/eb/394/

"How Can Biotechnology Be Applied to Agriculture?"
http://www.greenfacts.org/en/gmo/index.htm#il1

Microbial Genomics at the U.S. Department of Energy
http://microbialgenomics.energy.gov/index.shtml

Nitrogen Fixing Bacteria in Agriculture
http://www.mapletoninternational.com/content/documents/Nitrogen%20Fixing%20Bacteria%20in%20Agriculture.pdf

Regional Biofertilizer Development Center
http://dacnet.nic.in/RBDCImphal/FarmerInfo.htm

See also: Biopesticides; Genetic engineering; Genetic engineering: Agricultural applications; Genetically modified foods; High-yield crops.

Bioinformatics

CATEGORY: Bioinformatics; Molecular genetics; Techniques and methodologies

SIGNIFICANCE: Bioinformatics is the application of information technology to the management of biological information to organize data and extract meaning. It is a hybrid discipline that combines elements of computer science, information technology, mathematics, statistics, and molecular genetics.

KEY TERMS

algorithm: a mathematical rule or procedure for solving a specific problem; in bioinformatics, a computer program is built to implement an algorithm, but different algorithms may be used to achieve the same result—that is, to align two sequences

database: an organized collection of information within a computer system that can be used for storage and retrieval as well as for complex searches and analyses

GenBank: a comprehensive, annotated collection of publicly available DNA sequences maintained by the National Center for Biotechnology Information and available through its Web site

genomics: the use of high-throughput technology to analyze molecular events within cells at the whole genome scale (for example, all of the genes, all of the messenger RNA, or all of the proteins)

Human Genome Project: a publicly funded international project to determine the complete DNA sequence of human genomic (chromosomal) DNA and to map all of the genes, which produced a "final" sequence in April, 2003

microarray: a technology to measure gene expression using nucleic acid hybridization of messenger RNA to a miniature array of DNA probes for many genes; microarray analysis often involves processing data with a variety of statistical methods that identify genetic expression patterns

proteomics: a collection of technologies that examine proteins within a cell in a holistic fashion, identifying or quantitating a large number of proteins within a single sample, identifying many protein-protein or protein-DNA interactions, and so on

pharmacogenomics: the identification and study of genes that code for the enzymes used for drug metabolism, along with the influences that genetic variation plays on a patient's response to a drug

modern molecular genetics laboratory, and large, publicly funded genome projects have determined the complete genomic sequences for humans, mice, fruit flies, dozens of bacteria, and many other species of interest to geneticists. All of this information is now freely available in online databases. Computational molecular biology tools allow for the design of polymerase chain reaction (PCR) primers, restriction enzyme cloning strategies, and even entire *in silico* experiments. This greatly accelerates the work of researchers but also changes the daily lives of many biologists so that they spend more time working with computers and less time working with test tubes and pipettors. The rapid accumulation of enormous amounts of molecular sequence data and their cryptic and subtle patterns have created a need for computerized databases and analysis tools.

Bioinformatics provides essential support services to modern molecular genetics for organizing, analyzing, and distributing data. As DNA sequencing and other molecular genetic technologies become more automated, data are generated ever more rapidly, and computing systems must be de-

THE NEED FOR BIOINFORMATICS

While the discovery and identification of genome sequences improves understanding of biological systems, the ability to organize, categorize, and analyze these sequences has necessitated the development of important bioinformatic tools. Information derived from bioinformatics is becoming increasingly important for biological research in proteomics, microarray technology, oncology, pharmacogenomics, and other disciplines. Applications of bioinformatics include identification of the genetic contributions to an illness, which may be accomplished by cloning the gene for a particular disease. Once the contributing genes and their predisposing disease variants have been identified, diagnostic tests can be created to determine future risk.

Today, the ability to sequence cloned DNA molecules has become a routine, automated task in the

Steven Brenner, of the University of California at Berkeley, next to a computer running bioinformatics software in November, 2001. He advocates distributing information freely as "open source code," claiming that this is the best way to debug bioinformatics software and advance research. (AP/Wide World Photos)

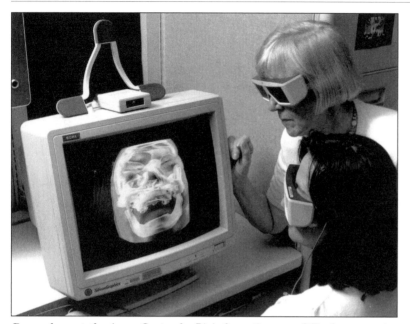

Researchers at the Ames Center for Bioinformatics wear 3-D glasses as they manipulate a high-resolution image of a skull and facial tissue. Such technology can help surgeons visualize the potential results of reconstructive surgery. (Getty Images)

signed to store the data and make them available to scientists in a useful fashion. The use of these vast quantities of data for the discovery of new genes and genetic principles relies on the development of sophisticated new data-mining tools. The challenge of bioinformatics is in finding new approaches to deal with the volume and complexity of the data, and in providing researchers with access both to the raw data and to sophisticated and flexible analysis tools in order to advance researchers' understanding of genetics and its role in health and disease.

DATABASE DESIGN

The DNA sequence data collected by automated sequencing equipment can be represented as a simple sequence of letters: G, A, T, and C—which stand for the four nucleotide bases on one strand of the DNA molecule (guanine, adenine, thymine, and cytosine). These letters can easily be stored as plain text files on a computer. Similarly, protein sequences can also be stored as text files using the twenty single-letter abbreviations for the amino acids.

There is a significant advantage to storing DNA and protein sequences as plain text files, also known as flat files. Text files take up minimal amounts of hard-drive space, can be used on any type of computer and operating system, and can easily be moved across the Internet. However, a text file with a bunch of letters representing a DNA or protein sequence is essentially meaningless without some basic descriptive information, such as the organism from which it comes, its location on the genome, the person or organization that produced the sequence, and a unique identification number (accession number) so that it can be referenced in scientific literature. This additional annotation information can also be stored as text—even in the same file with the sequence information—but there must be a consistent format, a standard.

In addition to maintaining basic flat-file structures for text data, it is useful to maintain sequence data in relational databases, which allow for much faster searching across multiple query terms and the linkage of sequence data files with other relevant information. The most sophisticated and widely used relational database system for bioinformatics is the Entrez system at the National Center for Biotechnology Information (NCBI). Entrez is a relational database that includes cross-links between all of the DNA sequences in GenBank. GenBank exchanges data with the DNA Data Bank of Japan (DDBJ) and the European Molecular Biology Laboratory (EMBL) on a daily basis to ensure that all three centers maintain the same set of data, and all peer-reviewed journals require the submission of sequence data to GenBank prior to publication of research articles; publicly funded sequencing projects, such as the Human Genome Project, submit new sequence data to GenBank as it is collected, so that the scientific community can have immediate access to it. Entrez also includes all the derived protein sequences (translations from cDNAs and predicted coding sequences in genomic DNA), the scientific literature in MedLine/PubMed, three-dimensional protein structures from the Protein Data Base (PDB), and human genetic information from the Online Mendelian Inheritance in Man (OMIM) database. Relational databases are even more important for more complex types of genomic data, such as gene expression microarrays and genetic variation and genotyping data sets.

KEY ALGORITHMS

Some of the key algorithms used in bioinformatics include sequence alignment (dynamic programming), sequence similarity (word matching from hash tables), assembly of overlapping fragments, clustering (hierarchical, self-organizing maps, principal components, and the like), pattern recognition, and protein three-dimensonal structure prediction. Bioinformatics is both eclectic and pragmatic: Algorithms are adopted from many different disciplines, including linguistics, statistics, artificial intelligence and machine learning, remote sensing, and information theory. There is no consistent set of theoretical rules at the core of bioinformatics; it is simply a collection of whatever algorithms and data structures have been found to work for the current data-management problems being faced by biologists. As new types of data become important in the work of molecular geneticists, new algorithms for bioinformatics will be invented or adopted.

NEW TYPES OF DATA

In addition to DNA and protein sequences, bioinformatics is being called upon to organize many other types of biological information that are being collected in ever greater amounts. Gene expression microarrays collect information on the amounts of mRNA produced from tens of thousands of different genes in a single tissue sample. Researchers realized that the technique of microarray analysis may identify new subclasses in disease states and establish biologic markers (biomarkers) that may be associated with diseases such as cancer. Studies are underway to examine how patients will respond to therapy when normal clinical predictors are inadequate. DNA microarrays or biochips can now be used to measure the functions of genes and proteins. DNA microarrays are microscopic slides containing cDNA (oligonucleotide) samples, which are fluorescently labeled probes used to quantitatively monitor quantities of transcripts (or mRNAs). Laser scanners are then used on the arrays to translate fluorescent emission into a numerical matrix of expression profiles. A number of clinical trials are

exploring the use of microarrays for prognosis or therapeutic guidance, and pharmaceutical firms have begun using microarray data to determine the success of their clinical trials for new drugs. Additionally, the technology is finding application in both forensics and food science.

Proteomics technologies are automating the process of mass spectroscopy, which allows investigators to identify and measure thousands of proteins in a single cell extract sample. Genes and proteins can also be organized into gene families based on sequence similarity, homology across organisms (comparative genomics), and function in metabolic or regulatory pathways. Many new technologies are being developed to measure genetic variation: genetic tests either for alleles of well-studied genes or for anonymous single nucleotide polymorphisms (SNPs) identified from genome sequence data. As these genotyping technologies are improved, it is becoming possible to collect data in an automated fashion for many genetic loci from a single DNA sample, or to test a single genetic locus on many thousands of DNA samples in parallel. These new data types require new database designs and the inclusion of new types of algorithms (from statistics, population genetics, and other disciplines) in bioinformatics data-management solutions.

INTEGRATION

In order to solve many biological problems, data from a variety of sources must be combined. Thus, despite advances in bioinformatics, a large challenge facing the discipline is the integration of various types of data in a form that allows scientists to extract meaningful insights into biology from the masses of information in molecular genetic databases. Successfully using multiple data sources remains complicated, however, and a lack of file formats and standardization is probably one reason why. These difficulties have prompted the development of the European Molecular Biology Open Software Suite (EMBOSS), which is software for multipurpose sequences analysis. EMBOSS automatically copes with data in a variety of formats, which has alleviated some of the challenges.

Genome browsers are yet another challenge. For example, it is extremely difficult to provide a display that allows someone to view all the relevant information about a gene or a chromosomal region, including the identity of encoded proteins; protein structure and functional information; involvement in metabolic and regulatory pathways; developmental and tissue-specific gene expression; evolutionary relationships to proteins in other organisms; DNA motifs bound by regulatory proteins; genetic synteny with other species (that is, having genes with loci on the same chromosome); phenotypes of mutations; and known alleles and SNPs and their frequency in various populations. Relatedly, in response to the needs in biomedical imagery, a model was developed that is a visual editor for profile hidden Markov models (HMMEditor), which can visualize the profile HMM architecture, transition probabilities, and emission probabilities for biological sequence analysis.

In clinical practice, tools that compare genomes between and among various species are also worthwhile. Someday, clinical bioinformaticians will look for available genomic data in electronic health records (EHRs) that will be aimed at considering the effects of genetic mutations.

Today there are a number of genomic databases—such as the National Center for Biotechnology Information, the Nucleotide Sequence Database, the DNA Data Bank of Japan, and the European Molecular Biology Laboratory—that are used for further analysis of genome and other data. The Sequence Retrieval System (SRS) of the European Bioinformatics Institute (EBI) integrates and links the main protein and nucleotide sequence databases and several other specialized molecular databases. The SRS has become an important integration system that allows both application and retrieval of data for analysis. The SRS contains more than 130 biological databases in total. Additionally, the rapid sequence Basic Local Alignment Search-Tool (BLAST) is now a vital tool in molecular biological research. The heart of NCBI's BLAST services is BLAST 2.0. In sum, BLAST is a set of five programs searching for similarity that allows researchers to perform sequence homology analysis with relative ease. The NCBI cross-references its own databases from DNA to proteins to three-dimensional structures to PubMed articles to genomes. Other special subject databases, such as those that focus on a particular species or on a particular type of molecule, link DNA and protein sequences back to the corresponding "reference" entries in GenBank. Entrez has continued to expand to access the entire National Library of Medicine Archives, and other indexed sequence databases have been developed to provide sequence entries (such as an index number or the source organism) in the major databases to access any additional information and share matches to one or more terms, which has greatly facilitated scientific searching and identification.

Despite the recent advances in data organization, genetic variants that are known reproducibly to influence diseases remain to be discovered, categorized, and manipulated to deliver this benchtop research to the bedside.

Stuart M. Brown, Ph.D.;
updated by Jesse Fishman, Pharm.D.

FURTHER READING

Baxevanis, Andreas D., and B. F. Francis Ouellette. *Bioinformatics: A Practical Guide to the Analysis of Genes and Proteins.* 2d ed. Hoboken, N.J.: John Wiley & Sons, 2003. This book provides a sound foundation of basic concepts of bioinformatics, with practical discussions and comparisons of both computational tools and databases relevant to biological research. The standard text for most graduate-level bioinformatics courses.

Bujnicki, J. *Practical Bioinformatics: Nucleic Acids and Molecular Biology.* New York: Springer Verlag, 2005. Bridges the gap between bioinformatics and molecular biology and provides numerous practical examples of the discipline that have lead to scientific advances.

Claverie, Jean-Michel, and Cedric Notredame. *Bioinformatics for Dummies.* Hoboken, N.J.: John Wiley & Sons, 2003. A practical introduction to bioinformatics: computer technologies that biochemical and pharmaceutical researchers use to analyze genetic and biological data. This reference addresses common biological questions, problems, and projects while providing a UNIX/Linux overview and tips on tweaking bioinformatic applications using Perl.

Krawetz, Stephen A., and David D. Womble. *Introduction to Bioinformatics: A Theoretical and Practical Approach.* Totowa, N.J.: Humana Press, 2003. Aimed at undergraduates, graduate students, and researchers. Includes four sections: "Biochemistry: Cell and Molecular Biology," "Molecular Genetics," "Unix Operating System," and "Computer Applications."

Mount, David W. *Bioinformatics: Sequence and Genome*

Analysis. Cold Spring Harbor, N.Y.: Cold Spring Harbor Laboratory Press, 2001. A textbook written for the biologist who wants to acquire a thorough understanding of popular bioinformatics programs and molecular databases. It does not teach programming but does explain the theory behind each of the algorithms.

Nucleic Acids Research 31, no. 1 (2003). This widely respected journal produces a special issue in January of each year devoted entirely to online bioinformatics databases. The articles represent the definitive statement by the directors of each of the major public databases of molecular biology data regarding the types of information and analysis tools in their databases and plans for development in the immediate future.

Quackenbush, J. "Microarray Analysis and Tumor Classification." *New England Journal of Medicine* 354 (2006): 2463-2472. Review article that details advances in microarray analysis and its clinical application.

WEB SITES OF INTEREST

Bioinformatics Organization
http://www.bioinformatics.org
Provides a helpful tutorial on bioinformatics.

European Bioinformatics Institute
http://www.ebi.ac.uk
Maintains databases concerning nucleic acids, protein sequences, and macromolecular structures, as well as postings of news and events and descriptions of ongoing scientific projects.

Human Genome Project Information: Bioinformatics
http://www.ornl.gov/sci/techresources/
Human_Genome/research/informatics.shtml
Details Human Genome Project bioinformatics research.

International Society of Intelligent Biological Medicine
http://www.isibm.org
Promotes research that is to be conducted toward the improvement of human health.

National Center for Biotechnology Information BLAST
http://blast.ncbi.nlm.nih.gov/Blast.cgi
Provides easy access to the most widely used sequence analysis searching.

See also: cDNA libraries; DNA fingerprinting; DNA sequencing technology; Forensic genetics; Genetic testing: Ethical and economic issues; Genetics: Historical development; Genomic libraries; Genomics; Human Genome Project; Icelandic Genetic Database; Linkage maps; Proteomics.

Biological clocks

CATEGORY: Human genetics and social issues

SIGNIFICANCE: Biological clocks control those periodic behaviors of living systems that are a part of their normal function. The rhythms may be of a daily, monthly, yearly, or even longer periodicity. In some cases, the clocks may be "programmed" to regulate processes that may occur at some point in the lifetime of the individual, such as those processes related to aging. Altered or disturbed rhythms may result in disease.

KEY TERMS

Alzheimer's disease: a disorder characterized by brain lesions leading to loss of memory, personality changes, and deterioration of higher mental functions

circadian rhythm: a cycle of behavior, approximately twenty-four hours long, that is expressed independent of environmental changes

free-running cycle: the rhythmic activity of an individual that operates in a constant environment

Huntington's disease: an autosomal dominant genetic disorder characterized by loss of mental and motor functions in which symptoms typically do not appear until after age thirty

suprachiasmatic nucleus (SCN): a cluster of several thousand nerve cells that contains a central clock mechanism that is active in the maintenance of circadian rhythms

TYPES OF CYCLES

Biological clocks control a number of physiological functions, including sexual behavior and reproduction, hormonal levels, periods of activity and rest, body temperature, and other activities. In humans, phenomena such as jet lag and shift-work disorders are thought to result from disturbances to the innate biological clock.

A squirrel hibernates in the hands of University of Minnesota biochemist Matt Andrews. Because the ground squirrel possesses the ability to put its body into this form of stasis, it is nearly immune to strokes. The genetics of such biological clocks may one day lead to better treatments for strokes in humans. (AP/Wide World Photos)

The most widely studied cycles are circadian rhythms. These rhythms have been observed in a variety of animals, plants, and microorganisms and are involved in regulating both complex and simple behaviors. Typically, circadian rhythms are innate, self-sustaining, and have a cyclicity of nearly, but not quite, twenty-four hours. Normal temperature ranges do not alter them, but bursts of light or temperature can change the rhythms to periods of more or less than twenty-four hours. Circadian rhythms are apparent in the activities of many species, including humans, flying squirrels, and rattlesnakes. They are also seen to control feeding behavior in honeybees, song calling in crickets, and hatching of lizard eggs.

What is known about the nature of the biological clock? The suprachiasmatic nucleus (SCN) consists of a few thousand neurons or specialized nerve cells that are found at the base of the hypothalamus, the part of the brain that controls the nervous and endocrine systems. The SCN appears to play a major role in the regulation of circadian rhythms in mammals and affects cycles of sleep, activity, and reproduction. The seasonal rhythm in the SCN appears to be related to the development of seasonal depression and bulimia nervosa. Light therapy is effective in these disorders. Blind people, whose biological clocks may lack the entraining effects of light, often show free-running rhythms.

Genetic control of circadian rhythms is indicated by the findings of single-gene mutations that alter or abolish circadian rhythms in several organisms, including the fruit fly (*Drosophila*) and the mouse. A mutation in *Drosophila* affects the normal twenty-four-hour activity pattern so that there is no activity pattern at all. Other mutations produce shortened (nineteen-hour) or lengthened (twenty-nine-hour) cycles. The molecular genetics of each of these mutations is known.

A semidominant autosomal mutation, CLOCK, in the mouse produces a circadian rhythm one hour longer than normal. Mice that are homozygous (have two copies) for the CLOCK mutation develop twenty-seven- to twenty-eight-hour rhythms when initially placed in darkness and lose circadian rhythmicity completely after being in darkness for two weeks. No anatomical defects have been seen in association with the CLOCK mutation.

Biological Clocks and Aging

Genes present in the fertilized egg direct and organize life processes from conception until death. There are genes whose first effects may not be evident until middle age or later. Huntington's disease (also known as Huntington's chorea) is such a disorder. An individual who inherits this autosomal dominant gene is "programmed" around midlife to develop involuntary muscle movement and signs of mental deterioration. Progressive deterioration of body functions leads to death, usually within fifteen years. It is possible to test individuals early in life before symptoms appear, but such tests, when no treatment for the disease is available, are controversial.

Alzheimer's disease (AD) is another disorder in which genes seem to program processes to occur after middle age. AD is a progressive, degenerative disease that results in a loss of intellectual function. Symptoms worsen until a person is no longer able to care for himself or herself, and death occurs on an average of eight years after the onset of symptoms. AD may appear as early as forty years of age, although most people are sixty-five or older when they are diagnosed. Age and a family history of AD are clear risk factors. Gene mutations associated with AD have been found on human chromosomes 1, 14, 19, and 21. Although these genes, especially the apolipoprotein *e4* gene, increase the likelihood of a person getting AD, the complex nature of the disorder is underscored when it is seen that the mutations account for less than half of the cases of AD and that some individuals with the mutation never get AD.

Impact and Applications

Evidence has accumulated that human activities are regulated by biological clocks. It has also become evident that many disorders and diseases, and even processes that are associated with aging, may be affected by abnormal clocks. As understanding of how genes control biological clocks develops, possibili-

ties for improved therapy and prevention should emerge. It may even become possible to slow some of the harmful processes associated with normal aging.

Donald J. Nash, Ph.D.

Further Reading

Finch, Caleb Ellicott. *Longevity, Senescence, and the Genome.* Reprint. Chicago: University of Chicago Press, 1994. Provides a comparative review of research on organisms from algae to primates, expanding traditional gerontological and geriatric issues to intersect with behavioral, developmental, evolutionary, and molecular biology. Illustrated.

Foster, Russell G., and Leon Kreitzman. *Rhythms of Life: The Biological Clocks That Control the Daily Lives of Every Living Thing.* London: Profile Books, 2004. Explains the workings of biological clocks in human beings and other creatures. Describes how these clocks are controlled by nerve cells and genes.

Hamer, Dean, and Peter Copeland. *Living with Our Genes: Why They Matter More than You Think.* New York: Doubleday, 1998. Links DNA and behavior and contains a good chapter on biological clocks and aging.

Koukkari, Willard L., and Robert B. Sothern. *Introducing Biological Rhythms: A Primer on the Temporal Organization of Life, with Implications for Health, Society, Reproduction, and the Natural Environment.* New York: Springer, 2006. Provides a comprehensive overview of circadian, tidal, lunar, and other biological rhythms and describes the application and implications of these rhythms in daily life.

Medina, John J. *The Clock of Ages: Why We Age, How We Age—Winding Back the Clock.* New York: Cambridge University Press, 1996. Designed for the general reader. Covers aging on a system-by-system basis and includes a large section on the genetics of aging.

Nelson, James Lindemann, and Hilde Lindemann Nelson. *Alzheimer's: Answers to Hard Questions for Families.* New York: Main Street Books, 1996. Reviews Alzheimer's disease for the general reader, guides caregivers through the difficult moral and ethical problems associated with the disease, and discusses support services.

Zallen, Doris Teichler. *Does It Run in the Family? A Consumer's Guide to DNA Testing for Genetic Disorders.* New Brunswick, N.J.: Rutgers University Press, 1997. Focuses on the practical aspects of obtaining genetic information, clearly explaining how genetic disorders are passed along in families. Provides useful information on genetic disorders, including Huntington's disease and Alzheimer's disease.

WEB SITES OF INTEREST

Learn Geneticism, The Time of Our Lives
http://learn.genetics.utah.edu/content/begin/DNA/clockgenes
Well-written and illustrated discussion of biological clocks and the mechanisms of their genes. Discusses the implications of these clocks for human sleep and health.

National Institute of General Medical Sciences
http://www.nigms.nih.gov/Publications/Factsheet_CircadianRhythms.htm
A fact sheet providing questions and answers about circadian rhythms.

Time Matters: Biological Clockworks
http://www.hhmi.org/biointeractive/clocks/museum.html
A virtual museum exhibit exploring the inputs, outputs, and mechanisms of biological clocks. Focuses on circadian rhythms, with information on the genetic sources of these rhythms.

Web MD, Sleep and Circadian Rhythm Disorders
http://www.webmd.com/sleep-disorders/guide/circadian-rhythm-disorders-cause
Discusses the causes of circadian rhythm disorders and their relationship to sleep.

See also: Aging; Alzheimer's disease; Biological determinism; Cancer; Developmental genetics; Huntington's disease; Inborn errors of metabolism; Telomeres.

Biological determinism

CATEGORY: Human genetics and social issues
SIGNIFICANCE: Biological determinists argue that there is a direct causal relationship between the biological properties of human beings and their behavior. From this perspective, social and eco-

nomic differences between human groups can be seen as a reflection of inherited and immutable genetic differences. This contention has been used by groups in power to claim that stratification in human society is based on innate biological differences. In particular, biological determinism has been used to assert that certain ethnic groups are biologically defective and thus intellectually, socially, and morally inferior to others.

KEY TERMS

determinism: the doctrine that everything, including one's choice of action, is determined by a sequence of causes rather than by free will
intelligence quotient (IQ): performance on a standardized test, often assumed to be indicative of an individual's level of intelligence
reductionism: the explanation of a complex system or phenomenon as merely the sum of its parts
reification: the oversimplification of an abstract concept such that it is treated as a concrete entity

THE USE OF INHERITANCE TO PROMOTE SOCIAL ORDER

The principle of biological determinism lies at the interface between biology and society. A philosophical extension of the use of determinism in other sciences, such as physics, biological determinists view human beings as a reflection of their biological makeup and hence simple extensions of the genes that code for these biological processes. Long before scientists had any knowledge of genetics and the mechanisms of inheritance, human societies considered certain groups to be innately superior by virtue of their family or bloodlines (nobility) while others were viewed as innately inferior (peasantry). Such views served to preserve the social order. According to evolutionary biologist Stephen Jay Gould, Plato himself circulated a myth that certain citizens were "framed differently" by God, with the ranking of groups in society based on their inborn worth.

As science began to take a more prominent role in society, scientists began to look for evidence that would justify the social order. Since mental ability is often considered to be the most distinctive feature of the human species, the quantification of intelligence was one of the main tactics used to demonstrate the inferiority of certain groups. In the mid-1800's, measurements of the size, shape, and

Psychologist Arthur Jensen grades IQ tests with students in 1970. He proposed that intelligence is an inherited trait. (Time & Life Pictures/Getty Images)

anatomy of the skull, brain, and other body features were compiled by physician Samuel George Morton and surgeon Paul Broca, among others. These measurements were used to depict races as separate species, to rank them by their mental and moral worth, and to document the subordinate status of various groups, including women. In the first decades of the twentieth century, such measurements were replaced by the intelligence quotient (IQ) test. Although its inventor, Alfred Binet, never intended it to be used in this way, psychologists such as Lewis M. Terman and Robert M. Yerkes promoted IQ as a single number that captured the complex, multifaceted, inborn intelligence of a person. IQ was soon used to restrict immigration, determine occupation, and limit access to higher education. Arthur Jensen, in 1979, and Richard Herrnstein and Charles Murray, in 1994, reasserted the claim that IQ is an inherited trait that differs among races and classes.

PROBLEMS WITH THE PRINCIPLE OF BIOLOGICAL DETERMINISM

Geneticists and sociobiologists (who study the biological basis of social behavior) have uncovered a variety of animal behaviors that are influenced by biology. However, the genetic makeup of an organism ("nature") is expressed only within the specific context of its environment ("nurture"). Thus genes that are correlated with behavior usually code for predispositions rather than inevitabilities. For such traits, the variation that occurs within a group is usually greater than the differences that occur between groups. In addition, the correlation between two entities (such as genes and behavior) does not necessarily imply a causal relationship (for example, the incidences of ice cream consumption and drowning are correlated only because both increase during the summer). Complex, multifaceted behaviors such as intelligence and violence are often reified, or treated as discrete concrete entities (as IQ and impulse control, respectively), in order to make claims about their genetic basis. Combined with the cultural and social bias of scientific researchers, reification has led to many misleading claims regarding the biological basis of social structure.

Biological and cultural evolution are governed by different mechanisms. Biological evolution occurs only between parents and offspring (vertically), while cultural evolution occurs through communication without regard to relationship (horizontally) and thus can occur quickly and without underlying genetic change. Moreover, the socially fit (those who are inclined to reproduce wealth) are not necessarily biologically fit (inclined to reproduce themselves). The reductionist attempt to gain an understanding of human culture through its biological components does not work well in a system (society) shaped by properties that emerge only when the parts (humans) are put together. Cultures cannot be understood as biological behaviors any more than biological behaviors can be understood as atomic interactions.

IMPACT AND APPLICATIONS

Throughout history, biological determinism has been used to justify or reinforce racism, genocide, and oppression, often in the name of achieving the genetic improvement of the human species (for example, the "racial health" of Nazi Germany). Gould has noted that claims of biological determinism tend to be revived during periods when it is politically expedient to do so. In times of economic hardship, many find it is useful to adopt an "us against them" attitude to find a group to blame for social and economic woes or to free themselves from the responsibility of caring for the "biologically inferior" underprivileged. As advances in molecular genetics lead to the identification of additional genes that influence behavior, society must guard against using this information as justification for the mistreatment or elimination of groups that are perceived as "inferior" or "undesirable" by the majority.

Lee Anne Martínez, Ph.D.

FURTHER READING

Begley, Sharon, and Andrew Murr. "Gray Matters." *Newsweek* 125, no. 13 (March 27, 1995): 48. Discusses the differences between the brains of males and females.

Carlson, Elof Axel. "The Blank Slate, the Human Nature, and the Biological Determinism Fallacies." In *Neither Gods nor Beasts: How Science Is Changing Who We Think We Are.* Cold Spring Harbor, N.Y.: Cold Spring Harbor Laboratory Press, 2008. A discussion of biological determinism is included in Carlson's book, in which he argues for an understanding of human beings based on their biology and their origins as a species.

Gould, Stephen Jay. *The Mismeasure of Man.* New York: Norton, 1996. Refutes Richard Herrnstein and Charles Murray's argument and presents an engaging historical overview of how pseudoscience has been used to support racism and bigotry.

Grossinger, Richard. "The Limits of Genetic Determinism." In *Embryos, Galaxies, and Sentient Beings: How the Universe Makes Life.* Berkeley, Calif.: North Atlantic Books, 2003. Describes the gap between science's description of life as random and mechanical and the depth of human experience.

Herrnstein, Richard, and Charles Murray. *The Bell Curve: Intelligence and Class Structure in American Life.* New York: Simon & Schuster, 1994. Asserts that IQ plays a statistically important role in the shaping of society by examining such sociological issues as school dropout rates, unemployment, work-related injury, births out of wedlock, and crime.

McDermott, Robyn. "Ethics, Epidemiology, and the Thrifty Gene: Biological Determinism as a Health Hazard." In *Health and Healing in Comparative Perspective*, edited by Elizabeth D. Whitaker. Upper Saddle River, N.J.: Pearson Prentice Hall, 2006. A discussion of late twentieth century ideas about the genetic nature of diabetes. Includes an examination of some of the ethical consequences of the biological deterministic paradigm, particularly the popular confusion of "genes" with "race."

Moore, David S. *Dependent Gene: The Fallacy of Nature vs. Nurture.* New York: W. H. Freeman, 2001. Few books examine how genes and the environment interact to produce everything from eye color to behavioral tendencies. This book lays to rest the popular myth that some traits are purely genetic and others purely a function of environment; rather, all traits are the result of complex, dependent interactions of both—interactions that occur at all stages of biological and psychological development. An informed argument against simplistic determinism.

Rose, Steven. "The Rise of Neurogenetic Determinism." *Nature* 373, no. 6513 (February 2, 1995): 380-382. Discusses how advances in neuroscience have led to a resurgence of the belief that genes are largely responsible for deviant human behavior.

Sussman, Robert, ed. *The Biological Basis of Human Behavior: A Critical Review.* 2d ed. New York: Simon & Schuster, 1998. Fifty-nine essays examine genetics, the various interpretations of the early evolution of human behavior, new attempts to link human physical variation to behavioral differences between people, evolutionary psychology, and the influences of hormones and the brain on behavior.

WEB SITE OF INTEREST

Nature or Nurture, Biological Determinism
http://www.nurture-or-nature.com/articles/biological-determinism/index.php

This Web site, which explores the longstanding debate over nurture versus nature, includes a section describing biological determinism, genetic determinism, and the biological approach to free will and determinism.

See also: Aggression; Aging; Alcoholism; Altruism; Behavior; Bioethics; Biological clocks; Cloning: Ethical issues; Criminality; Developmental genetics; Eugenics; Eugenics: Nazi Germany; Gender identity; Genetic engineering: Social and ethical issues; Genetic screening; Genetic testing: Ethical and economic issues; Heredity and environment; Human genetics; Intelligence; Miscegenation and antimiscegenation laws; Natural selection; Race; Sociobiology; Twin studies; XYY syndrome.

Biological weapons

CATEGORY: Genetic engineering and biotechnology; Human genetics

SIGNIFICANCE: Just as twentieth century discoveries in chemistry and physics led to such devastating weapons as poison gases and nuclear bombs, so humanity in the twenty-first century faces the prospect that the biotechnological revolution will lead to the development and use of extremely deadly biological weapons.

KEY TERMS

anthrax: an acute bacterial disease that affects animals and humans and that is especially deadly in its pulmonary form

biological weapon: the military or terrorist use of such organisms as bacteria and viruses to cause disease and death in people, animals, or plants

bioterrorist: an individual or group that coercively threatens or uses biological weapons, often for ideological reasons

ethnic weapons: genetic weapons that target certain racial groups

genetic engineering: the use of recombinant DNA to alter the genetic material in an organism

immune system: the biological defense mechanism that protects the body from disease-causing microorganisms

recombinant DNA: DNA prepared by transplanting and splicing genes from one species into the cells of another species

smallpox: an acute, highly infectious, often fatal disease characterized by fever followed by the eruption of pustules

EARLY HISTORY

Biological warfare antedates by several centuries the discovery of the gene. Just as the history of genetics did not begin with Gregor Mendel, whose pea-plant experiments eventually helped found modern genetics, the history of biological warfare began long before the Japanese dropped germ-filled bombs on several Chinese cities during World War II. For example, the Assyrians, six centuries before the common era, knew enough about rye ergot, a fungus disease, to poison their enemies' wells. The ancient Greeks also used disease as a military weapon, and the Romans catapulted diseased animals into enemy camps. A famous medieval use of biological weapons occurred during the Tatar siege of Kaffa, a fortified Black Sea port, then held by Christian Genoans. When Tatars started dying of the bubonic plague, the survivors catapulted cadavers into the walled city. Many Genoans consequently died of the plague, and the remnant who sailed back to Italy contributed to the spread of the Black Death into Europe.

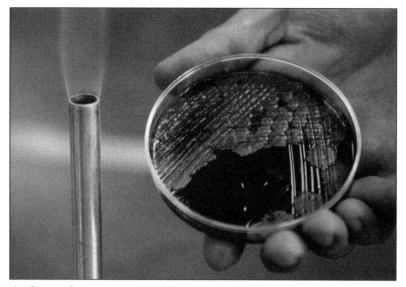

Anthrax colonies grow on culture in a petri dish in Mexico City, where in 2001 two germ banks housed dozens of these cultures virtually unguarded. (AP/Wide World Photos)

Once smallpox was recognized as a highly contagious disease, military men made use of it in war. For example, the conquistador Francisco Pizarro presented South American natives with smallpox-contaminated clothing, and, in an early case of ethnic cleansing, the British and Americans used deliberately induced smallpox epidemics to eliminate native tribes from desirable land.

As scientists in the nineteenth and twentieth centuries learned more about the nature and modes of reproduction of such diseases as anthrax and smallpox, germ warfare began to become part of such discussions as the First International Peace Conference in The Hague (1899). The worldwide revulsion against the chemical weapons used in World War I, along with a concern that biological weapons would be more horrendous, led to the Geneva Protocol (1925), which prohibited the first use of germ weapons, but not their development.

FROM GERM WARFARE TO GENETIC WEAPONS

With the accelerating knowledge about the genetics of various disease-causing microorganisms, several countries became concerned with the threat to their security posed by the weaponizing of these pathogens. Although several states signed the Geneva Protocol in the late 1920's, others signed only after assurances of their right to retaliate. The United States, which did not ratify the treaty until 1975, did extensive research on germ weapons during the 1950's and 1960's. American scientists were able to make dry infectious agents that could be packed into shells and bombs, and estimates were made that ten airplanes with such bombs could kill or seriously disable tens of millions of people. Unknown to Congress and the American people, tests using apparently harmless microbes were performed on such large communities as San Francisco. When news of these secret tests was made public, many questioned their morality. Extensive criticism of the research and development of these weapons, together with the realization that these weapons posed a threat to the attackers as well as the attacked, led President Richard Nixon to end the American biological weapons program formally in 1969.

Abhorrence of biological weapons extended to the world community, and in 1972 the Biological and Toxin Weapons Convention (BTWC)—a treaty that prohibited the development, production, and stockpiling of bacteriological weapons—was signed in Washington, D.C., London, and Moscow and was put into force in 1975. Although it was eventually signed by most members of the United Nations, the nations that signed the pact failed to reach agreement on an inspection system that would control the proliferation of these weapons. A pivotal irony of the BTWC is that while most of the world was renouncing germ warfare, biologists were learning how to manipulate DNA, the molecule that carries genetic information, in powerful new ways. This knowledge made possible the creation of "super-bugs," infectious agents for which there are no cures.

Some scientists warned the public and international agencies about these new germ weapons. Other investigators discovered that American researchers were creating infectious agents that would confuse diagnosticians and defeat vaccines. Similarly, Soviet researchers on an island in the Aral Sea, described as the world's largest BW test site, were producing germ weapons that could be loaded on missiles. When Boris Yeltsin became president of Russia, he discovered that the secret police and military officials had misinformed him about BW programs, in which deadly accidents had occurred. Also troubling was the spread of biological agents to such countries as Iraq. American and French companies legally shipped anthrax and botulinum bacteria to Iraq, whose scientists later acknowledged that they had used these microbes to make tons of biological weapons during the 1980's.

With the demise of the Soviet Union and increasing violence in the Balkans and Middle East, politicians became fearful that experts who had dedicated their careers to making biological weapons would now sell their knowledge to rogue nations or terrorist groups. Indeed, deadly pathogens were part of world trade, since the line separating legitimate and illegitimate research, defensive and offensive BWs, was fuzzy. In the 1980's members of a reli-

gious cult spread salmonella, a disease-causing bacterium, in an Oregon town, causing more than seven hundred people to become very ill. The same company that sold salmonella to this religious cult also sold pathogens to the University of Baghdad. Bioterrorism had become both a reality and a threat.

THE FUTURE OF GENETIC WEAPONS

Some scientists and politicians believe that a nation's best defense against bioterrorism is advanced genetic knowledge, so that vaccines can be tailored to respond to traditional and new BWs. For example, the Human Genome Project, which succeeded in mapping the human genetic material, has the potential for revealing both the vulnerabilities and defenses of the immune system. (The human genome sequence contains 3.2 billion bases and approximately 34,000 genes. These data freely are available

U.N. inspectors supervise the destruction of growth media for biological weapons in Iraq in 1996. (AP/Wide World Photos)

on the Internet in a variety of forms, including text files and graphical "genome browsers.") On the other hand, such knowledge could prove dangerous if the genetic vulnerabilities of certain ethnic groups could be targeted by bioengineered microbes. Some scientists find these speculations about genocidal BWs unevidenced and unsubstantial. Genetic similarities between different ethnic groups are more significant than their differences. Other scientists point out that dramatic genetic differences between ethnic groups are a reality. For example, milk is a poison for certain Southeast Asian populations. Other genetic differences could therefore be exploited to create BWs to attack group-specific gene clusters. Believers in ethnic BWs point to existing techniques for selectively killing certain cells and for inactivating certain DNA sequences. These techniques, developed with the hope of curing genetic diseases, could also be used to cause harm. Knowledge of the structure of the human genome will increasingly lead to knowledge of its function, and this knowledge will make it possible to manipulate, in both benign and malign ways, these information-laden molecules. Modern biotechnology thus presents humanity with both a great promise, to better health and life in peace, and a great peril, to multiply sickness and death in war. The hope of many scientists, politicians, and ordinary people is that humanity will choose the path of promise.

Robert J. Paradowski, Ph.D.

FURTHER READING

Alibeck, Ken, with Stephen Handelman. *Biohazard: The Chilling True Story of the Largest Covert Biological Weapons Program in the World, Told from the Inside by the Man Who Ran It.* New York: Random House, 1999. Alibeck was a Kazakh physician who helped create the Soviet Union's advanced biological warfare program. For some, this autobiographical account is the best and most comprehensive overview of the BW controversy; for others, Alibeck's treatment is tarnished by his long association with the Soviet system.

British Medical Association. *Biotechnology, Weapons, and Humanity.* Amsterdam: Harwood Academic, 1999. Concerned that genetic engineering and biotechnology will be used to produce germ weapons, the physicians of the British Medical Association produced this helpful guide to facilitate public debate. Contains a glossary of technical terms and an excellent set of references.

Cole, Leonard A. *The Eleventh Plague: The Politics of Biological and Chemical Warfare.* New York: W. H. Freeman, 1996. Cole, who has published other books on chemical and biological weapons, examines various programs in the United States and Iraq, while emphasizing that morality is important in responding to the challenges posed by these weapons.

Croddy, Eric. *Chemical and Biological Warfare: A Comprehensive Survey for the Concerned Citizen.* New York: Copernicus Books, 2002. Describes existing chemical and biological weapons, how they work, which countries own them, and the threat each weapon poses.

Guillemin, Jeanne. *Biological Weapons: From the Invention of State-Sponsored Programs to Contemporary Bioterrorism.* New York: Columbia University Press, 2005. A history of biological weapons, beginning with American, British, and Japanese weapons programs that predate World War II and continuing through the Cold War and twenty-first century bioterrorism.

Lutwick, Larry I., and Suzanne M. Lutwick, eds. *Beyond Anthrax: The Weaponization of Infectious Diseases.* New York: Springer, 2009. Collection of essays that provide a historical overview of bioterrorism and analyses of specific diseases and infections, such as smallpox, botulism, and viral hemorrhagic fevers, that can be used as biological weapons.

Miller, Judith, Stephen Engelberg, and William Broad. *Germs: Biological Weapons and America's Secret War.* New York: Simon & Schuster, 2001. This book, written by three *New York Times* reporters, explores the ideas and actions of scientists and politicians involved in the past, present, and future of germ warfare. Includes forty-two pages of notes and a select bibliography.

Piller, Charles, and Keith R. Yamamoto. *Gene Wars: Military Control over the New Genetic Technologies.* New York: William Morrow, 1988. A journalist teamed with a molecular biologist to write this book in order to demystify new biological technologies for the nonscientist and to alert scientists of their special responsibility to enlighten public debates about BW research. Contains appendixes on recombinant DNA and BW treaties and a select bibliography.

Web Sites of Interest

CBC News In Depth, Biological Weapons
http://www.cbc.ca/news/background/
bioweapons/#first

This Web site accompanied news reports about biological warfare that the Canadian Broadcasting Corporation (CBC) first aired in 2004. The site includes a biowar time line, a biowar dictionary, and a chart providing information about the various chemicals and organisms that can be used as weapons.

CBWInfo.com
http://www.cbwinfo.com

CBWI is an independent American organization that provides information about chemical and biological weapons to emergency, safety, and security personnel. Its Web site features news about and a history of these weapons, as well as the Biological Gateway, a collection of information about biological agent properties, their uses, defenses against them, and some background on the general concepts of biological warfare.

Federation of American Scientists, Biological and
Chemical Weapons
http://www.fas.org/programs/ssp/bio/index.html

The federation is dedicated to educating the public about biological weapons. Its Web site posts resources on biosecurity, information on chemical weapons, and an online compilation of documents related to the United States' ratification of the Chemical Weapons Convention.

See also: Anthrax; Bioethics; Biopesticides; Biopharmaceuticals; Emerging and reemerging infectious diseases; Eugenics; Eugenics: Nazi Germany; Genetic engineering: Risks; Genetic engineering: Social and ethical issues; Smallpox.

Biopesticides

CATEGORY: Genetic engineering and biotechnology
SIGNIFICANCE: As an alternative to chemical pesticides, agricultural scientists have begun using ecologically safer methods such as biopesticides to protect plants from insects.

Key terms

Agrobacterium tumefaciens: a species of bacteria that is able to transfer genetic information into plant cells
Bacillus thuringiensis: a species of bacteria that produces a toxin deadly to caterpillars, moths, beetles, and certain flies
baculovirus: a strain of virus that is capable of causing disease in a variety of insects
transformation: the process of transferring a foreign gene into an organism
transgenic organism: an organism synthesizing a foreign protein, the gene of which was obtained from a different species of organism

Bacillus thuringiensis

Hungry insects are the bane of gardeners, since their appetite results in defoliation of the crops. This problem is worsened for farmers, whose livelihoods depend on keeping fields free of destructive insects. Although effective, chemical pesticides have a variety of drawbacks that include contamination of surface and groundwater and adverse health effects on noninsect species, including humans. The increasing popularity of organically grown produce that is untreated by chemicals is indicative of the wariness of consumers concerning human-made pesticides. In response to consumer concern over the safety of chemical pesticides, agricultural biologists have turned to nature to solve pest problems. Biopesticides are naturally derived insecticides. The process of evolution has produced biopesticides that are very specific and effective in their activity. A biopesticide may be sprayed directly on crops or may be genetically engineered to be produced by a crop itself.

One of the oldest commercial biopesticides is a bacterium called *Bacillus thuringiensis* (*Bt*). Since the 1950's, *Bt* has been used on crops susceptible to destruction by insect larvae. *Bt* is a spore-forming bacterium, meaning it is capable of producing an environmentally hardy form that protects the genetic material from adverse conditions. When conditions improve, the spore can germinate to reestablish the normally growing and dividing form of the organism. The basis of *Bt*'s action is the ingestion of the spores by the feeding insect larvae. During the sporulation process, *Bt* produces a protein crystal. When the protein is synthesized by the bacteria, it is an inactive form known as a proenzyme. After being di-

Entomology professor Thomas Miller holds a petri dish containing genetically modified pink bollworm carrying a gene lethal to its offspring. (AP/Wide World Photos)

gested by a larva, enzymes in the larval gut cleave the proenzyme into an active version that kills the larva by binding to receptors in the insect's midgut cells and blocking those cells from functioning. Only caterpillars (tobacco hornworms and cotton bollworms), beetles, and certain flies have the gut biochemistry to activate the toxin. The toxin does not kill insects that are not susceptible, nor does it harm vertebrates in any way. This makes *Bt* a very specific pesticide.

Initially, *Bt* was expensive and remained active following spraying for only a relatively short time. These obstacles were overcome in the early 1990's, when scientists utilized genetic engineering technology to produced transgenic cotton plants that generated their own *Bt* toxin. The toxin gene was first isolated from *Bt* cells and ligated (enzymatically attached) into a Ti plasmid. A Ti plasmid is a circular string of double-stranded DNA that originates in the *Agrobacterium tumefaciens* bacteria. *A. tumefaciens* has the ability to take a portion of that Ti plasmid, called the T-DNA, and transfer it and whatever foreign gene is attached to it into a plant cell. Cotton plants were exposed to *A. tumefaciens* carrying the toxin gene and were transformed. The transgenic plants synthesized the *Bt* toxin and became resistant to many forms of larvae. This approach has become known as plant-incorporated bioprotectants.

Many crystal toxins have been isolated from various strains of *Bt*. These toxins make up a large collection of proteins active against pests from nematodes to aphids. Researchers are in the process of reengineering the toxin genes to improve upon their characteristics and to design better methods of transporting genes from one *Bt* strain to another.

OTHER BIOPESTICIDES

Several species of fungi are toxic to insects, including *Verticillium lecanii* and *Metarhizium anisopliae.* Natural fungicides that have been discovered include oils of tea tree, cinnamon, jojoba, neem, and rosemary.

In the mid-1990's, a viral biopesticide called baculovirus became widely popular. Baculoviruses are sprayed onto high-density pest populations just like chemical pesticides. Baculoviruses have several advantages over conventional pesticides. The most important advantage is their strong specificity against moths, sawflies, and beetles but not against beneficial insects. Also, viruses, unlike bacteria, tend to persist in the environment for a longer period. Finally, baculoviruses are ideal for use in developing countries because they can be produced cheaply and in great quantity with no health risks to workers. One limitation of baculovirus is that it must be administered at a certain time and location to be effective. Rather than spraying onto a crop and killing the insects that subsequently feed, baculovirus needs to be applied directly to the target insect population. Knowledge of insect behavior after hatching, the insect population's distribution within the crop canopy, and the volume of foliage ingested by each larva are essential. For example, moths usually do the most damage at the late larval stage. To minimize crop damage from moths, baculovirus needs to be sprayed as early as possible before the insects reach that late stage.

Another biopesticide approach has been to make transgenic plants that manufacture proteins isolated from insect-resistant plant species. Tomatoes naturally make an enzyme inhibitor that deters insects by keeping their digestive enzymes (trypsin and chymotrypsin) from functioning. These inhibitors were isolated by Clarence Ryan at the University of Washington. Ryan transformed tobacco plants with two different forms of inhibitor (inhibitors I and II from tomatoes). The tomato proteins were effectively produced in tobacco and made the transgenic plants resistant to tobacco hornworm larvae.

Biopesticides exist and have been refined for the control of insect pests even though they are not toxic to the insects. These biochemical pesticides include compounds that interfere with insect pheromones (chemicals that attract insects to a potential mate). Their use inhibits insect mating and so the production of the next generation of the particular insect.

Biopesticides are also being implemented to control the population of mosquitoes in some regions of the world that are susceptible to malaria. The idea is that by controlling the mosquito population, the spread of the microorganism responsible for the disease, which occurs when the mosquito takes a blood meal, will be lessened. As of 2009, this strategy was being tested in several malaria-prone countries in Africa.

Biopesticides and Nontarget Species

Researchers have long had a concern as to the effect of chemical insecticides on nontarget species. Target species frequently display resistance to chemical controls due to large effective population sizes and prior histories of exposure to chemical agents, which favors the increase in resistance alleles in a population. The exact opposite is typically true for nontarget species that occupy the treatment area. When biopesticides such as the CrylAb endotoxins, derived from the soil bacterium *Bacillus thuringiensis* (*Bt*), were first proposed as a control agent, many scientists believed that the collateral effects on nontarget species would be significantly limited.

Initially these toxins were sprayed on crops, thus potentially increasing the exposure of nontarget species. Even with the development of transgenic crops such as corn, it was possible for *Bt* to move from the treatment area to the feeding grounds of nontarget species through pollen dispersal. One of the first documented accounts of *Bt*-induced mortality in a nontarget species was provided in 1999 by researchers at Cornell University. They demonstrated that the pollen from *Bt*-treated corn increased mortality among monarch butterflies (*Danaus plexippus*) when applied to the surfaces of milkweed plants, the butterfly's primary food source. In this study, monarchs exposed to *Bt* had a slower rate of growth and increased mortality. It was suggested that field monarchs could also be exposed to corn pollen containing *Bt* endotoxins. Given the popularity of the monarch and the noticeable decline in North American populations during the 1990's, it appeared that the future of biopesticides was dim.

Since then, additional studies indicate a less significant effect of *Bt* toxins on nontarget species. The dispersal of *Bt* pollen is not believed to occur more than a few meters from the edges of the treatment area, and even at these distances the levels have been shown to be sublethal. Research involving monarchs and swallowtail butterflies (*Papilio* species) has indicated that lethality is not elevated at low-level *Bt* exposure, although there is evidence of reduced growth rates. Furthermore, only a fraction of the nontarget organism's population would be exposed at a given time, and frequently the larval periods of the target and nontarget organisms do not overlap. However, this evidence suggests that biopesticides are not producing the observed decrease in nontarget populations.

It is likely that there may be a limited effect of biopesticides on nontarget species, and most researchers agree that additional research needs to be conducted. The genetics of *Bt* resistance have been determined for a number of insects, although for others the exact mechanism has remained elusive. However, the greatest threat to the nontarget organisms rests with habitat destruction. A decrease in the population size due to reduced resources may serve to weaken the population and enhance the sublethal effects of biopesticide production. The physiological effect and population genetics of *Bt* susceptibility in nontarget species will need to be examined in some detail to prove to the public the value of biopesticides.

Michael Windelspecht, Ph.D.

At the University of Florida's Institute of Food and Agricultural Sciences, Dov Borovsky has developed a "diet pill" for mosquitoes that causes them to starve to death. It may help eradicate mosquitoes and the diseases they transmit. (AP/Wide World Photos)

BIOPESTICIDE RESISTANCE

As with chemical pesticides, over time insect populations grow resistant to biopesticides. *Bt*-resistant moths can now be found around the world. Resistance arises when pesticides are too effective and destroy more than 90 percent of a pest population. The few insects left are often very resistant to the pesticide, breed, and with succeeding generations create large, resistant populations.

Entomologists have suggested strategies for avoiding pesticide-resistant insect populations. One strategy suggests mixing biopesticide-producing and nonproducing plants in the same field, thereby giving the pesticide-susceptible part of the insect population places of refuge. These refuges would allow resistant and nonresistant insects to interbreed, making the overall species less resistant. Other strat-

egies include synthesizing multiple types of *Bt* toxin in a single plant to increase the toxicity range and reduce resistance, making other biological toxins besides *Bt* in a single plant, and reducing the overall exposure time of insects to the biopesticides.

Organizations including the U.S. Environmental Protection Agency (EPA) and Department of Agriculture recommend using biopesticides as part of what is termed an "integrated pest management approach" that uses a number of control and crop growth strategies. The aim is to decrease the use of conventional pesticides while maintaining or even increasing crop yield. This approach also helps lessen the development of resistance, since the same biopesticide is not used constantly.

James J. Campanella, Ph.D.;
updated by Brian D. Hoyle, Ph.D.

FURTHER READING

Copping, Leonard C., ed. *The Manual of Biocontrol Agents.* 3d ed. Surrey, England: British Crop Protection Council, 2004. An international expert in biological control of crops presents an overview of biological control agents, including biopesticides.

Copping, Leonard C., and J. J. Menn. "Biopesticides: A Review of Their Action, Applications, and Efficacy." *Pest Management Science* 56 (2002): 651-676. An academic review of biopesticides.

Khan, Mohammad Saghir, A. Zaidi, and J. Musarrat, eds. *Microbial Strategies for Crop Improvement.* New York: Springer, 2009. Multidisciplinary consideration of microbiological approaches for crop improvement, including the use of biopesticides.

Koul, Opender, and G. S. Dhaliwal, eds. *Microbial Biopesticides.* New York: Taylor & Francis, 2002. International experts on biopesticides explore developments in using those based on bacteria, fungi, viruses, and nematodes, discussing their advantages and disadvantages and their role in genetic engineering.

Regnault-Roger, Catherine, Bernard J. R. Philigène, and Charles Vincent, eds. *Biopesticides of Plant Origin.* Secaucus, N.J.: Intercept, 2005. Expert authors summarize the known plant biopesticides and aspects of their use.

Walters, Dale. *Disease Control in Crops: Biological and Environmentally-Friendly Approaches.* New York: Wiley-Blackwell, 2009. An expert on crop control discusses nonchemical approaches including biopesticides with an emphasis on integrated pest management.

WEB SITES OF INTEREST

U.S. Department of Agriculture
http://www.epa.gov/agriculture/tbio.html
Government site with a link to information on biopesticides.

U.S. Environmental Protection Agency
http://www.epa.gov/pesticides
Government site with a link to information on biopesticides, including the Federal Insecticide, Fungicide, and Rodenticide Act, enacted to monitor the harmful effects of toxic pesticides on humans and the environment and ensure industry compliance.

World Wildlife Federation
http:www.panda.org/about_our_earth/teacher_resources/webfieldtrips/sus_agriculture/bioipesticides
Organization Web site on the uses and agricultural advantages of biopestides, with links to sources of other biopesticide information.

See also: Biofertilizers; Genetic engineering: Agricultural applications; Genetic engineering: Risks; Genetic engineering: Social and ethical issues; Genetically modified foods; High-yield crops; Population genetics.

Biopharmaceuticals

CATEGORY: Diseases and syndromes; Genetic engineering and biotechnology

SIGNIFICANCE: Biopharmaceuticals are drugs that are designed through a combination of genetics and biotechnology. Biopharmaceuticals differ from pharmaceuticals, compounds usually produced by traditional chemical synthesis. Typically derived from proteins, such as enzymes or antibodies, biopharmaceuticals are genetically engineered to treat or target a specific disease.

KEY TERMS

clinical trial: an experimental research study that determines the safety and effectiveness of a medical treatment or drug

humanized antibody: a human antibody that has been engineered to contain a portion of a nonhuman variable region with known therapeutic activity

pharmacogenomics: the field of science that examines how variations in genes alter the metabolism and effectiveness of drugs

HISTORY OF BIOPHARMACEUTICALS

Drugs have been used by humans for thousands of years. More than three thousand years ago, the Sumerians were the first culture to compile written medical information that outlined symptoms and treatments for disease. Most ancient cultures used medicines derived from plants and animals. These drugs were different from modern biopharmaceuticals in many ways, but the most significant difference is that the drugs were not engineered to treat a particular disease. Since there was no real understanding of the underlying problem, a rational approach to drug selection and design was difficult, if not impossible. One philosophy of medicine that developed to address this problem was called the doctrine of similitudes, in which treatments were based on similarities of structure with disease manifestation. For example, the leaves of St. John's wort looked similar to damaged skin, so it was thought this plant extract could effectively treat cuts and burns.

It was not until the twentieth century that the underlying genetic basis for disease was discovered. The discovery that DNA is the genetic material that provides instructions to make proteins was revolutionary. In the mid-1900's, sickle-cell disease was shown to be caused by a single nucleotide mutation from an A (adenine) to a T (thymine) in the hemoglobin beta-chain gene. This small change alters the shape of a red blood cell from a biconcave disc to a sharply-pointed crescent. Although it was now possible to identify genetic mutations, there was still no way to manipulate or make changes to genetic information.

The advent of recombinant DNA technology in the 1970's provided the first chance to engineer, or manipulate, genes. Restriction enzymes became an important tool in this new technology. Restriction enzymes were first found in bacteria, where they function to protect the cell from foreign DNA by cutting it up at specific sequences. These sequences are usually palindromes of the letters that signify the four nucleotides that make up DNA: guanine (G), adenine (A), thymine (T), and cytosine (C). Most restriction enzymes cut the DNA in such a way that an overhang, called a sticky end, is created. Since excess unbound DNA, provided by the scientist, will readily bind its complementary base, these engineered sticky ends can be used to splice different pieces of DNA together in a laboratory. The resulting sequence is called recombinant DNA.

With the ability to engineer DNA now possible with restriction enzyme technology, scientists looked again to use bacteria as a host "factory" in order to convert known DNA sequences into protein. Bacteria are ideal for protein production because they reproduce quickly, are easy to genetically manipulate, and can be grown in large quantities. Many bacteria contain circular pieces of DNA that are separate from their genome, called plasmids. Plasmids can be readily transferred between bacteria and are also inherited by daughter cells when a bacterium divides. With the use of restriction enzymes, plasmids are isolated from bacteria and engineered to contain a foreign gene. The recombinant plasmid is re-inserted back into bacteria, which work nonstop to transcribe and translate the recombinant gene. The gene is then expressed as a fully functional protein. The first biopharmaceutical produced in bacteria was recombinant human insulin, which was marketed in 1982.

The future for biopharmaceuticals looks bright. In 1991, there were only fourteen biopharmaceuticals approved for use by the U.S. Food and Drug Administration (FDA). By 2001, nearly three hundred had been approved for use, with an additional fifty in phase III clinical trials. By 2003, more than 330 major companies in the United States were working to produce and develop biopharmaceuticals.

DESIGN OF BIOPHARMACEUTICALS

A popular method for the identification of disease-related genes is called genomics. Gene chip analysis is used to screen thousands of genes in a single experiment. This approach is drastically faster and more efficient than traditional methods and can be used for any disease, even those that are not hereditary.

Once the genomic information is obtained, it is used to build a broad understanding of how a disease gene functions and what role the gene plays in the cell. This information is gathered through the use of experimental models, genetic analysis, biochemical analysis, and structural analysis. Experimental models can range from cell culture to transgenic mice and can provide physiological information about the disease. Genetic analysis can provide information about where and when the gene is expressed. Biochemical analysis can provide information about protein-protein interactions, post-translational modifications of the protein, and its enzymatic activity. Structural analysis can yield extremely detailed information about the physical arrangement of the atoms that make up the protein. All these approaches can identify important potential targets for treatment of the disease. A better understanding of the disease at the genetic and molecular levels facilitates the design of a biopharmaceutical.

Once a disease is better understood, it becomes possible to target a key pathway or protein for biopharmaceutical intervention. The resultant drug

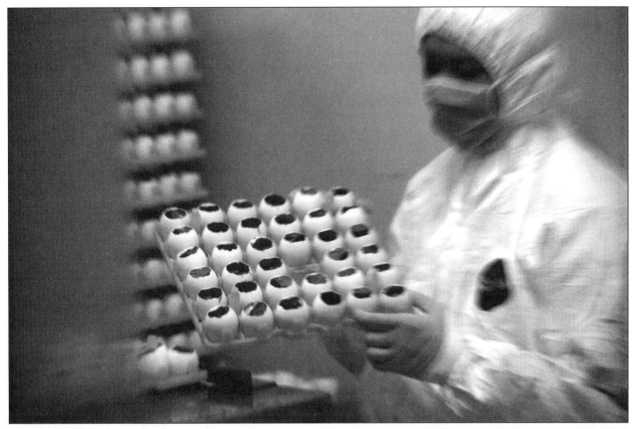

A technician in China checks flu virus growing in eggs to produce an inactivated H1N1 vaccine in September, 2009. (AP/ Wide World Photos)

and the way that it is used clinically will vary from disease to disease. For example, type I diabetes is caused by a deficiency in the hormone insulin. Without insulin, the body is not able to regulate the level of glucose in the blood. Lack of insulin was first corrected by an injection of the first biopharmaceutical, recombinant human insulin. It was developed by Genentech and marketed as Humulin by Eli Lilly & Company in 1982.

Another example of a biopharmaceutical is the enzyme tissue plasminogen activator (tPA). Most heart attacks are caused by a blood clot blocking the flow of blood through a coronary artery. Formation and removal of blood clots is a highly regulated and well-understood process. Tissue plasminogen activator is one of the key players in blood clot removal. This knowledge led to the development of recombinant tPA, which can be provided by injection or infusion to heart attack patients. Once in the bloodstream, tPA breaks up coronary artery clots and restores blood flow to the heart, preventing any further muscle damage.

CLINICAL TRIALS

Before a biopharmaceutical can be used to treat disease, it must undergo rigorous clinical trials that test its safety and effectiveness in humans. There are four phases of clinical trials. Phase I trials involve studies on a small number of patients (fewer than one hundred) in order to determine drug safety and dosage. Phase II trials involve more patients (up to five hundred) in order to determine effectiveness and additional safety information, such as side effects. Phase III trials are the most extensive and involve large numbers of people (between one thousand and three thousand). These trials establish risk-benefit information and are compared with other currently used treatments. Phase IV trials determine the drug's optimal use in a clinical setting.

In 2003, the entire process of drug design—from

discovery to clinical trials—cost approximately $802 million and took an average of twelve years. By 2006, there were 111 biopharmaceuticals in late-stage development that targeted thirty-eight different disease categories, the majority of which was cancer. Many years of research and millions of dollars are expended in an extraordinary effort that yields little success—only one in five thousand drugs makes it to market. By 2008, 64 therapeutic biological products were under review in the Center for Drug Evaluation and Research at the FDA.

BIOPHARMACEUTICALS TODAY

Biopharmaceuticals are classified into several categories, including blood factors, thrombolytic agents, hormones, hematopoietic growth factors, interferons, interleukin-based products, vaccines, monoclonal antibodies, and other products. Some FDA-approved biopharmaceuticals of particular interest include Aralast, Campath, Gardasil, and ATryn. Aralast is marketed by Baxter and was approved for use by the FDA in 2003. Aralast is the trade name for the recombinant human protein known as alpha-1 proteinase inhibitor (A1PI). A1PI deficiency, also called alpha-1-antitrypsin deficiency, results in the destruction of lung tissue, which can lead to emphysema. Aralast is given to patients intravenously each week, protecting against future lung damage.

Campath is marketed by Millennium Pharmaceuticals and was approved by the FDA in 2001. Campath is the trade name for a humanized antibody against the CD52 antigen found on lymphocytes. The antibody is used to treat chronic lymphocytic leukemia and works by destroying lymphocytes through agglutination and complement activation.

While many biopharmaceuticals are designed to target a specific disease, the popular vaccine Gardasil was designed to prevent genital human papillomavirus (HPV) infection. Gardasil is marketed by Merck & Company and was approved by the FDA for use in young women in 2006. HPV is the most commonly sexually transmitted disease in the United States and has been shown to cause cervical cancer, the second leading cause of cancer deaths among women worldwide. Gardasil vaccinates against the four most common strains of HPV. It is a quadrivalent vaccine that stimulates the immune system to make antibodies that recognize and destroy HPV 6, 11, 16, and 18, protecting against future infection.

ATryn is the trade name for recombinant antithrombin III, an anticoagulant manufactured by GTC Biotherapeutics and sold by Ovation Pharmaceuticals. It is produced from the milk of transgenic goats that have been genetically modified to produce human antithrombin. In 2009, the FDA approved the use of ATryn to treat patients with hereditary antithrombin deficiency who need an anticlotting agent to undergo procedures that involve blood loss, such as surgery and childbirth. It is the first biopharmaceutical to be produced from genetically engineered animals, goats being chosen for their high reproduction rate.

Currently, 100,000 people die each year because of adverse drug reactions. A trend in pharmaceutical research is the production of designer drugs through the new field of pharmacogenomics. These drugs are specifically matched to an individual patient's genetic profile and their particular form of disease. With the use of pharmacogenomics it will be possible to avoid adverse drug reactions. Research in pharmacogenomics will also increase the pool of drugs available to treat disease. While many drugs never make it to market because they work for only a small subset of patients, pharmacogenomic research will identify these specific patients as treatment successes.

Matthew J. F. Waterman, Ph.D.;
updated by Diana R. Lazzell

FURTHER READING

Barr, E., and H. L. Sings. "Prophylactic HPV Vaccines: New Interventions for Cancer Control." *Vaccine* 26, no. 49 (2008): 6244-6257. Overviews the clinical trials evaluating the long-term safety and efficacy of the quadrivalent HPV vaccine.

Collins, F., and V. McKusick. "Implications of the Human Genome Project for Medical Science." *JAMA* 285, no. 5 (2001): 540-541. Overviews the significant impact of the Human Genome Project on medical research, including specific examples of drug design.

Nagle, P. C., C. A. Nicita, L. A. Gerdes, and C. J. Schmeichel. "Characteristics of and Trends in the Late-Stage Biopharmaceutical Pipeline." *American Journal of Managed Care* 14, no. 4 (2008): 226-229. Provides a review of the drug development databases and analyzes the biopharmaceuticals in late-stage development in the United States.

Niemann, H., and W. A. Kues. "Transgenic Farm Animals: An Update." *Reproduction, Fertility, and Development* 19, no. 6 (2007): 762-770. A review of the history of genetic engineering in farm animals to date.

Wu-Pong, S., and Y. Rojanasakul. *Biopharmaceutical Drug Design and Development.* Totowa, N.J.: Humana Press, 1999. Outlines the process of biopharmaceutical design, including basic molecular biology, major classes of biopharmaceuticals, and clinical trials.

Web Sites of Interest

American Chemical Society, The Pharmaceutical Century
http://pubs.acs.org/journals/pharmcent
Posts articles about the science of biopharmaceuticals, including the role played by genetics and the Human Genome Project in the development of new drugs.

Clinical Today.com
http://www.clinicaltoday.com
Articles about the most recent developments in clinical trials.

ClinicalTrials.gov
http://www.clinicaltrials.gov
Search ongoing and completed FDA clinical trials using a variety of search criteria.

Food and Drug Administration (FDA)
http://www.fda.gov/cder/index.html
Information regarding all drugs regulated by the FDA.

International Biopharmaceutical Association-Alliance
http://www.ibpaalliance.org
Wikipedia-powered glossary of commonly encountered biopharmaceutical terms and common pharmaceutical companies that produce biopharmaceuticals.

Pharma Industry Today
http://www.pharmaindustrytoday.com
Up-to-date news from the pharmaceutical and biotechnology industries.

See also: Cloning; Cloning: Ethical issues; Cloning vectors; Gene therapy; Gene therapy: Ethical and economic issues; Genetic engineering; Genetic engineering: Medical applications; Genetic engineering: Risks; Genetic engineering: Social and ethical issues; Genetically modified foods; Molecular genetics; Synthetic genes; Transgenic organisms; Xenotransplants.

Bipolar affective disorder

Category: Diseases and syndromes
Also known as: Bipolar disorder; manic-depressive illness; manic depression; manic disorder; manic affective disorder

Definition

Bipolar affective disorder results in extreme swings in mood, energy, and ability to function. The mood changes of bipolar disorder are more dramatic than normal ups and downs; they can hurt relationships and cause poor job or school performance. Bipolar disorder can be treated. Individuals should contact their doctors if they think they may have this condition.

The two extremes of the illness are mania and depression. In mania, energy peaks; an individual's mood may be overly happy or irritable. In depression, lethargy takes over; the mood may be very blue.

There are three forms of this condition. Bipolar I disorder involves recurrent episodes of mania and depression. Bipolar II disorder involves milder episodes of mania (called hypomania) that alternate with episodes of depression. Cyclothymic disorder may be diagnosed in patients who experience frequent depressive symptoms and hypomania for at least two years and who have been without symptoms for no more than two months.

Risk Factors

A family history of bipolar disorder increases an individual's chance of developing this condition. Patients should tell their doctors if they have a family member with bipolar disorder.

Etiology and Genetics

Bipolar affective disorder is a complex condition in which the etiology most often depends on a combination of genetic and environmental factors.

While specific genetic factors and their interactions are not well understood, it is clear from twin, family, and adoption studies that there is a strong genetic component. One study suggests that children of individuals with bipolar affective disorder are seven times more likely to develop the disease than are children whose parents are unaffected, but there is otherwise no predictable pattern of inheritance. Twin studies suggest a concordance rate for identical twins of 43 percent as compared to only 6 percent for fraternal twins.

Although dozens of genes may play small roles in the development of bipolar affective disorder, two specific genes have been identified that may be more major contributors. The *ANK3* gene, found at position 10q21 on the long arm of chromosome 10, encodes a protein known as ankyrin G. This is an adapter protein that regulates the assembly of ion channels on neuronal axons. A second gene, *CACNA1C*, found on the short arm of chromosome 12 at position 12p13.3, specifies the alpha subunit of a specific calcium channel in neurons. Studies in mice suggest that the levels of the mouse versions of both of these proteins are depressed in the brains of mice in response to lithium treatment. These studies suggest that mutant alleles may cause an overproduction of the proteins and provide a possible therapeutic explanation for the effectiveness of lithium treatment in some patients.

SYMPTOMS

Symptoms of bipolar affective disorder include dramatic mood swings, ranging from elated excitability to hopeless despondency; periods of normal mood in between ups and downs; and extreme changes in energy and behavior.

Signs and symptoms of mania include persistent and inexplicable elevation in mood, increased energy and effort toward goal-directed activities, restlessness and agitation, racing thoughts, jumping from one idea to another, and rapid speech or pressure to keep talking. Trouble concentrating; a decreased need for sleep; overconfidence or inflated self-esteem; and poor judgment, often involving spending sprees and sexual indiscretions, are other signs and symptoms of mania.

Signs and symptoms of depression include a prolonged sad, hopeless, or empty mood; feelings of guilt, worthlessness, or helplessness; and a loss of interest or pleasure in activities once enjoyed, including sex. Decreased energy or fatigue; trouble concentrating, remembering, and/or making decisions; restlessness or diminished movements; agitation; sleeping too much or too little; unintended weight loss or gain; and thoughts of death or suicide with or without suicide attempts are other signs and symptoms of depression.

Severe episodes of mania or depression may sometimes be associated with psychotic symptoms, such as hallucinations, delusions, and disorders of thought.

SCREENING AND DIAGNOSIS

A doctor will ask a patient about his or her symptoms and medical history, and a physical exam may be done. In some cases, lab tests are ordered; they will help to rule out other causes of mood and behavior symptoms. Patients may also be referred to a mental health specialist.

Diagnosis of bipolar disorder is based on the presence of symptoms over time; the absence of other causes, such as some medications, thyroid disease, or Parkinson disease; and a family history of bipolar disorder.

Mania is diagnosed if a patient's mood is elevated and there are three or more of the mania symptoms listed above. If the mood is irritable, not elevated, four symptoms must be present for a diagnosis of mania. Symptoms last during most of the day, nearly every day, for one week or longer.

A depressive episode is diagnosed if there are five or more of the depressive symptoms listed above. Symptoms last for most of the day, nearly every day, for a period of two weeks or longer.

Some medicines and other medical issues may show similar features, such as corticosteroids, thyroid disease, and multiple sclerosis. The diagnosis is made only when none of these other causes are present.

TREATMENT AND THERAPY

Individuals should talk with their doctors about the best plans for them. Medications are among the treatment options. Many patients are treated with a combination of two or more of these medications: lithium, a mood stabilizer, often used as initial treatment, that helps prevent manic and depressive episodes from returning; antiseizure medications, also used as mood stabilizers instead or in combination with lithium; valproate (Depakote); carbamazepine

(Tegretol); lamotrigine (Lamictal); and topiramate (Topamax).

Benzodiazepines, including clonazepam (Klonopin) and lorazepam (Ativan), can be used to treat agitation or insomnia; zolpidem (Ambien) can be used to treat insomnia. Antidepressants, such as serotonin reuptake inhibitors or bupropion (Wellbutrin), are used to treat depression.

Antipsychotic medications are used for acute manic or mixed episodes and maintenance treatment. Classic antipsychotic medications, such as haloperidol (Haldol), are not often used because of risks of tardive dyskinesia (uncontrollable movements); atypical antipsychotic medications (risperidone, olanzapine, aripiprazole, ziprasidone, and quetiapine) are more effective with less risk of tardive dyskinesia.

The treatment plan is based on the pattern of the patient's illness. Treatment may need to be continued indefinitely, and it should prevent significant mood swings.

Psychotherapy is another treatment option and may include cognitive-behavioral therapy, counseling, family therapy, and interpersonal and social rhythm therapy—a form of therapy designed to treat bipolar disorder. Electroconvulsive therapy may be effective when medications fail; it can be used for both mania and depression.

PREVENTION AND OUTCOMES

There are no guidelines for preventing bipolar disorder.

Krisha McCoy, M.S.;
reviewed by Rosalyn Carson-DeWitt, M.D.
"Etiology and Genetics" by Jeffrey A. Knight, Ph.D.

FURTHER READING

Belmaker, R. H. "Bipolar Disorder." *New England Journal of Medicine* 351, no. 5 (July 29, 2004): 476-486.

EBSCO Publishing. *Health Library: Bipolar Affective Disorder.* Ipswich, Mass.: Author, 2009. Available through http://www.ebscohost.com.

El-Mallakh, Rif S., and S. Nassir Ghaemi, eds. *Bipolar Depression: A Comprehensive Guide.* Washington, D.C.: American Psychiatric, 2006.

Goodwin, Frederick K., and Kay Redfield Jamison. *Manic-Depressive Illness: Bipolar Disorders and Recurrent Depression.* 2d ed. New York: Oxford University Press, 2007.

Kasper, Siegfried, and R. M. A. Hirschfeld, eds. *Handbook of Bipolar Disorder: Diagnosis and Therapeutic Approaches.* New York: Taylor & Francis, 2005.

Kramlinger, Keith, ed. *Mayo Clinic on Depression.* Rochester, Minn.: Mayo Clinic, 2001.

WEB SITES OF INTEREST

Canadian Psychiatric Association
http://www.cpa-apc.org

Child and Adolescent Bipolar Foundation
http://www.bpkids.org

Depression and Bipolar Support Alliance
http://www.dbsalliance.org

Mood Disorders Society of Canada
http://www.mooddisorderscanada.ca

National Institute of Mental Health
http://www.nimh.nih.gov/index.shtml

Psychiatry Online
http://www.psychiatryonline.com

See also: Aarskog syndrome; Aggression; Behavior; Depression; Sociobiology.

Bloom syndrome

CATEGORY: Diseases and syndromes
ALSO KNOWN AS: Congenital telangiectatic erythema; Bloom's syndrome; Bloom-Torre-Machacek syndrome; BS

DEFINITION

Bloom syndrome (BS) is a rare disorder that is inherited in an autosomal recessive manner. Individuals with BS have an enormous predisposition to varied cancer types and are immunocompromised. BS is characterized by chromosome instability.

RISK FACTORS

BS is rare, with only several hundred confirmed cases, but it is least rare among the Ashkenazi Jewish population. In this population, approximately 1 percent are carriers of BS. Roughly one out of three people with the disease have one or both parents of Ashkenazi Jewish descent.

ETIOLOGY AND GENETICS

BS was first described by David Bloom in 1954 and belongs to a class of human diseases called the chromosome breakage syndromes. A person affected with BS carries two mutant *BLM* genes, one inherited from each parent. When both parents carry the mutant *BLM* gene, there is a 25 percent chance for each pregnancy that the offspring will have BS. The *BLM* gene is on chromosome 15 and has been traced to band q26.1. The normal *BLM* gene encodes a protein belonging to the RecQ DNA helicase family. DNA helicases are enzymes that unwind the two complementary spiral strands comprising a DNA molecule. These enzymes are crucial for unwinding before DNA can be replicated in cell division. However, *BLM* gene mutations lead to changes in helicase enzyme activity, affecting the unwinding and copying of DNA as it is replicated.

During normal replication, each chromosome—made up of tightly coiled DNA as well as proteins—makes two identical DNA structures, called sister chromatids. Sister chromatids normally exchange some small sections of DNA during replication in a process called sister chromatid exchange (SCE). When the *BLM* gene is mutated, more errors occur during replication. In BS, there is an increased level of spontaneous SCEs—as many as ten times the amount of SCEs seen in normal cells. In addition to excessive amounts of SCEs, increased breaks and rearrangements between nonhomologous chromosomes (chromosomes that are not of the same pair) are also observed. This chromosome instability and hyperrecombination is thought to contribute to the increased risk of cancer and other features of BS, although the mechanisms are not currently fully understood.

More than sixty BS-causing mutations in the *BLM* gene have been identified. However, one particular mutation causes almost all cases of BS among those of Ashkenazi Jewish descent. This founder deletion/insertion mutation, referred to as *blm^Ash^*, has reached a carrier frequency of roughly 1 percent in the Ashkenazi Jewish population.

SYMPTOMS

BS is physically characterized by proportional, but unusually small prenatal and postnatal size (although the brain and head are disproportionately small) and sun-sensitive skin lesions that are limited to the face and the back of the hands and forearms.

The other most important clinical characteristic, which is not physically observable, is a striking predisposition to cancer. Tumors can be benign or malignant. They may arise at an early age, and with great frequency in a large variety of body locations and cell types. Leukemias, lymphomas, and carcinomas are particularly common. Individuals with BS also usually experience decreased fertility (males are infertile), increased predisposition to multiple infections because of a compromised immune system, and sometimes diabetes.

SCREENING AND DIAGNOSIS

Diagnosis can be presumed based on clinical features associated with BS and is confirmed or ruled out by chromosome analysis. Cells from patients will show a significant (up to tenfold) increase in SCEs compared with cells from normal individuals, as well as increased chromosome breakage and rearrangements, and increased presence of chromosome structures called quadriradicals.

The identification of the *BLM* gene and the specific *BLM* gene mutation responsible for BS in Ashkenazi Jews have made carrier testing possible for this population. Prenatal diagnosis of BS is also available for couples who carry *BLM* gene mutations that have already been identified.

TREATMENT AND THERAPY

There is no direct effective treatment for BS. Mutations will continue to arise in excess, but the life span of affected individuals may be maximized by early diagnosis and systematic management. Avoiding the sun (especially in infancy and youth), treating bacterial infections promptly, treating diabetes if it arises, and avoiding environmental exposures or therapies that may further damage chromosomes (such as X rays) are all measures that may increase life span and improve the quality of life of affected individuals. Most important, those with BS should be in close communication with an experienced physician, who can develop a cancer surveillance program.

PREVENTION AND OUTCOMES

There is no effective means of prevention for BS. Genetic counseling should be available for parents of an affected child, and prenatal screening is an option for parents who are carriers. The Bloom Syndrome Registry comprises the files of 168 people with BS and shows cancer as the most common cause of death. Within the registry, the mean age at death was 23.6 years, with a range of less than 1 to 49.

Sabina Maria Borza, M.A.

Further Reading

Freivogel, Mary E. "Bloom Syndrome." In *Gale Encyclopedia of Genetic Disorders.* Farmington Hills, Mich.: Gale Group, 2002. This entry provides thorough, accessible information about the disorder.

Klug, William S., Michael R. Cummings, Charlotte Spencer, and Michael A. Palladino. *Concepts of Genetics.* 9th ed. San Francisco: Benjamin Cummings, 2008. A comprehensive introduction to various topics in genetics.

Parker, Philip M. *Bloom Syndrome: A Bibliography and Dictionary for Physicians, Patients, and Genome Researchers.* San Diego: ICON Group International, 2007. This guide helps students and medical professionals swiftly find accurate information about Bloom syndrome on the Internet.

Web Sites of Interest

Bloom's Syndrome Foundation
http://www.bloomssyndrome.org

eMedicine: Bloom Syndrome (Congenital Telangiectatic Erythema)
http://emedicine.medscape.com/article/1110271-overview

Genetics Home Reference: Bloom Syndrome
http://ghr.nlm.nih.gov/condition
=bloomsyndrome

OMIM: Online Mendelian Inheritance in Man
http://www.ncbi.nlm.nih.gov/entrez/
dispomim.cgi?id=210900

Orphanet Encyclopedia: Bloom Syndrome
http://www.orpha.net/data/patho/Pro/en/
BloomSyndrome-FRenPro430.pdf

See also: Cancer; DNA replication; Hereditary diseases; Mutagenesis and cancer; Mutation and mutagenesis.

Blotting
Southern, Northern, and Western

Category: Techniques and methodologies

Significance: Blotting is a technique that allows identification of a specific nucleic acid or amino acid sequence even when it is mixed in with all of the other material from a cell. This allows the rapid identification of the changes associated with mutant alleles.

Key terms

blotting: the transfer of nucleic acids or proteins separated by gel electrophoresis onto a filter paper, which allows access by molecules that will interact with only one specific sequence

hybridization: incubation of a target sequence with an identifying probe, which allows the formation of annealed hybrids

Northern blot: a blot designed to detect messenger RNA

probe: a nucleic acid sequence or antibody that can attach to a specific DNA or RNA sequence or protein; the probes are often labeled with radioactive compounds or enzymes so their position can be determined

Southern blot: a blot designed to detect specific DNA fragments

Western blot: a blot that uses antibodies to detect specific proteins

Limitations of Gel Electrophoresis

Using gel electrophoresis to separate proteins and nucleic acids has been an invaluable tool in analyzing living systems. Changes in these molecules—such as a mobility shift in a mutant protein or the change in the size of a plasmid that has received a DNA insert—can be easily detected using this technique. However, the ability to differentiate between types of molecules is quite limited. An extract of red blood cell proteins run through an acrylamide gel might show one major band for hemoglobin that can be discerned from the many other proteins in the cell. However, the hundreds of different proteins that might be produced in a liver extract will produce a tight ladder of bands that are impossible to tell apart.

The situation can be even worse with DNA. A restriction enzyme digest of a plasmid or simple virus might yield fewer than six pieces of DNA that could be easily separated on an agarose gel. If one were to digest the total genomic DNA of even a simple organism, such as *Escherichia coli*, with a typical restriction enzyme such as *Eco*RI, the result would be a thousand bands of numerous sizes (4×10^6 base pairs of DNA, since *Eco*RI recognizes a six-base-pair site, which should occur, on average, every 4^6 or 4,096 bp). After separation on a gel, the result would be a smear with no individual bands visible. Working with an even more complex genome, such as the human genome, would result in millions of bands. The only way to study a specific protein or nucleic acid sequence using gel electrophoresis, therefore, would be to find a way to label it specifically so that it could be differentiated from the general background.

BASIC BLOTTING TECHNIQUES

In 1975, Ed Southern developed a method that allowed the detection of specific DNA sequences after they had been separated by agarose gel electrophoresis. What makes a piece of DNA unique is the sequence of the nucleotides. This is most efficiently detected by the hybridization of the antiparallel strand. This can only occur if the two strands are separated into single strands. Therefore, the first step is to soak the agarose gel in a strong base, such as 1 molar sodium hydroxide, and high salt, which stabilizes the single-stranded form. The base is then neutralized with a strong buffer, such as tris-hydrochloride, again in high salt. The DNA can now be analyzed by its ability to hybridize to a radioactive piece of single-stranded DNA. Since this radioactive DNA can "explore" the different sequences to find the one matching sequence, it is also known as a probe (an instrument or device that can be used to explore and send back information).

Although the agarose is porous, it would be very slow and inefficient to try to perfuse the gel with radioactive probe and then remove the pieces that did not hybridize. Southern realized that he needed to move the DNA to a thin material to be able to probe it efficiently. The material chosen was nitrocellulose, consisting of a variant of paper (cellulose) with reactive nitro groups attached. The treated gel is placed on a sponge soaked with a high-salt solution. The nitrocellulose sheet is placed onto the gel and then a stack of dry paper towels is laid on top. The

salt solution is drawn through the gel to the dry towels and carries the DNA from the gel up into the paper. The positively charged nitro groups on the nitrocellulose stick to the negatively charged DNA, thereby holding the DNA in a pattern matching the band locations in the gel. The nitrocellulose is removed from the gel and baked at 80 degrees Celsius (176 degrees Fahrenheit) or treated with ultraviolet light, both of which covalently cross-link the DNA to the paper, locking it in its position. The filter is soaked in a solution that promotes reassociation of single-stranded DNA, and radioactive, single-stranded DNA is added. Since the added DNA could stick nonspecifically to the nitrocellulose, the paper is pretreated with unrelated DNA, such as sheared salmon DNA, which will bind the available nitro groups but not react with the probe.

A large molar excess of probe must be used to drive the hybridization reaction (reforming the "hybrid" of two matching antiparallel strands together), which means that it is necessary to make sure that enough probe is available in the solution to randomly run into the correct sequence on the paper and reanneal to it. The hybridization is done at an elevated temperature—often 50-65 degrees Celsius (122-149 degrees Fahrenheit), so that only strands that match exactly will stay together and those with short, random matches will come apart. After overnight hybridization, the paper is washed multiple times with a detergent-salt solution, which removes the DNA that did not hybridize. The paper is placed against a piece of X-ray film, and the radioactive emissions from the probe darken the film next to them. When the film is developed, a pattern of bands appears that corresponds to the position in the original gel of the DNA piece for which the researcher was probing.

EXPANDED TECHNIQUES TO STUDY RNA AND PROTEINS

The basic method of blotting has been expanded to include the study of RNA and proteins. James Alwine developed a very similar method to transfer messenger RNA (mRNA) that had been separated on an agarose gel. Since the mRNA started as single-stranded, there was no need to treat the gel with denaturant. However, to block the formation of internal double-stranded regions, which could alter the migration during electrophoresis, the gel contained an organic solvent. Other than that, the two methods are very similar. Although the DNA transfer system was named the Southern blot in honor of Ed Southern, Alwine decided to defer the credit and called his system the Northern blot to indicate that it was related but in a different direction.

Similarly, when W. N. Burnette developed a system for transferring and detecting specific proteins, he named the system Western blotting. This system of naming has been expanded: A technique for detecting viral DNA in tree leaves was named the Midwestern blot and a variant of the Northern blot developed in Israel was named the Middle Eastern blot.

Since proteins are generally smaller than DNA fragments, they are usually separated on polyacrylamide gels, which have a much smaller pore size than agarose gels. It is therefore necessary to use electrical current to pull the proteins out of the gel. The nitrocellulose is pressed onto the gel with a porous plastic pad. The gel is then placed in a buffer tank and electrodes are placed on either side. When a voltage is applied, the current that flows through the gel carries the proteins onto the nitrocellulose. The reactive side chains of the nitrocellulose also bind proteins very effectively, so they are all retained on the paper. The specific probe used to detect a protein is an antibody that either can be radioactively labeled or can have an enzymatic side chain attached, which will produce light or a colored dye when the appropriate chemicals are added. Since the antibody is a protein, it could also stick nonspecifically to the paper, so the blot is pretreated with a general protein such as serum albumin before the antibody is added.

BLOTTING IN GENETIC ANALYSIS

The ability to detect individual molecules in a large background has been very important for genetic analyses. For instance, restriction fragment length polymorphism (RFLP) analysis is a method that uses the change in the size of a DNA fragment in the genome, generated by restriction enzyme digestion as a genetic marker. The isolation of many disease genes, including the one causing Huntington's disease, depended on RFLP mapping to localize the gene. It would not be possible to detect the changes in a single DNA fragment out of the millions generated by digesting the human genome without having the Southern blot to pick out the correct piece. Many other mutations that change a

specific region of DNA—such as deletions, inversions, and duplications—are often detected by changes in a Southern blot pattern. The sensitivity of hybridization can be tuned to a level where probes that differ by only a single nucleotide will not attach efficiently. This allows the rapid identification of the positions of point mutations. When polymerase chain reaction (PCR) is used to amplify DNA from a crime scene or to detect human immunodeficiency virus (HIV) in the bloodstream, the presence of DNA pieces on a gel is not sufficient proof that the correct DNA has been found. The DNA must be blotted and probed with the expected sequence to confirm that it is the correct piece.

Northern blot analysis allows scientists to see how mRNA is altered in different mutants. Northern blots can indicate if a mutant allele is no longer transcribed or if the level of mRNA produced has been dramatically decreased or increased. Deletions or insertions will also show up as shortened or lengthened messages. Alternative splicing can be seen as multiple bands on a Northern blot which hybridize to the same probe. Point mutations that do not detectably alter the mRNA can still dramatically alter the protein product. Changes of a single amino acid can alter the electrophoretic mobility and the difference in apparent molecular weight can often only be detected by a Western blot. These changes can also alter protein stability, which can be detected as decreased protein levels showing up on the Western. The ability to detect changes at the DNA, RNA, and protein level through blotting techniques has greatly increased the ability of scientists to study genetic alterations.

FUTURE DIRECTIONS

Blotting techniques are the most generally efficient methods for detecting specific proteins or nucleic acids. Most improvements in the past years have been aimed at speeding up the transfer process using vacuums or pressure or the hybridization process by changing the conditions. The next step will be developing silicon chips that can interact with specific nucleic acid or amino acid sequences and produce an electrical output when they "hybridize" with the correct sequence. This will diminish the time required to confirm a sequence from several hours to minutes.

J. Aaron Cassill, Ph.D.

FURTHER READING

Alwine, J. C., D. J. Kemp, and G. R. Stark. "Method for Detection of Specific RNAs in Agarose Gels by Transfer to Diazobenyloxymethyl-Paper and Hybridization with DNA Probes." *Proceedings of the National Academy of Sciences* 74 (1977): 5350. The original description of Northern blotting.

Burnette, W. N. "Western Blotting: Remembrance of Past Things." *Methods in Molecular Biology* 536 (2009): 5-8. The scientist who devised the Western blotting technique recalls why the method was developed, how it works, and how its simplicity and relevance resulted in its expansive application as a research tool in biology and medicine.

Kurien, Biji T., and R. Hal Scofield. *Protein Blotting and Detection: Methods and Protocols.* New York: Humana Press, 2009. Researchers discuss numerous techniques based on the Western blot, providing advice for using these methods in a laboratory setting.

Southern, E. M. "Detection of Specific Sequences Among DNA Fragments Separated by Gel Electrophoresis." *Journal of Molecular Biology* 98, no. 3 (1975): 503-517. The original description of Southern blotting and of blotting in general. This is one of the most often cited articles in biology research.

Walker, John M., and Ralph Rapley, eds. *Molecular Biomethods Handbook.* 2d ed. Totowa, N.J.: Humana Press, 2008. Describes the technologies that are used to investigate and define cellular processes at the molecular level, focusing on various nucleic acid, protein, and cell-based methods.

WEB SITES OF INTEREST

Protocol Monkey.com
http://www.protocolmonkey.com/index.php
Protocol Monkey.com provides information about methods of laboratory research. The search engine will retrieve protocols about various types of blotting.

Westernblotting.com
http://www.westernblotting.org
A resource for Western blot methods, providing information about blotting protocols and research articles.

See also: Antibodies; DNA sequencing technology; Gel electrophoresis; Genetic testing; Huntington's disease; Immunogenetics; Model organisms; Polymerase chain reaction; Repetitive DNA; RFLP analysis.

Brachydactyly

CATEGORY: Diseases and syndromes
ALSO KNOWN AS: Brachymesophalangy; brachyphalangy; hypophalangia

DEFINITION

Brachydactyly describes both a morphologic feature and a group of congenital hand deformities characterized by short fingers. In most cases, the reduced finger length is attributed to an underdevelopment or absence of finger segments called phalanges, although shortening of other bones of the hands or feet (metacarpals or metatarsals) may also occur.

RISK FACTORS

The greatest risk factor for the development of brachydactyly is the inheritance of one of several disease-causing mutations. Brachydactyly can also result from embryologic disturbances and has been observed among infants exposed to drugs known to alter fetal development (teratogens); in this case, it is usually found in conjunction with other malformations.

ETIOLOGY AND GENETICS

Although brachydactyly was the first human trait to be ascribed an autosomal dominant Mendelian inheritance pattern, it has subsequently proven to be a genetic smorgasbord, demonstrating the concepts of incomplete penetrance (wherein carriers of causative mutations do not show evidence of disease), variable expressivity (differences in clinical presentation among individuals with the same mutation), and locus heterogeneity (multiple genes or gene combinations leading to the same phenotype). In addition, evidence suggests that certain forms of brachydactyly may be inherited as semidominant or autosomal recessive traits.

Brachydactyly can be inherited alone (isolated brachydactyly), in association with skeletal abnormalities, or as part of a syndrome. In 1951, Julia Bell developed a classification scheme for isolated brachydactyly based on the characteristic hand malformations found in family pedigrees. Categorization of brachydactyly still follows this general model of types A through E, with subtypes used to further delineate particular patterns of digit abnormalities. The majority of isolated brachydactyly types are very rare; however, brachydactyly types A3 and D are relatively common findings within certain populations.

Causative mutations have been identified for many, but not all, of the isolated brachydactyly types. Mutations in the Indian hedgehog (*IHH*) gene (2q35-q36) have been identified in families with brachydactyly type A1, although linkage has also been shown to a locus on 5p13.3-p13.2. Mutations in two separate genes have been associated with brachydactyly type A2: the bone morphogenetic protein receptor 1B (*BMPR1B*) gene (4q21-q20) and the growth and differentiation factor 5 (*GDF5*) gene (20q11.2). This divergence among families with a common phenotype exemplifies the genetic heterogeneity within brachydactyly.

The phenotype of patients with brachydactyly type B has been shown to correlate with the nature of the mutation in the receptor kinase-like orphan receptor 2 (*ROR2*) gene (9q22). Mutations in *ROR2* have also been identified in patients with autosomal recessive Robinow syndrome. More recently, mutations in the noggin (*NOG*) gene (17q22) have been identified in patients with brachydactyly type B for whom *ROR2* mutations were not detected.

The inheritance pattern of brachydactyly type C is not straightforward and has been suggested to be autosomal dominant, autosomal recessive, or semidominant. As observed for brachydactyly type A2, mutations in *GDF5* have been identified in families with brachydactyly type C.

Both brachydactyly types D and E have been linked to mutations in the homeobox-containing (*HOXD13*) gene (2q31-q32).

SYMPTOMS

Isolated brachydactyly is characterized by shortening of one or more digits and may affect the hands, feet, or both. Other finger abnormalities, including syndactyly (fused digits), clinodactyly (sideways deviation of the finger), or symphalangism (fused phalanges), may also be present. Syndromic forms of brachydactyly may be associated with skeletal defects (such as short stature, shortened limbs, and scoliosis), hypertension, cardiac malformations, mental retardation, or a host of other abnormalities.

SCREENING AND DIAGNOSIS

Family history is a strong predictor of disease. The benign nature of isolated brachydactyly makes prenatal screening unnecessary, although it may be valuable for syndromic forms of the disease. Prenatal ultrasound performed from twelve to seventeen weeks of gestation has been used to successfully diagnose brachydactyly. Diagnosis based on analysis of DNA from the fetus is possible if the familial mutation is known.

TREATMENT AND THERAPY

Plastic surgery is an option to enhance hand function but is not applicable in most cases. If needed, hand function may also be improved through physical therapy. For those with syndromic brachydactyly, treatment of associated conditions (such as blood pressure medication for patients with hypertension) may be indicated.

PREVENTION AND OUTCOMES

There is currently no method of preventing brachydactyly occurrence among individuals who inherit disease-causing mutations. The prognosis for patients with isolated brachydactyly is generally favorable; the ability to achieve normal hand function is reliant on the extent and severity of the defect. In cases of syndromic brachydactyly, prognosis is influenced by the nature of the associated conditions.

Crystal L. Murcia, Ph.D.

FURTHER READING

Everman, David B. "Hands and Feet." In *Human Malformations and Related Anomalies*, edited by Roger E. Stevenson and Judith G. Hall. 2d ed. Oxford, England: Oxford University Press, 2005. A source for in-depth descriptions of structural and genetic findings associated with the various brachydactyly types.

Firth, Helen V., Jane A. Hurst, and Judith G. Hall, eds. *Oxford Desk Reference: Clinical Genetics*. Oxford, England: Oxford University Press, 2005. A quick reference guide that provides an overview of clinical evaluation and classification of brachydactyly.

Temtamy, Samia A., and Mona S. Aglan. "Brachydactyly." *Orphanet Journal of Rare Diseases* 3 (2008): 15. This review provides insights into the commonalities and differences among the isolated brachydactyly types.

WEB SITES OF INTEREST

Birth Disorder Information Directory: Brachydactyly
http://www.bdid.com/brachy.htm

Genetics Home Reference
http://ghr.nlm.nih.gov

Geneva Foundation for Medical Education and Research
http://www.gfmer.ch/genetic_diseases_v2/
gendis_detail_list.php?cat3=650

Online Mendelian Inheritance in Man
http://www.ncbi.nlm.nih.gov/sites/
entrez?db=OMIM

See also: Apert syndrome; Carpenter syndrome; Congenital defects; Hereditary diseases; Polydactyly.

BRAF gene

CATEGORY: Genetic engineering and biotechnology; Molecular genetics

SIGNIFICANCE: The *BRAF* gene is part of the tightly regulated RAS/MAPK pathway and instructs protein development that transmits chemical signals from outside the cell to the nucleus of the cell. *BRAF* is an oncogene, and mutations in *BRAF* are evident in cancers and other syndromes. As part of the RAS/MAPK pathway, it contributes to cell activities related to growth, proliferation, differentiation, survival, and apoptosis.

KEY TERMS

amino acids: form the structure of proteins in the body

apoptosis: cell self-destruction

differentiation: a cell's ability to develop a more specialized function

gene mutation: a permanent change in the sequence of DNA

missense mutation: introduction of an incorrect amino acid into a protein

oncogene: mutations in normal cellular genes

proteins: made up of amino acids, provide or support almost all chemical processes in cells

RAS/MAPK pathway: controls cellular proliferation and differentiation

somatic mutation: a noninherited change in genetic structure that is not passed to offspring

THE *BRAF* GENE

A gene is a subunit of DNA that carries specific instructions for cells, with approximately 25,000 genes in the human body. *BRAF* is the gene symbol for v-raf murine sarcoma viral oncogene homolog B1. The *BRAF* gene is expressed in most tissues, especially neuronal tissue. It belongs to the oncogene class of genes, contains 766 amino acids, has a molecular weight of 84436 daltons, and is located on chromosome 7 at position 34, or more specifically, from base pair 140,080,750 to base pair 140,271,032 on chromosome 7. As a serine/threonine kinase in the RAS/RAF/MEK/ERK/MAPK pathway, *BRAF* relays mitogenic signals to the nucleus of the cell from the cell membrane. A normal *BRAF* gene switches on and off to deliver appropriate proteins necessary in the cell cycle of growth and development. Within the MEK/ERK pathway, it assumes an antiapoptotic role necessary to regulate cell death. A mutation in a gene may be caused by environmental exposures, or it can occur if a mistake is made during cell division. The most common *BRAF* mutation is *V600E*, a transversion mutation at T1799A that represents 80 percent of *BRAF* mutations.

IMPLICATIONS OF *BRAF* MUTATION

Somatic mutations in the *BRAF* gene, or noninherited changes in the body's cells during an individual's lifetime, are seen in several cancers. A somatic mutation of *BRAF* leads to an overexpression of protein that interferes with the normal cell cycle, which may lead to the overactive cell growth evident in the development of cancer. The most common cancers demonstrating a mutation in *BRAF* include malignant melanoma (70 to 80 percent), colorectal cancer (5 to 40 percent, depending on the mutation), ovarian cancer (approximately 30 percent of low-grade serous tumors), and thyroid cancer (a *V600E* mutation evident in half of cases). Breast and lung cancers show *BRAF* mutations, but in fewer cases. *BRAF* testing provides information related to prognosis, particularly in colorectal and metastatic colorectal cancers.

Cardiofaciocutaneous syndrome is an extremely rare, autosomal dominant condition that results from gene mutations, with approximately 75 percent of cases caused by mutation in the *BRAF* gene. According to the National Institutes of Health (NIH), there are less than two hundred cases in the United States, but that number may be an underestimation since mild cases are often not reported. The disorder is caused by overactive expression of protein resulting in alterations in cell communication during development of the fetus. Individuals with cardiofaciocutaneous syndrome have a variety of body changes including heart defects; distinctive facial features with a high forehead, a short nose, droopy eyelids and down-slanting eyes, and low ear placement on the head; a large head, known as macrocephaly; dry, rough skin with small bumps, known as keratosis pilaris; and thin or missing hair, eyebrows, and eyelashes. At birth, infants exhibit poor muscle tone and failure to thrive because of feeding problems, resulting in a lack of growth and weight gain, and they may have seizures. Cancer is not usually seen in relationship to this syndrome.

FUTURE EFFORTS RELATED TO *BRAF*

Because identifiable mutations of the *BRAF* gene may be associated with specific cancers, testing for *BRAF* mutations holds promise in cancer treatment, and commercial applications are being developed. Studies are being conducted that attempt to measure *BRAF* mutations and their relationship to clinical tumor stages in an attempt to define prognostic factors and, eventually, treatment strategies. Gene testing is reliable and low cost, and is likely to become a standard of care in some cancers.

Studies are currently being conducted that indicate that *BRAF* positive tumors may be more sensitive to certain drugs. Additional studies are addressing the potential for anticancer agents or immunotherapy applications targeting the *BRAF* gene, including the potential for antiangiogenic activity since *BRAF* inhibition causes cell cycle arrest and death through apoptosis. The missense mutation *V600E* has been shown to be sensitive to treatment in early laboratory studies through inhibition of the kinase in select melanoma cell lines leading to cell death. Investigation into the management of *BRAF* mutations holds promise for therapeutic interventions in multiple diseases and syndromes.

Patricia Stanfill Edens, Ph.D., R.N., FACHE

FURTHER READING

Kim, I. J., et al. "Development and Applications of a *BRAF* Oligonucleotide Microarray." *Journal of Molecular Diagnostics* 9, no. 1 (February, 2007): 55-63.

Pratilas, C. A., and D. B. Solit. "Therapeutic Strategies for Targeting *BRAF* in Human Cancer." *Reviews on Recent Clinical Trials* 2, no. 2 (May, 2007): 121-134.

Taube, J. M., et al. "Benign Nodal Nevi Frequently Harbor the Activating V600E *BRAF* Mutation." *American Journal of Surgical Pathology* 33, no. 4 (April, 2009): 568-571.

WEB SITES OF INTEREST

National Cancer Institute. Understanding Cancer Series: Gene Testing
http://www.cancer.gov/cancertopics/ understandingcancer/genetesting/allpages

National Center for Biotechnology Information
http://www.ncbi.nlm.nih.gov

See also: Cancer; Hereditary diseases; Mutagenesis and cancer; Mutation and mutagenesis; Oncogenes.

BRCA1 and *BRCA2* genes

CATEGORY: Classical transmission genetics; Molecular genetics

SIGNIFICANCE: Breast cancer, the most common cancer found in women in the United States, can be inherited. *BRCA1* and *BRCA2* are two genes linked directly with susceptibility to developing breast cancer, as well as to developing ovarian, prostate, and other types of cancer. Mutations in these genes can eliminate their ability to control cell growth.

KEY TERMS

chemoprevention: using natural or synthetic chemicals to reduce the risk of developing cancer

DNA: deoxyribonucleic acid, the carrier of genetic information in cell nuclei

gene therapy: repairing or manipulating genes by insertion of DNA to reduce the risk of cancer

mutation: alteration in the normal DNA pattern or chemical sequence along a gene

prophylactic surgery: removing tissue that enhances the risk of developing cancer

tumor-suppressor gene: genes that control cell growth and cell death

GENETICS

BRCA1 and *BRCA2* genes are tumor-suppressor genes that produce proteins which help repair any damage to the genetic information in a cell and halt abnormal cell growth. If these genes are mutated, then the DNA repair function is usually lost. Mutations in the *BRCA1* and *BRCA2* genes are transmitted in an autosomal dominant pattern in a family. In the early 1990's, it was determined that mutations in the *BRCA1* gene, located on chromosome 17, increase the risk of breast cancer. Shortly thereafter, breast cancer was also linked to mutations in the *BRCA2* gene, located on chromosome 13. Most breast cancers are not caused by inherited changes in genes, but for those that are, about a third are due to *BRCA1* mutations, another third to *BRCA2* mutations, and a third due to mutations in other genes that are being identified.

CANCER RISK

In addition to an increased risk for breast cancer in men and women, inherited mutations in either of the *BRCA* genes also significantly increase the risk for a woman to develop ovarian cancer. Mutations in the *BRCA1* gene also increase the risk for prostate cancer in men and for colon cancer, while mutations in the *BRCA2* gene have also been linked to increased risk of malignant melanoma; cancers of the pancreas, colon, gallbladder, and stomach; and prostate and breast cancer in men. Estimates of lifetime risk for breast cancer in women with an altered *BRCA1* or *BRCA2* gene is 45 to 85 percent, as compared to about 13 percent for women in the general population.

GENETIC TESTING

If a family has a strong history of breast and/or ovarian cancer, genetic testing that identifies mutations in the *BRCA* genes may be beneficial. A family member who has been diagnosed with breast or ovarian cancer provides a blood sample. DNA sequencing analyzes the DNA pattern of the *BRCA1* and *BRCA2* genes and compares it to the normal sequence in these genes. If a mutation is found in the DNA sequence of one of these genes, then it is likely that the tested person's cancer was caused by an altered *BRCA* gene. Other family members can then be tested for this particular gene change and the risk for that individual developing breast or ovarian cancer can be gauged.

MANAGEMENT AND THERAPY

Individuals with mutated *BRCA1* or *BRCA2* genes can manage their increased cancer risk through several approaches. Early diagnosis of breast cancer can be increased through periodic mammography, clinical breast exams, ultrasound, and breast self-exams. In some cases, prophylactic surgery is recommended to remove as much of the at-risk tissue as possible. To reduce the risk of developing cancer, or to reduce the risk of cancer reoccurring, chemoprevention (chemotherapy) is often employed. Progress is being made in using gene therapy to repair mutated genes that increase the risk for cancer.

IMPACT

Identification and isolation of the *BRCA1* and *BRCA2* genes proved that breast cancer, as well as some other types of cancer, can be inherited. DNA analysis to identify acquired *BRCA* gene mutations can help doctors more accurately predict survival of women with breast and/or ovarian cancer and implement proper treatment to help control the disease. *BRCA* gene abnormalities are found most commonly among younger women under the age of forty. Investigation of *BRCA* gene mutations has not only generated increased interest in breast and ovarian cancer research but also initiated research into using gene therapy as a treatment for cancer and motivated geneticists to escalate their search for genetic-related links as the source of other diseases.

Alvin K. Benson, Ph.D.

FURTHER READING

Bolin, Robert B. *Unwanted Inheritance.* Bloomington, Ind.: iUniverse, 2007. Case study of a mother and her daughters dealing with the adverse effects of *BRCA* gene mutations.

Greene, Diane Tropea. *Apron Strings: Inheriting Courage, Wisdom and . . . Breast Cancer.* Highland City, Fla.: Rainbow Books, 2007. True story of a family devastated by breast cancer caused by the *BRCA2* gene mutation.

Lu, Karen H., ed. *Hereditary Gynecologic Cancer: Risk, Prevention and Management.* London: Informa HealthCare, 2008. Contains an overview of hereditary gynecological cancers, as well as detailed information about hereditary breast cancer.

McPhee, Stephen, and Maxine Papadakis. *Current Medical Diagnosis and Treatment,* 48th ed. New York: McGraw-Hill Professional, 2008. Contains insights into the signs, symptoms, epidemiology, etiology, and treatment of breast cancer.

Zimmerman, Barbara T. *Understanding Breast Cancer*

Genetics. Jackson: University Press of Mississippi, 2004. A description of current breast cancer research that includes studies of the relationship between genetics and the environment.

WEB SITES OF INTEREST

Imaginis: The Women's Health Resource on the Web
http://www.imaginis.com/breasthealth/
genetic_risks.asp

Mayo Clinic: "Information on BRCA Gene Mutations"
http://www.mayoclinic.com

National Institutes of Health. National Cancer Institute
http://www.cancer.gov/cancertopics/prevention
-genetics-causes/genetics

See also: Breast cancer; Cancer; Genetic screening; Genetic testing; Genetic testing: Ethical and economic issues; Hereditary diseases; Human Genome Project; Mutagenesis and cancer; Mutation and mutagenesis; Oncogenes; Ovarian cancer; Tumor-suppressor genes.

Breast cancer

CATEGORY: Diseases and syndromes
ALSO KNOWN AS: Ductal carcinoma; lobular carcinoma

DEFINITION

Approximately one in eight women develops breast cancer over the course of her lifetime. In the United States there are approximately 180,000 new cases of breast cancer yearly. More than forty different genes have been found to be altered in breast cancers. It is estimated that about 5 to 10 percent of all breast cancers can be attributed to inherited gene mutations. Approximately 80 to 85 percent of these can be attributed to mutations in the *BRCA1* or *BRCA2* gene. Other gene mutations associated with a high risk of breast cancer include *TP53*, *PTEN*, *STKII/LKB1*, and *CDH1*. Genes associated with a low-to-moderate risk of breast cancer include *ATM* and *CHEK2*. Each of these gene mutations is associated with a different disease or syndrome. *BRCA1* and *BRCA2* are associated with hereditary breast and ovarian cancer; *TP53* and *CHEK2* with Li-

Fraumeni syndrome, *PTEN* with Cowden's disease, *STKII/LKB1* with Peutz-Jeghers syndrome, *CDH1* with hereditary diffuse gastric carcinoma syndrome, and *ATM* with ataxia telangiectasia.

RISK FACTORS

Female gender and increasing age are considered risk factors for breast cancer. Modifiable risks include lifestyle choices that effect exposure to endogenous estrogens or environmental toxins. Family history of breast cancer is also a risk factor.

ETIOLOGY AND GENETICS

The *BRCA1* gene is located on chromosome 17q21 and encodes a protein that is 1,863 amino acids long. Germ-line mutations of *BRCA1* are associated with 50 percent of hereditary breast cancers and with an increased risk of ovarian cancer.

The *BRCA2* gene is on chromosome 13q12-13 and encodes a protein of 3,418 amino acids. Germ-line mutations of *BRCA2* are thought to account for approximately 35 percent of families with multiple-case, early-onset female breast cancer. Mutations of *BRCA2* are also associated with an increased risk of male breast cancer, ovarian cancer, prostate cancer, and pancreatic cancer.

Although *BRCA1* was cloned in 1994 and *BRCA2* in 1995, the function of these genes has been difficult to identify. Part of the difficulty has been that the proteins coded by these genes do not resemble any proteins of known function. In 1997, David Livingston and coworkers of the Dana-Farber Cancer Institute found that the *BRCA1* gene product associates with repair protein RAD51. A few months later, Allan Bradley of Baylor College of Medicine and Paul Hasty of Lexicon Genetics reported that the BRCA2 protein binds to the RAD51 repair protein. This work suggests that both genes may be in the same DNA-repair pathway. Bradley and Hasty also showed that embryonic mouse cells with inactivated mouse *BRCA2* genes are unable to survive radiation damage, again suggesting that the *BRCA* genes are DNA-repair genes. Initially, it was thought that the breast cancer genes were typical tumor-suppressor genes that normally function to control cell growth. The 1997 work suggests that the breast cancer gene mutations act indirectly to disrupt DNA repair and allow cells to accumulate mutations, including mutations that allow cancer development. In 2002 the detailed structure of the

Discoveries of Breast Cancer Genes

Prior to the discoveries of *BRCA1* and *BRCA2*, there were many hints that susceptibility to at least some breast cancers was inherited. The time line below shows some of the discoveries leading up to the discoveries of *BRCA1* and *BRCA2* as well as later discoveries about breast cancer genes.

1966   Henry Lynch began the first studies on inherited cancers.

1970   The first cancer-causing gene (oncogene) was reported in chickens by Peter Vogt.

1976   J. Michael Bishop and Harold Varmus reported the discovery of oncogenes in the DNA of normal chromosomes.

1978   M. H. Bronstein et al. see a link between Cowden disease, an inherited tumorogenic syndrome, and breast cancer.

1979   Arnold Levine and David Baltimore discover *p53*, a gene mutated in approximately half of all known cancers, including breast cancer.

1985   The mutant *p53* gene is cloned by Arnold Levine.

1987   Michael Swift et al. report a hereditary link between ataxia telangiectasia mutated (*ATM*) and many cancers, including breast cancer.

1988   Dennis Slamon reports that the *HER-2/neu* growth factor gene is overexpressed in 30 percent of the most aggressive breast cancers.

1990   Mary-Claire King and coworkers report the discovery of *BRCA1* in Ashkenazi Jewish women and locate it on chromosome 17.

1990   David Malkin et al. report a link between the *p53* gene product and breast cancer.

1991   Elizabeth Claus et al. do a statistical analysis of familial breast cancer and predict a dominant breast cancer gene will be found.

1993   Theodore Krontiris et al. report an association between *HRAS1* (Harvey rat sarcoma oncogene 1) and breast cancer.

1994   Yoshio Miki et al. announce the cloning of *BRCA1* on chromosome 17.

1995   Richard Wooster et al. announce the discovery and cloning of *BRCA2* on chromosome 13.

1996   Prasanna Athma et al. report that heterozygotes for the recessive allele *ATM* are more susceptible to breast cancer.

1997   Danny Liaw et al. find that germ-line mutations in the *PTEN* gene lead to Cowden disease and associated breast cancer.

1998   Dennis Slamon tests Herceptin, a monoclonal antibody that targets the product of *HER-2/neu*, against aggressive breast cancers.

1999   François Ugolini et al. implicate *FGFR1* (fibroblast growth factor receptor gene 1) in some breast cancers.

2000   Tommi Kainu et al. propose a *BRCA3* gene to explain non-*BRCA1/BRCA2* hereditary breast cancers in several families.

2001   Paul Yaswen reports that multiple copies of the gene *ZNF217* are seen in 40 percent of breast cancers.

2001   Minna Allinen et al. find a mutation in the *CHEK2* gene that leads to hereditary breast cancers. This is proposed as *BRCA3*.

2002   Alan D'Andrea et al. report that the same inherited mutations in the six genes that cause Fanconi anemia also increase the susceptibility to breast cancer.

Richard W. Cheney, Jr., Ph.D.

BRCA2 protein was determined. It has structural motifs that show it to be capable of binding to DNA. Although the specific role of the BRCA2 protein is uncertain, it is now clear that it does play a role in repairing double-stranded breaks in DNA. The understanding of the function of *BRCA1* and *BRCA2* is incomplete, but what is known will encourage additional studies.

The *TP53* gene was the first gene identified associated with inherited breast cancer The gene is located on chromosome 17p13.1. It is a tumor-suppressor gene that encodes a protein transcription factor that stops the cell cycle until DNA repair has occurred; a defective p53 protein no longer stops cell division, and unrepaired DNA can be replicated, resulting in accumulated mutations in the cell. About 1 percent of women who develop breast cancer before the age of thirty have germ-line mutations in p53. Families with this syndrome have extremely high rates of brain tumors and other cancers in both children and adults.

The *PTEN* gene is located on 10q23.3 and is also a tumor-suppressor gene. It encodes an enzyme that modifies proteins and fats by removing phosphate groups. More than a hundred mutations in *PTEN* have been identified associated with Cowden syndrome. The mutation results in a defective phosphatase enzyme resulting in noncancerous growths (hamartomas) as well as cancerous tumors including breast cancer, prostate cancer, endometrial cancer, skin cancer, and brain tumors.

The *STKII* (serine/threonine kinase II) gene is located at chromosome 19p13.3 and encodes a tumor-suppressor enzyme. More than 140 mutations have been identified associated with Peutz-Jeghers syndrome. The loss of this enzyme function is associated with polyps in the gastrointestinal tract that can become cancerous. This same loss of tumor-suppression function is associated with increased risk for breast cancer.

The *CDH1* gene is located on chromosome 16q22.1 and encodes an epithelial cadherin protein. E-cadherin helps cells stick together. An inherited mutation in *CDH1* increases the risk of cancer of the milk-producing glands associated with hereditary diffuse gastric cancer (HDGC).

There is an increased incidence of breast cancer associated with the ataxia telangiectasia *AT* gene and the *HRAS1* gene. A mutated form of the gene, called *ATM* (ataxia telangiectasia mutated), is located on chromosome 11q22-23 and codes for a serine/threonine-specific protein kinase that plays a role in DNA damage repair. The *ATM* gene is found in the rare recessive hereditary disorder ataxia telangiectasia, which has a very wide range of symptoms, including cerebellar degeneration, immunodeficiency, balance disorder, high risk of blood can-

cers, extreme sensitivity to ionizing radiation, and an increased risk of breast cancer. Individuals with one mutated copy of the *ATM* gene have an increased risk of cancer. The *ATM* gene was identified as a phosphatidylinositol-3 kinase (an enzyme that adds a phosphate group to a lipid molecule) that transmits growth signals and other signals from the cell membrane to the cell interior. The *ATM* gene was found to be similar in sequence to other genes that are known to have a role in blocking the cell cycle in cells whose DNA is damaged by ultraviolet light or X rays. It is possible that the mutated *ATM* gene does not stop the cell from dividing, and the damaged DNA may lead to cancers. It is disturbing to note that individuals with a mutated *ATM* gene may be more sensitive to ionizing radiation and

Dr. Robert Cardiff of the University of California at Davis holds a genetically engineered mouse used to study breast cancer. (AP/ Wide World Photos)

should therefore avoid low X-ray doses, such as those received from a mammogram used to detect the early stages of breast cancer.

SCREENING AND DIAGNOSIS

A simple blood test can check for *BRCA1* and *BRCA2* mutations. Such testing has been controversial, however, raising a number of social and psychological issues. There is a concern that the technical ability to test for genetic conditions is ahead of the ability to predict outcomes or risks, prescribe the most effective treatment, or counsel individuals. Part of the dilemma about testing is the uncertainty

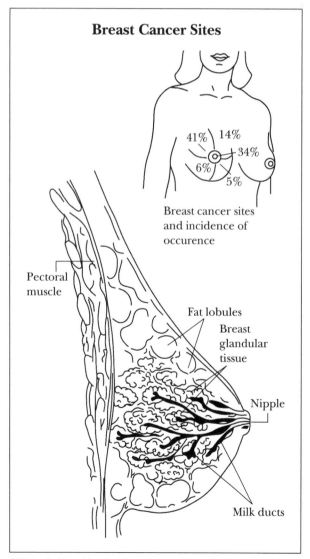

Breast Cancer Sites

Breast cancer sites and incidence of occurence

Pectoral muscle

Fat lobules

Breast glandular tissue

Nipple

Milk ducts

(Hans & Cassidy, Inc.)

about the meaning of the test results. If a test confirms the presence of a mutation in a breast cancer gene in a woman with a family history of breast cancer, then there is a high risk—but not a certainty—that the woman will develop breast cancer. Even if a test is negative, it does not mean the woman is not at risk for breast cancer, because the large majority of breast cancers are not inherited. If a test is positive, then it is not clear what the best course for the woman would be. Increased monitoring with mammography and even removal of both breasts as a preventive measure should reduce the chances of developing cancer but do not guarantee a cancer-free life. Even if a woman does not yet have cancer, she may feel the additional psychological stress of knowing she has a high risk of developing it.

There is also concern that test results may be misused by employers or insurers. A number of states have passed laws that prevent health insurance companies from using genetic test results to discriminate against patients. In 2008, the federal government passed the Genetic Information Nondiscrimination Act (GINA), which prohibits discrimination based on genetic information by employers and insurance companies with the exception of life insurance, disability insurance, and long-term care insurance. In 1996, the National Cancer Institute established the Cancer Genetics Network as a means for individuals with a family history of cancer to enroll in research studies and learn of their genetic status while receiving counseling.

SYMPTOMS

The symptoms of hereditary breast cancer are similar to somatic breast cancer: a lump or thickening in the breast or lymph nodes under the armpits; change in the size, shape, or feel of the breast or nipple; discharge from the nipple.

TREATMENT AND THERAPY

Treatment for hereditary breast cancer is similar to somatic breast cancer. It depends on the type and stage of the cancer, and if it is hormone-sensitive estrogen receptor (ER) positive or human epidermal growth factor receptor 2 (HER2) positive.

PREVENTION AND OUTCOMES

Some patients want to reduce their risk of breast cancer and choose to have preventive or prophylactic treatment. Cancer screening is a way to detect breast cancer early when it may be easier to treat. Mammography and clinical breast exams are two common screening methods. Magnetic resonance imaging (MRI) is also used. Prophylactic surgery is a preventive option. Bilateral prophylactic mastectomy (removal of healthy breasts) or prophylactic salpingo-oophorectomy (removal of healthy Fallopian tubes and ovaries) are two options. Some women choose to have both procedures. Chemoprevention is another strategy. Two drugs have been approved for this use: tamoxifen and raloxifene. Tamoxifen has been shown to reduce the risk of breast cancer in premenopausal and postmenopausal women. Raloxifene is approved for use in postmenopausal women.

Susan J. Karcher, Ph.D., and Bryan Ness, Ph.D.;
updated by Sandra Ripley Distelhorst

FURTHER READING

Abeloff, Martin D., et al., eds. *Abeloff's Clinical Oncology.* 4th ed. Philadelphia: Churchill Livingstone/Elsevier, 2008. See chapter 95, "Cancer of the Breast."

Bowcock, Anne M., ed. *Breast Cancer: Molecular Genetics, Pathogenesis, and Therapeutics.* Totowa, N.J.: Humana Press, 1999. Detailed information geared toward researchers and health professionals. The chapter "Hereditary Breast Cancer Genes" discusses *BRCA1* and *BRCA2* mutations among Ashkenazi women. Also addresses surgery, chemotherapy, drug resistance, and the *MDR* gene.

Dickson, Robert B., and Marc E. Lipman, eds. *Genes, Oncogenes, and Hormones: Advances in Cellular and Molecular Biology of Breast Cancer.* Boston: Kluwer Academic, 1992. Contains papers on the genetics and molecular biology of breast cancer, including the role of suppressor genes, the role of the retinoblastoma gene, oncogenes and stimulatory growth factors, and much more. Index.

Gronwald, J., N. Tung, W. D. Foulkes, et al. "Tamoxifen and Contralateral Breast Cancer in *BRCA1* and *BRCA2* Carriers: An Update." *International Journal of Cancer* 118, no. 9 (2006): 2281-2284. An article from the Hereditary Breast Cancer Clinical Study Group.

Kemeny, Mary Margaret, and Paula Dranov. *Beating the Odds Against Breast and Ovarian Cancer: Reducing Your Hereditary Risk.* Reading, Mass.: Addison-Wesley, 1992. Designed for women with a family history of breast or ovarian cancer who are motivated to evaluate risk factors, nutrition, warning signs, and options for treatment.

Love, Susan M., with Karen Lindsey. *Dr. Susan Love's Breast Book.* Illustrations by Marcia Williams. 3d ed. New York: Perseus, 2000. Perhaps the most comprehensive book on breast health, including breast cancer. One chapter discusses the genetic risks for breast cancer.

Lynch, Henry T. *Genetics and Breast Cancer.* New York: Van Nostrand Reinhold, 1981. One of the seminal works on genetic breast cancer, by a pioneer in the investigation of hereditary breast-ovarian cancer syndrome.

Lynch, Henry T., E. Silva, C. Synder, and J. F. Lynch. "Hereditary Breast Cancer: Part I—Diagnosing Hereditary Breast Cancer Syndromes." *The Breast Journal* 14, no. 1 (2008): 3-13. Discusses rates and trends in hereditary breast cancer.

National Cancer Institute. *Genetic Testing for Breast Cancer: It's Your Choice.* Bethesda, Md.: Author, 1997. One of the National Cancer Institute's large number of pamphlets and monographs on various cancers, including genetic risks for cancer, designed to provide responsible and detailed information to the public.

Yang, Haijuan, et al. "*BRCA2* Function in DNA Binding and Recombination from a *BRCA2*-DSS1-ssDNA Structure." *Science* 297 (September 13, 2002): 1837-1848. This study presents evidence that the failure of *BRCA2* in DNA repair through homologous recombination may account for unsuppressed tumor growth.

WEB SITES OF INTEREST

American Cancer Society, All About Breast Cancer
http://www.cancer.org
Searchable information on breast cancer, including an overview, a detailed guide, and practical resources.

National Cancer Institute, National Institutes of Health
http://www.nci.nih.gov/breast
Provides information on the genetics of breast cancer and useful links.

National Institutes of Health, National Library of Medicine. Genetics Home Reference
http://www.nlm.nih.gov
This site includes information on breast cancer genetics.

National Women's Health Information Center
http://womenshealth.gov
Searchable information on breast disease, including breast genetics.

See also: Aging; *BRCA1* and *BRCA2* genes; Cancer; Cell cycle; DNA repair; Genetic counseling; Genetic screening; Genetic testing; Genetic testing: Ethical and economic issues; Hereditary diseases; Human Genome Project; Model organism: *Mus musculus*; Mutagenesis and cancer; Mutation and mutagenesis; Oncogenes; Tumor-suppressor genes.

Burkitt's lymphoma

CATEGORY: Diseases and syndromes
ALSO KNOWN AS: Burkitt lymphoma; Burkitt's tumor; malignant lymphoma Burkitt's type

DEFINITION

Burkitt's lymphoma is a rapidly proliferating non-Hodgkin's lymphoma and the most common malignant tumor among children and young adults in Central Africa and New Guinea. Worldwide it accounts for about 30 to 40 percent of all childhood lymphomas. Three types are differentiated: endemic (African), sporadic (occurs throughout the world), and immunodeficiency-associated (most often seen in AIDS patients). Genetically, Burkitt's lymphoma is defined by a chromosomal translocation of the proto-oncogene *c-myc* to one of the immunoglobulin (Ig) heavy- or light-chain loci. In addition, most endemic and many sporadic and immunodeficient Burkitt's lymphoma cases are associated with the Epstein-Barr virus (EBV), and the African endemic Burkitt's lymphoma type is furthermore associated with malaria.

RISK FACTORS

The endemic Burkitt's lymphoma type usually affects children aged five to ten. A suggested risk factor for the endemic form is chronic infection with malaria, or some other infectious agent carried by mosquitoes, combined with inadequate medical care and late diagnosis and treatment of the disease. Pediatric patients with sporadic Burkitt's lymphoma are usually slightly older than those with the endemic form. Adults diagnosed with the disease are commonly between thirty and fifty years old. Burkitt's lymphoma is one of the most common types of lymphoma seen in AIDS patients. Risk factors for the sporadic or immunodeficiency-associated forms include lifestyle behaviors that increase the risk of HIV infection. All three Burkitt's lymphoma types are more prevalent in males.

SYMPTOMS

In sporadic cases in children, main symptoms of Burkitt's lymphoma include abdominal pain and vomiting along with the occurrence of a large abdominal tumor accompanied by fluid (ascites) buildup. Jaw and other facial bones besides abdominal sites are most commonly involved in the endemic form. Other sites of tumor development include the central nervous system and breast. Symptoms may appear as soon as four to six weeks after the lymphoma begins. Lymphoma starting in the blood marrow might induce easy bleeding and anemia. It is essential to see a doctor as soon as symptoms occur because of the aggressive (fast-growing) nature of the tumor.

SCREENING AND DIAGNOSIS

The diagnosis of Burkitt's lymphoma is usually made by a needle biopsy from a suspected disease site such as the jaw area, abdomen (ascites), bone marrow, or a lymph node. Microscopic analysis of cell morphology is used to determine if the disease is present and, if so, its stage of development. Early clinical and laboratory diagnosis spares the child any life-threatening complications from the rapid tumor growth. Other common diagnostic tests may include a complete blood count (CBC), a platelet count, and a lumbar puncture. Further tests may include specialized radiographic exams such as a computed tomography (CT) scan to look for hidden tumor masses, as well as a gallium scan. This scan requires injection of the radioactive isotope gallium, which concentrates in areas of rapid cell division and allows for visualization of tumor sites by nuclear scan techniques. Gene expression profiling is used to accurately distinguish between Burkitt's lymphoma and diffuse large B-cell lymphoma, another type of B-cell tumor.

ETIOLOGY AND GENETICS

Burkitt's lymphoma is a monoclonal proliferation of B-lymphocytes. The common cytogenetic hallmark of all types is the reciprocal translocation of the *c-myc* proto-oncogene located on the long arm of chromosome 8 to one of the immunoglobulin (Ig) heavy- or light-chain loci on chromosomes 14 (heavy chain, more than 80 percent of cases), 22 (lambda light chain), or 2 (kappa light chain).

Proto-oncogenes like *c-myc* normally help control the cell cycle by regulating the number of cell divisions. They are especially active when high rates of cell division are needed, as in embryonic development, wound healing, or regeneration. The proto-oncogene might be transformed into an oncogene when the chromosomes break and reunite, resulting in a reciprocal translocation. The localization of the chromosomal breakpoints with respect to *c-myc* vary between the different forms of Burkitt's lymphoma, suggesting a different time point of the translocation event at different stages of B-cell development. In endemic cases the chromosomal recombination likely occurs during VDJ rearrangement in early B-cell differentiation, while in the sporadic form the translocation probably takes place during Ig class-switch events. The chromosomal rearrangement in certain translocation events results in deregulation and subsequent continuous overexpression of the *c-myc* gene due to the new location that places the gene under control of genetic enhancer elements normally involved in Ig gene regulation. Abnormal activation of the basic helix-loop-helix-leucine zipper-transcription factor MYC leads to a multitude of events, including the regulation of a large number of genes involved in cell proliferation, differentiation, apoptosis, cell cycle control, and immune response. Abnormal activation of MYC is able to trigger most characteristics of Burkitt's lymphoma cells; however, it is not sufficient by itself. Associated with MYC activation in more than 30 percent of Burkitt's lymphoma cases are mutations in *TP53*, a transcription factor involved in cell cycle arrest, DNA repair, and apoptosis and functioning as a tumor suppressor. The *c-myc* gene also frequently accumulates mutations at mutational hotspots, which may lead to an increased transforming activity. Concurrent translocations of MYC and BCL2 were described in a small subset of cases, which were associated with an especially poor prognosis.

The occurrence of Burkitt's lymphoma in patients from equatorial Africa seems to have a close correlation with the prevalence of EBV. About 95

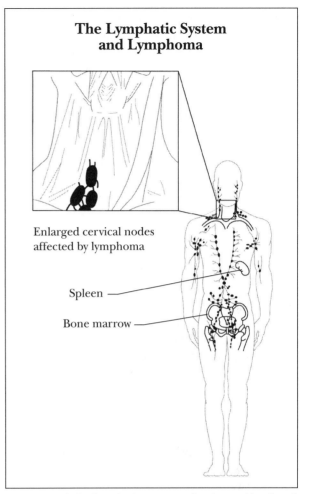

The Lymphatic System and Lymphoma

Enlarged cervical nodes affected by lymphoma

Spleen

Bone marrow

Anatomy of the lymphatic system, showing major lymph nodes. Enlarged lymph nodes may occur for a wide variety of reasons, including but not limited to lymphoma (cancer). (Hans & Cassidy, Inc.)

percent of lymphomas from equatorial Africa carry the EBV genome. By contrast, only 10 to 20 percent of sporadic cases of Burkitt's lymphoma in Europe and North America are positive for EBV, but 40 to 50 percent of HIV-infected individuals are. EBV has a single, linear, double-stranded DNA genome and was the first herpesvirus to be completely sequenced. However, EBV infection is not limited to areas where Burkitt's lymphoma is found; it infects people worldwide with a prevalance in about 95 percent of the adult population, mostly without producing symptoms. EBV is also the causative agent of infectious mononucleosis, a common disease in which B cells are infected, and is also highly associated with nasopharyngeal carcinoma, a lymphoepithelial tumor with high prevalence in South China. In spite of extensive ongoing research, the contribution of EBV to Burkitt's lymphoma tumor genesis remains an enigma. In a recently emerging model, cells infected with EBV accumulate genetic and epigenetic changes, predisposing them to tolerate the consequences of *c-myc* translocation events, even long after silencing of EBV gene expression.

The endemic Burkitt's lymphoma form affects children in equatorial Africa and New Guinea, areas that are characterized by holoendemic malaria. Chronic infection with malaria is believed to impair immune resistance in general and specifically to EBV. The sporadic or non-African form has a similar cellular appearance as the endemic form and is also putatively related to impaired immunity, allowing for development of EBV. The immunodeficiency-associated Burkitt's lymphoma variant is usually found in HIV-positive patients and can be the first tumor manifestation in AIDS. HIV infection, analogous to malaria, leads to polyclonal B-cell activation and allows poorly controlled proliferation of EBV+ B cells.

TREATMENT AND THERAPY

Burkitt's lymphoma, like all types of non-Hodgkin's lymphoma, are grouped in four stages according to the Cotswold Modification of the Ann Arbor staging system: stage I, a tumor in one lymph region only; stage II, lymphomas in at least two lymph regions on the same side of the diaphragm; stage III, lymphomas in lymph nodes and/or spleen, and on both sides of the diaphragm; stage IV, extranodal involvement (lung, liver, bone marrow). Non-Hodgkin's lymphoma in children is most commonly staged according to the system of the St. Jude's Children's Research Hospital. Stages I and II are identical to those described for adult patients. Stage III in children is defined by the presence of a large chest or abdominal tumor, whereas in stage IV patients the central nervous system and bone marrow are also involved.

Due to the aggressive nature of the tumor, prompt diagnosis and initiation of appropriate therapy is mandatory. Large chest or abdominal tumors may be removed surgically before starting chemotherapy.

Pediatric Burkitt's lymphoma patients are treated with chemotherapy and radiation therapy. Endemic

Burkitt's lymphoma is mostly treated with cyclophosphamide (Cytoxan), given either orally or intravenously. This drug acts by suppressing the immune system but also shows severe side effects. Tumors affecting the jaw and other facial bones are also treated with radiation therapy. Sporadic Burkitt's lymphoma in children is treated with high-dose chemotherapy, usually a combination of cyclophosphamide, methotrexate (MTX), vincristine (Oncovine), prednisone (Medicorten), and doxorubicin (Adriamycin) for a short course. Radiation therapy of the head and spine may be used to prevent infiltration of the tumor into the central nervous system. In addition, intrathecal (direct injection into the patient's spinal fluid) chemotherapy with the drug methorexate may be applied.

Adult patients with sporadic Burkitt's lymphoma are treated with a combination of radiation therapy and a high-dose chemotherapy regimen called CODOX-M/IVAC, which seems to render good results. CODOX-M/IVAC is a combination of cyclophosphamide, methotrexate, vincristine, doxorubicin, ifosfamide (Ifex), etoposide (VePesid), and cytarabine (ARA-C).

Adult AIDS patients are treated with low-dose chemotherapy because their immune system is already strongly impaired. Response to treatment is better for non-HIV patients than for HIV-positive individuals.

Newer methods of treatment have been applied including bone marrow or stem cell transplantation. Also, treatment with the new drug rituximab (Rituxan), a monoclonal antibody, used in combination with standard chemotherapy, shows promising results in clinical trials, indicating an improvement in the rates of remission and survival in high-risk patients.

PREVENTION AND OUTCOMES

Prevention of the endemic form of Burkitt's lymphoma is difficult as a result of the high incidence of malaria in central Africa, concurrent with inadequate medical care. Some risk factors for sporadic Burkitt's lymphoma can be lowered by a change of lifestyle behaviors that increase the risk of HIV infection.

Because of the good response of Burkitt's lymphoma to chemotherapy, the prognosis for children with the disease is generally good: 80 percent of children treated for early-stage Burkitt's lymphoma remain free from returning tumors three years after treatment. With combination chemotherapy and central nervous system prophylaxis, the survival rate is now at least 90 percent in both children and adults. Involvement of bone marrow and the central nervous system in tumor spread generally worsens the prognosis. Adults with the disease, especially those in the advanced stage, do more poorly than affected children. AIDS patients with Burkitt's lymphoma have an average length of survival of six months. After chemotherapy, patients should have regular follow-up examinations because of the possibility of long-term relapse.

Phillip A. Farber, Ph.D.;
updated by Nicola E. Wittekindt, Ph.D.

FURTHER READING

Aldoss, I. T., et al. "Adult Burkitt Lymphoma: Advances in Diagnosis and Treatment." *Oncology (Williston Park)* 22, no. 13 (2008): 1508-1517.

Bornkamm, G. W. "Epstein-Barr Virus and the Pathogenesis of Burkitt's Lymphoma: More Questions than Answers." *International Journal of Cancer* 124, no. 8 (2009): 1745-1755.

Burkitt, Denis Parsons, and D. H. Wright. *Burkitt's Lymphoma.* Edinburgh: Livingstone, 1970.

Carbone, A., et al. "HIV-Associated Lymphomas and Gamma-herpesviruses." *Blood* 113 (2009): 1213-1224.

Cotran, R. S., et al. *Robbins Pathologic Basis of Disease.* 8th ed. Philadelphia: Saunders, 2009.

Hartmann, E. M., G. Ott, and A. Rosenwald. "Molecular Biology and Genetics of Lymphomas." *Hematology/Oncology Clinics of North America* 22, no. 5 (2008): 807-823.

Heim, S., and Felix Mitelman. *Cancer Cytogenetics.* 3d ed. New York: J. Wiley, 2009.

Thorley-Lawson, D. A., and M. J. Allday. "The Curious Case of the Tumour Virus: Fifty Years of Burkitt's Lymphoma." *Nature Reviews Microbiology* 6, no. 12 (2008) : 913-924.

WEB SITES OF INTEREST

American Cancer Society
http://www.cancer.org./docroot/home/index.asp

Lymphoma Research Foundation
http://www.lymphoma.org/site/pp.asp?c=chKOI6PEImE&b=1573289

The Leukemia and Lymphoma Society
http://www.leukemia-lymphoma.org./hm_lls

See also: Cancer; Oncogenes; Tumor-suppressor genes.

C

Canavan disease

CATEGORY: Diseases and syndromes

ALSO KNOWN AS: Spongy degeneration of the brain; Canavan-Van Bogaert-Bertrand disease; aspartoacylase deficiency; ASPA deficiency

DEFINITION

Canavan disease is an inherited disorder that causes progressive, fatal degeneration of the brain. It is one of the neurologic disorders known as the leukodystrophies. Canavan disease is named for Myrtelle Canavan, the researcher who first described it in 1931.

RISK FACTORS

Parents who carry the Canavan gene may transmit it to their offspring in an autosomal recessive genetic pattern. For the disease to occur, both parents must carry the gene. Each child has a 1 in 4 risk of inheriting both abnormal genes and suffering from this tragic disease. The genetic mutation is more common in Jews of Eastern and Central European background (Ashkenazi) than in the population at large, though it has been found in all ethnic groups. It is estimated that 1 in 40 Ashkenazi Jews are carriers of the Canavan gene, resulting in a disease risk of about 1 in 6,400 births in this group.

ETIOLOGY AND GENETICS

Canavan disease is caused by a deficiency of the enzyme aspartoacylase, which is necessary for the breakdown of a substance in the brain called N-acetylaspartic acid. The lack of the enzyme leads to the accumulation of N-acetylaspartic acid, which in turn destroys myelin, a substance that acts as essential insulation in the brain. This destruction causes brain tissue to degenerate, resulting in a spongelike appearance and progressive disability and death.

The gene for aspartoacylase has been located on chromosome 17, and recent work in molecular genetics has identified three common alterations in DNA that are responsible for the disease. The first is a mutation in codon 285 (a section or location of DNA in the gene); the second is in codon 231; and the third is in codon 305. These three mutations are the cause of greater than 99 percent of cases of Canavan disease in Ashkenazi Jews and over 55 percent in other ethnic groups.

The disease is inherited as an autosomal recessive trait and must be carried by both parents for children to be affected. The inheritance of two altered genes leads to the deficiency of aspartoacylase. If only one parent carries the Canavan gene, then children will have a 50 percent chance of inheriting the altered gene and being carriers but will not have the disease. A single gene is sufficient for the production of the enzyme that prevents buildup of N-acetylaspartic acid. If both parents carry the Canavan gene mutation, then each pregnancy carries a 25 percent risk of disease and a 50 percent risk of carrier status.

SYMPTOMS

Newborns with Canavan disease appear normal. Symptoms of the disease begin to appear in infancy, most often by age six months. The hallmark is developmental delay, which is most often noted when infants fail to achieve early developmental milestones such as head control. The size of the head begins to increase (macrocephaly) and muscle tone and strength decrease. Motor skills are most severely affected. As damage to critical brain tissue continues, severe feeding problems, seizures, and blindness follow. Brain degeneration is progressive, and most children with the disease become severely disabled and die by age ten.

SCREENING AND DIAGNOSIS

Because of the increased prevalence of the Canavan gene in the Ashkenazi population, prepregnancy screening is recommended for couples in this group. Molecular diagnostic testing of blood samples is required to identify the three common gene mutations described above. If both parents are carriers of the gene, then prenatal diagnosis is available. Chorionic villus sampling may be performed in the first trimester of pregnancy, while amniocentesis is available in the second trimester.

Diagnosis of Canavan disease in at-risk infants is based on the presence of elevated amounts of N-acetylaspartic acid in the urine.

TREATMENT AND THERAPY

There is no cure or specific treatment for Canavan disease. Therapy is directed at modification of symptoms. Supportive measures such as the placement of feeding tubes, anticonvulsant medication, communication assistance, and physical therapy are often used. Antibiotics are used to treat common infections such as pneumonia.

Experimental studies for the treatment of Canavan disease are focused on gene therapy, including the introduction of functional aspartoacylase genes into an affected child's brain to attempt to increase the missing enzyme. Medications to reduce the amount of destructive N-acetylaspartic acid in the brain are also being studied.

PREVENTION AND OUTCOMES

Screening and genetic counseling are suggested for couples of Ashkenazi Jewish background and for those with a family history of Canavan disease. DNA testing is reliable and can identify whether one or both parents carry the Canavan gene in more than 99 percent of those screened. Parent screening and prenatal diagnosis are currently the only effective prevention for the disease.

Canavan disease is invariably fatal, with death often occurring by eighteen months of age, though there have been rare cases of children living into the teen years.

Rachel Zahn, M.D.

FURTHER READING

American College of Medical Genetics Board of Directors. *Position Statement on Carrier Testing for Canavan Disease.* Bethesda, Md.: Author, 1998. Screening recommendations for those at risk.

National Tay-Sachs & Allied Diseases Association. *A Genetics Primer for Understanding Tay-Sachs and the Allied Diseases.* Brookline, Mass.: Author, 1995. A closer look at the group of diseases that include Canavan.

Rezvani, I. "Defects in Metabolism of Amino Acids." In *Kliegman: Nelson Textbook of Pediatrics*, edited by R. M. Kliegman, R. E. Behrman, H. B. Jenson, and B. F. Stanton. 18th ed. Philadelphia: Saunders Elsevier, 2007. The classic medical text of pediatric diseases.

WEB SITES OF INTEREST

Canavan Foundation
http://canavanfoundation.org

Canavan Research Foundation
http://canavan.org

National Tay-Sachs & Allied Diseases Association, Inc.
http://ntsad.org

See also: Adrenoleukodystrophy; Alexander disease; Cerebrotendinous xanthomatosis; Hereditary diseases; Krabbé disease; Leukodystrophy; Metachromatic leukodystrophy; Pelizaeus-Merzbacher disease; Refsum disease; Vanishing white matter disease.

Cancer

CATEGORY: Diseases and syndromes

SIGNIFICANCE: At its root cancer is a genetic disease. It is characterized by unrestrained growth and reproduction of cells, loss of contact inhibition, and, eventually, metastasis (the wandering of cancer cells from a primary tumor to other parts of the body). All of these changes represent underlying mutations or inappropriate expression of genes involved in the control of the cell cycle and related processes.

KEY TERMS

carcinogen: a substance or other environmental factor that produces or encourages cancer

oncogenes: genes that cause cancer but that, in their normal form, called proto-oncogenes, are important in controlling the cell cycle and related processes

tumor: a mass formed by the uncontrolled growth of cells, which may be malignant (considered cancerous) or benign (nonmalignant)

tumor-suppressor genes: genes involved in regulating the cell cycle and preventing cell division until an appropriate time; when mutated, these genes can cause cancer

THE PROBLEM OF CANCER

Cancer is characterized by abnormal cell growth that leads to the invasion and destruction of healthy tissue by cells that meet certain criteria. Normal cells in the human body are continuously growing but are under normal cell control mechanisms. Cancer cells begin as normal cells that, as a result of genetic mutations, start to grow uncontrollably, escaping from the normal rules regulating cell growth and behavior. Cancer cells are characterized by six traits that differentiate them from normal healthy cells: cells that grow to abnormally large size; disregard for normal growth signals; lack of sensitivity to growth inhibition factors (antigrowth signals); avoidance of natural cell death (apoptosis); uninhibited replication; ability to form new vascular supply (angiogenesis); and ability for metastasis and invasion of neighboring tissues. Contact inhibition, in which cells contacting other cells prevent unrestrained growth, is lost in cancer cells. Normal cells also remain in one location, or at least in the same tissue, but malignant tumors, in their later stages, metastasize, allowing their cells to wander freely in the body, leading to the development of tumors in other organs. A final common feature is that cancer cells lose their normal cell shape.

The area where cancer begins to form a tumor is called the primary site. Most types of cancer begin in one place (the breast, lung, or bowel, for example) from which the cells invade neighboring areas and form secondary tumors. To make matters more complicated, some types of cancer, such as leukemia, lymphoma, and myeloma, begin in several places at the same time, usually in the bone marrow or lymph nodes. Primary tumors begin with one abnormal cell. This cell, as is true of all cells, is extremely small, no more than 0.002 or 0.003 millimeter across (about one-twentieth the width of a human hair). Therefore early cancer is very difficult to locate. Even if there are more than 100,000 cancer cells in a tumor, it is barely visible except under a microscope.

In general, cancer cells divide and reproduce about every two to six weeks, although different types of cancer grow at different rates. If they divide on the average of once per month, a single cell will multiply into approximately four thousand cells by the end of a year. After twenty months, there will be one million cells, which would form a tumor about the size of a pinhead and would still be undetectable. A tumor can be discovered only when a lump of approximately one billion cells is present. This would be about the size of a small grape. It would take about two and one-half years for a single cancer cell to reach this size. Within seven months, the one billion cells would grow to more than 100 billion cells, and the tumor would weigh about four ounces. By the fortieth month of growth, the lump of cancer cells would weigh about two pounds. By the time a tumor has reached this size, death often occurs. Death normally occurs about three and one-half years after the first cancer cell begins to grow. It takes about forty-two cell doublings to reach the lethal stage. The problem is that, in most cases, tumors are detectable only after thirty doublings. By this time, cancer cells may have invaded many other areas of the body beyond the primary site.

HOW CANCER CELLS GROW AND INVADE

Cancer cells are able to break down the barriers that normally keep cells from invading other groups of cells. With the aid of a microscope, cancer cells can be observed breaking through the boundary between cells, called the basement membrane. Cancer cells can make substances that break down the intercellular matrix, the "glue" that holds cells together. The intercellular matrix is a complex mixture of substances, including collagen, a strong, fibrous protein that gives strength to tissues. Cancer cells produce collagenase, an enzyme that breaks down collagen. Cancer cells also produce hyaluronidase, which further breaks down the intercellular matrix. This causes cancer cells to lose their normal shape and allows them to push through normal boundaries and establish themselves in surrounding tissues. Cancer cells have jagged edges, are irregular in shape, have large nuclei, and have hard-to-detect borders, making them relatively easy to identify microscopically. Normal cells, on the other hand, have a regular, smooth edge and shape.

There are many steps involved in the process of metastasizing, not all of which are understood by researchers. First is the entry into a blood vessel or lymph channel. Lymph channels, or lymphatics, comprise a network of vessels that carry lymph from the tissues to the bloodstream. Lymph is a colorless liquid that drains from spaces between cells. It consists mainly of water, salts, and proteins and eventually enters the bloodstream near the heart. The function of lymph is to filter out bacteria and other foreign particles that might enter the blood and cause infections. A mass of lymph vessels is called a lymph node. In the human body, lymph nodes are found in the neck, under the arms, and in several other places. Every body tissue has a network of lymph and blood vessels running through it.

Once a malignant tumor develops and metastasizes, the cells often travel through the body using the lymphatic system, a network of vessels that filter pathogens and transport lymph, a fluid similar to blood plasma. Cancer cells may gain entry into a nearby lymph vessel by breaking down defensive enzymes. Once in the lymph system, they can travel to nodes (glandlike masses of cells that produce white blood cells) and eventually into the bloodstream. Whatever route they take, groups of cancer cells can break away from the primary site of the tumor and float along whatever vessel they have invaded, forming numerous secondary tumors along the way. Because cancer cells are not considered foreign substances, such as bacteria or viruses, they are able to evade the body's immune system. Because of their overall resemblance to normal cells, cancer cells fool the body into thinking they are normal and therefore not dangerous.

Cancer cells eventually enter narrow blood vessels called capillaries and stay there for a brief period before they enter tissues such as lungs, bones, skin, and muscle. The secondary tumors then capture their own territory. As a tumor establishes itself, its cells often secrete signal proteins that stimulate new blood vessels to form (a process called angiogenesis) to increase blood supply to the growing tumor. The body thus not only fails to destroy developing tumors, but unwittingly helps establish them as well.

THE GENETICS OF CANCER

Cancer has been known since antiquity, but it was not until the twentieth century that the underlying causes of cancer began to be explored. In 1910, Peyton Rous discovered a type of cancer in chickens called a sarcoma (a cancer of connective tissue) that could be passed on to other chickens. He demonstrated this by removing tumors from affected chick-

ens, grinding the tumors up, filtering the grindate, and then injecting the filtrate into healthy chickens. Injected chickens invariably developed sarcoma tumors, suggesting that something smaller than the tumor cells was being passed on and was stimulating cancer development in otherwise normal cells. It is now known that the filtrate contained a cancer-causing virus, now called the Rous sarcoma virus. Similar types of viruses were discovered to be responsible for cancers in a variety of animals, but none was discovered in humans initially.

As the genetic material of some of the tumor viruses was later analyzed, all of them were discovered to contain genes called oncogenes, because they promoted oncogenesis (tumor development). Even more surprising was the discovery that humans have genes in their genome that are homologous (having a high degree of similarity) to viral oncogenes. The human genes did not seem to cause cancer under normal circumstances and were called proto-oncogenes. In cancer cells, some of these proto-oncogenes were discovered to have mutations or, in some cases, were simply overexpressed. In recognition of their abnormal state, these genes were called cellular oncogenes, to distinguish them from viral oncogenes. It is now known that proto-oncogenes are important in controlling the cell cycle by stimulating cell division only at the appropriate time. When they are transformed into oncogenes, uncontrolled cell growth and division occurs, two of the hallmarks of cancer.

A second type of cancer-causing gene, called a tumor-suppressor gene, was discovered to be the cause of retinoblastoma, a cancer of the retina, most often occurring in children. Tumor-suppressor genes have an effect opposite to that of proto-oncogenes; they suppress cell division and thus prevent unrestrained cell proliferation. If both alleles of a tumor-suppressor gene have a mutation that makes them nonfunctional, then cell division can occur unchecked. Retinoblastoma occurs in children when they inherit one faulty copy from a parent. If the other copy experiences a mutation, which frequently occurs, then retinoblastoma develops.

HOW CANCER DEVELOPS

The development of cancer is typically more complicated than implied above. Information gathered from the Human Genome Project helped improve our understanding of the role of genetics and genetic mutations not only in the development of cancer, but also in its treatment. The development of cancer generally requires mutations in more than a single proto-oncogene or tumor-suppressor gene. Any factors that increase mutation rates or decrease the ability of a cell to repair mutations will increase the likelihood that cancer will develop. Inheritance of already mutated genes can also greatly increase a person's chance of developing cancer, which accounts for the above-normal occurrence of certain types of cancer in some families.

One of the best-studied cases of oncogenesis involves colorectal cancer, which takes years to develop from a small cluster of abnormal cells into life-threatening cancer. It involves the loss or mutation of three tumor-suppressor genes and one proto-oncogene. Often colorectal cancer runs in families, because the loss of the first gene, the *APC* tumor-suppressor gene, is often inherited, resulting in an increased chance of developing colorectal cancer. Loss of this gene causes increased cell growth and some other genetic changes. In the next step, the *ras* oncogene is mutated, causing even more cell growth. Two more tumor-suppressor genes are lost, *DCC* and *p53*, at which point a tumor called a carcinoma has developed. Additional gene loss, which occurs much more easily in tumor cells, leads to metastasis, and the cancer then spreads to other organs and tissues.

The recent identification of specific genes associated with an increased risk of breast cancer has received a great deal of attention. Mutations in the *BRCA1* and *BRCA2* genes are responsible for up to half of all cases of breast cancer in women with a family history of the disease. Furthermore, the presence of such mutations helps guide treatment choices, with some women voluntarily undergoing prophylactic mastectomy if they have a family history of the genetic mutation. Although the identification of the *BRCA1* and *BRCA2* genes may help assess a woman's risk for developing breast cancer, it is important to note that it is not a definitive test. Women who have the mutation may never develop cancer, and breast cancer may develop in women who do not have these mutations.

Inheritance of a gene loss or mutation does not mean a person will get cancer; it simply means they have a higher chance of developing cancer. Although development of cancer is ultimately genetically based, environmental factors also play a part.

In the case of colorectal cancer, a diet low in roughage is often considered to increase colorectal cancer rates. Exposure to carcinogens, chemicals, or other factors, such as radiation, can also increase the likelihood of cancer. Exposure can occur in the diet, as a result of skin exposure, or inhalation. For example, smoking cigarettes is known to increase the occurrence of lung cancer, as well as a variety of other cancers. Excess exposure to damaging UV rays in sunlight or other sources is known to significantly increase the occurrence of skin cancer. Carcinogens promote cancer because they cause damage to DNA, and if the damage happens to occur to a tumor-suppressor gene or oncogene, then cancer may occur.

Inheritance of some mutations is particularly potent in increasing the chances of developing cancer. One example is the genetic disease xeroderma pigmentosa. Individuals with this disease develop skin cancer in response to even relatively brief exposure to UV radiation and must therefore avoid exposure to sunlight. In these types of highly heritable cancers, it appears that the mutations cause some kind of deficiency in the cellular DNA repair systems. As a result of a decreased ability to repair mutations, it is just a matter of time before mutations occur in proto-oncogenes or tumor-suppressor genes, so that the only way to prevent cancer is to control exposure to as many environmental carcinogens as possible and to aggressively screen for tumors.

CANCER TREATMENT

Cancers vary in their severity and rate of growth, which means that proper treatment depends on correctly diagnosing the type of cancer. For example, some forms of prostate cancer grow extremely slowly, and metastasis is rare until very late stages in the disease, sometimes many years after initial diagnosis. Treatment may comprise simply monitoring the tumor, avoiding carcinogenic exposure as much as possible, and possibly changing one's lifestyle. On the other hand, some types of skin cancer progress so rapidly that aggressive treatment may be required, unless it is caught very early. Although survival rates for many types of cancer have risen, treatment for most cancers is still only partially successful, and the later a tumor is detected, the greater chance that it will be untreatable.

New therapies are constantly being developed,

Leland H. Hartwell, winner of the 2001 Nobel Prize in Physiology or Medicine with R. Timothy Hunt and Paul M. Nurse, at the Fred Hutchinson Cancer Research Center in Seattle, Washington, shortly after the Nobel Foundation's announcement. The three men won for their work on cell division and its implications for cancer research. (AP/Wide World Photos)

but most cancers are still treated using surgery (removal of tumors), chemotherapy, and radiation therapy, either singly or, more often, in combination. More important than the specific treatment used is detecting tumors in their earliest stages, before they have extensively invaded surrounding tissues or metastasized. Survival rates are high for most cancers when treated very early.

The very nature of cancer makes treatment difficult. Because the cells involved are difficult for the immune system to recognize as dangerous, the body is typically inefficient at destroying them. Many of the treatments, other than surgical removal, rely on the fact that cancer cells divide much faster and more frequently than normal cells. Therefore, any agent that can cause higher mortality in rapidly dividing cells has potential as a cancer treatment. Chemotherapeutic agents are essentially toxic chemicals that are most toxic to dividing cells. Thus, they kill cancer cells much more readily than most other body cells, but any other body cells undergoing cell division are susceptible, so chemotherapy also kills some normal cells. Cancer patients often feel very ill during chemotherapy because of this.

Radiation therapy works similarly, being more damaging to dividing cells. An added advantage of radiation therapy, if the tumor has not yet metastasized, is that it can be focused more intensely in the vicinity of the tumor, preventing damage to other tissues. If the tumor has metastasized, then more widespread exposure to radiation may be used, with the obvious drawback that many other normal cells will also be damaged. Radiation therapy is often used to treat leukemia. Radiation is used to kill the patient's bone marrow, and then new bone marrow is transplanted from a compatible donor. The new bone marrow can then restore normal function to production of blood cells.

Genetics has played a part in improving chemotherapy. It has long been known that some people will respond better than others to certain chemo-therapeutic drugs. It is now known that some of these differences are genetic, and the underlying genetic differences have been uncovered in some cases. Therefore, as part of cancer treatment for some kinds of cancer, a person may be tested genetically to make more intelligent choices about which drugs to use. As more genetic data become available, it is anticipated that more effective and personalized treatments will be developed.

TARGETED THERAPY

The Human Genome Project opened a new avenue of cancer therapy called targeted therapy. The availability of gene and protein databases led to the identification of hundreds of human proteins and kinases that may harbor mutations and play a role in cancer development. Targeted therapies, which may be classified as either monoclonal antibodies or small molecule kinase inhibitors, act directly on these kinases. These products promise greater efficacy than blanket chemotherapy while keeping associated adverse effects to a minimum. Because kinase mutations tend to be found only in cancerous cells, normal healthy cells are largely unaffected by the targeted therapy.

Targeted therapies that are available are directed against a variety of proteins and are effective for a number of types of cancer. Agents targeting the epidermal growth factor receptor (EGFR-type I) pathway, for example, disrupt the signals that mediate cell growth. Several EGFR agents have been approved by the Food and Drug Administration (FDA) for cancer treatment, such as cetuximab (Erbitux) and panitumumab (Vectibix) for colorectal cancer; cetuximab for head and neck cancer; cetuximab and erlotinib (Tarceva) for pancreatic cancer; trastuzumab (Herceptin) and lapatinib (Tykerb) for breast cancer; erlotinib for hepatocellular carcinoma; and cetuximab, erlotinib, and gefitinib for lung cancer.

Other targeted therapies block the activity of ABL, which is a protein that controls cell proliferation in chronic myeloid leukemia (CML). CML cells have a genetic mutation that results in partial sequences of the *ABL* and *BRC* genes. The kinase inhibitor imatinib (Gleevec) targets these cells specifically, leading to complete remission in 75 percent of patients newly diagnosed with CML and 40 percent remission rates in patients with chronic CML who

have failed other therapies. Additional agents in this class are currently in development, and imatinib is being studied for the treatment of gastrointestinal stromal tumors.

Another approach of targeted therapies is to inhibit angiogenesis. Bevacizumab (Avastin) is an antivascular endothelial growth factor (anti-VEGF) agent that prevents cancer cells from building a vascular network, depriving them of their nutrient base and essentially starving them. Bevacizumab is used for treating colorectal cancer, breast cancer, renal cell cancer, hepatocellular cancer, and pancreatic cancer.

Sunitinib (Sutent) and sorafenib (Nexavar) target multiple kinases to prevent tumor growth, angiogenesis, and metastasis. Sunitinib has been approved for the treatment of renal cell cancer and gastrointestinal stromal tumors; sorafenib is approved for treating hepatocellular cancer and renal cell cancer.

Lastly, a class of targeted therapies called mTOR inhibitors inhibit a specific protein that disrupts the cascade signaling cell growth. Temsirolimus (Torisel) is used for treating renal cell cancer and is being studied for the treatment of breast cancer.

INNOVATIONS AND FUTURE TREATMENTS

Although the immune system cannot normally identify cancer cells accurately, there has been some success in immunological approaches. Research is progressing on development of vaccines against cancer, but so far this approach is still in its early experimental stages.

Photodynamic therapy also shows promise. It is based on the observation that certain chemicals, when ingested by single-celled organisms, release damaging oxygen radicals when exposed to light, thus killing the organisms. It has been observed that cancer cells retain these chemicals longer than normal cells. Treatment involves administering the chemical by injection, then waiting for a specified period for it to be retained by cancer cells and flushed out of normal cells. Then the tissue in which the cancer cells are located is exposed to laser light. This method works on any tissues that can be exposed to laser light, which includes any part of the body accessible to endoscopy.

Information from the Human Genome Project is being used not only to develop new, more specific therapies but also to control adverse events and to identify which patients will benefit most from a particular therapy. Researchers are exploring whether genes that predict cancer risk may also predict outcomes and susceptibility to symptoms such as fatigue and depression. Furthermore, biomarkers are being examined as predictors of cancer risk and treatment effectiveness. Many clinical trials include genetic assessments in an attempt to find specific markers that identify those patients most likely to respond to a particular therapy. For example, researchers know that women with breast cancer who have an overproduction of the protein HER-2 are much more likely to respond to trastuzumab therapy.

The ultimate treatment for cancer would be replacement or repair of the mutated genes responsible. Currently such treatment is not possible. There are many hurdles to overcome, including designing safe methods for inserting corrected gene copies. There is danger that improper gene therapy methods could actually make things worse, causing additional tumors or other diseases. A much better understanding of the genetics of cancer and future improvements in gene therapy techniques hold the promise of someday being able to cure or prevent most kinds of cancer.

Leslie V. Tischauser, Ph.D., and Bryan Ness, Ph.D.; updated by Cheryl Pokalo Jones

FURTHER READING

Bowcock, Anne M., ed. *Breast Cancer: Molecular Genetics, Pathogenesis, and Therapeutics.* Totowa, N.J.: Humana Press, 1999. Detailed information geared toward researchers and health professionals. The chapter "Hereditary Breast Cancer Genes" discusses *BRCA1* and *BRCA2* mutations among Ashkenazi women. Also addresses surgery, chemotherapy, drug resistance, and the *MDR* gene.

Bradlow, H. Leon, Jack Fishman, and Michael P. Osborne, eds. *Cancer Prevention: Novel Nutrient and Pharmaceutical Developments.* New York: New York Academy of Sciences, 1999. Examines several classes of nutrients and pharmaceutical agents believed to be important for tumor inhibition. Reviews novel preclinical models that facilitate analysis of chemopreventive agent efficacy and mechanisms of gene-nutrient interaction and provides information on clinical trials studying chemopreventive regimens.

Coleman, William B., and Gregory J. Tsongalis, eds. *The Molecular Basis of Human Cancer.* Totowa, N.J.: Humana Press, 2002. Surveys the known molecular mechanisms governing neoplastic transformation in the breast, prostate, lung, liver, colon, skin, leukemias, and lymphomas. Illuminates both recent developments and established concepts in epidemiology, molecular techniques, oncogenesis, and mutation mechanisms.

Cowell, J. K., ed. *Molecular Genetics of Cancer.* 2d ed. San Diego: Academic Press, 2001. Focuses on tumors of tissues such as liver, lung, bladder, and brain and surveys research in the cloning and analysis of genes central to the development and progression of human cancers.

Davies, Kevin, and Michael White. *Breakthrough: The Race to Find the Breast Cancer Gene.* New York: John Wiley, 1996. A history of the research into the genetic causes of breast cancer and other types of cancer.

Ehrlich, Melanie, ed. *DNA Alterations in Cancer: Genetic and Epigenetic Changes.* Natick, Mass.: Eaton, 2000. A comprehensive overview of the numerous and varied genetic alterations leading to the development and progression of cancer. Topics include oncogenes, tumor-suppressor genes, cancer predisposition, DNA repair, and epigenetic alteration such as methylation.

Feng, Qinghua, Mujun Yu, and Nancy B. Kiviat. "Molecular Biomarkers for Cancer Detection in Blood and Bodily Fluids." *Critical Reviews in Clinical Laboratory Sciences* 43, nos. 5/6 (2006): 497-560. Review of the approaches used to develop biomarkers of clinical interest and the reasons so few biomarkers are available for clinical use.

Fitzgerald, Patrick J. *From Demons and Evil Spirits to Cancer Genes: The Development of Concepts Concerning the Causes of Cancer and Carcinogenesis.* Washington, D.C.: American Registry of Pathology, Armed Forces Institute of Pathology, 2000. Traces the history, epidemiology, and genetics of neoplasms, cancer, and medical oncology.

Greaves, Mel F. *Cancer: The Evolutionary Legacy.* New York: Oxford University Press, 2000. Presents a Darwinian explanation for cancer that includes historical anecdotes and scientific findings.

Habib, Nagy A., ed. *Cancer Gene Therapy: Past Achievements and Future Challenges.* New York: Kluwer Academic/Plenum, 2000. Reviews forty-one preclinical and clinical studies in cancer gene therapy, organized into sections on the vectors available to carry genes into tumors, cell cycle control, apoptosis, tumor-suppressor genes, antisense and ribozymes, immunomodulation, suicidal genes, angiogenesis control, and matrix metallo proteinase.

Hanski, C., H. Scherübl, and B. Mann, eds. *Colorectal Cancer: New Aspects of Molecular Biology and Immunology and Their Clinical Applications.* New York: New York Academy of Sciences, 2000. Explores the immunological and molecular aspects of colon and rectal cancer.

Hodgson, Shirley V., and Eamonn R. Maher. *A Practical Guide to Human Cancer Genetics.* 2d ed. New York: Cambridge University Press, 1999. Gives a general overview of the underlying molecular genetic basis of cancer, the genetics of human cancers by site of origin, and a review of inherited cancer-predisposing syndromes.

Kornek, Gabriela, and Edgar Selzer. "Targeted Therapies in Solid Tumours: Pinpointing the Tumour's Achilles Heel." *Current Pharmaceutical Design* 15, no. 2 (2009): 207-242. Overview of targeted therapies approved and in development.

Liotta, L. A. "Cancer Cell Invasion and Metastasis." *Scientific American* (1992). Provides a basic description of cancer genetics.

Maruta, Hiroshi, ed. *Tumor-Suppressing Viruses, Genes, and Drugs: Innovative Cancer Therapy Approaches.* San Diego: Academic Press, 2002. An international field of experts addresses a number of innovative approaches to treating cancer, such as viral therapy using specific viral mutants, gene therapy using a variety of tumor-suppressor genes, and drug therapy targeted to block oncogenic signal pathways.

Mendelsohn, John, et al. *The Molecular Basis of Cancer.* 2d ed. Philadelphia: Saunders, 2001. Surveys the principles that constitute the scientific basis for understanding the pathogenesis of cancer and emphasizes clinical implications for treatment. Covers recent advances and current research, with descriptions of the basic mechanisms of malignant cells and molecular abnormalities, as well as new approaches to therapy.

Reyes-Gibby, Cielto C., et al. "Molecular Epidemiology, Cancer-Related Symptoms, and Cytokines Pathway." *Lancet Oncology* 9, no. 8 (2008): 777-785. Explores the use of genetic variations in predicting a patient's response to chemotherapy.

Schneider, Katherine A. *Counseling About Cancer: Strategies for Genetic Counseling.* 2d ed. New York: Wiley-Liss, 2002. A reference guide to assist genetic counselors and other health care providers help patients and families through the emotional difficulties of managing hereditary cancer.

Sung, Hye-Jin, and Je-Yeol Cho. "Biomarkers for the Lung Cancer Diagnosis and Their Advances in Proteomics." *Biochemistry and Molecular Biology Reports* 41, no. 9 (2008): 615-625. The promise of biomarkers for the diagnosis and treatment of lung cancer, and the clinical applicability.

Vogelstein, Bert, and Kenneth W. Kinzler, eds. *The Genetic Basis of Human Cancer.* 2d ed. New York: McGraw-Hill, 2002. Introduces the fundamentals of genetics and human phenotypes, gene mutation, the Human Genome Project, and gene imprinting and covers advances in the field.

Wilson, Samuel, et al. *Cancer and the Environment: Gene-Environment Interaction.* Washington, D.C.: National Academy Press, 2002. Includes "The Links Between Environmental Factors, Genetics, and the Development of Cancer," "Gene-Environment Interaction in Special Populations," and "Gene-Environment Interaction in Site-Specific Cancers."

WEB SITES OF INTEREST

American Cancer Society
http://www.cancer.org
Comprehensive and searchable site covering all aspects of cancer.

National Cancer Institute
http://www.cancer.gov
Site links to comprehensive information on genetics and cancer, including a cancer-basics tutorial.

See also: Aging; Bloom syndrome; *BRAF* gene; *BRCA1* and *BRCA2* genes; Breast cancer; Burkitt's lymphoma; Cell culture: Animal cells; Cell cycle; Cell division; Chemical mutagens; Chromosome mutation; Chronic myeloid leukemia; Colon cancer; Cowden syndrome; Developmental genetics; DNA repair; *DPC4* gene testing; Familial adenomatous polyposis; Gene therapy; Genetic engineering: Medical applications; Genetic testing: Ethical and economic issues; Hereditary diffuse gastric cancer; Hereditary diseases; Hereditary leiomyomatosis and renal cell cancer; Hereditary nonpolyposis colorectal cancer; Hereditary papillary renal cancer;

Homeotic genes; Human Genome Project; Huntington's disease; Hybridomas and monoclonal antibodies; Hypercholesterolemia; Insurance; Model organism: *Caenorhabditis elegans*; Model organism: *Mus musculus*; Mutagenesis and cancer; Mutation and mutagenesis; Nondisjunction and aneuploidy; Oncogenes; Ovarian cancer; Pancreatic cancer; Prostate cancer; Proteomics; Reverse transcriptase; RNA transcription and mRNA processing; Signal transduction; Stem cells; Steroid hormones; Telomeres; Tumor-suppressor genes; Wilms' tumor aniridia-genito-urinary anomalies-mental retardation (WAGR) syndrome.

Cardiomyopathy

CATEGORY: Diseases and syndromes

DEFINITION

Cardiomyopathy refers to heart muscle disease. The damaged heart does not effectively pump blood. The disease usually progresses to the point where patients develop life-threatening heart failure. In addition, patients with cardiomyopathy are more likely to have irregular heartbeats or arrhythmias.

There are two major categories of cardiomyopathy: ischemic and nonischemic cardiomyopathy. Ischemic cardiomyopathy occurs when the heart muscle is damaged from heart attacks due to coronary artery disease. Nonischemic cardiomyopathy, the less common category, includes types of cardiomyopathy that are not related to coronary artery disease.

There are three main types of nonischemic cardiomyopathy. In dilated cardiomyopathy, damaged heart muscles lead to an enlarged, floppy heart. The heart stretches as it tries to compensate for weakened pumping ability.

In hypertrophic cardiomyopathy, heart muscle fibers enlarge abnormally. The heart wall thickens, leaving less space for blood in the chambers. Since the heart does not relax correctly between beats, less blood fills the chamber and is pumped from the heart.

In restrictive cardiomyopathy, portions of the heart wall become rigid and lose their flexibility. Thickening often occurs due to abnormal tissue invading the heart muscle.

RISK FACTORS

Individuals whose family members have cardiomyopathy are at risk for the disease. Other risk factors include alcoholism, obesity, diabetes, hypertension, coronary artery disease, and certain drugs.

ETIOLOGY AND GENETICS

While there are several identifiable environmental factors that can trigger each of the various categories of cardiomyopathy, genetic factors play an important role as well, particularly in the cases of hypertrophic and restrictive cardiomyopathy. In hypertrophic cardiomyopathy, the cardiac muscle fibers often show abnormal growth and arrangement as a result of mutations in any of several genes that encode protein components of sarcomeres, the contractile units of heart muscle.

Two genes in particular have been identified in which mutations account for about 80 percent of cases of inherited hypertrophic cardiomyopathy. *MYH7*, found on the long arm of chromosome 14 at position 14q12, encodes the heavy chain of polypeptide 7 of cardiac myosin; and *MYBPC3*, found on the short arm of chromosome 11 at position 11p11.2, specifies the cardiac myosin binding protein C. In both of these genes the inheritance pattern seems to be autosomal dominant, meaning that a single copy of the mutation is sufficient to cause full expression of the disease. An affected individual has a 50 percent chance of transmitting the mutation to each of his or her children. Many cases, however, result from a spontaneous new mutation, so in these instances affected individuals will have unaffected parents.

Other genetic syndromes may include cardiomyopathy as one of many clinical manifestations. For example, Barth syndrome, a rare X-linked condition affecting male children in their first year of life, can include dilated cardiomyopathy. One type of hereditary hemochromatosis, a disease associated with abnormal iron absorption, can lead to restrictive cardiomyopathy, as well as additional abnormalities in the liver, pancreas, and pituitary gland.

SYMPTOMS

Symptoms vary depending on the type of cardiomyopathy and its severity. Patients with hypertrophic cardiomyopathy often do not notice any symptoms. Sudden cardiac death may be the first indication of the condition.

It may take years for symptoms of dilated cardiomyopathy to develop. Blood clots may form due to the abnormal pooling of blood in the heart. If a clot moves to another part of the body (embolism), symptoms associated with that organ (the brain, for example) may be the first sign of the heart disease.

Cardiomyopathy ultimately leads to heart failure and the following symptoms: fatigue; weakness; shortness of breath, often worse when lying down or with exertion; cough; swelling in the feet or legs; chest pain; and irregular heart rhythm.

SCREENING AND DIAGNOSIS

The doctor will ask about a patient's symptoms and medical history and will perform a physical exam. The doctor will also listen to a patient's heart with a stethoscope; cardiomyopathies often produce heart murmurs and other abnormal heart sounds.

Tests may include a chest X ray to look for heart enlargement; an electrocardiogram, a test that records the heart's activity by measuring electrical currents through the heart muscle; and an echocardiogram, a test that uses high-frequency sound waves (ultrasound) to examine the size, shape, and motion of the heart. Blood tests can check for damage to the heart and other organs and possibly the underlying cause or causes of the cardiomyopathy. Other tests include cardiac catheterization, in which a tube-like instrument is inserted into the heart through a vein or artery (usually in the arm or leg) in order to detect problems with the heart and its blood supply, and a heart biopsy, the removal of a sample of heart tissue for testing.

TREATMENT AND THERAPY

When heart failure is due to blockages in the coronary arteries, treatment directed at relieving these blockages through angioplasty, stent placement, or coronary artery bypass surgery may lead to improvements in heart function and symptoms. For certain genetic causes, other treatments may also lead to improvements in function. For many patients, however, treatment is aimed at relieving symptoms and preventing further damage.

Lifestyle modification aims to eliminate anything that contributes to the disease or worsens symptoms. These lifestyle changes may include avoiding alcohol, losing weight if the patient is overweight, eating a low-fat diet to minimize the risk and extent of coronary artery disease, and limiting salt intake to decrease fluid retention. Patients should follow their doctors' advice for exercise; they may need to limit their physical activity.

Medications for cardiomyopathy may include diuretics to eliminate extra fluid; ACE (angiotensin converting enzyme) inhibitors to help relax blood vessels, lower blood pressure, and decrease the heart's workload; digitalis to slow and regulate the heart rate and modestly increase its force of contractions; and calcium channel blockers to lower blood pressure and relax the heart. Other medications include beta blockers to slow the heart and limit disease progression; antiarrhythmia agents to prevent irregular heart rhythms; and immune system suppressants, including steroids (depending on the underlying cause).

Surgical options include implanting a pacemaker to improve the heart rate and pattern. For patients with hypertrophic disease, doctors may remove part of the thickened wall separating the heart's chambers. Surgery may also be needed to replace a heart valve.

A heart transplant may be possible for otherwise healthy patients who do not respond to medical treatment. Candidates often wait a long time for a new heart. Those waiting may temporarily receive a ventricular assist device—a mechanical pump that assumes some or most of the heart's pumping function.

PREVENTION AND OUTCOMES

Aggressively treating hypertension, coronary artery diseases, and their risk factors is the best way to prevent most cases of cardiomyopathy. Other, less common causes, however, are not preventable. Individuals with a family history of the disease should ask their doctors about screening tests, especially before starting an intense exercise program.

Debra Wood, R.N.; reviewed by Michael J. Fucci, D.O. "Etiology and Genetics" by Jeffrey A. Knight, Ph.D.

FURTHER READING

Domino, Frank J., ed. *Griffith's Five-Minute Clinical Consult, 2008.* 16th ed. Philadelphia: Lippincott Williams & Wilkins, 2007.

EBSCO Publishing. *Health Library: Cardiomyopathy.* Ipswich, Mass.: Author, 2009. Available through http://www.ebscohost.com.

Goldman, Lee, and Dennis Ausiello, eds. *Cecil Medicine.* 23d ed. Philadelphia: Saunders Elsevier, 2008.

Goroll, Allan H., and Albert G. Mulley, Jr., eds. *Primary Care Medicine: Office Evaluation and Management of the Adult Patient.* 6th ed. Philadelphia: Wolters Kluwer Health/Lippincott Williams & Wilkins, 2009.

Libby, Peter, et al., eds. *Braunwald's Heart Disease: A Textbook of Cardiovascular Medicine.* 8th ed. Philadelphia: Saunders/Elsevier, 2008.

Maron, Barry J., ed. *Diagnosis and Management of Hypertrophic Cardiomyopathy.* Malden, Mass.: Blackwell-Futura, 2004.

WEB SITES OF INTEREST

American Heart Association
http://www.americanheart.org

Canadian Cardiovascular Society
http://www.ccs.ca/home/index_e.aspx

Canadian Family Physician
http://www.cfpc.ca/cfp

The Cardiomyopathy Association
http://www.cardiomyopathy.org

Genetics Home Reference
http://ghr.nlm.nih.gov

National Heart, Lung, and Blood Institute
http://www.nhlbi.nih.gov

See also: Atherosclerosis; Barlow's syndrome; Heart disease.

Carpenter syndrome

CATEGORY: Diseases and syndromes
ALSO KNOWN AS: Acrocephalopolysyndactyly type II; ACPS II

DEFINITION

Carpenter syndrome is a rare genetic disorder that is inherited in an autosomal recessive manner.

The disorder is characterized by craniosynostosis (early closure of the cranial sutures of the skull), webbing of certain fingers or toes, and/or more than the normal number of fingers or toes, and sometimes congenital heart defects.

RISK FACTORS

It is not known whether particular risk factors lead to Carpenter syndrome. However, since the disorder is inherited in an autosomal recessive manner, a person will develop it if both parents are carriers of the defective *RAB23* gene. Both men and women have been diagnosed with Carpenter syndrome.

ETIOLOGY AND GENETICS

Carpenter syndrome was first identified in the early 1900's by George Carpenter. The disorder belongs to a group of rare genetic disorders known as acrocephalopolysyndactyly, which are characterized by craniosynostosis, webbing or fusion of digits, and more than the normal number of digits. Carpenter syndrome is a pleiotropic disorder, meaning that a single gene influences multiple phenotypic traits. Because it is inherited in an autosomal recessive manner, a person who shows the symptoms of the syndrome must carry two mutant genes (one inherited from each parent).

While the molecular basis of many craniosynostosis syndromes had been discovered, the cause of Carpenter syndrome was unknown until recently. The disorder is rare and estimated to occur in approximately one out of every one million live births. In 2007, scientists found linkage of Carpenter syndrome to chromosome 6, more specifically the *RAB23* gene. In fifteen independent families with Carpenter syndrome, five different *RAB23* mutations were identified, four of which were truncating (when a codon for one amino acid is changed into a stop codon) and one missense (when a codon for one amino acid is changed into a codon for another amino acid).

The *RAB23* gene encodes a member of the RAB guanosine triphosphatase (RAB GTPase) family of vesicle transport proteins and acts as a negative regulator of hedgehog (HH) signaling. The HH signaling pathway provides cells with the information that they need to develop properly; cells develop differently based on where they are in the embryo and eventually become part of different body parts. The discovery of *RAB23* mutations in patients with Carpenter syndrome did not surprise scientists, since other disorders associated with faulty HH signaling share some of the same physical characteristics as Carpenter syndrome, especially regarding superfluous, webbed, or shortened digits. However, researchers were surprised that finding *RAB23* mutations in those with Carpenter syndrome implies that HH signaling is involved in cranial sutures. There is much left to learn about HH signaling and craniosynostosis in Carpenter syndrome.

SYMPTOMS

The most common physical symptom of Carpenter syndrome is craniosynostosis, leading to either a short and broad head or a cone-shaped head (acrocephaly). Facial malformations are also usually present in those with Carpenter syndrome, as are digit abnormalities (webbing or fusion, extra digits, and/or shortened digits). Individuals with Carpenter syndrome may also have congenital heart defects, vision problems, hernias, undescended testes (in males), developmental delays, and a highly arched and narrow palate. They may be of short stature and tend toward obesity.

SCREENING AND DIAGNOSIS

There is no specific way to test for Carpenter syndrome, but the disorder may be ruled out as a diagnosis if a suspected patient tests positive for another genetic disorder that also has skull malformations. Diagnosis is usually made based on the skull malformations observed at birth, along with the presence of other Carpenter syndrome symptoms. X rays and CT scans may be key in ensuring that a Carpenter syndrome diagnosis is correct.

An ultrasound anatomy scan can be performed during pregnancy to look at skull development. If deformities are noticed, however, it may not always be possible to distinguish between Carpenter syndrome and other potential causes.

TREATMENT AND THERAPY

Treatment for Carpenter syndrome varies for each individual. Usually, the most crucial and immediate surgical interventions correct skull malformations during the first year of life in order to create room for the rapidly growing brain. Several procedures may be necessary. If serious heart defects are present, then heart surgery may also be necessary shortly after birth.

Hand and foot reconstruction, jaw surgery, and surgery to move undescended testes may also be part of the treatment plan for Carpenter syndrome-affected individuals. Diet, vision, and the highly arched and narrow palate of those with Carpenter syndrome may also need to be addressed.

PREVENTION AND OUTCOMES

There is no effective means of prevention for Carpenter syndrome. If the skull malformations and heart defects associated with Carpenter syndrome are treated appropriately and promptly, however, then affected individuals will live and benefit from an improved physical appearance following surgical intervention. Vision problems and developmental delay may persist, but many of the symptoms of Carpenter syndrome can be treated and greatly diminished.

Sabina Maria Borza, M.A.

FURTHER READING

Johnson, Paul A. "Carpenter Syndrome." In *Gale Encyclopedia of Genetic Disorders*. 2d ed. Detroit: Thomson Gale, 2005. This entry is available in eBook format and provides thorough, accessible information about the disorder.

Katzen, J. T., and J. G. McCarthy. "Syndromes Involving Craniosynostosis and Midface Hypoplasia." *Otolaryngologic Clinics of North America* 33, no. 6 (2000): 1257-1284. This technical article discusses surgical management of Carpenter syndrome and other similar syndromes.

Klug, William S., Michael R. Cummings, Charlotte Spencer, and Michael A. Palladino. *Concepts of Genetics*. 9th ed. San Francisco: Benjamin Cummings, 2008. A comprehensive introduction to various topics in genetics.

WEB SITES OF INTEREST

About.com: Carpenter Syndrome
http://rarediseases.about.com/od/acps/a/carpenter.htm

Carpenter Syndrome
http://carpentersyndrome.com/

National Organization for Rare Disorders
http://www.rarediseases.org/search/rdbdetail_abstract.html?disname=Carpenter%20'syndrome

Online Mendelian Inheritance in Man
http://www.ncbi.nlm.nih.gov/entrez/
dispomim.cgi?id=201000

See also: Brachydactyly; Congenital defects; Hereditary diseases; Polydactyly.

cDNA libraries

CATEGORY: Bioinformatics; Techniques and methodologies

SIGNIFICANCE: A cDNA library is a set of cloned DNA copies of the RNAs found in a specific cell type at a specific time. This library can be used to construct probes for mapping these genes, to study the changing expression of genes over time (during development, for example), or to clone genes into organisms for further study or production of proteins.

KEY TERMS

complementary DNA (cDNA): also known as copy DNA, a form of DNA synthesized by reverse transcribing RNAs (usually messenger RNAs) into DNA

DNA library: a collection of DNA fragments cloned from a single source, such as a genome, chromosome, or set of mRNAs

in situ hybridization: a technique that uses a molecular probe to determine the chromosomal location of a gene

introns: noncoding segments of DNA within a gene that are removed from mRNA copies of the gene before polypeptide translation

reverse transcriptase: an enzyme, isolated from retroviruses, that synthesizes a DNA strand from an RNA template

GENE CLONING AND DNA LIBRARIES

In order to study and map genes, researchers need to take potentially very large sections of DNA (such as a chromosome or whole genome), break them into smaller, manageable fragments, and clone these fragments to construct a DNA library. A genomic or chromosome library may contain many thousands of cloned fragments, many of which will represent stretches of noncoding DNA between genes. If the researcher is interested is studying the protein-coding regions, or genes, of the DNA, it is better to start with the messenger RNAs (mRNAs) of the cell, which represent the genes being actively transcribed in the cell at that time. By constructing and cloning complementary DNA (cDNA) copies of these mRNAs, researchers can create a library that contains copies of only the active genes.

cDNA LIBRARY CONSTRUCTION

DNA copies of mRNAs are synthesized using the enzyme reverse transcriptase. This enzyme was independently discovered by Howard Temin and David Baltimore in 1970 in retroviruses, which "reverse transcribe" their RNA genomes into DNA after infecting their host cells. In the late 1970's, researchers began using the enzyme to make DNA copies of mRNAs, and later to construct cDNA libraries.

To create a cDNA library from a sample of cells, mRNAs from the cells are isolated and purified. Reverse transcriptase is used to synthesize a complementary DNA strand using each mRNA strand as a template, resulting in a collection of double-stranded RNA-DNA hybrids. To obtain double-stranded cDNAs suitable for cloning, the enzyme RNase H is used to digest the RNA strand, and DNA polymerase I is used to synthesize the second DNA strand using the first as a template. If desired, "sticky ends" can be added to the cDNAs for cloning into a vector. The set of recombinant vectors are inserted into bacterial cells in the process of transformation, resulting in a cloned cDNA library. The library is maintained as a collection of bacterial colonies, each colony containing a different cloned DNA fragment.

APPLICATIONS

A cDNA library represents the coding sequences of genes that were actively expressed in the original cell sample at the time the sample was taken. In effect, it can represent a snapshot of active genes in the cells at that time. Comparing the cDNAs of different tissues from the same organism can reveal the differences in gene expression of these tissues. Also, comparing cDNAs of cells in the same tissue over time can show how gene expression changes in the same cells. This approach has been especially fruitful in developmental genetic research, because the developmental pattern of an organism can be correlated with the activity of specific genes.

Cloned cDNAs can also be used to find the chromosomal location of an expressed gene. One strand of the cDNA clone is labeled with a fluorescent tag and used as a molecular probe. In the technique of in situ hybridization, the probe will base pair, or hybridize, to the complementary sequence in a preparation of partially denatured chromosomes, and the chromosomal location of the original gene will be visible because of the fluorescent label. Such a probe can also be used to screen a chromosome or genomic library for the cloned fragment containing the target gene. Using the entire cDNA library to probe a genome will generate a cDNA map that suggests the most biologically and medically important parts of the genome, aiding researchers in the search for disease genes.

Genes of eukaryotes (nonbacterial organisms) usually contain introns, noncoding segments that are transcribed but removed from mRNAs before translation, but bacterial genes do not. Often, a eukaryotic gene put into a bacterial cell will not produce a functional polypeptide because the cell does not have the biochemical machinery for removing introns. If the goal of the research is to have a bacterium make the protein product of a gene, it may be necessary to clone a cDNA version of the gene, which lacks introns, using a special expression vector that allows the cell to transcribe the inserted gene and translate it to the proper polypeptide.

ADVANTAGES AND DISADVANTAGES

Because cDNA libraries contain only DNA of expressed genes, they are much smaller and more easily managed and studied than chromosome or genomic libraries that have all coding and noncoding regions. The cDNA versions of genes have only the protein-coding sequence, without introns, so that cloning them in bacteria allows expression of the protein products of the genes. In contrast to other DNA libraries, cDNA libraries can be used to study variable patterns of gene expression among cell types or over time. In eukaryotes, cDNA copies of genes are not identical to the original sequences of the genes and also lack the promoter region necessary for proper transcription of the gene. However, using cDNA as a molecular probe can lead to the identification of the original gene.

Stephen T. Kilpatrick, Ph.D.

FURTHER READING

Dale, Jeremy, and Malcolm von Schantz. "Genomic and cDNA Libraries." In *From Genes to Genomes: Concepts and Applications of DNA Technology.* 2d ed. Hoboken, N.J.: Wiley, 2007. This textbook introduces readers to significant techniques and concepts involved in cloning genes and in studying their expression and variation.

Sambrook, Joseph, and David Russell. *Molecular Cloning: A Laboratory Manual.* 3d ed. Cold Spring Harbor, N.Y.: Cold Spring Harbor Laboratory Press, 2001. Contains detailed protocols for mRNA isolation, cDNA synthesis, and library construction.

Watson, James D., et al. *Recombinant DNA—Genes and Genomes: A Short Course.* 3d ed. New York: W. H. Freeman, 2007. An introduction to techniques for cloning genes, including construction of cDNA libraries.

Ying, Shao-Yao. *Generation of cDNA Libraries: Methods and Protocols.* Totowa, N.J.: Humana Press, 2003. Designed for laboratory researchers, this book presents techniques for generating cDNA/mRNA libraries, including such methods as electrophoresis, Northern blotting, single-cell microarray analysis, subtractive cloning, and gene cloning.

WEB SITES OF INTEREST

Molecular Biology Web Book, Genomic and cDNA Libraries

http://www.web-books.com/MoBio/Free/Ch9B.htm

Discusses the construction of cDNA and genomic libraries, with links to an article about the subject published in a molecular biology textbook.

National Human Genome Research Institute

http://www.genome.gov/glossary.cfm?key=cDNA%20library

The Talking Library section of the institute's Web site provides a visual definition of the term "cDNA library," as well as a recorded definition by one of the institute's researchers. This oral explanation can be heard with the use of RealPlayer.

See also: Bioinformatics; DNA fingerprinting; DNA sequencing technology; Fluorescence in situ hybridization (FISH); Forensic genetics; Genetic testing: Ethical and economic issues; Genetics: Historical development; Genomic libraries; Genomics; Human Genome Project; Icelandic genetic database; Linkage maps; Proteomics; Reverse transcriptase.

Celiac disease

CATEGORY: Diseases and syndromes
ALSO KNOWN AS: Celiac sprue; nontropical sprue; gluten-sensitive enteropathy

DEFINITION

Celiac disease is an autoimmune disease of the digestive tract. For patients with celiac disease, eating food with gluten damages little protrusions in the small intestine. These protrusions, called villi, absorb nutrients from foods. The condition affects absorption of all nutrients. Untreated patients often become malnourished.

RISK FACTORS

Individuals whose family members have celiac disease are at risk for the illness. Individuals also are at risk if they have a history of another autoimmune disease, such as Type I diabetes, autoimmune thyroid disease, lupus, dermatitis herpetiformis (a skin condition associated with celiac disease), and rheumatoid arthritis.

ETIOLOGY AND GENETICS

Celiac disease is a complex disorder that is determined by an interaction between both genetic and environmental components. More than 97 percent of affected individuals have at least one allele of either of two closely linked predisposing genes found at the major histocompatibility locus located on the short arm of chromosome 6 (at position 6p21.3). These alleles, known as HLA-DQA1 and HLA-DQB1, are necessary but not sufficient to predispose the development of celiac disease, since some unaffected individuals also carry one or both of them.

Additionally, there are several other regions in the genome that are unlinked to the HLA region and that probably contain genes for celiac disease susceptibility (at positions 1q31, 2q11, 2q33, 3q21, 3q25, 3q28, 4q27, 5q31, 12q24, 15q11, and 19p13.1). Each of these has a relatively weak effect as compared with the HLA alleles, but they serve to complicate the inheritance patterns and make predictions of outcomes more unreliable.

Incidence of the disease in first-degree relatives of affected individuals is about ten times greater than for the general population, and concordance rates for celiac disease in identical twins have been reported to be about 70 percent.

The environmental factors that may serve to trigger the development of the disease are also not well understood, although a diet high in gluten is certainly a prerequisite. Other contributing environmental factors that have been reported include stress, pregnancy, traumatic injury, surgery, and systemic infections.

SYMPTOMS

Symptoms vary and may start in childhood or adulthood. Children often have different symptoms from adults. Symptoms may not develop if a large section of the intestine is undamaged. Malnutrition may produce the first signs of the condition, which are often the most serious.

Signs and symptoms in children may include ab-

A supermarket in Albany, N.Y., carries a selection of gluten-free products appropriate for people with celiac disease. (AP/Wide World Photos)

dominal pain; nausea or lack of appetite; vomiting, in later stages of the disease; diarrhea; malodorous, bulky stools; irritability; and failure to thrive (in infants). Other signs and symptoms in children may include short stature, delayed puberty, anemia, pale skin, seizures, hepatitis, angular cheilitis (cracked sores in the corners of the mouth), and aphthous ulcers (shallow sores in the mucous membranes of the mouth).

Signs and symptoms in adults include bloating, gas, diarrhea, and a foul-smelling, light-colored, oily stool. Additional signs and symptoms are weight loss, a hearty or a poor appetite, fatigue, abdominal pain, bone pain, behavior changes, muscle cramps and joint pain, seizures, dizziness, skin rash, dental problems, missed menstrual periods, infertility, altered sensation in the limbs, anemia, and osteopenia.

SCREENING AND DIAGNOSIS

The doctor will ask about an individual's symptoms and medical history and will conduct a physical exam. Symptoms of celiac disease are similar to those of other conditions. It may take a long time to get a diagnosis. Early diagnosis and treatment reduce the risk of complications.

Tests may include blood tests to detect the presence of gluten antibodies (produced by the immune system) and to look for evidence of malabsorption (anemia, vitamin and mineral deficiencies). Stool tests can check for evidence of malabsorption. Other tests include endoscopy, in which a thin, lighted tube is inserted down the throat to examine the intestine; biopsy, in which a small sample of tissue is removed during endoscopy to test for inflammation and tissue damage; and repeat biopsy, a biopsy performed several weeks after treatment begins to confirm the diagnosis.

TREATMENT AND THERAPY

A lifelong gluten-free diet is the only treatment for celiac disease; fortunately, it is very effective. Symptoms usually go away within days of starting the diet. Healing of the villi may take months or years. Additional intake of gluten can damage the intestine, even if the patient has no symptoms. Delayed growth and tooth discoloration may be permanent. Nutritional supplements, given through a vein, may be needed if the intestinal damage is significant and does not heal. Since gluten is added to many foods, the diet can be complicated and often frustrating. Some patients find support groups helpful.

Individuals with celiac disease must avoid all foods containing wheat, rye, or barley. This includes most bread, pasta, cereal, and processed foods. Special gluten-free breads and pastas are available; they are made with potato, rice, soy, or bean flour. Patients who are lactose intolerant before their small intestine heals need to avoid milk products. A dietitian can assist patients with meal planning.

Gluten is found in some unexpected foods and beverages; patients should carefully read all labels. Other foods with gluten include flavored coffee, beer, tuna in vegetable broth, packaged rice mixes, some frozen potatoes, creamed vegetables, commercially prepared vegetables, salads and salad dressings, pudding, some ice cream, and many other products. Ordering at restaurants can be especially challenging, since many foods on the menu may surprisingly contain gluten.

Patients with celiac disease should be tested for nutritional deficiencies. Bone density testing may also be needed. If vitamin or mineral deficiencies are found, the doctor may recommend taking supplements. Once the disease is under control with a gluten-free diet, however, this is often not necessary.

PREVENTION AND OUTCOMES

There are no guidelines for preventing celiac disease because the cause is not understood. If celiac disease runs in an individual's family, he or she should ask the doctor about a screening test. The earlier patients start the gluten-free diet, the less damage there will be to their intestines.

Debra Wood, R.N.; reviewed by Daus Mahnke, M.D.
"Etiology and Genetics" by Jeffrey A. Knight, Ph.D.

FURTHER READING

Conn, H. F., and R. E. Rakel. *Conn's Current Therapy.* 53d ed. Philadelphia: W. B. Saunders, 2001.

DiMarino, Anthony J., Jr. *Sleisenger and Fordtran's Gastrointestinal and Liver Disease: Review and Assessment.* 8th ed. Philadelphia: Saunders, 2007.

Domino, Frank J., ed. *Griffith's Five-Minute Clinical Consult, 2008.* 16th ed. Philadelphia: Lippincott Williams & Wilkins, 2007.

EBSCO Publishing. *Health Library: Celiac Disease.* Ipswich, Mass.: Author, 2009. Available through http://www.ebscohost.com.

Goldman, Lee, and Dennis Ausiello, eds. *Cecil Medi-*

cine. 23d ed. Philadelphia: Saunders Elsevier, 2008.

Green, Peter H. R., and Rory Jones. *Celiac Disease: A Hidden Epidemic*. New York: Collins, 2006.

Kliegman, Robert M., et al., eds. *Nelson Textbook of Pediatrics*. 18th ed. Philadelphia: Saunders, 2007.

Shepard, Jules E. Dowler. *The First Year—Celiac Disease and Living Gluten-Free: An Essential Guide for the Newly Diagnosed*. Cambridge, Mass.: Da Capo Press, 2008.

Web Sites of Interest

Canadian Celiac Association
http://www.celiac.ca

Celiac Disease Foundation
http://www.celiac.org

"Celiac Disease: What You Should Know." American Academy of Family Physicians
http://www.aafp.org/afp/20061201/1921ph.html

Celiac Sprue Association
http://www.csaceliacs.org

Health Canada
http://www.hc-sc.gc.ca/index-eng.php

"What I Need to Know About Celiac Disease." National Digestive Diseases Information Clearinghouse
http://digestive.niddk.nih.gov/ddiseases/pubs/celiac_ez

See also: Autoimmune disorders; Colon cancer; Crohn disease; Familial adenomatous polyposis.

Cell culture
Animal cells

Category: Cellular biology; Techniques and methodologies

Significance: The ability to grow and maintain cells or tissues in laboratory vessels has provided researchers with a means to study cell genetics and has contributed to the understanding of what differentiates "normal" cells from cancer cells. The technology involved in growing viruses in cell culture has proved vital both to under-standing virus replication and for development of viral vaccines.

Key terms

cell lines: cells maintained for an indeterminate time in culture

HeLa cells: the first human tumor cells shown to form a continuous cell line

micropropagation: removal of small pieces of plant tissue for growth in culture

primary cells: explants removed from an animal

transformation: any physical change to a cell, but generally the change of a normal cell into a cancer cell

Early History

Methodology for maintaining tissues in vitro (in laboratory vessels) began in 1907 with Ross Harrison at Yale College. Harrison placed tissue extracts from frog embryos on microscope slides in physiological fluids such as clotted frog lymph. The material was sealed with paraffin and observed; specimens could be maintained for several weeks. In 1912, Alexis Carrel began the maintenance of cardiac tissues from a warm-blooded organism, a chicken, in a similar manner. The term "tissue culture" was originally applied to the cells maintained in the laboratory in this manner, reflecting the origin of the technique. More appropriate to current techniques, the proper terminology is "cell culture," since it is actually individual cells which are grown, developing as explants from tissue. Nevertheless, the terms tend to be used interchangeably for convenience.

Types of Cell Culture

The most common form of mammalian cell culture is that of the primary explant. Cells are removed from the organism, preferably at the embryonic stage; treated with an enzyme such as trypsin, which serves to disperse the cells; and placed in a laboratory growth vessel. Most of these vessels are composed of polystyrene or similar forms of plastic.

Most forms of cells are anchorage-dependent, meaning they will attach and spread over a flat surface. Given sufficient time, such cells will cover the surface in a layer one cell thick, known as a monolayer.

A few forms of cells, mainly hematopoietic

(blood-forming) or transformed (cancer) cells, are anchorage-independent and will grow in suspension as long as proper nutrients are supplied.

Similar procedures are used in preparation of nonmammalian cell lines such as those from poikilotherms (cold-blooded organisms such as fish) or insects. Insect lines have become particularly important as techniques were developed for cloning genes in insect pathogens known as baculoviruses. Such cells can often be maintained at room temperature in suspension.

DEVELOPMENT OF CELL LINES

A characteristic of primary cells is that of a finite life span; normal cells will replicate approximately fifty times, exhibit symptoms of "aging," and die. When primary cells are removed from a culture and cultured separately, they become known as a cell strain.

A few rare cells may enter "crisis" and begin to exhibit characteristics of abnormal cells such as anchorage-independence or unusual chromosome numbers. If these cells survive, they represent what

is called a "cell line." Cell lines express characteristics of cancer cells and are often immortal.

During the first half century of work in cell culture, only nonhuman cells were grown in culture. In 1952, George Gey, a physician at Johns Hopkins Hospital, demonstrated that human cells could also be grown continuously in culture. Using cervical carcinoma explants from a woman named Henrietta Lacks, Gey prepared a continuous line from these cells. Known as HeLa cells, these cultures became standard in most laboratories studying the growth of animal viruses. Ironically, growth of HeLa cells was so convenient and routine that the cells frequently contaminated other cultures found in the same laboratories.

NUTRIENT REQUIREMENTS

Particular cells may have more stringent requirements for growth than other types of cells; in addition, primary cells have greater requirements than cell lines. However, certain generalities apply to the growth requirements for all cells. All cells must be maintained in a physiological salt solution. Re-

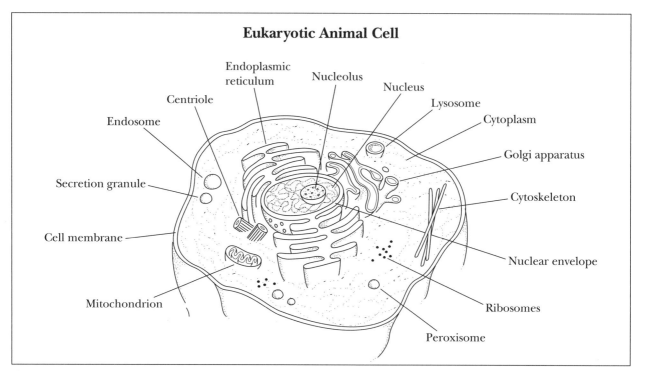

Eukaryotic Animal Cell

All animal cells are eukaryotic cells, which differ from more primitive prokaryotic cells in having a nucleus that houses the primary genetic material. This drawing depicts the basic features of a eukaryotic animal cell. (Electronic Illustrators Group)

quired vitamins and amino acids are included in the mixture. Antibiotics such as penicillin and streptomycin are routinely added to suppress the growth of unwanted microorganisms. Nevertheless, sterility is of utmost importance since some organisms are unaffected by these antibiotics. Depending upon the type of cell, the particular pH, or acid content, of the culture may be variable. Most mammalian cells grow best at a pH of 7.0-7.2. For this reason, cells are generally grown in special incubators which utilize a relatively high CO_2 atmosphere.

Replication of animal cells requires the presence of certain growth factors in the medium. Historically, the source of such factors has been serum, usually obtained from fetal bovines. Genetic engineering techniques have resulted in production of commercially available growth factors, eliminating the requirement for expensive serum for growth of some forms of cells in culture.

GENETICS OF CELLS IN CULTURE

Study of cultured animal cells has resulted in significant advancement in understanding many areas of cell regulation. For example, the role played by cell receptors in response to the presence of extracellular ligands such as hormones and other metabolites was clarified by studying the response of cells to such stimulation. Intracellular events, including the roles of enzymes in cell activities, were clarified and remain a primary area of research.

The ability to transform mammalian cells using isolated DNA has allowed for significant applications in genome analysis. Such genetic manipulation has led to a greater understanding of the role specific genes play in cell regulation. In particular, use of cultured cells was instrumental in clarifying the roles played by specific gene products in intracellular trafficking, the movement of molecules to specific sites within the cell. Similar techniques continue to be used to further understand the regulatory process.

MAMMALIAN CELLS AND ONCOGENESIS

During the 1960's, Leonard Hayflick at the Wistar Institute in Philadelphia, Pennsylvania, observed that primary cells in culture exhibit a finite life span; normal cells generally divide no more than approximately fifty times (a phenomenon now called the Hayflick limit). Any cells that survive generally take on the characteristics of cancer cells.

During the same period, Howard Temin at the University of Wisconsin, while studying the growth of RNA tumor viruses in cultured cells, reported the apparent requirement for DNA production by these viruses in transforming normal cells into cancer cells. Temin's and Hayflick's investigations contributed significantly to explaining how cancer cells differ from normal cells and the understanding of genes involved in development of cancer cells. Eventually, this led to the discovery of oncogenes.

The term "oncogene" is somewhat misleading. Its definition was originally based on the fact that mutations in such genes may contribute to transformation of cells from normal to cancerous. The study of these genes in cultured cells clarified their role: Most oncogene products can be classified as growth factors, which stimulate cell growth; receptors, which respond to such stimulation; or intracellular molecules, which transfer such signals to the cell DNA. In other words, the normal function of the oncogene is to regulate replication of normal cells; only when these proteins are inappropriately expressed do they result in transformation of the cell.

APPLICATION OF CELL CULTURE TO VIROLOGY

The use of mammalian cells for the study of viruses represented among the earliest, and arguably among the most important, applications of the technique of cell culture. Prior to the 1940's, study of most animal viruses, including those that cause disease in humans, was confined to in vivo studies in animals. For example, the study of poliovirus required inoculation of the virus directly into the brains of suitable monkeys.

In 1949, John Enders and his coworkers demonstrated the growth of poliovirus in human embryonic cells, eliminating the requirement for monkeys. Their work played a critical role in the later development of poliovirus vaccines by Jonas Salk and Albert Sabin. The ability to grow viruses in cells maintained in the laboratory opened the field to nearly all virologists and biochemists, rather than restricting such studies to those with access to animal facilities.

Richard Adler, Ph.D.

FURTHER READING

Butler, Michael. *Animal Cell Culture and Technology.* 2d ed. New York: BIOS Scientific, 2004. Designed as an introduction to animal cell cultures for

readers with minimal background knowledge of
the subject. Describes the basic requirements for
establishing and maintaining cell cultures in the
laboratory and in large-scale operations.

Castilho, Leda R., et al., eds. *Animal Cell Technology:
From Biopharmaceuticals to Gene Therapy.* New York:
Taylor and Francis Group, 2008. An overview of
the biological and engineering concepts related
to mammalian and insect cell technology, describing the workings of animal cell technology,
the science upon which it is based, and its numerous applications.

Freshney, R. Ian. *Culture of Animal Cells: A Manual of
Basic Technique.* 5th ed. Hoboken, N.J.: Wiley-Liss,
2005. Basically a how-to text on the science and
art of tissue culture. Useful as a source of recipes
and techniques, as well as an extensive bibliography.

Gold, Michael. *A Conspiracy of Cells: One Woman's Immortal Legacy and the Medical Scandal It Caused.* Albany: State University of New York Press, 1986. A
full account of the history behind development
of the HeLa cell line. Much of the account deals
with the (literal) spread of these cells throughout
the field of cell culture.

Hayflick, L., and P. Moorhead. "The Serial Cultivation of Human Diploid Cell Strains." *Experimental
Cell Research* 25 (December, 1961): 585-621. The
classic work that first reported the limited life
span of human cells in culture.

Pollack, Robert, ed. *Readings in Mammalian Cell Culture.* 2d ed. Cold Spring Harbor, N.Y.: Cold
Spring Harbor Press, 1981. A collection of reprints consisting of nearly all classic papers in the
field of cell culture.

WEB SITES OF INTEREST

Growth of Animal Cells in Culture
http://www.ncbi.nlm.nih.gov/books/bv.fcgi?rid
=mcb.section.1383

This page from an online textbook about molecular biology provides a detailed discussion, with
links to numerous illustrations, about animal cell
culture.

Introduction to Animal Cell Culture
http://catalog2.corning.com/lifesciences/media/
pdf/intro_animal_cell_culture.pdf

Although this eight-page technical bulletin was
written by an employee of Corning Incorporated

and includes some advertising for the company's
glass products, it provides straightforward and understandable information about cell culture techniques and applications.

See also: Cancer; Cell culture: Plant cells; Cell cycle;
Cell division; Gene regulation: Eukaryotes; Gene
regulation: Viruses; Mitosis and meiosis; Oncogenes; Stem cells; Totipotency; Tumor-suppressor
genes; Viral genetics.

Cell culture
Plant cells

CATEGORY: Cellular biology; Techniques and methodologies
SIGNIFICANCE: Plant cell culture is the establishment and subsequent growth of various plant
cells, tissues, or organs in vitro, using an artificial
nutritional medium usually supplemented by various plant growth regulators. It has become a
tool that plant geneticists use for purposes ranging from the basic study of plant development to
the genetic improvement of economically important agricultural plant species.

KEY TERMS

callus: a group of undifferentiated plant cells growing in a clump
morphogenesis: the induction and formation of organized plant parts or organs
plant growth regulators: hormonelike substances that
profoundly affect plant growth and development
somatic embryos: asexual embryoid structures derived
from somatic cells
totipotency: the ability of a plant cell or part to regenerate into a whole plant

CULTURING PLANT CELLS

Plant cell cultures are typically initiated by taking
explants—such as root, stem, leaf, or flower tissue—
from an intact plant. These explants are surface-
sterilized and then placed in vitro on a formulated,
artificial growth medium containing various inorganic salts, a carbon source (such as sucrose), vitamins, and various plant growth regulators, depend-

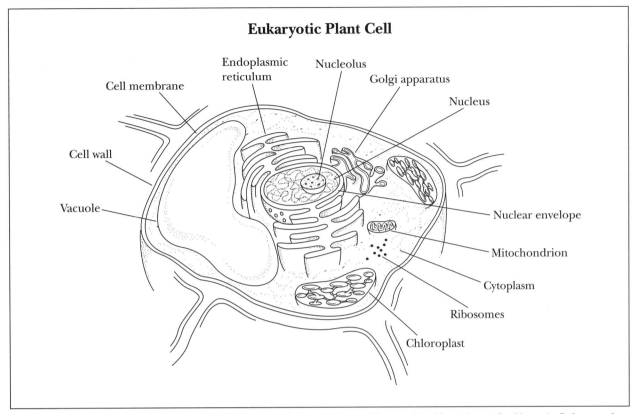

Eukaryotic Plant Cell

All plant cells, like animal cells, are eukaryotic cells. However, plant cells contain chloroplasts, the "factories" that produce chlorophyll during photosynthesis. This drawing depicts the basic features of a eukaryotic plant cell. (Electronic Illustrators Group)

ing on the desired outcome. There are many commercially available media formulations; the two most common include MS (murashige and skoog) and WPM (woody plant media). Alternatively, customized formulations may be necessary for culturing certain plant species. One of the most important uses of plant tissue culture has been for the mass propagation of economically important agricultural and horticultural crops. Since the 1980's, however, plant cell culture has become an important tool allowing for direct genetic manipulations of several important agricultural crops, including corn, soybeans, potatoes, cotton, and canola, to name only a few.

APPEARANCE IN CULTURE

The underlying basis for the prevalent and continued use of plant cell culture is the remarkable totipotent ability of plant cells and tissues. They are able to dedifferentiate in culture, essentially becom-

ing a nondifferentiated clump of meristematic, loosely connected cells termed callus. Callus tissue can be systematically subcultured and then, depending on exposure to various plant growth regulators incorporated in the growth media, induced to undergo morphogenesis. Morphogenesis refers to the redifferentiation of callus tissue to form specific plant organs, such as roots, shoots, or subsequent whole plants. Many plant species can also be manipulated in culture to form somatic embryos, which are asexual embryoid structures that can then develop into plantlets. The totipotency of plant cells thus allows for a single cell, such as a plant protoplast, to be able to regenerate into a complete, whole plant. An analogous comparison of the totipotency of plant cells would be that of stem cells in animals. Genetic manipulation of individual plant cells coupled with their totipotency makes plant cell culture a powerful tool for the plant geneticist.

ROLE OF PLANT GROWTH REGULATORS

Hormones or plant growth regulators (PGRs) are naturally occurring or synthetic compounds that, in small concentrations, have tremendous regulatory influence on the physiological and morphological growth and development of plants. There are several established classes of PGRs, including auxins, cytokinins, gibberellins, abscisic acid (ABA), and ethylene. Additionally, several other compounds, such as polyamines, oligosaccharides, and sterols, exert hormonelike activity in plant cell cultures. While each class has a demonstrative and unique effect on overall whole plant growth and development, auxins and cytokinins continue to be the most widely used in manipulating plant growth in vitro. Auxins (such as IAA, NAA, and 2,4-D) and cytokinins (such as zeatin, kinetin, and BAP) are frequently used in combination in plant tissue culture. Generally, a high auxin-to-cytokinin ratio results in the induction of root tissue from callus, while a high cytokinin-to-auxin ratio results in the induction of shoot formation. For many plant species, an intermediate ratio of auxin to cytokinin results in continued callus formation.

There are also specific uses of certain PGRs in plant cell culture. For example, 2,4-D is typically used to induce somatic embryogenesis in cultures but then must be removed for subsequent embryoid development. Gibberellins, such as GA_4 and GA_7, can be inhibitory to morphogenesis. Some PGRs may even elicit opposite morphogenic effects in two different plant species. Nevertheless, the use of PGRs remains essential in plant cell culture to direct morphological development.

APPLICATIONS AND POTENTIAL

Plant cell culture as a tool has greatly enhanced the ability of the plant geneticist in the area of crop improvement. Haploid cell cultures initiated from pollen can result in homozygous whole plants, which are very useful as pure lines in breeding programs. In such plants, recessive mutations are easily identified.

The enzymatic removal of the plant cell wall yields naked plant protoplasts, which are more amenable to genetic manipulation. Protoplasts of different species can be chemically or electrically fused to give somatic hybrids that may not be obtained through traditional sexual crossing due to various types of sexual incompatibility. As they divide and regenerate cell walls, these somatic hybrids can then be selected for desired agriculture characteristics, such as insect or disease resistance.

The isolation of plant protoplasts from leaves results in millions of individual cells. As they divide, grow, and differentiate into whole plants, some may contain spontaneous mutations or other changes which can be selected for. Screening for such characteristics, such as salt tolerance or disease resistance, can be done in vitro, thereby saving time and space.

Another use of plant cell culture in crop improvement involves directed genetic transformation. Genes from other species, including bacteria, animals, and other plants, have been introduced into cell cultures, resulting in genetically modified (GM) plants. The most common technique used to transfer desired genes uses the bacterium *Agrobacterium tumefaciens.* Other techniques include electroporation, microinjection, and particle bombardment with "gene guns." As genetic engineering of plants proceeds and is refined, plant cell culture will continue to play a vital role as a tool in this effort.

Thomas J. Montagno, Ph.D.

FURTHER READING

George, Edwin F., Michael A. Hall, and Geert-Jan De Klerk, eds. *Plant Propagation by Tissue Culture.* 3d ed. 2 vols. London: Springer, 2008. An exhaustive presentation of nutritional media components and discussion of PGR effects in culture. Provides specifics on the culture of several hundred species. Illustrations, photographs.

Neumann, Karl-Hermann Neumann, Ashwani Kumar, and Jafargholi Imani. *Plant Cell and Tissue Culture—a Tool in Biotechnology: Basics and Application.* Berlin: Springer, 2009. An introduction to the field of plant cell and tissue culture, including a detailed description of various techniques employed in laboratories worldwide, an account of applications in plant propagation and gene technology, and a survey of advances in the field.

Razdan, M. K. *Introduction to Plant Tissue Culture.* 2d ed. Enfield, N.H.: Science, 2003. Provides an introductory overview of plant tissue culture, including information about its history, basic aspects, and applications to plant breeding, horticulture, and forestry.

Trigiano, Robert, and Dennis Gray, eds. *Plant Development and Biotechnology.* Boca Raton, Fla.: CRC

Press, 2005. Includes a history of plant tissue and cell culture, descriptions of various methods of plant cell culture, and discussions of plant propagation and development concepts and crop improvement techniques. Designed for employees in tissue culture research laboratories.

_____. *Plant Tissue Culture Concepts and Laboratory Exercises.* 2d ed. Boca Raton, Fla.: CRC Press, 2000. A concise historical presentation of plant cell culture, along with current trends. Includes detailed student experiments and procedures. Illustrations, photographs.

WEB SITES OF INTEREST

Plant Tissue Culture
http://www.liv.ac.uk/~sd21/tisscult/introduction.htm
Prepared by the School of Biological Sciences at the University of Liverpool, this site provides an introduction to plant cell culture, describing what it is and its uses; three illustrated case studies; and a bibliography.

Plant Tissue Culture Exchange
http://aggie-horticulture.tamu.edu/tisscult
Texas A&M University maintains this Web site, which contains case studies, articles, resources, and other information about plant tissue culture.

See also: Cancer; Cell culture: Animal cells; Cell cycle; Cell division; Gene regulation: Eukaryotes; Genetic engineering: Agricultural applications; Genetically modified foods; High-yield crops; Mitosis and meiosis; Oncogenes; Shotgun cloning; Stem cells; Totipotency.

Cell cycle

CATEGORY: Cellular biology
SIGNIFICANCE: During the phases of the cell cycle, cells divide (mitosis and cytokinesis), grow (G_1), replicate their DNA (S), and prepare for another cell division (G_2). Protein signals regulate progress through these phases of the cell cycle. Mutations that alter signal structure, time of synthesis, or how signals are received can cause cancer.

KEY TERMS

checkpoint: the time in the cell cycle when molecular signals control entry to the next phase
cyclins: proteins whose levels rise and fall during the cell cycle
kinase: an enzyme that catalyzes phosphate addition to molecules
oncogene: a gene whose products stimulate inappropriate cell division, causing cancer
tumor suppressor: a gene whose product normally prevents or slows cell division; when mutated, these genes can lead to uncontrolled cell division

DEFINING CELL CYCLE PHASES

The eukaryotic cell cycle is defined by five phases. Two of these, mitosis and cytokinesis, do not last long. Mitosis itself has five phases:

(1) prophase, when duplicated attached chromatids with replicated DNA condense and become visible as chromosomes, each composed of two sister chromatids
(2) metaphase, when the chromosomes attach to spindle fibers and move to the middle of the cell
(3) anaphase, when sister chromatids separate
(4) telophase, when the separated sister chromatids, now chromosomes, move to opposite poles of the cell, during which cytokinesis often starts
(5) interphase, a time between successive mitoses when cells approximately double in size

An early experiment showed that DNA replicates long before mitosis. After a short exposure (pulse) of cells to radioactive thymidine to allow synthesis of radioactive DNA, the "hot" nucleotide was removed and cells were allowed to grow for different chase times in a medium containing nonradioactive nucleotides. (The term "chase" refers to the second part of what is called a radioactive pulse-chase experiment.) Autoradiography showed that after the pulse period, 40 percent of cells were labeled, but only interphase nuclei were radioactive. This established that DNA is not actually synthesized during mitosis. Labeled mitotic chromosomes were seen only in cells chased for 4 to 14 hours. After longer chase times, labeling was again confined to interphase nuclei. From this kind of study, the cell cycle could be divided into five major phases (times listed are typical of a cell in an adult organism):

(1) mitosis (M), one hour

(2) cytokinesis (C), thirty minutes

(3) gap 1 (G$_1$), a time of cell growth, which lasts the generation time minus the times of the other phases

(4) synthesis (S), nine hours of DNA synthesis

(5) gap 2 (G$_2$), four to five hours of preparation for the next mitosis

IDENTIFICATION OF CELL CYCLE SWITCHES

Cells reproduce at different rates. Embryonic cells divide hourly or more often, while neurons stop dividing altogether shortly after birth. Cells divide, or stop dividing, in response to chemical signals. When mitotic cells are fused with G$_1$, S, or G$_2$ cells, fused cells at first contain the chromosomes of the mitotic cell alongside the intact nucleus of the other cell. After a few minutes, the intact nucleus disintegrates and its chromosomes also condense, suggesting that some chemical signal is causing the intact nucleus to respond as if it were undergoing mitosis. This suggests that cells in mitosis contain a substance that induces nondividing cells to become "mitotic."

The first chemical signal controlling the cycle was discovered in studies of amphibian oogenesis. Oocyte maturation begins with the first meiotic

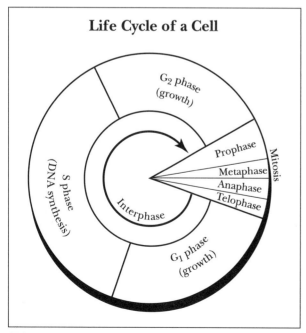

Life Cycle of a Cell

G$_2$ phase (growth)

Prophase

Metaphase

Anaphase

Telophase

Mitosis

Interphase

S phase (DNA synthesis)

G$_1$ phase (growth)

(Electronic Illustrators Group)

division, when the germinal vesicle (the oocyte nucleus) breaks down and chromosomes and spindle fibers first appear at one pole of the oocyte. In a key experiment, cytoplasm taken from oocytes during germinal vesicle breakdown was injected into immature oocytes. Condensed chromosomes quickly appeared in the injected oocytes. A protein called MPF (maturation promoting factor) was purified from the older oocytes. MPF was later found in developing frog embryos, where its levels fluctuated, peaking just before the embryonic cells began mitosis. Thus, MPF also controls mitosis as well as meiosis and is often called mitosis-promoting factor. MPF consists of cyclin and cyclin-dependent kinase (cdk). Cyclin-bound cdk catalyzes phosphorylation of other cellular proteins. Levels of cdk were shown to be constant in the cell, while cyclin levels rose and peaked late in G$_2$, explaining why MPF activity is highest during mitosis and why mitotic cells induce nuclear breakdown and chromosome condensation when fused to nonmitotic cells.

To study cell cycle regulation further, researchers turned to yeast, a model single-cell eukaryote easily subject to genetic manipulation. Mutagenized yeast was screened for temperature-sensitive mutations that reproduced at lower temperatures but were blocked at one or another point in the cell cycle when grown at higher temperatures. One such temperature-sensitive mutant was arrested in G$_2$ at the higher temperature. These cells had a defective cell-division-cycle-2 (*cdc2*) gene encoding a yeast version of the frog cdk in MPF. Cellular cdc2 levels are stable, but its kinase activity depends on a yeast cyclin whose levels peak at the end of G$_2$. The active yeast MPF triggers passage through the G$_2$ checkpoint, committing the cell to mitosis. Other mutants were found encoding separate G$_1$ cyclin and G$_1$ cdk proteins that together form an active kinase that triggers passage through a G$_1$ checkpoint into the S phase of the cell cycle. Among higher eukaryotes, different combinations of cyclins and cdk's act at still other checkpoints in the cell cycle.

HOW MPF AND G$_1$ CDK'S WORK

The proteins phosphorylated by yeast MPF and G$_1$ cyclin-cdk catalysis function in pathways that promote mitosis and cytokinesis, on one hand, and DNA replication, on the other. How are cdk's activated, what proteins do they phosphorylate, and what do these phosphorylated proteins do?

Yeast MPF made by joining late G_2 cyclin and cdk is not active until it is itself phosphorylated. MPF first receives two phosphates. Then the addition of a third phosphate causes the first two to come off in a peculiar MPF activation pathway. Infact, MPF remains unphosphorylated and inactive in cells experimentally prevented from replicating. In normal cells, blocking premature activation of MPF could prevent condensing chromosomes from damaging DNA that is still replicating. When properly activated, MPF phosphorylates (a) proteins that break down the nuclear membrane, (b) histones and other chromatin proteins thought to start chromosome condensation, and (c) microtubule-associated proteins associated with mitotic spindle formation.

G_1 cyclin and cdk production occurs when cells reach a suitable size during G_1 and when they are stimulated by a growth factor. For example, EGF (epidermal growth factor) stimulates embryonic cell growth by binding to cell membrane receptors. EGF-receptor binding converts the intracellular domain of the receptor into an active protein kinase that catalyzes self-phosphorylation. The autophosphorylated receptor activates a G-protein encoded by the *ras* gene, which binds GTP. Then, ras-GTP activates the first in a series of protein kinases, setting off an intracellular kinase cascade. Sequential phosphorylations finally stimulate synthesis of G_1 cyclin and G_1 cdk. Active cyclin-bound G_1 cdk then phosphorylates the Rb protein, causing it to detach from protein EF2, which becomes an active transcription factor that stimulates synthesis of proteins needed for replication in the S phase.

To summarize, MPF is activated by a phosphorylation pathway in which the kinase itself becomes phosphorylated, while G_1 cdk is made in response to growth factors like EGF that initiate phosphorylation cascades, resulting in the eventual synthesis of cyclin and cdk. MPF phosphorylates other proteins, permitting transition across the G_2 checkpoint, while the G_1 cdk allows progress through the G_1 checkpoint.

THE CELL CYCLE AND CANCER

With the discovery of the first MPF, scientists had already begun to suspect that mutations in genes encoding proteins involved in cell cycling might cause the uncontrollable cell divisions associated with cancer. Many cancers are associated with onco-

genes (called proto-oncogenes when they function and are expressed correctly) encoding proteins involved in cell cycle control. Some oncogenes are carried by viruses, but most arise by mutation of their normal counterparts, resulting in inappropriate activity of the protein encoded by the gene. Representative human oncogenes include *neu* (a growth-factor-receptor oncogene associated with breast and ovarian cancers), *trk* (a receptor oncogene associated with colon cancer), *ras* (a G-protein oncogene), *L-myc* (a transcription-factor oncogene causing small-cell lung cancer), *cdk-4* (a cyclin-dependent kinase oncogene causing a muscle sarcoma), and *CYCD1* (a cyclin oncogene associated with lymphoma). Each of these oncogenes produces proteins that promote unrestricted passage through the cell cycle. In contrast, retinoblastoma is a rare eye cancer in which the *Rb* oncogene product is not made, so that *EF2* transcription factor is always active and genes involved in replication are continuously on. Because *Rb* restrains unwanted cell divisions, it is called a tumor-suppressor gene. Unfortunately, the *Rb* oncogene is also associated with more common human adult lung, breast, and bladder cancers. Another tumor-suppressor gene, *p53*, is also implicated in several human cancers; a defective *p53* gene allows cells with damaged DNA to replicate, increasing the chances of cancer development.

In the brief history of cell-cycle studies, the discovery of an oncogene identifies the cause of a cancer while the newfound actor in a phosphorylation cascade is a candidate for an oncogene. The study of the cell cycle is an excellent example of the synergy between basic and applied science: The fundamental drive to know how cells grow and divide has merged with a fervent desire to conquer a group of human diseases increasingly prevalent in an aging population.

Gerald K. Bergtrom, Ph.D.

FURTHER READING

Alberts, Bruce, et al. "The Cell Cycle." In *Molecular Biology of the Cell*. 5th ed. New York: Garland Science, 2008. This cell biology textbook devotes an entire chapter to the cell cycle.

Becker, Wayne M., et al. *The World of the Cell*. 7th ed. San Francisco: Pearson/Benjamin Cummings, 2009. Provides an excellent overview of cell components regulating the cell cycle and how their dysfunction can cause cancer.

Campbell, Neil A., and Jane Reece. *Biology.* 8th ed. San Francisco: Pearson, Benjamin Cummings, 2008. A standard, introductory biology textbook for undergraduate majors that includes a detailed account of meiosis.

McCormick, F. "Signaling Networks That Cause Cancer." *Trends in Cell Biology* 9, no. 12 (December, 1999): M53-M56. A review of signaling pathway components whose inappropriate activity causes cancer.

Morgan, David O. *The Cell Cycle: Principles of Control.* London: New Science Press in association with Oxford University Press, 2007. Explains the mechanisms that control cell division, including a description of the phases and main events of the cell cycle, the main model organisms in cell-cycle analysis, cell-cycle control in development, and the failure of controls in cancer.

Murray, A. W., and Tim Hunt. *The Cell Cycle: An Introduction.* New York: W. H. Freeman, 1993. An informative overview for both students and general readers, without too much scientific jargon. Bibliographical references, index.

Orr-Weaver, T. L., and R. A. Weinberg. "A Checkpoint on the Road to Cancer." *Nature* 392, no. 6673 (March 19, 1998): 223-224. Describes a mutation in a gene regulating the cell cycle in cancer cells.

Stein, Gary S., and Arthur B. Pardee, eds. *Cell Cycle and Growth Control: Biomolecular Regulation and Cancer.* 2d ed. Hoboken, N.J.: Wiley-Liss, 2004. Explains the cell cycle and cell growth control, with an emphasis on aberrations accompanying the onset and progression of cancer.

WEB SITES OF INTEREST

Cells Alive!
http://www.cellsalive.com

This site provides interactive visuals that enable users to learn about the structure and function of eukaryotic cells. One of the pages contains text and an animation that explain the cell cycle.

Kimball's Biology Pages, The Cell Cycle
http://users.rcn.com/jkimball.ma.ultranet/
BiologyPages/C/Cell cycle.html

John Kimball, a retired Harvard University biology professor, includes a page about the cell cycle in his online cell biology text.

See also: Cancer; Cell division; Chemical mutagens; Chromosome mutation; Cytokinesis; DNA repair; Gene regulation: Eukaryotes; Mitosis and meiosis; Mutation and mutagenesis; Oncogenes; Stem cells; Telomeres; Totipotency; Tumor-suppressor genes.

Cell division

CATEGORY: Cellular biology

SIGNIFICANCE: Eukaryotic cell division (mitosis and cytokinesis) is a short part of the cell cycle. In the longer time between successive cell divisions cells grow and replicate their DNA. Molecular signals tell cells when to enter each stage of the cycle.

KEY TERMS

asexual reproduction: a form of reproduction wherein an organism's cell DNA doubles and is distributed equally to progeny cells

binary fission: cell division in prokaryotes in which the plasma membrane and cell wall grow inward and divide the cell in two

chromatid: one-half of a replicated chromosome

chromatin: the material that makes up chromosomes; a complex of fibers composed of DNA, histone proteins, and nonhistone proteins

chromosome: a self-replicating structure, consisting of DNA and protein, that contains part of the nuclear genome of a eukaryote; also used to describe the DNA molecules constituting the prokaryotic genome

cyclin-dependent kinases (cdk's): proteins that regulate progress through the eukaryotic cell cycle

cyclins: proteins whose levels rise and fall during the cell cycle

cytokinesis: movements of and in a cell resulting in the division of one eukaryotic cell into two

DNA replication: synthesis of new DNA strands complementary to parental DNA

genome: the species-specific, total DNA content of a single cell

meiosis: a type of cell division that leads to production of gametes (sperm and egg) during sexual reproduction

mitosis: nuclear division, a process of allotting a complete set of chromosomes to two daughter nuclei

phases of mitosis and meiosis: periods—including prophase, metaphase, anaphase, and telophase—characterized by specific chromosomal events during cell division

phases of the cell cycle: mitosis, cytokinesis, G_1 (gap 1), S (DNA synthesis), and G_2 (gap 2)

phosphorylation: a chemical reaction in which a phosphate is added to a molecule, common in the control of cell activity, including the regulation of passage through different stages of the cell cycle

ASEXUAL VS. SEXUAL REPRODUCTION

A cell's genetic blueprint is encoded in genes written in the four-letter alphabet of DNA, which stands for the four nucleotides that make up the strands of DNA: guanine (G), adenine (A), thymine (T), and cytosine (C). Reproduction of this blueprint is an essential property of life. Prokaryotes (cells without nuclei) contain a single chromosome in the form of a circular double helix. They replicate their DNA and reproduce asexually by binary fission. Eukaryotic cells, with two or more pairs of linear, homologous chromosomes in a nucleus, replicate their DNA and reproduce asexually by mitosis. In sexual reproduction in higher organisms, special cells called germ cells are set aside to form gametes by meiosis. During meiosis, the germ cells duplicate their chromosomes and separate the homologs into gametes. After mitosis, new cells have a copy of all of the chromosomes originally present in the parent cell; after meiosis, gametes (sperm or egg) contain only one of each homologous chromosome originally present in the parent cell. Though their chromosomal outcomes are quite different, the cellular events of mitosis and meiosis share many similar features, discussed below mostly in the context of mitosis. The focus here is on when cells replicate their DNA, when they physically divide, and how they partition duplicate sets of genetic information into progeny cells.

BINARY FISSION VS. MEIOSIS, MITOSIS, AND CYTOKINESIS

During binary fission, which occurs in prokaryotic cells (cells that have no nucleus—primary bacteria), these small cells grow larger, become pinched in the middle, and eventually produce two new cells. A specific base sequence in the circular bacterial DNA molecule attaches to the cell membrane. When this sequence replicates during DNA synthesis it also attaches to the cell membrane, but

on the opposite side of the cell. As the bacterial cell grows and divides, the two DNA attachment points become separated into the progeny cells, ensuring that each gets a copy of the original circular DNA molecule. DNA replication and cell division in prokaryotes are therefore simultaneous processes.

Mitosis (and meiosis) and cytokinesis, by contrast, are processes well separated in time from DNA replication. When first observed in the microscope in the 1880's, mitosis seemed to be a busy time in the life of a cell. During prophase (the initial phase of mitosis), nuclei seem to disintegrate in a matter of minutes at the same time that chromosomes take shape from nondescript nuclear substance. Spindle fibers form at opposite poles and grow toward the center of the cell. After about thirty minutes, cells are in metaphase. The spindle fibers extend across the cell, attaching to fully formed chromosomes lined up at the metaphase plate in the middle of the cell. Each chromosome is actually composed of two attached strands, or chromatids.

During anaphase the chromatids of each chromosome pull apart and move toward opposite poles of the cell. Telophase is characterized by the reformation of nuclei around the chromosomes and the de-condensation of the chromosomes back to the shapeless substance now called chromatin.

Cytokinesis, meaning "cell movement," begins during telophase, lasts about thirty minutes, and is the actual division of the parent cell into two cells, each of which gets one of the newly forming nuclei. The processes of mitosis and cytokinesis, which together typically last about 1.5 hours, ensure that duplicated pairs of chromosomes are partitioned correctly into progeny cells.

Meiosis actually consists of two cell divisions, each progressing through prophase, metaphase, anaphase, and telophase. In the first division, homologous chromosomes with their chromatids are separated into progeny cells; in the second, chromatids are pulled apart into the cells that will become gametes. The result is to produce haploid eggs or sperm, rather than the diploid progeny with paired homologous chromosomes that result from mitosis.

THE CELL CYCLE

Early histologists studying mitosis noted that it often took cells about twenty hours to double, implying a long period between successive cell divisions. This period was called interphase, meaning simply "between" the mitotic phases. An interphase also separates the first meiotic division from a prior mitosis, though there is not always an interphase between the first and second meiotic divisions. One might have suspected that cells were not just biding their time between mitoses, but it was only in the middle of the twentieth century that the cell cycle was fully characterized, showing interphase to be a long and very productive time in the life of a cell.

In an elegant experiment, cultured cells were exposed to radioactive thymidine, a DNA precursor. After a few minutes, radioactive DNA was detected in the nuclei of some cells. However, no cells actually in mitosis were radioactive. This meant that DNA is not synthesized during mitosis. Radioactive condensed mitotic chromosomes were detected only four to five hours after cells had been exposed to the radioactive DNA precursor, suggesting that replication had ended four to five hours before the beginning of mitosis. Studies like this eventually revealed the five major intervals of the cell cycle: mitosis, cytokinesis, gap 1 (the G_1 phase, a time of cell growth), DNA synthesis (the S phase of DNA synthesis), and gap 2 (the G_2 phase, during which a cell continues growing and prepares for the next mitosis).

The overall length of the cell cycle differs for different cell types. Human neurons stop dividing shortly after birth, never to be replaced. Many other differentiated cells do not divide but are replaced periodically by stem cells that have the capacity to continue to divide and differentiate. Clearly, human genes must issue instructions telling cells when and when not to reproduce.

CONTROLLING THE CELL CYCLE

Sometimes cells receive faulty instructions (for example, from environmental carcinogens) or respond inappropriately to otherwise normal commands from other cells. Cancer is a group of diseases in which normal regulation of the cell cycle has been lost and cells divide out of control. In research published in the 1970's, cells synchronized in mitosis were mixed with others synchronized in other phases of the cell cycle in the presence of polyethylene glycol (the main ingredient in automobile antifreeze). The antifreeze caused cells to fuse. Right after mixing, chromosomes and a mitotic spindle could be seen alongside an intact nucleus in the fused cells. Later, the intact nucleus broke down and chromosomes condensed. The conclusion from studies like this is that mitosing cells contain a substance that causes nuclear breakdown and chromosome condensation in nonmitosing cells. Similar results were seen when cells in meiosis were fused with nonmeiotic cells. When purified, the substances from meiotic and mitotic cells could be injected into nonmitosing cells, where they caused nuclear breakdown and the appearance of chromosomes from chromatin. The substance was called maturation (or mitosis) promoting factor (MPF). MPF contains one polypetide called cyclin and another called cyclin-dependent kinase (cdk). The kinase enzyme catalyzes transfer of a phosphate to other proteins; it is active only when bound to cyclin—hence the name. The kinase is always present in cells, while cyclin concentrations peak at mitosis and then fall. This explains why MPF activity is highest during mitosis and why mitotic cells fused to G_1 cells, for example, can cause the G_1 cell nucleus to disappear and chromosomes to emerge from chromatin.

Since the initial discovery of MPF, studies of eukaryotic cells, from yeast cells to human cells, have revealed many different cyclin-dependent kinases and other regulatory proteins that exert control at different checkpoints on the cell cycle, determining whether or not cells progress from one stage to another. Scientists remain ignorant of the exact causes of most cancers, but because of the compelling need to know, researchers are beginning to understand the normal controls on cellular reproduction.

A final word on the cyclin-dependent protein kinase: This enzyme is one of a large number of kinases that participate in regulating cell chemistry and behavior in response to many extracellular signals (such as hormones). The phosphorylation of cellular proteins has emerged as a major theme in the regulation of many cellular activities, including cell division.

Gerald K. Bergtrom, Ph.D.

FURTHER READING

Alberts, Bruce, et al. *Molecular Biology of the Cell.* 5th ed. New York: Garland Science, 2008. This cell biology textbook contains numerous references to cell division.

Baringa, M. "A New Twist to the Cell Cycle." *Science* 269, no. 5524 (August 4, 1995): 631-632. Addresses how periodic changes in cyclin concentrations regulate the cell cycle.

Campbell, Neil A., and Jane Reece. *Biology.* 8th ed. San Francisco: Pearson, Benjamin Cummings, 2008. Includes a detailed account of meiosis in a standard textbook for undergraduate majors.

Karp, Gerald. *Cell and Molecular Biology: Concepts and Experiments.* 5th ed. Chichester, England: John Wiley and Sons, 2008. Detailed accounts of mitosis and events and regulation of the cell by cyclins and kinases are included in this standard textbook for professionals and undergraduate majors.

Morgan, David O. *The Cell Cycle: Principles of Control.* London: New Science Press in association with Oxford University Press, 2007. Explains the mechanisms that control cell division, including a description of the phases and main events of the cell cycle, the main model organisms in cell-cycle analysis, cell-cycle control in development, and the failure of controls in cancer.

Murray, A. W., and Tim Hunt. *The Cell Cycle: An Introduction.* New York: W. H. Freeman, 1993. An informative overview for both students and general readers, without too much scientific jargon. Bibliographical references, index.

Orr-Weaver, T. L., and R. A. Weinberg. "A Checkpoint on the Road to Cancer." *Nature* 392, no. 6673 (March 9, 1998): 223-224. Describes a mutation in a gene regulating the cell cycle in cancer cells.

Stein, Gary S., and Arthur B. Pardee, eds. *Cell Cycle and Growth Control: Biomolecular Regulation and Cancer.* 2d ed. Hoboken, N.J.: Wiley-Liss, 2004. Explains the cell cycle and cell growth control, with an emphasis on aberrations accompanying the onset and progression of cancer.

WEB SITES OF INTEREST

The American Society for Cell Biology, Image and Video Library
http://cellimages.ascb.org

The society's library has a section with more than twenty images and videos that demonstrate the processes of cell division, growth, and death.

Cells Alive!
http://www.cellsalive.com

Cells Alive! provides interactive visuals that enable users to learn about the structure and function of eukaryotic cells. The site contains individual

pages with text and animation that explain the cell cycle, animal cell meiosis, and animal cell mitosis.

Kimball's Biology Pages
http://users.rcn.com/jkimball.ma.ultranet/BiologyPages

John Kimball, a retired Harvard University biology professor, includes pages about the cell cycle, meiosis, and mitosis in his online cell biology text.

Nova Online, How Cells Divide: Mitosis Versus Meiosis
http://www.pbs.org/wgbh/nova/baby/divide.html

The process of cell division is explained in several formats, including one that uses flash animation technology.

See also: Cell culture: Animal cells; Cell culture: Plant cells; Cell cycle; Cytokinesis; Gene regulation: Eukaryotes; Mitosis and meiosis; Polyploidy; Totipotency.

Central dogma of molecular biology

CATEGORY: Molecular genetics

SIGNIFICANCE: The central dogma states precisely how DNA is processed to produce proteins. Originally thought to be a unidirectional process proceeding from DNA to RNA and then to protein, it is now known to include reverse transcription and the enzymatic activity of certain RNA molecules. The central dogma lies at the core of molecular genetics, and understanding it, and particularly reverse transcription, is key to comprehending both the way viruses cause disease and methods that have revolutionized biology.

KEY TERMS

codon: three nucleotides in DNA or RNA that correspond with a particular amino acid or stop signal

colinearity: the exact correspondence between DNA or RNA codons and a protein amino acid sequence

complementary bases: the nucleic acid bases in different strands of nucleic acid in RNA and DNA that pair together through hydrogen bonds: guanine-cytosine and adenine-thymine (in DNA and RNA) and adenine-uracil (in RNA)

exon: the part of the coding sequence of mRNA that specifies the amino acid sequence of a protein

hydrogen bond: a weak chemical bond that forms between atoms of hydrogen and atoms of other elements, including oxygen and nitrogen

intron: a noncoding intervening sequence present in many eukaryotic genes that is transcribed but removed before translation

retrovirus: a virus that carries reverse transcriptase that converts its RNA genome into a DNA copy that integrates into the host chromosome

reverse transcription: the conversion of RNA into DNA catalyzed by the enzyme reverse transcriptase

ribozyme: catalytic RNA

subunit: a polypeptide chain of a protein

ORIGINAL CENTRAL DOGMA

Nobel Prize winner Francis Crick, who was codiscoverer with James Watson of the double helical structure of DNA, coined the term "central dogma" in 1958 to describe the fact that the processing of genetic information contained in DNA proceeded unidirectionally by its conversion first to an RNA copy, called messenger RNA (mRNA), in a molecular process called transcription. Then the genetic information contained in the sequence of bases in the mRNA was read in the ribosome, and the appropriate amino acids carried by transfer RNAs (tRNAs) were assembled into protein according to the genetic code in a process called translation. The basis of these reactions stemmed from the properties of DNA, particularly its double helical structure. The fact that the two strands of DNA were held together by hydrogen bonds between specific nucleic acid bases (guanine-cytosine, adenine-thymine) on the two strands clearly suggested how the molecule could be duplicated. Watson and Crick postulated that if they split the double-stranded structure at the hydrogen bonds, attached new complementary bases, and reformed the hydrogen bonds, precise copies identical to the original DNA would result. In an analogous manner, RNA was produced by using one DNA strand as a template and adding the correct complementary bases according to what came to be called Watson Crick base pairing. Thus the original dogma stated that transfer of genetic information proceeded unidirectionally, that is, only from DNA to RNA to protein. The only exception was the duplication of DNA in a process called replication.

MODIFIED CENTRAL DOGMA

Several discoveries made it necessary to change the central dogma. The first and most heretical information came from the study of retroviruses, including the human immunodeficiency virus (HIV). Howard Temin reported that viruses of this group contained an enzyme called reverse transcriptase, which was capable of converting RNA to DNA and thus challenging the whole basis of molecular reactions and the central dogma. Temin and David Baltimore were subsequently awarded Nobel Prizes for their work describing this new enzyme. They were able to show that it synthesizes a DNA strand complementary to the RNA template, and then the DNA-RNA hybrid is converted to a DNA-DNA molecule, which inserts into the host chromosome. Only then can transcription and translation take place.

The second significant change was finding that RNA can act as a template for its own synthesis. This situation occurs in RNA bacteriophages such as MS2 and QB. These phages are very simple, with genomes specifying only three proteins, a coat and attachment proteins and an RNA replicase subunit. This subunit combines with three host proteins to form the mature RNA replicase that catalyzes the replication of the single-stranded RNA. Thus translation to form the protein subunit of RNA replicase occurs using the RNA genome as mRNA upon viral infection without transcription taking place. Only then is the RNA template successfully replicated.

The third natural modification of the original dogma also concerned the properties of RNA. Thomas Cech in 1982 discovered that introns could be spliced out of eukaryotic genes without proteins catalyzing the process. For the discovery and characterization of catalytic RNA, Cech and Sidney Altman were awarded Nobel Prizes for their work in 1989. Their experiments demonstrated that RNA introns, also called ribozymes, had enzymatic activity that could produce a functional mRNA. This process occurred by excising the introns and combining the exons, thus restoring colinearity of DNA and amino acid sequence. RNA processing thus demonstrates another needed modification of the central dogma: The colinearity of gene and protein in prokaryotes predicts that gene expression results directly from the sequence of bases in its DNA. In the case of eukaryotic genes with multiple introns, however, colinearity does not result until the RNA processing has taken place. Therefore, the correspondence of the codons in the original DNA sequence containing the introns does not correspond to the order of amino acids in the protein product.

Numerous examples also exist of DNA rearrangements occurring before final gene expression takes place. Examples include the formation of antibodies, the expression of different mating types in yeast, and the expression of different surface antigens in parasites, such as the trypanosome protozoan parasite, which causes sleeping sickness. All of these gene products are produced as a result of gene rearrangements, and the original DNA sequences are not colinear with the amino acid sequences in the protein.

Francis Crick, who with James Watson won the 1962 Nobel Prize in Physiology or Medicine for their discovery of the double helix structure of DNA. Crick articulated the "central dogma" of molecular biology and coined the term. (AP/Wide World Photos)

IMPORTANCE AND APPLICATIONS

The theoretical importance of the central dogma is unquestioned. For example, one modern-day scourge, the human immunodeficiency virus (HIV), replicates its genetic material by reverse transcription (central dogma modification), and one of the drugs shown to contain this virus, azidothymidine (AZT), targets the reverse transcriptase enzyme. Perhaps even more important is the use of the reverse transcription polymerase chain reaction (RT-PCR), one application of the polymerase chain reaction originally devised in 1983 by Kary B. Mullis, formerly of Cetus Corporation. RT-PCR employs reverse transcriptase to form a double-stranded molecule from RNA, resulting in a revolutionary technique that can generate usable amounts of DNA from extremely small quantities of DNA or from poor-quality DNA. Also of practical importance is the laboratory modification of hammerhead ribozymes (central dogma modification), found naturally in plant pathogens, for clinical uses, such as to target RNA viruses infecting patients, including HIV and papillomavirus.

Steven A. Kuhl, Ph.D.

FURTHER READING

Allison, Lizabeth A. "From Gene to Protein." In *Fundamental Molecular Biology.* Malden, Mass.: Blackwell, 2007. This chapter's explanation of how DNA produces proteins includes information about the central dogma.

Cech, T. R. "RNA as an Enzyme." *Scientific American* 255, no. 5 (November, 1986): 64-75. A Nobel Prize winner describes his revolutionary discovery that RNA can catalyze reactions. Includes both charts and color illustrations.

Crick, F. "Central Dogma of Molecular Biology." *Nature* 227, no. 5258 (August 8, 1970): 561-563. The seminal paper in which Nobel laureate Crick, a codiscoverer of DNA's double helical structure, proposed his theory of how molecular reactions occur.

O'Connell, Joe, ed. *RT-PCR Protocols.* Totowa, N.J.: Humana Press, 2002. Collects several papers on the use of reverse transcription polymerase chain reaction in analysis of mRNA, quantitative methodologies, detection of RNA viruses, genetic analysis, and immunology. Tables, charts, index.

Ohtsuki, Takashi, and Masahiko Sisido. "The Central Dogma: From DNA to RNA, and to Protein." In *Automation in Proteomics and Genomics: An Engineering Case-Based Approach,* edited by Gil Alterovitz, Roseann Benson, and Marco Ramoni. Hoboken, N.J.: John Wiley, 2009. Focuses on the molecules and bioprocesses that are related to protein biosynthesis.

Varmus, H. "Retroviruses." *Science* 240, no. 4858 (June 10, 1988): 1427-1435. Describes properties of different retroviruses, including the mechanism of reverse transcription.

Watson, James D., et al. *Molecular Biology of the Gene.* 6th ed. San Francisco: Pearson/Benjamin Cummings, 2008. An eminently readable discussion of the subject by a codiscoverer of DNA's double helical structure. Contains numerous illustrations.

WEB SITES OF INTEREST

Eastern Michigan University, Biology 301: Genetics
http://www.emunix.emich.edu/~rwinning/ genetics/transcr.htm

This online course about genetics includes three pages about genes and transcription, including an illustrated discussion of the central dogma.

Genetic Science Learning Center, University of Utah
http://learn.genetics.utah.edu/content/ begin/dna

Offers information and activities about DNA, proteins, and protein synthesis.

See also: DNA structure and function; Gene regulation: Viruses; Genetic code; Genetic code, cracking of; Molecular genetics; Protein synthesis; Reverse transcriptase; RNA structure and function; RNA world.

Cerebrotendinous xanthomatosis

CATEGORY: Diseases and syndromes

ALSO KNOWN AS: Van Bogaert-Scherer-Epstein disease; xanthomatosis, cerebrotendinous; cerebral cholesterinosis; CTX

DEFINITION

Cerebrotendinous xanthomatosis (CTX) is a slowly progressive genetic disorder caused by muta-

tions in the *CYP27A1* gene, resulting in deficiency of the mitochondrial enzyme sterol 27-hydroxylase (CYP27A1) and deposition of cholestanol (5-alpha-dihydro derivative of cholesterol) in nerve cells and body tissues. Excess cholestanol eventually damages the brain, spinal cord, tendons, lens of the eye, and arteries, leading to paralysis, ataxia, dementia, and/ or coronary heart disease by adulthood.

RISK FACTORS

CTX is rare, occurring in only 3 to 5 per 100,000 births worldwide and 1 per 50,000 births in the United States. The incidence is much higher in the Moroccan Jewish population, occurring in 1 per 108 births. More than three hundred patients have been diagnosed with CTX. Most experts believe that the prevalence is higher; however, no epidemiologic studies exist to verify this.

CTX is an autosomal recessive disorder and can be inherited only if the parents each carry the defective *CYP27A1* gene. For every pregnancy, the risk of the child inheriting both genes and being affected is 25 percent and the risk of the child inheriting one gene and being a carrier is 50 percent. Prenatal testing is available for at-risk populations and for those with a family history of CTX.

ETIOLOGY AND GENETICS

CTX is caused by defects in the *CYP27A1* gene located on the long arm of chromosome 2 (2q33-qter). *CYP27A1* belongs to the Cytochrome P450 superfamily, contains nine exons and eight introns, and consists of 33,515 bases spanning 18.6 kilobase pairs (kpb) of DNA. Mutations of the gene result in deficient CYP27A1, causing truncated versions of the enzyme, and involve mainly adrenodoxin-binding or heme-binding sites. Most alleles are point mutations (for example, amino acid substitution, frameshift, splice-junction variants). More than fifty mutations of the *CYP27A1* gene have been identified, but no genotype-phenotype relationship has been established.

CTX is a lipid storage disease of bile acid synthesis. Bile acids play a major role in the body's homeostasis, are necessary for the absorption of fats and fat-soluble vitamins, and serve as the main catabolic pathway for cholesterol. Cholesterol is the precursor for the synthesis of bile acids, and CYP27A1 is the enzyme initiating the first step in oxidizing the side chain of the sterol which metabolizes it to the primary bile acids, cholic acid, and chenodeoxy-cholic acid (CDCA).

In affected individuals who have the faulty *CYP27A1* gene and lack the CYP27A1 enzyme, cholesterol metabolism is impaired, bile acid production is disrupted, and little or virtually no CDCA is formed. Instead, large amounts of bile alcohols are produced and excreted in urine, and the cholesterol-to-cholestanol ratio is altered, causing pathological storage of cholestanol in tissues and plasma, and eventually in the circulation and brain. Cholestanol is usually present in minute amounts but increases manyfold in the absence of CDCA and becomes toxic. Over time, as cholestanol deposits accumulate, lipid-filled nodules called xanthomas form, first in the Achilles and other tendons and then in the brain, most often in the cerebellar hemispheres and cerebellum. These lesions contain cholesterol and cholestanol in the form of lipid crystals and foamy cell granulomata, respectively, and are responsible for the characteristic ataxia and neurodegeneration associated with CTX.

SYMPTOMS

CTX is characterized by these hallmark conditions: chronic, often intractable, diarrhea in infancy; cataracts in early childhood; and tendon xanthomas in adolescence, with progressive neurologic dysfunction occurring simultaneously or in early adulthood. Such symptoms include dementia and seizures, ataxia and motor disturbances, and peripheral neuropathy. Patients with CTX are also at greater risk for osteoporosis and coronary artery disease.

SCREENING AND DIAGNOSIS

CTX can be diagnosed by clinical features and biochemical testing. Despite the hallmark signs of the disorder, CTX can be difficult to diagnose; clinical presentation and time of onset can vary considerably, and tendon xanthomas may or may not be present. CTX should be suspected whenever there are complaints of chronic diarrhea and evidence of cataracts in the same patient. High cholestanol levels in tissues, plasma, and cerebrospinal fluid; increased concentrations of bile alcohols in urine; and greatly reduced or absent CDCA are diagnostic for CTX. Magnetic resonance imaging (MRI) can detect evidence of cerebral and cerebellar atrophy and xanthomatous lesions in the brain of affected patients. The definitive diagnosis is made by mutation analysis of the *CYP27A1* gene coding for the CYP27A1 enzyme.

TREATMENT AND THERAPY

Unlike most genetic diseases, CTX is a treatable disorder, provided that it is diagnosed before irreversible damage to the nervous system occurs. First-line treatment for CTX is long-term administration of CDCA, which helps prevent further neurological degeneration. In most cases, treatment with exogenous CDCA normalizes cholestanol levels and bile acid synthesis, compensating for the lack of enzyme activity.

PREVENTION AND OUTCOMES

Left untreated, CTX is a slowly progressive and fatal disease. Greater awareness and early diagnosis is crucial since treatment with CDCA can resolve the metabolic abnormalities, stabilize neurologic function, and stop progression of the disease. With genetic testing and a known treatment available, the prognosis for CTX is becoming less grim.

Barbara Woldin

FURTHER READING

"Ataxias." In *Textbook of Clinical Neurology*, edited by Christopher Goetz. 3d ed. Philadelphia: Elsevier/Saunders, 2007. Discusses clinical features, evaluation guidelines, genetic testing, and aspects of ataxia in CTX.

Feldman, Mark, Lawrence Friedman, and Lawrence Brandt. "Other Inherited Metabolic Disorders of the Liver." In *Sleisenger & Fordtran's Gastrointestinal and Liver Disease*. 8th ed. Philadelphia: Elsevier/Saunders, 2006. Discusses CTX in relation to bile acid synthesis.

Moore, David, and James Jefferson. "Cerebrotendinous Xanthomatosis." In *Moore and Jefferson: Handbook of Medical Psychiatry*. 2d ed. Philadelphia: Elsevier/Mosby, 2004. In-depth chapter on CTX, including symptoms, clinical features, etiology, course of disease, and treatment.

WEB SITES OF INTEREST

GeneReviews
http://www.ncbi.nlm.nih.gov/bookshelf/br.fcgi?book=gene&part=ctx

NORD
http://www.rarediseases.org/search/
rdbdetail_abstract.html?disname
=Cerebrotendinous%20Xanthomatosis

Orphanet
http://www.orpha.net/consor/cgi-bin/
OC_Exp.php?Lng=GB&Expert=909

See also: Adrenoleukodystrophy; Alexander disease; Canavan disease; Hereditary diseases; Krabbé disease; Leukodystrophy; Metachromatic leukodystrophy; Pelizaeus-Merzbacher disease; Refsum disease; Vanishing white matter disease.

Charcot-Marie-Tooth syndrome

CATEGORY: Diseases and syndromes
ALSO KNOWN AS: Charcot-Marie-Tooth disease; hereditary motor and sensory neuropathies

DEFINITION

Charcot-Marie-Tooth (CMT) disease is a group of genetic disorders that affects movement and sensation in the limbs. The disease progresses slowly and causes damage to the peripheral nerves that control muscles and transmit sensation.

RISK FACTORS

The primary risk factor for developing CMT is having family members with this disease.

ETIOLOGY AND GENETICS

There are many variations of Charcot-Marie-Tooth syndrome, and at least fifteen different genes have been identified in which mutations leading to the condition might occur. The most common group, known as CMT1, includes all those with identifiable abnormalities in the myelin sheath that surrounds nerve cells. CMT1A disease results from a duplication of the *PMP-22* gene, found at position 17p11.2 on the short arm of chromosome 17. Since the PMP-22 protein is an integral part of the myelin sheath, an excess of this protein causes an abnormal sheath to develop. CMT1B disease results from a mutation in a different gene, *MPZ*, found on the long arm of chromosome 1 (at position 1q22). The

myelin zero protein, encoded by this gene, is also a critical component of the myelin sheath. Both of these disease variants are inherited in an autosomal dominant manner, meaning that a single copy of the mutation is sufficient to cause full expression of the disease. An affected individual has a 50 percent chance of transmitting the mutation to each of his or her children. Many cases, however, result from a spontaneous new mutation, so in these instances affected individuals will have unaffected parents. The much rarer variants, CMT1C and CMT1D, result from mutations in the *LITAF* gene, at position 16p13.3, and the *EGR2* gene, at position 10q21.1, respectively.

CMT2 disease (several different subtypes, designated A-L) is also inherited in an autosomal dominant manner. The molecular defect in this case always involves an abnormality in the axons themselves, rather than in the surrounding myelin sheath.

There are at least six different subtypes of CMT4 disease, resulting from mutations in several different genes. Most of these are identifiable as demyelinating neuropathies, and they are distinguished by an autosomal recessive pattern of inheritance, meaning that both copies of the particular gene must be deficient in order for the individual to be afflicted. Typically, an affected child is born to two unaffected parents, both of whom are carriers of the recessive mutant allele. The probable outcomes for children whose parents are both carriers are 75 percent unaffected and 25 percent affected. If one parent has CMT4 disease and the other is a carrier, there is a 50 percent probability that each child will be affected.

Finally, CMTX disease can result from mutations in at least three distinct genes found on the X chromosome. CMTX1 is a sex-linked dominant disease due to a mutation in the *GJB1* gene, at position Xq13.1, while CMTX2 and CMTX3 are sex-linked recessive diseases resulting from mutations in genes found at positions Xq22.2 and Xq26, respectively.

SYMPTOMS

Symptom onset and type vary depending on the type of CMT. Usually, symptoms first appear in children and young adults. The first sign of CMT is often a high-arched foot or difficulty walking.

Other symptoms may include hammertoes, decreased sensation in the feet and legs, muscle cramping in legs and forearms, difficulty holding the foot up in a horizontal position, frequent sprained ankles and ankle fractures, and problems with balance.

Patients may also experience muscle weakness and atrophy in the lower extremities, which can spread to the upper extremities later in life; foot drop; a diminished ability to detect hot and cold, vibration, and position; difficulty writing, fastening buttons and zippers, and manipulating small objects; and scoliosis. Delay in learning how to walk is a symptom of CMT3; congenital glaucoma is a symptom only of CMT4.

SCREENING AND DIAGNOSIS

The doctor will ask about a patient's symptoms and medical history and will perform a physical exam. Tests may include a nerve conduction study, a test that measures the speed and amplitude of nerve impulses in the extremities; an electromyogram (EMG), a test that records the electrical activity of muscle cells; and a deoxyribonucleic acid (DNA) blood test to confirm certain types of CMT, even if there are no symptoms.

TREATMENT AND THERAPY

Although there is no cure for CMT, treatment may help to improve function, coordination, and mobility. Treatment is also essential to protect against injury due to muscle weakness and diminished sensation. Treatment may include physical and occupational therapy, moderate exercise, braces of lower legs, shoe inserts to correct foot deformity, foot care and routine exams with a specialist (podiatrist), and orthopedic surgery.

PREVENTION AND OUTCOMES

There are no known ways to prevent CMT once a person is born with the condition. Individuals who have CMT or have risk factors may want to talk to a genetic counselor before deciding to have children.

Michelle Badash, M.S.;
reviewed by J. Thomas Megerian, M.D., Ph.D., F.A.A.P.
"Etiology and Genetics" by Jeffrey A. Knight, Ph.D.

FURTHER READING

EBSCO Publishing. *DynaMed: Peroneal Muscular Atrophy.* Ipswich, Mass.: Author, 2009. Available through http://www.ebscohost.com/dynamed.
_____. *Health Library: Charcot-Marie-Tooth Syndrome.* Ipswich, Mass.: Author, 2009. Available through http://www.ebscohost.com.
Haratai, Y., and E. P. Bosch. "Disorders of Peripheral Nerve." In *Neurology in Clinical Practice*, edited by Walter G. Bradley et al. 5th ed. 2 vols. Philadelphia: Butterworth-Heinemann/Elsevier, 2008.
Nave, K. A., M. W. Sereda, and H. Ehrenreich. "Mechanisms of Disease: Inherited Demyelinating Neuropathies—from Basic to Clinical Research." *Nature Reviews. Neurology* 3, no. 8 (August, 2007): 453-464.
Pareyson, D. "Differential Diagnosis of Charcot-Marie-Tooth Disease and Related Neuropathies." *Neurological Sciences* 25, no. 2 (June, 2004): 72-82.
Shy, M. E. "Charcot-Marie-Tooth Disease: An Update." *Current Opinion in Neurology* 17, no. 5 (October, 2004): 579-585.

WEB SITES OF INTEREST

About Kids Health
http://www.aboutkidshealth.ca

Charcot-Marie-Tooth Association
http://www.charcot-marie-tooth.org

Genetics Home Reference
http://ghr.nlm.nih.gov

Health Canada
http://www.hc-sc.gc.ca/index-eng.php

Muscular Dystrophy Association
http://www.mda.org

National Institute of Neurological Disorders and Stroke
http://www.ninds.nih.gov

See also: Amyotrophic lateral sclerosis; Congenital muscular dystrophy; Duchenne muscular dystrophy; Hereditary diseases.

Chediak-Higashi syndrome

CATEGORY: Diseases and syndromes
ALSO KNOWN AS: CHS

DEFINITION

Chediak-Higashi syndrome is a rare autosomal recessive disorder that is associated with a defect in many cells of the body which leads to recurrent infections, prolonged bleeding, increased bruising, and neurologic damage. Affected individuals also have a defect in melanin production that causes decreased pigmentation (partial albinism) of the skin, eyes, and hair.

RISK FACTORS

The primary risk for developing Chediak-Higashi syndrome is a family history of the disease. Consanguinity, or the close familial relationship between parents, is commonly found in affected individuals. The incidence of the disease is unknown, but there have been fewer than five hundred reported cases.

ETIOLOGY AND GENETICS

The main genetic defect in Chediak-Higashi syndrome is of the *LYST* gene, also known as the *CHS1* gene, located on the long arm of chromosome 1 and localized on band 42 (chromosome 1, q42.1-2). Most mutations of this gene result in the absence of lysosomal trafficking protein. Milder forms of the defect encode a partially functioning protein. There are also atypical forms of the syndrome found in adults that do not map to this gene location.

The protein expressed by this gene is important in the fusion of vesicles within the cell. These vesicles include lysosomes of leukocytes and fibroblasts, dense bodies of platelets, and melanosomes of melanocytes. A defect or lack of this protein causes these structures to become larger in size and irregular in shape, which changes how they function.

One of the important features of this syndrome is the effect on the immune system. The lysosome is unable to fuse with the phagosome and cannot release the toxic granules important for killing ingested bacteria. The larger shape and rigid structure of the lysosome inside the cell make it difficult for the leukocytes to leave the blood circulation to fight infection in the tissues. This increases the vulnerability to skin infections and abscess formation. Targeted cell killing by natural killer cells and cytotoxic T cells is abnormal as well. In fibroblasts, the enlarged lysosomes inhibit wound healing. These defects of the cells in the immune system are responsible for the increased infection risk in patients with this syndrome.

Because other cells in the body require lysosomal trafficking protein to function properly, a number of defects are associated with Chediak-Higashi syndrome. Platelets perform poorly during clot formation because they are unable to access important substances from the dense bodies. In epidermal melanocytes, the oversized melanosomes do not transfer melanin to the surrounding keratinocytes, which could explain the lack of pigment seen in the skin, hair, and eyes in this syndrome.

A complication of the syndrome, seen in more than 80 percent of affected individuals, is an accelerated phase of a nonmalignant infiltration of lymphocytic cells in the bone marrow. This lymphoma phase of the disease often results in death. People with milder atypical disease and those children who have successfully been treated by stem cell transplantation often develop neurologic symptoms in early adulthood.

SYMPTOMS

The primary symptoms of Chediak-Higashi syndrome are an increased susceptibility to infection, particularly of the skin, lungs, and mucous membranes. Outwardly, affected individuals may have pale eyes and skin, with a metallic sheen to their hair. Other physical characteristics can include an enlarged liver and spleen. Later in life, they also have increasing signs of neurological problems, including visual problems, muscle weakness, and difficulty with sensation in the lower extremities. They often have difficulty walking and can have seizures. The cause of the progressive peripheral neuropathy is not understood.

SCREENING AND DIAGNOSIS

The presence of giant inclusions within white blood cells on a peripheral blood smear is the only clinical diagnostic test currently available for this disease. Giant granules can also be found in cells from skin, muscle, and nerves. The numbers of both platelets and white cells in the blood are low, and there are other signs that the hematological system is not working properly, such as an enlarged spleen and liver. Neurologically, the brain may appear smaller on MRI or CT scans due to atrophy.

TREATMENT AND THERAPY

There are no specific treatments for Chediak-Higashi syndrome. Allogenic hematopoietic stem cell transplantation appears to have been successful in some patients. In addition, great care must be taken in preventing and treating infections with early diagnosis, aggressive antibiotic therapy, and surgical drainage of abscesses. Nonsteroidal anti-inflammatory drugs (NSAIDs) should not be used because they interfere with normal clotting function.

PREVENTION AND OUTCOMES

Prenatal testing is currently not available on a routine basis. However, it may be of some value for families in which the mutation has been identified. Genetic counseling is recommended for prospective parents with a family history of Chediak-Higashi syndrome.

Death often occurs in the first ten years of life either from chronic infections or as the result of the lymphoma-like complication. However, some persons have survived longer.

Jane Blood-Siegfried, D.N.Sc.

FURTHER READING

Nussbaum, Robert L., Roderick, R. McInnes, and Huntington F. Willard. *Thompson and Thompson Genetics in Medicine.* 7th ed. New York: Saunders, 2007. This is a classic textbook that is very easy to read and understand.

Pritchard, Dorian J., and Bruce R. Korf. *Medical Genetics at a Glance.* 2d ed. Malden, Mass.: Blackwell, 2008. This book is another text on genetics that is easy to understand.

Stevenson, Roger E., et al., eds. *Human Malformations and Related Anomalies.* 2d ed. New York: Oxford University Press, 2006. This book provides descriptions of conditions that can be diagnosed at birth.

WEB SITES OF INTEREST

Bookshelf Gene Reviews
http://www.ncbi.nlm.nih.gov/bookshelf/br.fcgi
?book=gene&part=chediak-higashi

Chediak-Higashi Syndrome Association
http://www.chediak-higashi.org

Genetics Home Reference
http://ghr.nlm.nih.gov/condition
=chediakhigashisyndrome

Online Mendelian Inheritance in Man (OMIM)
http://www.ncbi.nlm.nih.gov/entrez/
dispomim.cgi?id=214500

See also: Albinism; Autoimmune disorders; Hemophilia; Hereditary diseases; Immunogenetics.

Chemical mutagens

CATEGORY: Molecular genetics-

SIGNIFICANCE: Mutagens are naturally occurring or human-made chemicals that can directly or indirectly create mutations or changes in the information carried by the DNA. Mutations may cause birth defects or lead to the development of cancer.

KEY TERMS

deamination: the removal of an amino group from an organic molecule

tautomerization: a spontaneous internal rearrangement of atoms in a complex biological molecule which often causes the molecule to change its shape or its chemical properties

THE DISCOVERY OF CHEMICAL MUTAGENS

The first report of mutagenic action of a chemical occurred in 1946, when Charlotte Auerbach showed that nitrogen mustard (a component of the poisonous "mustard" gas widely used in World War I) could cause mutations in fruit flies (*Drosophila melanogaster*). Since that time, it has been discovered that many other chemicals are also able to induce mutations in a variety of organisms. This led to the birth of genetic toxicology during the last half of the twentieth century, dedicated to identifying potentially mutagenic chemicals in food, water, air, and consumer products. Continued research has identified two modes by which mutagens cause mutations in DNA: (1) by interacting directly with DNA and (2) indirectly, by tricking the cell into mutating its own DNA.

CHEMICAL MUTAGENS WITH DIRECT ACTION ON DNA

Base analogs are chemicals that structurally resemble the organic bases purine and pyrimidine and may be incorporated into DNA in place of the normal bases during DNA replication. An example is bromouracil, an artificially created compound extensively used in research. It resembles the normal base thymine and differs only by having a bromine atom instead of a methyl (CH_3) group. Bromouracil is incorporated into DNA by DNA polymerase, which pairs it with an adenine base just as it would thymine. However, bromouracil is more unstable than thymine and is more likely to change its structure slightly in a process called tautomerization. After the tautomerization process, the new form of bromouracil pairs better with guanine rather than adenine. If this happens to a DNA molecule being replicated, DNA polymerase will insert guanine opposite bromouracil, thus changing an adenine-thymine pair to guanine-cytosine by way of the two intermediates involving bromouracil. This type of mutation is referred to as a transition, in which a purine is replaced by another purine and a pyrimidine is replaced by another pyrimidine.

Another class of chemical mutagens are those that alter the structure and the pairing properties of bases by reacting chemically with them. An example is nitrous acid, which is formed by digestion of nitrite preservatives found in some foods. Nitrous acid removes an amino (NH_3) group from the bases cytosine and adenine. When cytosine is deaminated, it becomes the base uracil, which is not a normal component of DNA but is found in RNA. It is able to pair with adenine. Therefore, the action of nitrous acid on DNA will convert what was a cytosine-guanine base pair to uracil-guanine, which, if replicated, will give rise to a thymine-adenine pair. This is also a transition type of mutation.

Alkylating agents are a large class of chemical mutagens that act by causing an alkyl group (which may be methyl, ethyl, or a larger hydrocarbon group) to be added to the bases of DNA. Some types of alkylation cause the base to become unstable, resulting in a single-strand break in the DNA; this type of event can cause a mutation if the DNA is replicated with no base present or can lead to more serious breaks in the DNA strand. Other alkylation products will change the pairing specificity of the base and create mutations when the DNA is replicated.

Intercalating agents such as acridine orange, proflavin, and ethidium bromide (which are used in labs as dyes and mutagens) have a unique mode of action. These are flat, multiple-ring molecules that interact with bases of DNA and insert themselves between them. This insertion causes a "stretching" of the DNA duplex, and the DNA polymerase is "fooled" into inserting an extra base opposite an intercalated molecule. The result is that intercalating agents cause frame-shift mutations in which the "sense" of the DNA message is lost, just as if an extra letter were inserted into the phrase "the fat cat ate the hat" to make it "the ffa tca tat eth eha t." This occurs because genes are read in groups of three bases during the process of translation. This type of mutation always results in production of a nonfunctional protein.

CHEMICAL MUTAGENS WITH INDIRECT ACTION

Aromatic amines are large molecules that bind to bases in DNA and cause them to be unrecognizable to DNA polymerase or RNA polymerase. An example is N-2-acetyl-2-aminofluorine (AAF), which was originally used as an insecticide. This compound and other aromatic amines are relatively inactive on DNA until they react with certain cellular enzymes, after which they react readily with guanine. Mutagens of this type and all others with indirect action work by triggering cells to induce mutagenic DNA repair pathways, which results in a loss of accuracy in DNA replication.

One of the oldest known environmental carcinogens is the chemical benzo(α)pyrene, a hydrocarbon found in coal tar, cigarette smoke, and automobile exhaust. An English surgeon, Percivall Pott, observed that chimney sweeps had a high incidence of cancer of the scrotum. The reason for this was later found to be their exposure to benzo(α)pyrene in the coal tar and soot of the chimneys. Like the aromatic amines, benzo(α)pyrene is activated by cellular enzymes and causes mutations indirectly.

Another important class of chemical mutagens with indirect action are agents causing cross-links between the strands of DNA. Such cross-links prevent DNA from being separated into individual strands as is needed during DNA replication and transcription. Examples of cross-linking agents are psoralens (compounds found in some vegetables and used in treatments of skin conditions such as psoriasis) and cis-platinum (a chemotherapeutic agent used to fight cancer).

Another important class of chemical mutagens are those that result in the formation of active species of oxygen (oxidizing agents). Some of these are actually created in the body by oxidative respiration (endogenous mutagens), while others are the result of the action of chemicals such as peroxides and radiation. Reactive oxygen species cause a wide variety of damage to the bases and the backbone of DNA and may have both direct and indirect effects.

DETECTION OF CHEMICAL MUTAGENS

The Ames test, developed by biochemistry professor Bruce Ames and his colleagues, is one of the most widely used screening methods for chemical mutagens. It employs particular strains of the bacterium *Salmonella typhimurium* that require the amino acid histidine because of mutations in one of the genes controlling histidine production. The bacteria are exposed to the potential mutagen and then spread on an agar medium lacking histidine. The strains can grow only if they develop a mutation restoring function to the mutated gene required for histidine synthesis. The degree of growth indicates the strength of the mutagen; mutagens of different types are detected by using bacterial strains with different mutations. Mutagens requiring metabolic activation are detected by adding extracts of rat liver cells (capable of mutagen activation) to the tested substance prior to exposure of the bacteria. The Ames test and others like it involving microorganisms are rapid, safe, and relatively inexpensive ways to detect mutagenic chemicals, but it is not always clear how the results of the Ames test should be interpreted when determining the degree of mutagenicity predicted in humans.

IMPACT AND APPLICATIONS

Mutations can have serious consequences for cells of all types. If they occur in gametes, they can cause genetic diseases or birth defects. If they occur in somatic (body) cells of multicellular organisms, they may alter a growth-controlling gene in such a way that the mutated cell begins to grow out of control and forms a cancer. DNA is subject to a variety of types of damage by interaction with a wide array of chemical agents, some of which are ubiquitous in the environment, while others are the result of human intervention. Methods of detection of chemicals with mutagenic ability have made it possible to reduce the exposure of humans to some of these mutagenic and potentially carcinogenic chemicals.

Beth A. Montelone, Ph.D.

FURTHER READING

Ahern, Holly. "How Bacteria Cope with Oxidatively Damaged DNA." *ASM News* 59 (March, 1993): 119-122. Discusses oxidative damage to DNA.

Connell, D. W. "Mutagens and Mutagenesis." In *Basic Concepts of Environmental Chemistry.* 2d ed. Boca Raton, Fla.: CRC/Taylor and Francis, 2005. Discusses mutagens and mutagenesis within the broader context of genotoxicity.

Frickel, Scott. *Chemical Consequences: Environmental Mutagens, Scientist Activism, and the Rise of Genetic Toxicology.* New Brunswick, N.J.: Rutgers University Press, 2004. A historical and sociological account of the birth of the new field of genetic toxicology. Describes how scientist-activists transformed chemical mutagens into a greater understanding of environmental problems.

Friedberg, Errol C., et al. *DNA Repair and Mutagenesis.* Washington, D.C.: ASM Press, 1995. Provides extensive descriptions of the mechanisms of chemical mutagenesis.

Hollaender, Alexander, and Frederick J. De Serres, eds. *Chemical Mutagens: Principles and Methods for Their Detection.* 10 vols. New York: Kluwer Academic, 1971-1986. This illustrated, multivolume work is the most comprehensive set of volumes on chemical mutagenesis.

Kuroda, Yukioki, et al., eds. *Antimutagenesis and Anticarcinogenesis Mechanisms II.* New York: Plenum Press, 1990. Part of the proceedings of the Second International Conference on Mechanisms of Antimutagenesis and Anticarcinogenesis, held December, 1988, in Ohito, Japan. Addresses topics such as antimutagens in food, environmental toxicology, free radicals, aspects of mammalian and human genetics, and molecular aspects of mutagenesis and antimutagenesis.

Neumann, David, et al. *Human Variability in Response to Chemical Exposures: Measures, Modeling, and Risk Assessment.* Boca Raton, Fla.: CRC Press, 1998. Addresses genetic evidence for variability in the human response to chemicals associated with reproductive and developmental effects, the nervous system and lungs, and cancer.

WEB SITES OF INTEREST

Genetic Toxicology Association
http://www.gta-us.org
The search engine enables users to retrieve information about chemical mutagens.

Mutation, Mutagens, and DNA Repair
http://www-personal.ksu.edu/~bethmont/
mutdes.html
Beth A. Montelone, a professor of biology at Kansas State University, created this site as a supplement for a class on human genetics. The site contains a

page providing information about chemical mutagens.

See also: Biochemical mutations; Cancer; DNA repair; DNA replication; Mutagenesis and cancer; Mutation and mutagenesis; Oncogenes; Repetitive DNA; Tumor-suppressor genes.

Chloroplast genes

CATEGORY: Molecular genetics

SIGNIFICANCE: Plants are unique among higher organisms in that they meet their energy needs through photosynthesis. The specific location for photosynthesis in plant cells is the chloroplast, which also contains a single, circular chromosome composed of DNA. Chloroplast DNA (cpDNA) contains many of the genes necessary for proper chloroplast functioning. A better understanding of the genes in cpDNA has improved the understanding of photosynthesis, and analysis of the DNA sequence of these genes has also been useful in studying the evolutionary history of plants.

KEY TERMS

chloroplast: the cell structure in plants responsible for photosynthesis

genome: all of the DNA in the nucleus or in one of the organelles such as a chloroplast

open reading frames: DNA sequences that contain all the components found in active genes, but whose functions have not yet been identified

photosynthesis: the process in which sunlight is used to take carbon dioxide from the air and convert it into sugar

THE DISCOVERY OF CHLOROPLAST GENES

The work of nineteenth century Austrian botanist Gregor Mendel showed that the inheritance of genetic traits follows a predictable pattern and that the traits of offspring are determined by the traits of the parents. For example, if the pollen from a tall pea plant is used to pollinate the flowers of a short pea plant, all the offspring are tall. If one of these tall offspring is allowed to self-pollinate, it produces a mixture of tall and short offspring, three-quarters of them tall and one-quarter of them short. Similar patterns are observed for large numbers of traits from pea plants to oak trees. Because of the widespread application of Mendel's work, the study of genetic traits by controlled mating is often referred to as Mendelian genetics.

In 1909, German botanist Carl Erich Correns discovered a trait in four-o'clock plants (*Mirabilis jalapa*) that appeared to be inconsistent with Mendelian inheritance patterns. He discovered four-o'clock plants that had a mixture of leaf colors on the same plant: Some were all green, many were partly green and partly white (variegated), and some were all white. If he took pollen from a flower on a branch with all-green leaves and used it to pollinate a flower on a branch with all-white leaves, all the resulting seeds developed into plants with white leaves. Likewise, if he took pollen from a flower on a branch with all-white leaves and used it to pollinate a flower on a branch with all-green leaves, all the resulting seeds developed into plants with green leaves. Repeated pollen transfers in any combination always resulted in offspring whose leaves resembled those on the branch containing the flower that received the pollen—that is, the maternal parent. These results could not be explained by Mendelian genetics.

Since Correns's discovery, many other such traits have been discovered. It is now known that the reason these traits do not follow Mendelian inheritance patterns is that their genes are not on the chromosomes in the nucleus of the cell where most genes are located. Instead, the gene for the four-o'clock leaf color trait is located on the single, circular chromosome found in chloroplasts. Because chloroplasts are specialized for photosynthesis, many of the genes on the single chromosome produce proteins or RNA that either directly or indirectly affect synthesis of chlorophyll, the pigment primarily responsible for trapping energy from light. Because chlorophyll is green and because mutations in many chloroplast genes cause chloroplasts to be unable to make chlorophyll, most mutations result in partially or completely white or yellow leaves.

IDENTITY OF CHLOROPLAST GENES

Advances in molecular genetics have allowed scientists to take a much closer look at the chloroplast genome. The size of the genome has been determined for a number of plants and algae and ranges

Chloroplasts and Other Parts of a Plant Cell

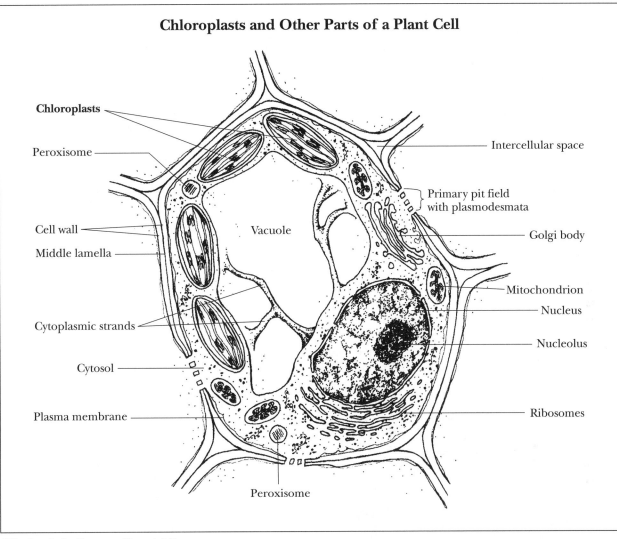

Chloroplasts

Peroxisome

Cell wall

Middle lamella

Cytoplasmic strands

Cytosol

Plasma membrane

Vacuole

Peroxisome

Intercellular space

Primary pit field with plasmodesmata

Golgi body

Mitochondrion

Nucleus

Nucleolus

Ribosomes

(Kimberly L. Dawson Kurnizki)

from 85 to 292 kilobase pairs (kb, or one thousand base pairs), with most being between 120 kb and 160 kb. The complete DNA sequence for many different chloroplast genomes of plants and algae have been determined. Although a simple sequence does not necessarily identify the role of each gene, it has allowed the identity of a number of genes to be determined, and it has allowed scientists to estimate the total number of genes. In terms of genome size, chloroplast genomes are relatively small and contain a little more than one hundred genes.

Roughly half of the chloroplast genes produce either RNA molecules or polypeptides that are important for protein synthesis. Some of the RNA genes

occur twice in the chloroplast genomes of almost all land plants and some groups of algae. The products of these genes represent all the ingredients needed for chloroplasts to carry out transcription and translation of their own genes. Half of the remaining genes produce polypeptides directly required for the biochemical reactions of photosynthesis. What is unusual about these genes is that their products represent only a portion of the polypeptides required for photosynthesis. For example, the very important enzyme ATPase—the enzyme that uses proton gradient energy to produce the important energy molecule adenosine triphosphate (ATP)—comprises nine different polypeptides. Six of these polypeptides are products of chloroplast genes, but the other three are products of nuclear genes that must be transported into the chloroplast to join with the other six polypeptides to make active ATPase. Another notable example is the enzyme ribulose biphosphate carboxylase (RuBP carboxylase, or Rubisco), which is composed of two polypeptides. The larger polypeptide, called *rbcL*, is a product of a chloroplast gene, whereas the smaller polypeptide is the product of a nuclear gene.

The last thirty or so genes remain unidentified. Their presence is inferred because they have DNA sequences that contain all the components found in active genes. These kinds of genes are often called open reading frames (ORFs) until the functions of their polypeptide products are identified.

Impact and Applications

The discovery that chloroplasts have their own DNA and the further elucidation of their genes have had some impact on horticulture and agriculture. Several unusual, variegated leaf patterns and certain mysterious genetic diseases of plants are now better understood. The discovery of some of the genes that code for polypeptides required for photosynthesis has helped increase understanding of the biochemistry of photosynthesis. The discovery that certain key chloroplast proteins such as ATPase and Rubisco carboxylase are composed of a combination of polypeptides coded by chloroplast and nuclear genes also raises some as yet unanswered questions. For example, why would an important plant structure like the chloroplast have only part of the genes it needs to function? Moreover, if chloroplasts, as evolutionary theory suggests, were once free-living bacteria-like cells, which must have had all the genes needed for photosynthesis, why and how did they transfer some of their genes into the nuclei of the cells in which they are now found?

Of greater importance has been the discovery that the DNA sequences of many chloroplast genes are highly conserved—that is, they have changed very little during their evolutionary history. This fact has led to the use of chloroplast gene DNA sequences for reconstructing the evolutionary history of various groups of plants. Traditionally, plant systematists (scientists who study the classification and evolutionary history of plants) have used structural traits of plants such as leaf shape and flower anatomy to try to trace the evolutionary history of plants. Unfortunately, there are a limited number of structural traits, and many of them are uninformative or even misleading when used in evolutionary studies. These limitations are overcome when gene DNA sequences are used.

A DNA sequence several hundred base pairs in length provides the equivalent of several hundred traits, many more than the limited number of structural traits available (typically much fewer than one hundred). One of the most widely used sequences is the *rbcL* gene. It is one of the most conserved genes in the chloroplast genome, which in evolutionary terms means that even distantly related plants will have a similar base sequence. Therefore, *rbcL* can be used to retrace the evolutionary history of groups of plants that are very divergent from one another. The *rbcL* gene, along with a few other very conservative chloroplast genes,

Carl Erich Correns, whose experiments with four-o'clock plants led to the discovery of chloroplast genes. (National Library of Medicine)

has already been used in attempts to answer some basic questions about the origins and evolution of some of the major flowering plant groups. Less conservative genes and ORFs show too much evolutionary change to be used at higher classification levels but are extremely useful in answering questions about the origins of closely related species, genera, or even families. As analytical techniques are improved, chloroplast genes show promise of providing even better insights into plant evolution.

Bryan Ness, Ph.D.

FURTHER READING

Blankenship, Robert E. "Origin and Evolution of Photosynthesis." In *Molecular Mechanisms of Photosynthesis.* Malden, Mass.: Blackwell Science, 2002. A concise overview of photosynthesis, including a chapter with information on the genetic origins of the process.

Daniell, Henry, and Christine Chase, eds. *Molecular Biology and Biotechnology of Plant Organelles: Chloroplasts and Mitochondria.* Dordrecht, the Netherlands: Springer, 2004. Describes basic concepts and current understanding of plant organelle genetics and molecular biology, chloroplasts, and nuclear genetic engineering.

Doyle, Jeff J. "DNA, Phylogeny, and the Flowering of Plant Systematics." *Bioscience* 43, no. 6 (June, 1993): 380-389. Introduces the reader to the basics of using DNA to construct plant phylogenies and discusses the future of using DNA in evolutionary studies in plants.

Palevitz, Barry A. "'Deep Gene' and 'Deep Time': Evolving Collaborations Parse the Plant Family Tree." *The Scientist* 15, no. 5 (2001): 12. Describes the Deep Green Project, an attempt to use DNA sequence data to trace the evolutionary history of all plants.

Palmer, Jeffry D. "Comparative Organization of Chloroplast Genomes." *Annual Review of Genetics* 19 (1985): 325-354. One of the best overviews of chloroplast genome structure, from algae to flowering plants.

Svetlik, John. "The Power of Green." *Arizona State University Research Magazine*, Winter, 1997. Provides a review of research at the Arizona State University Photosynthesis Center and offers good background for understanding the genetics of chloroplasts.

WEB SITES OF INTEREST

Chloroplast Genome Database
http://chloroplast.cbio.psu.edu/link.html

This database contains annotated chloroplast/plastid genomes from the National Center for Biotechnology Information. Users can search for genes by their annotated names, download protein and nucleotide sequences extracted from a selected chloroplast genome, and retrieve other information.

Kimball's Biology Pages
http://users.rcn.com/jkimball.ma.ultranet/BiologyPages/C/Chloroplasts.html

John Kimball, a retired Harvard University biology professor, includes a page about chloroplasts in his online cell biology text.

See also: Cell culture: Plant cells; DNA isolation; Extrachromosomal inheritance; Genomics; Hybridization and introgression; Model organism: *Chlamydomonas reinhardtii*.

Cholera

CATEGORY: Diseases and syndromes

DEFINITION

Cholera is an infection of the small intestine caused by the comma-shaped bacterium *Vibrio cholerae*. Cholera arose centuries ago in India and was disseminated throughout Asia and Europe by trade and pilgrimage. It was devastating, causing epidemics that resulted in countless deaths. By the early twentieth century, cholera had been confined mostly to Asia. In 1961, however, a cholera pandemic beginning in Indonesia spread to Africa, the Mediterranean nations, and North America. Because cholera has a 50 to 60 percent fatality rate when its symptoms are not treated quickly, occasional cases cannot be ignored; both the consequences to afflicted people and the potential for the outbreak of epidemics are great.

RISK FACTORS

In the poorer nations of the world, cholera is still widespread and occurs where sanitation is inade-

Cholera in Marine Plankton

Outbreaks of cholera can occur in nonendemic areas when an infected person travels to another country or when infected water is carried in the ballast of ships to another country. These two processes alone, however, could not explain all of the outbreaks of cholera observed worldwide. In the late 1960's, *Vibrio cholerae* was found in the ocean associated with marine plankton. This association, along with climate change, helps to explain the spread of cholera.

Plankton are the small organisms suspended in the ocean's upper layers. Plankton can be divided into two groups, phytoplankton (small plants) and zooplankton (small animals). *Vibrio cholerae* is found associated with the surface and gut of copepods, which are members of the zooplankton group. These small crustaceans act as a reservoir for the cholera bacteria, allowing them to survive in the ocean for long periods of time. Then, a change in weather that causes the ocean temperature to rise could also cause currents that stir up nutrients from lower layers of the ocean to the upper layers. Numbers of phytoplankton, which live in the upper layers of ocean waters, increase in these periods as a result of the warmer temperatures and greater availability of nutrients. Zooplankton numbers increase as well, because of the increase in their main food source, the phytoplankton. Consequently, the number of cholera bacteria increase to a level that can cause the disease. Thus, climate change can result in an outbreak of cholera in a region where cholera is endemic, or, if currents move the plankton to other coastal areas, in a new,

nonendemic region. This scenario is believed to explain the 1991 cholera epidemic in Peru, when the oceanic oscillation known as El Niño would have caused a warming of ocean temperature.

Because of the association of *Vibrio cholerae* with plankton, scientists believe they may be able to track or identify future epidemics by the use of satellite imagery. Increases in phytoplankton turn the ocean color from blue to green. Thus, changes in green areas in the ocean on satellite pictures show where the phytoplankton and, by association, zooplankton and cholera bacteria are relocating or increasing in number.

The association of cholera with zooplankton has also helped reveal a new way to prevent the disease. People get cholera by ingesting several thousand cholera bacteria at one time. A single copepod can harbor ten thousand bacteria; therefore, the ingestion of one infected copepod can cause disease in a person. Researchers have found a simple and inexpensive way to reduce this risk from copepods dramatically. Filtering water through four layers of fabric used to make saris, which are commonly worn in regions plagued by cholera, removes 99 percent of copepods from water containing high levels of plankton.

Now that the entire genetic sequence of *Vibrio cholerae* has been determined, scientists are armed with additional genetic data to elucidate the relationship of the bacterium with copepods, which may help them find more ways of controlling the spread of the disease.

Vicki J. Isola, Ph.D.

quate. In the United States and other industrialized nations, where sanitation is generally good, only a few cases occur each year. These usually result from the return of afflicted travelers from regions where cholera is endemic.

ETIOLOGY AND GENETICS

The disease occurs when cholera toxin binds to intestinal cells and stimulates the passage of water from the blood into the intestine. This water depletion and resultant cardiovascular collapse are major causes of cholera mortality. Study of the genetics and the biochemistry of cholera has shown that the toxin is a protein composed of portions called A and B subunits, each produced by a separate gene. When a bacterium secretes a molecule of cholera toxin, it binds to a cell of the intestinal lining (an in-

testinal mucosa cell) via B subunits. Then the A subunits cause the mucosal cell to stimulate the secretion of water and salts from the blood to produce diarrhea. Lesser amounts of the watery mix are vomited and exacerbate dehydration.

The use of bacterial genetics to compare virulent *V. cholerae* and strains that did not cause the disease helped in the discovery of the nature of the cholera toxin and enabled production of vaccines against the protein. These vaccines are useful to those individuals who visit areas where cholera is endemic, ensuring that they do not become infected with it during these travels. Unfortunately, the vaccines are effective only for about six months.

The basis for the operation of cholera toxin is production of a hormone substance called cyclic adenosine monophosphate (cAMP). The presence of excess cAMP in intestinal mucosa cells causes movement of water and other tissue components into the intestine and then out of the body. The accumulation of cAMP is caused by the ability of the cholera toxin to modify an enzyme protein, adenyl cyclase, to make it produce excess cAMP via modification of a control substance called a G-protein. This modification, called adenine ribosylation, is a mechanism similar to that causing diphtheria, another dangerous disease that can be fatal, although in diphtheria other tissues and processes are affected.

SYMPTOMS

Infection is almost always caused by consumption of food or water contaminated with the bacterium. It is followed in one to five days by watery diarrhea that may be accompanied by vomiting. The diarrhea and vomiting may cause the loss of as much as a pint of body water per hour. This fluid loss depletes the blood water and other tissues so severely that, if left unchecked, it can cause death within a day.

SCREENING AND DIAGNOSIS

The signs and symptoms of cholera are usually evident in areas where the disease is endemic. However, the only way to confirm this diagnosis is to test a patient's stool sample for *V. cholerae*. Health care providers in remote parts of the world conduct rapid cholera dipstick tests that enable them to confirm quickly if a patient has the disease. Polymerase chain reaction (PCR) assays or other genetic tests can also to confirm a diagnosis.

TREATMENT AND THERAPY

Treatment of cholera combines oral or intravenous rehydration of afflicted individuals with saline-nutrient solutions and chemotherapy with antibiotics, especially tetracycline. The two-pronged therapy replaces lost body water and destroys all *V. cholerae* in infected individuals. Antibiotic prophylaxis, which destroys the bacteria, leads to the cessation of production of cholera toxin, the substance that causes diarrhea, vomiting, and death.

PREVENTION AND OUTCOMES

Cholera has, for centuries, been a serious threat to humans throughout the world. During the twentieth century, its consequences to industrialized nations diminished significantly with the advent of sound sanitation practices that almost entirely pre-

vented the entry of *V. cholerae* into the food and water supply. In poorer nations with less adequate sanitation, the disease flourishes and is still a severe threat.

It must be remembered that dealing with cholera occurs at three levels. The isolation and identification of cholera toxin, as well as development of current short-term cholera vaccines, were highly dependent on genetic methodology. Vaccine protects most travelers from the disease. However, wherever the disease afflicts individuals, its treatment depends solely upon rehydration and use of antibiotics. Finally, current cholera prevention focuses solely on adequate sanitation. Medicine seeks to produce a long-lasting vaccine for treatment of cholera to enable prolonged immunization at least at the ten-year level of tetanus shots. Efforts aimed at this goal are ongoing and utilize molecular genetics to define more clearly why long-term vaccination has so far been unsuccessful. Particularly useful will be fine genetic sequence analysis and the use of gene amplification followed by DNA fingerprinting.

Sanford S. Singer, Ph.D.

FURTHER READING

Chadhuri, Keya, and S. N. Chatterjee. *Cholera Toxins.* Berlin: Springer, 2009. Offers comprehensive information about the *Vibrio cholerae* toxins, including their physical and chemical structures, biosynthesis and genetic regulation, physiology, and role in the development of a cholera vaccine.

Colwell, Rita R. "Global Climate and Infectious Disease: The Cholera Paradigm." *Science* 274, no. 5295 (1996): 2025-2031. An analysis of the role climate change might play in the spread of cholera. Includes a good overview of the history of cholera.

Gotuzzo, E., and C. Seas. "Cholera and Other *Vibrio* Infections." In *Cecil Medicine*, edited by Lee Goldman and Dennis Ausiello. 23d ed. Philadelphia: Saunders Elsevier, 2008. Describes these pathogens and the diseases they cause.

Heidelberg, John F., et al. "DNA Sequence of Both Chromosomes of the Cholera Pathogen *Vibrio cholerae*." *Nature* 406, no. 6795 (August 3, 2000): 477-483. Presents the results of a report revealing the complete genome sequence of the bacteria responsible for cholera and describes how this sequencing may improve treatment of the disease.

Holmgren, John. "Action of Cholera Toxin and Prevention and Treatment of Cholera." *Nature* 292 (1981): 413-417. Clearly describes both the composition and bioaction of the cholera toxin.

Keusch, Gerald, and Masanobu Kawakami, eds. *Cytokines, Cholera, and the Gut.* Amsterdam: IOS Press, 1997. Surveys the role of peptide mediators in the intestinal responses to infectious and inflammatory challenges presented by diverse disease states, including inflammatory bowel disease and infectious diarrheas and dysenteries. Examines the epidemiology and pathogenesis of cholera and related diarrheal diseases.

Pennisi, Elizabeth. "Cholera Strengthened by Trip Through Gut." *Science* 296, no. 5575 (June 7, 2002): 1783-1784. Examines the effect that passing through a host's gut has on cholera bacteria.

Rakel, Robert E., Edward T. Bope, and Howard F. Conn. *Conn's Current Therapy 2008: Text with Online Reference.* Philadelphia: Elsevier Saunders, 2008. Provides general readers with a succinct overview of cholera and its treatment.

Sherman, Irwin W. "King Cholera." In *The Power of Plagues.* Washington, D.C.: ASM Press, 2006. Describes the nature and evolution of cholera and other diseases and explains how scientists discovered the causes of and controls for infectious diseases.

Wachsmuth, Kate, et al., eds. *Vibrio Cholerae and Cholera: Molecular to Global Perspectives.* Washington, D.C.: ASM Press, 1994. A comprehensive guide to the disease and its genetics.

WEB SITES OF INTEREST

Centers for Disease Control and Prevention
http://www.cdc.gov/nczved/dfbmd/
disease_listing/cholera_gi.html
Offers information about cholera and what the United States is doing to treat it.

Food and Drug Administration
http://vm.cfsan.fda.gov
The FDA's "Bad Bug Book" provides information on *Vibrio cholerae*, the bacterium that causes cholera.

Mayo Clinic.com
http://www.mayoclinic.com/health/cholera/
DS00579
An overview of the causes, symptoms, treatment, and prevention of cholera.

Medline Plus
http://www.nlm.nih.gov/MEDLINEPLUS/ency/
article/000303.htm
A concise, "user-friendly" article about cholera, describing its causes, symptoms, diagnosis, and treatment.

See also: Anthrax; Archaea; Bacterial genetics and cell structure; Bacterial resistance and super bacteria; Diphtheria; Emerging and reemerging infectious diseases; Gene regulation: Bacteria; Gene regulation: *Lac* operon; Transgenic organisms.

Chorionic villus sampling

CATEGORY: Techniques and methodologies; Bioethics
ALSO KNOWN AS: CVS
SIGNIFICANCE: Chorionic villus sampling (CVS) is an important method of intrauterine diagnosis of genetic disorders. First conceptualized in 1968 by Jan Mohr, it is used to diagnose certain genetic disorders of the fetus in the prenatal period, allowing for awareness and planning by the parents and early pregnancy termination, if determined to be appropriate.

KEY TERMS

amniocentesis: a procedure generally performed in the second trimester in which amniotic fluid surrounding the fetus is sampled and fetal DNA is analyzed for genetic abnormalities

anaphase: a stage in mitosis and meiosis where sister chromosomes separate and move toward opposite poles of the cell

chromosome: an organized structure containing DNA and protein found within cells

karyotype: a profile of an individual's chromosomes, which indicates the number, size, and shape of chromosomes present

meiosis: a process where chromosomal segregation into daughter cells leads to gamete formation in which each gamete possesses half the number of chromosomes as did the parent cell

nuchal translucency: clear space in the folds of the developing neck of the fetus that can be detected via ultrasound and can indicate a potential genetic disorder or malformation

placental mosaicism: an occurrence where there exists a discrepancy between the chromosomal makeup of the cells of the placenta and the cells of the fetus

trisomy: a genetic abnormality in which there are three copies of a given chromosome in a cell instead of the normal two

INDICATIONS FOR CVS

Many indications exist for performing CVS, including advanced maternal age, abnormal results from prenatal screening tests, and known or suspected family history of genetic abnormalities. Advanced maternal age, generally considered to be greater than thirty-give years, is one of the most important indications for CVS because a woman's eggs age along with her, which can increase the chance for errors during cell division. The most common error is called nondisjunction and occurs in anaphase during meiosis. Nondisjunction is the failure of two sister chromosomes to separate in one parent and thus the inheritance of two copies of that particular chromosome from that parent and one copy from the other parent, resulting in three copies of a chromosome instead of the usual two. This resulting trisomy is generally sporadic in nature and can be detected through CVS.

Prenatal screening tests involve quantifying maternal levels of certain pregnancy-related hormones, including AFP, PAPP-A, hCG, E3, and inhibin A, and performing an ultrasound to evaluate the fetus for nuchal translucency. Combining the results of these tests, a risk of fetal abnormality can be calculated, and CVS may then be indicated for further evaluation of the fetus. The prenatal screening tests can predict only the likelihood of genetic disorders, while CVS is able to provide an actual diagnosis. In cases where the parents are known carriers of a genetic abnormality detectable by CVS or have had a previous child affected by a genetic disorder, a family history exists of such genetic disorders, or the parents are of a specific ethnic background considered to be at high risk, they may choose to have CVS performed directly without prior screening tests because of the increased probability of having a fetus with a genetic disorder.

DIAGNOSTIC CAPABILITIES AND LIMITATIONS

Chorionic villus sampling can detect a variety of genetic syndromes. Chromosomal disorders including trisomy 21 (Down syndrome), trisomy 18 (Edwards syndrome), trisomy 13 (Patau syndrome), Turner syndrome, and Klinefelter syndrome can be diagnosed. Numerous other genetic diseases such as cystic fibrosis, Tay-Sachs disease, congenital adrenal hyperplasia (CAH), and sickle-cell disease are also diagnosed through CVS. Neural tube defects such as spina bifida and anencephaly cannot be diagnosed with CVS testing and require amniocentesis if suspected.

Confusion may occur when placental mosaicism is present. This occurs when the cells of the placenta sampled through CVS indicate an abnormality while others are unaffected, leading to potentially false positive results. Placental mosaicism is rare, occurring in only about 1-2 percent of pregnancies. In this special case, amniocentesis will be performed to confirm the suspected mosaicism. It is

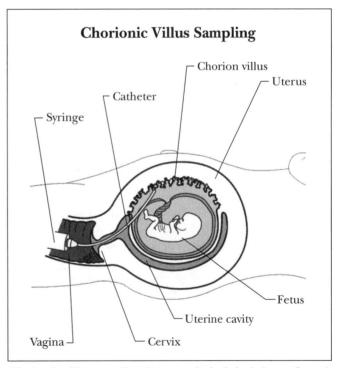

Chorionic villus sampling is one method of obtaining embryonic cells from a pregnant woman. Examination of these cells helps physicians determine fetal irregularities or defects, which allows time to assess the problem and make recommendations for treatment. (Hans & Cassidy, Inc.)

also possible to have a negative CVS result and still have a baby with a genetic disorder. The rate of false positive or false negative results can vary widely with each genetic disorder.

LABORATORY PROCESS

Chorionic villus sampling is performed either transabdominally or transcervically to sample cells of the chorionic villus of the placenta; the cells are analyzed in a laboratory. The cells are subsequently halted in the middle of cell division and examined microscopically for karyotype abnormalities such as too many or too few chromosomes and abnormal variations in size of the chromosomes.

EARLY INTERVENTION

At this point in time, the scope of genetic disorders able to be diagnosed through chorionic villus sampling for which in utero treatment is available is small. One such disorder is congenital adrenal hyperplasia, a genetic disorder caused by lack of an enzyme that leads to a hormonal deficiency and possibly ambiguous genitalia. If the fetus is treated with these vital hormones within a narrow, critical period of time, then the need for postnatal intervention can be avoided along with the gender assignment confusion that often results for parents. Currently, for most other genetic disorders diagnosed through CVS, the primary intervention available is termination of the pregnancy, which can pose a considerable emotional dilemma for the parents.

IMPACT

Chorionic villus sampling allows for earlier diagnosis of a wide array of genetic disorders than does amniocentesis. It is performed in the first trimester, between ten and twelve weeks of gestation, whereas amniocentesis is generally performed in the second trimester, between fifteen and seventeen weeks of gestation. If termination of the pregnancy is indicated, it can then be performed in the first trimester, potentially allowing for a safer, less physically traumatic procedure. It is also likely to be less psychologically damaging to the parents to have termination performed at an earlier time.

CVS is currently widely available, and increasing numbers of women are undergoing the procedure. The Centers for Disease Control and Prevention (CDC) reported a large increase in the amount of CVS procedures performed in the late 1980's and early 1990's, especially among women greater than thirty-five years of age. This increase in procedure rate has resulted in a large percentage of the pregnant female population becoming aware of their child's genetic condition prior to birth and allows more time to prepare for the care of the neonate. The future impact of CVS is likely to be wide as prenatal interventions for commonly diagnosed genetic conditions are discovered and the survival rate of the affected offspring is greatly increased.

Jennifer Birkhauser, M.D.

FURTHER READING

Creasy, Robert K., Robert Resnick, and J. D. Iams. *Creasy and Resnick's Maternal-Fetal Medicine: Principles and Practice.* 6th ed. Philadelphia: Saunders/ Elsevier, 2009. A popular reference text for genetic disorders from a maternal-fetal medicine viewpoint.

Harms, Roger W., ed. *Mayo Clinic Guide to a Healthy Pregnancy.* New York: HarperCollins, 2004. A resource for non-medical professionals regarding pregnancy and prenatal testing options.

Miller, Orlando J., and Eeva Therman. *Human Chromosomes.* 4th ed. New York: Springer, 2001. An understandable, concise introduction to chromosomes for all readers.

Pierce, Benjamin. *Genetics: A Conceptual Approach.* 3d ed. New York: Macmillan, 2007. An excellent, comprehensive approach to the field of genetics.

Queenan, John T., John C. Hobbins, and Catherine Y. Spong. *Protocols for High-Risk Pregnancies.* 4th ed. Hoboken, N.J.: Wiley-Blackwell, 2005. A comprehensive description of chorionic villus sampling.

WEB SITES OF INTEREST

March of Dimes: Chorionic Villus Sampling (CVS)
http://www.marchofdimes.com/pnhec/159_521.asp

Mayo Clinic: Chorionic Villus Sampling
http://www.mayoclinic.com/health/chorionic -villus-sampling/MY00154

National Library of Medicine and the National Institutes of Health. MedlinePlus
http://www.nlm.nih.gov/medlineplus/ency/ article/003345.htm

See also: Amniocentesis; Bioethics; Genetic screening; Genetic testing; Genetic testing: Ethical and economic issues; Prenatal diagnosis.

Choroideremia

CATEGORY: Diseases and syndromes

ALSO KNOWN AS: Diffuse total choroidal vascular atrophy of X-linked inheritance; progressive tapetochoroidal dystrophy; tapetochoroidal dystrophy (TCD)

DEFINITION

Choroideremia is an X-linked recessive eye disorder causing progressive degeneration of the choriocapillaries, retinal pigment epithelium, and photoreceptors of the retina. Choroideremia causes progressive loss of night vision, peripheral vision, and finally central vision in affected males in the first or second decade of life.

The prevalence of choroideremia is estimated to be 1 in 50,000, although choroideremia may be misdiagnosed as other retinal degenerative conditions such as retinitis pigmentosa or gyrate dystrophy, which makes the estimation of prevalence inexact. Thioridizine (Mellaril) toxicity and Bietti crystalline dystrophy must also be considered in the differential diagnosis.

RISK FACTORS

Family history of choroideremia with a usual pattern of X-linked recessive inheritance with female carriers will pose a risk for the children of the carrier. Consistent with X-linked recessive inheritance, choroideremia will predominantly affect males; however, female carriers may exhibit mild findings of retinal dysfunction.

ETIOLOGY AND GENETICS

Choroideremia was reported by Clement McCulloch in 1969 in a family with 1,600 descendants of an Irish man who immigrated to Canada in 1850. His determination that choroideremia was an X-linked recessive disorder was controversial due to the fact that female carriers in some cases showed mild clinical forms of the disease. Subsequently, other lineages including a large Finnish family have been identified with the X-linked recessive pattern of inheritance of choroideremia.

The Xq21.2 locus of the long arm of the X chromosome has been identified in choroideremia. The Finnish lineage, which comprise a large proportion of all choroideremia patients worldwide, all carry the same mutation, while J. A. Van Den Hurk and colleagues described different point mutations in other lineages. All these mutations result in a termination of a codon, which results in a truncated protein. The gene product REP1 of the *CHM* gene is nonfunctional in choroideremia patients. REP1 is a Rab escort protein geranylgeranyl transferase involved in trafficking of Rab proteins in the cell. This protein attaches isoprenoids to TAB 27. A result of a nonfunctional REP1 product is degenerations of the layers of the retina.

There has been some debate about whether the primary degenerative effect occurs in the vascular basement membranes of the choriocapillaries, leading in turn to degeneration of the other layers of the retina. More recent research points to the retinal pigment epithelium as the site of the primary defect. In the end stages of severe choroideremia, the retinal layers thin to reveal the underlying sclera.

Additional syndromes can involve degeneration of the choroid in association with mental retardation, deafness, and cleft lip and palate. These syndromes are considered distinct from isolated choroideremia.

SYMPTOMS

The age of onset is variable; however, most patients present between age ten and thirty. Initial symptoms may be nonspecific and can include glare as well as night vision disturbance. As the disorder progresses, there is an increasing loss of peripheral vision and in the later stages a loss of central vision.

Anyone with a known family history of choroideremia should have regular ophthalmic examinations including visual field testing. In a case of symptoms with no known history of choroideremia, a baseline eye examination from an ophthalmologist is recommended. In the case of persistent symptoms, repeat examinations over a period of time may detect a progressive condition.

SCREENING AND DIAGNOSIS

Affected males and female carriers should undergo a complete dilated eye examination by a qualified ophthalmologist. Visual field testing and multifocal electroretinogram (ERG) testing is helpful to establish a diagnosis. A retinal specialist may be consulted to differentiate between choroideremia and other retinal disorders such as gyrate dystrophy and retinitis pigmentosis.

TREATMENT AND THERAPY

There is no treatment for choroideremia at this time. Support of patients with choroideremia includes genetic counseling and visual aids.

PREVENTION AND OUTCOMES

Carriers of choroideremia and affected individuals may be detected through genetic counseling. Families with a history of choroideremia should have a high index of suspicion and consult with genetic counseling and ophthalmology specialists to access tests that may identify affected individuals and carriers.

Ellen Anderson Penno, M.D., M.S., FRCSC

FURTHER READING

McCulloch, C. "Choroideremia." *Transactions of the American Ophthalmological Society* 67 (1969): 142-195. An early report of clinical observations of choroideremia in a Canadian immigrant family.

Merin, Saul. *Inherited Eye Diseases: Diagnosis and Management.* 2d ed. Boca Raton, Fla.: Taylor and Francis, 2005. A comprehensive text which includes summaries of inherited retinal disease.

Regillo, Carl. *2009-2010 Basic and Clinical Science Course (BCSC) Section 12: Retina and Vitreous.* San Francisco: American Academy of Ophthalmology, 2009. A concise text that covers a variety of retinal disorders.

WEB SITES OF INTEREST

American Academy of Ophthalmology: Choroideremia
http://one.aao.org/CE/EducationalProducts/
snippet.aspx?F=bcsccontent\bcscsection12\bcsc2007
section12_2007-07-12_010755\hereditaryretinaland
choroidaldystrophies\bcsc12090079.xml

Choroideremia Research Foundation
http://www.choroideremia.org

National Eye Institute, National Institutes of Health: Choroideremia
http://www.ncbi.nlm.nih.gov/entrez/
dispomim.cgi?id=303100

See also: Aniridia; Best disease; Corneal dystrophies; Hereditary diseases.

Chromatin packaging

CATEGORY: Molecular genetics

SIGNIFICANCE: The huge quantity of DNA present in each cell must be organized and highly condensed in order to fit into the discrete units of genetic material known as chromosomes. Gene expression can be regulated by the nature and extent of this DNA packaging in the chromosome, and errors in the packaging process can lead to genetic disease.

KEY TERMS

chromatin: the material that makes up chromosomes; a complex of fibers composed of DNA, histone proteins, and nonhistone proteins

histone proteins: small, basic proteins that are complexed with DNA in chromosomes and that are essential for chromosomal structure and chromatin packaging

nonhistone proteins: a heterogeneous group of acidic or neutral proteins found in chromatin that may be involved with chromosome structure, chromatin packaging, or the control of gene expression

nucleosome: the basic structural unit of chromosomes, consisting of 146 base pairs of DNA wrapped around a core of eight histone proteins

CHROMOSOMES AND CHROMATIN

Scientists have known for many years that an organism's hereditary information is encrypted in molecules of DNA that are themselves organized into discrete hereditary units called genes and that these genes are organized into larger subcellular structures called chromosomes. James Watson and Francis Crick elucidated the basic chemical structure of the DNA molecule in 1953, and much has been learned since that time concerning its replication and expression. At the molecular level, DNA is composed of two parallel chains of building blocks called nucleotides, and these chains are coiled around a central axis to form the well-known "double helix." Each nucleotide on each chain attracts and pairs with a complementary nucleotide on the opposite chain, so a DNA molecule can be described as consisting of a certain number of these nucleotide base pairs. The entire human genome consists of more than six billion base pairs of DNA, which, if completely unraveled, would extend for more than 2 meters (6.5 feet). It is a remarkable feat of engineering that in each human cell this much DNA is condensed, compacted, and tightly packaged into chromosomes within a nucleus that is less than 10^{-5} meters in diameter. What is even more astounding is the frequency and fidelity with which this DNA must be condensed and relaxed, packaged and unpackaged, for replication and expression in each individual cell at the appropriate time and place during both development and adult life. The essential processes of DNA replication or gene expression (transcription) cannot occur unless the DNA is in a more open or relaxed configuration.

Chemical analysis of mammalian chromosomes reveals that they consist of DNA and two distinct classes of proteins, known as histone and nonhistone proteins. This nucleoprotein complex is called chromatin, and each chromosome consists of one linear, unbroken, double-stranded DNA molecule that is surrounded in predictable ways by these histone and nonhistone proteins. The histones are relatively small, basic proteins (having a net positive charge), and their function is to bind directly to the negatively charged DNA molecule in the chromosome. Five major varieties of histone proteins are found in chromosomes, and these are known as H1, H2A, H2B, H3, and H4. Chromatin contains about equal amounts of histones and DNA, and the amount and proportion of histone proteins are constant from cell to cell in all higher organisms. In fact, the histones as a class are among the most highly conserved of all known proteins. For example, for histone H3, which is a protein consisting of 135 amino acid "building blocks," there is only a single amino acid difference in the protein found in sea urchins as compared with the one found in cattle. This is compelling evidence that histones play the same essential role in chromatin packaging in all higher organisms and that evolution has been quite intolerant of even minor sequence variations between vastly different species.

Nonhistones as a class of proteins are much more heterogeneous than the histones. They are usually acidic (carrying a net negative charge), so they will most readily attract and bind with the positively charged histones rather than the negatively charged DNA. Each cell has many different kinds of nonhistone proteins, some of which play a structural role in chromosome organization and some of which are more directly involved with the regulation

of gene expression. Weight for weight, there is often as much nonhistone protein present in chromatin as histone protein and DNA combined.

NUCLEOSOMES AND SOLENOIDS

The fundamental structural subunit of chromatin is an association of DNA and histone proteins called a "nucleosome." First discovered in the 1970's by Ada and Donald Olins and Chris Woodcock, each nucleosome consists of a core of eight histone proteins: two each of the histones H2A, H2B, H3, and H4. Around this histone octamer are wound 146 base pairs of DNA in one and three-quarter turns (approximately eighty base pairs per turn). The overall shape of each nucleosome is similar to that of a lemon or a football. Each nucleosome is separated from its adjacent neighbor by about 55 base pairs of "linker DNA," so that in its most unraveled state they appear under the electron microscope to be like tiny beads on a string. Portions of each core histone protein protrude outside the wound DNA and interact with the DNA that links adjacent nucleosomes.

The next level of chromatin packaging involves a coiling and stacking of nucleosomes into a ribbon-like arrangement, which is twisted to form a chromatin fiber about 30 nanometers (nm) in diameter commonly called a "solenoid." Formation of solenoid fibers requires the interaction of histone H1, which binds to the linker DNA between nucleosomes. Each turn of the chromatin fiber contains about 1,200 base pairs (six nucleosomes), and the DNA has now been compacted by about a factor of fifty. The coiled solenoid fiber is organized into large domains of 40,000 to 100,000 base pairs, and these domains are separated by attached nonhistone proteins that serve both to organize and to control their packaging and unpackaging.

LONG DNA LOOPS AND THE CHROMOSOME SCAFFOLD

Physical studies using the techniques of X-ray crystallography and neutron diffraction have suggested that solenoid fibers may be further organized into giant supercoiled loops. The extent of this additional looping, coiling, and stacking of solenoid fibers varies, depending on the cell cycle. The most relaxed and extended chromosomes are found at interphase, the period of time between cell divisions. Interphase chromosomes typically have a

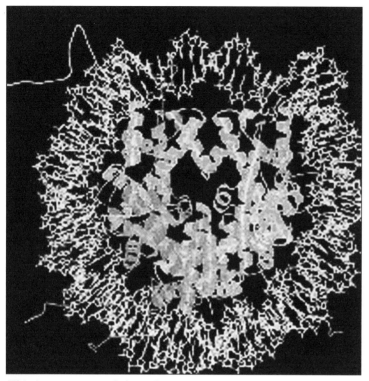

This image, captured through neutron crystallography, shows the molecular structure of the fundamental repeating unit of the chromosome, the nucleosome core complex: 146 base pairs of DNA wrapped around a core of eight histone proteins. (U.S. Department of Energy Genomes to Life Program, http://doegenomestolife.org)

venting tangles in the DNA. Apparently this same enzyme activity is necessary for the coiling and looping of solenoid fibers along the chromosome scaffold that occurs during the transition between interphase and metaphase chromosome structure. In the most highly condensed metaphase chromosomes, the DNA has been further compacted by an additional factor of one hundred.

IMPACT AND APPLICATIONS

Studies of chromatin packaging continue to reveal the details of the precise chromosomal architecture that results from the progressive coiling of the single DNA molecule into increasingly compact structures. Evidence suggests that the regulation of this coiling and packaging within the chromosome has a significant effect on the properties of the genes themselves. In fact, errors in DNA packaging can lead to inappropriate gene expression and developmental abnormalities. In humans, the blood disease thalassemia, several neuromuscular diseases, and even male sex determination can all be explained by the altered assembly of chromosomal structures.

Chromatin domains, composed of coiled solenoid fibers, may contain several genes, or the boundary of a domain can lie within a gene. These domains have the capacity to influence gene expression, and this property is mediated by specific DNA sequences known as locus control regions (LCRs). An LCR is like a powerful enhancer that activates transcription, thereby turning on gene expression. The existence of such sequences was first recognized from a study of patients with beta-thalassemia and a related condition known as hereditary persistence of fetal hemoglobin. In these disorders, there is an error in the expression of a cluster of genes, known as the beta-globin genes, that prevents the appearance of adult type hemoglobin. The beta-globin genes are linearly arrayed over a 50-kilobase-pair chromatin domain, and the LCR is found upstream from this cluster. Affected patients were found to have normal beta-globin genes, but there was a deletion of the upstream LCR that led to failure to activate the genes appropriately. Further

diameter of about 300 nm. Chromosomes that are getting ready to divide (metaphase chromosomes) have the most highly condensed chromatin, and these structures may have a diameter of up to 700 nm. One major study on the structure of metaphase chromosomes has shown that a skeleton of nonhistone proteins in the shape of the metaphase chromosome remains even after all of the histone proteins and the DNA have been removed by enzymatic digestion. If the DNA is not digested, it remains in long loops (10 to 90 kilobase pairs) anchored to this nonhistone protein scaffolding.

In the purest preparations of metaphase chromosomes, only two scaffold proteins are found. One of these forms the latticework of the scaffold, while the other has been identified as toposiomerase II, an enzyme that is critical in DNA replication. This enzyme cleaves double-stranded DNA and then rapidly reseals the cut after some of the supercoiling has been relaxed, thus relieving torsional stress and pre-investigation led to the conclusion that the variation in expression of these genes observed in different patients was caused by differences in the assembly of the genes into higher-order chromatin structures. In some cases, gene expression was repressed, while in others it was facilitated. Under normal circumstances, a nonhistone protein complex was found to bind to the LCR, causing the chromatin domain to unravel and making the DNA more accessible to transcription factors, thus enhancing gene expression.

DNA sequencing studies have demonstrated a common feature in several genes whose altered expression leads to severe human genetic disease. For example, the gene that causes myotonic dystrophy has a large number of repeating nucleotide triplets in the DNA region immediately adjacent to the protein-encoding segment. Physical studies have shown that this results in the formation of unusually stable nucleosomes, since these repeated sequences create the strongest naturally occurring sites for association with the core histones. It has been suggested that these highly stable nucleosomes are unusually resistant to the unwinding and denaturation of the DNA that must occur in order for gene expression to begin. RNA polymerase is the enzyme that makes an RNA transcript of the gene, and its movement through the protein-coding portion of the gene is inhibited if the DNA is unable to dissociate from the nucleosomes. Thus, although the necessary protein product would be normal and functional if it could be made, it is a problem with chromatin unpackaging that leads to reduced gene expression that ultimately leads to clinical symptoms of the disease. Both mild and severe forms of myotonic dystrophy are known, and an increase in the clinical severity correlates exactly with an increased number of nucleotide triplet repeats in the gene. Similar triplet repeats have been found in the genes responsible for Kennedy disease, Huntington's disease (Huntington's chorea), spinocerebellar ataxia type I, fragile X syndrome, and dentatorubral-pallidoluysian atrophy.

Fascinating and unexpected research results have suggested that a central event in the determination of gender in mammals depends on local folding of DNA within the chromosome. Molecular biologists Peter Goodfellow and Robin Lovell-Badge successfully cloned a human gene from the Y chromosome that determines maleness. This *SRY* gene (named from the sex-determining region of the Y chromosome) encodes a protein that selectively recognizes a specific DNA sequence and helps assemble a chromatin complex that activates other male-specific genes. More specifically, binding of the SRY protein causes the DNA to bend at a specific angle and causes conformation that facilitates the assembly of a protein complex to initiate the cascade of gene activation leading to male development. If the bend is too tight or too wide, gene expression will not occur, and the embryo will develop as a female.

The unifying lesson to be learned from these examples of DNA packaging and disease is that DNA sequencing studies and the construction of human genetic maps will not by themselves provide all the answers to questions concerning human variation and genetic disease. An understanding of human genetics at the molecular level depends not only on the primary DNA sequence but also on the three-dimensional organization of that DNA within the chromosome. Compelling genetic and biochemical evidence has left no doubt that the packaging process is an essential component of regulated gene expression.

Jeffrey A. Knight, Ph.D.

FURTHER READING

Becker, Peter B. *Chromatin Protocols.* Totowa, N.J.: Humana Press, 1999. Western scientists provide step-by-step instructions for analyzing the relationship between chromatin structure and function and for elucidating the molecular mechanisms that control such vital cellular functions as transcription, replication, recombination, and DNA repair.

Elgin, Sarah C. R., and Jerry L. Workman, eds. *Chromatin Structure and Gene Expression.* 2d ed. New York: Oxford University Press, 2000. Examines numerous facets of chromatin structure, including its histones, nucleosomes, and fiber elements; its relationship to DNA structure; its replication and assembly; and its initiation of expression.

Kornberg, Roger, and Anthony Klug. "The Nucleosome." *Scientific American* 244, no. 2 (February, 1981): 52-64. Provides a somewhat dated but highly readable summary of the primary association of DNA with histone proteins. Kornberg received the 2006 Nobel Prize in Chemistry for his research on the process by which DNA is copied to RNA.

Krebs, Jocelyn E., Elliott S. Goldstein, and Stephen T. Kilpatrick. "Chromatin." In *Lewin's Essential Genes.* 2d ed. Sudbury, Mass.: Jones and Bartlett, 2010. A chapter on chromatin is included in this textbook on genetics.

Lodish, Harvey, et al. *Molecular Cell Biology.* 6th ed. New York: W. H. Freeman, 2008. Covers chromatin structure from a cellular and biochemical perspective.

Russell, Peter. *Genetics.* 5th ed. Menlo Park, Calif.: Benjamin Cummings, 1998. A college-level textbook with an excellent discussion of chromatin structure and organization.

Turner, Bryan. *Chromatin and Gene Regulation: Mechanisms in Epigenetics.* Malden, Mass.: Blackwell, 2001. Explores the relationship between gene expression and DNA packaging by explaining the chromatin-based control mechanisms. Provides an overview of transcription in bacteria, covers refined structures and control mechanisms, and covers dosage compensation.

Van Holde, Kensal. *Chromatin.* New York: Springer, 1988. Contemporary views on chromatin's functions and structure, addressing structures of DNA, proteins of chromatin (both histone and nonhistone), the nucleosome, higher-order structures, transcription, and replication. Bibliography.

Wolffe, Alan P. "Genetic Effects of DNA Packaging." *Science and Medicine* 2, no. 6 (December, 1995): 68-77. Excellent summary for the general reader of the relationship between gene expression and DNA packaging.

WEB SITES OF INTEREST

Chromatin Structure and Function Page
http://www.chromatin.us/chrom.html

A wide range of information on the impact of chromatin biology, histones, and epigenetics on the biological processes and on the researchers who study them. Links to numerous resources.

Waterborg's Chromatin Home Page
http://sbs.umkc.edu/waterborg/chromat/chromatn.html

Jakob Waterborg, a professor of biological sciences at the University of Missouri, Kansas City, has compiled this collection of slides and papers on the nuclear/nucleosome packaging of DNA, gene transcription in chromatin, and dynamic histone acetylation.

See also: Cell division; Central dogma of molecular biology; Chromosome structure; Developmental genetics; Fragile X syndrome; Gene regulation: Eukaryotes; Huntington's disease; Mitosis and meiosis; Molecular genetics.

Chromosome mutation

CATEGORY: Cellular biology; Molecular genetics

SIGNIFICANCE: Unlike gene mutations, which alter individual genes, chromosome mutations delete, duplicate, or rearrange chromosome segments. Chromosome mutations may create gene mutations if they delete genes or if the breakpoints of rearranged segments disrupt gene structure or alter gene expression. Even when they do not create gene mutations, chromosome mutations may reduce fertility and are an important cause of inherited infertility in humans. They also play important roles in the evolution of species.

KEY TERMS

deletion: a missing chromosome segment

duplication: a chromosome segment repeated in the same or in a different chromosome

fission: separation of a single chromosome into two chromosomes

fusion: joining of two chromosomes to become a single chromosome

inversion: a chromosome segment with reversed orientation when compared to the original chromosome structure

translocation: a chromosome segment transferred from one chromosome to a nonhomologous chromosome

DISCOVERY

As the fruit fly *Drosophila melanogaster* became a premier organism for genetic research in the early years of the twentieth century, geneticists who worked with it were the first to discover chromosome mutations. Calvin Bridges proposed deletions in 1917, duplications in 1919, and translocations in 1923 as explanations of phenomena he had observed in genetic experiments. Alfred Sturtevant proposed inversions in 1926 to explain experimental genetic data. Their proposals were directly confirmed as chromosome mutations when methods for microscopic examination of chromosomes were developed in the 1920's and 1930's.

DELETIONS

A deletion results when a chromosomal segment is lost. A deletion creates an imbalance in the genetic material because a relatively large segment of it is missing. Most deletions are lethal, even when heterozygous. Some small deletions persist in the heterozygous state but are usually lethal when homozygous. These small deletions are usually characterized by deleted portions of only one or two genes and behave genetically as recessive alleles when paired with a typical recessive allele of the affected gene.

DUPLICATIONS

A duplication arises when a chromosomal segment is duplicated and inserted either into the same chromosome, as its parent segment, or into another chromosome. Duplications are present in most genomes. Genome projects (including the Human Genome Project) have revealed large duplicated segments containing multiple genes dispersed throughout the chromosomes in most species. Some duplications are repeated in tandem in the same chromosome and are subject to unequal crossing over, a process in which duplicated segments mispair with one another and a crossover takes place within the mispaired segment. Unequal crossing over increases the number of tandem duplications in one chromosome and decreases that number in the other.

INVERSIONS

Two breaks within the same chromosome may liberate a chromosome segment. If the segment is reinserted into the same chromosome, but in reverse orientation, an inversion results. Also, rare crossing over between duplicated segments in the same chromosome may produce an inversion. If a breakpoint of the inversion lies within a gene, it disrupts the gene, causing a gene mutation. Additionally, an inversion may place a gene in another location in the chromosome, removing the gene from its regulatory elements and altering its expression, a phenomenon known as the position effect.

When one chromosome carries an inversion and its homologous partner does not, the individual carrying these two chromosomes is said to be an inversion heterozygote. The two homologous chromosomes in an inversion heterozygote cannot pair properly in meiosis; one of them must form a loop in the inverted region. A crossover within the inversion loop results in chromosomes that carry large deletions and duplications. Because of the imbalance of chromosomal material created by the deletions and duplications, progeny resulting from such crossovers usually do not survive. In genetic experiments, crossing over appears to be suppressed within an inversion, whereas, in reality, crossing over does take place within the inversion but crossover-type progeny fail to survive. For this reason, inversion heterozygotes may suffer a reduction in fertility that is proportional to the size of the inversion. An individual who is homozygous for an inversion, however, suffers no loss of fertility, because the chromosomes pair normally.

TRANSLOCATIONS

A break in a chromosome may liberate a chromosome fragment, which if reattached to a different chromosome is called a translocation. Most translocations are reciprocal: Two chromosome breaks, each in a different chromosome, liberate two fragments, and each fragment reattaches to the site where the other fragment was originally attached; in other words, the two fragments exchange places. If the breakpoint of a translocation is within a gene, a gene mutation may result. Also, a gene at or near the breakpoint may undergo a change in its expression because of position effect.

Translocations alter chromosome pairing in meiosis. During meiosis in a reciprocal translocation heterozygote, the two chromosomes with translocated segments pair with two other chromosomes without translocations. The pairing of these four chromosomes forms an X-shaped structure called a quadrivalent, so named because it contains four chromosomes paired with one another, instead of the usual two. Depending on the orientation of the quadrivalent during meiosis, some gametes may receive a balanced complement of chromosomes and others an unbalanced complement with large duplications and deletions. Typically, about half of all gametes in a reciprocal translocation heterozygote carry an unbalanced chromosome complement, a situation that significantly reduces the individual's fertility. However, translocation homozygotes suffer no loss of fertility, because the chromosomes pair normally with no quadrivalent.

FUSIONS

Very rarely, two chromosomes may fuse with one another to form a single chromosome. Chromosomes with centromeres at or very near the ends of the chromosomes may undergo breakage at the centromeres and fuse with each other in the centromeric region, resulting in a single chromosome with the long arms of the original chromosomes on either side of the fused centromere. Such a chromosome fusion is called a Robertsonian translocation. In other cases, two chromosomes may fuse with one another producing a dicentric chromosome (a chromosome with two centromeres). For the fused chromosome to persist, one of the centromeres ceases to function, leaving the other centromere as a single, functional centromere for the fused chromosome.

FISSIONS

A chromosome break produces two fragments, which may function as individual chromosomes if each has telomeres on both ends and a functional centromere. Typically, chromosome breakage produces one fragment with a telomere on one end and a centromere, and another fragment with a telomere on one end and no centromere. For both fragments to function as chromosomes, one must acquire a telomere and the other a centromere and a telomere. These events are highly unlikely, so fissions are rarer than fusions. However, complex translocations with other chromosomes may rarely produce functional chromosomes from a fission event, and cases of functional chromosomes arising from fissions have been documented.

IMPACT ON HUMAN GENETICS AND MEDICINE

Chromosome mutations are responsible for several human genetic disorders. For example, about 20 percent of hemophilia A cases result from a gene mutation caused by an inversion with a breakpoint in the *F8C* gene, which encodes blood clotting factor VIII. Cri du chat syndrome, a severe disorder characterized by severe mental retardation and distinctive physical features, is usually caused by deletion of a small chromosomal region near the end of chromosome 5. A few cases of this syndrome are associated with deletions that result from a translocation with a breakpoint near the end of chromosome 5 or crossovers within a small inversion in that chromosome region. Robertsonian translocations that

fuse the long arm of chromosome 21 with the long arm of another chromosome (usually chromosome 14) are responsible for some inherited cases of Down syndrome. A reciprocal translocation between chromosomes 9 and 22, called the Philadelphia chromosome, causes increased susceptibility to certain types of cancer by altering the expression of a gene located at the breakpoint of the translocation. Other translocations are likewise associated with certain cancers. Chromosome mutations may also cause infertility in humans. Reciprocal translocations are especially notorious, although certain inversions are also associated with infertility.

IMPLICATIONS FOR EVOLUTION

Heterozygous carriers of inversions, translocations, fusions, and fissions often suffer losses of fertility, but homozogotes do not. Thus, natural selection may disfavor heterozygotes while favoring homozygotes either for the original chromosome structure or for the mutation. Accumulation of different chromosome mutations in isolated populations of a species may eventually differentiate the chromosomes to such a degree that the isolated populations diverge into separate species. Their members can no longer produce fertile offspring when hybridized with members of another population because the chromosomes cannot properly pair with one another. Indeed, accumulated chromosome mutations are often evident when geneticists compare the chromosomes of closely related species. For example, the chromosomes of different *Drosophila* species are differentiated mostly by translocations and fusions. Comparison of human, chimpanzee, gorilla, and orangutan chromosomes reveals numerous inversions that distinguish the chromosomes of these species. One of the most striking cases of chromosome evolution is the origin of human chromosome 2. This chromosome matches two separate chromosomes in the great apes and apparently arose from a fusion of these two chromosomes after the divergence of the human and chimpanzee lineages. The presence in human chromosome 2 of DNA sequences corresponding to a nonfunctional centromere and telomere at sites corresponding to these structures in the great ape chromosomes is strong evidence of a chromosome fusion during evolution of the human lineage.

Daniel J. Fairbanks, Ph.D.

FURTHER READING

Burnham, Charles R. *Discussions in Cytogenetics.* Minneapolis, Minn.: Burgess, 1962. A classic book on chromosome mutations written by one of the pioneers in the field. Though out of print, this book remains available in many libraries.

Calos, Michele. *Molecular Evolution of Chromosomes.* New York: Oxford University Press, 2003. Describes the role of chromosome mutations in evolution.

Gersen, Steven L., and Martha B. Keagle, eds. *The Principles of Clinical Cytogenetics.* 2d ed. Totowa, N.J.: Humana Press, 2005. Includes basic information about cytogenetics, including chapters on DNA, chromosomes, and cell division; human chromosome nomenclature; and examining and analyzing chromosomes.

Leyden, Guy T., ed. *Genetic Translocations and Other Chromosome Aberrations.* New York: Nova Biomedical Books, 2008. Contains research from numerous scientists who have studied the relationship of genetic translocations and chromosome aberration on disease.

Lima-de-Faria, A. *Praise of Chromosome "Folly": Confessions of an Untamed Molecular Structure.* Hackensack, N.J.: World Scientific, 2008. Argues that the chromosome appears to be an independent molecular structure that follows its own path, without obedience to gravity, randomness, selection, or magnetism.

Miller, Orlando J., and Eeva Therman. *Human Chromosomes.* 4th ed. New York: Springer Verlag, 2001. A good textbook on human chromosomes, including common chromosome mutations.

WEB SITES OF INTEREST

Genetics Home Reference
http://ghr.nlm.nih.gov/handbook/mutationsanddisorders
Features several pages on Mutations and Health, with information on gene mutations, chromosomal changes, and health conditions that run in families.

National Center for Biotechnology Information
http://www.ncbi.nlm.nih.gov/Class/MLACourse/Original8Hour/Genetics/mutations.html
Features illustrations and text describing the various kinds of mutations.

Scitable
http://www.nature.com/scitable/topic/Chromosomes-and-Cytogenetics-7
Scitable, a library of science-related articles compiled by the Nature Publishing Group, contains a section on chromosomes and cytogenetics that features information about chromosome mutation.

See also: Cell cycle; Cell division; Central dogma of molecular biology; Chemical mutagens; Chromosome structure; Chromosome theory of heredity; Congenital disorders; Cystic fibrosis; Down syndrome; Epistasis; Evolutionary biology; Hemophilia; Hereditary diseases; Huntington's disease; Inborn errors of metabolism; Infertility; Mitosis and meiosis; Molecular genetics; Mutation and mutagenesis; Punctuated equilibrium.

Chromosome structure

CATEGORY: Cellular biology; Classical transmission genetics

SIGNIFICANCE: The separation of the alleles in the production of the reproductive cells is a central feature of the model of inheritance. The realization that the genes are located on chromosomes and that chromosomes occur as pairs that separate during meiosis provides the physical explanation for the basic model of inheritance. When chromosome structure is modified, changes in information transmission produce abnormal developmental conditions, most of which contribute to early miscarriages and spontaneous abortions.

KEY TERMS

histones: a class of proteins associated with DNA

homologous chromosomes: chromosomes that have identical physical structure and contain the same genes; humans have twenty-two pairs of homologous chromosomes and a pair of sex chromosomes that are only partially homologous

karyotyping: an analysis or physical description of all the chromosomes found in an organism's cells; often includes either a drawing or photograph of the chromosomes

spindle fibers: minute fibers composed of the protein tubulin that are involved in distributing the chromosomes during cell division

DISCOVERY OF CHROMOSOMES' ROLE IN INHERITANCE

The development of the microscope made it possible to study what became recognized as the central unit of living organisms, the cell. One of the most obvious structures within the cell is the nucleus. As study continued, dyes were used to stain cell structures to make them more visible. It became possible to see colored structures called chromosomes ("color bodies") within the nucleus that became visible when they condensed as the cell prepared to divide.

The association of the condensed, visible state of chromosomes with cell division caused investigators to speculate that the chromosomes played a role in the transmission of information. Chromosome counts made before and after cell division showed that the chromosome number remained constant from generation to generation. When it was observed that the nuclei of two cells (the egg and the sperm) fused during sexual reproduction, the association between information transport and chromosome composition was further strengthened. German biologist August Weismann, noting that the chromosome number remained constant from generation to generation despite the fusing of cells, predicted that there must be a cell division that reduced the chromosome number in the egg and sperm cells. The reductional division, meiosis, was described in 1900.

Following the rediscovery of Gregor Mendel's rules of inheritance in 1900, the work of Theodor Boveri and Walter Sutton led to the 1903 proposal that the character-determining factors (genes) proposed by Mendel were located on the chromosomes and that the factor segregation that was a central part of the model occurred because the like chromosomes of each pair separated during the reductional division that occurs in meiosis. This hypothesis, the "chromosome theory of heredity," was confirmed in 1916 by the observations of the unusual behavior of chromosomes and the determining factors located on them by Calvin Bridges.

CHROMOSOME STRUCTURE AND RELATION TO INHERITANCE

With the discovery of the nucleic acids came speculation about the roles of DNA and the associated proteins. During the early 1900's, it was generally accepted that DNA formed a structural support system to hold critical information-carrying proteins on the chromosomes. The identification of the structure of DNA in 1953 by American biologist James Watson and English physicist Francis Crick and the recognition that DNA, not the proteins, contained the genetic information led to study of chromosome structure and the relationships of the DNA and protein components.

It is now recognized that each chromosome contains one DNA molecule. Each plant and animal species has a specific number of chromosomes. Humans have twenty-three kinds of chromosomes, present as twenty-three pairs. Each chromosome can be recognized by its overall length and the position of constrictions, called centromeres, that are visible only when the cell is reproducing. At all other stages of the cell's life, the chromosome material is diffuse and is seen only as a general color within the nucleus. When the cell prepares for division, the fibrous DNA molecule tightly coils and condenses into the visible structures. Since there must be information for the two cells that result from the process of division, the chromosomes are present in a duplicated condition when they first become visible.

A major feature of the visible, copied chromosomes is the centromere. This constriction may be located anywhere along the chromosome, so its position is useful for identifying chromosomes. In karyotyping, the standard system used to identify human chromosomes, the numbering begins with the longest chromosome with the constriction nearest the center (chromosome 1), referred to as having a metacentric centromere placement. Chromosomes with nearly the same length but with the centromere constriction removed from the center position have higher numbers (chromosomes 2 and 3) and are referred to as acrocentric. Shorter chromosomes with a centromere near the middle are next, and the numbering proceeds based on the distance the centromere is removed from the central position. Short chromosomes with a centromere near one end have the highest numbers and are referred to as telocentric.

Most of the chromosomes have a centromere that is not centrally located, which results in arms of unequal length. The short arm is referred to as "petite" and is designated the *p* arm. The long arm is designated the *q* arm. This nomenclature is useful for referring to features of the chromosome. For example, when a portion of the long arm of chromosome 15 has been lost, the arm is shorter than normal. The loss, a deletion, is designated 15*q̄* (chromosome 15, long arm, deletion). Prader-Willi syndrome, in which an infant has poor sucking ability and poor growth, and later becomes a compulsive eater, results from this deletion. Cri du chat ("cry of the cat") syndrome results from 5*p̄*, a deletion of a portion of the short arm of chromosome 5. The cry of these individuals is like that of a cat, and they are severely mentally retarded and have numerous physical defects.

Some chromosomes have additional constrictions referred to as secondary constrictions. The primary centromere constrictions are located where the spindle fibers attach to the chromosomes to move them to the appropriate poles during cell division. The secondary constrictions are sites of specific gene activity. Both of these regions contain DNA base sequence information that is specific to their functions.

HISTONES

The DNA of the chromosomes is wound around special proteins called histones. This results in an orderly structure that condenses the DNA mass so that the bulky DNA does not require as much storage space. The wrapped DNA units then fold into additional levels of compaction, by means of a process called condensation. The exact processes involved in these higher levels of folding are not fully understood, but the overall condensation reduces the bulk of the DNA nearly one thousandfold. If the DNA is removed from a condensed chromosome, the proteins remain and have nearly the same shape as the chromosome, indicating that it is the proteins that form the chromosome shape. The presence of these proteins and the fact that the DNA is wrapped around them raises many questions about how the DNA is copied in preparation for cell division and how the DNA information is read for gene activity. These are areas of active research.

The histone proteins form a structure called a "nucleosome" ("nuclear body"). There are four

kinds of histones, and two of each kind join together to form a cylinder-shaped nucleosome structure. The fibrous DNA molecule wraps around each nucleosome approximately two and one-half times with a sequence of unwound DNA between each nucleosome along the entire length of the DNA molecule. The structure, called chromatin, looks like a string of beads when isolated sections are viewed with an electron microscope. When the chromatin is digested with enzymes that break the DNA backbone in the unwound regions, repeated lengths of chromatin are recovered, showing that the nucleosome wrapping is very regular. These nucleosome regions join together to form the additional folding as the chromosome condenses when the cell prepares for division.

In addition to the histone proteins, nonhistone proteins attach to the chromatin. With an electron microscope, chromatin loops can be seen extending from a protein matrix. There is evidence that these loops represent replication units along the chromosome, but how the DNA molecule is freed from the

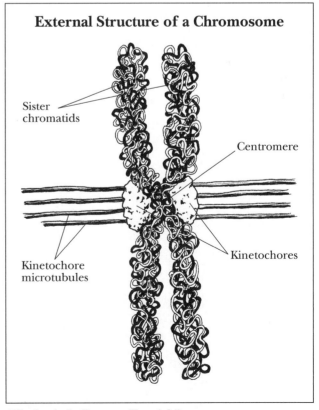

External Structure of a Chromosome

Sister chromatids

Centromere

Kinetochore microtubules

Kinetochores

(Kimberly L. Dawson Kurnizki)

Internal Structure of a Chromosome

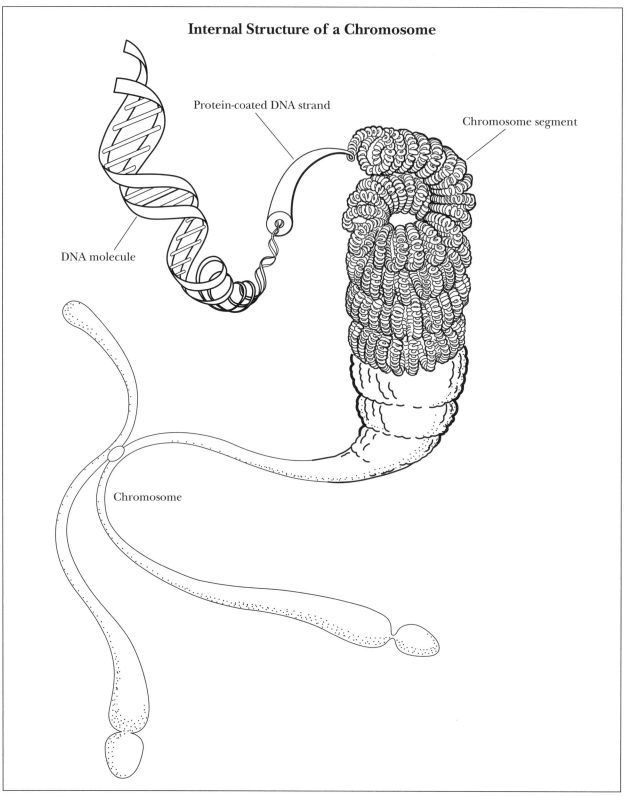

Protein-coated DNA strand

Chromosome segment

DNA molecule

Chromosome

histone proteins to be replicated is a major unsolved puzzle.

The condensation of the chromatin is not uniform over the entire chromosome. In the regions immediately adjacent to the centromere, the chromatin is tightly condensed and remains that way throughout the visible cycle. All of the available evidence indicates that this chromatin does not contain actively expressed genes. It also replicates later than the remaining DNA. This more highly condensed chromatin is called heterochromatin ("the other chromatin"). The remaining chromatin is referred to as euchromatin ("true chromatin") because it contains actively expressed genes and it replicates as a unit.

GIEMSA STAIN AND CHROMOSOME PAINTING

When chromosomes are treated with a dye called Giemsa stain, regular banding patterns appear. The bands vary in width, but their positions on the individual chromosomes are consistent. This makes the bands useful in identifying specific chromosome regions. When a chromosome has a structural modification, such as an inversion—which results when two breaks occur and the region is reversed when the fragments are rejoined—the change in the banding pattern makes it possible to recognize where it has occurred. When a loss of a chromosomal region produces a deficiency disorder, changes in the banding patterns of a chromosome can identify the missing region. Karyotype analysis is a useful tool in genetic counseling because disorders caused by chromosome structure modifications can be identified. Associations between disorders and missing chromosome regions are useful in identifying which functions are associated with specific regions. Other stains produce different banding patterns and, when used in combination with the Giemsa banding patterns, allow diagnosis of structure modifications that can be quite complex.

It is also possible to use fluorescent dyes, in a process called chromosome painting, to identify the DNA of individual chromosomes, which allows the recognition of small regions that have been exchanged between chromosomes that are too small to be recognized otherwise. Color differences within chromosomes or at their tips clearly show which chromosomes have exchanged DNA, how much DNA each has exchanged, and where on the chromosomes the exchanges have taken place.

Many cancer cells, for example, have multiple chromosome modifications, with DNA from two or three chromosomes associated in one highly modified chromosome structure.

CHROMOSOME DISORDERS

At the ends of the chromosomes are structures called telomeres, which are composed of specific repetitive DNA sequences that help protect the ends of chromosomes from damage and prevent DNA molecules from sticking together. Research that began in the early 1990's led to the discovery that the telomere regions of the chromosomes are shortened at each DNA replication. When the telomeres have been reduced to some critical point, the cell is no longer able to divide and often dies not long after, a phenomenon called apoptosis. Other observations indicate that the telomere is returned to its normal length in tumor cells, suggesting that this might contribute to the long life of tumor cells, possibly making them immortal. The relationship of cell age to telomere length and the mechanisms that lead to telomere shortening are not understood clearly, but this is an area of active research because it has implications for aging and cancer treatment.

The DNA of each chromosome carries a unique part of the information code in the sequence of the bases. The specific sequences are in linear order along the chromosome and form linked sequences of genes called linkage groups. When the like chromosomes pair and separate during meiosis, one copy of each chromosome is transmitted to the offspring. During meiosis, there may be an exchange of material between the paired chromosomes, but this does not change the information content because the information is basically the same for both chromosomes in any region. There may be differences in the coding sequences, but functionally the same informational content is transmitted. Extreme changes in chromosome structure that result in the moving of information to another chromosome may have consequences for how specific information is expressed; a change in position might result in different regulation or in changes in how the information is transmitted during meiosis.

Each chromosome has a specific arrangement of genes. Although homologous chromosomes exchange DNA during meiosis, as long as this process occurs normally, the gene arrangement on the chromosomes remains unchanged. Position affects result when genes are moved to different regions of the same chromosome or to another chromosome. A normal allele may show a mutant phenotype expression in a new position in the chromosome set. The best-known case occurs when a gene is placed adjacent to a heterochromatic region. The relocated DNA is condensed like the heterochromatic-region DNA and normally active genes now remain inactive. Ninety percent of patients with the disorder chronic myeloid leukemia have an exchange of material, called a translocation, between chromosomes 9 and 22. Chromosome 22 is shorter than normal and is called the Philadelphia chromosome, after the city in which it was discovered. The placing of a specific gene from chromosome 9 within the broken region adjacent to a gene on chromosome 22 causes the uncontrolled expression of both of the genes and uncontrolled cell reproduction, the hallmark of leukemia.

The separation of like chromosomes during meiosis occurs because the two chromosome arms are attached to a specific centromere. When the centromere is moved to one of the poles, the arms are pulled along, ensuring movement of all of the material of the paired chromosomes to the opposite poles and inclusion in the newly formed cells. Translocations occur when chromosomes are broken and material is placed in the wrong position by the repair system, causing a chromosome region to become attached to a different centromere. This leads to an inability to properly separate the regions of the arm, which can result in duplication of some of the chromosomal regions (when two copies of the same arm move to one cell) or deficiencies (when none of the material from a chromosome arm moves into a cell). This is a common outcome with translocation heterozygotes (individuals with both normal chromosomes and translocated chromosomes in the same cells). Pairing of like chromosome regions occurs, but rather than two chromosomes paired along their entire lengths, the arms of the two translocated chromosomes are paired with the arms of their normal pairing partners. The separation of the chromosomes produces duplications of material from one chromosome arm or a deficiency of that material 50 percent of the time. If these cells are involved in fertilization, the offspring will show duplication or deficiency disorders.

D. B. Benner, Ph.D.; updated by Bryan Ness, Ph.D.

FURTHER READING

Adolph, Kenneth W., ed. *Gene and Chromosome Analysis.* 3 vols. San Diego: Academic Press, 1993-1994. Covers a range of topics, including cDNA cloning methods, mammalian embryogenesis, microcell hybrids, tumor-suppressor genes, prenatal cytogenetics, and the polymerase chain reaction.

Bickmore, Wendy A. *Chromosome Structural Analysis: A Practical Approach.* New York: Oxford University Press, 1999. Describes research on chromatin and chromosome structure, specifically examining the mapping of protein, a study of whole chromosome structure and biochemical techniques for analyzing the substructure of mammalian nuclei, and the experimental manipulation of chromosome structure.

Campbell, Neil A., and Jane Reece. *Biology.* 8th ed. San Francisco: Pearson, Benjamin Cummings, 2008. A college-level biology textbook that provides introductory explanations of chromosomes.

Greider, Carol, and Elizabeth Blackburn. "Telomeres, Telomerase, and Cancer." *Scientific American* 274, no. 2 (February, 1996): 92. Contains a review of the nature of telomeres and their importance in the lives of cells.

Lima-de-Faria, A. *Praise of Chromosome "Folly": Confessions of an Untamed Molecular Structure.* Hackensack, N.J.: World Scientific, 2008. Argues that the chromosome appears to be an independent molecular structure that follows its own path, without obedience to gravity, randomness, selection, or magnetism.

Russell, Peter. *Genetics.* 5th ed. Menlo Park, Calif.: Benjamin Cummings, 1998. A college-level textbook with an excellent discussion of chromosome structure and function.

Sharma, Archana, and Sumitra Sen. *Chromosome Botany.* Enfield, N.H.: Science Publishers, 2002. Focuses on the chromosome as a vehicle of hereditary transmission, covering topics such as structural details, identification of gene sequences at the chromosome level, specific and genetic diversity in evolution, and the genome as affected by environmental agents.

Sumner, Adrian T. *Chromosomes: Organization and Function.* Malden, Mass.: Blackwell, 2003. Textbook providing an overview of both the structure and the behavior of chromosomes, including information on cell division, the assembly of chromatin, sex chromosomes, and chromosomes and disease.

WEB SITES OF INTEREST

Biology Corner
http://www.biologycorner.com/bio1/celldivision
-chromosomes.html
This Web site, designed as a resource for students and teachers, contains an illustration and textual description of chromosomal structure.

Genetics Home Reference, Chromosomes
http://ghr.nlm.nih.gov/chromosomes
Provides basic information about chromosomes in general and about specific chromosome pairs.

Scitable
http://www.nature.com/scitable/topic/
Chromosomes-and-Cytogenetics-7
Scitable, a library of science-related articles compiled by the Nature Publishing Group, contains a section on chromosomes and cytogenetics that features information about chromosome structure.

See also: Chromatin packaging; Cell cycle; Cell division; Central dogma of molecular biology; Chromosome mutation; Chromosome theory of heredity; Chromosome walking and jumping; Classical transmission genetics; Dihybrid inheritance; DNA replication; DNA structure and function; Epistasis; Extrachromosomal inheritance; Incomplete dominance; Mendelian genetics; Mitosis and meiosis; Molecular genetics; Monohybrid inheritance; Multiple alleles; Mutation and mutagenesis; Nondisjunction and aneuploidy; Parthenogenesis; Penetrance; Polygenic inheritance; RNA structure and function; Transposable elements.

Chromosome theory of heredity

CATEGORY: Classical transmission genetics; History of genetics

SIGNIFICANCE: The chromosome theory of heredity originated with American geneticist Walter Sutton, who first suggested that genes were lo-

cated on chromosomes. This theory guided much of genetic research in the early twentieth century, including development of the earliest genetic maps based on linkage. In 1931, several experiments confirmed the chromosome theory by demonstrating that certain rearrangements of the heritable traits (or genes) were always accompanied by corresponding rearrangements of the microscopically observable chromosomes.

KEY TERMS

crossing over: the breakage of chromosomes followed by the interchange of the resulting fragments; also, the recombination of genes that results from the chromosomal rearrangement

genetic mapping: the locating of gene positions along chromosomes

independent assortment: the inheritance of genes independently of one another when they are located on separate chromosomes

linkage: the frequent inheritance of two or more genes together as a unit if they are located close together on the same chromosome

linkage mapping: a form of genetic mapping that uses recombination frequencies to estimate the relative distances between linked genes

physical mapping: a form of genetic mapping that associates a gene with a microscopically observable chromosome location

MENDEL'S LAW OF INDEPENDENT ASSORTMENT

In a series of experiments first reported in 1865, Austrian botanist Gregor Mendel established the first principles of genetics. Mendel showed that the units of heredity were inherited as particles that maintained their identity across the generations; these units of heredity are now known as genes. These genes exist as pairs in all the body's cells except for the egg and sperm cells. When Mendel studied two traits at a time (dihybrid inheritance), he discovered that different genes were inherited independently of one another, a principle that came to be called the law of independent assortment. For example, if an individual inherits genes *A* and *B* from one parent and genes *a* and *b* from the other parent, in subsequent generations the combinations *AB*, *Ab*, *aB*, and *ab* would all occur with equal frequency. Gene *A* would go together with *B* just as often as with *b*, and gene *B* would go with *A* just as often as with *a*. Mendel's results were ignored for many years after he published his findings, but his principles were rediscovered in 1900 by Erich Tschermak von Seysenegg in Vienna, Austria, Carl Erich Correns in Tübingen, Germany, and Hugo de Vries in Amsterdam, Holland. Organized research in genetics soon began in various countries in Europe and also in the United States.

SUTTON'S HYPOTHESIS

Mendel's findings had left certain important questions unanswered: Why do the genes exist in pairs? Why do different genes assort independently? Where are the genes located? Answers to these questions were first suggested in 1903 by a young American scientist, Walter Sutton, who had read about the rediscovery of Mendel's work. By this time, it was already well known that all animal and plant cells contain a central portion called the nucleus and a surrounding portion called the cytoplasm. Division of the cytoplasm is a very simple affair: The cytoplasm simply squeezes in two. The nucleus, however, undergoes mitosis, a complex rearrangement of the rod-shaped bodies called chromosomes, which exist in pairs. Sex cells (eggs or sperm) are "haploid," with one chromosome from each pair. All other body cells, called somatic cells, have a "diploid" chromosome number in which all chromosomes are paired. During mitosis, each chromosome becomes duplicated; then the two strands (or chromatids) split apart and separate. One result of mitosis is that the chromosome number of each cell is always preserved. Sutton also noticed that eggs in most species are many times larger than sperm because of a great difference in the amount of cytoplasm. The nuclei of egg and sperm are approximately equal in size, and these nuclei fuse during fertilization, a process in which two haploid sets of chromosomes combine to make a complete diploid set. From these facts, Sutton concluded that the genes are probably in the nucleus, not the cytoplasm, because the nucleus divides carefully and exactly while the cytoplasm divides inexactly. Also, if genes were in the cytoplasm, one would expect the mother's contribution to be much greater than the father's, contrary to the repeated observation that the parental contributions to heredity are usually equal.

Of all the parts of diploid cells, only the chromosomes were known to exist in pairs. If genes were located on the chromosomes, it would explain why they existed in pairs (except singly in eggs and sperm cells). In fact, the known behavior of chromosomes exactly paralleled the postulated behavior of Mendel's genes. Sutton's hypothesis that genes were located on chromosomes came to be called the chromosome theory of heredity. According to Sutton's hypothesis, Mendel's genes assorted independently because they were located on different chromosomes. However, there were only a limited number of chromosomes (eight in fruit flies, fourteen in garden peas, and forty-six in humans), while there were hundreds or thousands of genes. Sutton therefore predicted that Mendel's law of independent assortment would apply only to genes located on different chromosomes. Genes located on the same chromosome would be inherited together as a unit, a phenomenon now known as linkage.

In 1903, Sutton outlined his chromosomal theory of heredity in a paper entitled "The Chromosomes in Heredity." Many aspects of this theory were independently proposed by Theodor Boveri, a German researcher who had worked with sea urchin embryos at the Naples Marine Station in Italy.

LINKAGE AND CROSSING OVER

Sutton had predicted the existence of linked genes before other investigators had adequately described the phenomenon. The subsequent discovery of linked genes lent strong support to Sutton's hypothesis. English geneticists William Bateson and Reginald C. Punnett described crosses involving linked genes in both poultry and garden peas, while American geneticist Thomas Hunt Morgan made similar discoveries in the fruit fly (*Drosophila melanogaster*). Instead of assorting independently, linked genes most often remain in the same combinations in which they were transmitted from prior generations: If two genes on the same chromosome both come from one parent, they tend to stay together through several generations and to be inherited as a unit. On occasion, these combinations of linked genes do break apart, and these rearrangements were attributed to "crossing over" of the chromosomes, a phenomenon in which chromosomes were thought to break apart and then recombine. Some microscopists thought they had observed X-shaped arrangements of the chromosomes that looked like the result of crossing over, but many other scientists were skeptical about this claim because there was no proof of breakage and recombination of the chromosomes in these X-shaped arrangements.

GENETIC MAPPING

Sutton had been a student of Thomas Hunt Morgan at Columbia University in New York City. When Morgan began his experiments with fruit flies around 1909, he quickly became convinced that Sutton's chromosome theory would lead to a fruitful line of research. Morgan and his students soon discovered many new mutations in fruit flies, representing many new genes. Some of these mutations were linked to one another, and the linked genes fell into four linkage groups corresponding to the four chromosome pairs of fruit flies. In fruit flies as well as other species, the number of linkage groups always corresponds to the number of chromosome pairs.

One of Morgan's students, Alfred H. Sturtevant, reasoned that the frequency of recombination of linked genes should be small for genes located close together and higher for genes located far apart. In fact, the frequency of crossing over between linked genes could serve as a rough measure of the distance between them along the chromosome. Sturtevant assumed that the frequency of recombination would be roughly proportional to the distance along the chromosome; recombination between closely linked genes would be a rare event, while recombination between genes further apart would be more common. Sturtevant first used this technique in 1913 to determine the relative positions of six genes on one of the chromosomes of *Drosophila*. For example, the genes for white eyes and vermilion eyes recombined about 30 percent of the time, and the genes for vermilion eyes and miniature wings recombined about 3 percent of the time. Recombination between white eyes and miniature wings took place 34 percent of the time, close to the sum of the two previously mentioned frequencies (30 percent plus 3 percent). Therefore, the order of arrangement of the genes was:

white ← 30 units → vermilion ← 3 units → miniature

Since the distances were approximately additive (the smaller distances added up to the larger distances), Sturtevant concluded that the genes were arranged along each chromosome in a straight line like beads on a string. In all, Sturtevant was able to determine such a linear arrangement among six genes in his initial study (an outgrowth of his doctoral thesis) and many more genes subsequently.

Calvin Bridges, another one of Morgan's students, worked closely with Sturtevant. Over the next several years, Sturtevant and Bridges conducted numerous genetic crosses involving linked genes. They used recombination frequencies to determine the arrangement of genes along chromosomes and the approximate distances between these genes, thus producing increasingly detailed genetic maps of several *Drosophila* species.

The use of Sturtevant's technique of making linkage maps was widely copied. As each new gene was discovered, geneticists were able to find another gene to which it was linked, and the new gene was then fitted into a genetic map based on its linkage distance to other genes. In this way, geneticists began to make linkage maps of genes along the chromosomes of many different species. There are now more than one thousand genes in *Drosophila* whose locations have been mapped using linkage mapping. Extensive linkage maps have also been developed for mice (*Mus musculus*), humans (*Homo sapiens*), corn or maize (*Zea mays*), and bread mold (*Neurospora crassa*). In bacteria such as *Escherichia coli*, other methods of genetic mapping were developed based on the order in which genes were transferred during bacterial conjugation. These mapping techniques reveal that the genes in bacteria are arranged in a circle or, more precisely, in a closed loop resembling a necklace. This loop can break at

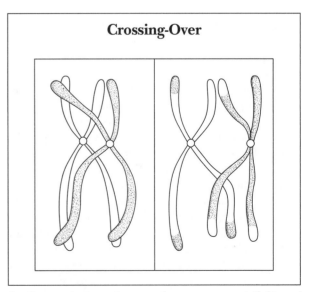

Crossing-Over

In the crossing-over process, chromosomes meet (left) and recombine (right). (Electronic Illustrators Group)

any of several locations, after which the genes are transferred from one individual to another in the order of their location along the chromosome. The order can be determined by interrupting the process and testing to see which genes had been transferred before the interruption.

CONFIRMATION OF THE CHROMOSOME THEORY

The first confirmation of the chromosome theory was published in 1916 by Bridges, who studied the results of a type of abnormal cell division. When egg or sperm cells are produced by meiosis, only one chromosome of each chromosome pair is normally included in each of the resultant cells. In a very small proportion of cases, one pair of chromosomes fails to separate (or "disjoin"), so that one of the resultant cells has an extra chromosome while the other cell is missing that chromosome. This abnormal type of meiosis is called nondisjunction. In fruit flies, as in humans and many other species, females normally have two X chromosomes (XX) and males have two unequal chromosomes (XY). Bridges discovered some female fruit flies that had the unusual chromosome formula XXY; he suspected that these unusual females had originated from nondisjunction, in which two X chromosomes had failed to separate during meiosis. Bridges studied one cross using a white-eyed XXY female mated to a normal, red-eyed male. (The gene for white eyes was known to be sex-linked; it was carried on the X chromosome.) Bridges was able to predict both the genetic and chromosomal anomalies that would occur as a result of this cross. Among the unusual predictions that were verified experimentally was the existence of a chromosome configuration (XYY) that had never been observed before. Using the assumption that the gene for white eyes was carried on the X chromosome in this and other crosses, Bridges was able to make unusual predictions of both genetic and chromosomal results. These studies greatly strengthened the case for the chromosomal theory.

In 1931, Harriet Creighton and Barbara McClintock were able to confirm the chromosomal theory of inheritance much more directly. Creighton and McClintock used corn plants whose chromosomes had structural abnormalities on either end, enabling them to recognize the chromosomes under the microscope. One chromosome, for example, had a knob at one end and an attached portion of another chromosome at the other end, as shown in the figure headed "Creighton and McClintock's Cross." Creighton and McClintock then crossed plants differing in two genes located along this chromosome. One gene controlled the color of the seed coat while the other produced either a starchy or waxy kernel. The parental gene combinations (*C* with *wx* on the abnormal chromosome and *c* with *Wx* on the other chromosome) were always preserved in noncrossovers. However, a crossover between the two genes produced two new gene combinations: *C* with *Wx* and *c* with *wx*.

In this cross, Creighton and McClintock observed that the chromosomal appearance in the offspring could always be predicted from the phenotypic appearance: Seeds with colorless seed coats and starchy kernels had normal chromosomes, seeds with colored seed coats and waxy kernels had chromosomes with the knob at one end and the extra interchanged chromosome segment at the other end, seeds with colorless seed coats and waxy kernels had the interchanged segment but no knob, and seeds with colored coats and starchy kernels had the knob but not the interchanged segment. In other words, whenever the two genes showed rearrangement of the parental combinations, a corresponding switch of the chromosomes could be observed under the microscope. The interchange of chromosome segments was always accompanied by the recombination of genes, or, in the words of the original paper,

> cytological crossing-over . . . is accompanied by the expected types of genetic crossing-over. . . . Chromosomes . . . have been shown to exchange parts at the same time they exchange genes assigned to these regions.

In short, genetic recombination (the rearranging of genes) was always accompanied by crossing over (the rearranging of chromosomes). This historic finding established firm evidence for the chromosomal theory of heredity. Later that same year, Curt Stern published a paper describing a similar experiment using fruit flies.

PHYSICAL MAPPING AND FURTHER CONFIRMATION

Other evidence that helped confirm the chromosome theory came from the study of rare chromosome abnormalities. In 1933, Thomas S. Painter called attention to the large salivary gland chromo-

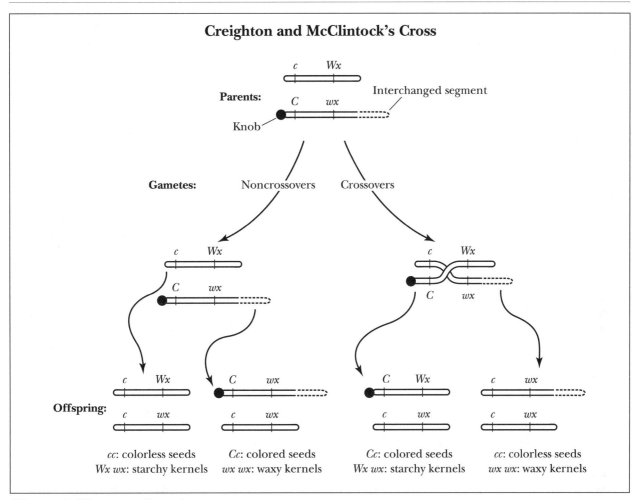

Creighton and McClintock's Cross

(Electronic Illustrators Group)

somes of *Drosophila.* Examination of these large chromosomes made structural abnormalities in the chromosomes easier to identify. When small segments of a chromosome were missing, a gene was often found to be missing also. These abnormalities, called chromosomal deletions, allowed the first physical maps of genes to be drawn. In all cases, the physical maps were found to be consistent with the earlier genetic maps (or linkage maps) based on the frequency of crossing over.

When Bridges turned his attention to the "bar eyes" trait in fruit flies, he discovered that the gene for this trait was actually another kind of chromosome abnormality called a "duplication." Again, a chromosome abnormality that could be seen under the microscope could be related to a genetic map based on linkage. Larger chromosome abnormali-

ties included "inversions," in which a segment of a chromosome was turned end-to-end, and "translocations," in which a piece of one chromosome became attached to another. There were also abnormalities in which entire chromosomes were missing or extra chromosomes were present. Each of these chromosomal abnormalities was accompanied by corresponding changes in the genetic maps based on the frequency of recombination between linked genes. In cases in which the location of a chromosomal abnormality could be identified microscopically, this permitted an anchoring of the genetic map to a physical location along the chromosome. The correspondence between genetic maps and chromosomal abnormalities provided important additional evidence in support of the chromosomal theory. Other forms of physical mapping were developed decades later in mammals and bacteria. The increasingly precise mapping of gene locations led the way to the development of modern molecular genetics, including techniques for isolating and sequencing individual genes.

The discovery of restriction endonuclease enzymes during the 1970's allowed geneticists to cut DNA molecules into small fragments. In 1980, a team headed by David Botstein measured the sizes of these "restriction fragments" and found many cases in which the length of the fragment varied from person to person because of changes in the DNA sequence. This type of variation is generally called a "polymorphism." In this case, it was a polymorphism in the length of the restriction fragments (known as a restriction fragment length polymorphism, or RFLP). The use of the RFLP technique has allowed rapid discovery of the location of many human genes. The Human Genome Project (an effort by scientists worldwide to determine the location and sequence of every human gene) would never have been proposed had it not been for the existence of this mapping technique.

Eli C. Minkoff, Ph.D.

FURTHER READING

Botstein D., R. L. White, M. Skolnick, and R. W. Davis. "Construction of a Genetic Linkage Map in Man Using Restriction Fragment Length Polymorphisms." *American Journal of Human Genetics* 32, no. 3 (1980): 314-331. Botstein's initial paper on the RFLP technique.

Carlson, E. A. *The Gene: A Critical History.* 1966. Reprint. Ames: Iowa State University Press, 1989. A classic text that examines the major theories from the early to mid-twentieth century concerning the structure of the gene.

Cummings, Michael R. *Human Heredity: Principles and Issues.* 8th ed. Florence, Ky.: Brooks/Cole/Cengage Learning, 2009. College text that surveys topics such as genetics as a human endeavor; cells, chromosomes, and cell division; transmission of genes from generation to generation; cytogenics; the source of genetic variation; cloning and recombinant DNA; genes and cancer; genetics of behavior; and genes in populations.

Griffiths, Anthony J. F., et al. *An Introduction to Genetic Analysis.* 9th ed. New York: W. H. Freeman, 2008. A classic text that includes discussions of advances in genetics research.

Hartl, Daniel L., and Elizabeth W. Jones. "The Chromosomal Basis of Heredity." In *Essential Genetics: A Genomics Perspective.* 4th ed. Boston: Jones and Bartlett, 2006. This textbook on genetics contains a chapter about the chromosomal theory.

Hartwell, Leland, et al. "The Chromosome Theory of Inheritance." In *Genetics: From Genes to Genomes.* 3d ed. Boston: McGraw-Hill Higher Education, 2008. This introductory genetics textbook devotes a chapter to the chromosomal theory of inheritance.

Lewin, B. *Genes IX.* Sudbury, Mass.: Jones and Bartlett, 2007. Provides an integrated account of the structure and function of genes and incorporates research in the field, including topics such as accessory proteins (chaperones), the role of the proteasome, reverse translocation, and the process of X chromosome inactivation.

Peters, James A., ed. *Classic Papers in Genetics.* Englewood Cliffs, N.J.: Prentice-Hall, 1959. Includes many of the classic papers that contributed to chromosomal theory, including those in which Mendel established the principles of genetics, Sutton first proposed the chromosomal theory of heredity, Sturtevant produced the first genetic map based on linkage, and Creighton and McClintock confirmed that the recombination of linked genes always took place by a process that also rearranged the chromosomes.

Scriver, Charles, et al., eds. *The Metabolic and Molecular Bases of Inherited Disease.* 8th ed. 4 vols. New York: McGraw-Hill, 2001. An authority on heredity of disease and genetic inheritance, covering genetic perspectives, basic concepts, how inherited diseases occur, diagnostic approaches, and the effects of hormones.

WEB SITES OF INTEREST

Genome News Network, Genetic and Genomics Timeline
http://www.genomenewsnetwork.org/resources/timeline/1902_Boveri_Sutton.php
Discusses Theodor Boveri and Walter Sutton's theories about chromosomes and heredity.

Scitable
http://www.nature.com/scitable/topic/Chromosomes-and-Cytogenetics-7
Scitable, a library of science-related articles compiled by the Nature Publishing Group, contains a section on chromosomes and cytogenetics that fea-

tures information about chromosome theory and cell division. One page, "Developing the Chromosome Theory," discusses how research by Walther Flemming, Theodore Boveri, and Walter Sutton connected chromosomes to heredity.

See also: Cell division; Chromosome mutation; Chromosome structure; Classical transmission genetics; Genetic code; Genetic code, cracking of; Linkage maps; Mendelian genetics; Mitosis and meiosis; Model organism: *Drosophila melanogaster*; Restriction enzymes; RFLP analysis; Transposable elements.

Chromosome walking and jumping

CATEGORY: Genetic engineering and biotechnology; Techniques and methodologies

SIGNIFICANCE: Chromosome walking and jumping were once used as mapping methods to find defective genes that cause hereditary diseases. Although these techniques have been rendered obsolete by the completion of the Human Genome Project, they have assisted in curing diseases, seeking preventive measures, and detecting genetic carriers.

KEY TERMS

genomic library: a group of cloned DNA fragments representative of an organism's genome

kilobase pairs (kb): a measurement of 1,000 base pairs in DNA

marker: a unique DNA sequence with a known location with respect to other markers or genes

repetitive DNA: nucleotide sequences, usually noncoding, that are present in many copies in a eukaryotic genome

GENE HUNTING

The science of molecular genetics began in the early 1950's when Alfred Hershey and Martha Chase conducted a series of experiments that proved that DNA did indeed carry life's hereditary information. This discovery was soon followed by James Watson and Francis Crick's determination that the structure of DNA was that of a double helix that could "unzip" and make copies of itself. By the late 1960's, researchers began to actively seek the knowledge to identify, isolate, and manipulate certain sections of DNA within the human genome.

Around this time, several geneticists autonomously recognized the possibilities of chromosome walking and jumping to locate genes. Hans Lehrach suggested such techniques at the European Molecular Biology Laboratory, and Sherman Weissman proposed similar methods at Yale University. Weissman's student Francis S. Collins elaborated his mentor's chromosome-jumping concepts. Interested in identifying disease-causing genes, Collins sought to examine sizable areas of genetic material for unknown genes believed to be responsible for triggering erratic biochemical behavior. As a result of Collins's work, investigators began to adopt the chromosome-jumping procedure as a reliable, efficient molecular biology tool. This novel exploratory method enabled researchers to span chromosomes expeditiously and bypass repetitive or insignificant genetic information. Based on Collins's chromosome-jumping technology, gene searching became less time-consuming and resulted in the identification of defective genes that code for abnormal proteins and cause such diseases as cystic fibrosis. Understanding the nature of such mutations makes the development of treatments and cures more likely and can lead to the ability to detect the presence of the mutated gene in carriers.

PROCEDURE

Geneticists initiate chromosome walking and jumping by collecting genetic samples from people who have a specific disease and from their close relatives. For walking, researchers select a cloned DNA fragment from a genomic library that contains the marker closest to the gene being sought. A small part of the cloned DNA fragment that is on the end nearest the gene being sought is subcloned. The subcloned fragment is then used to screen the genomic library for a clone with a fragment closer to the gene. Then a small part of this new cloned fragment is subcloned to be used to screen for the next closer fragment. This series of steps is repeated as many times as needed, until a fragment is found that appears to contain a gene. This fragment is carefully analyzed, and if it does contain the gene of interest, the process is halted; if not, chromosome walking is continued. Chromosome walking is slow, and repetitive DNA sequences or regions that do not appear in the library can halt the process.

Another method used to maneuver to genes more quickly and to bypass troublesome regions of DNA that cannot be easily mapped by chromosome walking, such as those containing repetitive DNA, is chromosome jumping. Using chromosome jumping, researchers can travel the same distance they can using chromosome walking but they are able to advance farther along the chromosome in less time because this method uses much larger fragments. Chromosome jumping is achieved by selecting a large DNA segment from the area where geneticists believe the desired gene is located and joining the ends to form a circle. This moves DNA sequences together that naturally would occur at distances of several kilobases. Researchers cut out and clone these junctions into a phage vector, and the various junction segments are then used to form libraries. Researchers then use probes from the DNA sample to seek clones with matching start and end sequences and jump along the chromosome. After each jump, bidirectional walking is often done in the new region. A combination of chromosome jumping and walking can be done until the gene is found.

GENE DISCOVERY

Collaborating with Lap-Chee Tsui and researchers at Toronto's Hospital for Sick Children, Collins examined DNA from patients suffering from cystic fibrosis. Tsui realized that the *CF* gene was located on chromosome 7. Since that chromosome consists of 150 million DNA base pairs, chromosome walking toward the *CF* gene would be a very slow process that would take approximately 18 years to complete. After Tsui contacted Collins, his colleague at the University of Michigan, the two researchers devised a technique for jumping along the chromosome. Tsui and Collins estimated that jumping along the chromosome would be five to ten times faster than walking because it would allow researchers to cover 100,000 to 200,000 DNA bases at one time. In addition, areas on the chromosome that might otherwise be difficult to cross could simply be jumped over. Using markers Tsui made from chromosome 7 library fragments, they applied the chromosome jumping technique and scanned the genetic material to target where they should use chromosome walking to find the *CF* gene.

They discovered the *CF* gene in 1989. Analysis revealed that the mutation is a deletion of DNA base pairs. This gene codes the cystic fibrosis transmembrane conductance regulator (CFTR) protein. Tsui determined that the shape of CFTR and how it functions are affected by the mutated gene's coding. The abnormal CFTR is unable to create a release channel to remove chloride and sodium from cells. Mucus builds up, adhering to lungs and organs, and bacteria proliferate. Cystic fibrosis is the most frequent fatal hereditary disease in Caucasians. Geneticists estimate that one in twenty-five white Americans has a recessive *CF* gene, and one in two thousand white babies are born with cystic fibrosis. Internationally, researchers associated with Tsui's Toronto-based consortium continue to study DNA fragments for additional *CF* gene mutations and have detected at least one thousand distinct mutations.

IMPACT

Chromosome walking and jumping have been utilized to find other disease-causing genes. Collins and his team identified the tumor-producing neurofibromatosis gene in 1990. Three years later, they located the gene for Huntington's disease (Huntington's chorea), an extreme neurological disorder. This method also detected the location on the X chromosome of the choroideremia gene, which causes gradual blindness, mostly in males, as the retina and choroid coat degenerate. Investigating Duchenne muscular dystrophy, Louis Kunkel at the Harvard Medical School used chromosome walking to detect the absence of a gene on the X chromosome that codes the dystrophin protein for muscles. Not all genes found by these methods are linked to diseases. Andrew Sinclair and his team in London applied chromosome walking to seek the gene that signals development of testes in many embryonic mammals. Although these techniques are useful, they raise ethical concerns. As genes with disease-causing mutations are identified, people can undergo testing to determine whether they carry the mutations. This information can affect reproductive choices, particularly if both partners have a recessive allele for a potentially lethal disease. Fetal material can be genetically analyzed, resulting in complex decisions to continue or terminate a pregnancy if the fetus has the mutation.

Once the mapping of the human genome was completed, however, geneticists arrived at a time when they no longer needed to depend on chromosome walking and jumping as tools to seek human genes. Investigators continue to use walking and jumping, however, to locate genes of other organisms, particularly such agricultural plants as rice and wheat. The Human Genome Project, an international, collaborative scientific research program with primary end goals of identifying, mapping, and understanding the entire human genome, was completed in April, 2003, fifty years after the discovery of the double helix structure. Due to the development of improved technologies for accelerating the elucidation of the genome, this project was completed years earlier than what was originally anticipated. Approximately 20,500 human genes have been identified and mapped through the Human Genome Project. This project has provided the entire world with a resource of information that has revolutionized the field of medicine and biological research.

The completion of the Human Genome Project has significantly eased the task of locating and analyzing the mechanism of actions of genes involved in complex diseases. Understanding the molecular basis of a disease can ultimately lead to new ways to diagnose and treat patients. The ability to understand the pathophysiology of a disease on a molecular level has led to the development of more specific and effective drug treatments. More recent developments in novel detection methods have led to significant advances in the ability to provide more rapid, efficient, and less expensive methods of DNA sequencing.

Elizabeth D. Schafer, Ph.D.; updated by Kimberly Lynch

Further Reading

Adams, Jill. "Sequencing Human Genome: The Contributions of Francis Collins and Craig Venter." *Nature Education* 1, no. 1 (2008). Collins and other researchers master gene mapping.

Gelehrter, Thomas D., Francis S. Collins, and David Ginsburg. *Principles of Medical Genetics.* 2d ed. Baltimore: Williams & Wilkins, 1998. Collins and his University of Michigan colleagues explore basic concepts and advances in genetics, including positional cloning, molecular genetics, genome mapping, and ethics, in a text comprehensible by readers unfamiliar with genetics. Glossary and illustrations.

Metzker, M. L. "Emerging Techniques in DNA Sequencing." *Genome Research* 15 (2005): 1767-1776. This article outlines the major technological advances that led to the completion of the Human Genome Project.

Rommens, Johanna M., Michael C. Iannuzzi, et al. "Identification of the Cystic Fibrosis Gene: Chromosome Walking and Jumping." *Science* 245, no. 4922 (September 8, 1989): 1059-1065. This issue's cover story, cowritten by Francis S. Collins and his research team, announced one of chromosome jumping's first major discoveries.

Tsui, Lap-Chee, et al., eds. *The Identification of the CF (Cystic Fibrosis) Gene: Recent Progress and New Research Strategies.* New York: Plenum Press, 1991. This collection of technical papers represents the work of notable researchers on chromosome jumping who attended an international workshop seven months after the *CF* gene was identified.

Web Sites of Interest

National Human Genome Research Institute. "All About the Human Genome Project (HGP)"
http://www.genome.gov/10001772

Wellcome Trust Sanger Institute. "Human Genome Project"
http://www.sanger.ac.uk/HGP

See also: Cystic fibrosis; Genetic screening; Genetic testing; Genomic libraries; Linkage maps.

Chronic granulomatous disease

Category: Diseases and syndromes
Also known as: CGD; fatal granulomatosis of childhood; chronic granulomatous disease of childhood; progressive septic granulomatosis

Definition

Chronic granulomatous disease develops when a specific gene from both parents passes to the child. This gene causes abnormal cells to develop in the immune system called phagocytic cells. They normally kill bacteria. With this disease, these cells are impaired. As a result, the body cannot fight some types of bacteria. It also makes infections likely to recur.

The increased risk of infections can lead to premature death. Repeated lung infections are often the cause of death with this disease. Preventive care and treatment can help to reduce and temporarily control infections. CGD is a rare condition, occurring in about 1 in 200,000 births in the United States.

RISK FACTORS

Individuals whose parents have the recessive trait and females have an increased chance of developing CGD.

ETIOLOGY AND GENETICS

Mutations in at least four different genes have been identified that can result in clinical symptoms that are recognized as CGD. The most common form (50-70 percent of all cases) involves a mutation in the *CYBB* gene, found on the short arm of the X chromosome at position Xp21.1. This gene encodes the b subunit of cytochrome B, one of the essential protein complexes of the electron transport system found in mitochondria. The inheritance pattern of this disease is typical of all sex-linked recessive mutations (those found on the X chromosome). Mothers who carry the mutated gene on one of their two X chromosomes face a 50 percent chance of transmitting this disorder to each of their male children. Female children have a 50 percent chance of inheriting the gene and becoming carriers like their mothers. Although females rarely express the syndrome fully, female carriers may occasionally show minor manifestations. Affected males will pass the mutation on to all of their daughters but to none of their sons.

An autosomal recessive form of the disease accounts for approximately 20-40 percent of all cases, and this results from a mutation in the *NCF1* gene, found on the long arm of chromosome 7 at position 7q11.23. The protein product of this gene is known as neutrophil cytosolic factor-1, an important component of the NADPH oxidase complex. In autosomal recessive inheritance, both copies of a particular gene must be deficient in order for the individual to be afflicted. Typically, an affected child is born to two unaffected parents, both of whom are carriers of the recessive mutant allele. The probable outcomes for children whose parents are both carriers are 75 percent unaffected and 25 percent affected. If one parent has CGD and the other is a carrier, there is a 50 percent probability that each child will be affected.

The remaining 10 percent of cases of CGD result from mutations in either of two additional autosomal genes, and both of these are also inherited in a classic autosomal recessive manner. The *NCF2* gene, which encodes neutrophil cytosolic factor-2, is found at position 1q25 on the long arm of chromosome 1, and the *CYBA* gene, which specifies a subunit of cytochrome B, is located on the long arm of chromosome 16 at position 16q24.

SYMPTOMS

Symptoms typically begin to appear in childhood. In some patients, they may not appear until the teen years. Symptoms include swollen lymph nodes in the neck and abscesses in the neck's lymph nodes or liver. Other symptoms of CGD are frequent skin infections that are resistant to treatment, such as chronic infections inside the nose; impetigo (a bacterial skin infection); abscesses; furuncles (boils); eczema worsened by an infection; and abscesses near the anus. Frequent pneumonia that is resistant to treatment, persistent diarrhea, infections of the bones, infections of the joints, and fungal infections are additional symptoms.

SCREENING AND DIAGNOSIS

The doctor will ask about a patient's symptoms and medical history, and a physical exam will be done. Tests may include a biopsy, the removal of a sample of tissue to test for the condition; a dihydrorhodamine reduction (DHR) flow cytometry test, a blood test that looks at the ability of phagocytes to make chemicals that can destroy bacteria; an erythrocyte sedimentation rate (ESR) to test for inflammation; a chest X ray, a test that uses radiation to take pictures of structures inside the chest cavity; a bone scan; a liver scan; and a complete blood count (CBC).

TREATMENT AND THERAPY

Patients should talk with their doctors about the best plans for them. Treatment options include medications, such as antibiotics and interferon gamma. Antibiotics are used for preventive and fungal treatments and for new infections. Interferon gamma reduces the number of infections in patients, but it is not useful in acute (newly active) infections.

A bone marrow transplantation may be an option. A suitable donor will need to be found. It is a definitive cure. Surgery to treat CGD may involve the debridement or removal of abscesses.

Patients should avoid some live viral vaccines and should consult with an immunologist before receiving one.

PREVENTION AND OUTCOMES

CGD is an inherited disease. There are no preventive steps to reduce the risk of being born with the disease. Genetic counseling may be helpful; it can be used to detect carrier status in women. Early diagnosis is essential and will allow for early treatment. The bone marrow transplant donor search can also be started.

Diana Kohnle; reviewed by Julie D. K. McNairn, M.D.
"Etiology and Genetics" by Jeffrey A. Knight, Ph.D.

FURTHER READING

Bernhisel-Broadbent, J., et al. "Recombinant Human Interferon-Gamma as Adjunct Therapy for Aspergillus Infection in a Patient with Chronic Granulomatous Disease." *Journal of Infectious Diseases* 163, no. 4 (April, 1991): 908-911.

Dinauer, M. C., J. A. Lekstrom-Himes, and D. C. Dale. "Inherited Neutrophil Disorders: Molecular Basis and New Therapies." *Hematology: The Education Program of the American Society of Hematology* (2000): 303-318.

EBSCO Publishing. *Health Library: Chronic Granulomatous Disease.* Ipswich, Mass.: Author, 2009. Available through http://www.ebscohost.com.

Lekstrom-Himes, J. A., and J. I. Gallin. "Immunodeficiency Diseases Caused by Defects in Phagocytes." *New England Journal of Medicine* 343, no. 23 (December 7, 2000): 1703-1714.

Pogrebniak, H. W., et al. "Surgical Management of Pulmonary Infections in Chronic Granulomatous Disease of Childhood." *Annals of Thoracic Surgery* 55, no. 4 (April, 1993): 844-849.

WEB SITES OF INTEREST

HealthLink B. C. (British Columbia)
http://www.healthlinkbc.ca/kbaltindex.asp

IDF Patient and Family Handbook, Immune Deficiency Foundation
http://www.primaryimmune.org/publications/book_pats/book_pats.htm

"Chronic Granulomatous Disease." Medline Plus
http://www.nlm.nih.gov/medlineplus/ency/article/001239.htm

National Institute of Allergy and Infectious Diseases
http://www3.niaid.nih.gov

National Organization of Rare Disorders
http://www.rarediseases.org

Primary Immunodeficiency Resource Center
http://www.jmfworld.com

University of Maryland Medical Center
http://www.umm.edu

See also: Autoimmune disorders; Farber disease; Hereditary diseases; Immunogenetics.

Chronic myeloid leukemia

CATEGORY: Diseases and syndromes
ALSO KNOWN AS: Chronic myelocytic leukemia; CML; chronic myelogenous leukemia; chronic granulocytic leukemia

DEFINITION

Chronic myeloid leukemia (CML) is a cancer of the blood and bone marrow. With CML, the bone marrow makes abnormal blood cells, including myeloblasts, a type of white blood cell that fights infection; red blood cells (RBCs) that carry oxygen; and platelets, which make blood clot and stop bleeding in cuts or bruises.

CML progresses gradually. It is often slow growing for many years. Eventually, it may transform itself into acute myeloid leukemia (AML). This is a more aggressive type of leukemia; it progresses much more rapidly and is more serious.

Cancer occurs when cells in the body become abnormal. They divide without control or order. Leukemia is cancer of the white blood cells and their parent cells. Leukemia cells do not function normally; they cannot do what normal blood cells do. In this case they cannot fight infections, which means that the patient is more likely to become infected with viruses or bacteria. The cancerous cells also overgrow the bone marrow. This forces out other normal components, like platelets. Platelets are needed to help the blood clot. As a result, patients with leukemia may bleed more easily.

RISK FACTORS

Males and individuals who are middle-aged or older are at risk of developing CML. Other risk factors include exposure to atomic bomb radiation, exposure to a nuclear reactor accident, and smoking. Smoking is the only lifestyle factor that has been linked to leukemia, and its association with CML is still unclear.

ETIOLOGY AND GENETICS

Chronic myeloid leukemia is not an inherited disease, and it is not passed from parent to child. Yet it has a very characteristic and well understood genetic basis. It results from an acquired mutation, a mutational event that occurs usually in one adult life and that involves only blood stem cells and white blood cells. It is generally diagnosed in cytogenetic studies by the appearance of a specific chromosomal rearrangement that results in a perceptibly shorter chromosome 22, known as the Philadelphia chromosome, named after the city in which it was first discovered. Segments of the long arms of chromosomes 9 and 22 break off, and they reattach to the wrong chromosome, yielding two translocated or hybrid chromosomes. The break on chromosome 22 occurs at the *BCR* locus (breakpoint cluster region), while chromosome 9 is cleaved in the middle of the *ABL* (Abelson leukemia virus) gene. One of the fusion products results in a new gene, *BCR-ABL*, that is expressed on chromosome 22. This hybrid gene encodes a protein with tyrosine kinase activity, and it is the overexpression of tyrosine kinase in these blood stem cells which activates certain signal transduction pathways and leads to uncontrolled cell growth. All of the resulting cells will have the Philadelphia chromosome, and this can be detected in a simple blood test.

SYMPTOMS

Symptoms of CML may also be caused by other, less serious health conditions. A patient should see a doctor if he or she has any of the following symptoms: tiredness, lack of energy, fatigue, unexplained weight loss, night sweats, fever, pain or a feeling of fullness below the ribs, bone pain, joint pain, reduced exercise tolerance, enlargement of the liver or spleen, and unexplained bleeding or unusual bruising.

SCREENING AND DIAGNOSIS

The doctor will ask about a patient's symptoms and medical history and will conduct a physical exam. The doctor will look for swelling of the liver or the spleen, as well as swelling in lymph nodes in the armpits, groin, or neck. A patient will likely be referred to an oncologist, a doctor who focuses on treating cancer.

Tests may include blood tests to check for changes in the number or appearance of different types of blood cells; a bone marrow aspiration, in which a sample of liquid bone marrow is removed to test for cancer cells; a bone marrow biopsy, in which a sample of liquid bone marrow and a small piece of bone are removed to test for cancer cells; and a routine microscopic exam to examine a sample of blood, bone marrow, lymph node tissue, or cerebrospinal fluid. Bone, blood marrow, lymph node tissue, or cerebrospinal fluid tests can distinguish among types of leukemia. A cytogenetic analysis can look for certain changes of the chromosomes (most often to test for the Philadelphia chromosome). Additional tests include a chest X ray, which may detect signs of lung infection; a computed tomography (CT) scan, a type of X ray that uses a computer to make pictures of structures inside the body; a magnetic resonance imaging (MRI) scan, which uses magnetic waves to make pictures of structures inside the body; and an ultrasound, which uses sound waves to examine masses and organs inside the body.

TREATMENT AND THERAPY

Patients should talk to their doctors about the best plans for them. Treatment options include targeted drug therapy. Three drugs work to inhibit the molecule that triggers the development of leukemia and the gene that is associated with it. This medication is often used in the early stages of CML, and it has replaced chemotherapy and biologic therapy as a treatment of choice. The drugs for this therapy are imatinib (Gleevec), dasatinib (Sprycel), and nilotinib (Tasigna).

Chemotherapy is the use of drugs to kill cancer cells. It may be given in many forms, including by pill, injection, and catheter. The drugs enter the bloodstream and travel through the body. While these drugs will focus on cancer cells, some healthy cells will also be killed.

Biologic therapy is the use of medications or substances made by the body. The substance is used to increase or restore the body's natural defenses against cancer. This type of therapy is also called biological response modifier therapy. It is still being tested in clinical trials.

High-dose chemotherapy with stem cell transplant is another treatment option. In this treatment, high doses of chemotherapy are followed by a transplantation of stem cells (immature blood cells); these will replace blood-forming cells destroyed by cancer treatment. Stem cells are removed from the blood or bone marrow of the patient or a donor. They are then infused into the patient.

Donor lymphocyte infusion is another form of treatment. Lymphocytes are a type of white blood cell. A donor's cells are infused into the patient. The cancer cells do not recognize these cells, and they do not attack them.

Splenectomy, a surgery to remove the spleen, may be done if the spleen has become enlarged from the leukemia. It may also be done if other complications develop.

PREVENTION AND OUTCOMES

There are no guidelines for preventing CML. It is possible that smoking is associated with CML, so patients may reduce their risk by not smoking.

Krisha McCoy, M.S.; reviewed by Igor Puzanov, M.D.
"Etiology and Genetics" by Jeffrey A. Knight, Ph.D.

FURTHER READING

Carella, Angelo M., et al., eds. *Chronic Myeloid Leukaemia: Biology and Treatment.* London: M. Dunitz, 2001.

Cortes, Jorge, and Michael Deininger, eds. *Chronic Myeloid Leukemia.* New York: Informa Healthcare, 2007.

EBSCO Publishing. *Health Library: Chronic Myeloid Leukemia.* Ipswich, Mass.: Author, 2009. Available through http://www.ebscohost.com.

Fioretos, Thoas, and Bertil Johansson. "Chronic Myeloid Leukemia." In *Cancer Cytogenetics*, edited by Sverre Heim and Felix Mitelman. 3d ed. Hoboken, N.J.: Wiley-Blackwell, 2009.

Mughal, Tariq I., and John M. Goldman, eds. *Chronic Myeloproliferative Disorders.* London: Informa Healthcare, 2008.

WEB SITES OF INTEREST

American Cancer Society
http://www.cancer.org

B. C. (British Columbia) Cancer Agency
http://www.bccancer.bc.ca/default.htm

Canadian Cancer Society
http://www.cancer.ca

The Leukemia and Lymphoma Society
http://www.leukemia-lymphoma.org/hm_lls

National Cancer Institute
http://www.cancer.gov

See also: Cancer; Hereditary diseases; Mutagenesis and cancer; Mutation and mutagenesis.

Classical transmission genetics

CATEGORY: Classical transmission genetics; History of genetics

SIGNIFICANCE: In sexual reproduction, parents produce specialized cells (eggs and sperm) that fuse to produce a new individual. Each of these cells contains one copy of each of the required units of information, or genes, which provide the blueprint necessary for the offspring to develop into individual, functioning organisms. Transmission genetics refers to the passing of the information needed for the proper function of an organism from parents to their offspring as a result of reproduction.

KEY TERMS

chromosomes: structures in haploid cells (eggs and sperm) that carry genetic information from each parent

cross: the mating of parents to produce offspring during sexual reproduction

gene: a sequence of base pairs that specifies a product (either RNA or protein); the average gene in bacteria is one thousand base pairs long

linkage: a relation of gene loci on the same chromosome; the more closely linked two loci are, the more often the specific traits controlled by these loci are expressed together

meiosis: the process of nuclear division during sexual

reproduction that produces cells that contain half the number of chromosomes as the original cell

sexual reproduction: reproduction that requires fusion of haploid gametes, each of which contains one copy of the respective parent's genes, as a first step

DISCOVERY OF TRANSMISSION GENETICS

The desire to improve plant and animal production is as old as agriculture. For centuries, humans have been using selective breeding programs that have resulted in the production of thousands of varieties of plants and breeds of animals. The Greek philosopher-scientist Hippocrates suggested that small bits of the body of the parent were passed to the offspring during reproduction. These small bits of arms, heads, stomachs, and livers were thought to develop into a new individual. Following the development of the microscope, it became possible to see the cells, the small building blocks of living organisms. Study of the cell during the 1800's showed that sexual reproduction was the result of the fusion of specialized cells from two parents (eggs and sperm). It was also observed that these cells contained chromosomes ("color bodies" visible when the cells reproduced) and that the number and kind of these chromosomes was the same in both the parents and the offspring. This suggested that the chromosomes carried the genetic information and that each parent transmitted the same number and kinds of chromosomes. For example, humans have twenty-three kinds of chromosomes. The offspring receives one of each kind from each parent and so has twenty-three chromosome pairs. Since the parents and the offspring have the same number and kinds of chromosomes, and since each parent transmits one complete set of the chromosomes, it was thought that there must be a process of cell division that reduces the parent number from two sets of chromosomes to one set in the production of the egg or sperm cells. The parents would each have twenty-three pairs (forty-six) chromosomes, but their reproductive cells would each contain only one of each chromosome (twenty-three).

In the 1860's, the Austrian botanist Gregor Mendel repeated studies of inheritance in the garden pea and, using the results, developed a model of genetic transmission. The significance of Mendel's work was not recognized during his lifetime, but it was rediscovered in 1900. In that same year, the pre-dicted reductional cell division during reproduction was fully described, and the science of genetics was born.

A STUDY OF VARIATION

In many respects, genetics is the study of variation. It is recognized that a particular feature of an animal or plant is inherited because there is variation in the expression of that feature, and variation in expression follows a recognizable inheritance pattern. For example, it is known that blood types are inherited, both because there is variation (blood types A, B, and O) and because examination of family histories reveals patterns that show transmission of blood-type information from parents to children.

Variation in character expression may have one of two sources: environmental conditions or inherited factors. If a plant is grown on poor soil, it might be short. The same plant grown on good soil might be tall. A plant that is short because of an inherited factor cannot grow tall even if it is placed on richer soil. From this example, it can be seen that there may be two different ways to determine whether a specific character expression is environmentally or genetically determined: testing for environmental influences and testing for inherited factors. Many conditions are not so easily resolved as this example; there may be many complex environmental factors involved in producing a condition, and it would be impossible to test them all. Knowledge of inheritance patterns can, however, help in determining whether inherited factors play a role in a condition. Cancer-associated genes have been located using family studies that show patterns consistent with a genetic contribution to the disease. There are certainly environmental factors that influence cancer production, but those factors are not as easily recognized.

The patterns of transmission genetics were discovered because the experimenters focused their attention on single, easily recognized characteristics. Mendel carefully selected seven simple characteristics of the pea plant, such as height of the plant, color of the flower, and color of the seeds. The second reason for success was the use of carefully controlled crosses. The original parents were selected from varieties that did not show variations in the characteristic of interest. For example, plants from a pure tall variety were crossed with plants from a pure short variety. Control of the information passed by the parents allowed the experimenter to follow the variation of expression from parents to offspring through a number of generations.

TRANSMISSION PATTERNS

The classic genetic transmission pattern is the passing of information for each characteristic from each parent to each offspring. The offspring receives two copies of each gene. (The term "gene" is used to refer to a character-determining factor; Mendel's original terminology was "factor.") Each parent also had two copies of each gene, so in the production of the specialized reproductive cells, the number must be reduced. Consider the following example. A tall pea plant has two copies of the information for height, and both copies are for tall height (tall/tall). This plant is crossed with a plant with two genes for short height (short/short). The information content of each plant is reduced to one copy: The tall plant transmits one tall gene, and the short plant transmits one short gene. The offspring receive both genes and have the information content tall/short.

The situation becomes more complex and more interesting when one or both of the parents in a cross have two different versions of the gene for the same characteristic. If, for example, one parent has the height genes tall/short and the other has the genes short/short, the cells produced by the tall/short parent will be of two kinds: ½ carry the tall gene and ½ carry the short gene. The other parent has only one kind of gene for height (short), so all of its reproductive cells will contain that gene. The offspring will be of two kinds: ½ will have both genes (tall/short), and ½ will have only one kind of gene (short/short). Had it been known that the one parent had one copy for each version of the gene, it could have been predicted that the offspring would have been of two kinds and that each would have an equal chance of appearing. Had it not been known that one of the parents had the two versions of the gene, the appearance of two kinds of offspring would have revealed the presence of both genes. The patterns are repeatable and are therefore useful in predicting what might happen or revealing what did happen in a particular cross. For example, blood-type patterns or DNA variation patterns can be used to identify the children that belong to parents in kidnapping cases or in cases in which children are mixed up in a hospital.

In a second example, the pattern is more complex, because both parents carry both versions of the gene: a tall/short to tall/short cross. Each parent will produce ½ tall-gene-carrying cells and ½ short-gene-carrying cells. Any cell from one parent may randomly join with any cell from the other parent, which leads to the following patterns: ½ tall × ½ tall = ¼ tall/tall; ½ tall × ½ short = ¼ tall/short; ½ short × ½ tall = ¼ short/tall; ½ short × ½ short = ¼ short/short. Tall/short and short/tall are the same, yielding totals of ¼ tall/tall; ½ tall/short; and ¼ short/short, or a 1:2:1 ratio.

This was the ratio that Mendel recognized and used to develop his model of transmission genetics. Mendel used pure parents (selected to breed true for the one characteristic), so he knew when he had a generation in which all of the individuals had one of each gene.

As in the previous example, if it had been known that each of the parents had one of each gene, the ratio could have been predicted; conversely, by using the observed ratio, the information content of the parents could be deduced. Using a blood-type example, if one parent has blood-type genes AO and the other parent has the genes BO, the possible combinations observed in their offspring would be AB, AO, BO, and OO, each with the same probability of occurrence (½ A gene-bearing and ½ O gene-bearing cells in one parent × ½ B gene-bearing and ½ O gene-bearing cells in the other parent).

REDUCTIONAL DIVISION

Transmission genetics allows researchers to make predictions about specific crosses and explains the occurrence of characteristic expressions in the offspring. In genetic counseling, probabilities of the appearance of a genetic disease can be made when there is an affected child in the family or a family history of the condition. This is possible because, for most inherited characteristics, the pattern is established by the reduction of chromosome numbers that occurs when the reproductive cells are produced and by the random union of reproductive cells from the two parents. The recognition that the genes are located on the chromosomes and the description of the reductional division in which the like chromosomes separate, carrying the two copies of each gene into different cells during the reductional division of meiosis, provide the basis of the regularity of the transmission pattern. It is this regularity that allows the application of mathematical treatments to genetics. Two genes are present for each character in each individual, but only one is passed to each offspring by each parent; therefore, the 50 percent (or ½) probability becomes the basis for making predictions about the outcome of a cross for any single character.

The classical pattern of transmission genetics occurs because specialized reproductive cells, eggs and sperm, are produced by a special cell reproduction process (meiosis) in which the chromosome number is reduced from two complete sets to one set in each of the cells that result from the process. This reduction results because each member of a pair of chromosomes recognizes its partner, and the chromosomes come together. This joining (pairing) appears to specify that each chromosome in the pair will become attached to a "motor" unit from an opposite side of the cell that will move the chromosomes to opposite sides of the cell during cell division. The result is two new cells, each with only one of the chromosomes of the original pair. This process is repeated for each pair of chromosomes in the set.

INDEPENDENT GENES

Humans have practiced selective breeding of plants and animals for centuries, but it was only during the nineteenth and twentieth centuries that the patterns of transmission of inherited characters were understood. This change occurred because the experimenters focused on a single characteristic and could understand the pattern for that characteristic. Previous attempts had been unsuccessful because the observers attempted to explain a large number of character patterns at the same time. Mendel expanded his model of transmission to show how observations become more complex as the number of characteristics examined is expanded. Consider a plant with three chromosomes and one simple character gene located on each chromosome. In the first parent, chromosome 1 contains the gene for tall expression, chromosome 2 contains the gene for expression of yellow seed color, and chromosome 3 contains the gene for purple flower color. In the other parent, chromosome 1 contains a gene for short height, chromosome 2 contains a gene for green seed color, and chromosome 3 contains a gene for white flower color. Each parent will transmit these genes to their offspring,

who will have the genes tall/short, yellow/green, and purple/white. In the production of reproductive cells, the reductional division of meiosis will pass on one of the character expression genes for each of the three characters. (It is important to remember that the products of the reductional cell division have one of each chromosome. If this did not occur, information would be lost, and the offspring would not develop normally.) The characteristics are located on different chromosomes, and during the division process, these chromosome pairs act independently. This means that the genes that came from any one parent (for example, the tall height, yellow seed, and purple flower expression genes from the one parent) do not have to go together during the division process. Since chromosome pairs act independently, different segregation patterns occur in different cells. The results from one meiosis may be a cell with the tall, green, and purple genes and one with the short, yellow, and white genes. In the same plant, another meiosis might produce a cell with the short, yellow, and purple genes, and the second cell would have the tall, green, and white genes.

Since these genes are independent, height does not influence seed color or flower color, nor does flower color influence seed color or height. The determining gene for each characteristic is located on a different chromosome, so the basic transmission model can be applied to each gene independently, and then the independent patterns can be combined. The tall/short height genes will segregate so that ½ of the cells will contain the tall gene and ½ will contain the short gene. Likewise, the yellow/green seed color genes will separate so that ½ of the cells will contain the yellow gene and ½ will contain a green gene. Finally, ½ of the cells will contain a purple flower gene and ½ will contain a white gene. These independent probabilities can be combined because the probability of any combination is the product of the independent probabilities. For example, the combination tall, purple, white will occur with a probability of $\frac{1}{2} \times \frac{1}{2} \times \frac{1}{2} = \frac{1}{8}$. This means that one should expect eight different combinations of these characters. The possible number of combinations for n chromosome pairs is 2^n. For humans, this means that any individual may produce 2^{23} different chromosome combinations. This is the same idea as tossing three coins simultaneously. Each coin may land with a head or a tail up, but how each coin lands is independent of how the other coins land. Knowledge of transmission patterns based on chromosome separation during meiosis allows researchers to explain the basic pattern for a single genetic character, but it also allows researchers to explain the great variation that is observed among individuals within a population in which genes for thousands of different characters are being transmitted.

Continuous Variation

The principles of transmission genetics were established by studying characters with discrete expressions—plants were tall or dwarf, seeds were yellow or green. In 1903, Danish geneticist Wilhelm Johannsen observed that characteristics that showed continuous variation, such as weight of plant seeds, fell into recognizable groups that formed a normal distribution. These patterns could also be explained by applying the principles of transmission genetics.

Assume a plant has two genes that influence its height and that these genes are on two different chromosomes (for example, 1 and 3). Each gene has two versions. A tall gene stimulates growth (increases the height), but a short gene makes no contribution to growth. A plant with the composition tall-1/tall-1, tall-3/tall-3 would have a maximum height because four genes would be adding to the plant's height. A short-1/short-1, short-3/short-3 plant would have minimum height because there would be no contribution to its height by these genes. Plants could have two contributing genes (tall-1/short-1, tall-3/short-3) or three contributing genes (tall-1/short-1, tall-3/tall-3). The number of offspring with each pattern would be determined by the composition of the parents and would be the result of gene segregation and transmission patterns. Many genes contributing to a single character expression apply to many interesting human characteristics, such as height, intelligence, amount of skin pigmentation, hair color, and eye color.

Linkage Groups

Mendel's model of the transmission of genes was supported by the observations of chromosome pair separation during the reductional division, but early in the twentieth century, it was recognized that some genes did not separate independently. Work in American geneticist Thomas Hunt Morgan's lab-

oratory, especially by an undergraduate student, Alfred Sturtevant, showed that each chromosome contained determining genes for more than one characteristic and established that genes located close together on the same chromosome stayed together during the separation of the paired chromosomes during meiosis. If a pea plant had a chromosome with the tall height gene and, immediately adjacent to it, a gene for high sugar production, and if the other version of this chromosome had a gene for short height and a gene that limited the sugar production, the most likely products from meiosis would be two kinds of cells: one with the genes for tall height and high sugar production and one with the genes for short height and limited sugar production. These genes are said to be "linked," or closely associated on the same chromosome, because they go together as the chromosomes in the pair separate. It is generally accepted that humans contain approximately 21,000 genes, but there are only twenty-three kinds of chromosomes. This means that each chromosome contains many different genes. Each chromosome is considered a linkage group, and one of the goals of genetic study is to locate the gene responsible for each known characteristic to its proper chromosome.

A common problem in medical genetics is locating the gene for a specific genetic disease. Family studies may show that the disease is transmitted in a pattern consistent with the gene being on one of the chromosomes, but there is no way of knowing its location. Variations in DNA structure are also inherited in the classic pattern, and these DNA pattern modifications can be determined using current molecular procedures. DNA variation patterns are analyzed for linkage to the disease condition. If a specific DNA pattern always occurs in individuals with the disease condition, it indicates that the DNA variation is on the same chromosome and close to the gene of interest because it is transmitted along with the disease-producing gene. This information locates the chromosome position of the gene, allowing further work to be done to study its structure. With the completion of the Human Genome Project, it is predicted that tracking down the genes responsible for genetic defects will be a much faster process than before. Many more genetic markers have now been identified, which, in theory, should greatly enhance the techniques used to locate a faulty gene.

D. B. Benner, Ph.D.

FURTHER READING

Carlson, Elof Axel. *Mendel's Legacy: The Origin of Classical Genetics.* Cold Spring Harbor, N.Y.: Cold Spring Harbor Laboratory Press, 2004. Traces how the major principles of classic genetics emerged from Gregor Mendel's discoveries in 1865 through other scientists' concepts of reproductive cell biology in the early twentieth century.

Cummings, Michael R. *Human Heredity: Principles and Issues.* 8th ed. Florence, Ky.: Brooks/Cole/ Cengage Learning, 2009. College text that surveys topics such as genetics as a human endeavor; cells, chromosomes, and cell division; transmission of genes from generation to generation; cytogenics; the source of genetic variation; cloning and recombinant DNA; genes and cancer; genetics of behavior; and genes in populations.

Gonick, Larry, and Mark Wheelis. *The Cartoon Guide to Genetics.* New York: Harper Perennial, 1991. An easy-to-read presentation of the basic concepts of transmission genetics.

Lewis, Ricki. *Human Genetics: Concepts and Applications.* 9th ed. Dubuque, Iowa: McGraw-Hill, 2009. An introductory text for undergraduates with sections on fundamentals, transmission genetics, DNA and chromosomes, population genetics, immunity and cancer, and genetic technology.

Moore, John A. *Science as a Way of Knowing.* Reprint. Cambridge, Mass.: Harvard University Press, 1999. Traces the development of scientific thinking with an emphasis on understanding hereditary mechanisms.

Rheinberg, Hans-Jörg, and Jean-Paul Gaudillière, eds. *Classical Genetic Research and Its Legacy: The Mapping Cultures of Twentieth-Century Genetics.* New York: Routledge, 2004. Traces the history of classical genetics and linkage mapping procedures.

Stansfield, William D. *Schaum's Outline of Theory and Problems of Genetics.* 4th ed. New York: McGraw-Hill, 2002. Provides explanations of basic genetics concepts and an introduction to problem solving.

WEB SITES OF INTEREST

MendelWeb
http://www.mendelweb.org

This site, designed for teachers and students, revolves around Mendel's 1865 paper and includes educational activities, images, interactive learning, and other resources.

Scitable
http://www.nature.com/scitable/topic/Gene -Inheritance-and-Transmission-23

Scitable, a library of science-related articles compiled by the Nature Publishing Group, features a Gene Inheritance and Transmission Topic Room with numerous articles and other resources about this subject.

The Virtually Biology Course, Principle of Segregation
http://staff.jccc.net/pdecell/transgenetics/ monohybrid1.html

Paul Decelles, a professor at Johnson Community College in Overland Park, Kansas, has included a page about Mendelian genetics in his online biology course.

See also: Cell division; Chromosome mutation; Chromosome structure; Chromosome theory of heredity; Dihybrid inheritance; Epistasis; Extrachromosomal inheritance; Genetic code; Genetic code, cracking of; Hybridization and introgression; Incomplete dominance; Lamarckianism; Linkage maps; Mendelian genetics; Mitochondrial genes; Mitosis and meiosis; Monohybrid inheritance; Multiple alleles; Nondisjunction and aneuploidy; Parthenogenesis; Penetrance; Polygenic inheritance.

Cleft lip and palate

CATEGORY: Diseases and syndromes
ALSO KNOWN AS: Oral-facial clefts

DEFINITION

An oral-facial cleft is a birth defect. A cleft lip is an opening in the upper lip, usually just below the nose. A cleft palate is an opening in the roof of the mouth (hard palate) or in the soft tissue at the back of the mouth (soft palate). In the majority of cases, a cleft lip and cleft palate occur together.

RISK FACTORS

Male infants are at risk for having cleft lip and palate, as are infants who have other birth defects, and infants who have a sibling, parent, or other close relative born with an oral-facial cleft. A geneticist can best define the actual risk, which can vary

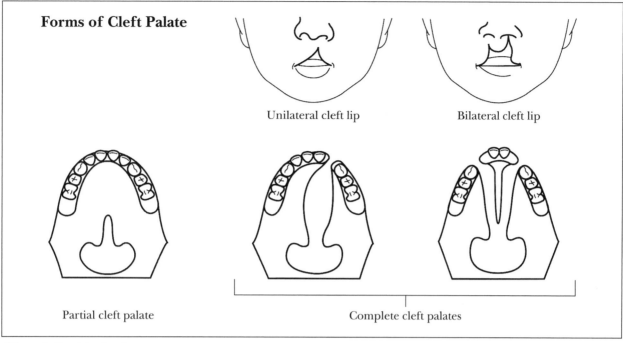

Forms of Cleft Palate

Unilateral cleft lip

Bilateral cleft lip

Partial cleft palate

Complete cleft palates

(Hans & Cassidy, Inc.)

greatly among families. In general, if one child in a family has a cleft palate, the next child has about a 4 percent chance of also having a cleft palate. If only the lip has a cleft, the risk of this occurring in a second child is about 2 percent.

Risk factors for the mother during pregnancy include taking certain drugs, such as antiseizure drugs (especially phenytoin) or retinoic acid (used for dermatologic conditions, such as acne), and consuming alcohol (especially in the development of a cleft lip). Other risk factors for pregnant women are having an illness or infection and having folic acid deficiency at conception or during early pregnancy.

ETIOLOGY AND GENETICS

Cleft lip and/or palate is a birth defect that typically involves a complex etiology that includes both environmental and genetic factors. It is important to first distinguish isolated clefts from those associated with other syndromes. In fact, clefts have been linked to between three hundred and four hundred different syndromes, many of which are exceedingly rare. It is estimated that these syndromes account for about 15 percent of the total number of clefts diagnosed at birth, and in most cases the ge-

netic basis for inheritance of the syndrome is well understood. The remaining 85 percent of clefts, however, are isolated, and it is usually not possible to assign specific genetic or environmental causation factors.

It is most likely that several genes may act to increase susceptibility for clefts, and this potential is realized only when particular environmental factors trigger the event at the appropriate developmental stage. Three specific genes have so far been identified that appear to have an association with isolated clefts. The *IRF* gene (interferon regulatory factor), found on the long arm of chromosome 1 at position 1q32-q34, has been linked to a rare autosomal dominant form of isolated cleft. A single copy of the mutation is sufficient to cause full expression of the trait. An affected individual has a 50 percent chance of transmitting the mutation to each of his or her children. Most cases of this type of isolated cleft lip or palate, however, result from a spontaneous new mutation, so in these instances affected individuals will have unaffected parents. The other two genes, *UBB* (at position 17p12-p11.2) and *SATB2* (at position 2q33), are associated with clefts, but the specific pattern of inheritance remains unclear.

SYMPTOMS

The major symptom of a cleft lip and/or cleft palate is a visible opening in the lip or palate. Other symptoms that can occur as a result of an oral-facial cleft include feeding problems (especially with cleft palate); problems with speech development; dental problems, including missing teeth, especially when cleft lip extends to the upper gum area; recurrent middle ear infections; and hearing problems.

SCREENING AND DIAGNOSIS

A doctor can diagnose cleft lip or cleft palate by examining a newborn baby. A newborn with an oral-facial cleft may be referred to a team of medical specialists soon after birth. Rarely, a partial or "submucous" cleft palate may not be diagnosed for several months or even years.

Cleft lip and palate are sometimes associated with other medical conditions. A doctor should be able to tell parents if their child's clefting is part of a syndrome. Some syndromes may require treatment in addition to taking care of a cleft lip or palate.

Prenatal diagnosis (diagnosis prior to birth) can also be accomplished using ultrasound examination. Cleft lip is more easily diagnosed via prenatal ultrasound than is cleft palate. Diagnosis can be made as early as eighteen weeks of pregnancy. Prenatal diagnosis gives the parents and the medical team the advantage of advanced planning for the baby's care.

TREATMENT AND THERAPY

The main treatment of cleft lip and palate is surgery to close the opening in a lip and/or palate. Additional surgical treatment for oral-facial clefts may include bite alignment surgery (if the jaw is not aligned properly) and plastic and/or nasal surgery to improve facial appearance and function. Prior to surgery for cleft palate, a dental plate may temporarily be placed in the roof of the mouth to make eating and drinking easier.

Treatment for middle ear infections and fluid buildup include medications to treat infection or prevent fluid buildup and surgery to drain built-up fluid and prevent future infections. Most children with cleft palate are at risk for hearing loss, which may interfere with learning language.

The hearing of these children should be tested regularly; rarely, children with cleft palate may benefit from hearing aids.

PREVENTION AND OUTCOMES

In order to help prevent oral-facial clefts in their unborn children, pregnant women and women who are likely to become pregnant can consume four hundred micrograms of folic acid daily by taking a multivitamin or eating foods containing folic acid, such as fruits and orange juice, green leafy vegetables, dried beans and peas, pasta, rice, bread, flour, and cereals. Women can avoid smoking or drinking alcohol during pregnancy, and they should take medications during pregnancy only as directed by their doctors. Pregnant women can also get early and regular prenatal care. Women who are thinking about having a child and have risk factors for oral-facial cleft can seek medical advice on additional ways to prevent the disorder and can consider genetic counseling.

Rick Alan; reviewed by Kari Kassir, M.D.
"Etiology and Genetics" by Jeffrey A. Knight, Ph.D.

FURTHER READING

Beers, Mark H., ed. *The Merck Manual of Medical Information.* 2d home ed., new and rev. Whitehouse Station, N.J.: Merck Research Laboratories, 2003.

Berkowitz, Samuel, ed. *Cleft Lip and Palate: Diagnosis and Management.* 2d ed. New York: Springer, 2006.

Cummings, C. W., et al., eds. *Cummings Otolaryngology Head and Neck Surgery.* 4th ed. St. Louis: Mosby, 2005.

EBSCO Publishing. *Health Library: Cleft Lip and Palate.* Ipswich, Mass.: Author, 2009. Available through http://www.ebscohost.com.

Kleigman, Robert M., et al., eds. *Nelson Textbook of Pediatrics.* 18th ed. Philadelphia: Saunders Elsevier; 2007.

Losee, Joseph E., and Richard E. Kirschner, eds. *Comprehensive Cleft Care.* New York: McGraw-Hill Medical, 2009.

WEB SITES OF INTEREST

About Kids Health
http://www.aboutkidshealth.ca

Cleft Lip and Palate Association
http://www.clapa.com

Cleft Palate Foundation
http://www.cleftline.org

Genetics Home Reference
http://ghr.nlm.nih.gov

Medline Plus
http://medlineplus.gov

Women's Health Matters
http://www.womenshealthmatters.ca/index.cfm

See also: Congenital defects; Hereditary diseases.

Cloning

CATEGORY: Genetic engineering and biotechnology

SIGNIFICANCE: Cloning includes both gene cloning and the cloning of entire organisms. Gene cloning, an important technique for understanding how cells work, has produced many useful products, including human medicines. Organ cloning includes reproductive cloning and therapeutic cloning. Ethical and safety concerns have led to a consensus that human cloning should be banned. Therapeutic cloning could lead to treatments for many human diseases, but ethical concerns related to human genetic manipulation raises much debate.

KEY TERMS

cloning vector: a plasmid or virus into which foreign DNA can be inserted to amplify the number of copies of the foreign DNA in the host cell or organism

DNA: dexoyribonucleic acid, a long-chain macromolecule, made of units called nucleotides and structured as a double helix joined by weak hydrogen bonds, that forms genetic material for most organisms

DNA hybridization: formation of a double-stranded nucleic acid molecule from single-stranded nucleic acid molecules that have complementary base sequences

ligase: an enzyme that joins recombinant DNA molecules together

plasmid: a DNA molecule that replicates independently of chromosomes

recombinant DNA technology: methods used to splice a DNA fragment from one organism into DNA from another organism and then clone the new recombinant DNA molecule

reproductive cloning: cloning to produce individual organisms

restriction enzyme: a protein (an enzyme) that recognizes a specific nucleotide sequence in a piece of DNA and causes a sequence-specific cleavage of the DNA

stem cells: cells that are able to divide indefinitely in culture and to give rise to specialized cells

therapeutic cloning: cloning to produce a treatment for a disease

TYPES OF CLONING

There are three different definitions of a clone. One is a group of genetically identical cells descended from a single common ancestor. This type of clone is often made by plant cell tissue culture in which a whole line of cells is made from a single cell ancestor. A second type of clone is a gene clone, or recombinant DNA clone, in which copies of a DNA sequence are made by genetic engineering. A third type of clone is an organism that is descended asexually from a single ancestor. A much-celebrated example of an organismal clone is the sheep Dolly (1997-2003), produced by placing the nucleus of a cell from a ewe's udder, with its genetic material (DNA), into an unfertilized egg from which the nucleus had been removed.

DNA CLONING

DNA is cloned to obtain specific pieces of DNA that are free from other DNA fragments. Clones of specific pieces of DNA are important for basic research. DNA is made up of four different compounds known as nucleotide bases. Once a piece of DNA is cloned, the specific DNA bases can be identified. This is called sequencing. Once this specific pattern of DNA sequencing is accomplished, the DNA is said to be "sequenced," revealing the genetic code detailed by the nucleotide bases. This valuable information helps answer the following questions and can be used in a variety of ways. Where does the gene begin and end? What type of control regions does the gene have? Cloned DNAs can be used as hybridization probes, where sequences that are complementary to the cloned DNA can be detected. Such DNA hybridization is useful to detect similarities between genes from different organisms, to detect the presence of specific disease genes, and to determine in what tissues that gene is expressed. The gene is expressed when a messenger RNA (mRNA) is made from the gene and the mRNA is translated into a protein product. A DNA clone is also used to produce the protein product for which that gene codes. When a clone is expressed, the protein made by that gene can be studied or an antibody against that protein can be made. An antibody is used to show in which tissues of an organism that protein is found. Also, a DNA clone may be expressed because the gene codes for a useful product. This is a way to obtain large amounts of the specific protein.

PRODUCTS OF RECOMBINANT DNA TECHNOLOGY

Recombinant DNA technology has produced clones put to use for a wide variety of human purposes. For example, rennin and chymosin are used in cheese making. One of the most important applications, however, is in medicine. Numerous recombinant DNA products are useful in treating human diseases, including the production of human insulin (Humalin) for diabetics. Other human pharmaceuticals produced by gene cloning include clotting factor VIII to treat hemophilia A, clotting factor IX to treat hemophilia B, human growth hormone, erythropoietin to treat certain anemias, interferon to treat certain cancers and hepatitis, tissue plasminogen activator to dissolve blood clots after a heart attack or stroke, prolastin to treat genetic emphysemas, thrombate III to correct a genetic antithrombin III deficiency, and parathyroid hormone. The advantages of the cloned products are their high purity, greater consistency from batch to batch, and the steady supply they offer.

HOW TO CLONE DNA

DNA is cloned by first isolating it from its organism. Vector DNA must also be isolated from bacte-

These five cows on an Iowa farm in April, 2000, were cloned by Robert Lanza and colleagues of Advanced Cell Technologies in Worcester, Massachusetts. The cows' cells—unlike those of the first cloned vertebrate, Dolly the sheep—appeared to have a prolonged youth. (AP/Wide World Photos)

ria. (A vector is a plasmid or virus into which DNA is inserted.) Both the DNA to be cloned and the vector DNA are cut with a restriction enzyme that makes sequence-specific cuts in the DNAs. The ends of DNA molecules cut with restriction enzymes are then joined together with an enzyme called ligase. In this way the DNA to be cloned is inserted into the vector. These recombinant DNA molecules (vector plus random pieces of the DNA to be cloned) are then introduced into a host, such as bacteria or yeast, where the vector can replicate. The recombinant molecules are analyzed to find the ones that contain the cloned DNA of interest.

REGULATION OF DNA CLONING

In the 1970's the tools to permit cloning of specific pieces of DNA were developed. There was great concern among scientists about the potential hazards of some combinations of DNA from different sources. Concerns included creating new bacterial plasmids with new drug resistances and putting DNA from cancer-causing viruses into plasmids. In February, 1975, scientists met at a conference center in Asilomar, California, to discuss the need to regulate recombinant DNA research. The result of this conference was the formation of the Recombinant DNA Molecule Program Advisory Committee at the National Institutes of Health, and guidelines for recombinant DNA work were established.

GENETICALLY MODIFIED ORGANISMS

Numerous cloned genes have been introduced into different organisms to produce genetically modified organisms (GMOs). Genes for resistance to herbicides and insects have been introduced into soybean, corn, cotton, and canola, and these genetically engineered plants are in cultivation in fields in the United States and other countries. Fish and fruit and nut trees that mature more rapidly have been created by genetic engineering. Edible vaccines have been made—for example, a vaccine for hepatitis B in bananas. A tomato called the Flavr Savr is genetically engineered to delay softening. Plants that aid in bioremediation by taking up heavy metals such as cadmium and lead are possible.

Concerns about genetically modified organisms include safety issues—for example, concerns that foreign genes introduced into food plants may contain allergens and that the antibiotic resistance markers used in creating the GMOs might be transferred to other organisms. There are concerns about the environmental impact of GMOs; for example, if these foreign genes are transferred to other plants by unintended crossing of a GMO with a weed plant, weeds may become difficult or impossible to eradicate and jeopardize crop growth. There is a concern about the use of genetically modified organisms as food. There is a concern about loss of biodiversity if only one, genetically modified, variety of a crop plant is cultivated. There are also ethical concerns surrounding whether certain GMOs might be made available only in rich countries, and there are concerns about careful labeling of GMOs so that consumers will be aware when they are using products from GMOs. All of these questions remain in flux as the marketing of GMOs proceeds.

According to the U.S. government's Human Genome Project, 252 million acres of genetically modified crops were planted in 22 countries by 10.3 million farmers in 2006. U.S. government statistics state that in 2006, countries growing 97 percent of the global transgenic crops were the United States (53 percent), Argentina (17 percent), Brazil (11 percent), Canada (6 percent), India (4 percent), China (3 percent), Paraguay (2 percent), and South Africa (1 percent). Corn, soybeans, cotton, alfalfa, and canola were the major crops, often modified for insect resistance. Rice has been genetically enhanced for more iron and vitamins to alleviate malnutrition in Asia. Other plants have been modified to survive weather variances.

Genetically modified organisms may soon include cows resistant to mad cow disease and nut and fruit trees that yield bounties faster. Plants producing new plastics and fish that grow faster are potential genetically modified organisms. It is expected that the next decade will see huge increases in genetically modified organisms as worldwide researchers gain more access to genomic resources.

ORGANISMAL CLONING

A goal of organismal cloning is to develop ways of efficiently altering animals genetically in order to reproduce certain animals that are economically valuable. Animals have been altered by the introduction of specific genes, such as human proteins that will create drug-producing animals. Some genes have been inactivated in organisms to create animal models of human diseases. For example, "knockout mice" are used as models for diabetes research. Another goal is to conduct research that might lead to the development of human organs for transplant produced from single cells. Similarly, animals might be genetically engineered to make their organs better suited for transplantation to humans. Finally, the cloning of a human might be a solution to human infertility.

ARE ORGANISMAL CLONES NORMAL?

There is, however, a concern about the health of cloned animals. First of all, when inserting a new nucleus into an egg from which the nucleus has been removed, and then implanting such eggs into surrogate mothers, only very few of the eggs develop properly. There are suggestions of other abnormalities in cloned animals that might be due to the cloning process. The first vertebrate to be successfully cloned, the sheep Dolly, developed first arthritis and then a lung disease when six years old; although neither condition was unusual in sheep, both appeared years earlier than normal, and Dolly was euthanized. Was she genetically older than her chronological age?

STEM CELLS

Stem cells are unspecialized cells that are able to divide continuously and, with the proper conditions, be induced to give rise to specialized cell types. In the developing embryo they give rise to the hundreds of types of specialized cells that make up the adult body. Embryonic stem cells can be isolated from three- to five-day-old embryos. Some tissues in the adult, such as bone marrow, brain, and muscle, contain adult stem cells that can give rise to cell types of the tissue in which they reside.

A goal of research on stem cells is to learn how stem cells become specialized cells. Human stem cells could be used to generate tissues or organs for transplantation and to generate specific cells to replace those damaged as a result of spinal cord injury, stroke, burns, heart disease, diabetes, osteoarthritis, rheumatoid arthritis, and other conditions.

A 2009 study demonstrated that human corneal stem cells can repair cloudy corneas in mice. The cornea is the outermost portion of the eye and provides protection along with 70 percent of the eye's focusing power. Deep corneal scratches can cause scarring that impairs vision. Mice treated with human stem cells cleared their corneas. Further study and investigation of this type of stem cell therapy could develop a potential stem cell therapy for corneal scarring in humans.

REGULATION OF ORGANISMAL CLONING

Until the cloning of the sheep Dolly in 1997, it was thought that adult specialized cells could not be made to revert to nonspecialized cells that can give rise to any type of cell. However, Dolly was created from a specialized adult cell from a ewe's udder. After the publicity about Dolly, U.S. president Bill Clinton asked the National Bioethics Advisory Commission to form recommendations about the ethical, religious, and legal implications of human cloning. In June, 1997, that commission concluded that attempts to clone humans are "morally unacceptable" for safety and ethical reasons. There was a moratorium on using federal funds for human cloning. In January, 1998, the U.S. Food and Drug Administration (FDA) declared that it had the authority to regulate human cloning and that any human cloning must have FDA approval.

While there is general agreement in the United States and in many other countries that reproductive human cloning should be banned because of ethical and safety concerns, there is ongoing debate about whether or not to allow therapeutic cloning to treat human disease or research cloning to study how stem cells develop. The Human Cloning Prohibition Act of 2001 to ban both reproductive and therapeutic cloning passed in the U.S. House of Representatives, but the Senate did not support the ban. The ban was again considered by the lawmakers in 2002. In the meantime, individual states such as California and New Jersey have passed bills that approve of embryonic stem cell research. Such research might lead to treatments for diseases such as Parkinson's, diabetes, and Alzheimer's. The research is controversial because embryos must be destroyed to obtain the stem cells, and some groups believe that constitutes taking a human life. The embryos used are generally extra embryos left over from in vitro fertilizations. In December, 2002, and January, 2003, a company called Clonaid announced the births of several babies it claimed were the result of human cloning but then failed to produce any scientific evidence that the babies were clones. In February, 2003, the U.S. Congress considered a ban on both reproductive and therapeutic cloning. In late February, the House passed the Human Prohibition Cloning Act of 2003, banning the cloning of human beings but allowing limited research on some existing stem cell lines.

In May, 2008, President George W. Bush signed into law the Genetic Information Nondiscrimination Act (GINA). GINA prohibits U.S. employers and insurance companies from discriminating on the basis of genetic test information. GINA protects Americans from discrimination based on genetic testing information. Insurance companies may not discriminate with reduced coverage or increased pricing based on information derived from genetic testing. Employers are prohibited from making adverse employment decisions based on an individual's genetic code. Under GINA law, insurers and employers may not demand or request a genetic test.

GINA protections are meant to encourage increased genetic testing without the fear of job loss or insurance complications. It is hoped that more genetic testing will enable researchers to devise therapies for a wide range of hereditary diseases. Genetic testing may also enable earlier treatments with better outcomes and decreased health care costs.

Executive order 13505, "Removing Barriers to Responsible Scientific Research Involving Human Stem Cells," was issued by President Barack Obama in March, 2009. This executive order requires the Health and Human Services secretary and the National Institutes of Health (NIH) director to review and issue new NIH guidelines regarding scientific research and human stem cells.

The tension between scientific possibility, public policy, and societal values continues in the arena of cloning. Through therapeutic cloning there is great potential for the treatment of human diseases, but the ethical concerns about such procedures must be carefully considered as well.

Susan J. Karcher, Ph.D.;
updated by Richard P. Capriccioso, M.D.

FURTHER READING

Boylan, Michael. "Genetic Engineering." In *Medical Ethics*, edited by Boylan. Upper Saddle River, N.J.: Prentice Hall, 2000. Considers the ethical concerns of gene therapy and organismal cloning. Tables, list for further reading.

Cibelli, Jose B., Robert P. Lanza, Michael D. West, and Carol Ezzell. "The First Human Cloned Embryo." *Scientific American* 286, no. 1 (2002): 44-48. Describes the production of cloned early-stage human embryos and embryos generated only from eggs, not embryos.

Espejo, Roman, ed. *Biomedical Ethics: Opposing Viewpoints*. San Diego: Greenhaven Press, 2003. Presents debates about many aspects of organismal cloning. Illustrations, bibliography, index.

Fredrickson, Donald S. *The Recombinant DNA Controversy, a Memoir: Science, Politics, and the Public Interest, 1974-1981*. Washington, D.C.: ASM Press, 2001. An overview of the initial concerns about potential hazards of recombinant DNA cloning.

Klotzko, Arlene Judith, ed. *The Cloning Sourcebook*. New York: Oxford University Press, 2001. A collection of twenty-seven essays on the science, context, ethics, and policy issues surrounding cloning.

Kreuzer, Helen, and Adrianne Massey. *Recombinant DNA and Biotechnology: A Guide for Teachers*. Washington, D.C.: ASM Press, 2001. Descriptions of recombinant DNA cloning methods and applications. Illustrations, index.

Lauritzen, Paul, ed. *Cloning and the Future of Human Embryo Research*. New York: Oxford University Press, 2001. Discusses cloning from the perspective of human embryo research and reproductive technology, seeing it as an extension of work that began with in vitro fertilization.

Schatten, G., R. Prather, and I. Wilmut. "Cloning Claim Is Science Fiction, Not Science." *Science* 299 (2003): 344. Letter to the editor written by prominent scientists expressing concern that evidence to support the claims of cloned humans has not been produced.

WEB SITES OF INTEREST

Genetic Information Nondiscrimination Act of 2007
http://www.ornl.gov/sci/techresources/Human _Genome/publicat/GINAMay2008.pdf
The actual GINA law—protection meant to encourage increased genetic testing without fear of job loss or insurance complications.

Human Genome Project Information
http://www.ornl.gov/sci/techresources/Human _Genome/graphics/slides/talks.shtml
Includes links to two PowerPoint presentations. "Genomics and Its Impact on Science and Society:

The Human Genome Project and Beyond" covers basic science, the Human Genome Project, what is known so far, next steps in genomic research, medicine, and benefits. "Beyond the Human Genome Project" covers what scientists have learned from the human genome sequence, what the next steps are in scientific discovery in genomics, and the diverse future applications of genomics.

1 Federal Register, Presidential Document Executive Order 13505 of March 9, 2009: "Removing Barriers to Responsible Scientific Research Involving Human Stem Cells"
http://edocket.access.gpo.gov/2009/pdf/E9 -5441.pdf
Presidential directive to review and issue new NIH guidelines regarding scientific research and human stem cells.

Stem Cell Information
http://stemcells.nih.gov/info/basics
Comprehensive source of information from the National Institutes of Health on the biological properties of stem cells, important questions about stem cell scientific research, and potential stem cell use in research and disease treatments.

Stem Cells, AlphaMed Press. "Stem Cell Therapy Restores Transparency to Defective Murine Corneas"
http://www3.interscience.wiley.com/journal/ 122318105/abstract?CRETRY=1&SRETRY=0
Research conducted by Du Yiqin et al. An example of current stem cell research that could contribute to effective human stem cell therapies.

See also: Animal cloning; Biopharmaceuticals; Cloning: Ethical issues; Cloning vectors; DNA replication; DNA sequencing technology; Gene therapy; Gene therapy: Ethical and economic issues; Genetic engineering; Genetic engineering: Agricultural applications; Genetic engineering: Historical development; Genetic engineering: Industrial applications; Genetic engineering: Medical applications; Genetic engineering: Risks; Genetic engineering: Social and ethical issues; Genetically modified foods; Knockout genetics and knockout mice; Plasmids; Polymerase chain reaction; Restriction enzymes; Reverse transcriptase; Shotgun cloning; Stem cells; Synthetic genes; Transgenic organisms; Xenotransplants.

Cloning
Ethical issues

CATEGORY: Bioethics; Genetic engineering and biotechnology; Human genetics

SIGNIFICANCE: Although cloning of plants has been performed for hundreds of years and cloning from embryonic mammalian cells became commonplace in the early 1990's, the cloning of the sheep Dolly from adult cells raised concerns that cloning might be used in a dangerous or unethical manner.

KEY TERMS

bioethics: the study of human actions and goals in a framework of moral standards relating to use and abuse of biological systems

clone: an identical genetic twin of any organism or DNA sequence; clones can occur naturally or experimentally

cloning: the process of producing a genetic twin in the laboratory by experimental means

reproductive cloning: production of an embryo by somatic cell nuclear transfer followed by implantation into a uterus in order to obtain the live birth of a cloned individual

somatic cell nuclear transfer: a method of cloning adult mammals by transplanting the nucleus of an adult somatic cell into an egg cell from which the original haploid nucleus has been removed

therapeutic cloning: production of an embryo by somatic cell nuclear transfer to obtain therapeutic embryonic stem cells from a blastocyst that is not allowed to develop further

BIOETHICS AND CLONING

Bioethics was founded as a discipline by ethicist Van Rensselaer Potter (1911-2001) in the early 1970's as the formal study and application of ethics to biology and biotechnology. The discipline was initially created as an ethical values system to help guide scientists and others in making decisions that could affect the environment. The world has become even more complex since Potter's original vision of a planet challenged by ecological catastrophe. Humans have developed the ability to take genes from one organism and transfer them to another, creating something entirely new to nature, with unknown consequences. Moreover, humans have the ability to make endless genetic copies of these organisms by cloning. Bioethics now includes asking hard ethical questions about biotechnology, and, as Potter suggests, "promot[ing] the evolution of a better world for future generations."

Cloning involves making a genetic twin of an organism or of a DNA sequence. The process of cloning has actually been performed with plants for centuries.

Cuttings can be removed from many species and induced to make roots. These cuttings are then grown into full-size, genetically identical copies of the parent plant. The emergence of crops that cannot be propagated in the standard fashion, such as seedless navel oranges, has led to whole groves of cloned siblings. Few would suggest that such cloning is inherently wrong or unethical. Animal cloning has been quietly occurring since the early 1990's. Eggs fertilized in vitro are allowed to develop to the eight-cell stage, at which point the cells are separated. Each individual cell then develops into an embryo that is implanted in a female. Thus, a single zygote can be used to make eight identical individuals. This type of cloning has been used routinely in animal husbandry to propagate desirable genetic traits.

In 1996, a team of scientists in Scotland headed by Ian Wilmut used somatic cell nuclear transfer to clone a mammal—a sheep named Dolly—from adult cells for the first time. While bioethicists had seen no wrong in cloning orange trees and embryonic mammals, they were troubled by the cloning of a sheep. It is important to realize that the cloning of Dolly was not the key bioethical issue. Rather, the issue that worried the ethicists was the implication of the clone's existence: If scientists were able to clone one mammal in this manner, then they were only a small step away from cloning other species, including humans.

In fact, in subsequent years, other laboratories around the world were able to reproductively clone a wide range of mammals using somatic cell nuclear transfer: cow and mouse in 1997; goat and pig in 2000; housecat, mouflon (an endangered wild sheep), and gaur (an endangered wild ox) in 2001; deer, rabbit, rat, horse, and mule in 2003; water buffalo, dog, and wolf in 2005, and camel and ferret in 2009.

Meanwhile, in 2004 Korean scientist Hwang Woo-Suk and collaborators reported success in so-called therapeutic cloning, in which cloned human embryos are allowed to develop only to the blastocyst stage in order to harvest embryonic stem cells. In a series of sensational revelations, these results—published in the highest profile journals—were demonstrated to be fabricated and Hwang was dismissed in disgrace, forfeiting his status as a national hero in South Korea. Geneticists in several laboratories have carried out human cloning through the very early stages of embryogenesis, but there has been no published case where cloned human embryos have been used to create embryonic stem cell lines.

In 2007, a research group led by Shoukhrat Mitalipov at the Oregon Health and Science University produced cloned rhesus monkey blastocysts

Chief executive of Clonaid Brigitte Boisselier (left) and the founder of the Raelian movement, Claude Vorilhon, announced in January, 2003, the birth of a human clone, as well as imminent births of other cloned children. Physical evidence and independent confirmation of the cloning were never offered and the announcement was concluded to be a hoax, but in the wake of media attention the issue of human cloning became the focus of renewed public debate. (AP/Wide World Photos)

from which they successfully obtained embryonic stem cells. This latter result suggests that therapeutic cloning may indeed become a viable technology in humans as well.

If bioethics is concerned with protecting future generations of humans, do these accomplishments represent a potential threat? What would be the social ramifications of human cloning? Would it have the potential to change humanity as it is now known forever? Was cloning simply wrong? Christian bioethicists, for example, were troubled by the implications of humans being able to manipulate themselves in this way, many considering it morally wrong.

Many scientists, including Wilmut, were quick to point out that they would never support human cloning but did not believe that cloning itself was unethical. Most ethicists agreed that cloning animals could help human society in many ways. Genetically engineered animals had the potential to be used to create vast quantities of protein-based therapeutic drugs. Commercial animals that are top producers, such as cows with high milk yields, could also be cloned. Human replacement organs could be grown in precisely controlled environments.

However, cloning, if misapplied, has frightening possibilities in the minds of many. Although only science fiction now, it is possible to envision a future world of human clones designed to fill certain roles, as genetically programmed soldiers, workers, or even an elite society of "perfect" cloned individuals. Others have envisioned the possibility of cloning an extra copy of themselves as donors of perfectly matched organs during old age. Even the possibility that individuals might be cloned without their knowledge or permission has been anticipated.

HUMAN CLONING

There is an apparent widespread consensus that human reproductive cloning is legally, socially, and ethically unappealing. The United Kingdom and Australia both passed legislation banning reproductive cloning in humans. In the United States, legislation has been proposed on numerous occasions, to ban both reproductive cloning and any type of human cloning, but it has never been passed into law. The United Nations adopted a nonbinding declaration against reproductive cloning, and the European Union's Convention on Human Rights and Biomedicine bans reproductive cloning (although the convention had not been ratified by all member states as of 2009).

Therapeutic cloning, on the other hand, occupies a much more controversial position in public opinion. Production of large numbers of embryonic stem cells of a patient's own genotype promises to deliver, in the relatively near future, staggering therapeutic benefits for an enormous variety of diseases. Patient advocates and others have lobbied strongly to proceed with therapeutic cloning research in humans. Many other groups, however, consider it profoundly immoral to destroy human embryos in order to obtain these stem cells. As a result, this technology is a highly contentious political topic around the world.

Bioethicist Karen Rothenberg, in statements delivered to the U.S. Senate's Public Health and Safety Subcommittee of the Labor and Human Resources Committee in the 1990's, suggested why society is made uneasy by the potential implications of human reproductive cloning. She broke her argument down into three *I*s. The first *I* is "interdependence." Cloning makes humans uneasy because it requires only one parent. People are humbled because it takes two humans to produce a baby. If part of the definition of humanity is the interdependence upon one another to reproduce, then a cloned human begs the question of just what is human. Rothenberg's second *I* is "indeterminateness." Cloning removes all randomness from human reproduction. With cloning, people predetermine whether they want to reproduce any physical or mental type available. They can control all possible genetic variables in cloning with a predicted outcome. However, does the same genetic variability that decides one's hereditary fate at conception also define some part of humanity? The last *I* is "individuality." It is disconcerting for people to imagine ten or one hundred copies of themselves walking around. Twins and triplets are common now, but what would such a vast change mean to individuality and the concept of the human soul? In closing, Rothenberg asked whether "the potential benefits of any scientific innovation [are] outweighed by its potential injury to our very concept of what it means to be human."

Andrew Scott of the Urban Institute takes a different view. He believes that bioethics does not apply to cloning but only to what happens after cloning. Cloning does not present a moral dilemma to

Scott, assuming that the process does not purposely create "abnormalities." Scott states that "the clone [would] simply be another, autonomous human being . . . carrying the same genes as the donor, and [living] life in a normal, functional way." He suggests that as long as clones are not programmed to be "human drones" and are not used in an unethical way, cloning should not be a bioethical worry. Many nonscientists miss the point that a clone is simply a genetically identical copy, not a copy in every aspect. If someone were to have cloned Albert Einstein, the cloned Einstein would not be identical behaviorally or in other ways to the original. What made Einstein who he was involved not merely his genes but also his many life experiences, which are impossible to duplicate in a clone. The same would be true of a cloned child brought to life by grieving parents who have lost their original child in an accident. The clone would be like a twin, not the same child.

Perhaps the right questions are not being asked. Better questions may be: Can humans be trusted not to abuse the technology of cloning? Can those in positions of power be trusted not to use cloning to their advantage and the endangerment of humanity? Probably the most basic question is, What compelling reason is there to clone a human in the first place? Carl B. Feldbaum, the president of the Biotechnology Industry Organization, believes that people should be wary of anyone who asks them to allow human cloning and states:

> In the future, society may determine that there are sound reasons to clone certain animals to improve the food supply, produce biopharmaceuticals, provide organs for transplantation and aid in research. I can think of no ethical reason to apply this technique to human beings, if in fact it can be applied.

The ethical issues are even more complicated than they first appear. Is the actual process of cloning, as performed by Wilmut, ethical if applied to humans? Wilmut's cloning process produced many failures before Dolly was conceived; only she survived of her 277 cloned sisters. Her early death at the age of six was also potentially precipitated by the cloning process. Bioethicists question whether manipulating human embryos to produce clones with only a 0.4 percent success rate is moral; to someone who believes that human life begins at conception, the cloning procedure as performed by Wilmut would almost certainly be unacceptable. The low success rate of cloning by somatic cell nuclear transfer, as well as health problems in cloned individuals, have consistently been observed in each vertebrate species that has been cloned.

Cloning offers a new and perhaps frightening view of life and the biological universe. If almost any cell in the body can be used as the basis to clone an entirely new organism, this makes each cell the potential equivalent of a fertilized egg. Does this insight lead to a renewed respect for life, or does it render life paradoxically cheapened? If each cell contains all the genetic information needed to create a new individual, then what is a single cell worth among millions of copies? When one million or one hundred million potential copies exist, what is one copy alone worth?

James J. Campanella, Ph.D., and Bryan Ness, Ph.D.;
updated by Carina Endres Howell, Ph.D.

FURTHER READING

Andrews, Lori B. *The Clone Age: Adventures in the New World of Reproductive Technology.* New York: Henry Holt, 1999. A lawyer specializing in reproductive technology, Andrews examines the legal ramifications of human cloning, from privacy to property rights.

Baudrillard, Jean. *The Vital Illusion.* Edited by Julia Witwer. New York: Columbia University Press, 2000. A sociological perspective on what human cloning means to the idea of what it means to be human.

Bonnicksen, Andrea L. *Crafting a Cloning Policy: From Dolly to Stem Cells.* Washington, D.C.: Georgetown University Press, 2002. Political and policy issues surrounding human cloning.

Brannigan, Michael C., ed. *Ethical Issues in Human Cloning: Cross-Disciplinary Perspectives.* New York: Seven Bridges Press, 2001. A collection of writings from a broad variety of Western and non-Western traditions and perspectives—philosophical, religious, scientific, and legal—good for sparking debate.

Harris, John. *On Cloning.* London: Routledge, 2004. A frank discussion of the myths of human cloning and a presentation of the benefits that cloning could have for humans.

Klotzko, Arlene Judith. *A Clone of Your Own? The Science and Ethics of Cloning.* London: Cambridge University Press, 2006. A bioethicist who considers reproductive cloning "inevitable" makes a strong argument against genetic determinism.

_____, ed. *The Cloning Sourcebook.* New York: Oxford University Press, 2001. A collection of twenty-seven essays on the science, context, ethics, and policy issues surrounding cloning.

Lauritzen, Paul, ed. *Cloning and the Future of Human Embryo Research.* New York: Oxford University Press, 2001. Places the ethical debate on human cloning in the larger context of reproductive technology.

MacKinnon, Barbara, ed. *Human Cloning: Science, Ethics, and Public Policy.* Urbana: University of Illinois Press, 2000. Experts from a variety of perspectives argue both for and against human cloning.

Pence, Gregory E. *Cloning After Dolly: Who's Still Afraid?* Lanham, Md.: Rowman & Littlefield, 2004. A strong advocate of therapeutic, and even reproductive, cloning advances outspoken arguments.

Rantala, M. L., and Arthur J. Milgram, eds. *Cloning: For and Against.* Chicago: Open Court, 1999. Scientists, journalists, ethicists, religious leaders, and legal experts represent all viewpoints, presenting all sides of the human cloning debate.

Shostak, Stanley. *Becoming Immortal: Combining Cloning and Stem-Cell Therapy.* Albany: State University of New York Press, 2002. Examines the question of whether human beings are equipped for potential immortality.

Wilmut, Ian, and Roger Highfield. *After Dolly: The Uses and Misuses of Human Cloning.* New York: W. W. Norton, 2006. Written by the scientist who produced the cloned sheep, Dolly, this book describes the scientific process that produced Dolly and addresses its implications.

Yount, Lisa, ed. *The Ethics of Genetic Engineering.* San Diego: Greenhaven Press, 2002. Essays written by scientists, science writers, ethicists, and consumer advocates present the growing controversy over genetically modifying plants and animals, altering human genes, and cloning humans.

WEB SITES OF INTEREST

ActionBioScience.org
http://www.actionbio science.org/biotech/mcgee.html

"Cloning." American Journal of Bioethics
http://www.bioethics.net/topics.php?catId=4

Human Genome Project Information. "Cloning Fact Sheet"
http://www.ornl.gov/sci/techresources/Human_Genome/elsi/cloning.shtml

President's Council on Bioethics
http://bioethics.gov

World Health Organization. "A Dozen Questions and Answers on Human Cloning"
http://www.who.int/ethics/topics/cloning

See also: Animal cloning; Bioethics; Biological weapons; Cloning; Cloning vectors; Eugenics; Eugenics: Nazi Germany; Gene therapy; Gene therapy: Ethical and economic issues; Genetic engineering: Medical applications; Genetic engineering: Risks; Genetic engineering: Social and ethical issues; Genetics in television and films; Knockout genetics and knockout mice; Polymerase chain reaction; Restriction enzymes; Reverse transcriptase; Shotgun cloning; Stem cells; Synthetic genes; Transgenic organisms; Xenotransplants.

Cloning vectors

CATEGORY: Genetic engineering and biotechnology

SIGNIFICANCE: Cloning vectors are one of the key tools required for propagating (cloning) foreign DNA sequences in cells. Cloning vectors are vehicles for the replication of DNA sequences that cannot otherwise replicate. Expression vectors are cloning vectors that provide not only the means for replication but also the regulatory signals for protein synthesis.

KEY TERMS

bacteriophage: a virus that infects bacterial cells, often simply called a phage

foreign DNA: DNA taken from a source other than the host cell that is joined to the DNA of the cloning vector; also known as insert DNA

plasmid: a small, circular DNA molecule that replicates independently of the host cell chromosome

recombinant DNA molecule: a molecule of DNA created by joining DNA molecules from different sources, most often vector DNA joined to insert DNA

restriction enzyme: an enzyme capable of cutting DNA

at specific base pair sequences, produced by a variety of bacteria as a protection against bacteriophage infection

THE BASIC PROPERTIES OF A CLONING VECTOR

Cloning vectors were developed in the early 1970's from naturally occurring DNA molecules found in some cells of the bacteria *Escherichia coli* (*E. coli*). These replicating molecules, called plasmids, were first used by the American scientists Stanley Cohen and Herbert Boyer as vehicles, or vectors, to replicate other pieces of DNA (insert DNA) that were joined to them. Thus the first two essential features of cloning vectors are their ability to replicate in an appropriate host cell and their ability to join to foreign DNA sequences to make recombinant molecules. Plasmid replication requires host-cell-specified enzymes, such as DNA polymerases that act at a plasmid sequence called the "origin of replication." Insert DNA is joined (ligated) to plasmid DNA through the use of two kinds of enzymes: restriction enzymes and DNA ligases. The plasmid DNA sequence must have unique sites for restriction enzymes to cut. Cutting the double-stranded circular DNA at more than one site would cut the plasmid into pieces and would separate important functional parts from one another. However, when a restriction enzyme cuts the circular plasmid at one unique site, it converts it to a linear molecule. Linear, insert DNA molecules, produced by cutting DNA with the same restriction enzyme as was used to cut the plasmid vector, can be joined to cut plasmid molecules using the enzyme DNA ligase. This catalyzes the covalent joining of the insert DNA and plasmid DNA ends to create a circular, recombinant plasmid molecule. Most cloning vectors have been designed to have many unique restriction enzyme cutting sites all in one stretch of the vector sequence. This part of the vector is referred to as the multiple cloning site.

In addition to an origin of replication and a multiple cloning site, most vectors have a third element: a selective marker. In order for the vector to replicate, it must be present inside an appropriate host cell. Introducing the vector into cells is often a very inefficient process. Therefore, it is very useful to be able to select, from a large population of host cells, those rare cells that have taken up a vector. This is the role of the selectable marker. The selectable marker is usually a gene that encodes resistance to an antibiotic to which the host is normally sensitive.

For example, if a plasmid vector has a gene that encodes resistance to the antibiotic ampicillin, only those *E. coli* cells that harbor a plasmid will be able to grow on media containing ampicillin.

Many vectors have an additional selective marker that is rendered inactive when a plasmid is recombinant. A commonly used marker gene of this kind is the *lacZ* gene, which encodes the enzyme beta-galactosidase. This enzyme breaks the disaccharide lactose into two monosaccharides. The pUC plasmid vector has a copy of the *lacZ* gene which has been carefully engineered to contain a multiple cloning site within it, while maintaining the functionality of the expressed enzyme. When a DNA fragment is inserted into the multiple cloning site, the *lacZ* gene is no longer capable of making functional beta-galactosidase. This loss of function can be detected by putting X-gal into the growth media. X-gal has a structure similar to lactose but cannot be broken down by beta-galactosidase. Rather, beta-galactosidase modifies X-gal and produces a blue color. Thus, colonies of the bacterium *E. coli* containing recombinant plasmids will be normal colored, whereas those that have normal, nonrecombinant plasmids will be blue. Typical selection media then contain ampicillin and X-gal. The ampicillin only allows *E. coli* that contain a plasmid to grow, and the X-gal identifies which colonies have recombinant plasmids.

There are a number of procedures for introducing the plasmid vector into the host cell. Transformation is a procedure in which the host cells are chemically treated so that they will allow small DNA molecules to pass through the cell membrane. Electroporation is a procedure that uses an electric field to create pores in the host cell membrane to let small DNA molecules pass through.

VIRUSES AND CLONING VECTORS

In addition to plasmid cloning vectors, some bacteriophages (or phages) have been modified to serve as cloning vectors. Bacteriophages, like other viruses, are infectious agents that are made of a genome, either DNA or RNA, that is surrounded by a protective protein coat. Phage vectors are used similarly to the way plasmid vectors are used. The vector and insert DNAs are cut by restriction enzymes so that they subsequently can be joined by DNA ligase. The newly formed recombinant DNA molecules must enter an appropriate host cell to replicate. In

The Ti Plasmid of *Agrobacterium*

Two species of naturally occurring plant pathogenic bacteria, *Agrobacterium tumefaciens* and *Agrobacterium rhizogenes*, infect many plant species and have been harnessed through biotechnology to effect permanent genetic transformation of plants. Virulent (disease-causing) *Agrobacterium* species can infect plants and transfer a small portion of their own bacterial DNA, called T-DNA (transferred DNA), into the plant. The T-DNA is actually a small fragment of a large (approximately 200-kilobase-pair) plasmid called the Ti (tumor-inducing) plasmid in *A. tumefaciens* and the Ri (root-inducing) plasmid in *A. rhizogenes*.

The T-DNA fragment of the Ti plasmid is defined on both ends by 24-base-pair direct repeat sequences called the left-hand and right-hand border sequences. The T-DNA fragment is released from the plasmid by the action of endonucleases, which cut the DNA at specific points within the right-hand and left-hand border sequences. The endonucleases are two of the *Vir* (virulence) genes encoded on the Ti plasmid adjacent to the T-DNA. Several other *Vir* genes are produced when *Agrobacterium* cells are introduced into plant tissue, usually through a wound. Following infection of a plant, *Agrobacterium* cells sense the presence of phenolic wound compounds and the acidic environment within wounded plant tissues. These conditions trigger a series of several *Vir* genes to produce Vir proteins that direct excision of the T-DNA and facilitate transport and incorporation of the T-DNA segment into the plant's genome. Once the T-DNA is incorporated into the plant genome, expression of the T-DNA-encoded genes causes the plant to produce unusual quantities of plant hormones and other compounds that cause the plant cells to grow abnormally near the infection site, producing characteristic tumors.

Purposeful genetic transformation of plants requires a tool that can be used to insert new genes into a plant. This tool, regardless of its derivation, is called a vector. To date, the most common means for stable genetic transformation of plants involves the use of vectors derived from bacteria of the genus *Agrobacterium*. Biotechnologists have harnessed *Agrobacterium* to insert new genes of interest into plants by modifying the T-DNA segment of the bacterial DNA using standard recombinant methods. By deleting the genes on the T-DNA that cause tumors and then inserting desirable genes in their place, a wide variety of vectors can be produced to transfer desirable genes into plants. The genes transferred by way of *Agrobacterium* vectors become a permanent part of the plant's genome. DNA from plants, animals, bacteria, and viruses can be introduced into plants in this way.

One major drawback of *Agrobacterium* transformation is that insertion of T-DNA into the plant genome is essentially random. The genes on the T-DNA segment may not be efficiently transcribed at their location or the insertion of T-DNA may knock out an important plant gene by inserting in the middle of it. Therefore, a plant genetically transformed using an *Agrobacterium* vector is not necessarily guaranteed to perform as desired. A final drawback is that the vector works only with dicots, while many of the world's most important crops are monocots, such as wheat, rice, corn, and many other grain crops.

Robert A. Sinnott, Ph.D.

order to introduce the phage DNA into cells, a whole phage particle must be built. This is referred to as "packaging" the DNA. The protein elements of the phage are mixed with the recombinant phage DNA and packaging enzymes to create an infectious phage particle. Appropriate host cells are then infected with it. The infected cells then make many copies of each recombinant molecule, along with the proteins needed to make a completed phage particle. In many cases, the final step of viral infection is the lysis of the host cell. This releases the mature phage particles to infect nearby host cells. Phage vectors have two advantages relative to plasmid vectors: First, viral delivery of recombinant DNA to host cells is much more efficient than the transformation or electroporation procedures used to introduce plasmid DNA into host cells, and second, phage vectors can be used to clone larger fragments of insert DNA.

Viruses that infect cells other than bacteria have been modified to serve as cloning vectors. This permits cloning experiments using many different kinds of host cells, including human cells. Viral vectors, just like the natural viruses from which they are derived, have specific host and tissue ranges. A particular viral vector will be limited for use in specific species and cell types. The fundamental practice of all virally based cloning vectors involves the covalent joining of the insert DNA to the viral DNA to make a recombinant DNA molecule, introduction of the recombinant DNA into the appropriate host cell, and then propagation of the vector through the natural mechanism of viral replication. There are two fundamentally different ways that viruses propagate in cells. Many viruses, such as the phages already described, enter the host cell and subvert the cell's biosynthetic machinery to its own reproduction, which ultimately leads to lysis and thereby kills the host cell as the progeny viruses are released. The second viral life strategy is to enter the host cell and integrate the viral DNA into the host cell chromosome so that the virus replicates along with the host DNA. Such integrating viruses can be stably maintained in the host cell for long periods. The retroviruses, of which the human immunodeficiency virus (HIV) is an example, are a group of integrating viruses that are potentially useful vectors for certain gene therapy applications. Using cloning vectors and host cells other than bacteria allows scientists to produce some proteins that bacteria cannot properly make, permits experiments to determine the function of cloned genes, and is important for the development of gene therapy.

EXPRESSION VECTORS

Expression vectors are cloning vectors designed to express the gene contained in the recombinant vector. In order to accomplish this, they must also provide the appropriate regulatory signals for the transcription and translation of the foreign gene. Regulatory sequences, which direct the cellular transcription machinery, are very different in bacteria and higher organisms. Thus, unless the vector provides the appropriate host regulatory sequences, foreign genes will not normally be expressed.

Expression vectors make it possible to produce proteins encoded by eukaryotic genes (that is, genes from higher organisms) in bacterial cells. Furthermore, producing proteins in this way often results in higher production rates than in the cells from which the gene was obtained. This

technology not only is of immense benefit to scientists who study proteins but also is used by industry (particularly the pharmaceutical industry) to make valuable proteins. Proteins such as human insulin,

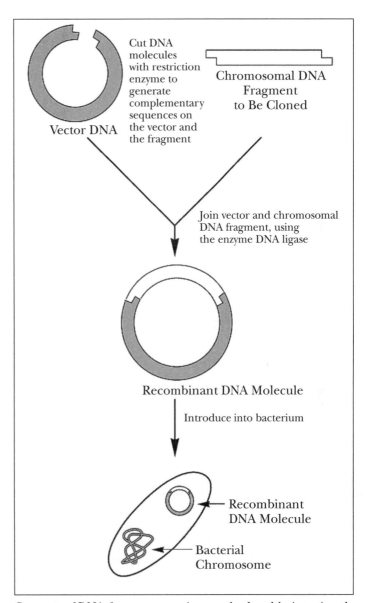

Segments of DNA from any organism can be cloned by inserting the DNA segment into a plasmid—a small, self-replicating circular molecule of DNA separate from chromosomal DNA. The plasmid can then act as a "cloning vector" when it is introduced into bacterial cells, which replicate the plasmid and its foreign DNA. This diagram from the Department of Energy's Human Genome Program site illustrates the process. (U.S. Department of Energy Human Genome Program, http://www.ornl.gov/hgmis.)

growth hormone, and clotting factors that are difficult and extremely expensive to isolate from their natural sources are readily available because they can be produced much more cheaply in bacteria. An added benefit of expression vectors is that actual human proteins are produced by bacteria and therefore do not provoke allergic reactions as frequently as insulin that is isolated from other species.

ARTIFICIAL CHROMOSOMES

In 1987, a new type of cloning vector was developed by David Burke, Maynard Olson, and their colleagues. These new vectors, artificial chromosomes, filled the need created by the Human Genome Project (HGP) to clone very large insert DNAs (hundreds of thousands to millions of base pairs in length). One of the goals of the HGP—to map and ultimately sequence all the chromosomes of humans, as well as a number of other "model" organisms' genomic sequences—required a vector capable of propagating much larger DNA fragments than plasmid or phage vectors could propagate. The first artificial chromosome vector was developed in the yeast *Saccharomyces cerevisiae*. All the critical DNA sequence elements of a yeast chromosome were identified and isolated, and these were put together to create a yeast artificial chromosome (YAC). The elements of a YAC vector are an origin of replication, a centromere, telomeres, and a selectable marker suitable for yeast cells. A yeast origin of replication (similar to the origin of replication of bacterial plasmids) is a short DNA sequence that the host's replicative enzymes, such as DNA polymerase, recognize as a site to initiate DNA replication. In addition to replicating, the new copies of a chromosome must be faithfully partitioned into daughter cells during mitosis. The centromere sequence mediates the partitioning of the chromosomes during cell division because it serves as the site of attachment for the spindle fibers in mitosis. Telomeres are the DNA sequences at the ends of chromosomes. They are required to prevent degradation of the chromosome and for accurate replication of DNA at the ends of chromosomes.

YACs are used much as plasmid vectors are. Very large insert DNAs are joined to the YAC vector, and the recombinant molecules are introduced into host yeast cells in which the artificial chromosome is replicated just as the host's natural chromosomes are. YAC cloning technology allows very large chromosomes to be subdivided into a manageable number of pieces that can be organized (mapped) and studied. YACs also provide the opportunity to study DNA sequences that interact over very long distances. Since the development of YACs, artificial chromosome vectors for a number of different host cells have been created.

IMPACT AND APPLICATIONS

Cloning vectors are one of the key tools of recombinant DNA technology. Cloning vectors make it possible to isolate particular DNA sequences from an organism and make many identical copies of this one sequence in order to study the structure and function of that sequence apart from all other DNA sequences. Until the development of the polymerase chain reaction (PCR), cloning vectors and their host cells were the only means to collect many copies of one particular DNA sequence. For long DNA sequences (those over approximately ten thousand base pairs), cloning vectors are still the only means to do this.

Gene therapy is a new approach to treating and perhaps curing genetic disease. Many common diseases are the result of defective genes. Gene therapy aims to replace or supplement the defective gene with a normal, therapeutic gene. One of the difficulties faced in gene therapy is the delivery of the therapeutic gene to the appropriate cells. Viruses have evolved to enter cells, sometimes only a very specific subset of cells, and deliver their DNA or RNA genome into the cell for expression. Thus viruses make attractive vectors for gene therapy. An ideal vector for gene therapy would replace viral genes associated with pathogenesis with therapeutic genes; the viral vector would then target the therapeutic genes to just the right cells. One of the concerns related to the use of viral vectors for gene therapy is the random nature of the viral insertion into the target cell's chromosomes. Insertion of the vector DNA into or near certain genes associated with increased risk of cancer could theoretically alter their normal expression and induce tumor formation.

Plasmid DNA vectors encoding immunogenic proteins from pathogenic organisms are being tested for use as vaccines. DNA immunization offers several potential advantages over traditional vaccine strategies in terms of safety, stability, and effectiveness. Genes from disease-causing organisms are cloned into plasmid expression vectors that provide the regulatory signals for efficient protein production in humans. The plasmid DNA is inoculated intramuscularly or intradermally, and the muscle or skin cells take up some of the plasmid DNA and express the immunogenic proteins. The immune system then generates a protective immune response. There are two traditional vaccination strategies: One uses live, attenuated pathogenic organisms, and the other uses killed organisms. The disadvantage of the former is that, in rare cases, the live vaccine can cause disease. The disadvantage of the latter strategy is that the killed organism does not enter the patient's cells and make proteins like the normal pathogen. Therefore, one part of the immune response, the cell-mediated response, is usually not activated, and the protection is not as good. In DNA immunization, the plasmids enter the patient's cells, and the immunogenic proteins produced there result in a complete immune response. At the same time, there is no chance that DNA immunization will cause disease, because the plasmid vector does not carry all of the disease-causing organism's genes.

Craig S. Laufer, Ph.D.; updated by Bryan Ness, Ph.D.

FURTHER READING

Anderson, W. French. "Gene Therapy." *Scientific American* 273, no. 3 (September, 1995): 124. Provides a good review of the promises and problems of gene therapy.

Brown, T. A. "Vectors for Gene Cloning: Plasmids and Bacteriophages." In *Gene Cloning and DNA Analysis: An Introduction.* 5th ed. Malden, Mass.: Blackwell, 2006. Describes the principles and methods of gene cloning and DNA analysis for readers with little knowledge of these subjects. Contains more than 250 two-color illustrations.

Cohen, Philip. "Creators of the Forty-seventh Chromosome." *New Scientist* 148, no. 2003 (November 11, 1995): 34. Describes the efforts to develop human artificial chromosomes.

Friedmann, Theodore. "Overcoming the Obstacles to Gene Therapy." *Scientific American* 276, no. 6 (June, 1997): 96. Elaborates on the relative merits of different delivery systems for gene therapy.

Hassett, Daniel E., and J. Lindsay Whitton. "DNA Immunization." *Trends in Microbiology* 4, no. 8 (August, 1996): 307-312. Reviews the process of DNA immunization and compares it to traditional vaccination strategies.

Jones, P., and D. Ramji. *Vectors: Cloning Applications and Essential Techniques.* New York: J. Wiley, 1998. A laboratory manual that allows quick and easy access to the key protocols required by those working with vectors.

Lodge, Julia, Peter A. Lund, and Steve Minchin. *Gene Cloning: Principles and Applications.* New York: Taylor & Francis, 2007. Describes the many available gene-cloning techniques and how they can be used in the research laboratory for numerous applications, including biotechnology, medicine, agriculture, and pharmaceuticals.

Lu, Quinn, and Michael P. Weiner, eds. *Cloning and Expression Vectors for Gene Function Analysis.* Natick, Mass.: Eaton, 2001. Reprints forty-three articles from the journal *BioTechnique* to provide an overview of cloning vectors and strategies, protein expression and purification, gene tagging and epitope tagging strategies, and special purpose vectors.

Watson, James D., et al. *Recombinant DNA—Genes and Genomes: A Short Course.* 3d ed. New York: W. H. Freeman, 2007. Nobel laureate Watson uses accessible language and diagrams to address the methods, underlying concepts, and far-reaching applications of recombinant DNA technology. An excellent reference for details on how the different cloning vectors work and to what purposes each is particularly suited.

WEB SITES OF INTEREST

Molecular Biology, Cloning Vectors
http://www.web-books.com/MoBio/Free/Ch9A4.htm

This page in an online book about molecular biology uses text and illustrations to describe cloning vectors.

Waksman Student Scholars, Genetic Engineering Vectors
http://dwb.unl.edu/Teacher/NSF/C08/C08Links/mbclserver.rutgers.edu/~sofer/cloningvectors.html

The Waksman Student Scholars site was designed by professors at Rutgers University to be a resource about molecular biology for high school students and teachers. The site includes a page providing information about cloning vectors.

See also: Animal cloning; Biopharmaceuticals; Cloning; Cloning: Ethical issues; DNA replication;

DNA sequencing technology; Gene therapy; Genetic engineering; Genetic engineering: Medical applications; Genetic engineering: Risks; Genomic libraries; Knockout genetics and knockout mice; Plasmids; Polymerase chain reaction; Protein synthesis; Restriction enzymes; Reverse transcriptase; Shotgun cloning; Stem cells; Synthetic genes; Telomeres; Transgenic organisms; Xenotransplants.

Cockayne syndrome

CATEGORY: Diseases and syndromes
ALSO KNOWN AS: CS; dwarfism-retinal atrophy-deafness syndrome; Neill-Dingwall syndrome; progeroid nanism; progeria-like syndrome

DEFINITION

Cockayne syndrome (CS) is a rare, heterogeneous, multisystem disease that is typically apparent at birth or during childhood. The disorder is characterized by cachetic dwarfism, neurological deficits, sensitivity to sunlight (photosensitivity), and premature aging (progeria).

RISK FACTORS

The incidence of CS has been estimated at about 1 in 560,000 live births in Western Europe; a similar incidence is likely in other parts of the world. Individuals with a family history of CS or whose parents carry known mutations in the *ERCC6/CSB* or *ERCC8* (*CSA; CKN1*) gene are at increased risk. No predispositions based upon sex or ethnicity have been noted.

ETIOLOGY AND GENETICS

CS is caused by mutations in the *ERCC6/CSB* gene on chromosome 10q11 or in the *ERCC8/CSA* gene on chromosome 5q12.1. Roughly 75 percent of cases are attributable to mutations in the *ERCC6/CSB* gene, with the remaining 25 percent due to mutations in the *ERCC8/CSA* gene. CS is an autosomal recessive disorder, meaning that the affected individual carries two copies of the defective gene in each cell. The parents are heterozygous, each carrying one copy of the defective gene. Heterozygous individuals do not typically show any symptoms of the disease.

The CSB and CSA proteins are believed to play a role in the repair of transcriptionally active genes (transcription-coupled DNA repair). The defect in DNA repair is thought to result in the accumulation of DNA damage and cell death. There is currently no clear association between type of mutation and the severity of symptoms observed.

CS is a heterogeneous disease. Three distinct major subtypes have been identified that vary in severity and prognosis. CS type I (Type A) is the classic form of the disease in which growth and developmental abnormalities are noted in the first few years of life. CS type II (Type B) is the most severe form of the disease, with growth failure present at birth and little postnatal neurological development. CS type III (Type C) is a less common, milder form of the disease that typically is seen later in life. A fourth subtype called xeroderma pigmentosum-Cockayne syndrome (XP-CS) has been recognized that combines features of both diseases but is caused by mutations in genes other than *ERCC6/CSB* or *ERCC8/CSA*.

SYMPTOMS

Individuals with CS display physical features of cachectic dwarfism, with thinning of the skin and hair, sunken eyes, and a stooped standing posture. Symptoms may also include disproportionately long arms and legs, large low-set ears, bird-like facies (facial expressions), microcephaly (small head), joint contractures, gait disturbances, and sensitivity to sunlight. Neurodevelopmental delays, which vary in severity, are characteristic of the disease. Hearing loss (deafness), eye problems (retinal atrophy and cataracts), and dental caries are also common and become progressively worse with age. Other observed symptoms, including hypertension, early atherosclerosis, intracranial calcification, and glomerulosclerosis, appear to be related to premature aging.

SCREENING AND DIAGNOSIS

Clinical diagnosis of CS is usually made when an infant or child fails to grow properly (postnatal growth failure) and has signs of neurologic dysfunction. In addition, affected children often have a characteristic appearance, are sensitive to sunlight, and show signs of premature aging (progeria). Diagnosis of CS may be delayed or missed due to heterogeneity in symptoms among patients as well as to the rarity and progressive nature of the disease. Diagnosis can be confirmed by sequencing of the *ERCC8/CSA* and *ERCC6/CSB* genes in the affected individual.

TREATMENT AND THERAPY

No specific treatment for CS currently exists. Patients are treated based on their individual symptoms. Infants with the most severe form of CS may require tube feeding to prevent malnutrition. Physical therapy may be performed to minimize effects of joint contractures and to maintain mobility. Speech, vision, hearing, and occupational therapy may also be provided.

PREVENTION AND OUTCOMES

For individuals with a family history of CS, genetic counseling is recommended. Genetic testing can be performed on the unborn child to determine if he or she carries a mutation in the *ERCC8/CSB* or *ERCC6/CSA* gene. Parents who are heterozygous for the causative mutation (carriers) have a 25 percent chance of having a normal child, a 50 percent chance of having a heterozygous (carrier) child, and a 25 percent chance of having a child with CS.

Typical life expectancy is dependent on the subtype of the disease but is generally shorter than normal. Individuals with CS type I usually live one or two decades. Individuals with CS type II usually die before age seven. Individuals with CS type III can survive into their thirties or forties. Individuals who do survive longer do not have an increased incidence of skin cancer despite their photosensitivity.

Valerie L. Gerlach, Ph.D.

FURTHER READING

De Boer, J., and J. H. Hoeijmakers. "Nucleotide Excision Repair and Human Syndromes." *Carcinogenesis* 21 (2000): 453–460.

Friedberg, E. C., G. C. Walker, and W. Siede. *DNA Repair and Mutagenesis.* Washington, D.C.: ASM Press, 1995.

Nance, M. A., and S. A. Berry. "Cockayne Syndrome: Review of 140 Cases." *American Journal of Medical Genetics* 42 (1992): 68-84.

Tan, W. H., H. Baris, C. D. Robson, and V. E. Kimonis. "Cockayne Syndrome: The Developing Phenotype." *American Journal of Medical Genetics A* 135 (2005): 214-216.

WEB SITES OF INTEREST

Amy and Friends Cockayne Syndrome UK
http://www.amyandfriends.org

Share and Care Cockayne Syndrome Network
http://cockaynesyndrome.net

See also: Developmental genetics; DNA repair; Dwarfism; Hereditary diseases.

Colon cancer

CATEGORY: Diseases and syndromes
ALSO KNOWN AS: Colorectal cancer; cancer of the colon and rectum

DEFINITION

Colorectal cancer is a disease in which cancer cells grow in the colon and/or rectum. The colon and the rectum are parts of the large intestine.

Cancer occurs when cells in the body divide out of control or order. If cells keep dividing, a mass of tissue, called a growth or tumor, forms. The term "cancer" refers to malignant tumors. They can invade nearby tissue and spread to other parts of the body. A benign tumor does not invade or spread.

RISK FACTORS

Individuals who are fifty years of age or older are at risk for colon cancer. Other risk factors include eating a diet that is high in beef and fat and low in fiber; having polyps (benign growths) in the colon and rectum (especially due to familial polyposis, an inherited condition); having a personal history of colorectal cancer; and having a family history of colorectal cancer, especially in a parent, sibling, or child. Individuals who have ulcerative colitis (inflammation of the lining of the colon) or Crohn disease are also at risk. Long-term insulin use in people with Type II diabetes is associated with an increased risk. Obesity, a high body mass index (BMI), physical inactivity, having diabetes, smoking, and alcohol intake are also risk factors.

ETIOLOGY AND GENETICS

Colorectal cancer is a complex condition that can be triggered by either environmental or genetic events. Approximately 75 percent of affected individuals have sporadic disease (no clear evidence of having inherited the disorder), while the remaining 25 percent have family histories that suggest an inherited predisposition to the development of colon cancer. In only about 5 percent of the total cases, however, can a specific mutation in a known gene be identified as the causative agent.

Two quite different classes of molecular events are known to lead to colon cancer, and these are referred to as chromosomal instability and microsatellite instability. About 85 percent of all colon cancers show evidence of chromosomal instability in tumor tissue, as detected by chromosomal deletions or loss of whole chromosomes. Most commonly, the deleted regions involve either the short arm of chromosome 17 or the long arms of either chromosome 5 or 18. Three different tumor suppressor genes have been located to these regions (*APC*, at position 5q21-q22; *AXIN2*, at position 17q24; and *SMAD4*, at position 18q21.1), and it has been suggested that it is the loss of these genes that is the genetic trigger for development of the disease. The remaining 15 percent of cancers showing microsatellite instability result in tumors that have intact chromosomal complements but appear to be deficient in the deoxyribonucleic acid (DNA) mismatch repair system. This inability to fix mutations by this pathway presumably increases the likelihood that deleterious mutations might accumulate in cancer-associated genes. Genes that have been identified that affect mismatch repair include *PMS2*, at position 7p22, *BLM*, at position 15q26.1, and *MYH* on the short arm of chromosome 1, at position 1p34.3.

Most of the genes noted above that can cause a predisposition to colon cancer are inherited in an autosomal dominant fashion, meaning that a single copy of the mutation is sufficient for expression. An affected individual has a 50 percent chance of transmitting the mutation to each of his or her children. Many cases of familial colorectal cancer, however, result from a spontaneous new mutation, so in these instances affected individuals will have unaffected parents. Mutations in the *BLM* and *MYH* genes show autosomal recessive inheritance, which means that both copies of the gene must be deficient in order for the individual to be afflicted. Typically, an affected child is born to two unaffected parents, both of whom are carriers of the recessive mutant allele. The probable outcomes for children whose parents are both carriers are 75 percent unaffected and 25 percent affected. If one parent is affected and the other is a carrier, there is a 50 percent probability that each child will be affected.

SYMPTOMS

Colorectal cancer often does not have any symptoms. However, some symptoms associated with the disease include a change in bowel habits, such as diarrhea, constipation, or a feeling that the bowel does not empty completely, lasting for more than a few days in people aged fifty and older; blood (either bright red or very dark) in the stool; stools that are narrower than usual; abdominal discomfort (frequent gas pains, bloating, fullness, and/or cramps); unexplained weight loss; and constant fatigue. These symptoms may also be caused by other, less serious health conditions. Individuals experiencing these symptoms should see a doctor.

SCREENING AND DIAGNOSIS

The doctor will ask about a patient's symptoms and medical history, and a physical exam will be done. Tests include a digital rectal exam, the use of a doctor's gloved finger to examine the rectum for lumps or growths; a fecal occult blood test, a test to check for hidden blood in the stool; and a barium enema, a rectal injection of barium given to coat the lining of the colon and rectum, which is done before X rays in order to create a better image of the lower intestine. Other tests include sigmoidoscopy, an examination of the lower colon using a lighted tube called a sigmoidoscope; colonoscopy, an examination of the rectum and entire colon using a lighted tube called a colonoscope; polypectomy, the removal of a polyp during a sigmoidoscopy or colonoscopy; biopsy, the removal of colon or rectal tissue to be tested for cancer cells; a computed tomography (CT) scan, a type of X ray that uses a computer to make pictures of structures inside the body, used to identify the spread of the tumor outside the colon; a magnetic resonance imaging (MRI) scan, a test that uses magnetic waves to make pictures of structures inside the body; and a positron emission tomography (PET) scan, a test that produces images showing the amount of functional activity in tissue being studied.

TREATMENT AND THERAPY

Once colon cancer is found, staging tests are performed to find out if the cancer has spread and, if so, to what extent. Treatment depends on the stage of the cancer.

Surgery is the main treatment for colon cancer. It requires removal of the cancerous tumor and nearby colon or rectum tissue, and it may also involve nearby lymph nodes. In most cases, the healthy portions of the colon or rectum are reconnected; sometimes they cannot be joined. In this case, a temporary or permanent colostomy is necessary. This is a surgical opening through the abdomen into the colon; body waste can exit here into a special bag.

Radiation therapy is the use of radiation to kill cancer cells and shrink tumors. It is directed at the site of the tumor from a source outside the body. It is used alone or in combination with chemotherapy in rectal cancer.

Chemotherapy uses drugs to kill cancer cells. It may be given in many forms, including pill, injection, and via a catheter. The drugs enter the bloodstream and travel through the body, killing mostly cancer cells. Some healthy cells can also be killed.

PREVENTION AND OUTCOMES

The cause of most colorectal cancer is not known. However, it is possible to prevent many colon cancers by finding and removing polyps that could become cancerous. Beginning at age fifty, both men and women at average risk for the development of colorectal cancer should follow one of these five screening options: a yearly fecal occult blood test or fecal immunochemical test, a flexible sigmoidoscopy every three to five years, a yearly fecal occult blood test or fecal immunochemical test plus flexible sigmoidoscopy every five years, a double contrast barium enema (X rays of the colon and rectum) every five years, or a colonoscopy every ten years. Patients should be sure to discuss these cancer screening tools with their doctors to see which options are best for them.

Individuals with any of the following risk factors should begin colorectal cancer screening earlier, at age forty, and/or undergo screening more often: a strong family history of colorectal cancer or polyps, a known family history of hereditary colorectal cancer syndromes, a personal history of colorectal cancer or adenomatous polyps, and a personal history of chronic inflammatory bowel disease.

Laurie LaRusso, M.S., ELS;
reviewed by Daus Mahnke, M.D.
"Etiology and Genetics" by Jeffrey A. Knight, Ph.D.

FURTHER READING

Casciato, Dennis A., ed. *Manual of Clinical Oncology.* 6th ed. Philadelphia: Wolters Kluwer Health/ Lippincott Williams & Wilkins, 2009.

EBSCO Publishing. *DynaMed: Colon Carcinoma.* Ipswich, Mass.: Author, 2009. Available through http://www.ebscohost.com/dynamed.

_____. *Health Library: Colon Cancer.* Ipswich, Mass.: Author, 2009. Available through http://www.ebscohost.com.

Levin, Bernard, et al., eds. *American Cancer Society's Complete Guide to Colorectal Cancer.* Atlanta: American Cancer Society, 2006.

Potter, John D., and Noralane M. Lindor, eds. *Genetics of Colorectal Cancer.* New York: Springer, 2009.

Skarin, Arthur T., Jeffrey Meyerhardt, and Mark P. Saunders, eds. *Colorectal Cancer: Dana-Farber Cancer Institute Handbook.* Maryland Heights, Mo.: Elsevier/Mosby, 2007.

U.S. Preventive Services Task Force. "Screening for Colorectal Cancer: U.S. Preventive Services Task Force Recommendation Statement." *Annals of Internal Medicine* 149, no. 9 (November 4, 2008): 627-637.

WEB SITES OF INTEREST

American Cancer Society
http://www.cancer.org

Canadian Cancer Society
http://www.cancer.ca

CancerCare
http://www.cancercare.org

Colorectal Cancer Association of Canada
http://www.colorectal-cancer.ca

Genetics Home Reference
http://ghr.nlm.nih.gov

National Cancer Institute
http://www.cancer.gov

See also: *APC* gene testing; Cancer; Familial adenomatous polyposis; Genetic screening; Hereditary diseases; Mutagenesis and cancer; Mutation and mutagenesis; Oncogenes.

Color blindness

CATEGORY: Diseases and syndromes

DEFINITION

Color blindness is a condition in people whose eyes lack one or more of the three color receptors present in most human eyes. It is an important condition to understand because so many people experience it to some degree. It is also a window into the inner workings of the eye and a marvelous example of the workings of Mendelian genetics.

RISK FACTORS

Approximately 1.2 percent of males and 0.02 percent of females are protanopes (lack L cones); 1.5 percent of males and 0.01 percent of females are deuteranopes (lack M cones); but only 0.001 percent of males and females are tritanopes (lack S cones).

ETIOLOGY AND GENETICS

Light-sensitive structures in the retina called cones are the basis for color vision. A person with normal vision can distinguish seven pure hues (colors) in the rainbow: violet, blue, cyan, green, yellow, orange, and red. People with normal vision are trichromats, meaning that they have three types of cones: L, M, and S, named for particular sensitivities to light of long, medium, and short wavelengths. The human vision system detects color by comparing the relative rates at which the L, M, and S cones react to light. For example, yellow light causes the M and L cones to signal at about the same rate, and the person "sees" yellow. Strangely, the right amounts of green and red stimulate these cones in the same fashion, and the person will again see the color yellow even though there is no yellow light present. Since people have only three types of color receptors, it takes the proper mix of intensities of only three primary colors to cause a person to "see" all the colors of the rainbow. A tiny droplet of water on the screen of a color television or computer monitor will act like a magnifying lens and reveal that the myriad colors that are displayed are formed from tiny dots of only blue, green, and red.

People are referred to as "color blind" if they are dichromats, that is, if they have only two of the three types of cones. Tritanopes cannot distinguish between blue (especially greenish shades) and yellow. The genetic code for the S pigment lies on chromosome 7. The fact that the S pigment gene lies on an autosome explains why yellow-blue color blindness is manifested equally in males and females. The inheritance pattern is that of an autosomal dominant trait: Only one arm of the two arms of chromosome 7 has the defective allele in the affected parent, and since there is a 50 percent chance a child will receive the defective arm, 50 percent of the children will inherit the defect. In fact, the trait is often incompletely expressed, so that the majority of affected individuals retain some reduced S-cone function.

Anomalous trichromats are more common than dichromats. They need three primary colors to match the hues of the rainbow, but they match them with different intensities than normal trichromats do because the peak sensitivities of their cones occur at wavelengths slightly different from normal. Their color confusion is similar to that of the dichromats, but less severe. About 1 percent of males and 0.03 percent of females have anomalous L cones, while 4.5 percent of males and 0.4 percent of females have anomalous M cones. The fact that far more males than females have some degree of red-green color blindness implies that the genetic information for the pigments in L and M cones lies on the X chromosome.

The gene structures for M-cone and L-cone pigments are 96 percent the same, so it is likely that one began as a mutation of the other. Small mutations in either gene can slightly shift the color of peak absorption in the cones and produce an anomalous trichromat. Generally these mutations make M and L cones more alike. The similarity between the genes and the fact that they are adjacent to each other on the X chromosome can lead to a variety of copying errors during meiosis. People with normal color vision have one L-cone gene and one to three M-cone genes. The complete omission of either type of gene will result in severe red-green color blindness: protanopia or deuteranopia. Hybrid genes that are a combination of L-cone and M-cone genes lead to less severe types of red-green color blindness, especially if there is also a normal copy of the gene present.

Red-green color blindness follows an X-gene recessive inheritance pattern. Suppose that a man has a defective X gene (and is therefore color blind) and a woman is normal. Their male children are normal because they inherited their X genes from their mother, but their female children will be carriers because they had to inherit one X gene from their father. If the daughters married normal men, 50 percent of the grandsons got the defective gene from their mothers and were color blind, and 50 percent of the grandsons were normal. Likewise, 50 percent of the granddaughters were normal and 50 percent inherited the defective gene from their mothers and became carriers.

SYMPTOMS

Dichromats can match all of the colors they see in the rainbow by mixing only two primary colors of light, but they see fewer (and different) hues in the rainbow than a person with normal vision. Protanopes and deuteranopes cannot distinguish between red and green. More exactly, protanopes tend to confuse reds, grays, and bluish blue-greens, while deuteranopes tend to confuse purples, grays, and greenish blue-greens.

SCREENING AND DIAGNOSIS

The test most often used to diagnose red-green color blindness is called the Ishihara color test. It consists of a series of pictures of colored spots in which a figure (such as a number or symbol) is embedded using in a slightly different color. Those with normal color vision can easily distinguish the figure within the image, but those with color deficiencies cannot.

TREATMENT AND THERAPY

While color blindness cannot be cured, devices such as tinted filters or contact lenses can help an individual distinguish between different colors. Their practical use is somewhat limited, however. Computer software and cybernetic devices can also aid those born with this condition.

PREVENTION AND OUTCOMES

Color blindness cannot be prevented, but those born with it may manage so well that they may be unaware of the condition if not tested for it.

Charles W. Rogers, Ph.D.

FURTHER READING

Hsia, Yun, and C. H. Graham. "Color Blindness." In *The Science of Color.* Vol. 2 in *Readings on Color,* edited by Alex Byrne and David R. Hilbert. Cambridge, Mass.: MIT Press, 1997. A description of the genetics of color blindness aimed at students with a good science background. Includes a series of color plates.

McIntyre, D. A. *Colour Blindness: Causes and Effects.* Chester, England: Dalton, 2002. McIntyre, who is color blind, provides an introductory overview of this vision deficiency. The first part of the book explains the different types of color vision deficiency and the workings of color blindness tests; the second part focuses on the effects of color blindness, describing how color-blind people see their world, live their everyday lives, the professions for which they are best suited, and the techniques that may improve their vision.

Medeiros, John A. *Cone Shape and Color Vision: Unification of Structure and Perception.* Blountsville, Ala.: Fifth Estate, 2006. Provides a detailed description of the role of cones in human color vision that challenges accepted ideas about the subject.

Nathans, Jeremy. "The Genes for Color Vision." In *The Science of Color.* Vol. 2 in *Readings on Color,* edited by Alex Byrne and David R. Hilbert. Cambridge, Mass.: MIT Press, 1997. An account of how the genes for color blindness were discovered that is understandable to students with a good science background. Includes a series of color plates.

Rosenthal, Odeda, and Robert H. Phillips. *Coping with Color Blindness.* Garden City Park, N.Y.: Avery, 1997. A description of color blindness aimed at nonspecialists and covering causes, testing, and coping strategies.

Wagner, Robert P. "Understanding Inheritance: An Introduction to Classical and Molecular Genetics." In *The Human Genome Project: Deciphering the Blueprint of Heredity,* edited by Necia Grant Cooper. Mill Valley, Calif.: University Science Books, 1994. A superb, well-illustrated discussion of Mendelian genetics and disorders.

WEB SITES OF INTEREST

Causes of Color
http://webexhibits.org/causesofcolor

A good introduction to light and color, including the genetics of color blindness. Includes demonstrations of how a scene looks to people with different types of color blindness and how color blindness tests are constructed.

Colblindor

http://www.colblindor.com

A site about color blindness that features facts and articles about the color vision deficiency and tests for color blindness, among other information.

Genetics Home Reference, Genetics Home Reference

http://ghr.nlm.nih.gov/condition =colorvisiondeficiency

A fact sheet about color vision deficiency that provides information about the genes related to it, how people inherit it, and links to additional sources of information.

Howard Hughes Medical Institute, Seeing, Hearing, and Smelling the World

http://www.hhmi.org/senses

This site includes the articles "Color Blindness: More Prevalent Among Males" and "How Do We See Colors?"

Medline Plus, Color Blindness

http://www.nlm.nih.gov/medlineplus/ency/ article/001002.htm

This page from the Medline Plus encyclopedia provides an overview of the deficiency.

See also: Classical transmission genetics; Congenital defects; Dihybrid inheritance; Hereditary diseases; Monohybrid inheritance.

Complementation testing

CATEGORY: Techniques and methodologies
SIGNIFICANCE: Complementation testing is used to determine whether or not two mutations occur within the same gene.

KEY TERMS

allele: a form of a gene; each gene (locus) in most organisms occurs as two copies called alleles

cistron: a unit of DNA that is equivalent to a gene; it encodes a single polypeptide

inborn errors of metabolism: conditions that result from defective activity of an enzyme or enzymes involved in the synthesis, conversion, or breakdown of important molecules within cells

locus (pl. loci): the location of a gene, often used as a more precise way to refer to a gene; each locus occurs as two copies called alleles

FINDING MUTATIONS

Most traits are the result of products from several genes. Mutations at any one of these genes may produce the same mutant phenotype. If the same mutant phenotype is observed in two different strains of an organism, there is no way, using simple observation, to determine whether this shared mutant phenotype represents a mutation in the same or different genes, or loci. One way of solving this problem is through complementation testing. If the mutations are alleles of the same locus, then a cross between mutant individuals from the two strains will only produce offspring with the mutant phenotype. In genetic terms, they fail to complement each other and are therefore members of the same complementation group. If from the same cross, all the offspring are normal; the two mutations are at the same locus and they are said to complement each other. Researchers often want to define multiple alleles of a single gene in order to understand the gene's function better.

Often a researcher is interested in the genetic control of a particular biological process, such as the biochemistry of eye color in fruit flies. As a first step, researchers often screen large numbers of individuals to find abnormal phenotypes involving the process in which they are interested. For instance, researchers studying eye color in fruit flies may screen hundreds of thousands of fruit flies for abnormal eye colors. Complementation testing is then used to organize the mutations into complementation groups.

COMPLEMENTATION TESTING AND INBORN ERRORS OF METABOLISM

Human genetic diseases that affect the function of cellular enzymes are known as inborn errors of metabolism and were defined by Sir Archibald Garrod long before DNA was determined to be the hereditary material. Garrod studied families with alkaptonuria, a disease that causes urine to turn dark upon exposure to air. He determined that this biochemical defect was inherited in a simple Mendelian fashion.

George Beadle and Edward Tatum studied mutant strains of *Neurospora* and expanded on Garrod's work. They used radiation to generate random mu-

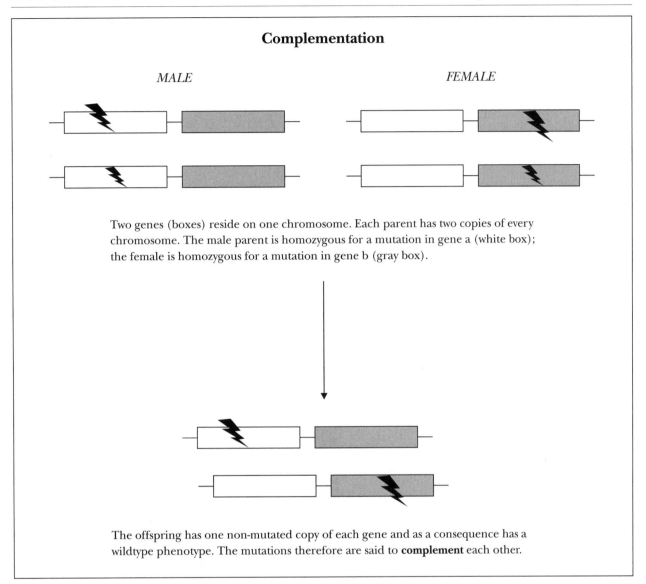

Complementation

MALE   *FEMALE*

Two genes (boxes) reside on one chromosome. Each parent has two copies of every chromosome. The male parent is homozygous for a mutation in gene a (white box); the female is homozygous for a mutation in gene b (gray box).

The offspring has one non-mutated copy of each gene and as a consequence has a wildtype phenotype. The mutations therefore are said to **complement** each other.

tations that resulted in strains of *Neurospora* that could not grow without specific nutritional supplements (essentially creating yeast with inborn errors of metabolism). Some of the mutant strains required the addition of a specific amino acid to the media. Each mutant strain had its own specific requirements for growth, and each strain was shown to have a single defective step in a metabolic pathway. When strains that had different defects were grown together, they were able to correct each other's metabolic defect. This correction was termed metabolic complementation. Using complementation tests, Beadle and Tatum were able to es-

tablish the number of genes required for a particular pathway. These studies formed the basis for the "one gene-one enzyme hypothesis": Each gene encodes a single enzyme required for a single step in a metabolic pathway. This hypothesis has since been renamed the "one gene-one polypeptide hypothesis" because some enzymes consist of multiple polypeptides, each of which is encoded by a single gene.

The Biochemical Basis for Complementation Testing

Complementation testing is useful for locating and identifying the genes affected by recessive or loss-of-function alleles. A researcher crosses two organisms that are each homozygous for a recessive mutation. If these two alleles affect the same gene, they will not complement each other, because the first-generation (F_1) offspring will inherit one mutant copy of the gene from one parent and a second mutant copy of the gene from the other parent, thus having no normal copies of the gene. If the mutations are alleles of two different genes, genes *A* and *B*, the F_1 offspring will receive a normal copy of *A* and a mutant copy of *B* from one parent and a mutant copy of *A* and a normal copy of *B* from the other, thus having one normal copy of each of the two genes and having a wild-type (normal) phenotype. The mutations are said to complement each other.

If a scientist is interested in a particular gene, obtaining as many alleles of that gene as possible will lead to a better understanding of how the gene works and what parts of the gene are essential for function. One way to identify new alleles of a gene is through an F_1 noncomplementation screen. In this type of screen, the researcher treats the model organism with radiation or chemicals to increase the rate of mutation. Any individuals from the screen that segregate the desired phenotype (white eyes, for example) in a Mendelian fashion are crossed with individuals carrying a known mutation in the gene of interest. If the progeny of this cross have white eyes (the mutant phenotype), then the two mutations have failed to complement each other and are most likely alleles of the same gene. Such noncomplementation screens have been used to identify genes involved in a wide variety of processes ranging from embryo development in fruit flies to spermatogenesis in *Caenorhabditis elegans*.

Michele Arduengo, Ph.D., ELS, and Bryan Ness, Ph.D.

FURTHER READING

Griffiths, Anthony J. F., et al. "Defining the Set of Genes by Using the Complementation Test." *Introduction to Genetic Analysis*. 9th ed. New York: W. H. Freeman, 2008. Text and illustrations describe how the complementation test is used in gene analysis.

Hartl, D. L., and Elizabeth W. Jones. *Genetics: Analysis of Genes and Genomes*. 7th ed. Sudbury, Mass.: Jones and Bartlett, 2009. An excellent introductory genetics textbook.

Hawley, R. Scott, and Michelle Y. Walker. "The Complementation Test." In *Advanced Genetic Analysis: Finding Meaning in a Genome*. Malden, Mass.: Blackwell, 2003. Illustrated textbook providing analytical tools for gene mutation and function, gene mapping, and chromosome segregation.

Lewin, Benjamin. *Genes IX*. Sudbury, Mass.: Jones and Bartlett, 2007. Includes a summary of complementation and a discussion of complementation in bacterial systems.

Tropp, Burton E., and David Freifelder. "Genetic Analysis in Molecular Biology." In *Molecular Biology: Genes to Proteins*. 3d ed. Sudbury, Mass.: Jones and Bartlett, 2008. Includes a description of complementation as a form of genetic analysis.

WEB SITE OF INTEREST

Genes and Mutations, Complementation Testing
http://www.ndsu.nodak.edu/instruct/mcclean/plsc431/mutation/mutation5.htm

Philip McClean, a professor in the department of plant science at North Dakota State University, provides a page describing complementation testing in his online explanation of genes and mutations.

See also: Biochemical mutations; Chemical mutagens; Chromosome mutation; Inborn errors of metabolism; Linkage maps; Mutation and mutagenesis; Model organism: *Caenorhabditis elegans*; Model organism: *Neurospora crassa*.

Complete dominance

CATEGORY: Classical transmission genetics

SIGNIFICANCE: Complete dominance represents one of the classic Mendelian forms of inheritance. In an individual that is heterozygous for a trait, the allele that displays complete dominance will determine the phenotype of the individual. Knowing whether the pattern of expression of a trait is dominant or recessive helps in making predictions concerning the inheritance of a particular genetic condition or disorder in a family's history.

KEY TERMS

alleles: different forms of a gene at a specific locus; for each genetic trait there are typically two alleles in most organisms, including humans

genotype: a description of the alleles at a gene locus

heterozygote: an individual with two different alleles at a gene locus

homozygote: an individual with two like alleles at a gene locus

incomplete dominance: the expression of a trait that results when one allele can only partly dominate or mask the other

locus (pl. loci): a gene, located at a specific location on a chromosome, which in humans and many other organisms occurs in the same location on homologous chromosomes

phenotype: the observed expression of a gene locus in an individual

The Discovery and Definition of Dominance

Early theories of inheritance were based on the idea that fluids carrying materials for the production of a new individual were transmitted to offspring from the parents. It was assumed that substances in these fluids from the two parents mixed and that the children would therefore show a blend of the parents' characteristics. For instance, individuals with dark hair mated to individuals with very light hair were expected to produce offspring with medium-colored hair. The carefully controlled breeding studies carried out in the 1700's and 1800's did not produce the expected blended phenotypes, but no other explanation was suggested until Gregor Mendel proposed his model of inheritance. In the 1860's, Mendel repeated studies using the garden pea and obtained the same results seen by other investigators, but he counted the numbers of each type produced from each mating and developed his theory based on those observations.

One of the first observations Mendel dealt with was the appearance of only one of the parental traits in the first generation of offspring (the first filial, or F_1, generation). For example, a cross of tall plants and dwarf plants resulted in offspring that were all tall. Mendel proposed that the character expression (in this case height) was controlled by a determining "factor," later called the "gene." He then proposed that there were different forms of this controlling factor corresponding to the different expressions of the characteristic and termed these "alleles." In the case of plant height, one allele produced tall individuals and the other produced dwarf individuals. He further proposed that in the cross of a tall (*D*) plant and a dwarf (*d*) plant, each parent

contributed one factor for height, so the offspring were *Dd*. (Uppercase letters denote dominant alleles, while lowercase letters denote recessive alleles.) These plants contained a factor for both the tall expression and the dwarf expression, but the plants were all tall, so "tall" was designated the dominant phenotype for the height trait.

Mendel recognized from his studies that the determining factors occurred in pairs—each sexually reproducing individual contains two alleles for each inherited characteristic. When he made his crosses, he carefully selected pure breeding parents that would have two copies of the same allele. In Mendel's terminology, the parents would be homozygous: A pure tall parent would be designated *DD*, while a dwarf parent would be designated *dd*. His model also proposed that each parent would contribute one factor for each characteristic to each offspring, so the offspring of such a mating should be *Dd* (heterozygous). The tall appearance of the heterozygote defines the character expression (the phenotype) as dominant. Dominance of expression for any characteristic cannot be guessed but must be determined by observation. When variation is observed in the phenotype, heterozygous individuals must be examined to determine which expression is observed. For phenotypes that are not visible, such as blood types or enzyme activity variations, a test of some kind must be used to determine which phenotype expression is present in any individual.

Mendel's model and the appearance of the dominant phenotype also explains the classic 3:1 ratio observed in the second (F_2) generation. The crossing of two heterozygous individuals (*Dd × Dd*) produces a progeny that is ¼ *DD*, ½ *Dd*, and ¼ *dd*. Because there is a dominant phenotype expression, the ¼ *DD* and the ½ *Dd* progeny all have the same phenotype, so ¾ of the individuals are tall. It was this numerical relation that Mendel used to establish his model of inheritance.

The Functional Basis of Dominance

The development of knowledge about the molecular activity of genes through the 1950's and 1960's provided information on the nature of the synthesis of proteins using the genetic code passed on in the DNA molecules. This knowledge has allowed researchers to explain variations in phenotype expression and to explain why a dominant allele behaves the way it does at the functional level. An enzyme's function is determined by its structure, and that structure is coded for in the genetic information. The simplest situation is one in which the gene product is an enzyme that acts on a specific chemical reaction that results in a specific chemical product, the phenotype. If that enzyme is not present or if its structure is modified so that it cannot properly perform its function, then the chemical action will not be carried out. The result will be an absence of the normal product and a phenotype expression that varies from the normal expression. For example, melanin is a brown pigment produced by most animals. It is the product of a number of chemical reactions, but one enzyme early in the process is known to be defective in albino animals. Lacking normal enzyme activity, these animals cannot produce melanin, so there is no color in the skin, eyes, or hair. When an animal has the genetic composition *cc* (*c* designates colorless, or albino), it has two alleles that are the same, and neither can produce a copy of the normal enzyme. Animals with the genetic composition *CC* (*C* designates colored, or normal) have two copies of the allele that produces normal enzymes and are therefore pigmented. When homozygous normal (*CC*) and albino (*cc*) animals are crossed, heterozygous (*Cc*) animals are produced. The *c* allele codes for production of an inactive enzyme, while the *C* allele codes for production of the normal, active enzyme. The presence of the normal enzyme promotes pigment production, and the animal displays the pigmented phenotype. The presence of pigment in the heterozygote leads to the designation that the pigmented phenotype is dominant to albinism or, conversely, that albinism is a recessive phenotype because it is seen only in the homozygous (*cc*) state.

The same absence or presence of an active copy of an enzyme explains why blood types A and B are both dominant to blood type O. When an *A* allele or a *B* allele is present, an active enzyme promotes the production of a substance that is identified in a blood test; the blood type A expression or the blood type B expression is seen. When neither of these alleles is present, the individual is homozygous *OO*. There is no detectable product present, and the blood test is negative; therefore, the individual has blood type O. When the *A* allele and the *B* allele are both present in a heterozygous individual, each produces an active enzyme, so both the *A* and the *B* product are detected in blood tests; such an individual has blood type AB. The two phenotypes are both expressed in the heterozygote, a mode of gene expression called codominance.

When there are a number of alleles present for the expression of a characteristic, a dominance relation among the phenotype expressions can be established. In some animal coats, very light colors result from enzymes produced by a specific allele that is capable of producing melanin but at a much less efficient rate than the normal version of the enzyme. In the rabbit, chinchilla (c^{ch}) is such an allele. In the Cc^{ch} heterozygote, the normal allele (*C*) produces a normal, rapidly acting enzyme, and the animal has normal levels of melanin. The normal pigment phenotype expression is observed because the animals are dark in color, so this expression is dominant to the chinchilla phenotype expression. In the heterozygote $c^{ch}c$, the slow-acting enzyme produced by the c^{ch} allele is present and produces pigment, in a reduced amount, so the chinchilla phenotype expression is observed and is dominant to the albino phenotype expression. The result is a dominance hierarchy in which the normal pigment phenotype is dominant to both the chinchilla and the albino phenotypes, and the chinchilla expression is dominant to the albino expression.

It is important to note that the dominant phenotype is the result of the protein produced by each allele. In the previous examples, both the albino allele and the chinchilla allele produce a product—a version of the encoded enzyme—but the normal allele produces a version of the enzyme that produces more pigment. The relative ability of the enzymes to carry out the function determines the observed phenotype expression and therefore the dominance association. The *C* allele does not inhibit the activity of either of the other two alleles or their enzyme products, and the allele does not, therefore, show dominance; rather, its enzyme expression does.

DOMINANT MUTANT ALLELES

Dominance of a normal phenotype is fairly easy to explain at the level of the functioning protein because the action of the normal product is seen, but dominance of mutant phenotypes is more difficult. Polydactyly, the presence of extra fingers on one hand or extra toes on one foot, is a dominant phenotype. The mechanism that leads to this expression and numerous other developmental abnormalities is not yet understood. One insight comes from the genetic expression of enzymes that are composed of two identical polypeptide subunits. In this situation, the gene locus codes for one polypeptide, but it takes two polypeptide molecules joined together to form a functional enzyme molecule. In order to function normally, both of the polypeptide subunits must be normal. A heterozygote can have one allele coding for a normal polypeptide and the other allele coding for a mutant, nonfunctional polypeptide. These polypeptides will join together at random to form the enzyme. The possible combinations will be defective-defective, which results in a nonfunctional enzyme; defective-normal, which also results in a nonfunctional enzyme; and normal-normal, which is a normal, functional enzyme. The majority of the enzyme molecules will be nonfunctional, and their presence will interfere with the action of the few normal units. The normal function will be, at best, greatly reduced, and the overall phenotype will be abnormal. One form of hereditary blindness is dominant because the presence of abnormal proteins interferes with the transport of both protein types across a membrane to their proper location in the cells that react to light. The abnormal phenotype appears in the heterozygote, so the abnormal phenotype is dominant. A number of human disease conditions, including some forms of cancer, display a dominant mode of inheritance.

Sometimes a trait that appears to be dominant is actually more complex. The Manx trait in cats, which results in a very short, stubby tail, occurs only in heterozygous individuals. On the surface, this would appear to be a simple case of dominance, where the Manx allele, *T*, is dominant to the normal tail allele, *t*. Recall that when two heterozygotes are mated, the expected phenotype ratio in the offspring is 3:1, dominant:recessive. If two Manx cats are mated, the phenotype ratio in the offspring is 2:1, Manx:normal, because kittens that are homozygous for the Manx allele (*TT*) die very early in development and are reabsorbed by the mother cat. Therefore, the Manx allele does not display complete dominance, but rather incomplete dominance. The Manx allele is lethal in the homozygous state and causes a short, stubby tail in the heterozygous state. This occurs because the Manx allele causes a developmental defect that affects spinal development. If one normal allele is present, the spine develops enough for the cat to survive, although it will display the Manx trait. In mutant homozygotes (*TT*) the spine is unable to develop, which proves lethal to the developing fetus.

IMPACT AND APPLICATIONS

One of the aims of human genetic research is to find cures for inherited conditions. When a condition shows the recessive phenotype expression, treatment may be effective. The individual lacks a normal gene product, so supplying that product can have a beneficial effect. This is the reason for the successful treatment of diabetes using insulin. There are many technical issues to be considered in such treatments, but current successes give hope for the treatment of other recessive genetic conditions.

Dominant disorders, on the other hand, will be much more difficult to treat. An affected heterozygous individual has a normal allele that produces normal gene product. The nature of the interactions between the products results in the defective phenotype. Supplying more normal product may not improve the situation. A great deal more knowledge about the nature of the underlying mechanisms will be needed to make treatment effective.

D. B. Benner, Ph.D.; updated by Bryan Ness, Ph.D.

FURTHER READING

Campbell, Neil A., and Jane Reece. *Biology*. 8th ed. San Francisco: Pearson, Benjamin Cummings, 2008. A college-level biology text that provides an introduction to many topics relating to genetics.

Ringo, John. "Genes, Environment, and Interactions." In *Fundamental Genetics*. New York: Cambridge University Press, 2004. The concept of genetic dominance, including complete dominance, is discussed in this chapter.

Russell, Peter J. *Fundamentals of Genetics*. 2d ed. San Francisco: Benjamin Cummings, 2000. Introduces the three main areas of genetics: transmission genetics, molecular genetics, and population and quantitative genetics. Reflects advances in the field, such as the structure of eukaryotic chromosomes, alternative splicing in the production of mRNAs, and molecular screens for the isolation of mutants.

Snustad, D. Peter, and Michael J. Simmons. "Incomplete and Complete Dominance." In *Principles of Genetics*. 5th ed. Hoboken, N.J.: John Wiley & Sons, 2009. This textbook provides an explanation of complete dominance within the broader context of allelic variation and gene function.

Strachan, Tom, and Andrew Read. *Human Molecular Genetics 3.* 3d ed. New York: Garland Press, 2004. Provides introductory material on DNA and chromosomes and describes principles and applications of cloning and molecular hybridization. Surveys the structure, evolution, and mutational instability of the human genome and human genes. Examines mapping of the human genome, study of genetic diseases, and dissection and manipulation of genes.

WEB SITES OF INTEREST

Scitable
http://www.nature.com/scitable/topicpage/
Genetic-Dominance-Genotype-Phenotype
-Relationships-489

Scitable, a library of science-related articles compiled by the Nature Publishing Group, features the article "Genetic Dominance: Genotype-Phenotype Relationships," which explains complete dominance and other aspects of the genetic concept of dominance.

Suite 101.com, DNA and Complete Genetic Dominance
http://humangenetics.suite101.com/article.cfm/
complete_genetic_dominance

Suite 101.com, an online compilation of articles, contains an article describing complete genetic dominance with links to additional information.

See also: Biochemical mutations; Dihybrid inheritance; Epistasis; Incomplete dominance; Mendelian genetics; Monohybrid inheritance; Multiple alleles.

Congenital adrenal hyperplasia

CATEGORY: Diseases and syndromes
ALSO KNOWN AS: Adrenal hyperplasia; CAH

DEFINITION

Congenital adrenal hyperplasia (CAH) causes the body to have low levels of certain hormones. The disorder can be life-threatening. With proper treatment, however, people with CAH can lead normal, healthy lives.

Parents who suspect their child may have CAH can talk to their child's doctor about treatment op-

tions. Women who are pregnant and suspect their children may be affected by CAH can ask their doctors about testing. Prenatal treatment may also be available; this can correct problems before the child is born.

RISK FACTORS

CAH is an inherited disorder. Most people who carry the gene for CAH do not have the disorder. Individuals who have someone in their immediate family with CAH can talk to their doctors about genetic testing; this is important if they are expecting or planning to have a child.

ETIOLOGY AND GENETICS

"CAH" is the term applied to a family of diseases that may result from mutations in any one of several genes on five different chromosomes. In each case, the mutant gene causes a deficiency in one of the enzymes involved in the biosynthesis of cortisol and aldosterone. All these mutant genes are inherited in an autosomal recessive fashion, which means that both copies of a particular gene must be deficient in order for the individual to be afflicted.

Typically, an affected child is born to two unaffected parents, both of whom are carriers of the recessive mutant allele. The probable outcomes for children whose parents are both carriers are 75 percent unaffected and 25 percent affected. If one parent has CAH and the other is a carrier, there is a 50 percent probability that each child will be affected. A simple blood test is available to screen for and identify the most common carrier phenotype.

Approximately 90 percent of cases of CAH result from mutations in the gene that encodes the enzyme 21-hydroxylase, which is found on the short arm of chromosome 6 at position 6p21.3. An additional 5-8 percent of cases are caused by 11-beta-hydroxylase deficiency, resulting from mutations in a gene on the long arm of chromosome 8 at position 8q21-22. The remaining cases result from mutant genes located on chromosomes 1 (at position 1p13), 10 (at position 10q24.3), or 15 (at position 15q23-24).

In all cases, the mutant genes encode enzymes that either partially or completely block the conversion of adrenal precursors into cortisol and aldosterone. As a result, there are markedly increased levels of these precursor hormones, which include progesterone and dehydroepiandrosterone (DHEA).

It is the presence of excess DHEA that causes the masculinization of female infants during development.

SYMPTOMS

A parent whose child has any of the symptoms of CAH should not assume it is due to CAH. These symptoms may be caused by other, less serious health conditions, or by a reaction to medication. Parents should see their child's doctor if their child experiences any one of the symptoms.

The most obvious symptom in newborn girls is the unusual appearance of genitalia. Parts of the vagina may be enlarged, and it may resemble a penis.

There are no obvious visual symptoms in newborn boys. Boys as young as two or three years old may begin to show signs of puberty. They may become very muscular, experience penis growth, develop pubic hair, and have a deepening voice.

Both boys and girls may have excessive facial and/or body hair. They may also grow very fast compared to other children their age; most will stop growing sooner than their peers and are often relatively short as adults. Boys and girls may also have difficulty fighting respiratory infections and illnesses; have high blood pressure; and develop dehydration, low blood pressure, a low sodium level, and a high potassium level in the blood due to a condition known as "salt wasting," caused by lack of aldosterone. They may also exhibit high blood pressure with low blood potassium, poor feeding and vomiting, a failure to gain weight, short stature, and severe acne.

SCREENING AND DIAGNOSIS

A pregnant woman's doctor will ask about her pregnancy and medical history. The doctor may perform blood or urine tests to check hormone levels, in particular cortisol and aldosterone, and/or amniocentesis, in which a sample of the fluid that surrounds the baby in the womb is collected and examined.

If a child has already been born, the doctor will ask about the child's symptoms and medical history, and a physical exam will be conducted. The doctor may take a small amount of blood and urine to test for hormone levels.

In borderline cases, genetic testing is done with blood tests. A child may also be referred to a specialist. An endocrinologist focuses on hormones; a pediatric urologist focuses on the urinary system in children.

TREATMENT AND THERAPY

Parents should talk with their doctors about the best plans for them and/or their children. Prenatal treatment with dexamethasone is often given when CAH is found before childbirth. The drug is administered early in the pregnancy, and it is usually taken as a pill or liquid. The doctor will determine the correct dosage.

Most children born with CAH need to take hormone replacement all of their lives. With constant monitoring, no side effects are expected. The goal of the treatment is to keep the body's normal balance of hormones. It is important to know that during stressful situations the dose of cortisol needs to be increased. Individuals should follow their doctors' directions.

Salt wasting illness will often require additional table salt in a patient's diet. Cortisol can increase the appetite, but this can lead to excess weight gain. Calorie intake should be followed closely.

In most cases, surgery can correct unusually formed genitalia, if desired. Surgery is often done when the child is between one and three years old.

PREVENTION AND OUTCOMES

CAH is an inherited disorder. There are no preventive measures, except the use of dexamethasone.

Diane Stresing; reviewed by Rosalyn Carson-DeWitt, M.D.
"Etiology and Genetics" by Jeffrey A. Knight, Ph.D.

FURTHER READING

Bachelot, A., et al. "Hormonal Treatment of Congenital Adrenal Hyperplasia Due to 21-Hydroxylase Deficiency." *Annales D'endocrinologie* 68, no. 4 (September, 2007): 274-280.

Carlson, A. D., et al. "Congenital Adrenal Hyperplasia: Update on Prenatal Diagnosis and Treatment." *Journal of Steroid Biochemistry and Molecular Biology* 69, nos. 1-6 (April-June, 1999): 19-29.

Charmandari, Evangelia, George Chrousos, and Deborah P. Merke. "Classic Congenital Adrenal Hyperplasia." In *Adrenal Glands: Diagnostic Aspects and Surgical Therapy*, edited by Dimitrios Linos and Jon A. van Heerden. New York: Springer, 2005.

EBSCO Publishing. *Health Library: Adrenal Hyperplasia.* Ipswich, Mass.: Author, 2009. Available through http://www.ebscohost.com.

Homma, K., et al. "Elevated Urine Pregnanetriolone Definitively Establishes the Diagnosis of Classical 21-Hydroxylase Deficiency in Term and Preterm Neonates." *Journal of Clinical Endocrinology and Metabolism* 89, no. 12 (December, 2004): 6087-6091.

Lajic, S., et al. "Prenatal Treatment of Congenital Adrenal Hyperplasia." *European Journal of Endocrinology* 151, no. 3 (November, 2004): 63-69.

Meyer-Bahlburg, H. F., et al. "Cognitive and Motor Development of Children with and Without Congenital Adrenal Hyperplasia After Early-Prenatal Dexamethasone." *Journal of Clinical Endocrinology and Metabolism* 89, no. 2 (February, 2004): 610-614.

New, M. I., et al. "Prenatal Diagnosis for Congenital Adrenal Hyperplasia in 532 Pregnancies." *Journal of Clinical Endocrinology and Metabolism* 86, no. 12 (December, 2001): 5651-5657.

Ogilvie, C. M., et al. "Congenital Adrenal Hyperplasia in Adults: A Review of Medical, Surgical, and Psychological Issues." *Clinical Endocrinology* 64, no. 1 (January, 2006): 2-11.

Speiser, Phyllis W., ed. *Congenital Adrenal Hyperplasia.* Philadelphia: W. B. Saunders, 2001.

WEB SITES OF INTEREST

Cares Foundation: Congenital Adrenal Research Education and Support
http://www.caresfoundation.org

Congenital Adrenal Hyperplasia Education and Support Network
http://www.congenitaladrenalhyperplasia.org

Genetics Home Reference
http://ghr.nlm.nih.gov

The MAGIC Foundation
http://www.magicfoundation.org

Save Babies Through Screening Foundation of Canada
http://www.savebabiescanada.org/ehome.htm

See also: Adrenoleukodystrophy; Adrenomyelopathy; Chorionic villus sampling; Hereditary diseases.

Congenital defects

CATEGORY: Diseases and syndromes

SIGNIFICANCE: Congenital defects are malformations caused by abnormalities in embryonic or fetal development that may interfere with normal life functions or cause a less severe health problem. The defect may be morphological or biochemical in nature. Understanding the causes of birth defects has led to improved means of detection and treatment.

KEY TERMS

sensitive period: a critical time during development when organs are most susceptible to teratogens

teratogen: any agent that is capable of causing an increase in the incidence of birth defects

teratology: the science or study of birth defects

NORMAL DEVELOPMENT

In order to understand the causes of birth defects, it is necessary to have some understanding of the stages of normal development. If the time and sequence of development of each organ are not correct, an abnormality may result. It has been useful to divide human pregnancy into three major periods: the preembryonic stage, the embryonic stage, and the fetal stage.

The preembryonic stage is the first two weeks after fertilization. During this stage, the fertilized egg undergoes cell division, passes down the Fallopian tube, and implants in the uterine wall, making a physical connection with the mother. It is of interest to note that perhaps as many as one-half of fertilized eggs fail to implant, while half of those which do implant do not survive the second week. The second stage, the embryonic stage, runs from the beginning of the third week through the end of the eighth week. There is tremendous growth and specialization of cells during this period, as all of the body's organs are formed. The embryonic stage is the time during which most birth defects are initiated.

The fetal stage runs from the beginning of the ninth week to birth. Most organs continue their rapid growth and development during this final period of gestation leading up to birth. By the end of the eighth week, the embryo, although it has features of a human being, is only about 1 inch (2.54

centimeters) long. Its growth is amazing during this period, reaching 12 inches (30 centimeters) by the end of the fifth month and somewhere around 20 inches (50 centimeters) by birth. It is evident from the description of normal development that the changes the embryo and fetus undergo are very rapid and complicated. It is not unexpected that mistakes can happen, leading to congenital disorders.

CAUSES OF BIRTH DEFECTS

Throughout history, examples of birth defects have been described by all cultures and ethnic groups. Although the incidence of specific malformations may vary from group to group, the overall incidence of birth defects is probably similar in all people on earth. It is estimated that three out of every hundred newborns have some sort of major or minor disorder. An additional 2 to 3 percent have malformations that fully develop sometime after birth. When it is also realized that perhaps another 5 percent of all fertilized eggs have severe enough malformations to lead to an early, spontaneous abortion, the overall impact of birth defects is considerable.

Humans have long sought an explanation for why some couples have babies afflicted with serious birth defects. Such children were long regarded as "omens" or warnings of a bad event to come. The word "teratology" (Greek for "monster causing") was coined by scientists to reflect the connection of "monster" births with warnings. Frequently, ancient people sacrificed such babies. It was thought that such pregnancies resulted from women mating with animals or evil spirits. Maternal impression has long been invoked as an explanation for birth defects, and from early Greek times until more recent times, stories and superstitions abounded.

Of the birth defects in which a specific cause has been identified, it has been found that some are caused by genetic abnormalities, including gene mutations and chromosomal changes, while others are caused by exposure of the pregnant woman and her embryo or fetus to some sort of environmental toxin such as radiation, viruses, drugs, or chemicals.

EXAMPLES OF BIRTH DEFECTS

Many birth defects are caused by changes in the number or structure of chromosomes. The best-known chromosomal disorder is Down syndrome, which results from individuals having an extra chromosome 21, giving them forty-seven chromosomes rather than the normal forty-six. A person with Down syndrome characteristically has a flattened face, square-shaped ears, epicanthal folds of the eye, a short neck, poor muscle tone, slow development, and subnormal intelligence. Cystic fibrosis is an example of a defect caused by a single gene. Affected people inherit a recessive gene from each parent. The disorder is physiological in nature and results in a lack of digestive juices and the production of thick and sticky mucus that tends to clog the lungs, pancreas, and liver. Respiratory infections are common, and death typically occurs by the age of thirty. Cleft lip, or cleft palate, is multifactorial in inheritance (some cases are caused by chromosomal abnormalities or by single-gene mutations). Multifactorial traits are caused by many pairs of genes, each having a small effect, and are usually influenced by factors in the environment. The result is that such traits do not follow precise, predictable patterns in a family.

Genetic factors account for the great majority (perhaps 85 to 90 percent) of the birth defects in which there is a known cause. The remaining cases of known cause are attributed to maternal illness; congenital infections; exposure to chemicals, drugs, and medicines; and physical factors such as X rays, carbon dioxide, and low temperature. The "government warning" on liquor bottles informs pregnant women that if they drink alcohol during a sensitive period of prenatal development, they run the risk of having children with fetal alcohol syndrome. There is a wide variation in the effects of alcohol on a developing fetus. Alcohol exposure can lead to an increased frequency of spontaneous abortion, and it depresses growth rates, both before and after birth. Facial features of a child exposed to alcohol may include eye folds, a short nose, small mid-face, a thin upper lip, a flat face, and a small head. These characteristics are likely to be associated with mental retardation. Frequently, however, otherwise normal children have learning disorders and only a mild growth deficiency. Variation in the symptoms of prenatal alcohol exposure has made it difficult to estimate the true incidence of fetal alcohol syndrome. Estimates for the United States range from 1 to 3 per 1,000 newborns.

In 50 to 60 percent of babies born with a major birth disorder, no specific cause can be identified.

Because of this rather large gap in knowledge, nonscientific explanations about the causes of birth defects flourish. What is known is that most congenital defects, whether caused by a genetic factor or an environmental factor, are initiated during the embryonic period. It is also known that some disorders, such as learning disorders, frequently result from damage to the fetus during the last three months of pregnancy. Knowledge about what can be done by parents to avoid toxic exposure and activity that could cause birth defects is critical.

Donald J. Nash, Ph.D.

FURTHER READING

Berul, Charles I., and Jeffrey A. Towbin, eds. *Molecular Genetics of Cardiac Electrophysiology.* Boston: Kluwer Academic, 2000. Reviews research regarding single-cell electrophysiology, animal models, and hereditary diseases, including structural anomalies.

Edwards, Jesse E. *Jesse E. Edwards' Synopsis of Congenital Heart Disease.* Edited by Brooks S. Edwards. Armonk, N.Y.: Futura, 2000. A comprehensive review of more than thirty-five categories of congenital cardiac lesions. Aimed at students, clinicians, and health care providers.

Ferretti, Patrizia, et al., eds. *Embryos, Genes, and Birth Defects.* 2d ed. Hoboken, N.J.: Wiley, 2006. The first six chapters discuss concepts of genetics and other information that explains the developmental anomalies leading to birth defects, while the remainder of the book focuses on genetic and environmental factors that cause birth defects in specific organs.

Harvey, Richard P., and Nadia Rosenthal, eds. *Heart Development.* San Diego: Academic Press, 1999. A broad discussion of the molecular basis of cardiovascular development, including the lineage origins and morphogenesis of the developing cardiovascular system, the genetic dissection of cardiovascular development in a variety of model organisms, and the molecular basis of congenital heart defects.

Judd, Sandra J., ed. *Congenital Disorders Sourcebook.* 2d ed. Detroit: Omnigraphics, 2007. Describes the most common types of nonhereditary birth defects and disorders related to premature birth, gestational injuries, congenital infections, and birth complications, including disorders of the heart, brain, gastrointestinal tract, musculoskel-

etal system, urinary tract, and reproductive organs. Cerebral palsy, spina bifida, and fetal alcohol syndrome and their related complications are also discussed in detail.

Kramer, Gerri Freid, and Shari Maurer. *The Parent's Guide to Children's Congenital Heart Defects: What They Are, How to Treat Them, How to Cope with Them.* Foreword by Sylvester Stallone and Jennifer Flavin-Stallone. New York: Three Rivers Press, 2001. Collects the expertise of more than thirty leading experts in pediatric cardiology—cardiologists, surgeons, nurses, nutritionists, counselors, and social workers—to give detailed answers to parents' concerns about managing a child's heart defect.

Riccitiello, Robina, and Jerry Adler. "Your Baby Has a Problem." *Newsweek* 129, no. 9 (Spring/Summer, 1997): 46. Discusses how advances in medicine have reduced the number of birth defects and how surgeries have been designed to correct some birth defects before babies are born.

Rossen, Anne E. "Understanding Congenital Disorders." *Current Health* 18, no. 9 (May, 1992): 26. A useful article describing some congenital disorders, including some that are not apparent early in life, such as Huntington's disease. Some of the environmental factors causing congenital defects are also covered.

Tomanek, Robert J., and Raymond B. Runyan, eds. *Formation of the Heart and Its Regulation.* Foreword by Edward B. Clark. Boston: Birkhauser, 2001. Details the major events in heart development and their control via genes, cell-cell interactions, growth factors, and other contributing elements.

Wynbrandt, James, and Mark D. Ludman. *The Encyclopedia of Genetic Disorders and Birth Defects.* 3d ed. New York: Facts On File, 2008. Written for the general public, this guidebook offers clinical and research information on hereditary conditions and birth defects. More than six hundred alphabetically arranged and cross-referenced entries cover genetic anomalies, diagnostic procedures, causes of mutations, and high risk groups. Also includes an essay on the basics of genetic science and its medical applications.

WEB SITES OF INTEREST

Centers for Disease Control and Prevention
http://cdc.gov/ncbddd/bd
The CDC's Web site provides a page with links to

numerous resources about birth defects, including information about genetics, monitoring for and prevention of birth defects, specific birth defects, and the impact of smoking, alcohol, and medications on pregnancy.

Kids Health.org
http://kidshealth.org/parent/system/ill/birth_defects.html
Kids Health.org, a site with information about children's health development designed for both parents and children, contains an article about birth defects, with links to related articles.

March of Dimes Foundation
http://www.marchofdimes.com
Includes fact sheets and links to resources on birth defects.

Medline Plus
http://medlineplus.gov
Medline, sponsored by the National Institutes of Health, is one of the first stops for any medical question, and it offers information and references on most genetic diseases, birth defects, and disorders.

National Birth Defects Network
http://www.nbdpn.org
The Web site for this support group features information and other resources about birth defects.

National Institutes of Health, National Library of Medicine
http://www.nlm.nih.gov/medlineplus/birthdefects.html
Government site that includes dozens of links to resources on birth defects, with information on genetics, treatments, and statistics.

See also: Albinism; Androgen insensitivity syndrome; Apert syndrome; Brachydactyly; Carpenter syndrome; Cleft lip and palate; Color blindness; Consanguinity and genetic disease; Cornelia de Lange syndrome; Cri du chat syndrome; Crouzon syndrome; Cystic fibrosis; Developmental genetics; Down syndrome; Dwarfism; Edwards syndrome; Ellis-van Creveld syndrome; Fragile X syndrome; Hemophilia; Hereditary diseases; Hermaphrodites; Huntington's disease; Inborn errors of metabolism; Klinefelter syndrome; Metafemales; Mitochondrial diseases; Neural tube defects; Phenylketonuria (PKU); Prader-Willi and Angelman syndromes; Prion diseases: Kuru and Creutzfeldt-Jakob syndrome; Pseudohermaphrodites; Sickle-cell disease; Tay-Sachs disease; Thalidomide and other teratogens; Turner syndrome; XYY syndrome.

Congenital hypothyroidism

CATEGORY: Diseases and syndromes
ALSO KNOWN AS: Cretinism

DEFINITION

The thyroid is a gland in the lower neck. It makes hormones that regulate growth, brain development, and metabolism. Hypothyroidism is a low or absent production of these hormones. Congenital means the condition is present since birth. If this condition is not treated it can cause damage to the brain; this can lead to mental retardation and abnormal growth.

RISK FACTORS

Risk factors for congenital hypothyroidism may include medication during pregnancy, such as radioactive iodine therapy; having maternal autoimmune disease; having too much iodine during pregnancy; and an inborn error of metabolism. Some babies are born early (before forty weeks). This may cause a temporary shortage in the thyroid hormones.

ETIOLOGY AND GENETICS

Most cases of congenital hypothyroidism are sporadic, meaning that they occur in families with no history of the disorder and there is no identifiable genetic basis for the condition. About 15-20 percent of cases, however, are inherited, and at least seven different genes have been identified in which mutations can occur that will lead to expression of the disease. These genes can cause loss of thyroid function either by adversely affecting the growth and development of the thyroid gland itself or by disrupting the production of thyroid hormones in an otherwise normal appearing gland.

The *PAX8* gene, found on the long arm of chromosome 2 at position 2q12-q14, encodes a transcription factor that is essential for the proper formation of thyroxine-producing follicular cells. The *TSHR* gene (at position 14q31) specifies the thyroid-stimulating hormone receptor protein. Mutations in either of these genes affect thyroid gland development, and they are inherited in an autosomal dominant fashion, meaning that a single copy of the mutation is sufficient to cause full expression of the condition. An affected individual has a 50 percent chance of transmitting the mutation to each of his or her children.

Most cases of inherited congenital hypothyroidism, however, result from mutations in any one of five genes that reduce or eliminate the production of specific thyroid hormones. These genes are *TG, TPO, TSHB, DUOX2,* and *SLC5A5,* and they are found on chromosomes 8, 3, 1, 15 and 19, respectively. The inheritance pattern for mutations in each of these genes is autosomal recessive, meaning that both copies of the gene must be deficient in order for the individual to be afflicted. Typically, an affected child is born to two unaffected parents, both of whom are carriers of the recessive mutant allele. The probable outcomes for children whose parents are both carriers are 75 percent unaffected and 25 percent affected. If one parent has congenital hypothyroidism of this type and the other is a carrier, there is a 50 percent probability that each child will be affected.

SYMPTOMS

Symptoms or signs take time to develop. The symptoms of congenital hypothyroidism may include a puffy face, coarse facial features, a dull look, a thick protruding tongue, poor feeding, choking episodes, constipation or reduced stooling, prolonged jaundice, short stature, and a swollen and protuberant belly button. Other symptoms may include decreased activity, sleeping a lot, rarely crying or a hoarse cry, dry and brittle hair, a low hairline, poor muscle tone, cool and pale skin, goiter (enlarged thyroid), birth defects (such as a heart valve abnormality), poor weight gain due to poor appetite, poor growth, difficult breathing, slow pulse, low temperature, and swollen hands, feet, and genitals.

SCREENING AND DIAGNOSIS

At birth, most infants are screened for this condition. Tests may include a measurement of free (un-

bound) thyroxine (T_4) levels in the blood, a measurement of thyroid stimulating hormone (TSH) in the blood, and a thyroid scan (technetium). Nuclear imaging (scintigraphy) may help determine the cause of congenital hypothyroidism, which can guide treatment and prognosis.

TREATMENT AND THERAPY

The outcome is best if the condition is caught early. It is important to start treatment before the brain and nervous system are fully developed. If treatment is given early, it can prevent damage; left untreated, the condition can lead to mental and growth retardation.

Hormone replacement therapy is often done with the hormone thyroxine, given in the form of either levothyroxine, Levothroid, Levoxyl, or Synthroid. These tablets should be given at least thirty minutes before a meal or feeding.

Once medication starts, the levels of thyroid hormones are checked often, which will help to keep the values within normal range. If values are kept within a normal range, there are no side effects or complications.

PREVENTION AND OUTCOMES

Most cases of congenital hypothyrodism cannot be prevented. A mother can do some things during her pregnancy to reduce the risk. She should not have radioactive iodine treatment or use iodine as an antiseptic. Mothers should also consume enough, but not too much, iodine.

Dianne Scheinberg, M.S., RD, LDN;
reviewed by Rosalyn Carson-DeWitt, M.D.
"Etiology and Genetics" by Jeffrey A. Knight, Ph.D.

FURTHER READING

Bongers-Schokking, J. J., et al. "Influence of Timing and Dose of Thyroid Hormone Replacement on Development in Infants with Congenital Hypothyroidism." *Journal of Pediatrics* 136, no. 3 (March, 2000): 292-297.

Castanet, M., M. Polak, and J. Léger. "Familial Forms of Thyroid Dysgenesis." *Endocrine Development* 10 (2007): 15-28.

EBSCO Publishing. *Health Library: Congenital Hypothyroidism.* Ipswich, Mass.: Author, 2009. Available through http://www.ebscohost.com.

Grüters, A., H. Krude, and H. Biebermann. "Molecular Genetic Defects in Congenital Hypothyroidism." *European Journal of Endocrinology* 151, supplement 3 (November, 2004): U39-44.

LaFranchi, S. H., and J. Austin. "How Should We Be Treating Children with Congenital Hypothyroidism?" *Journal of Pediatric Endocrinology and Metabolism* 20, no. 5 (May, 2007): 559-578.

Rose, S. R., et al. "Update of Newborn Screening and Therapy for Congenital Hypothyroidism." *Pediatrics* 117, no. 6 (June, 2006): 2290-2303.

WEB SITES OF INTEREST

All Thyroid.org
http://www.allthyroid.org

Genetics Home Reference
http://ghr.nlm.nih.gov

"Health Guides on Thyroid Disease." Thyroid Foundation of Canada
http://www.thyroid.ca/Guides/HG09.html

"Neonatal Hypothyroidism." Medline Plus
http://www.nlm.nih.gov/medlineplus/ency/article/001193.htm

"Screening for Congenital Hypothyroidism." Canadian Task Force on Preventive Health Care
http://www.ctfphc.org/Full_Text/Ch18full.htm

See also: Autoimmune polyglandular syndrome; Hereditary diseases; Pseudohypoparathyroidism.

Congenital muscular dystrophy

CATEGORY: Diseases and syndromes

ALSO KNOWN AS: Hereditary progressive muscular dystrophy; Fukuyama congenital muscular dystrophy; Ullrich congenital muscular dystrophy; rigid spine syndrome; Walker-Warburg syndrome; muscle-eye-brain disease

DEFINITION

The term congenital muscular dystrophy (CMD) refers to a group of inherited, genetically and clinically heterogeneous disorders. Their common denominator is muscular weakness, evident at birth or in the first year of life, with dystrophic changes on skeletal muscle biopsy. Two main clinical categories can be defined. Syndromic CMD comprises disease forms with multisystem involvement and developmental delay. Nonsyndromic CMD includes subtypes characterized by muscular disease only. After a century of clinical characterization efforts, molecular and genetic advances have improved diagnostic precision and suggested potential therapeutic strategies. This complex spectrum of disorders results from defects in genes needed for normal muscle function, as well as eye and brain development.

RISK FACTORS

Family history is the only known risk factor. No ethnic group is selectively affected, except the Japanese population in the Fukuyama form.

ETIOLOGY AND GENETICS

The most common CMDs have autosomal recessive inheritance, with the exception of Ullrich CMD (UCMD), for which cases of autosomal dominant transmission have been reported. In autosomal recessive forms, both copies of the gene in each cell display the mutation. Each sibling of a proband (subject) has a 25 percent chance of having the disease and a 50 percent chance of being an asymptomatic carrier (heterozygote). Each unaffected sibling of a patient has a 66 percent chance of carrying the mutated gene.

The many CMD phenotypes are caused by overlapping genetic defects affecting essential muscle proteins required early in life for proper motor development. In the mid-1990's, a deficiency in merosin (skeletal muscle laminin) was discovered in a number of patients. This protein is the backbone of muscle cell basal membrane and is essential for cell adhesion, migration, and survival. Subsequent studies localized the genetic defect to the region coding for the laminin alpha 2 chain of merosin. Soon thereafter, a defect was found in a subunit of the protein integrin, which bridges laminin and intracellular proteins. Merosin also binds the "sugar" branches of dystroglycan, an important glycoprotein that stabilizes the muscle cell and is also expressed in the developing nervous system. Perturbations in the glycosylation ("sugarcoating") of dystroglycan lead to its inability to function and bind merosin. Therefore, deficiencies in glycosyltransferases (enzymes catalyzing the transfer of sugar groups) result in severe, syndromic disease forms, with eye and brain involvement: Fukuyama CMD (FCMD), muscle-eye-brain disease (MEBD), and Walker-Warburg syndrome

(WWS). Inadequate neuronal migration results in lissencephaly ("cobblestone" cortex, broad or absent folds), enlarged ventricles, and brain stem and cerebellar developmental defects.

Disturbances in cell adhesion and cell cycle signaling also contribute to UCMD pathogenesis, which is caused by the deficit in collagen VI, a ropelike extracellular matrix molecule.

All these findings, together with the need for accurate clinical diagnosis and genetic counseling, prompted a CMD classification emphasizing a combination of clinical, biochemical and genetic criteria. Based on the categories proposed by Thomas Voit and Fernando Tomé in 2004, four disease groups are distinguished: defects of laminin alpha 2, primarily affecting the basement membrane (congenital muscular dystrophy type 1A: MDC1A); defects due to abnormal glycosylation of dystroglycan (FCMD, MEB, WWS, and other rare CMD types: MDC1B, MDC1C, MDC1D); disorders with marked contractures (UCMD and rigid spine syndrome, or RSS); and alpha-7 integrin deficiency.

In the first group, classic CMD (MDC1A) is caused by mutations in the laminin alpha 2 gene (on chromosome 6q22-q23), spanning all protein domains. Many missense, nonsense, splice-site, and deletion mutations have been described. Complete lack of expression accounts for approximately half of all CMD cases and usually leads to a more severe phenotype. Molecular diagnosis is not a priority in most patients, given the relatively homogeneous clinical presentation, neuroimaging findings, and immunohistochemical analysis. However, it serves to ascertain the status of a second fetus for the parents of an affected child.

The second group includes overlapping, heterogeneous phenotypes caused by mutations affecting glycosyltransferases and related proteins involved in posttranslational modification of dystroglycan. Fukuyama CMD is characterized by mutations (mainly insertion) in the fukutin gene on chromosome 9q31-q33, resulting in a complete loss of glycosylated dystroglycan. The main gene affected in MEBD is *POMGNT1* (O-linked mannose beta 1,2- N-acetyl-glucosaminyltransferase, on locus 1p34.1), but additional glycosyltransferases have been implicated. A Walker-Warburg phenotype is most severe and can be caused by mutations in all six transferases: fukutin (9q31-q33), protein-O-mannosyltransferase 1 (POMT1, on 9q34.1), protein-O-mannosyltransferase 2 (POMT2,

14q24.3), fukutin-related protein (FKRP, 19q13.33), POMGNT1 (1p34.1) and "L-acetylglucaminyl-transferase-like" (LARGE, 22q12.3-q13.1). Congenital muscular dystrophy type 1B (MDC1B) is associated with mutations at the locus 1q42, MDC1C with mutations in FKRP, and MDC1D with LARGE alterations.

In the third group, Ullrich CMD is caused by deficiencies in collagen VI, related to mutations in one of three genes: *COL6A1* (21q22.3), *COL6A2* (21q22.3), and *COL6A3* (2q37). RSS is due to a mutation in the selenoprotein *N1* gene (locus 1p36.13). This glycoprotein has uncertain functions but seems to be implicated in reduction-oxidation reactions.

In the fourth group, a mutation on chromosome 12, at the 12q13 locus, leads to CMD with integrin alpha 7 deficiency.

In addition, several rare disease forms of unknown genetic causality have been described.

SYMPTOMS

Symptoms vary according to the type of disease. The classical clinical description is centered on congenital hypotonia (diminished muscle tone, "floppy" appearance), muscle weakness, contractures, and joint deformities. All are evident before age two. The clinical course is variable. Depending on disease type, infants may display weak sucking, failure to meet motor milestones, or seizures. Spinal rigidity and scoliosis are RSS characteristics. Eye abnormalities include strabismus, myopia, retinal detachment, microphthalmos (small eyes), and cataracts. Learning disabilities or mental retardation can be present (such as in FCMD and MEBD), although in many forms the IQ is normal.

SCREENING AND DIAGNOSIS

Carrier testing is clinically available for some forms. Prenatal testing is available for classic CMD, FCMD, MEBD, WWS, MDC1C, MDC1D, and RSS. Laboratories may offer custom prenatal testing for other types of CMD.

Muscle biopsies show dystrophic changes in all CMDs. The dystrophic pattern is characterized by variation in fiber size, fibrosis, and sometimes fatty infiltration. Inflammation, necrosis (death), and regeneration of muscle fibers are less prominent. Immunostaining of muscle tissue reveals merosin deficiency in 50 percent of CMDs. Serum levels of creatine kinase (an enzyme released by damaged muscle) are often high. In syndromic CMD, brain magnetic resonance imaging (MRI) reveals developmental anomalies, such as lissencephaly and pontocerebellar underdevelopment. Abnormal white matter signal and dysmyelination are also noted. Classic CMD shows white matter changes, mostly around ventricles, after age six months.

Molecular genetic testing confirms the diagnosis in some forms, such as MDC1A, syndromic CMD, UCMD, and RSS.

TREATMENT AND THERAPY

No specific treatment exists for any CMD. The management is tailored to specific disease subtypes and patients. Symptomatic antiepileptic and antispastic medication may be necessary. Physical therapy helps preserve muscle function and prevent contractures. Surgical intervention for orthopedic complications and ventilatory assistance may be needed. Occupational and speech therapy are often undertaken.

PREVENTION AND OUTCOMES

Genetic counseling and carrier testing should be considered by individuals with affected family members. Morbidity and mortality are mainly connected to respiratory insufficiency, muscle weakness and contractures, seizures, feeding difficulty, and ocular and cardiac complications. Some patients die in infancy, while others can live into adulthood. Weakness is static or minimally progressive in classic CMD, with survival up to thirty years after diagnosis. With severe disease, such as WWS, patients die within the first years of life.

Mihaela Avramut, M.D., Ph.D.

FURTHER READING

Muntoni, Francesco, and Thomas Voit. "The Congenital Muscular Dystrophies in 2004: A Century of Exciting Progress." *Neuromuscular Disorders* 14 (2004): 635-649. Excellent, accessible review written by experts.

Nussbaum, Robert L., Roderick R. McInnes, and Huntington F. Willard. *Thompson and Thompson Genetics in Medicine.* 7th ed. New York: Saunders, 2007. A classic medical school textbook.

Voit, Thomas, and Fernando M. S. Tomé. "The Congenital Muscular Dystrophies." In *Myology: Basic and clinical*, edited by Andrew G. Engel and Clara Franzini-Armstrong. 3d ed. New York: McGraw-

Hill Medical, 2004. Authoritative work on muscle disorders, including CMD.

WEB SITES OF INTEREST

Muscular Dystrophy Association (MDA)
http://www.mda.org/disease/cmd.html

NIH GeneReviews
http://www.ncbi.nlm.nih.gov/bookshelf/br.fcgi?book=gene&part=cmd-overview

See also: Duchenne muscular dystrophy; Hereditary diseases.

Consanguinity and genetic disease

CATEGORY: Diseases and syndromes; Population genetics

SIGNIFICANCE: The late onset of sexual maturity and the random mating habits of most humans make studying rare mutations in human populations especially difficult. Small, isolated communities in which mates are chosen only from within the population lead to consanguineous populations that can serve as natural laboratories for the study of human genetics, especially in the area of human disease.

KEY TERMS

alleles: genetic variants of a particular gene

consanguineous: literally, "of the same blood," or sharing a common genetic ancestry; members of the same family are consanguineous to varying degrees

isolate: a community in which mates are chosen from within the local population rather than from outside populations

THE IMPORTANCE OF ISOLATES

When studying the genetics of the fruit fly or any other organism commonly used in the laboratory, a researcher can choose the genotypes of the flies that will be mated and can observe the next few generations in a reasonable amount of time. Experimenters can also choose to mate offspring flies with their siblings or with their parents. As one might expect, this is not possible when studying the inheritance of human characteristics. Thus, progress in human genetics most often relies on the observation of the phenotypes of progeny that already exist and matings that have already occurred. Many genetic diseases only appear when a person is homozygous for two recessive alleles; thus a person must inherit the same recessive allele from both parents. Since most recessive alleles are rare in the general population, the chance that both parents carry the same recessive allele is small. This makes the study of these diseases very difficult. The chance that both parents carry the same recessive allele is increased whenever mating occurs between individuals who share some of the same genetic background. These consanguineous matings produce measurably higher numbers of offspring with genetic diseases, especially when the degree of consanguinity is at the level of second cousin or closer.

In small religious communities in which marriage outside the religion is forbidden, and in small, geographically isolated populations in which migration into the population from the outside is at or near zero, marriages often occur between two people who share some common ancestry; therefore, the level of consanguinity can be quite high. These communities thus serve as natural laboratories in which to study genetic diseases. Geographically isolated mountain and island communities are found in many areas of the world, including the Caucasus Mountains of Eurasia, the Appalachian Mountains of North America, and many areas in the South Pacific. Culturally isolated communities are also of worldwide distribution. Among the Druse, a small Islamic sect, first-cousin marriages approach 50 percent of all marriages. The Amish, Hutterites, and Dunkers in the United States are each descended from small groups of original settlers who immigrated in the eighteenth and nineteenth centuries and rarely mated with people from outside their religions.

THE AMISH

There are many reasons the Amish serve as a good example of an isolate. The original immigration of Amish to America consisted of approximately two hundred settlers. In subsequent generations, the available mates came from the descendants of the original settlers. With mate choice this limited, it is inevitable that some of the marriages will be consanguineous. Consanguinity increases as further marriages take place between the offspring of consanguineous marriages. Current estimates are that the average degree of consanguinity of Amish marriages in Lancaster County, Pennsylvania, is at the level of marriages between second cousins.

Other factors that make the Amish good subjects for genetic research are their high fertility and their high level of marital fidelity. Thus, if both parents happen to be heterozygous for a particular genetic disease, the chance that at least one of the offspring will show the disease is high. In families of two children, there is a 44 percent chance that at least one child will show the trait. This increases to 70 percent of the families with four children and to more than 91 percent of the families with eight children, a common number among the Amish. Because

Amish children head to class in a one-room schoolhouse in 2008. The Amish are good subjects for the study of consanguinity because of their limited gene pool. (AP/Wide World Photos)

of the high marital fidelity, researchers do not have to worry about illegitimacy when making these estimates.

Many genetic diseases that are nearly nonexistent in the general population are found among the Amish. The allele for a type of dwarfism known as the Ellis-van Creveld syndrome is found in less than 0.1 percent of the general population; among the Lancaster Amish, however, the allele exists in approximately 7 percent of the population. Other genetic diseases at higher levels among the Amish include cystic fibrosis, limb-girdle muscular dystrophy, pyruvate kinase-deficient hemolytic anemia, and several inherited psychological disorders. Having more families and individuals with these diseases to study helps geneticists and physicians discover ways to treat the problems and even prevent them from occurring.

Richard W. Cheney, Jr., Ph.D.

Further Reading

Bittles, Alan H. "Genetic Aspects of Inbreeding and Incest." In *Inbreeding, Incest, and the Incest Taboo: The State of Knowledge at the Turn of the Century,* edited by Arthur P. Wolf and William H. Durham. Stanford, Calif.: Stanford University Press, 2005. The genetic consequences of inbreeding are discussed in this chapter, and there are many other references to consanguinity throughout the book.

Cavalli-Sforza, Luigi Luca, Antonio Moroni, and Gianna Zei. *Consanguinity, Inbreeding, and Genetic Drift in Italy.* Princeton, N.J.: Princeton University Press, 2004. Detailed study of consanguineous marriages and inbreeding in Italy and their genetic impact on the population.

Cross, Harold. "Population Studies of the Old Order Amish." *Nature* 262, no. 5563 (July 1, 1976): 17-20. Describes the advantages of isolates and some of the genetic characteristics seen in Amish populations.

Hartl, D. L., and Elizabeth W. Jones. "Inbreeding." In *Genetics: Analysis of Genes and Genomes.* 7th ed. Sudbury, Mass.: Jones and Bartlett, 2009. This excellent introductory genetics textbook devotes a section of chapter 17 to a discussion of the genetic impact of inbreeding.

McKusick, Victor, et al. "Medical Genetic Studies of the Amish with Comparison to Other Populations." In *Population Structure and Genetic Disorders,* edited by A. W. Eriksson et al. New York: Academic Press, 1981. Describes many of the inherited conditions seen among the Amish.

Shaw, Alison. *Negotiating Risk: British Pakistani Experiences of Genetics.* New York: Berghahn Books, 2009. Based on her study of British Pakistanis with an elevated risk of genetic disorders, Shaw examines the personal and social implications of genetic diagnosis. Includes discussion of consanguineous marriage, close kin marriages, and British Pakistani cousin marriages and genetic risk.

Web Site of Interest

Health Scout Network
http://www.healthscout.com/ency/68/219/main.html
This consumer health site includes a health encyclopedia entry on consanguinity and inbreeding.

See also: Cystic fibrosis; Dwarfism; Genetic load; Hardy-Weinberg law; Hereditary diseases; Heredity and environment; Inbreeding and assortative mating; Lateral gene transfer; Mendelian genetics; Natural selection; Polyploidy; Population genetics; Punctuated equilibrium; Quantitative inheritance; Sociobiology; Tay-Sachs disease.

Corneal dystrophies

CATEGORY: Diseases and syndromes

Definition

The cornea is the most anterior clear structure of the eye, analogous to a windshield. From anterior to posterior, the cornea contains epithelium, Bowman's membrane, stroma, Decemet's membrane, and endothelium. Corneal dystrophy describes primarily bilateral, often inherited, noninflammatory corneal disorders that are not related to systemic disease. The ICD 3 classification of corneal dystrophies was published in 2008 in the journal *Cornea.* Older classification schemes use anatomic location, whereas newer classification uses genetics to categorize corneal dystrophies into four categories.

Category 1 includes well defined dystrophies with known genes. Category 2 disorders are well defined with a known gene locus but no specific gene identified. Category 3 encompasses well-defined dystrophies with no known gene or locus. Dystrophies in Category 4 are poorly defined disorders with no known gene or locus. As the genetics of corneal dystrophies become more defined, these poorly defined disorders may be identified as variants of known dystrophies. Ophthalmologists will likely continue using anatomic classifications, however, as more genes are identified a new classification scheme may develop to include both anatomic and genetic features.

RISK FACTORS

Family history is the risk factor for corneal dystrophy. Unfortunately, symptoms are often subtle and may appear later in life or may be misdiagnosed because of the rarity of corneal dystrophies.

ETIOLOGY AND GENETICS

Autosomal dominant conditions in which 50 percent of children may inherit the disorder include epithelial basement membrane dystrophy (map-dot-fingerprint, Cogan's microcystic), Meesman's dystrophy (juvenile hereditary epithelial dystrophy), and Thiel-Behnke dystrophy (Vogt's anterior crocodile-shagreen, corneal dystrophy of Bowman layer type II). Epithelial basement membrane disorders may be found in up to 2 percent of the population.

The autosomal dominant stromal dystrophies include Granular corneal dystrophy (Groenouw's type 1), Lattice corneal dystrophy (Biber-Haab-Dimmer), Avellino corneal dystrophy, and superficial granular corneal dystrophy (Bowman type 1 corneal dystrophy, Reis-Bucklers). These disorders are associated with *TGFBI* gene mutations. There are several variants of lattice type corneal dystrophy that are autosomal dominant and are associated with the *TGFBI* gene.

Additional autosomal dominant corneal dystrophies affecting the stroma are Meretoja's, central corneal crystalline dystrophy (Schnyder's), Fleck corneal dystrophy (Francois and Neetens), central cloudy corneal dystrophy (posterior crocodile-shagreen), congenital hereditary stromal dystrophy, and posterior amorphous corneal dystrophy. Many of these conditions are not linked to a specific gene or locus, and some may be variants. Posterior autosomal dominant disorders that involve the epithelium are congenital hereditary corneal edema type 1, polymorphous deep corneal dystrophy of Schlichting, and Fuchs' endothelial corneal dystrophy.

Autosomal recessive corneal dystrophies include the anterior stromal dystrophies macular corneal dystrophy, gelatinous droplike dystrophy, and lattice corneal dystrophy type 3. In the anatomic area of the endothelium, the congenital hereditary corneal edema (CHED type 2) is also autosomal recessive. There are several other corneal dystrophies that are not yet defined as dominant or recessive. The penetrance (degree) of both the dominant and recessive dystrophies can be variable.

Disorders such as keratoconus and pellucid degeneration may be hereditary and can cause progressive corneal irregularity. These disorders are not generally considered corneal dystrophies. There are also systemic disorders that can affect the cornea. By definition, corneal dystrophies not associated with systemic disorders.

There has been a rapid increase in the understanding of the genetics of corneal dystrophies in recent years, and the understanding of these disorders from a genetics standpoint will likely change in the next decade as more research is undertaken. The first major breakthrough came in the 1990's when Robert Folberg and associates identified mutations of the 5q31 chromosome as the cause for some forms of lattice degeneration. In recent years, the majority of the classic autosomal dominant stromal corneal dystrophies have been shown to involve this chromosome. Several genes have been identified in these disorders.

In the epithelial dystrophies, the *KRT12* gene at 17q12, the *KRT3* gene at 12q13, and the *CDB2* gene at 10q24 locations have been identified. Some form of Messman's dystrophy may also be X-linked. Several stromal mutations have been associated with the *TGFBI* gene mutations. For autosomal recessive dystrophies, the 16q22 with mutations at the *CHST6* gene and 1p32 with mutations at the *M1S1* gene has been identified. Additional chromosomes that have been identified in corneal dystrophies include 9p34, 1p34.1,2q35, 20p12-q13.1, 20p13, 20p11.2, and 1p34.2-p32.

SYMPTOMS

The main symptoms of corneal dystrophies include blurred vision, pain, or spontaneous corneal abrasions also known as recurrent erosions. Symptoms may have a sudden onset such as in the case of an epithelial dystrophy, where a sudden spontaneous abrasion may occur upon awakening. Others may have a more subtle onset and can be mistaken for other corneal disorders such as herpes simplex infection. Dystrophies such as Fuch's corneal endothelial dystrophy may cause fluctuation of vision with a worsening of vision in the morning and improvement in vision later in the day.

SCREENING AND DIAGNOSIS

Unfortunately, family history of corneal dystrophies may be unkown due to either advanced age of onset or misdiagnosis. The most important screening test is a careful slit-lamp examination with a qualified ophthalmologist. An ophthalmologist trains as a medical doctor with a residency in ophthalmology. An optometrist is also qualified to do a slit-lamp examination and can refer the patient to an ophthalmologist for further diagnosis and treatment as needed.

SCREENING AND DIAGNOSIS

Discoveries of specific genes and mutations responsible for corneal dystrophies are very recent, so genetic screening is not yet standard of care in the diagnosis and treatment of most corneal dystrophies.

TREATMENT AND THERAPY

Depending on the type of corneal dystrophy, anatomic area affected, and severity of symptoms, treatment will range from observation to treatment with drops or ointments to corneal transplant. In severe cases, a fellowship-trained corneal specialist, an ophthalmologist with specialized corneal training, will be involved in diagnosis and treatment. Many corneal dystrophies have symptoms that are so mild that the dystrophy is never diagnosed. For those who are diagnosed, many retain functional vision throughout their lives. Severe cases may require corneal transplant for vision improvement.

PREVENTION AND OUTCOMES

No specific preventive measures are available for corneal dystrophies. For dystrophies that appear in infancy and childhood, treatment of the corneal disorder is important to avoid amblyopia, in which the visual system does not develop properly. Recognizing early-onset corneal dystrophy requires vigi-

lance from parents, family physicians, obstetricians, or pediatricians. Warning signs might include white opacities on the cornea or apparent white pupil (leukocoria), family history of corneal disorders, failure of the infant to fixate properly, or deviation of the eyes such as turning in or turning out. Older patients may seek care for eye discomfort or blurred vision.

Ellen Anderson Penno, M.D., M.S., FRCSC

FURTHER READING

American Academy of Ophthalmology. *2009-2010 Basic and Clinical Science Course (BCSC) Section 8: External Disease and Cornea.* San Francisco: Author, 2004.

Krachmer, Jay H., Mark J. Mannis, and Edward J. Holland. *Cornea.* St. Louis: Elsevier Mosby, 2004.

Merin, Saul. *Inherited Eye Diseases Diagnosis and Management.* 2d ed. Boca Raton, Fla.: Taylor & Francis Group, 2005.

WEB SITES OF INTEREST

American Academy of Ophthalmology
http://www.aao.org

Cornea Society
http://www.corneasociety.org

National Library of Medicine and the National Institutes of Health. MedlinePlus
http://www.nlm.nih.gov

See also: Aniridia; Best disease; Choroideremia; Hereditary diseases.

Cornelia de Lange syndrome

CATEGORY: Diseases and syndromes

ALSO KNOWN AS: Brachmann-de Lange syndrome; Amsterdam dwarfism; typus degenerativus amstelodamensis

DEFINITION

Cornelia de Lange syndrome (CdLS) is a developmental disorder of variable severity. Mutations of genes associated with the cohesin protein complex are present in more than half of individuals with CdLS. Cohesin regulates gene expression. Changes in gene expression affect the developing embryo and are currently believed to underlie the deficits seen in CdLS.

RISK FACTORS

CdLS is mostly a sporadic disorder, the result of spontaneous (de novo) mutations. The syndrome is inherited in an autosomal dominant and X-linked dominant pattern. Mildly affected individuals can have children, but 50 percent of their offspring will have CdLS. Germline mosaicism explains several affected children born to normal parents.

ETIOLOGY AND GENETICS

CdLS is estimated to occur in 1 out of 10,000 live births. It is a cohesinopathy. Cohesin is a multiprotein complex that holds sister chromatids together after DNA replication. It is also involved in the repair of double-stranded DNA and plays a regulatory role in gene expression. Disruption of cohesin function interferes with the normal course of embryonic development. Alterations in three different proteins associated with cohesin's ringlike structure have been found in CdLS.

Approximately 50 percent of individuals with CdLS have mutations in the Nipped-B-Like (*NIPBL*) gene. *NIPBL* is located on the long arm of chromosome 5. Its product, the NIPBL protein, also known as delangin, mediates the binding of cohesin to chromosomes. A variety of *NIPBL* mutations have been identified, ranging from point mutations to small insertions and deletions. Individuals with truncations and deletions are more profoundly affected than those with missense mutations.

Rare cases of CdLS have been associated with chromosomal rearrangements affecting the area on chromosome 5 where *NIPBL* is located.

The cohesion complex consists of four subunit proteins. Mutations in the structural maintenance of chromosome 1A and 3 (*SMC1A* and *SMC3*) have been linked to CdLS. Four percent of individuals with CdLS have mutations in the *SMC1A* gene. *SMC1A* is located on the short arms of the X chromosome. The gene is not subject to X inactivation, so both male and female patients can have *SMC1A* mutations. Individuals with *SMC1A* mutations tend to have a milder form of CdLS: Growth is less impaired, abnormalities of the arms and legs are rarely if at all present, and organ systems generally are spared. The principal finding in individuals with

SMC1A mutations is mental retardation, which does not tend to be severe.

SMC3 is located on the long arm of chromosome 10. Mutations in the *SMC3* gene are infrequent and mainly associated with mild-to-moderate mental retardation.

About 40 percent of individuals whose clinical features evoke CdLS have no identified mutations in the three aforementioned genes. Mutations in other structural genes associated with the cohesin complex may lie at the origin of the syndrome. Mechanisms other than gene mutations could hypothetically lead to the malfunction of *NIPBL*, *SMC1A*, and *SMC3*.

SYMPTOMS

Characteristic facial features include arched, well-defined eyebrows, long eyelashes, and a short, upturned nose. The head is small with low-set ears and an underdeveloped mandible. Hair growth often is abundant. Arm defects include absent forearms, single fingers, and small hands. Growth is impaired. Mental retardation varies from mild to severe. Other potential findings include eye and kidney anomalies, hearing loss, heart malformations, a dysfunctional gastrointestinal tract, underdeveloped genitalia, autism, and self-destructive behavior.

SCREENING AND DIAGNOSIS

Pregnancy-associated plasma protein-A (PAPP-A) levels are low in early gestation. High-resolution ultrasound can reveal increased nuchal translucency (cystic hygroma), dysmorphic facial features, and organ and possibly limb anomalies in a growth-retarded fetus. Otherwise, at birth, the infant's striking facial features and clinical findings will lead to the diagnosis of CdLS. The initial genetic evaluation consists of chromosomal analysis. When the results of this analysis are negative, mutational analysis of the *NIPBL* gene should be obtained. If there are no mutations in the *NIPBL* gene, then analysis of the *SMC1A* and *SMC3* genes can be pursued.

TREATMENT AND THERAPY

Gastroesophageal reflux disease requires surgery when medical treatment has failed. Surgical correction is needed for intestinal malrotation. High-calorie formula and feeding tube placement will support weight gain. Surgery may be necessary to correct heart defects, cleft palate, and undescended testicles or to improve mobility of the limbs. Vesicoureteral reflux warrants prophylactic antibiotics. Regular hearing screens are needed as sensorineural hearing loss may develop over time. Visits with the ophthalmologist facilitate the detection and treatment of common problems such as narrowing of the tear ducts, myopia, and cataracts. Seizure disorders require medication. Early intervention directed at optimizing developmental outcome includes speech, occupational, and physical therapy.

PREVENTION AND OUTCOMES

Prevention is possible in families in which a parent or child has been diagnosed with CdLS. In early pregnancy, fetal cells obtained by chorionic villus sampling or amniocentesis can be analyzed for a known CdLS-causing mutation. High-resolution fetal scans during pregnancy will monitor growth and detect anomalies associated with CdLS in families with no identified mutation. Long-term survival is influenced by the severity of organ and system involvement.

Elisabeth Faase, M.D.

FURTHER READING

Cassidy, Suzanne B., and Judith E. Allenson. *Management of Genetic Syndromes*. 2d ed. New York: Wiley-Liss, 2005. Comprehensive chapter on CdLS.

Dorsett, Dale, and Ian D. Krantz. "On the Molecular Etiology of Cornelia de Lange Syndrome." *Annals of the New York Academy of Sciences* 1151 (January, 2009): 22-37. Analyzes cohesin's role in CdLS.

Liu, Jinglan, and Ian D. Krantz. "Cohesin and Human Disease." *Annual Review of Genomics and Human Genetics* 9 (September, 2008): 303-320. Discusses cohesin's function in genetic disorders.

WEB SITES OF INTEREST

Cornelia de Lange Syndrome Foundation USA
www.cdlsusa.org

World Cornelia de Lange Syndrome Federation
www.cdlsworld.org

See also: Congenital defects; Developmental genetics; Hereditary diseases.

Cowden syndrome

CATEGORY: Diseases and syndromes
ALSO KNOWN AS: CS; Cowden disease; *PTEN* hamartoma tumor syndrome; multiple hamartoma syndrome

DEFINITION

Cowden syndrome (CS), one of the syndromes making up the *PTEN* hamartoma tumor syndrome, and is a hamartoma syndrome with cancer predisposition. Affected persons with CS may have macrocephaly, trichilemmomas, and papillomatous papules. Cancers most frequently seen with CS include benign and malignant forms of thyroid, breast, and endometrial cancers.

RISK FACTORS

Persons at risk for CS are identified through personal and/or family history of clinical manifestations of mucocutaneous lesions, macrocephaly, mental retardation, and benign and malignant cancers. A mutation in the *PTEN* gene can be identified in approximately 85 percent of persons meeting the diagnostic criteria for CS.

ETIOLOGY AND GENETICS

Although thought to be underdiagnosed, CS is a rare syndrome. Most disease is associated with a mutation in the *PTEN* gene, located on chromosome 10. No other gene has been associated with CS.

The *PTEN* gene encodes for a major lipid phosphatase, which can cause cell arrest and apoptosis, thereby suppressing tumor formation. The protein phosphatase may also inhibit cell migration and spreading.

When the *PTEN* gene is mutated, there is less ability to activate cell cycle arrest and apoptosis, with abnormal cell survival. While some recurrent mutations associated with CS have been identified, more than one hundred unique mutations in the *PTEN* gene have been reported. The described mutations in the *PTEN* gene associated with CS may be inherited or arise de novo. The inherited mutations are transmitted by autosomal dominant inheritance. The mutation may be passed from either the maternal or paternal lineage, with a 50 percent chance of transmission with each offspring. However, many people with CS are simplex cases, meaning that there is no obvious family history of the syndrome. Simplex cases may be related to underdiagnosis in the family, early death of family members before the clinical onset of the syndrome, late-onset disease in family members, or de novo mutations. An estimated 40 percent of persons with CS have de novo mutations.

SYMPTOMS

CS is a predisposition syndrome, yet nearly all persons with CS will have some clinical manifestation by their late twenties. Those who do develop disease have symptoms respective of the sporadic forms—that is, the symptoms of the diseases associated with CS are not unique to CS.

SCREENING AND DIAGNOSIS

Screening for CS includes clinical examination and evaluation of family history, with genetic testing for *PTEN* mutations to confirm the diagnosis. The criteria for testing for CS diagnosis includes the following: presence of any single pathognomonic criterion, presence of two or more major criteria, one major and three or more minor criteria, or four or more minor criteria. Patognomonic criteria include adult Lermitte-Duclos disease or mucocutaneous lesions. Major criteria are breast cancer, nonmedullary thyroid cancer, macrocephaly, or endometrial cancer. Minor criteria are other thyroid lesions, mental retardation, gastrointestinal (GI) harmartomas, fibrocystic disease of the breast, lipomas, fibromas, genitourinary tumors or structural manifestations, or uterine fibroids.

While sequencing the *PTEN* gene with attention to large deletions and rearrangements is performed to confirm the diagnosis of CS, such testing will not identify a mutation in all persons meeting the criteria of CS. Failure to identify a mutation, however, does not exclude a clinical diagnosis of CS.

TREATMENT AND THERAPY

For individuals with CS who are affected with disease, treatment and therapy will be similar to the clinical management of the respective disease—that is, the diseases associated with CS also occur sporadically and there is no special treatment based on an accompanying diagnosis of CS.

PREVENTION AND OUTCOMES

Medically, the most serious outcomes of CS relate to the increased cancer risks. The associated breast cancer is of early age onset, generally before the age of fifty, with the lifetime risk of 25 to 50 percent. The lifetime risk for thyroid cancer is 10 percent and for endometrial cancer is 5 to 10 percent. Skin cancer, renal cell carcinoma, and brain tumors may also be seen in CS. Although hamartomatous polyps of the GI tract may occur, the risk of colorectal cancer is not thought to be increased.

For breast cancer surveillance, mastectomy or intensified surveillance may be indicated. The later may include semi-annual or annual clinical breast examinations beginning at age twenty-five years with annual mammography and breast MRI beginning at age thirty to thirty-five years, or five to ten years earlier than the youngest age at diagnosis of breast cancer in the family, in addition to monthly self breast examination. Surveillance for thyroid cancer may include baseline thyroid ultrasound examination at age eighteen years and annual thyroid examination. Endometrial cancer surveillance may include annual suction biopsies beginning at age thirty-five to forty years for premenopausal women and annual transvaginal ultrasound examination for postmenopausal women. However, all cancer surveillance needs to be individualized and account for current research findings.

Other general surveillance for persons with CS may include annual physical examination starting at age eighteen, annual urinalysis, annual dermatologic examination, general cancer screening such as colonoscopy starting at age fifty, and a heightened awareness of the signs and symptoms of CS disease.

For persons with CS, genetic counseling and possible *PTEN* testing of other family members, including those who are asymptomatic, may be indicated to guide cancer prevention and improve outcomes.

Judy Mouchawar, M.D.

FURTHER READING

Offit, Kenneth. *Clinical Cancer Genetics*. New York: Wiley-Liss, 1998. A clinically oriented text of cancer genetic syndromes.

Schottenfeld, David, and Joseph F. Fraumeni, Jr. *Cancer Epidemiology and Prevention*. 2d ed. New York: Oxford University Press, 1996. A comprehensive text on cancer.

Vogel, Victor G. *Management of Patients at High Risk for Breast Cancer*. Malden, Mass.: Blackwell Science, 2001.

See also: Cancer; Hereditary diseases; Mutagenesis and cancer; Mutation and mutagenesis; Tuberous sclerosis.

Cri du chat syndrome

CATEGORY: Diseases and syndromes
ALSO KNOWN AS: 5p– syndrome; 5p deletion syndrome; chromosome 5p deletion syndrome; cat cry syndrome; Lejeune's syndrome

DEFINITION

Cri du chat syndrome is a genetic disease caused by a deletion (loss) of genetic material in the short (p) arm of chromosome number 5. The syndrome was given its name (French for "cat's cry") because the sounds these infants make are like those of a meowing cat.

RISK FACTORS

Most cases are spontaneous, meaning there are no known risk factors. Recent reviews have noted, however, that parents who carry translocations involving the 5p region have a risk of producing a child with cri du chat syndrome; this may account for 10 to 15 percent of all cases. A balanced translocation involves no loss of genetic material and no symptoms in the carrier; however, these translocations can become unbalanced when passed on to the next generation. The risk for male and female carriers is similar.

ETIOLOGY AND GENETICS

Identified by Jérôme Lejeune in 1963, cri du chat syndrome is rare, presenting in 1 in 20,000 to 1 in 100,000 babies, and accounts for approximately 1 percent of all severe mental retardation. It is slightly more common in females than in males and affects all ethnic groups. It is an autosomal disorder, involving breakage of the short arm of chromosome

number 5. Specific points of breakage, and the extent of the deletion, vary from patient to patient. However, the critical region, recently identified as chromosome band 5p15, is missing in all patients with the cri du chat phenotype. It is believed that breakage of chromosome 5 occurs in the process of meiosis, during development of the egg or the sperm, with about 70 to 80 percent of breaks being paternal (sperm) in origin.

A number of specific genes have been identified as deleted in these chromosomes, and the clinical phenotype of the patients appears to be related to the monosomic condition (presence of only one chromosome copy) in this region, rather than the presence of mutated genes as in many genetic disorders. Through correlation of phenotypic changes and breakpoints, two specific bands have been identified within the 5p15 region. The lack of genes within band 5p15.2 appears to account for many of the neurological findings, including mental retardation, while the distinctive cat cry appears to be the result of a deletion in band 5p15.3. Recent work suggests the presence of two critical but discontinuous regions of the chromosome are involved in producing the etiology of the syndrome. Two genes have been mapped to these regions that may be involved in brain development—*SEMA5A* (Semaphorine F) and *CTNND2* (δ-catenin)—and their presence in only one copy may account for much of the mental disability seen. The cri du chat deletion area is also known to contain the gene for telomerase reverse transcriptase enzyme (also known as TERT, and found at band 5p15.33), which keeps telomeres, the critical ends of chromosomes, long in cells that divide often. The presence of only one copy of the gene may be related to shortened life span of the chromosomes, as well as other control features of cell growth, and therefore the shortened life span that these patients experience.

SYMPTOMS

Symptoms usually include the distinctive cry (which some children lose by age two); difficulty in sucking and swallowing (including breast-feeding); severe delays in motor function, speech, and cognitive function; low birth weight and slow growth; behavioral problems, including aggression, tantrums, hyperactivity; and, in some cases, repetitive motions. Common facial appearance includes microcephaly (small head), hypertelorism, epicanthal folds, low-set ears, and micrognathia (small teeth). Other symptoms may include hypotonia, constipation, short fingers, heart defects (including patent ductus arteriosis, septal defects, and Tetralogy of Fallot), strabismus, down-turned mouth, and a round face. The distinction between symptoms and secondary complications of the syndrome is very unclear.

SCREENING AND DIAGNOSIS

Prenatal screening (cytogenetic and molecular) is useful in some cases, particularly where one child with cri du chat syndrome already exists in the family. On occasion such screening may be warranted by a combination of ultrasound findings. Diagnosis after birth results from a combination of clinical features that, when taken as a whole, create a distinct phenotype. This phenotype, along with the distinctive cry of the infant, leads to a suspicion of cri du chat syndrome. Diagnosis is confirmed on the basis of karyotype analysis of the chromosomes. Occasionally a mild form of the syndrome goes undiagnosed at birth, but the continued abnormal voice of the infant coupled with retardation of psychomotor skills will lead to diagnosis via karyotype analysis.

TREATMENT AND THERAPY

There is no treatment available for the underlying genetic disorder. Therapy, therefore, is focused on the individual symptoms. Surgeries may be required to correct heart defects or other structural errors. Some patients require gastrostomy tubes for feeding, due to laryngeal irregularities or other developmental problems. The degree of mental retardation will dictate the course of therapy needed for daily functionality.

PREVENTION AND OUTCOMES

There is no known prevention for cri du chat syndrome. Those with a family history of the syndrome should seek genetic counseling before attempting to become pregnant. Outcomes vary, but often mental retardation is severe enough that the individual is unable to care for himself or herself or to function constructively in society. About half of affected children learn sufficient verbal skills to be able to communicate.

Kerry L. Cheesman, Ph.D.

FURTHER READING

Parker, Philip M. *Cri-du-chat Syndrome: A Bibliography and Dictionary for Physicians, Patients, and Genome Researchers.* San Diego: ICON Group International, 2007.

WEB SITES OF INTEREST

Five P Minus Society
www.fivepminus.org

Genetics Home Reference
http://ghr.nlm.nih.gov/condition
=criduchatsyndrome

National Human Genome Research Institute
http://www.genome.gov/19517558

See also: Congenital defects; Hereditary diseases.

Criminality

CATEGORY: Human genetics and social issues
SIGNIFICANCE: The pursuit of genetic causes of criminality is a controversial field of study that has produced intriguing examples of the apparent contribution of genetic defects to criminal behavior. However, the nature of human criminality defies simple and straightforward explanations and instead likely involves a combination of genetic, psychological, and environmental influences. Research into the causes of human criminality have also come under strong criticism by opponents who fear that such discoveries may be used to identify and implicate certain ethnic or racial groups as genetically predisposed to deviant behavior. In short, experts disagree on the degree to which genetics determine human criminal behavior but generally acknowledge its critical role in shaping behavior.

KEY TERMS

metabolic pathway: a biochemical process that converts specific chemicals in the body to other, often more useful, chemicals with the help of proteins called enzymes.

neurotransmitter: a neurochemical that transmits messages between neurons.

BIOCHEMICAL ABNORMALITIES

Scientists have long sought an answer to the heritability of criminality. Early attempts to identify the roots of human criminal behavior were based on the concept of biological determinism, which explains and justifies human behavior as strictly a reflection of inborn human traits, with little or no attention paid to psychological or environmental influences. For example, Italian physician Cesare Lombroso reported in *L'uomo delinquente* (1876; criminal man) that certain "inferior" groups, by virtue of their "apish" appearance, were in actuality evolutionary throwbacks with criminal tendencies. Since that time, however, more sophisticated scientific theories and methods have been developed to identify the multiple etiologies of human behavior, including criminality.

Among the best-known theories of human behavior to find support in the scientific community are those suggesting certain biochemical imbalances, particularly involving neurochemicals, potentially play a role in generating a wide range of abnormality. Neurotransmitters are responsible for activating behavioral tendencies and patterns in explicit areas of the brain, so it makes sense that imbalances in these chemicals might also negatively affect behavior.

In some research studies, decreased levels of the neurotransmitter serotonin have been discovered in people who are depressed or aggressive, have attempted suicide, or have poor impulse control, such as impulsive arsonists and children who torture animals. In other studies, though, normal levels of serotonin have been found in these same groups, as well as abnormal levels in normal groups. As well, abnormalities in the brain's levels of dopamine (another primary neurotransmitter) have also been implicated in aggressive and antisocial behaviors, although studies have yielded mixed results. Put simply, the role of neurotransmitters, including serotonin and dopamine, in abnormal behavior remains controversial and likely does not adequately explain criminality without taking into account social and psychological influences.

Perhaps the most widely researched theories of criminality have addressed potential genetic influences. The majority of early investigations in this area examined the role of an abnormality of the sex chromosomes—47,XYY—involving the presence of an additional Y chromosome in an otherwise normal male karyotype. Beginning in the 1960's, Dr. Patricia Jacobs proposed that those males who possess this extra Y chromosome were overrepresented in prisons and mental institutions. She studied nine males (out of more than three hundred males in a maximum security prison) who had an XYY karyotype. These XYY males had above-average height (generally over six feet tall) and below-average intelligence, exhibited personality disorders, and were more prone to have engaged in antisocial acts leading to their incarceration. A number of studies also supported these early findings, which understandably generated considerable interest—and debate—into the abnormal behaviors potentially associated with a XYY condition. By the 1970's, multiple investigations into XYY males in various settings, not just prisons, yielded inconsistent findings with respect to behavior. In fact, the only dependable feature of XYY males, whether incarcerated or not, appears to be that of increased height. No definitive associations between XYY males and criminal behaviors have ever been absolutely demonstrated.

Another proposed genetic explanation for criminal behavior involves an abnormality in the enzyme monoamine oxidase A (MAOA). This important enzyme is responsible for degrading certain neurotransmitters, including dopamine and epinephrine. Theoretically, criminal behavior is more liable when the normal levels of neurotransmitters in the brain are disrupted, which in turn leads to behavioral alterations. To date, no definitive causal link to criminal behavior has ever been established in individuals with a MAOA abnormality.

The biology of criminality is comparable to the biology of aggression, with testosterone (or similar androgens) typically being referenced in order to explain belligerent male behavior. Yet defining male criminal behavior in terms of excessive testosterone, or another biochemical entity, has almost become a cliché in recent years—and one without solid scientific merit. A multimodal approach is instead preferable. Therefore, the roles of psychology and environment in criminal behavior must also be considered. When physiological dysfunction exists secondary to genetic dysfunction, cognitive deficits and impulsiveness may also coexist, which sets the stage for criminal tendencies to be acted out. First, a neural defect in almost any form is frequently associated with impatience, irritability, and impulsiveness. Next, misperceptions and ideation, symptoms associated with many different kinds of antisocial behavior, increase anxiety and the tendency to "act out" or "retaliate" for both real and imagined reasons. Finally, intellectual deficits not only diminish judgment but also lessen the person's ability to acknowledge feelings and describe them verbally rather than through inappropriate actions.

Overall, genetic abnormalities clearly play a role in affecting numerous human characteristics, including mental capabilities and behavior, but to ignore psychology and environment in human characteristics is to be simplistic. After all, criminality refers to a violation of the law and since there are numerous types of crimes and motivations for them (anger, revenge, financial gain), it is difficult to make claims of definitive, nonenvironmental links between biochemical disorders and criminal behavior without exploring all potential variables. In other words, the nature of human criminal behavior defies simple and straightforward explanations. The exact causes of aberrant behavior are complex and involve multiple influences, of which is genetics is one critical component.

IMPACT AND APPLICATIONS

Research into the biological and genetic causes of criminality entered the public spotlight starting in the early 1990's as part of the U.S. government's Violence Initiative, championed by Secretary of Health and Human Services Louis Sullivan. The uproar began when Frederick Goodwin, then director of the Alcohol, Drug Abuse, and Mental Health Administration, made comments comparing urban youth to aggressive jungle primates. The public feared that research on genetic links to criminality would be used to justify the disproportionate numbers of African Americans and Hispanics in the penal system. Psychiatrist Peter Breggin also warned that unproved genetic links would be used as an excuse to screen minority children and give them sedating drugs to intervene in their impending aggression and criminality. After all, forced sterilization laws had been enacted in thirty U.S. states in the 1920's to prevent reproduction by the "feebleminded" and "moral degenerate." Today, the general public remains highly suspicious of any medical or genetic research that might be used to target and marginalize minority or disadvantaged groups as predisposed to "criminal" behavior. This is all the more

the case as the Human Genome Project continues to discover genetic links to diseases and pathological behaviors.

In an era in which genes have been implicated in everything from bipolar disorders to the propensity to change jobs, the belief that genes are responsible for criminal behavior is very enticing. However, this belief may have severe ramifications. To the extent that society accepts the view that crime is the result of pathological and biologically deviant behavior, it is possible to ignore the necessity to change social conditions such as poverty and oppression that are also linked to criminal behavior. Moreover, this view may promote the claim by criminals themselves that their "genes" made them do it. While biochemical diagnosis and treatment with medications may be simpler and therefore more appealing than social interventions, this is perhaps reminiscent of the days when frontal lobotomy was the preferred method of biological intervention for aggressive mental patients. In the future, pharmacological solutions to social problems may be viewed as similarly questionable.

Criminality as a specific form of human behavior has been studied by scientists, psychiatrists, psychologists, sociologists, and others who ultimately seek to understand its causes, primarily in the hopes of lessening the occurrence and impact of its more deleterious manifestations. Those researchers who look for solutions in genetics sometimes lose sight of the roles that psychology and environment play in the various expressions of criminal behavior. The same can be said of social scientists who sometimes ignore the roles that genetics and neurochemistry play. It makes better sense to conclude that a combination of genetic, psychological, and environmental influences work in different ways for different individuals leading to the development of criminal behavior in some but not all.

Lee Anne Martínez, Ph.D.;
updated by George D. Zgourides, M.D., Psy.D.

FURTHER READING

Andreasen, Nancy C. *Brave New Brain: Conquering Mental Illness in the Era of the Genome.* New York: Oxford University Press, 2001. Surveys the way in which advances in the understanding of the human brain and the human genome are coming together in an ambitious effort to conquer mental illness.

Faraone, Stephen V., Ming T. Tsuang, and Debby W. Tsuang. *Genetics of Mental Disorders: A Guide for Students, Clinicians, and Researchers.* New York: Guilford Press, 1999. Introduces the investigative methods of human genetics as applied to mental disorders, their clinical applications, and some of the biological, ethical, and legal implications of the investigative processes and conclusions.

Gilbert, Paul, and Kent G. Bailey Hove, eds. *Genes on the Couch: Explorations in Evolutionary Psychotherapy.* Philadelphia: Brunner-Routledge, 2000. Examines models and interventions in psychotherapy based on evolutionary findings and includes topics such as psychotherapy in the context of Darwinian psychiatry, Jungian analysis, gender, and the syndrome of rejection sensitivity.

Glenn, Andrea L., and Adrian Raine. "The Neurobiology of Psychopathy." *Psychiatric Clinics of North America* 31 (2008): 463-475. Reviews current thinking regarding the roles of neurotransmitters and neuroendocrine functioning in abnormal behavior. Concludes the current state of research does not yet support a strictly physiological model of psychopathy.

Hare, Robert D. "Psychopathy: A Clinical and Forensics Overview." *Psychiatric Clinics of North America* 29 (2006): 709-724. An overview of current thinking into the role of various biopsychosocial factors influencing antisocial behavior.

Livesley, W. John. "Research Trends and Directions in the Study of Personality Disorders." *Psychiatric Clinics of North America* 31 (2008): 545-559. Updated information, from a psychiatric perspective, concerning the etiologies and typical courses of common personality disorders, including those involving criminal behavior.

Rose, Steven. "The Rise of Neurogenetic Determinism." *Nature* 373 (February, 1995). Comments on how technological advances have revived genetic explanations for behavior.

Sapolsky, Robert. "A Gene for Nothing." *Discover* 18 (October, 1997). An entertaining account of the complex interaction between genes and the environment.

Walsh, Anthony, and Kevin M. Beaver, eds. *Contemporary Biosocial Criminology: New Directions in Theory and Research.* New York: Taylor & Francis, 2008. Excellent resource describing how leading criminologists have integrated aspects of the biological sciences, including genetics, into their forensics work.

Wasserman, David, and Robert Wachbroit, eds. *Genetics and Criminal Behavior.* New York: Cambridge University Press, 2001. Explores issues surrounding causation and responsibility in the debate over genetic research into criminal behavior. Chapters include "Understanding the Genetics of Violence Controversy," "Separating Nature and Nurture," "Genetic Explanations of Behavior," "On the Explanatory Limits of Behavioral Genetics," "Degeneracy, Criminal Behavior, and Looping," "Genetic Plans, Genetic Differences, and Violence," "Crime, Genes, and Responsibility," "Genes, Statistics, and Desert," "Genes, Electrotransmitters, and Free Will," "Moral Responsibility Without Free Will," "Strong Genetic Influence and the New 'Optimism,'" and "Genetic Predispositions to Violent and Antisocial Behavior."

Williams, Juan. "Violence, Genes, and Prejudice." *Discover* 15 (November, 1994). Gives an excellent account of the controversy and debate that accompanied the U.S. government's funding of research on genetic links to violence and crime.

Wright, Robert. "The Biology of Violence." *The New Yorker* 71 (March 15, 1995). Discusses evolutionary psychology's view that violent responses to oppressive environments may be adaptive rather than genetically inflexible.

WEB SITES OF INTEREST

DNA Forensics and Genetics Links
http://www.lhup.edu/tnuttall/dna_forensics_and_genetics_links.htm
Site links and resources from the Director of the DNA Forensics Program at Lock Haven University of Pennsylvania.

Genetics Education Center, University of Kansas Medical Center
http://www.kumc.edu/gec/forensic.html
A list of forensics resources with an emphasis on genetics.

See also: Aggression; Alcoholism; Altruism; Behavior; Biological determinism; Developmental genetics; DNA fingerprinting; Eugenics; Eugenics: Nazi Germany; Forensic genetics; Hardy-Weinberg law; Heredity and environment; RFLP analysis; Sociobiology; Sterilization laws; XYY syndrome.

Crohn disease

CATEGORY: Diseases and syndromes
ALSO KNOWN AS: Crohn's disease; CD; regional enteritis

DEFINITION

Crohn disease is a severe, chronic inflammatory bowel disease. It causes inflammation, ulcers, and bleeding in the digestive tract. It usually affects the end portion of the small intestine called the ileum. However, any part of the digestive tract can be affected, from the mouth to the anus.

RISK FACTORS

Factors that increase an individual's chance of getting Crohn include having family members with inflammatory bowel disease and being of Jewish heritage.

ETIOLOGY AND GENETICS

Crohn disease is a complex condition whose expression very likely involves many different environmental and genetic factors. A genetic component has been suspected for many years, since the disease has been known to run in families and it is more common in people of Ashkenazi Jewish descent. By 2009, researchers had identified thirty-two different genes thought to be associated with Crohn disease, and some suggested that this number could rise to one hundred or more. Because of the complexity of gene interaction and variable environmental factors, there is no clear predictability with regard to the inheritance pattern within families.

The single most important gene associated with an increased risk of developing Crohn disease is called *NOD2*, found on the long arm of chromosome 16 at position 16q12. This gene specifies a protein known as nucleotide-binding oligomerization domain protein 2, and it functions as an intracellular receptor for bacterial products in white blood cells lining the intestinal epithelium. Disruption of this function is thought to alter the response to the bacteria naturally found in the intestines, causing the immune system to attack the intestinal epithelium and result in inflammation and possibly necrosis.

The *DLG5* gene on the long arm of chromosome 10 (at position 10q23) encodes a scaffolding protein that helps maintain structural integrity of the intestinal epithelium, and mutations in this gene may result in a weakened scaffold that can increase disease susceptibility. On chromosome 9, the *TLR4* gene (at position 9q32-q33) specifies a protein called toll-like receptor 4 that is involved with recognition of the lipopolysaccharide layer of bacterial cell walls, and disruption of this function could result in an altered immune response to the normal intestinal flora and subsequent inflammation. Other genes thought to have a significant impact on the predisposition to develop Crohn disease include *ITCH* (at position 20q11.22), *OCTN1* and *IBD5* (at position 5q31), *IRGM* (at position 5q33.1), *ATG16L1* (at position 2q37.1), and *IL23R* (at position 1p31.3).

SYMPTOMS

Symptoms include diarrhea, abdominal cramps and pain, rectal bleeding, anemia, weight loss, fatigue, weakness, nausea, fever, mouth sores, and sores and abscesses in the anal area.

SCREENING AND DIAGNOSIS

The doctor will ask about a patient's symptoms and medical history, and a physical exam will be done. Tests may include blood tests; a stool examination; barium swallow, a series of X rays of structures inside the throat that are taken after drinking a barium-containing liquid; and a barium enema X ray, the insertion of fluid into the rectum that makes the colon show up on an X ray. Other tests include a flexible sigmoidoscopy, in which a thin, lighted tube is inserted into the rectum to examine the rectum and the lower colon; a colonoscopy, in which a thin, lighted tube is inserted through the rectum and into the colon to examine the lining of the colon; and biopsy, the removal of a sample of colon tissue for testing (may be performed as part of a flexible sigmoidoscopy or colonoscopy). If patients are diagnosed with Crohn disease, their doctors will give them guidelines to follow.

TREATMENT AND THERAPY

Treatment may include avoiding foods that provoke symptoms. These foods are different for each individual and may include dairy foods (due to lactose intolerance), highly seasoned foods, and high-fiber foods.

Aminosalicylate medications are another treatment option and include sulfasalazine, mesalamine, and olsalazine. Anti-inflammatory medications used to treat Crohn disease include prednisone, methyl-prednisolone, and budesonide. Other medications include immune modifiers, such as azathioprine, 6-mercaptopurine, and methotrexate; biologic therapy, including infliximab and adalimumab; and antibiotic medications, such as metronidazole, ampicillin, and ciprofloxacin.

Very severe Crohn disease may not improve with medications. A patient may be advised to have the severely diseased section of his or her intestine removed; the two remaining healthier ends of the intestine are then joined together. The patient is still at high risk for recurrence of the disease elsewhere. Surgery may also be done if a patient has an obstruction or fistulas.

Untreated Crohn disease may lead to fistulas, or abnormal connections between the intestine and other organs or tissues, such as the bladder, vagina, or skin. The untreated disease may also lead to intestinal obstruction, arthritis, eye inflammation, liver disease, kidney stones, gallstones, skin rashes, and osteoporosis.

PREVENTION AND OUTCOMES

There are no guidelines for preventing Crohn disease because the cause is unknown.

Rosalyn Carson-DeWitt, M.D.;
reviewed by Jill D. Landis, M.D.
"Etiology and Genetics" by Jeffrey A. Knight, Ph.D.

FURTHER READING

EBSCO Publishing. *Health Library: Crohn Disease.* Ipswich, Mass.: Author, 2009. Available through http://www.ebscohost.com.

Goroll, Allan H., and Albert G. Mulley, Jr., eds. *Primary Care Medicine: Office Evaluation and Management of the Adult Patient.* 6th ed. Philadelphia: Wolters Kluwer Health/Lippincott Williams & Wilkins, 2009.

Sklar, Jill. *Crohn's Disease and Ulcerative Colitis: An Essential Guide for the Newly Diagnosed.* 2d ed., rev. and updated. New York: Marlowe, 2007.

Warner, Andrew S., and Amy E. Barto. *One Hundred Questions and Answers About Crohn's Disease and Ulcerative Colitis: A Lahey Clinic Guide.* 2d ed. Sudbury, Mass.: Jones & Bartlett, 2001.

WEB SITES OF INTEREST

American Gastroenterological Association
http://www.gastro.org

Crohn's and Colitis Foundation of America
http://www.ccfa.org

Crohn's and Colitis Foundation of Canada
http://www.ccfc.ca/English/index.html

Genetics Home Reference
http://ghr.nlm.nih.gov

Health Canada
http://www.hc-sc.gc.ca/index-eng.php

National Institute of Diabetes and Digestive and Kidney Diseases
http://www.niddk.nih.gov

See also: Autoimmune disorders; Celiac disease; Colon cancer; Familial adenomatous polyposis; Hereditary diseases.

Crouzon syndrome

CATEGORY: Diseases and syndromes
ALSO KNOWN AS: Craniofacial dysotosis

DEFINITION

Crouzon syndrome is a genetic disorder. It is one of many birth defects that results in abnormal fusion between bones in the skull and face. Normally, as an infant's brain grows, open sutures between the bones allow the skull to develop normally. When sutures fuse too early, the skull grows in the direction of the remaining open sutures. In Crouzon syndrome, the bones in the skull and face fuse too early. This results in an abnormally shaped head, face, and teeth. Crouzon disease is believed to affect 1 in 60,000 people.

RISK FACTORS

Those most at risk for Crouzon syndrome are children of parents with the disorder; children whose parents do not have the disorder, but who carry the gene that causes the disorder; and children whose fathers are at an older age at the time of conception.

ETIOLOGY AND GENETICS

Crouzon syndrome is inherited as an autosomal dominant disorder, meaning that a single copy of the mutation is sufficient to cause full expression of the syndrome. An affected individual has a 50 percent chance of transmitting the mutation to each of his or her children. Many cases of Crouzon syndrome, however, result from a spontaneous new mutation, so in these instances affected individuals will have unaffected parents.

Most individuals with Crouzon syndrome carry a mutation in a gene called *FGFR2*, which is found on the long arm of chromosome 10 at position 10q26. *FGFR2* encodes a protein known as fibroblast growth factor receptor 2. Like many similar receptors, this protein has multiple functions, but one particularly important effect is to signal the appropriate population of stem cells to develop into bone cells during embryonic and fetal development. It is believed that mutations in this gene result in an overstimulation by the receptor protein that can cause the bones of the skull to prematurely fuse.

Mutations in a second gene, *FGFR3*, found on the short arm of chromosome 4 at position 4p16.3, have also been known to be associated with Crouzon syndrome, although patients with these mutations also exhibit a characteristic skin condition known as acanthosis nigricans (dark, thick, velvety skin in body folds, often in the neck and armpit area). The gene product is another fibroblast growth factor receptor, one that is more commonly associated with achondroplastic dwarfism.

SYMPTOMS

The main signs and symptoms of Crouzon syndrome include flattened top and back of head; flattened forehead and temples; midface that is small and located further back in the face than normal; beaklike nose; compression of nasal passages, often causing reduced airflow through the nose; a large, protruding lower jaw; misalignment of teeth; and a high-arched, narrow palate, or cleft palate. Other symptoms and complications that can result from Crouzon syndrome include hearing loss, deformity of middle ears, absence of ear canals, Ménière's disease (dizziness, vertigo, or ringing in the ears), vision problems, crossed eyes or involuntary eye movement, curvature of the spine, headaches, fused joints (in some cases), and acanthosis nigricans (small, dark, velvety patches of skin).

SCREENING AND DIAGNOSIS

A doctor can usually diagnosis Crouzon syndrome at birth or in early childhood based on the patient's physical signs and symptoms. Tests are taken to confirm the diagnosis. They may include X rays, a test that uses radiation to take a picture of structures inside the body, especially bones; a magnetic resonance imaging (MRI) scan, a test that uses magnetic waves to make pictures of the inside of the body; and a computed tomography (CT) scan, a type of X ray that uses a computer to make pictures of the inside of the body. Genetic testing to confirm mutations in the *FGFR2* or *FGFR3* gene may be used if the clinical findings are not sufficient to make a diagnosis.

TREATMENT AND THERAPY

There is no cure yet for Crouzon syndrome. Because the molecular cause is now known, scientists are exploring ways to block the processes that lead to early fusion of the sutures without affecting other important growth processes. These efforts are currently restricted to experimental animals, but human advances may be on the horizon.

Currently, many of the symptoms can be treated with surgery. In addition, orthodontic treatment, eye and ear treatment, and supportive treatment are usually needed. Good dental care is also an important aspect of managing the care of children with Crouzon syndrome.

There are a number of surgeries used to treat the symptoms of Crouzon syndrome. They include craniectomy, which involves removal and replacement of portions of the cranial bone. This surgery is done as early as possible after birth to prevent pressure on and damage to the brain and to maintain a skull shape that is as normal as possible. Surgery to treat exophthalmos (protrusion of one or both eyeballs) is done directly on the eye sockets or on the bones surrounding the eye sockets to help minimize exophthalmos. Surgery to treat protruding lower jaw is often very successful in normalizing the appearance of the jaw by removing a portion of the jawbone. Surgery can also be done to repair a cleft palate.

Braces and other orthodontic treatments are usually necessary to help correct misalignment of teeth. An ophthalmologist (eye specialist) and otolaryngologist (ear, nose, and throat specialist) should monitor infants and children with Crouzon syndrome. These specialists can check for problems and provide corrective treatment as necessary. Supportive treatment for the disease includes special education for children with a mental deficiency or mental retardation.

PREVENTION AND OUTCOMES

There is no known way to prevent Crouzon syndrome. If a patient has Crouzon syndrome or has a family history of the disorder, he or she can talk to a genetic counselor when deciding to have children.

Rick Alan; reviewed by Rosalyn Carson-DeWitt, M.D.
"Etiology and Genetics" by Jeffrey A. Knight, Ph.D.

FURTHER READING

Dalben, Gda S., B. Costa, and M. R. Gomide. "Oral Health Status of Children with Syndromic Craniosynostosis." *Oral Health and Preventive Dentistry* 4, no. 3 (2006); 173-179.

EBSCO Publishing. *Health Library: Crouzon Syndrome.* Ipswich, Mass.: Author, 2009. Available through http://www.ebscohost.com.

Kjaer, I., et al. "Abnormal Timing in the Prenatal Ossification of Vertebral Column and Hand in Crouzon Syndrome." *American Journal of Medical Genetics* 90, no. 5 (February 28, 2000): 386-389.

Perlyn C. A., et al. "A Model for the Pharmacological Treatment of Crouzon Syndrome." *Neurosurgery* 59, no. 1 (July, 2006): 210-215.

WEB SITES OF INTEREST

The Centre for Craniofacial Care and Research, Sick Kids
http://www.sickkids.ca/craniofacial

Cleft Palate Foundation
http://www.cleftline.org

Crouzon Syndrome
http://www.familyvillage.wisc.edu/lib_crouz.htm

Faces: The National Craniofacial Association
http://www.faces-cranio.org

"FGFR-Related Craniosynostosis Syndromes." Gene Reviews
http://www.ncbi.nlm.nih.gov/bookshelf/br.fcgi ?book=gene&part=craniosynostosis

Genetics Home Reference
http://ghr.nlm.nih.gov

National Institute of Dental and Craniofacial Research
http://www.nidcr.nih.gov

See also: Congenital defects; Diastrophic dysplasia; Fibrodysplasia ossificans progressiva; Hereditary diseases.

Cystic fibrosis

CATEGORY: Diseases and syndromes
ALSO KNOWN AS: CFTR-related disorders, mucoviscidosis

DEFINITION

Cystic fibrosis is a life-limiting, multisystem, autosomal recessive disorder that results from a defective channel in the epithelial cell membrane that is responsible for chloride transport. Progressive, chronic lung problems, pancreatic insufficiency, endocrine abnormalities, and infertility are the major health problems associated with this disease.

RISK FACTORS

The primary risk factor for cystic fibrosis is having two abnormal copies of the cystic fibrosis transmembrane conductance regulator (*CFTR*) gene. However, both genetic and nongenetic modifiers exist that can affect the course of the disease. Cystic fibrosis most commonly occurs in the Caucasian population with an incidence of approximately 1 in 3,200; it occurs in all other populations, but with less frequency.

ETIOLOGY AND GENETICS

The *CFTR* gene is located on chromosome 7 at band q31.2. The gene is large, containing 180,000 base pairs and 1,480 amino acids. More than 1,500 disease-associated mutations have been detected, which are classified according to their effect on the function of the CFTR protein. Class I mutations cause no protein to be made, class II mutations prevent the protein from reaching its location in the cell membrane, class III mutations result in problems with the function of the protein, class IV mutations result in the reduced ability of the channel to transport chloride across the membrane, and class V mutations cause a reduced amount of functioning CFTR protein to be produced. Some genotype-phenotype correlation is possible, primarily in relation to determining whether an affected individual with be pancreatic sufficient or insufficient. Unfortunately, genotype-phenotype correlation remains poor for determining the severity of lung disease.

Cystic fibrosis follows an autosomal recessive inheritance pattern. Everyone has two copies of the *CFTR* gene; individuals with cystic fibrosis have a mutation in each of their copies of the gene that causes it to malfunction. Usually an affected individual has inherited one malfunctioning copy of the gene from each parent, who are both carriers of the disease. Carriers of cystic fibrosis have one functioning and one malfunctioning copy of the gene. Their functioning copy allows for enough normal transport of chloride across cell membranes that they do not develop cystic fibrosis. If two parents are carriers, they have, in each pregnancy, a 25 percent chance to have an affected child. In addition, they have a 50 percent chance to have a child who is a carrier and a 25 percent chance to have a child who is neither a carrier of nor affected with cystic fibrosis.

The effects of abnormal transport of chloride across the cell membrane are best understood in the sweat gland. Secretion of sweat across the gland is modified before it reaches the skin. Usually, sodium, followed by chloride, is reabsorbed through the gland through both sodium channels and the CFTR protein. When the CFTR protein is not functioning, limited amounts of chloride are reabsorbed, which in turn limits the amount of reabsorption of sodium. The sweat contains large amounts of sodium, which can lead to salt loss syndromes.

The problem with chloride transport in lung epithelial cells is less clear. The low-volume model contends that malfunctioning CFTR causes increased sodium, chloride (through other pathways), and fluid absorption. Airway surfaces become dehydrated; mucus becomes thick and viscous and cannot be eliminated from the lungs. The high-salt model contends that with chloride unable to be reabsorbed efficiently, more sodium chloride will be present in the airway surface liquid. The high salt content disables some of the body's immune defense mechanisms, causing affected individuals to be more susceptible to bacterial infection. It is also possible that defective chloride transport plays a role in the inflammatory process in the lungs, a theory that is supported by the increased inflammatory response in affected individuals prior to, or in the absence of, infection. The last hypothesis suggests that a natural immune response of normally functioning CFTR protein helps eliminate bacteria from the lung, and when the CFTR protein malfunctions this immune response is disabled.

In general, all organ systems affected in cystic fibrosis have problems with chloride and fluid secretion. The thick, sticky mucus blocks pancreatic ducts (pancreatic insufficiency) and prevents the release of enzymes into the intestine, which help us digest and obtain nutrients from our food. Fibrosis and replacement of pancreatic tissue with fatty deposits also interfere with pancreatic functioning and can lead to additional problems. Obstruction of the liver bile duct can lead to cirrhosis. Thickening of cervical mucus in women and absent or abnormal formation of the vas deferens in men occurs as well.

Current research is focusing on eradication of bacterial infection in the lung and maximizing and prolonging lung functioning. In addition, researchers are exploring other routes for chloride reabsorption in epithelial cells and developing therapies that correspond to the effect on protein function of the cystic fibrosis mutation classes.

SYMPTOMS

Symptoms and their age of onset vary greatly in cystic fibrosis, even between individuals with the same genotype. Prenatally, an echogenic bowel is a sign of cystic fibrosis. About 15 percent of babies with cystic fibrosis are born with meconium ileus, a fatal condition if not treated. Infants and children may have poor weight gain, frequent loose and greasy stools, and recurrent respiratory infections with colonization in the lung of specific bacterial pathogens. In addition, pancreatitis, diabetes developing in adolescence, nasal polyps, and cirrhosis of the liver can occur. About 95 percent of men are infertile because of abnormalities of the vas deferens, and women's fertility may be reduced. Any individual, even an adult with only some of the above symptoms, should be referred for clinical evaluation.

SCREENING AND DIAGNOSIS

Cystic fibrosis is diagnosed by either a positive sweat chloride test, genetic testing identifying two known mutations in the *CFTR* gene, and/or an abnormal nasal transepithelial potential difference (NPD). Genetic testing should always be confirmed by a sweat chloride test or NPD because, with the variability in clinical symptoms that can be associated with *CFTR* gene mutations, it is becoming more difficult to determine exactly what health problems should constitute a diagnosis of cystic fibrosis. *CFTR* mutations have been found in healthy men who are infertile as a result of congenital bilateral absence of the vas deferens (CBVAD) and in adults who have only pancreatitis. Interpretation of genetic test results should be performed by a genetics professional that is familiar with cystic fibrosis.

Sequencing of the *CFTR* gene remains labor intensive, and clinical significance of novel mutations is often difficult to determine. Therefore, genetic screening of common, clearly understood mutations has been developed, and two types are available to the general population. The first is genetic carrier screening. Uncommon mutations are not detected, and a negative screen does not eliminate a person's risk to be a carrier of cystic fibrosis, but reduces it. The amount of risk reduction depends on the number of mutations that are screened and the person's ethnicity. Genetic screening is available to all couples considering pregnancy but has primarily been marketed to the Caucasian population. In addition, newborn screening is performed in many states. The level of immunoreactive trypsinogen (IRT) is measured and when abnormal, is followed by either a repeat measurement of IRT in one to three weeks or with genetic screening of commonly occurring *CFTR* mutations. Follow-up diagnostic testing, usually via a sweat test, is required, as false-positive results can occur on the newborn screen.

TREATMENT AND THERAPY

Respiratory problems and infections that are associated with cystic fibrosis are treated with antibiotics and medication to dilate air passages. Chest physiotherapy, in which drainage of mucus from the

---

Gene Therapy for Cystic Fibrosis

Once scientists discovered the cystic fibrosis gene, *CF*, and its protein product, cystic fibrosis transmembrane regulator (CFTR), attempts at gene therapy were quickly initiated. Since most of the life-threatening complications of this disease are seen in the respiratory system, that system became the main target for gene replacement therapy.

Early attempts at gene therapy involved the attachment of a functional *CF* gene to a virus that acts as a vector and the subsequent introduction of this virus to the respiratory epithelium in an aerosol. Several problems arose. Although a cystic fibrosis patient's immune system does not function properly, especially in the respiratory system, the immune system is active enough to prevent many of the viruses from entering the target cells. Those that did penetrate and inserted the normal *CF* gene induced only a transient benefit. This most likely occurred because of the high turnover rate of surface epithelial cells. The epithelial cells could incorporate the gene that codes for normal CFTR, but cells that had not been repaired would soon replace the repaired cells. Continued aerosol applications were also not helpful, because the body began producing antibodies to the viral vector, which further reduced the virus's ability to enter cells and introduce an active *CF* gene. Another problem was the inflammation caused by the virus itself.

To surmount some of these difficulties, other approaches have been tried. A team of Australian researchers has looked at preconditioning the respiratory epithelium with a detergent-like substance found in normal lungs as a way of increasing virus uptake by the epithelium. This system has had success in mice and has led to longer-term improvements of lung function. These researchers speculate that long-tem improvement occurs when some epithelial stem cells have had defective DNA replaced by the DNA for functional CFTR. In Cleveland, researchers have tried to insert the *CF* gene directly into cells without a viral vector. They have accomplished this by compacting the DNA into a particle small enough to enter the cell.

Another novel gene therapy has been labeled SMaRT by its proponents. This therapy takes advantage of the need to remove introns (noncoding intervening sequences) from pre-messenger RNA (pre-mRNA) in eukaryotes and then to splice the exons (coding sequences) together to form functional mRNA. In this procedure, multiple copies of a "minigene" that contain a good copy of the exon that normally contains the defect in the *CF* gene are introduced to the epithelial cells. When the pre-mRNA is processed, there are so many more copies of the corrected exon that it is usually spliced into the CFTR mRNA. This technique has the advantage of not disrupting the cells' normal regulation of the CFTR protein. However, the viruses involved in the transfer of the minigenes face the same barriers that all viral vectors face in cystic fibrosis gene therapy.

Richard W. Cheney, Jr., Ph.D.

At the National Heart, Lung, and Blood Institute in Bethesda, Maryland, Dr. Ronald Crystal works on a nasal spray of proteins that have been genetically engineered to break down mucus in the lungs of individuals with cystic fibrosis. (AP/ Wide World Photos)

lungs is assisted by percussion of the chest, is performed one or more times per day. Lung transplantation may be an option for individuals whose disease has become severe. Weight gain is optimized by nutritional supplementation through medication, diet, and vitamins. Artificial reproductive technologies may be helpful in managing fertility problems.

PREVENTION AND OUTCOMES

Newborn screening has provided one significant area of prevention of some of the major problems associated with cystic fibrosis. Prompt nutritional supplementation can prevent many of the secondary problems that occur because of malnutrition and has been associated with improved lung functioning later in life. Strict adherence to the type and amount of chest physiotherapy and compliance with taking medications necessary to open lung airways

can minimize the occurrence of airway obstruction and recurrence of infection. The first isolation of typical cystic fibrosis-associated bacteria in airway secretions is treated aggressively with antibiotics in hopes of eradicating the bacteria from the lung. Early detection of diabetes and liver problems allows for prompt management of these complications, which play a significant role in the mortality and morbidity of cystic fibrosis.

Life expectancy for an individual with the typical health problems of cystic fibrosis has now extended into the mid-thirties; however, there is significant variation. With the onset of newborn screening, improvements in the understanding of the function of the *CFTR* gene, what factors modify its functioning, and the underlying pathophysiology of cystic fibrosis, life expectancy will continue to increase and quality of life will continue to improve.

Heather F. Mikesell, M.S.

FURTHER READING

Cutting, Gary. *Emory and Rimoin's Principles and Practice of Medical Genetics.* 4th ed. Vol 2. New York: Churchill Livingstone, 2002. A comprehensive, detailed summary of human genetics and genetic disorders.

Farrell, Philip M., et al. "Guidelines for Diagnosis of Cystic Fibrosis in Newborns Through Older Adults: Cystic Fibrosis Foundation Consensus Report." *Journal of Pediatrics* 153 (2008): S4-S14. A review of the protocols for diagnosing cystic fibrosis at various times in life and under different types of clinical presentations.

Nussbaum, Robert L., Roderick R. McInnes, and Huntington F. Willard. *Thompson and Thompson Genetics in Medicine.* 7th ed. New York: Saunders, 2007. A nice introduction to basic genetic principles.

O'Sullivan, Brian P., and Steven D. Freedman. "Cystic Fibrosis." *Lancet* 373 (2009): 1891-1904. A comprehensive, current review of the genetics, diagnosis, symptoms, treatment, and future research of cystic fibrosis.

WEB SITES OF INTEREST

Canadian Cystic Fibrosis Foundation
www.ccff.ca

Cystic Fibrosis Foundation
www.cff.org

GeneTests
www.genetests.org

See also: Amniocentesis and chorionic villus sampling; Biochemical mutations; Chromosome mutation; Chromosome walking and jumping; Congenital defects; Gene therapy; Gene therapy: Ethical and economic issues; Genetic counseling; Genetic engineering; Genetic screening; Genetic testing: Ethical and economic issues; Hereditary diseases; Human genetics; Inborn errors of metabolism; Multiple alleles.

Cytokinesis

CATEGORY: Cellular biology

SIGNIFICANCE: Cytokinesis is a process, usually occurring concurrent with mitosis, in which the cytoplasm and organelles are divided into two new cells. In eukaryotes, mitosis and meiosis involve division of the nucleus, while cytokinesis is the division of the cytoplasm.

KEY TERMS

binary fission: cell division in prokaryotes in which the plasma membrane and cell wall grow inward and divide the cell in two

cell cycle: a regular and repeated sequence of events during the life of a cell; it ends when a cell completes dividing

daughter cells: cells that result from cell division

interphase: the phase that precedes mitosis in the cell cycle, a period of intense cellular activities that include DNA replication

meiosis: a type of cell division that leads to production of gametes (sperm and egg) during sexual reproduction

mitosis: nuclear division, a process of allotting a complete set of chromosomes to two daughter nuclei

EVENTS LEADING TO CYTOKINESIS

Cytokinesis is the division or partitioning of the cytoplasm during the equal division of genetic material into the daughter cells. Before a cell can divide, its genetic material, DNA, has to be duplicated through DNA replication. The identical copies of DNA are then separated into one of the two daughter cells through a multistep process, which varies among prokaryotes, plants, and animals. With a single chromosome and no nucleus, prokaryotes (such as bacteria) utilize a simple method of cell division called binary fission (meaning "splitting in two"). The single circular DNA molecule is replicated rapidly and split into two. Each of the two circular DNAs then migrates to the opposite pole of the bacterial cell. Eventually, one bacterial cell splits into two through binary fission. On average, a bacterial cell can go through the whole process of cell division within twenty minutes.

In eukaryotes, cell division is a more complex process given the presence of a nucleus and multiple DNA molecules (chromosomes). Each chromosome needs to be replicated in preparation for the division. The replication process is completed during the interphase. Once replicated, the copies of each chromosome, called sister chromatids, are connected together in a region called the centromere. The chromosomes then go through a process of shortening, condensing, and packing with proteins that make them visible using a light microscope. Chromosomes then migrate and line up at the equator of the parent cell. Then the sister chromatids are separated and pulled to opposite poles. These multiple steps include interphase (cell growth and DNA replication), prophase (disintegration of nuclear envelope, formation of spindle fibers, condensation of chromosomes), metaphase (lining up of chromosomes at equator plate), anaphase (split of two sister chromatids), and telophase (completion of migration of chromatids to opposite poles). Although animal and plant cells share many common features in DNA replication and mitosis, some noticeable differences in interphase and cytokinesis exist. Even within the animal kingdom, cytokinesis may vary with the type of cell division. Particularly during oogenesis (the process of forming egg), both meiosis I and meiosis II engage in unequal partitioning of cytoplasm that is distinct from normal mitosis of animal and plant cells. In some cases, a cell will complete mitosis without cytokinesis, resulting in a multinucleate cell.

CYTOKINESIS IN ANIMALS

In animal cells, cytokinesis normally begins during anaphase or telophase and is completed following the completion of chromosome segregation. First, microfilaments attached to the plasma membrane and form a ring around the equator of the cell. This ring then contracts and constricts the cell's equator, forming a cleavage furrow, much like pulling the drawstring around the waist of a pair of sweatpants. Eventually the "waist" is pinched through and contracts down to nothing, partitioning the cytoplasm equally into two daughter cells. Partitioning the cytoplasm includes distributing cellular organelles so each daughter cell has what is needed for cellular processes.

CYTOKINESIS IN PLANTS

Cytokinesis in plant cells is different from that in animal cells. The presence of a tough cell wall (made up of cellulose and other materials) makes it nearly impossible to divide plant cells in the same manner as animal cells. Instead, it begins with formation of a cell plate. In early telophase, an initially barrel-shaped system of microtubules called a phragmoplast forms between the two daughter nuclei. The cell plate is then initiated as a disk suspended in the phragmoplast.

The cell plate is formed by fusion of secretory vesicles derived from the Golgi apparatus. Apparently, the carbohydrate-filled vesicles are directed to the division plane by the phragmoplast microtubules, possibly with the help of motor proteins. The vesicles contain matrix molecules, hemicelluloses, and/or pectins. As the vesicles fuse, their membranes contribute to the formation of the plasma membrane on either side of the cell plate. When enough vesicles have fused, the edges of the cell plate merge with the original plasma membrane around the circumference of the cell, completing the separation of the two daughter cells. In between the two plasma membranes is the middle lamella. Each of the two daughter cells then deposits a primary wall next to the middle lamella and a new layer of primary wall around the entire protoplast. This new wall is continuous with the wall at the cell plate. The original wall of the parent cell stretches and ruptures as the daughter cells grow and expand.

CYTOKINESIS IN SEXUAL REPRODUCTION

In animal oogenesis, the formation of ova, or eggs, occurs in the ovaries. Although the daughter cells resulting from the two meiotic divisions receive equal amounts of genetic material, they do not receive equal amounts of cytoplasm. Instead, during each division, almost all the cytoplasm is concentrated in one of the two daughter cells. In meiosis I, unequal partitioning of cytoplasm during cytokinesis produces the first polar body almost void of cytoplasm, and the secondary oocyte with almost all cytoplasm from the mother cell. During meiosis II, cytokinesis again partitions almost all cytoplasm to one of the two daughter cells, which will eventually grow and differentiate into a mature ovum, or egg. Another daughter cell, the secondary polar body, receives almost no cytoplasm. This concentration of cytoplasm is necessary for the success of sexual reproduction because a major function of the mature ovum is to nourish the developing embryo following fertilization.

Ming Y. Zheng, Ph.D.

FURTHER READING

Grant, Michael C. "The Trembling Giant." *Discover* 14, no. 10 (October, 1993): 82. Excellent illustrations on asexual reproduction (by reference to the aspen tree) through mitosis of plant cells and tissues.

Karp, Gerald. "M Phase: Mitosis and Cytokinesis." In *Cell and Molecular Biology: Concepts and Experiments.* 5th ed. Chichester, England: John Wiley and Sons, 2008. Detailed accounts of cellular reproduction in a standard textbook for professionals and undergraduate majors.

Murray, A. W., and Tim Hunt. *The Cell Cycle: An Introduction.* New York: W. H. Freeman, 1993. An informative overview for both students and general readers, without too much scientific jargon. Bibliographical references, index.

Murray, A. W., and M. W. Kirschner. "What Controls the Cell Cycle." *Scientific American* 264, no. 3 (March, 1991): 56. An illuminating description of a group of proteins that are involved in cell cycle control. The synthesis, processing, and degradation of these proteins seems to regulate the progression of a cell through various stages of the cell cycle.

Rappaport, R. *Cytokinesis in Animal Cells.* New York: Cambridge University Press, 1996. Reprint. 2005. Describes division in different kinds of cells and explains the mechanisms underlying this division.

Shaul, Orit, Marc van Montagu, and Dirk Inze. "Regulation of Cell Divisions in *Arabidopsis.*" *Critical Reviews in Plant Sciences* 15 (1996): 97-112. A review of what is known about plant cell cycle regulation and cell divisions. For serious students.

Staiger, Chris, and John Doonan. "Cell Divisions in Plants." *Current Opinion in Cell Biology* 5 (1993): 226-231. A condensed version on plant cell divisions. Provides a quick overview.

WEB SITES OF INTEREST

The Cytokinetic Mafia
http://www.bio.unc.edu/faculty/salmon/lab/mafia/mafia.html
The "mafia," a group of scientists who are "passionate" about cytokinesis, compiled this site on the subject. The site's "Cytokinesis Movies" section enables users to watch films demonstrating the cytokinetic process in selected plants and animals.

Scitable
http://www.nature.com/scitable/topicpage/Mitosis-and-Cell-Division-205
Scitable, a library of science-related articles compiled by the Nature Publishing Group, features a the article "Mitosis and Cell Division," which uses text and illustrations to describe cytokinesis and other aspects of the cell division process.

See also: Cell cycle; Cell division; Mitosis and meiosis; Polyploidy; Totipotency.

D

Dandy-Walker syndrome

CATEGORY: Diseases and syndromes

ALSO KNOWN AS: Dandy-Walker malformation; familial Dandy Walker; Dandy Walker malformation

DEFINITION

Dandy-Walker syndrome is a brain deformity present at birth consisting of a deformity of the cerebellum and the presence of a cyst in the lower portion of the brain. The deformity involves an area in the back of the brain that controls movement and cognitive learning. In many cases, there is also an abnormal accumulation of cerebrospinal fluid within the ventricles of the vein. The symptoms of this syndrome may develop suddenly or may go unnoticed.

RISK FACTORS

Dandy-Walker syndrome may be inherited; therefore, having a parent with Dandy-Walker syndrome may increase the risk of occurrence in his or her children. Aside from association with certain inherited genetic conditions, there are no known risk factors. The following factors are associated with Dandy-Walker syndrome, but do not increase the risk of its occurrence: absence of the corpus callosum, which connects the brain's hemispheres; and malformations of the heart, face, limbs, fingers, and toes.

ETIOLOGY AND GENETICS

Dandy-Walker syndrome is a complex condition with highly variable expression, and there appear to be multiple genetic and environmental factors that can contribute to its manifestation. The best evidence for genetic involvement centers around two adjacent genes found on the long arm of chromosome 3 at position 3q24. These two genes, known as *ZIC1* and *ZIC4*, appear to play a role in the development of the cerebellum. The proteins encoded by these genes are known as zinc finger protein of cerebellum 1 and 4. Cytogenetic analysis indicates that several patients exhibit a small deletion of part of the long arm of chromosome 3 that includes the 3q24 band. Researchers have identified homologous genes and proteins in mice and shown that mice with similar chromosomal deletions present with Dandy-Walker-like symptoms. There is considerable optimism that the establishment of an effective mouse model system will hasten the understanding of the genetic factors contributing to this malformation. Sporadic reports in the literature suggest that deletions of the 2q36.1 chromosomal region or deletions on chromosome 9 and the X chromosome may also be associated with Dandy-Walker syndrome.

In all cases where deletion of a chromosomal segment on chromosomes 2, 3, or 9 have been identified, the inheritance pattern will follow a classic autosomal dominant pattern. An affected individual has a 50 percent chance of transmitting the mutation to each of his or her children. In the majority of cases, however, where the genetic and environmental contributing factors are unknown, no predictable pattern of inheritance is possible.

SYMPTOMS

Symptoms of Dandy-Walker syndrome often occur in infancy but can also occur in older children. Eighty percent of cases are diagnosed within the first year of life. Symptoms may include impaired development of normal speech and language, slow motor development, irritability, vomiting, convulsions, unsteadiness, lack of muscle coordination, and jerky eye movements. Other symptoms may include an increased head circumference; bulging of the back of the skull; problems with the nerves that control the eyes, face, and neck; and abnormal breathing. Children with this condition may have problems with other organs, including heart malformations, kidney and urinary tract abnormalities, cleft lip, and extra digits.

Screening and Diagnosis

The doctor will ask about a patient's symptoms and medical history and will perform a physical exam. The doctor will also likely do a computed tomography (CT) scan or magnetic resonance imaging (MRI) scan to view the inside of the brain.

Treatment and Therapy

Patients should talk with their doctors about the best plans. Treatment will depend on the problems caused by the syndrome. This may involve placing a special tube called a shunt inside the skull to drain excess fluid in order to reduce pressure and help control swelling.

Prevention and Outcomes

There is no known way to prevent Dandy-Walker syndrome.

Krisha McCoy, M.S.;
reviewed by J. Thomas Megerian, M.D., Ph.D., F.A.A.P.
"Etiology and Genetics" by Jeffrey A. Knight, Ph.D.

Further Reading

Burton, Barbara K. "Dandy-Walker Malformation." In *Congenital Malformations: Evidence-Based Evaluation and Management*, edited by Praveen Kumar and Burton. New York: McGraw-Hill Medical, 2008.

EBSCO Publishing. *Health Library: Dandy-Walker Syndrome.* Ipswich, Mass.: Author, 2009. Available through http://www.ebscohost.com.

Sarnat, H. B., and L. Flores-Sarnat. "Developmental Disorders of the Nervous System." In *Neurology in Clinical Practice*, edited by Walter G. Bradley et al. 5th ed. 2 vols. Philadelphia: Butterworth-Heinemann/Elsevier, 2008.

Web Sites of Interest

Canadian Neurological Sciences Federation
http://www.ccns.org

Children's Craniofacial Association
http://www.ccakids.com

Health Canada
http://www.hc-sc.gc.ca/index-eng.php

National Institute of Neurological Disorders and Stroke
http://www.ninds.nih.gov

See also: Amyotrophic lateral sclerosis; Congenital defects; Hereditary diseases.

Deafness

CATEGORY: Diseases and syndromes
ALSO KNOWN AS: Hearing loss; hearing impairment

Definition

Deafness means a lack or loss of the sense of hearing, which may be partial or complete. Partial loss of hearing is often called hearing loss rather than deafness. Deafness can occur in one or both ears.

There are three primary types of hearing loss. Conductive loss is hearing loss caused by the inability of the sound to reach the inner ear. This can result from outer or middle ear problems, such as ear infection, excess wax, or swelling. This type of hearing loss is most likely to respond to medical or surgical treatment.

Sensorineural loss is hearing loss caused by disorders of the inner ear or auditory nerve. This type of loss is usually permanent. It can be caused by heredity or congenital problems; excess noise; old age; medications; infections, such as ear infections and meningitis; or from tumors compressing the nerve of hearing, such as an acoustic neuroma.

Mixed loss is a combination of both conductive and sensorineural loss.

Risk Factors

Risk factors for deafness include premature birth; increasing age for age-related hearing loss (presbycusis); taking ototoxic medications; and exposure to loud noise on the job, such as loud industrial noise, use of heavy equipment, or being a musician. Exposure to recreational loud noise, such as guns (target practice) and loud music, and a family history of deafness are also risk factors.

Etiology and Genetics

Genetic factors are responsible for about half of the children born each year with profound or partial hearing loss. Of these, about 75 percent present with deafness as the only or the major symptom, while the remaining 25 percent exhibit any one of several specific genetic syndromes. So many anatomical and neurological factors are essential for normal hearing that it is not surprising that more than thirty-five genes have been identified that, when mutated, can result in nonsyndromic deafness.

One major cause of genetic deafness is a mutation in either of two adjacent genes on the long arm of chromosome 13, known as *GJB2* and *GJB6* (at position 13q11-q12). These genes encode proteins called connexins, which are components of the gap junction channels which allow communication between neighboring cells in the inner ear. Another gene, known as *POU3F4*, is found on the X chromosome (at position Xq21.1) and helps specify shape and function of the stapes, one of the tiny bones in the middle ear. Mutations in this gene are known that cause the stapes to be fixed in place and unable to move with sound vibrations, thereby resulting in deafness. Two recently discovered genes associated with deafness, *MT-RNR1* and *MT-TS1*, are found on mitochondrial deoxyribonucleic acid (DNA) and therefore lack a chromosomal location entirely.

Several different patterns of inheritance are associated with nonsyndromic deafness. An autosomal recessive pattern accounts for about 75 percent of cases, which means that both copies of a particular gene must be deficient in order for the individual to be afflicted. Typically, an affected child is born to two unaffected parents, both of whom are carriers of the recessive mutant allele. About 20 percent of cases of inherited deafness show an autosomal dominant pattern, in which a single copy of the mutation is sufficient to cause full expression of the trait. An affected individual has a 50 percent chance of transmitting the mutation to each of his or her children. Most of the remaining 5 percent of cases exhibit a sex-linked recessive pattern in which mothers who carry the mutated gene on one of their two X chromosomes face a 50 percent chance of transmitting this disorder to each of their male children. Female children have a 50 percent chance of inheriting the gene and becoming carriers like their mothers. Fewer than 1 percent of cases result from mutations in mitochondrial genes, but in these cases deafness is transmitted from a mother to all of her children (fathers do not transmit mitochondrial genes to their offspring).

SYMPTOMS

Hearing loss usually comes on gradually but may come on suddenly. Symptoms may include difficulty hearing, ringing in the ears (tinnitus), dizziness, ear pain in case of an infection, and feeling of ear fullness (as in earwax or fluid).

A symptom of deafness in infants who are one to four months old may be the lack of response to sounds or voices. Symptoms noted at four to eight months may include disinterest in musical toys and lack of verbalization, such as babbling, cooing, and making sounds. A symptom for infants from eight to twelve months old may be the lack of recognition of the child's own name; lack of speech may be a symptom for children between twelve and sixteen months. According to the American Academy of Pediatrics, all children (including newborns) should be screened for hearing loss so that loss occurring before birth can be uniformly detected prior to three months of age.

SCREENING AND DIAGNOSIS

The doctor will ask about a patient's symptoms and medical history, and will perform a physical exam. As part of the diagnosis, the doctor may try to determine the location of the problem, the degree of loss, and the cause; it is not always possible to identify the exact cause of hearing loss. This information can help guide treatment.

Depending on the type of hearing loss, the doctor may order tests to confirm the patient's diagnosis. Tests may include an otoscopy, the examination of the structures inside the ear; a bone vibrator (also called a tuning fork test), which helps to determine the type of hearing loss; an audiogram (also called a hearing test), which measures the degree of hearing loss; and tympanometry, which measures middle ear fluid and pressures. A brain-stem auditory evoked response measures the electrical response in the brain to sounds in order to help determine the exact location of certain hearing problems. Other tests include a computed tomography (CT) scan, a type of X ray that uses a computer to make pictures of the inside of the body, in this case the head; and a magnetic resonance imaging (MRI) scan, which uses magnetic waves to make pictures of the inside of the body, in this case the head.

TREATMENT AND THERAPY

Treatment for deafness depends on the type of hearing loss. Options may include medical treatment, such as removal of earwax or use of antibiotics to treat an ear infection. In selected cases of sudden hearing loss, medical treatment with intratympanic steroids may be effective.

Treatment may also include the use of hearing aids—small devices that are worn in or behind the ear to help amplify sounds.

In some cases, surgery may be recommended to help improve hearing. Types of surgery include stapedectomy for treatment of otosclerosis; tympanoplasty for a perforated eardrum; tympanoplasty tubes for persistent middle ear infections or fluid; and cochlear implant, a surgically implanted electronic device that helps provide sound to a person with severe sensorineural hearing loss (although the devices do not completely restore hearing, improvements in implant technology continue to be made).

PREVENTION AND OUTCOMES

To help prevent deafness, individuals should avoid loud noises. In cases when loud noises cannot be avoided, individuals can reduce exposure to loud noises by wearing earplugs, earmuffs, or ear protectors. In addition, taking steps to reduce injuries or disease may prevent certain types of deafness.

There is currently no effective way to prevent congenital or genetic deafness. Hearing screening for newborns can help insure that hearing loss in young babies is detected and treated at the earliest possible stage.

Michelle Badash, M.S.;
reviewed by Elie Edmond Rebeiz, M.D., FACS
"Etiology and Genetics" by Jeffrey A. Knight, Ph.D.

FURTHER READING

Burkey, John M. *Baby Boomers and Hearing Loss: A Guide to Prevention and Care.* New Brunswick, N.J.: Rutgers University Press, 2006.

EBSCO Publishing. *Health Library: Deafness.* Ipswich, Mass.: Author, 2009. Available through http://www.ebscohost.com.

Martini, Alessandro, Dafydd Stephens, and Andrew P. Read, eds. *Genes, Hearing, and Deafness: From Molecular Biology to Clinical Practice.* Boca Raton, Fla.: Taylor & Frances, 2007.

Myers, David G. *A Quiet World: Living with Hearing Loss.* New Haven, Conn.: Yale University Press, 2000.

Olsen, Wayne, ed. *Mayo Clinic on Hearing.* New York: Kensington, 2003.

Plaza, G., and C. Herráiz. "Intratympanic Steroids for Treatment of Sudden Hearing Loss After Failure of Intravenous Therapy." *Otolaryngology—Head and Neck Surgery* 137, no. 1 (July, 2007): 74-78.

WEB SITES OF INTEREST

Alexander Graham Bell Association for the Deaf and Hard of Hearing
http://www.agbell.org

American Academy of Audiology
http://www.audiology.org

Canadian Academy of Audiology
http://www.canadianaudiology.ca

Canadian Association of the Deaf
http://www.cad.ca

Genetics Home Reference
http://ghr.nlm.nih.gov

MayoClinic.com
http://www.mayoclinic.com/health/hearing-loss/DS00172

National Institute on Deafness and Other Communication Disorders
http://www.nidcd.nih.gov

See also: Alport syndrome.

Depression

CATEGORY: Diseases and syndromes
ALSO KNOWN AS: Major depressive affective disorder; unipolar disorder; unipolar mood disorder

DEFINITION

Depression is a mental illness marked by feelings of profound sadness and lack of interest in activities. Depression is not the same as a blue mood. It is a persistent low mood that interferes with the ability to function and appreciate things in life. It may cause a wide range of symptoms, both physical and emotional. It can last for weeks, months, or years. People with depression rarely recover without treatment.

RISK FACTORS

Females and the elderly are at risk for depression. Other risk factors include chronic physical or mental illness, including thyroid disease, headaches, chronic pain, and stroke; a previous episode of depression; major life changes or stressful life events, such as bereavement or trauma; postpartum depression; the winter season for seasonal affective disorder; little or no social support; low self-esteem; and lack of personal control over an individual's circumstances. Additional risk factors are a family history of depression (parent or sibling); feelings of helplessness; using certain medications, including medications used to treat asthma, high blood pressure, arthritis, high cholesterol, and heart problems; smoking; anxiety; insomnia; personality disorders; and hypothyroidism.

ETIOLOGY AND GENETICS

Major depressive disorder is a condition in which multiple environmental and genetic factors play a contributing part. Some individuals are genetically predisposed to develop the condition, yet a detailed genetic analysis and prediction of inheritance patterns are not possible, since so many different genes seem to be implicated. It has been known for decades that depression tends to run in some families, and twin studies have confirmed that genetics plays a critical role. One study reports that in fraternal twins (who share approximately 50 percent of the same genes), if one twin develops depression the other will also be diagnosed with the condition about 20 percent of the time. In identical twins (who share 100 percent of the same genes), however, the rate of concordant diagnoses of depression rises to 76 percent.

Molecular genetics studies conducted during and since the completion of the Human Genome Project have identified several candidate genes that may play a role in the predisposition for or development of depression. There is considerable disagreement among the researchers, however, so the candidate genes in the following list must be considered as only possible contributing factors. The *TPH1* gene, located on the short arm of chromosome 11 (at position 11p15.3-p14) encodes the enzyme tryptophan hydroxylase 1, which is important for the synthesis of serotonin (a neurotransmitter produced in the brain that may affect mood). The *SLC6A4* gene on the long arm of chromosome 17 (at position 17q11.1-q12) specifies the serotonin transporter protein, and the gene *5HTR2A* (at position 13q14-q21) codes for the serotonin 2A receptor protein. The *COMT* gene on chromosome 22 (at position 22q11.2) encodes the enzyme catechol-o-methyltransferase, which is important for the metabolism of dopamine (another brain neurotransmitter). Finally, the gene *BDNF* (at position 11p13), which specifies the brain-derived neurotrophic factor, may be involved in the etiology of several different neuropsychiatric behaviors

SYMPTOMS

Depression can differ from person to person. Some people have only a few symptoms, while others have many. Symptoms can change over time and may include persistent feelings of sadness, anxiety, or emptiness; hopelessness; feelings of guilt, worthlessness, or helplessness; loss of interest in hobbies and activities; loss of interest in sex; tiredness; trouble concentrating, remembering, or making decisions; and trouble sleeping, waking up too early, or oversleeping. Other symptoms may include eating more or less than usual; weight gain or weight loss; thoughts of death or suicide, with or without suicide attempts; restlessness or irritability; and physical symptoms that defy standard diagnosis and do not respond well to medical treatments.

SCREENING AND DIAGNOSIS

There is no blood test or diagnostic test for depression. The doctor will ask about a patient's symptoms and medical history, giving special attention to alcohol and drug use, thoughts of death or suicide, family members who have or have had depression, sleep patterns, and previous episodes of depression.

The doctor may also perform specific mental health exams; this will help get detailed information about the patient's speech, thoughts, memory, and mood. A physical exam and other tests can help rule out other causes.

TREATMENT AND THERAPY

Treatment may involve the use of medicine, psychotherapy, or both. Severe depression usually requires hospital care and the use of drugs, such as olanzapine.

Up to 70 percent of depressed patients find relief from their symptoms with antidepressant medications, which can take two to six weeks to reach their maximum effectiveness. These medications include selective serotonin reuptake inhibitors (SSRIs), such as fluoxetine (Prozac), sertraline (Zoloft), paroxetine (Paxil), citalopram (Celexa), fluvoxamine (Luvox), and escitalopram (Lexapro).

The U.S. Food and Drug Administration advises that people taking antidepressants should be closely observed. For some, the medications have been linked to worsening symptoms and suicidal thoughts. These adverse effects are most common in young adults. These effects tend to occur at the beginning of treatment or when there is an increase or decrease in the dose.

Although the warning is for all antidepressants, of most concern are the SSRI class, such as Prozac (fluoxetine), Zoloft (sertraline), Paxil (paroxetine), Luvox (fluvoxamine), Celexa (citalopram), and Lexapro (escitalopram).

Another form of treatment is the use of tricyclic antidepressants, such as imipramine (Tofranil), doxepin (Adapin, Sinequan), clomipramine (Anafranil), nortriptyline (Pamelor), and mitriptyline (Elavil); and the use of monoamine oxidase inhibitors (MAOIs), such as phenelzine (Nardil) and tranylcypromine (Parnate). Other antidepressants include venlafaxine (Effexor), nefazodone (Serzone), mirtazapine (Remeron), bupropion (Wellbutrin), and duloxetine (Cymbalta).

Short-term psychotherapy (ten to twenty weeks) can help some people. Psychotherapy is designed to help patients cope with difficulties in relationships, change negative thinking and behavior patterns, and resolve difficult feelings.

Electroconvulsive therapy (ECT) is the use of an electric stimulus to produce a generalized seizure. It may be used in people with severe or life-threatening depression. ECT is also used for people who cannot take or do not respond to medicine. It is considered a safe and effective procedure.

A regular exercise program has been shown to relieve some of the symptoms of depression. It should play a large role in the overall management of depression. Phototherapy treatment is done by sitting under special lights; it usually lasts about thirty minutes every morning.

St. John's wort is an herb that is available without prescription. It is widely used in Europe for the treatment of mild to moderate depression. Studies have shown that the herb is as effective as standard antidepressants and has fewer side effects although it can interfere with some medications.

There is also recent evidence that dehydroepiandrosterone (DHEA), a dietary supplement, may help some people. DHEA is an ingredient in fish oil. Some experts disagree with these findings. Patients always should discuss the use of dietary and herbal supplements with their doctors.

Research suggests that diets high in tryptophan, certain B vitamins, and fish oil may be helpful. They have shown promise in both relieving and preventing depression. Patients should always discuss the use of such supplements with their doctors.

Vagal nerve stimulation (VNS) is used as a therapy for depression when multiple trials of medicine do not work. A pacemaker-like device stimulates the vagus nerve in the neck.

Prevention and Outcomes

Individuals can reduce their chances of becoming depressed by being aware of their personal risks; having psychiatric evaluations and psychotherapy, if needed; developing social supports; learning stress management techniques; exercising regularly; avoiding the abuse of alcohol or drugs; and getting adequate sleep, rest, and recreation.

Amy Scholten, M.P.H.;
reviewed by Rosalyn Carson-DeWitt, M.D.
"Etiology and Genetics" by Jeffrey A. Knight, Ph.D.

Further Reading

Aguirre, Blaise A. *Depression.* Westport, Conn.: Greenwood Press, 2008.

Beck, Aaron T., and Brad A. Alford. *Depression: Causes and Treatments.* 2d ed. Philadelphia: University of Pennsylvania Press, 2009.

EBSCO Publishing. *Health Library: Depression.* Ipswich, Mass.: Author, 2009. Available through http://www.ebscohost.com.

Groves, D. A., and V. J. Brown. "Vagal Nerve Stimulation: A Review of Its Applications and Potential Mechanisms That Mediate Its Clinical Effects." *Neuroscience and Biobehavioral Reviews* 29, no. 3 (May, 2005): 493-500.

Kramer, Peter D. *Against Depression.* New York: Penguin Books, 2005.

Linde, K., M. Berner, and L. Kriston. "St. John's Wort for Major Depression." Available through *EBSCO DynaMed Systematic Literature Surveillance* at http://www.ebscohost.com/dynamed.

See also: Aarskog syndrome; Aggression; Behavior;
Bipolar affective disorder; Sociobiology.

Developmental genetics

CATEGORY: Developmental genetics

SIGNIFICANCE: The discovery of the genes responsible for the conversion of a single egg cell into a fully formed organism has greatly increased our understanding of development. Common developmental mechanisms exist for diverse organisms and experimental manipulation of particular genes could potentially lead to treatments or cures for cancers and developmental abnormalities in humans.

KEY TERMS

differentiation: the process in which a cell establishes an identity that is distinct from its parent cell, usually involving alterations in gene expression

epigenesis: the formation of differentiated cell types and specialized organs from a single, homogeneous fertilized egg cell without any preexisting structural elements

fate mapping: following the movements of a cell and its descendants during development, often through introduction of a temporary or permanent marker into the cell

gene expression: the combined biochemical processes, called transcription and translation, that convert the linearly encoded information in the bases of DNA into the three-dimensional structures of proteins

induction: the process by which a cell or group of cells signals an adjacent cell to pursue a different developmental pathway and so become differentiated from its neighboring cells

morphogen: a chemical compound or protein that influences the developmental fate of surrounding cells by altering their gene expression or their ability to respond to other morphogens

EARLY HYPOTHESES OF DEVELOPMENT IN DIVERSE ORGANISMS

From the earliest times, people have noted that a particular organism produced offspring very much like itself in structure and function, and the fully formed adult consisted of numerous cell types and other highly specialized organs and structures, yet it came from one simple egg cell. How could such simplicity, observed in the egg cell, give rise to such complexity in the adult and always reproduce the same structures?

In the seventeenth century, the "preformationism" hypothesis was advanced to answer these questions by asserting that a miniature organism existed in the sperm or eggs. After fertilization, this miniature creature simply grew into the fully formed adult. Some microscopists of the time claimed to see a "homunculus," or little man, inside each sperm cell. That the preformationism hypothesis was ill-conceived became apparent when others noted that developmental abnormalities could not be explained satisfactorily, and it became clear that another, more explanatory hypothesis was needed to account for these inconsistencies.

In 1767, Kaspar Friedrich Wolff published his "epigenesis" hypothesis, in which he stated that the complex structures of chickens developed from initially homogeneous, structureless areas of the embryo. Many questions remained before this new hypothesis could be validated, and it became clear that the chick embryo was not the best experimental system for answering them. Other investigators focused their efforts on the sea squirt, a simpler organism with fewer differentiated tissues.

Work with the sea squirt, a tiny sessile marine animal often seen stuck to submerged rocks, led to the notion that development followed a mosaic pattern. The key property of mosaic development was that any cell of the early embryo, once removed from its surroundings, grew only into the structure for which it was destined or determined. Thus the early embryo consisted of a mosaic of cell types, each determined to become a particular body part. The determinants for each embryonic cell were found in the cell's cytoplasm, the membrane-bound fluid surrounding the nucleus. Other scientists, most notably Hans Driesch in 1892 and Theodor Boveri (working with sea urchin embryos) in 1907, noted that a two-cell-stage embryo could be teased apart into separate cells, each of which grew into a fully formed sea urchin. These results appeared to disagree with the mosaic developmental mechanism. Working from an earlier theory, the "germ-plasm" theory of August Weismann (1883), Driesch and Boveri proposed a new mechanism called regulative development.

The key property of regulative development was that any cell separated from its embryo could regulate its own development into a complete organism. In contrast to mosaic development, the determinants for regulative development were found in the nuclei of embryonic cells, and Boveri hypothesized that gradients of these determinants, or morphogens, controlled the expression of certain genes. Chromosomes were assumed to play a major role in controlling development; however, how they accomplished this was not known, and Weismann mistakenly implied that genes were lost from differentiated cells as more and more specific structures formed.

In spite of the inconsistencies among the several hypotheses, a grand synthesis was soon formed. Working with roundworm, mollusk, sea urchin, and frog embryos, investigators realized that both mosaic and regulative mechanisms operate during development, with some organisms favoring one mechanism over the other. The most important conclusion coming from these early experiments suggested that certain genes on the chromosomes interacted with both the cytoplasmic and nuclear morphogenetic determinants to control the proliferation and differentiation of embryonic cells. What exactly were these morphogens, where did they originate, and how did they form gradients in the embryo? How did they interact with genes?

THE MORPHOLOGY OF DEVELOPMENT

Before the "how and why" mechanistic questions of morphogens could be answered, more answers to the "what happens when" questions were needed. Using new, powerful microscopes in conjunction with cell-specific stains, many biologists were able to precisely map the movements of cells during embryogenesis and to create "fate maps" of such cell migrations. Fate maps were constructed for sea squirt, roundworm, mollusk, sea urchin, and frog embryos, which showed that specific, undifferentiated cells in the early embryo gave rise to complex body structures in the adult.

In addition, biologists observed an entire stepwise progression of intervening cell types and struc-

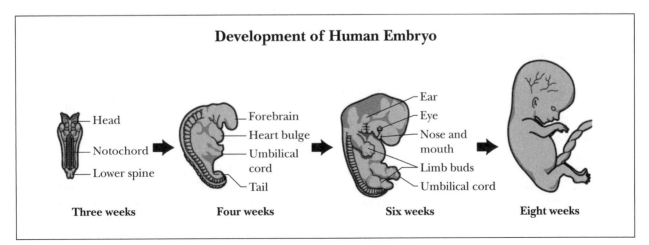

Development of Human Embryo

Three weeks: Head, Notochord, Lower spine

Four weeks: Forebrain, Heart bulge, Umbilical cord, Tail

Six weeks: Ear, Eye, Nose and mouth, Limb buds, Umbilical cord

Eight weeks

(Hans & Cassidy, Inc.)

tures that could be grouped into various stages and that were more or less consistent from one organism to another. Soon after fertilization, during the very start of embryogenesis, specific zones with defining, yet structureless, characteristics were observed. These zones consisted of gradients of different biochemical compounds, some of which were morphogens, and they seemed to function by an induction process. Some of these morphogen gradients existed in the egg before fertilization; thus it became evident that the egg was not an entirely amorphous, homogeneous cell, but one with some amount of preformation. This preformation took the form of specific morphogen gradients.

After these early embryonic events and more cell divisions, in which loosely structured patterns of morphogen gradients were established to form the embryo's polar axes, the cells aggregated into a structure called a "blastula," a hollow sphere of cells. The next stage involved the migration of cells from the surface of the blastula to its interior, a process called gastrulation. This stage is important because it forms three tissue types: the ectoderm (for skin and nerves), the mesoderm (for muscle and heart), and the endoderm (for other internal organs). Continued morphogenesis generates a "neurula," an embryo with a developing nervous system and backbone. During axis formation and cell migrations, the embryonic cells are continually dividing to form more cells that are undergoing differentiation into specialized tissue types such as skin or muscle. Eventually, processes referred to as "organogenesis" transform a highly differentiated embryo into one with distinct body structures that will grow into a fully formed adult.

EXPERIMENTAL SYSTEMS FOR STUDYING DEVELOPMENTAL GENES

In order to understand the details of development, biologists have traditionally studied organisms with the simplest developmental program, those with the fewest differentiated cell types that will still allow them to answer fundamental questions about the underlying processes. Sea squirts and roundworms have been valuable, but they exhibit a predominantly mosaic form of development and are not the best systems for studying morphogen-dependent induction. Frog embryogenesis, with both mosaic and regulative processes, was well described and contributed greatly to answering the "what and when" questions of sequential events, but, at the time, no effective genetic system existed for examining the role of genes in differentiation necessary for answering the "why and how" questions.

Historically, the issue was resolved by focusing once again on the morphogens. These mediators of cellular differentiation were found only in trace amounts in developing embryos and thus were difficult, if not impossible, to isolate in pure form for experimental investigation. An alternative to direct isolation of morphogens was to isolate the genes that make the morphogens. The organism deemed most suitable for such an approach was the fruit fly *Drosophila melanogaster*, even though its development was more complex than that of the roundworm. Fruit flies could be easily grown in large numbers in the laboratory, and many mutants could be generated quickly; most important, an effective genetic system already existed in *Drosophila*, making it easier to create and analyze mutants. The person who best used the fruit fly system and greatly contributed to the understanding of developmental genetics was Christiane Nüsslein-Volhard, who shared a 1995 Nobel Prize in Physiology or Medicine with Edward B. Lewis and Eric Wieschaus.

THE GENES OF DEVELOPMENT

The first important developmental genes discovered in *Drosophila* were the latest acting in morphogenesis, which led to the isolation of the gene for one of the morphogens controlling the anterior-posterior axis of the embryo, the *bicoid* gene. The study of mutants, such as those with legs in place of antennae, allowed the discovery of many other developmental genes, referred to generally as "homeotic" genes.

The *bicoid* gene's discovery validated the gradient hypothesis originally proposed by Boveri because its gene product functioned as a "typical" morphogen. It was a protein that existed in the highest concentration at the egg's anterior pole and diffused to lower concentrations toward the posterior pole, thus forming a gradient. Through the use of more fruit fly mutants, geneticists showed that the bicoid protein stimulated the gene expression of another early gene, called *hunchback*, which in turn affected the expression of other genes: *Krüppel* and *knirps*. The bicoid protein controls the *hunchback* gene by binding to the gene's control region.

Since these initial discoveries, a plethora of new

Caenorhabditis Studies Tracing Cell Fates

Caenorhabditis elegans, a free-living soil nematode (a type of worm) 1 millimeter in length, has proved invaluable as a model organism for studying development. In addition to its small size, it has a rapid life cycle, going from egg to sexual maturity in three and a half days and living only two to three weeks. The presence of rudimentary physiological systems, including digestive, nervous, muscular, and reproductive systems, enables comparative studies between *Caenorhabditis* and "higher" organisms, such as mice and humans. Because the animal is transparent, the formation of every cell in the 959-celled adult can be observed microscopically and manipulated to illuminate its developmental program.

In 1963 Sydney Brennern set out to learn everything there was to know about *Caenorhabditis elegans*. In a 1974 publication he demonstrated how specific mutations could be induced in the *C. elegans* genome through chemical mutagenesis and showed how these mutations could be linked to specific genes and specific effects on organ development. Proving the utility of the organism as a genetic model encouraged a cadre of researchers to pursue research with *C. elegans*.

One of Brenner's students, John Sulston, developed techniques to track and study cell divisions in the nematode, from fertilized egg through adult. Microscopic examination of individual cell nuclei of the animal as it developed, along with electron microscopy of serial sections of the animal, enabled scientists to trace each of the adult worm's 959 cells back to a single fertilized egg. This "lineage map" was then used to track the fates of cells in animals that had been experimentally manipulated. Using a fine laser beam, scientists could kill a single cell at some point in development of the animal, then determine what changes, if any, awaited the remaining cells. These studies proved that the *C. elegans* cell lineage is invariant; that is, every worm underwent exactly the same sequence of cell divisions and differentiation.

Studies on cell fate and lineage mapping also led to the discovery that specific cells in the lineage always die through programmed cell death. Robert Horvitz, another of Brenner's students, discovered two "death genes" in *C. elegans* as well as genes that protect against cell death and direct the elimination of the dead cell. He also identified the first counterparts of the death gene in humans.

The characterization of the invariant cell lineage of *C. elegans* and the genetic linkages have been of great value to understanding basic principles of development, including signaling pathways in multicellular organisms and pathways controlling cell death. This knowledge has been invaluable to medicine, where it has helped researchers to understand mechanisms by which bacteria and viruses invade cells and has provided insights into the cellular mechanisms involved in neurodegenerative diseases, autoimmune disorders, and cancer. For their pioneering work in the "genetic regulation of organ development and programmed cell death" Brenner, Sulston, and Horvitz were awarded the 2002 Nobel Prize in Physiology or Medicine.

Karen E. Kalumuck, Ph.D.;
updated by Crystal L. Muncia, Ph.D.

developmental genes have been discovered. It is now clear that some fifty genes are involved in development of a fruit fly larva from an egg, with yet more genes responsible for development of the larva into an adult fly. These genes are grouped into three major categories: maternal effect genes, segmentation genes, and homeotic genes. Maternal effect genes include the *bicoid* gene. These genes, produced by special "nurse" cells of the mother, make proteins that contribute to the initial morphogen gradients along the egg's axes before fertilization. Segmentation genes comprise three subgroups: gap, pair-rule, and segment polarity genes. Each of these types of segmentation genes determines a different aspect of the segments that make up a developing fruit fly. The *hunchback*, *Krüppel*, and *knirps* genes are all gap genes. Homeotic genes ultimately determine the segment identity of previously differentiated cell groups.

PATTERN FORMATION

Through the use of highly specific stains to track the morphogens in normal and mutant fruit fly embryos, a fascinating picture of the interactions among developmental genes has emerged. Even before fertilization, shallow, poorly defined gradients are established by genes of the mother, such as the *bicoid* gene and related genes. These morphogen gradients establish the anterior-to-posterior and dorsal-to-ventral axes. After fertilization, these morphogens

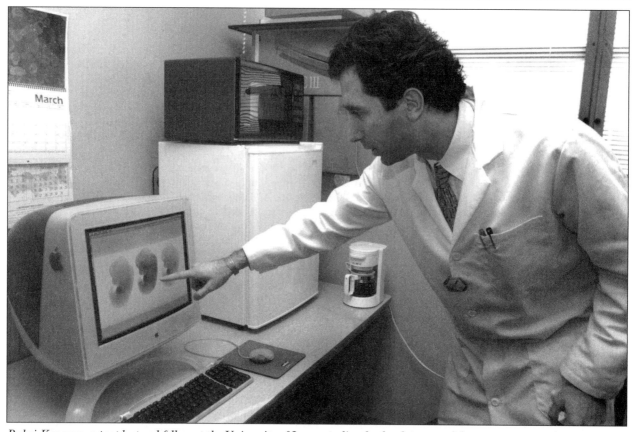

Bahri Karacay, a postdoctoral fellow at the University of Iowa, studies the development of the nervous system in a mouse embryo as part of a project that seeks to treat brain tumors in humans. (AP/Wide World Photos)

bind to the control regions of gap genes, whose protein products direct the formation of broadly defined zones that will later develop into several specific segments. The gap proteins then bind to the control regions of pair-rule genes, whose protein products direct further refinements in the segmentation process. The last group of segmentation genes, the segment polarity genes, direct the completion of the segmentation patterns observable in the embryo and adult fly, including definition of the anterior-posterior orientation of each segment. Homeotic genes then define the specific functions of the segments, including what appendages will develop from each one. Mutations in any of these developmental genes cause distinct and easily observed changes in the developing segment patterns. Genes such as *hunchback, giant, gooseberry,* and *hedgehog* were all named with reference to the specific phenotypic changes that result from improper control of segmentation.

Homeotic genes are often called the "master" genes because they control large numbers of other genes required to make a whole wing or leg. Several clusters of homeotic genes have been discovered in *Drosophila*. Mutations in a certain group of genes of the *bithorax* complex result in adult fruit flies with two sets of wings. Similarly, mutations in some of the genes in the *antennapedia* complex can result in adult fruit flies with legs, rather than antennae, on the head.

A general principle applying to developmental processes in all organisms has emerged from the elegant work with *Drosophila* mutants: Finer and finer patterns of differentiated cells are progressively formed in the embryo along its major axes by morphogens acting on genes in a cascading manner, in which one gene set controls the next in the sequence until a highly complex pattern of differentiated cells results. Each cell within its own patterned zone then responds to the homeotic gene products and contributes to the formation of distinct, identifiable body parts.

Another important corollary principle was substantiated by the genetic analysis of development in *Drosophila* and other organisms: in direct contrast to Weismann's implication about gene loss during differentiation, convincing evidence showed that genes were not systematically lost as egg cells divided and acquired distinguishing features. Even though a muscle cell was highly differentiated from a skin cell or a blood cell, each cell type retained the same numbers of chromosomes and genes as the original, undifferentiated, but fertilized egg cell. What changed in each cell was the pattern of gene expression, so that some genes were actively transcribed, whereas other genes were turned off. The morphogens, working in complex combinatorial patterns during the course of development, determined which genes would stay "on" and which would be turned "off."

MODERN TOOLS FOR STUDYING DEVELOPMENTAL GENETICS

Innovations in genetic manipulation technologies have transformed the study of developmental genetics. Creation of specific DNA alterations—ranging from single base-pair changes (called point mutations) to large-scale deletions or rearrangements of chromosome segments—provides a unique forum for researchers to assess gene function during development. Moreover, conditional mutations, wherein altered gene products are only produced within certain tissues or at specific times during development, allow for the study of gene function within the context of one or more systems of interest. The ability to create designer mutations was made possible through the work of Mario R. Capecchi and Oliver Smithies, who conceptualized and studied the use of homologous recombination for gene modification. Their work, combined with murine embryonic stem cell technologies pioneered by Martin J. Evans, gave birth to the first mice with targeted gene mutations. For these discoveries, Capecchi, Evans, and Smithies were awarded the 2007 Nobel Prize in Physiology or Medicine. With genome sequences complete for a number of model organisms, the possibilities for targeted mutations are limitless.

The zebra fish (*Danio rerio*) is a model organism uniquely suited to studies of developmental genetics. Genetic manipulations are possible with zebra fish that are not possible with higher model organisms. The effects of gene overexpression can be studied in zebra fish through direct injection of mRNA or DNA constructs into one- to four-cell stage embryos. Underexpression or "knock down" approaches to studying gene function in zebra fish include the use of morpholinos, antisense oligonucleotides that bind to complementary mRNA transcripts and either prevent their translation or inhibit proper splicing. Morpholinos have become a popular technique for assessing developmental gene function in zebra fish.

The study of developmental genetics has also benefited from improvements in the ability to visualize cell lineage and movements. Embryonic cells can be tagged with constructs that express markers continuously, at a desired time point, or within a target tissue type. Unlike injectable dyes, labeling is noninvasive and is not diluted by successive cell divisions.

The next frontier for the study of developmental genetics is the integration of the vast resources of molecular, cellular, and genetic data through a systems approach. Rather than assess gene function, cell division, and cell movements in isolation, these data can be integrated through computational methodology to generate graphic models. This technology is still in its infancy, but promises to yield a better understanding of the networks that underpin development.

IMPACT AND APPLICATIONS

The discovery and identification of the developmental genes in *Drosophila* and other lower organisms led to the discovery of similarly functioning genes in higher organisms, including humans. The base-pair sequences of many of the developmental genes, especially shorter subregions coding for sections of the morphogen that bind to the control regions of target genes, are conserved, or remain the same, across diverse organisms. This conservation of gene sequences has allowed researchers to find similar genes in humans. For example, thirty-nine *Hox* genes located in four clusters have been found in mice and humans, even though only eight homeobox genes localized in a single cluster were initially discovered in *Drosophila*. Some of the late-acting human homeobox genes are responsible for such developmental abnormalities as fused fingers and extra digits on the hands and feet. One of the most interesting abnormalities is craniosynostosis, a premature fusion of an infant's skull bones that can cause mental retardation. In 1993, developmental biologist Robert Maxson and his research group at the University of Southern California's Norris Cancer Center were the first to demonstrate that a mutation in a human homeobox gene *MSX2* was directly responsible for craniosynostosis and other bone/limb abnormalities requiring corrective surgeries. Maxson made extensive use of "knockout" mice, genetically engineered mice lacking particular genes, to test his human gene isolates. He and his research group made great progress in understanding the role of the *MSX2* gene as an inducer of surrounding cells in the developing embryo. When this induction process fails because of defective *MSX2* genes, the fate of cells destined to participate in skull and bone formation and fusion changes, and craniosynostosis occurs.

A clear indication of the powerful cloning methods developed in the late 1980's was the discovery and isolation in 1990 of an important mouse developmental gene called *brachyury* ("short tails"). The gene's existence in mutant mice had been inferred from classical genetic studies sixty years prior to its isolation. In 1997, Craig Basson, Quan Yi Li, and a team of coworkers isolated a similar gene from humans and named it T-box brachyury (*TBX5*). Discovered first in mice, the "T-box" is one of those highly conserved subregions of a gene, and it allowed Basson and Li to find the human gene. When mutated or defective in humans, *TBX5* causes a variety of heart and upper limb malformations referred to as Holt-Oram syndrome. *TBX5* codes for an important morphogen affecting the differentiation of embryonic cells into mesoderm, beginning in the gastrulation phase of embryonic development. These differentiated mesodermal cells are destined to form the heart and upper limbs.

One of the important realizations emerging from the explosive research into developmental genetics in the 1990's was the connection between genes that function normally in the developing embryo but abnormally in an adult, causing cancer. Cancer cells often display properties of embryonic cells, suggesting that cancer cells are reverting to a state of uncontrolled division. Some evidence indicates that mutated developmental genes participate in causing cancer. Taken together, the collected data from many isolated human developmental genes, along with powerful reproductive and cloning technologies, promise to lead to cures and preventions for a variety of human developmental abnormalities and cancers.

Chet S. Fornari, Ph.D., and Bryan Ness, Ph.D.;
updated by Crystal L. Murcia, Ph.D.

FURTHER READING

Beurton, Peter, Raphael Falk, and Hans-Jorg Rheinberger, eds. *The Concept of the Gene in Development and Evolution: Historical and Epistemological Perspectives.* New York: Cambridge University Press, 2000. A collection of essays that examines the question of what genes actually are; for philosophers and historians of science.

Bier, Ethan. *The Coiled Spring: How Life Begins.* Cold Spring Harbor, N.Y.: Cold Spring Harbor Laboratory Press, 2000. A basic overview of the development of embryos in both plants and animals.

Cronk, Quentin C. B., Richard M. Bateman, and Julie A. Hawkins, eds. *Developmental Genetics and Plant Evolution.* New York: Taylor & Francis, 2002. Developmental genetics for botanists.

DePamphilis, Melvin L., ed. *Gene Expression at the Beginning of Animal Development.* New York: Elsevier, 2002. Developmental genetics for zoologists.

Gilbert, Scott F. *Developmental Biology.* Sunderland, Mass.: Sinauer Associates, 2003. Presents a detailed description of all aspects of development.

Lewin, Benjamin. *Genes VII.* New York: Oxford University Press, 2001. Includes a comprehensive, clear discussion of genes and development, with excellent illustrations.

Nüsslein-Volhard, Christiane. "Gradients That Organize Embryo Development." *Scientific American,* August, 1996. The Nobel laureate reports on findings from the *Drosophila* studies.

Nüsslein-Volhard, Christiane, and J. Kratzschmar, eds. *Of Fish, Fly, Worm, and Man: Lessons from Developmental Biology for Human Gene Function and Disease.* New York: Springer, 2000. Designed for researchers, a consideration of the next phase of biology following the sequencing of several large genomes accomplished at the turn of the millennium: determining the functions of genes and the interplay between them and their protein products.

Skromne, Isaac, and Victoria E. Prince. "Current Perspectives in Zebrafish Reverse Genetics: Moving Forward." *Developmental Dynamics* 237 (2008): 861-

882. A comprehensive review of techniques used to study developmental genetics in zebra fish.

Slack, Jonathan Michael Wyndham. *Essential Developmental Biology.* 2d ed. Malden, Mass.: Blackwell, 2006. Easy-to-read guide to developmental biology, with discussions of individual model organisms and the development of organ systems.

WEB SITES OF INTEREST

Caltech MRI Atlases
http://atlasserv.caltech.edu/

DevBio.net
http://www.developmentalbiology.net

Embryo Images: Normal and Abnormal Mammalian Development
http://www.med.unc.edu/embryo_images/

Society for Developmental Biology
www.sdbonline.org

Virtual Library of Developmental Biology, Society for Developmental Biology
http://www.sdbonline.org/index.php?option=com
_content&task=view&id=23&Itemid=34

See also: Aging; Animal cloning; Cell cycle; Cell division; Congenital defects; Cytokinesis; DNA structure and function; Evolutionary biology; Genetic engineering; Hereditary diseases; Homeotic genes; In vitro fertilization and embryo transfer; Model organism: *Caenorhabditis elegans*; Model organism: *Drosophila melanogaster*; Model organism: *Mus musculus*; RNA structure and function; Stem cells; Telomeres; Totipotency; X chromosome inactivation.

Diabetes

CATEGORY: Diseases and syndromes

ALSO KNOWN AS: Diabetes mellitus; juvenile, insulin-dependent diabetes, Type I diabetes; adult onset, non-insulin-dependent diabetes, Type II diabetes; gestational diabetes; diabetes insipidus; unspecified diabetes mellitus; prediabetes; "sugar."

DEFINITION

Diabetes mellitus is a syndrome in which the body cannot metabolize glucose (sugar) appropri-

ately. The subsequent sustained elevated levels cause significant damage to the eyes, heart, kidneys, and other organs. Diabetes is a significant public health problem with 26.3 million persons affected in the United States, or 8 percent of the population. An additional 57 million have prediabetes that will often result in Type II diabetes in the future. About 95 percent have adult-onset or Type II diabetes, with around 5 percent with Type I. Diabetes is a disease related to both genetics and environmental or lifestyle factors.

RISK FACTORS

The primary risk factor for Type I diabetes is having a parent or sibling with the disease. The most common type of diabetes, Type II, has multiple risk factors, both genetic and environmental. These include excessive food intake or unhealthy eating habits that result in obesity especially around the waist area, an inactive or sedentary lifestyle, increased age (over forty-five years old), high blood pressure (140/90 mmHg or greater), family history, gestational (during pregnancy) diabetes, and high cholesterol (HDL under thirty-five and triglycerides over 250 mg/dL). African Americans, Hispanic Americans, Pacific Islanders, and Native Americans have a higher incidence of diabetes.

ETIOLOGY AND GENETICS

Diabetes mellitus comprises a number of different diseases, primarily Type I and Type II diabetes. Genetics plays a role in both types of diabetes, although both are thought to result from the interaction between genetics and the environment. In both, the body's ability to process sugars is impaired, with consequences that can lead to death if untreated. Glucose is a simple sugar required by all cells for normal functioning. Most of the body's glucose initially comes from carbohydrates broken down during digestion. Normally, blood glucose rises when carbohydrates are ingested. At a certain level, the blood glucose triggers the pancreas to release insulin, causing the blood glucose level to drop by increasing the uptake in muscle, fat, the liver, and the gut.

Patients with either type of diabetes have difficulty metabolizing glucose, with a subsequent rise in fasting and postprandial (after meals) blood sugar levels. In Type I diabetes, also called juvenile-onset or insulin-dependent diabetes, this is caused by destruction of the insulin-secreting cells in the pancreas. In Type II (adult-onset, maturity-onset, or non-insulin-dependent diabetes), cells become resistant to the effects of insulin even though the pancreas is still producing some insulin.

Genetics plays a significant role in the development of diabetes. Type I diabetes mellitus is a chronic autoimmune disease that results from a combination of genetic and environmental factors. Certain persons are born with a genetic susceptibility to the disease. The genetic basis for developing Type I diabetes appears to involve not so much mutant genes, but rather a bad combination of particular alleles. Some eighteen regions, labeled *IDDM1* to *IDMM18*, of the genome (the complete set of DNA with genes in the nucleus of each cell) are suspect for linking to Type I diabetes. Under primary investigation is *IDDM1*, containing human leukocyte antigen (HLA) complex genes related to immune response proteins. These HLA genes may increase susceptibility to Type I diabetes, but not always. *IDMM2* is the non-HLA insulin gene. Research on the remaining *IDDM3-IDDM18* continues for links to Type I diabetes.

The HLA genes on chromosome 6 assist the body in differentiating its own immune cells from external substances. These immune cells continually watch for small chained amino acids such as those found in tumor cells or infectious bacteria. Under normal circumstances, the immune cells will attack these chained amino acids to protect the body. The *CTLA4* gene that hinders this action has been associated with a number of diseases including Type I diabetes.

In addition, a rare type of autoimmune diabetes, resembling Type I, occurs as part of a syndrome called autoimmune polyendocrinopathy-candidiasis-ectodermal dystrophy (APECED), which is caused by mutation in *AIRE*, an autoimmune regulator gene. Although the function of *AIRE* is not known, expression of the gene has been detected in the thymus, pancreas, and adrenal cortex, and developmental studies suggest that mutations in *AIRE* might cause the thymus (which is integral to proper immune system function) to develop incorrectly.

Diabetes mellitus Type II is the more common type of diabetes. Type II diabetes appears to be a group of diseases, rather than a single disease, in which there are two defects: beta-cell dysfunction, leading to somewhat decreased production of insulin (although elevated levels of insulin also occur); and tissue resistance to insulin. As with Type I, peo-

ple who develop Type II are born with a genetic susceptibility but the development of actual disease may be dependent upon an environmental trigger. Some possible triggers include aging, sedentary lifestyle, and abdominal obesity. Obesity plays a significant role in the development of Type II diabetes. Among North Americans, Europeans, and Africans with Type II diabetes, between 60 and 70 percent are obese.

As with Type I, epidemiologic evidence suggests a strong genetic component to Type II diabetes. In identical twins over forty years of age, the likelihood is about 70 percent that the second twin will develop Type II diabetes once the first twin has developed the disease.

Mutant alleles for a number of genes have been implicated in susceptibility and development of Type II diabetes. The first genes to be implicated were the insulin gene, genes encoding important components of the insulin secretion pathways, and other genes involved in glucose homeostasis. Mutations are diverse and can include not only the genes themselves but also the transcription factors and control sequences. In March, 2008, the National Institutes of Health (NIH) announced that international scientists had confirmed six additional genetic variants connected to Type II diabetes, bringing the total genetic risk factors to sixteen. As more genes and their mutant alleles are discovered, better treatment options should become available, possibly even some tailored to specific types of mutations.

One way to discover susceptibility for Type II is through whole-genome linkage studies of families. Researchers have found the genes calpain 10 (*CAPN10*) and hepatocyte nuclear factor 4 alpha (*HNF4A*) are suspect for Type II diabetes. The *CAPN10* gene has been linked to high rates of Type II diabetes in Mexican Americans. A mutated *CAP10* may in some way alter insulin secretion as well as affect liver glucose production. Likewise the *HNF4A* gene transcription factor found around chromosome 20 is linked to Type II diabetes. *HNF4A* located in the liver is related to embryo development, and *HNF4A* found in the beta cells of the pancreas is related to insulin secretion. Many other genes are under study for their impact on Type II diabetes.

SYMPTOMS

In Type I, the first recognizable symptom is a condition called prediabetes in which the usual insulin release in response to elevated blood sugar levels in the blood is diminished. At a certain point, commonly between the ages of ten and fourteen, the person develops full-blown diabetes, with excessive thirst and urination, as well as weight loss despite adequate or increased caloric intake. In Type II diabetes, symptoms may develop slowly over time and include excessive thirst and hunger, frequent urination, unexplained fatigue or weight loss, impaired healing of sores, higher incidence of infections, and blurred vision.

SCREENING AND DIAGNOSIS

Screening people at high risk but without symptoms can lead to early diagnosis and avert long-term chronic disease resulting from lack of therapeutic intervention. The American Diabetes Association recommends screening based on risks such as advanced age, family history, personal gestational history, and central obesity (apple-shaped body type with fat around the waist and upper body). The practice of screening is controversial but diabetes often goes undetected in the early stages and therefore untreated. Some research shows that screening is not cost-effective, while others state that this method of prevention can save the healthcare system the high cost of treatment for complications from untreated diabetes.

The methods of screening for diabetes generally begin with a random plasma glucose test. If this yields abnormal values, either the fasting plasma glucose test (FPG) or the fasting two-hour oral glucose tolerance test (GTT) is used. Values greater than 140 mg/dL for the FPG or greater than 200 mg/dL on the GTT require further assessment and intervention.

TREATMENT AND THERAPY

Treatment for Type I diabetes includes regular blood glucose monitoring and management with insulin. The person with Type I may need lifestyle changes to optimize self-care and minimize the possibility of other complications from the disease such as ketoacidosis. Choosing a healthy diet with regular meals, balanced with adequate activity and insulin, is essential for disease management. Consultation with a registered dietitian may be useful to choose meals and snacks with the proper amounts of carbohydrates and fats.

Type II diabetes treatment requires similar approaches, but the patient may try initial control with diet and exercise. If that approach is ineffective, therapy can progress to oral medications that increase tissue sensitivity to circulating insulin, stimulate increased insulin secretion, or alter insulin action. Later, insulin therapy may be necessary. Even with medication, successful therapy must include weight control through regular physical activity and diet modification.

Once the genetic factors related to diabetes have been completely elucidated for all types of diabetes, treatments to modify the genes may become a reality. Genome technology could remove the risks of side effects currently caused by treatment with medications.

PREVENTION AND OUTCOMES

Although genetics has a definite role in the development of diabetes, personal choice can also impact the prevention of this disease. The primary prevention approaches for diabetes include choosing a healthy lifestyle and maintaining normal weight. Regular physical activity, balanced diet with adequate fiber and whole grains, weight loss to optimal level for the person's height and build, not smoking, and early screening for those at high risk are important. The Centers for Disease Control and Prevention (CDC) recommends that people eat right and be active.

Both types of diabetes lead to increased risk of heart and vascular disease, kidney problems, blindness, neurological problems, and other serious medical consequences. Related health concerns include increased infections, delayed healing, foot and skin problems, depression, neuropathy (nerve damage), impaired vision, gingivitis, and dental disease.

Rebecca Lovell Scott, Ph.D., PA-C,
and Bryan Ness, Ph.D.;
updated by Marylane Wade Koch, M.S.N., R.N.

FURTHER READING

Gibson, Greg. *It Takes a Genome: How a Clash Between Our Genes and Modern Life Is Making Us Sick.* Upper Saddle River, N.J.: Pearson Education, 2009. The author investigates the connection between human genes and the stressed culture that results in illness and disease.

Lowe, William L., Jr., ed. *Genetics of Diabetes Mellitus.* Boston: Kluwer Academic, 2001. An in-depth, scientifically based book written by multiple experts in the field of diabetes research.

McConkey, Edwin H. *How the Human Genome Works.* Sudbury, Mass.: Jones and Bartlett, 2004. Although the target audience are professionals in the health sciences, mature lay readers who want more information on the concept of the human genome can benefit from this book.

Milchovich, Sue K., and Barbara Dunn-Long. *Diabetes Mellitus: A Practical Handbook.* 8th ed. Boulder, Colo.: Bull, 2003. This basic reference book contains comprehensive information on living with diabetes.

Notkins, Abner Louis. "Immunologic and Genetic Factors in Type I Diabetes." *The Journal of Biological Chemistry* 277, no. 46 (2002): 43, 545-543, 548. An overview of the major lines of evidence used to consider Type I diabetes primarily an autoimmune disease provides specifics about the gene defects involved in Type I diabetes.

Pavenec, Michal, et al. "Direct Linkage of Mitochondrial Genome Variation to Risk Factors for Type 2 Diabetes in Conplastic Strains." *Genome Research* 17 (August, 2007): 1319-1326. This explores variation in mitochondrial DNA (mtDNA) and risk factors for diabetes.

Roep, B. O. "News and Views: Diabetes—Missing Links." *Nature* 450 (December 6, 2007): 799. This update provides specific confirmation about genes connected to risk factors for Type I diabetes in the major histocompatibility complex (MHC) genome region.

Silander, Kaisa, et al. "Genetic Variation Near the Hepatocyte Nuclear Factor-4α Gene Predicts Susceptibility to Type 2 Diabetes." *Diabetes* 53, no. 4 (April, 2004): 1141-1149. This study of Finish sibling pair families with Type II diabetes establishes *HNF4A* as the likely common gene and recommends further study.

WEB SITES OF INTEREST

American Diabetes Association
http://www.diabetes.org

Diabetes Genome Anatomy Project
http://www.diabetesgenome.org/

Genetic Landscape of Diabetes
http://www.ncbi.nlm.nih.gov/books/bv.fcgi?call=bv.View..ShowTOC&rid=diabetes.TOC&depth=1

Human Genome Project Information. "Medicine and Genetics"
http://www.ornl.gov/sci/techresources/Human_Genome/medicine/medicine.shtml

National Human Genome Research Institute
http://www.genome.gov/

National Institute of Diabetes & Digestive & Kidney Diseases
http://www.niddk.nih.gov

Your Genes, Your Choice
http://www.ornl.gov/sci/techresources/Human_Genome/publicat/genechoice/yourgenes.pdf

See also: Autoimmune disorders; Bacterial genetics and cell structure; Biopharmaceuticals; Cloning; Diabetes insipidus; Gene therapy: Ethical and economic issues; Genetic engineering; Heart disease; Hereditary diseases; Inborn errors of metabolism.

Diabetes insipidus

CATEGORY: Diseases and syndromes

DEFINITION

There are two forms of diabetes insipidus (DI). Central diabetes insipidus (central DI) is caused by inadequate antidiuretic hormone (ADH). Nephrogenic diabetes insipidus (NDI) is due to renal cells in the kidneys that do not respond to ADH.

RISK FACTORS

Among the factors that increase an individual's chance of developing diabetes insipidus are damage to the hypothalamus due to surgery, infection, tumor, or head injury; having polycystic kidney disease or another kidney disease that may affect the filtration process; and the use of certain medications, such as lithium, amphotericin B, or demeclocycline. Other risk factors include having high blood levels of calcium and low blood levels of potassium.

ETIOLOGY AND GENETICS

Nephrogenic diabetes insipidus can be either inherited or acquired. In cases of inherited disease, mutations in either of two different genes may be responsible. The great majority of inherited cases (perhaps as many as 90 percent) result from mutations in a gene called *AVPR2*, found on the long arm of the X chromosome at position Xq28. The product of this gene is the arginine vasopressin receptor-2 protein, and its normal function is to recognize and bind antidiuretic hormone in the kidneys. When the gene is mutated, the altered protein fails to respond to the hormone, and normal fluid balance regulation in the kidneys is disrupted. The inheritance pattern of this form of the disease is typical of all sex-linked recessive mutations (those found on the X chromosome). Mothers who carry the mutated gene on one of their two X chromosomes face a 50 percent chance of transmitting this disorder to each of their male children. Female children have a 50 percent chance of inheriting the gene and becoming carriers like their mothers. Although females rarely express the condition fully, female carriers may occasionally show minor manifestations. Affected males will pass the mutation on to all of their daughters but to none of their sons.

Mutations in a different gene, *AQP2*, are responsible for the remaining 10 percent of cases of inherited diabetes insipidus. Located on the long arm of chromosome 12 at position 12q13, this gene encodes a protein known as aquaporin-2, which is a pore-forming protein that permits the passage of water through the plasma membranes of certain kidney cells. It normally functions to reabsorb water from the urine, but mutations in the gene can result in an inactive pore protein that absorbs too little water, yielding an abundance of dilute urine and resulting dehydration. This form of the disease is inherited as an autosomal recessive, meaning that both copies of the *APQ2* gene must be deficient in order for the individual to be afflicted. Typically, an affected child is born to two unaffected parents, both of whom are carriers of the recessive mutant allele. The probable outcomes for children whose parents are both carriers are 75 percent unaffected and 25 percent affected. If one parent has diabetes insipidus of this type and the other is a carrier, there is a 50 percent probability that each child will be affected.

SYMPTOMS

Individuals who have any of the symptoms of diabetes insipidus should not assume that it is due to the disease. These symptoms may be caused by other health conditions. Individuals should see their doctors if they experience any one of the symptoms. Extreme thirst with preference for cold drinks is a symptom of central DI. Other symptoms of diabetes insipidus include muscle weakness; headache; fever; blurred vision; low blood pressure; rapid pulse; frequent urination, especially during the night (nocturia); and dehydration.

SCREENING AND DIAGNOSIS

The doctor will ask about a patient's symptoms and medical history. A physical exam may be done. Tests may include blood tests to check for electrolytes, ADH levels, or blood sugar. Urinalysis tests the urine specific gravity and/or osmolality (which measures how concentrated or dilute the urine is). A water deprivation test is done only under a doctor's supervision and measures urine output for a twenty-four-hour period. Diabetes insipidus can cause as much as 4 to 10 liters of urine to be excreted per day. In central DI, urine output is suppressed by a dose of vasopressin/ADH; in NDI, urine output is not suppressed by a dose of vasopressin/ADH. Other tests include a magnetic resonance imaging (MRI) scan of the head if central DI is suspected.

TREATMENT AND THERAPY

Individuals should talk with their doctors about the best plans for them. Treatment options for central DI include a synthetic form of ADH; this drug can be taken by mouth, inhaled through the nose, or injected. A diuretic "water pill" or an antidiabetic medication may also be taken in mild cases to boost the ADH effect on the renal cells in the kidney.

A diuretic "water pill" can be used to treat NDI. If lithium is causing the problem, another diuretic, amiloride, can be used.

In both central DI and NDI, symptoms can often be reduced by decreasing the amount of sodium in the diet and by medications called thiazide diuretics (diuretics that conserve water loss and decrease urine output in people with diabetes insipidus).

PREVENTION AND OUTCOMES

There are no known ways to prevent diabetes insipidus. It is wise for patients to seek medical attention promptly if they have excessive urination and thirst.

Diane Voyatzis Norwood, M.S., RD, CDE;
reviewed by Rosalyn Carson-DeWitt, M.D.
"Etiology and Genetics" by Jeffrey A. Knight, Ph.D.

FURTHER READING

Alpern, Robert J., and Steven C. Hebert. *Seldin and Giebisch's The Kidney: Physiology and Pathophysiology.* 4th ed. 2 vols. Boston: Elsevier Academic Press, 2008.

Beers, Mark H., et al. *The Merck Manual of Diagnosis and Therapy.* 18th ed. Whitehouse Station, N.J.: Merck Research Laboratories, 2006.

EBSCO Publishing. *DynaMed: Diabetes Insipidus.* Ipswich, Mass.: Author, 2009. Available through http://www.ebscohost.com/dynamed.

_____. *Health Library: Diabetes Insipidus.* Ipswich, Mass.: Author, 2009. Available through http://www.ebscohost.com.

Garofeanu, C. G., et al. "Causes of Reversible Nephrogenic Diabetes Insipidus: A Systematic Review." *American Journal of Kidney Diseases* 45, no. 4 (April, 2005): 626-637.

Majzoub, J. A., and A. Srivatsa. "Diabetes Insipidus: Clinical and Basic Aspects." *Pediatric Endocrinology Reviews* 4, supp. 1 (December, 2006): 60-65.

Rivkees, S. A., N. Dunbar, and T. A. Wilson. "The Management of Central Diabetes Insipidus in Infancy: Desmopressin, Low Renal Solute Load Formula, Thiazide Diuretics." *Journal of Pediatric Endocrinology and Metabolism* 20, no. 4 (April, 2007): 459-469.

Sands, J. M., and D. G. Bichet. "Nephrogenic Diabetes Insipidus." *Annals of Internal Medicine* 144, no. 3 (February 7, 2006): 186-194.

Toumba, M., and R. Stanhope. "Morbidity and Mortality Associated with Vasopressin Analogue Treatment." *Journal of Pediatric Endocrinology and Metabolism* 19, no. 3 (March, 2006): 197-201.

WEB SITES OF INTEREST

Canadian Diabetes Association
http://www.diabetes.ca

Diabetes Insipidus Foundation
http://www.diabetesinsipidus.org

"Diabetes Insipidus." Medline Plus
http://www.nlm.nih.gov/medlineplus/ency/article/000377.htm

Health Canada
http://www.hc-sc.gc.ca/index-eng.php

National Kidney and Urologic Diseases Information Clearinghouse
http://kidney.niddk.nih.gov

NDI Foundation
http://www.ndif.org

See also: Autoimmune disorders; Diabetes; Hereditary diseases; Inborn errors of metabolism.

Diastrophic dysplasia

CATEGORY: Diseases and Syndromes
ALSO KNOWN AS: Diastrophic dwarfism; broad bone-platyspondylic variant; DTD; DD

DEFINITION

Diastrophic dysplasia is an inherited disorder of bone and cartilage development. It causes bone, cartilage, and joint abnormalities throughout the body.

RISK FACTORS

Those whose biological parents both carry a mutant copy of the *SLC26A2* gene are at risk of inheriting diastrophic dysplasia. This disease is found in all populations and occurs equally in men and women. It is particularly prevalent in Finland.

ETIOLOGY AND GENETICS

Diastrophic dysplasia is an autosomal recessive genetic disease and is caused by mutations in the *SLC26A2* gene. Someone must possess two copies of the mutant form of this gene to have this disorder. The *SLC26A2* gene is located on the long arm of chromosome 5, in band regions 32-33. If both parents carry one mutant copy of *SLC26A2*, then each sibling has a 25 percent chance of receiving two mutant copies and contracting diastrophic dysplasia.

The *SLC26A2* gene encodes the information for the synthesis of a protein called solute carrier family 26, member 2. This protein is embedded in the cell membrane. The cell membrane is composed of phosphate-containing lipids that border the cell and delimit the cell interior from the exterior. The structure of the cell membrane prevents charged and large polar molecules from entering or exiting the cell. If the cell needs such molecules, then specific transport proteins inserted into the membrane facilitate the import or export of particular molecules. Solute carrier family 26, member 2 is a transport protein that allows the entrance of sulfate ions into cells.

Sulfate ions are essential for the production of normal cartilage. A major component of cartilage is a group of proteins called proteoglycans. Proteoglycans are proteins with long sugar chains attached to them, but these sugar molecules also have sulfate ions linked to them. The cells that produce cartilage are called chondrocytes, and if these cells do not possess a normal solute carrier family 26, member 2, then they cannot properly import sulfate ions for the synthesis of proteoglycans and they will make abnormal cartilage.

Cartilage also establishes the pattern for bone development. The long bones of the arms, legs, and other structures initially develop as long cartilage rods that are replaced later by bone (endochondral ossification). Because the cartilage precursors act as templates for the bones, if the original cartilage rods are abnormal, then the bones that replace them will also be abnormal. Chondrocytes in individuals with diastrophic dysplasia lack the ability to import sulfate ions and make normal proteoglycans, and therefore, they make structurally abnormal cartilage that is replaced by abnormal bone. The ends of the cartilage rod make the joint-specific cartilage, which is also abnormal in diastrophic dysplasia.

In developing humans, not all cartilage is used to make bone. Many structures, like joints, the voice box (larynx), external ears, and the windpipe (trachea) are made, largely, from cartilage. These structures are also abnormal in individuals with diastrophic dysplasia and often do not function properly.

SYMPTOMS

The main characteristics of this disease include short stature and short arms and legs. The joints show permanent shortening (contractures). The feet tend to turn downward and inward (club feet). The thumbs are placed farther back on the hand (hitchhiker thumbs). The spine is abnormally curved (kyphosis). About one-third of babies with diastrophic dysplasia are born with a hole in the roof of the mouth (cleft palate). Two-thirds of newborn children also show swollen ears.

SCREENING AND DIAGNOSIS

Ultrasound can detect clinical features such as shortened limbs with a normal-sized skull, a small chest, hitchhicker thumbs, and joint contractures as early as twelve weeks of gestation. X-ray analysis reveals poorly developed and malformed bones. Tissue (histopathological) analysis reveals abnormal cartilage that contains too few sulfate-containing proteoglycans. Molecular genetic testing can confirm the diagnosis.

TREATMENT AND THERAPY

In children, physical therapy and casting can help joint problems, and surgery can correct club feet. In young adults, a surgical technique called arthroplasty, which replaces abnormal joints with synthetic articulations made from cobalt chromoly and high molecular weight polyethylene, relieves pain and increases hip and knee joint mobility. Spinal surgery can fix excessive curvature of the spine. One caveat with surgical therapies is that deformities tend to reoccur after orthopedic surgery.

PREVENTION AND OUTCOMES

Rib and windpipe abnormalities can prevent proper breathing, which causes increased death rates in newly born babies with diastrophic dysplasia. If the baby survives, then surgical corrections probably will be required to allow the child to walk and to reduce the abnormal curvature of the spine. Spinal curvature and the health of the joints should be checked annually.

Obesity tends to place too great a load on the joints and should be avoided. If they survive early childhood, then children with diastrophic dysplasia have normal intelligence and can excel in academic, social, and artistic endeavors.

Michael A. Buratovich, Ph.D.

FURTHER READING

Moore, L. Keith, and T. V. N. Persuad. *Before We Are Born.* 7th ed. Philadelphia: Saunders, 2008. A very readable and beautifully illustrated summary of human development. The section on the formation of bones and cartilage, and their developmental abnormalities is particularly useful.

Read, Andrew, and Dian Donnai. *New Clinical Genetics.* Bloxham, Oxfordshire, England: Scion, 2007. Excellent introduction to the basic and advanced aspects of medical genetics.

Schwartz, Nancy B. "Carbohydrate Metabolism II: Special Pathways and Glycoconjugates." In *Textbook of Biochemistry with Clinical Correlations,* edited by Thomas M. Devlin. 5th ed. New York: Wiley-

Liss, 2002. A well-written but technical chapter that discusses the function, synthesis, and structure of proteoglycans and other components of cartilage.

WEB SITES OF INTEREST

GeneReviews: Diastrophic Dysplasia
http://www.ncbi.nlm.nih.gov/bookshelf/br.fcgi
?book=gene&part=diastrophic-d

Genetics Home Reference: Diastrophic Dysplasia
http://ghr.nlm.nih.gov/condition
=diastrophicdysplasia

The MAGIC Foundation
http://www.magicfoundation.org

See also: Crouzon syndrome; Fibrodysplasia ossificans progressiva; Hereditary diseases.

DiGeorge syndrome

CATEGORY: Diseases and syndromes
ALSO KNOWN AS: Velocardiofacial syndrome; chromosome 22q11 deletion syndrome

DEFINITION

DiGeorge syndrome is a rare genetic disease present at birth and is associated with recurrent infection, heart defects, and characteristic facial features. People with complete DiGeorge syndrome have no thymus or parathyroid glands.

RISK FACTORS

Maternal diabetes is thought to increase the risk of DiGeorge syndrome.

ETIOLOGY AND GENETICS

DiGeorge syndrome results from the deletion (loss) of a small part of the long arm of chromosome 22 at position 22q11.2. Most people with this deletion are missing about two million base pairs of deoxyribonucleic acid (DNA), which corresponds to about forty genes, although some patients have shorter deletions in the same region. The degree of phenotypic expression in different individuals seems to be related to the size of the deletion. There have been sporadic reports of patients with

similar symptoms who have deletions on other chromosomes (at positions 10p13, 17p13, and 18q21).

Efforts to identify a single one or two genes in the 22q11.2 band that are most critical to the expression of the syndrome have largely been unsuccessful, and it appears likely that the diminished expression of several related genes affecting common developmental processes is responsible. Three genes, however, have been identified that may be particularly important. The *TBX1* gene, which encodes the T-box transcription factor 1, is probably responsible for the heart defects characteristic of the syndrome. A gene called *HIRA* specifies another transcriptional regulator that acts early in the development of the nervous system, and the gene *UFD1L* codes for a protein involved in the degradation pathway of ubiquinated compounds.

The 22q11.2 deletion characteristic of DiGeorge syndrome can be inherited, but this is only rarely the case in new diagnoses. More than 90 percent of affected individuals have a de novo deletion, and fewer than 10 percent have an affected parent. Apparently the 22q11.2 region has a chromosomal structure that makes it occasionally susceptible to breakage during genetic recombination events that occur naturally during sperm and egg development.

SYMPTOMS

Features of DiGeorge syndrome are present at birth and do not worsen with age. Features may include immune deficiency leading to increased infections, cleft palate, heart defects, failure to thrive, small head, increased incidence of psychiatric disorders, and characteristic facial features, including elongated face, almond-shaped eyes, wide nose, and small ears. Other features of the disease may be learning difficulties; hypoparathyroidism, a disorder in which insufficient parathyroid hormone is secreted from the parathyroid glands, resulting in abnormally low levels of calcium in the blood; weak muscles; and short height. Occasional abnormalities include structural brain defects, scoliosis, umbilical or inguinal hernias, kidney abnormalities, anogenital abnormalities, eye abnormalities, thyroid problems, and tapered and hyperextensible fingers.

SCREENING AND DIAGNOSIS

The doctor will ask about a patient's symptoms and medical history and will perform a physical exam. Other tests may include blood tests to rule out other conditions, detect parathyroid hormone levels, and discover immune problems; genetic tests to look for deletions in chromosome 22; and a chest X ray, a test that uses radiation to take pictures of structures inside the body, especially to determine if the thymus is present.

TREATMENT AND THERAPY

Patients should talk with their doctors about the best plans. In infants, thymic tissue transplantation or bone marrow transplantation may help restore immune function, but the risks and benefits of these procedures must be carefully considered.

PREVENTION AND OUTCOMES

There is no known way to prevent DiGeorge syndrome.

Krisha McCoy, M.S.; reviewed by Kari Kassir, M.D.
"Etiology and Genetics" by Jeffrey A. Knight, Ph.D.

FURTHER READING

Baker-Gomez, Sherry. *Missing Genetic Pieces: Strategies for Living with VCFS, the Chromosome 22q11 Deletion.* Phoenix, Ariz.: Desert Pearl, 2004.

Cutler-Landsman, Donna, ed. *Educating Children with Velo-Cardio-Facial Syndrome.* San Diego: Plural, 2007.

EBSCO Publishing. *DynaMed: DiGeorge Syndrome.* Ipswich, Mass.: Author, 2009. Available through http://www.ebscohost.com/dynamed.

_____. *Health Library: DiGeorge Syndrome.* Ipswich, Mass.: Author, 2009. Available through http://www.ebscohost.com.

Jones, Kenneth L. *Smith's Recognizable Patterns of Human Malformation.* 6th ed. Philadelphia: Elsevier Saunders, 2006.

Kleigman, Robert M., et al., eds. *Nelson Textbook of Pediatrics.* 18th ed. Philadelphia: Saunders Elsevier; 2007.

Shprintzen, Robert J., and Karen J. Golding-Kushner. *Velo-Cardio-Facial Syndrome.* 2 vols. San Diego: Plural, 2008-2009.

WEB SITES OF INTEREST

Canadian Directory of Genetic Support Groups
http://www.lhsc.on.ca/programs/medgenet/c_sup.htm

"DiGeorge Syndrome." Genes and Disease, National Center for Biotechnology Information
http://www.ncbi.nlm.nih.gov/disease/DGS.html

"DiGeorge Syndrome." MayoClinic.com
http://www.mayoclinic.com/health/digeorge-syndrome/DS00998

Genetics Home Reference
http://ghr.nlm.nih.gov

Health Canada
http://www.hc-sc.gc.ca/index-eng.php

Immune Deficiency Foundation
http://www.primaryimmune.org

See also: Hereditary diseases.

Dihybrid inheritance

CATEGORY: Classical transmission genetics

SIGNIFICANCE: The simultaneous analysis of two different hereditary traits may produce more information than the analysis of each trait separately. In addition, many important hereditary traits are controlled by more than one gene. Traits controlled by two genes serve as an introduction to the more complex topic of traits controlled by many genes.

KEY TERMS

alleles: different forms of the same gene; any gene may exist in several forms having very similar but not identical DNA sequences

dihybrid: an organism that is heterozygous for both of two different genes

heterozygous: a condition in which the two copies of a gene in an individual (one inherited from each parent) are different alleles

homozygous: a condition in which the two copies of a gene in an individual are the same allele; synonymous with "purebred"

MENDEL'S DISCOVERY OF DIHYBRID INHERITANCE

Austrian botanist Johann Gregor Mendel was the first person to describe both monohybrid and dihybrid inheritance. When he crossed purebred round-seed garden peas with purebred wrinkled-seed plants, they produced only monohybrid round seeds. He planted the monohybrid round seeds and allowed them to fertilize themselves; they subsequently produced ¾ round and ¼ wrinkled seeds. He concluded correctly that the monohybrid generation was heterozygous for an allele that produces round seeds and another allele that produces wrinkled seeds. Since the monohybrid seeds were round, the round allele must be dominant to the wrinkled allele. He was able to explain the 3:1 ratio in the second generation by assuming that each parent contributes only one copy of a gene to its progeny. If *W* represents the round allele and *w* the wrinkled allele, then the original true-breeding parents are *WW* and *ww*. When eggs and pollen are produced, they each contain only one copy of the gene. Therefore the monohybrid seeds are heterozygous *Ww*. Since these two alleles will separate during meiosis when pollen and eggs are produced, ½ of the eggs and pollen will be *W* and ½ will be *w*. Mendel called this "segregation." When the eggs and pollen combine randomly during fertilization, ¼ will produce *WW* seeds, ½ will produce *Ww* seeds, and ¼ will produce *ww* seeds. Since *W* is dominant to *w*, both the *WW* and *Ww* seeds will be round, producing ¾ round and ¼ wrinkled seeds. When Mendel crossed a purebred yellow-seed plant with a purebred green-seed plant, he observed an entirely analogous result in which the yellow allele (*G*) was dominant to the green allele (*g*).

Once Mendel was certain about the nature of monohybrid inheritance, he began to experiment with two traits at a time. He crossed purebred round, yellow pea plants with purebred wrinkled, green plants. As expected, the dihybrid seeds that were produced were all round and yellow, the dominant form of each trait. He planted the dihybrid seeds and allowed them to fertilize themselves. They produced ⁹⁄₁₆ round, yellow seeds; ³⁄₁₆ round, green seeds; ³⁄₁₆ wrinkled, yellow seeds; and ¹⁄₁₆ wrinkled, green seeds. Mendel was able to explain this dihybrid ratio by assuming that in the dihybrid flowers, the segregation of *W* and *w* was independent of the segregation of *G* and *g*. Mendel called this "independent assortment." Thus, of the ¾ of the seeds that are round, ¾ should be yellow and ¼ should be green, so that ¾ × ¾ = ⁹⁄₁₆ should be round and yellow, and ¾ × ¼ = ³⁄₁₆ should be round and green. Of the ¼ of the seeds that are wrinkled, ¾ should be yellow and ¼ green, so that ¼ × ¾ = ³⁄₁₆ should be wrinkled and yellow, and ¼ × ¼ = ¹⁄₁₆ should be wrinkled and green. This relationship can be seen in the table headed "Dihybrid Inheritance and Sex Linkage."

SEX CHROMOSOMES

Humans and many other species have sex chromosomes. In humans, normal females have two X chromosomes and normal males have one X and one Y chromosome. Therefore, sex-linked traits,

Dihybrid Inheritance and Sex Linkage

Eggs \ Pollen	W;G	W;g	w;G	w;g
W;G	W W;G G round, yellow	W W;G g round, yellow	W w;G G round, yellow	W w;G g round, yellow
W;g	W W;G g round, yellow	W W;g g round, green	W w;G g round, yellow	W w;g g round, green
w;G	W w;G G round, yellow	W w;G g round, yellow	w w;G G wrinkled, yellow	w w;G g wrinkled, yellow
w;g	W w;G g round, yellow	W w;g g round, green	w w;G g wrinkled, yellow	w w;g g wrinkled, green

Note: Semicolons indicate that the two genes are on different chromosomes.

Mixed Sex-Linked and Autosomal Traits

	Sperm			
	A;R	*a;R*	*A;*Y	*a;*Y
A;R	*A A;R R* normal female	*A a;R R* normal female	*A A;R* Y normal male	*A a;R* Y normal male
A;r	*A A;R r* normal female	*A a;R r* normal female	*A A;r* Y red-green color-blind male	*A a;r* Y red-green color-blind male
a;R	*A a;R R* normal female	*a a;R R* albino female	*A a;R* Y normal male	*a a;R* Y albino male
a;r	*A a;R r* normal female	*a a;r r* albino female	*A a;r* Y red-green color-blind male	*a a;r* Y albino, red-green color-blind male

(Eggs — row labels at left)

Note: Semicolons indicate that the two genes are on different chromosomes.

which are controlled by genes on the X or Y chromosome, are inherited in a different pattern than the genes that have already been described. Since there are few genes on the Y chromosome, most sex-linked traits are controlled by genes on the X chromosome.

Every daughter gets an X chromosome from each parent, and every son gets an X from his mother and a Y from his father. Human red-green color blindness is controlled by the recessive allele (*r*) of an X-linked gene. A red-green color-blind woman (*rr*) and a normal man (*R*Y) will have normal daughters (all heterozygous *Rr*) and red-green color-blind sons (*r*Y). Conversely, a homozygous normal woman (*RR*) and a red-green color-blind man (*r*Y) will have only normal children, since their sons will get a normal X from the mother (*R*Y) and the daughters will all be heterozygous (*Rr*). A heterozygous woman (*Rr*) and a red-green color-blind man (*r* Y) will have red-green color-blind sons (*r* Y) and daughters (*rr*), and normal sons (*R*Y) and daughters (*Rr*) in equal numbers.

A dihybrid woman who is heterozygous for red-green color blindness and albinism (a recessive trait that is not sex linked) can make four kinds of eggs with equal probability: *R;A*, *R;a*, *r;A*, and *r;a*. A normal, monohybrid man who is heterozygous for al-

binism can make four kinds of sperm with equal probability: *R;A*, *R;a*, Y;*A*, and Y;*a*. By looking at the table headed "Mixed Sex-Linked and Autosomal Traits," it is easy to predict the probability of each possible kind of child from this mating.

The probabilities are $6/16$ normal female, $2/16$ albino female, $3/16$ normal male, $3/16$ red-green color-blind male, $1/16$ albino male, and $1/16$ albino, red-green color-blind male. Note that the probability of normal coloring is $1/4$ and the probability of albinism is $1/4$ in both sexes. There is no change in the inheritance pattern for the gene that is not sex linked.

OTHER EXAMPLES OF DIHYBRID INHERITANCE

A hereditary trait may be controlled by more than one gene. To one degree or another, almost every hereditary trait is controlled by many different genes, but often one or two genes have a major effect compared with all the others, so they are called single-gene or two-gene traits. Dihybrid inheritance can produce traits in various ratios, depending on what the gene products do. A number of examples will be presented, but they do not exhaust all of the possibilities.

The comb of a chicken is the fleshy protuberance that lies on top of the head. There are four forms of the comb, each controlled by a different combination of the two genes that control this trait. The first gene exists in two forms (*R* and *r*), as does the second (*P* and *p*). In each case, the form represented by the uppercase letter is dominant to the other form. Since there are two copies of each gene (with the exception of genes on sex chromosomes), the first gene can be present in three possible combinations: *RR*, *Rr*, and *rr*. Since *R* is dominant, the first two combinations produce the same trait, so the symbols *R_* and *P_* can be used to represent either of the two combinations. Chickens with *R_;P_* genes have what is called a walnut comb, which looks very much like the meat of a walnut. The gene combinations *R_;pp*, *rr;P_*, and *rr;pp* produce combs that are called rose, pea, and single, respectively. If two chickens that both have the gene combination *Rr;Pp* mate, they will produce progeny that are 9/16 walnut, 3/16 rose, 3/16 pea, and 1/16 single.

White clover synthesizes small amounts of cyanide, which gives clover a bitter taste. There are some varieties that produce very little cyanide (sweet clover). When purebred bitter clover is crossed with some varieties of purebred sweet clover, the progeny are all bitter. However, when the hybrid progeny is allowed to fertilize itself, the next generation is 9/16 bitter and 7/16 sweet. This is easy to explain if it is assumed that bitter/sweet is a dihybrid trait. The bitter parent would have the gene combination *AA;BB* and the sweet parent *aa;bb*, where *A* and *B* are dominant to *a* and *b*, respectively. The bitter dihybrid would have the gene combination *Aa;Bb*. When it fertilized itself, it would produce 9/16 *A_;B_*, which would be bitter, and 3/16 *A_;bb*, 3/16 *aa;B_*, and 1/16 *aa;bb*, all of which would be sweet. Clearly, both the *A* allele and the *B* allele are needed in order to synthesize cyanide. If either is missing, the clover will be sweet.

ABSENCE OF DOMINANCE

In all of the previous examples, there was one dominant allele and one recessive allele. Not all genes have dominant and recessive alleles. If a purebred snapdragon with red flowers (*RR*) is crossed with a purebred snapdragon with white flowers (*rr*), all the monohybrid progeny plants will have pink flowers (*Rr*). The color depends on the number of *R* alleles present: two *R*s will produce a red flower, one *R* will produce a pink flower, and no *R*s will produce a white flower. This is an example of partial dominance or additive inheritance.

Consider a purebred red wheat kernel (*AA;BB*) and a purebred white wheat kernel (*aa;bb*) (see the table headed "Partial Dominance"). If the two kernels are planted and the resulting plants are crossed with each other, the progeny dihybrid kernels will

Partial Dominance

		Pollen			
		A;B	*A;b*	*a;B*	*a;b*
Eggs	*A;B*	A A;B B red	A A;B b medium red	A a;B B medium red	A a;B b light red
	A;b	A A;B b medium red	A A;b b light red	A a;B b light red	A a;b b very light red
	a;B	A a;B B medium red	A a;B b light red	a a;B B light red	a a;B b very light red
	a;b	A a;B b light red	A a;b b very light red	a a;B b very light red	a a;b b white

Note: Semicolons indicate that the two genes are on different chromosomes. Dihybrid ratios may change if both genes are on the same chromosome.

be light red (*Aa;Bb*). If the dihybrid plants grown from the dihybrid kernels are allowed to self-fertilize, they will produce ¹⁄₁₆ red (*AA;BB*), ⁴⁄₁₆ medium red (*AA;Bb* and *Aa;BB*), ⁶⁄₁₆ light red (*AA;bb, Aa;Bb*, and *aa;BB*), ⁴⁄₁₆ very light red (*Aa;bb* and *aa;Bb*), and ¹⁄₁₆ white (*aa;bb*). The amount of red pigment depends on the number of alleles (*A* and *B*) that control pigment production. Although it may appear that this is very different than the example in the first table, they are in fact very similar.

All of the inheritance patterns that have been discussed are examples of "independent assortment," in which the segregation of the alleles of one gene is independent of the segregation of the alleles of the other gene. That is exactly what would be expected from meiosis if the two genes are not on the same chromosome. If two genes are on the same chromosome and sufficiently close together, they will not assort independently and the progeny ratios will not be like any of those described. In that case, the genes are referred to as "linked" genes.

James L. Farmer, Ph.D.

FURTHER READING

Madigan, Michael M., et al., eds. *Brock Biology of Microorganisms.* 12th ed. San Francisco: Pearson/ Benjamin Cummings, 2009. A college-level text organized into units on the principles of microbiology, evolutionary microbiology and microbial diversity, immunology and pathogenicity, microbial diseases, and microorganisms as tools for industry and research.

Mauseth, James D. "Genetics." In *Botany: An Introduction to Plant Biology.* 3d ed. Sudbury, Mass.: Jones and Bartlett, 2003. Includes an explanation of dihybrid crosses.

Thomas, Alison. "Dihybrid Inheritance." In *Introducing Genetics: From Mendel to Molecule.* Cheltenham, England: Nelson Thornes, 2003. This introductory genetics textbook devotes a chapter to an explanation of dihybrid inheritance.

Tortora, Gerard J., Berdell R. Funke, and Christine L. Case. *Microbiology: An Introduction.* 10th ed. San Francisco: Pearson Benjamin Cummings, 2010. An accessible introduction to the basic principles of microbiology, the interaction between microbe and host, and human diseases caused by microorganisms. Provides a general overview of antibiotics and how bacteria become resistant to them.

Wolf, Jason B., Edmund D. Brodie III, and Michael J. Wadel. *Epistasis and the Evolutionary Process.* New York: Oxford University Press, 2000. Primary focus is on the role of gene interactions (epistasis) in evolution. Leading researchers examine how epistasis impacts the evolutionary processes in overview, theoretical, and empirical chapters.

WEB SITE OF INTEREST

Scitable
http://www.nature.com/scitable/topicpage/Gregor-Mendel-and-the-Principles-of-Inheritance-593

Scitable, a library of science-related articles compiled by the Nature Publishing Group, features an article about Gregor Mendel's principles of inheritance that includes an illustrated section explaining dihybrid crosses. Another article in the site, "Epistasis: Gene Interaction and Phenotype Effects," also discusses Mendel's experiments with dihybrid crosses in pea plants and other information about epistatic relationships involving two genes.

See also: Chromosome theory of heredity; Classical transmission genetics; Complete dominance; Dihybrid inheritance; Epistasis; Incomplete dominance; Linkage maps; Mendelian genetics; Monohybrid inheritance; Multiple alleles.

DNA fingerprinting

CATEGORY: Human genetics and social issues

SIGNIFICANCE: DNA fingerprinting includes a variety of techniques in which individuals are uniquely identified through examination of specific DNA sequences that are expected to vary widely among individuals. Uses for these technologies include not only practical applications in forensic analysis and paternity tests but also basic research in paternity, breeding systems, and ecological genetics for many nonhuman species.

KEY TERMS

microsatellite: a type of VNTR in which the repeated motif is 1 to 6 base pairs; synonyms include simple sequence repeat (SSR) and short tandem repeat (STR)

minisatellite: a type of VNTR in which the repeated motif is 12 to 500 base pairs in length *polymerase chain reaction (PCR):* a laboratory procedure for making millions of identical copies of a short DNA sequence

variable number tandem repeat (VNTR): a type of DNA sequence in which a short sequence is repeated over and over; chromosomes from different individuals frequently have different numbers of the basic repeat, and if many of these variants are known, the sequence is termed a hypervariable

GENETIC DIFFERENCES AMONG INDIVIDUALS

All individuals, with the exception of twins and other clones, are genetically unique. Theoretically it is therefore possible to use these genetic differences, in the form of DNA sequences, to identify individuals or link samples of blood, hair, and other features to a single individual. In practice, individuals of the same species typically share the vast majority of their DNA sequences; in humans, for example, well over 99 percent of all of the DNA is identical. For individual identification, this poses a problem: Most of the sequences that might be examined are identical (or nearly so) among randomly selected individuals. The solution to this problem is to focus only on the small regions of the DNA that are known to vary widely among individuals. These regions, termed hypervariable, are typically based on repeat sequences in the DNA.

Imagine a simple DNA base sequence, such AAC (adenine-adenine-cytosine), which is repeated at a particular place (or locus) on a human chromosome. One chromosome may have eleven of these AAC repeats, while another might have twelve or thirteen, and so on. If one could count the number of repeats on each chromosome, it would be possible to specify a diploid genotype for this chromosomal locus: An individual might have one chromosome with twelve repeats, and the other with fifteen. If there are many different chromosomal variants in the population, most individuals will have different genotypes. This is the conceptual basis for most DNA fingerprinting.

DNA fingerprint data allow researchers or investigators to exclude certain individuals: If, for instance, a blood sample does not match an individual, that individual is excluded from further consideration. However, if a sample and an individual match, this is not proof that the sample came from that individual; other individuals might have the same genoytpe. If a second locus is examined, it

becomes less likely that two individuals will share the same genotype. In practice, investigators use enough independent loci that it is extremely unlikely that two individuals will have the same genotypes over all of the loci, making it possible to identify individuals within a degree of probability expressed as a percentage, and very high percentages are possible.

THE FIRST DNA FINGERPRINTS

Alec Jeffreys, at the University of Leicester in England, produced the first DNA fingerprints in the mid-1980's. His method examined a twelve-base sequence that was repeated one right after another, at many different loci in the human genome. Once collected from an individual, the DNA was cut using restriction enzymes to create DNA fragments that contained the repeat sequences. If the twelve-base sequence was represented by more repeats, the fragment containing it was that much longer. Jeffreys used agarose gel electrophoresis to separate his fragments by size, and he then used a specialized staining technique to view only the fragments containing the twelve-base repeat. For two samples from the same individual, each fragment, appearing as a band on the gel, should match. This method was used successfully in a highly publicized rape and murder case in England, both to exonerate one suspect and to incriminate the perpetrator.

While very successful, this method had certain drawbacks. First, a relatively large quantity of DNA was required for each sample, and results were most reliable when each sample compared was run on the same gel. This meant that small samples, such as individual hairs or tiny blood stains, could not be used, and also that it was difficult to store DNA fingerprints for use in future investigations.

VARIABLE NUMBER TANDEM REPEAT LOCI

The type of sequence Jeffreys used is now included in the category of variable number tandem repeats (VNTRs). This type of DNA sequence is characterized, as the name implies, by a DNA sequence which is repeated, one copy right after another, at a particular locus on a chromosome. Chromosomes vary in the number of repeats present.

VNTRs are often subcategorized based on the length of the repeated sequence. Minisatellites, like the Jeffreys repeat, include repeat units ranging from about twelve to several hundred bases in length. The total length of the tandemly repeated sequences may be several hundred to several thousand bases. Many different examples have since been discovered, and they occur in virtually all eukaryotes. In fact, the Jeffreys repeat first discovered in humans was found to occur in a wide variety of other species.

Shorter repeat sequences, typically 1 to 6 bases in length, were subsequently termed microsatellites. In humans, AC (adenine-cytosine) and AT (adenine-thymine) repeats are most common; an estimate for the number of AC repeat loci derived from the Human Genome Project suggests between eighty thousand and ninety thousand different AC repeat loci spread across the genome. Every eukaryote studied to date has had large numbers of microsatellite loci, but they are much less common in prokaryotes.

THE POLYMERASE CHAIN REACTION

The development of the polymerase chain reaction (PCR) in the mid-1980's, and its widespread use and optimization in DNA labs a few years later offered an alternative approach to DNA fingerprinting. The PCR technique makes millions of copies of short segments of DNA, with the chromosomal location of the fragments produced under the precise control of the investigator. PCR is extremely powerful and can be used with extremely small amounts of DNA. Because the fragments amplified are small, PCR can also be used on partially degraded samples. The size and chromosomal location of the fragments produced depend on the DNA primers used in the reaction. These are short, single-stranded DNA molecules that are complementary to sequences that flank the region to be amplified.

With this approach, an investigator must find and determine the DNA sequence of a region containing a VNTR. Primers are designed to amplify the VTNR region, together with some flanking DNA sequences on both ends. The fragments produced in the reaction are then separated by length using gel electrophoresis so that differences

in length, attributable to different numbers of the repeat, become apparent. For a dinucleotide repeat like AC, fragments representing different numbers of repeats, and hence different alleles, differ by a multiple of two. For instance, a researcher might survey a number of individuals and find fragments of 120, 122, 124, 128, and 130 base pairs in length.

CURRENT APPROACHES

Most current approaches to DNA fingerprinting use data collected simultaneously from a number of different VNTR loci, most commonly microsatellites. Preferably, the loci are PCR amplified using primers with fluorescent dyes attached, so that fragments from different loci are uniquely tagged with different colors. The fragments are then loaded in polyacrylamide DNA gels of the type used for DNA sequencing and separated by size. The fluorescent colors and sizes of the fragments are determined automatically, using the same automated machines typically used for DNA sequencing.

DNA fingerprint data generated in this way are easily stored and saved for future comparisons. Since each allelic variant is represented by a specific DNA fragment length, and because these are mea-

A criminalist at the Phoenix Police Department prepares samples of DNA taken from a crime scene for comparison to the DNA fingerprints of suspects. (AP/Wide World Photos)

sured very precisely, the initial constraint of running samples for comparison on the same gel is avoided.

Analyzing genes from cellular DNA can be limited if biological samples are limited. This occurred with many victims from the World Trade Center terrorist attack. Extensive burning and decomposition of victims found months later resulted in little biological tissue for genetic testing. Cells contain two copies of every gene, but cells contain thousands of mitochondria (organelles that provide energy for cells) that have their own DNA. Mitochondrial DNA is a source of more DNA analysis where the biological tissue supply is severely limited.

HUMAN FORENSIC AND PATERNITY TESTING

Although several different systems have been developed and used, a widely employed current standard comprises the Federal Bureau of Investigation's Combined DNA Index System (CODIS), with thirteen core loci. These thirteen are tetranucleotide (TCTA) microsatellite repeat loci, located on autosomes. Each locus has many known alleles, in some cases more than forty; the genetic variation is well characterized, and databases of variation within a variety of ethnic groups are available.

In addition to its role in criminal cases, this technique has seen widespread use to establish or exclude paternity, in immigration law to prove relatedness, and to identify the remains of casualties resulting from military combat and large disasters.

For validity concerns, it is important to consider false positives and false negatives. A false-positive genetic test identifies a genetic match when none exists, whereas a false-negative genetic test declares no match when a genetic match actually exists. How can false positives and false negatives occur in genetic testing? One way relates to the laboratory mechanics of genetic testing. Electrophoresis is used in genetic testing. Electrophoresis separates genetic fragments along a path that is around 8 inches in length. If a genetic fragment travels the exact same distance as another genetic fragment—the two genetic fragments are identical in chemical and genetic composition—then the two fragments are a "match." Small variations in migration can occur as a result of experimental error. If only fragments migrating the exact same length are accepted as matches, then false negatives will result since it is known that a genetically identical fragment may not

travel the exact same length as its twin as a result of experimental variability. Very small variances need to be accepted in order to minimize false negatives. On the other hand, if a very large variance of 2 inches is accepted as experimental, this unreasonably large range (compared to an 8-inch total path) will result in many false positives. Thus, minimizing false positives will increase false negatives, and vice versa.

OTHER USES FOR VNTR GENOTYPING

Soon after VNTRs were discovered in humans and used for DNA fingerprinting, researchers demonstrated that the same or similar types of sequences were found in all animals, plants, and other eukaryotes. The method pioneered by Jeffreys was, only a few years later, used for studies of paternity in wild bird populations. Since then, microsatellite analysis has come to dominate studies of relatedness, paternity, breeding systems, and other questions of individual identification in wild species of all kinds, including plants, insects, fungi, and vertebrates. Researchers now know, for example, that among the majority of birds which appear monogamous, between 10 and 15 percent of all progeny are fathered by males other than the recognized mate.

VNTR typing has been used to study the epidemiology of disease transmission. A 2008 study published in *Tuberculosis* genotyped forty-one *Mycobacterium tuberculosis* isolates from the Warao people, an indigenous population with a high tuberculosis (TB) incidence living in a geographically isolated area in Venezuela. This genetic study showed that 78 percent of the TB strains were in clusters, suggesting a very high transmission rate. VNTR typing is a useful tool to study the molecular epidemiology of tuberculosis, and this type of genetic analysis promises to yield more valuable information in the treatment and prevention of disease.

Paul R. Cabe, Ph.D.;
updated by Richard P. Capriccioso, M.D.

FURTHER READING

Burke, Terry, R., Wolf, G. Dolf, and A. Jeffreys, eds. *DNA Fingerprinting: Approaches and Applications.* Boston: Birkhauser, 2001. Describes repetitive DNA and the broad variety of practical applications to law, medicine, politics, policy, and more. Aimed at the layperson.

Fridell, Ron. *DNA Fingerprinting: The Ultimate Identity.* New York: Scholastic, 2001. The history of the technique, from its discovery to early uses. Aimed at younger readers and nonspecialists.

Herrmann, Bernd, and Susanne Hummel, eds. *Ancient DNA: Recovery and Analysis of Genetic Material from Paleographic, Archaeological, Museum, Medical, and Forensic Speciments.* New York: Springer-Verlag, 1994. Written when DNA fingerprinting was just coming to the fore and films such as *Jurassic Park* were in theaters, this collection of papers by first-generation researchers reflects the broad applications of the technology, including paleontological investigations.

Hummel, Susanne. *Fingerprinting the Past: Research on Highly Degraded DNA and Its Applications.* New York: Springer-Verlag, 2002. Manual about typing ancient DNA.

Maes, M., et al. "24-Locus MIRU-VNTR Genotyping Is a Useful Tool to Study the Molecular Epidemiology of Tuberculosis Among Warao Amerindians in Venezuela." *Tuberculosis* 88, no. 5 (2008): 490-494. A study that shows the value of DNA fingerprinting.

Rudin, Norah, and Keith Inman. *An Introduction to Forensic DNA Analysis.* Boca Raton, Fla.: CRC Press, 2002. An overview of many DNA typing techniques, along with numerous examples and a discussion of legal implications.

Varsha. "DNA Fingerprinting in the Criminal Justice System: An Overview." *DNA and Cell Biology* 25, no. 3 (March, 2006): 181-188. Also available at www.liebertonline.com/doi/pdf/10.1089/dna.2006.25.181. Overview of recent DNA fingerprinting applications in the court system.

Wambaugh, Joseph. *The Blooding.* New York: Bantam Books, 1989. The policeman-turned-writer offers a fascinating account of the British rape and murder case in which DNA fingerprinting was first used.

Web Sites of Interest

DNA Fingerprinting, Genetics and Crime: DNA Testing and the Courtroom
http://www.fathom.com/course/21701758/index.html

An online "seminar" from the University of Michigan explaining principles, procedures, and issues related to DNA fingerprinting.

Human Genome Project
http://www.ornl.gov/sci/techresources/Human_Genome/elsi/gmfood.shtml

Comprehensive Web site with information on the Human Genome Project, medicine and genetics, ethical, legal and social issues, and educational resources.

Iowa State University Extension and Office of Biotechnology, DNA Fingerprinting in Agricultural Genetics Programs
http://www.biotech.iastate.edu/biotech_info_series

Site links to a comprehensive and illustrative article on the role of DNA fingerprinting in agriculture.

See also: Criminality; Forensic genetics; Genetic testing; Genetics: Historical development; Human genetics; Paternity tests; Repetitive DNA.

DNA isolation

Category: Genetic engineering and biotechnology; Molecular genetics

Significance: Before it can be manipulated and studied, DNA must be isolated from other substances such as complex carbohydrates, proteins, and RNA. The isolation process is central to biotechnology and genetic engineering.

Key Terms

chloroform/isoamyl alcohol (CIA): a mixture of two chemicals used in DNA isolation to rid the extract of the contaminating compound phenol

lysis: the breaking open of a cell

osmotic shock: the lysing of cells by moving them from a liquid environment with a high solute concentration to an environment with a very low solute concentration

phenol: a simple chemical used in DNA extraction to precipitate proteins and aid in their removal

DNA Discovery and Extraction

Deoxyribonucleic acid (DNA) was discovered in 1869 by the Swiss physician Friedrich Miescher, who studied white blood cells in pus obtained from a surgical clinic. Miescher found that when bandages that

Differential Isolation of Organelle DNA

Discussions of DNA isolation usually concern isolation of DNA from the nucleus. While the nucleus is the location of most of the genetic information in the cell, DNA molecules also exist in other organelles, such as mitochondria and chloroplasts. Chromosomes of these organelles, referred to as nonnuclear or cytoplasmic DNA, contain a small subset of genes, mostly encoding proteins needed by these organelles.

Most standard DNA isolation techniques isolate both nuclear and nonnuclear DNA together. For a person working with nuclear DNA, this is usually not a concern because the amount of nuclear DNA is much greater than the amount of nonnuclear DNA. In working with nonnuclear DNA however, the presence of nuclear DNA can often cause problems. Some techniques used to examine nonnuclear DNA, such as the polymerase chain reaction (PCR), are not affected by the presence of nuclear DNA, but for other techniques, pure nonnuclear DNA is required.

Isolation strategies for nonnuclear DNA usually involve two steps. The first step is the isolation of intact mitochondria or chloroplasts from the cells, followed by the lysing of the mitochondria or chloroplasts to release the DNA so it can be purified. The process is the same for isolation of both mitochondrial and chloroplast DNA. Isolation of intact mitochondria (for example) requires that the membranes of the cells be lysed in a way that does not rupture the mitochondria. To achieve this goal, gentle mechanical, chemical, or enzy-matic methods (depending on the nature of the cell membrane and whether there is a cell wall) are employed to break open the cells and release the cytoplasmic contents. The lysis of the cells is usually done in an osmotically stabilized buffer. The solutes in this buffer match the concentration of the solutes inside the mitochondria, which prevents the mitochondria from bursting when the cells are lysed.

Once the cells are lysed, the lysate is centrifuged at low speed (usually between one thousand and three thousand times the force of gravity) to remove nuclei, membrane fragments, and other debris. The resulting supernatant contains the mitochondria in suspension. To concentrate the mitochondria, the supernatant is centrifuged at high speed (twelve thousand times the force of gravity). The pellet formed by this centrifugation will contain mitochondria and can be suspended in a small volume of liquid to create a concentrated suspension of mitochondria. This suspension may be treated with the enzyme DNase, which will degrade any nuclear DNA that remains without crossing the intact mitochondrial membrane. The enzyme will then be deactivated, and the mitochondria will be lysed. Lysis of the mitochondria is achieved by adding a strong detergent to the suspension of the mitochondria. Once the mitochondria have been lysed, the free mitochondrial DNA can be purified just as nuclear DNA would be, using phenol extraction and ethanol precipitation.

Douglas H. Brown, Ph.D.

had been removed from the postoperative wounds of injured soldiers were washed in a saline solution, the cells on the bandages swelled into a gelatinous mass that consisted largely of DNA. Miescher had isolated a denatured form of DNA—that is, DNA not in the normal double-stranded conformation. After a series of experiments, Miescher concluded that the substance he had isolated originated in the nuclei of the blood cells; he first called the substance nuclein and later nucleic acid.

The first problem when extracting DNA is lysing, or breaking open, the cell. Bacteria, yeast, and plant cells usually have a thick cell wall protecting them, which makes lysis more difficult. Bacteria, such as *Escherichia coli*, are the easiest of these cells to open by a process called alkaline lysis, in which cells are treated with a solution of sodium hydroxide and detergent that degrades both the cell wall and the cell membrane. Yeast cells are often broken open with enzymes such as lysozyme that degrade cell walls or by using a "French press," a piston in an enclosed chamber that forces cells open under high pressure. Plant tissue is usually mechanically broken into a fine cell suspension before extraction by grinding frozen tissue in a mortar and pestle. Once the suspension of cells is obtained, the tissue may be treated with a variety of enzymes to break down cell walls or with strong detergents, such as sodium lauryl sarcosine, that disrupt and dissolve both cell walls and cell membranes. Animal cells, such as white blood cells, do not have cell walls and can generally be opened by osmotic shock, the lysing of cells by moving them from a liquid environment with a high solute concentration to an environment with a very low solute concentration.

ISOLATION AND PURIFICATION

Although lysis methods differ according to cell type, the process of DNA isolation and purification is more standardized. The isolation process may be imagined as a series of steps designed to remove either naturally occurring biological contaminants from the DNA or contaminants added by the scientist during the extraction process. The biological contaminants already present in cells are proteins, fat, and ribonucleic acid (RNA); additionally, plant cells have high levels of complex carbohydrates. Contaminants intentionally added by scientists may include salts and various chemicals.

After cells are lysed, a high-speed centrifugation is performed to form large-scale, insoluble cellular debris, such as membranes and organelles, into a pellet. The liquid extract remaining still contains dissolved proteins, RNA, and DNA. If salts are not already present in the extract, they are added; salt must be present later for the DNA to precipitate efficiently. Proteins must be removed from the extract since some not only degrade DNA but also inhibit enzymatic reactions with DNA that would be involved in further DNA manipulations used in cloning, for example. Proteins are precipitated by mixing the extract with a chemical called phenol. When phenol and the extract are mixed in a test tube, they separate into two parts like oil and water. If these fluids are centrifuged, precipitated proteins will actually collect between the two liquids at a spot called the interphase. The liquid layer containing the dissolved DNA is then drawn up and away from the precipitated protein.

The protein-free solution still contains DNA, RNA, salts, and traces of phenol dissolved into the extract. To remove the contaminating phenol, the extract is mixed with a chloroform/isoamyl alcohol solution (CIA). Again like oil and water, the DNA extract and CIA separate into two layers. If the two layers are mixed vigorously and separated by centrifugation, the phenol will move from the DNA extract into the CIA layer. At this point the extract—removed to a new test tube—contains RNA, DNA, and salt.

The extract is next mixed with 100 percent ethanol, inducing the DNA to precipitate out in long strands. The DNA strands may be isolated by either spooling the sticky DNA around a glass rod or by centrifugation. If spooled, the DNA is placed in a new test tube; if centrifuged, the liquid is decanted from the pellet of DNA. The precipitated DNA, with salt and RNA present, is still not pure. It is washed for a final time with 70 percent ethanol, which does not dissolve the DNA but forces salts present to go into solution. The DNA is then reisolated by spooling or centrifugation and dried to remove all traces of ethanol. At this point, only DNA and RNA are left; this mixture can be dissolved in a low-salt buffer containing the enzyme RNase, which degrades any RNA present, leaving pure DNA.

Technological advances have allowed deproteinization by the use of "spin columns" without the employment of toxic phenol. The raw DNA extract is placed on top of a column containing a chemical matrix that binds proteins but not DNA; the column is then centrifuged in a test tube. The raw extract passes through the chemical matrix and exits protein-free into the collection tube. These newer methods not only increase safety and reduce the production of toxic waste; they are also much faster.

For the isolation of DNA for cloning, DNA is typically broken into fragments using enzymes called restriction endonucleases. The fragment of interest is typically separated from other DNA using gel electrophoresis in an agarose gel. The DNA is stained with ethinium bromide, which permits visualization using UV light. The fragment of DNA is cut out of the agarose gel and purified using spin columns, which contain silica to which DNA binds in the presence of chaotropic salts. The chaotropic salt, such as guanidium chloride, denatures biomolecules by disrupting the shell of hydration around them. This allows a positively charged ion to form a salt bridge between the negatively charged silica and the negatively charged DNA backbone when the salt concentration is high. After the DNA is adsorbed to the silica surface, all other molecules pass through the column. The DNA is then washed with high salt and ethanol, and ultimately eluted with low salt.

Plasmids are used as vectors to clone DNA of interest. Plasmids are extrachromosomal DNA that replicate independent of the chromosome and occur naturally in bacteria. To isolate plasmids independent of chromosomal bacterial DNA, holes are punctured in the bacterial cell wall by gently mixing a bacterial cell suspension with alkali, which is then neutralized. The holes that are generated are of a size that permits the plasmids to leak out of the cell while the chromosomal DNA remains trapped in the bacteria and is separated from the plasmid DNA and RNA by differential centrifugation. The cell debris forms a pellet, which is discarded. Proteins, RNA, and plasmid DNA are present in the supernatant. RNA is removed with RNAse. Plasmid DNA can be purified using either phenol/chloroform extraction and ethanol precipitation or silica column chromatography.

James J. Campanella, Ph.D.;
updated by Dervla Mellerick, Ph.D.

FURTHER READING

Gjerde, Douglas T., Christopher P. Hanna, and David Hornby. *DNA Chromatography.* Weinheim, Germany: Wiley-VCH, 2002. In chapters about instrumentation and operation, chromatographic principles, size-based separations, purification of nucleic acids, RNA chromatography, and special techniques, among others, this book bridges the chasm between the work of analytic chemists and molecular biologists. Illustrated.

Mirsky, Alfred. "The Discovery of DNA." *Scientific American,* June, 1968. The fascinating story of Friedrich Miescher's work.

Roe, Bruce A., Judy S. Crabtree, and Akbar S. Khan, eds. *DNA Isolation and Sequencing.* New York: John Wiley & Sons, 1996. Focus is on protocol, describing the most commonly used methods for DNA isolation, DNA sequencing, sequence analysis, and allied molecular biology techniques. Illustrated.

Sambrook, Joseph, and David W. Russell, eds. *Molecular Cloning: A Laboratory Manual.* 3d ed. 3 vols. Cold Spring Harbor, N.Y.: Cold Spring Harbor Laboratory Press, 2001. A standard for researchers, covering plasmids, bacteriophage, high-capacity vectors, gel electrophoresis, eukaryotic genomic DNA preparation and analysis, eukaryotic mRNA, polymerase chain reaction techniques, and more. Bibliographical references and index.

Trevors, J. T., and J. D. van Elsas, eds. *Nucleic Acids in the Environment.* New York: Springer, 1995. A laboratory manual that details molecular biological techniques such as DNA/RNA extraction and purification, and polymerase chain reaction methods. Illustrated.

Watson, James, et al. *Recombinant DNA.* New York: W. H. Freeman, 1992. Uses accessible language and exceptional diagrams to give a concise background on the methods, underlying concepts, and far-reaching applications of recombinant DNA technology.

Weissman, Sherman M., ed. *cDNA Preparation and Characterization.* San Diego: Academic Press, 1999. Examines the analysis and mapping of messenger RNA, gene mapping DNA, complementary isolation and purification DNA, and chromosome-mapping methods. Includes six pages of plates, illustrations.

WEB SITES OF INTEREST

DNA Extraction Virtual Lab
http://learn.genetics.utah.edu/content/labs/extraction/

Protocol Online: "DNA"
http://www.protocol-online.org/prot/Molecular_Biology/DNA/index.html

See also: Ancient DNA; DNA replication; DNA sequencing technology; DNA structure and function; RFLP analysis; RNA isolation; RNA structure and function.

DNA repair

CATEGORY: Molecular genetics

SIGNIFICANCE: To protect the integrity of their genetic material, cells are able to correct damage to DNA. Many of these mechanisms are found in organisms ranging from bacteria to humans, and there is a high degree of homology, or "sameness," between species. Disruption of DNA repair mechanisms in humans has been associated with the development of cancers.

KEY TERMS

base: the component of a nucleotide that gives it its identity and special properties

nucleotide: the basic unit of DNA, consisting of a five-carbon sugar, a nitrogen-containing base, and a phosphate group

DNA STRUCTURE AND DNA DAMAGE

All living things are continually exposed to agents that can damage their genetic material. Damage to or mutations in DNA can occur as a result of ionizing radiation, from assault by mutagenic chemicals, or as a by-product of other cellular processes, such as failure of the DNA mismatch repair (DMMR) pathway. As DNA is the blueprint for directing the functions of the cell, it must be accurately maintained. The integrity of DNA is also important because daughter cells receive copies of a parent cell's DNA during mitosis. DNA damage can include a break in a DNA molecule, the abnormal bonding of two nucleotides, or by the attaching or removal of a chemical group to or from a nucleotide. Mutations typically occur as a result of a copying error and can follow from DNA damage. To a bacterial cell, DNA damage may mean death. To a multicellular organism, damaged DNA in some of its cells may mean loss of function of organs or tissues or it may lead to cancer.

DNA is assembled from nucleotides, each defined by the base it contains. If the DNA double helix is pictured as a twisted ladder, the outside supports, sometimes referred to as the "backbone" of the DNA, are composed of alternating phosphates and ribose sugars. The "rungs" of the ladder are bases. Four bases are found in DNA: the double-ring purines, adenine and guanine, and the single-ring pyrimidines, cytosine and thymine. The structure of each base makes two base pairings most likely. In James Watson and Francis Crick's model of DNA, adenine pairs with thymine, and cytosine pairs with guanine. This base pairing holds the two strands of the double helix together and is essential for the synthesis of new DNA molecules (DNA replication) and for the transfer of information from DNA to RNA in the process of transcription. DNA replication is carried out by an enzyme called DNA polymerase, which reads the information (the sequence of bases) on a single strand of DNA, brings the appropriate nucleotide to pair with the template strand one nucleotide at a time, and joins it to the end of the new DNA chain. Transcription occurs through a process similar to DNA replication, except that a RNA polymerase copies only the portion of one DNA strand which codes for a gene, making an RNA copy. The RNA can be used as a template for synthesizing a particular protein, which is the final product of most genes.

One frequent form of DNA damage is the loss of a base. Purines are particularly unstable, and many are lost each day in human cells. If a base is absent, the DNA cannot be copied correctly during DNA replication. Another common type of DNA damage is a pyrimidine dimer, an abnormal linkage between two cytosines, two thymines, or a cytosine and an adjacent thymine in a DNA strand. These are caused by the absorption of ultraviolet light by the two bases. A pyrimidine dimer creates a distortion in the double helix that interferes with the processes of DNA replication and transcription. Another form of DNA damage is a break in the backbone of one or both strands of the double helix. Breaks can block DNA replication, create problems during cell division, or cause rearrangements in the chromosomes. DNA replication itself can cause problems by inserting an incorrect base or an additional or too few bases in a new strand. While DNA replication errors are not DNA damage as such, they can also lead to mutations and are subject to repair.

DNA REPAIR SYSTEMS

DNA repair systems are found in most organisms. Even some viruses, such as bacteriophages (viruses which infect bacteria) and herpes viruses (which infect animals), are capable of repairing some damage to their genetic material. The DNA repair systems of single-celled organisms, including bacteria and yeasts, have been extensively studied for many years. Techniques including the use of recombinant DNA methods revealed that DNA repair systems of multicellular organisms such as humans, animals, and plants are quite similar to those of microorganisms.

Scientists generally classify DNA repair systems into three categories on the basis of complexity, mechanism, and the fate of the damaged DNA. "Damage reversal" systems are the simplest: They usually require only a single enzyme to directly act on the damage and restore it to normal, usually in a single step. "Damage removal" systems are somewhat more complicated: These involve cutting out and replacing a damaged or inappropriate base or section of nucleotides and require several proteins to act together in a series of steps. "Damage tolerance" systems are those that respond to and act on damaged DNA but do not actually repair the original damage. Instead, they are ways for cells to cope with DNA damage in order to continue growth and division.

DAMAGE REVERSAL SYSTEMS

Photoreactivation is one of the simplest and perhaps oldest known repair systems: It consists of a single enzyme that can split pyrimidine dimers in the presence of light. An enzyme called photolyase catalyzes this reaction; it is found in many bacteria, lower eukaryotes, insects, and plants but seems to be absent in mammals (including humans). A similar gene is present in mammals but may code for a protein that functions in another type of repair.

X rays and some chemicals, such as peroxides, can cause breaks in the backbone of DNA. Simple breaks in one strand are rapidly repaired by the enzyme DNA ligase. Mutant strains of microorganisms with reduced DNA ligase activity tend to have high levels of recombination since DNA ends are very "sticky" and readily join with any other fragment of DNA. While recombination is important in generating genetic diversity during sexual reproduction, it can also be dangerous if DNA molecules are joined inappropriately. The result can be aberrant chromosomes that do not function properly.

DAMAGE REMOVAL SYSTEMS

Damage removal systems are accurate and efficient but require the action of several enzymes and are more energetically "expensive" to the cell. There are three types of damage removal systems that work in the same general way but act on different forms of DNA damage. In "base excision" repair, an enzyme called a DNA glycosylase recognizes a specific damaged or inappropriate base and cuts the base-sugar linkage to remove the base. The backbone then is cut by another protein (an endonuclease) that removes the baseless sugar, and a new nucleotide is inserted to replace the damaged one by a DNA polymerase enzyme. The remaining break in the backbone is reconnected by DNA ligase. There are a number of specific glycosylases for particular types of DNA damage caused by radiation and chemicals.

The "nucleotide excision" repair system works on DNA damage that is "bulky" and that creates a block to DNA replication and transcription, such as ultraviolet-induced pyrimidine dimers and some kinds of DNA damage created by chemicals. It probably does not recognize a specific abnormal structure but sees a distortion in the double helix. Several proteins joined in a complex scan the DNA for helix distortions. When one is found, the complex binds to the damage and creates two cuts in the DNA strand containing the damaged bases on either side of the damage. The short segment with the damaged bases

(around thirty nucleotides in humans) is removed from the double helix, leaving a short gap that can be filled by DNA polymerase using the intact nucleotides in the other DNA strand as a guide. In the last step, DNA ligase rejoins the strand. Mutants that are defective in nucleotide excision repair have been isolated in many organisms and are extremely sensitive to mutation by ultraviolet light and similar-acting chemical mutagens.

"Mismatch repair" occurs during DNA replication as a last "spell check" on its accuracy. By comparing mutation rates in *Escherichia coli* bacteria that either have or lack mismatch repair systems, scientists have estimated that this process adds between one hundred and one thousand times more accuracy to the replication process. It is carried out by a group of proteins that can scan DNA and look for incorrectly paired bases (or unpaired bases). The incorrect nucleotide is removed as part of a short stretch, and then a DNA polymerase attempts to insert the correct sequence. In 1993, Richard Fishel, Bert Vogelstein, and their colleagues isolated the first genes for human mismatch repair proteins and showed that they are very similar to those of the bacterium *Escherichia coli* and the simple eukaryote baker's yeast. Further studies in the 1990's revealed that mismatch repair genes are defective in people with hereditary forms of colon cancer.

NUCLEOTIDE EXCISION REPAIR: XERODERMA PIGMENTOSUM

Humans with the hereditary disease xeroderma pigmentosum (XP) are extremely sensitive to ultraviolet light and are at nearly a 100 percent risk of skin cancer in their lifetime. XP results when a child inherits genetic defects in the nucleotide excision repair system (NER) from both parents. These children often begin to exhibit symptoms of XP between the ages of one and two. The affected are often hypersensitive to light and are prone to sunburn, skin, and eye defects, such as cataracts. Eight different forms of the disease, labeled A through G and V, correspond to mutations in different components of the NER system. Forms A and C are the two most common, accounting for approximately 50 percent of cases. Rates of XP presentation vary, but have been estimated at 1 in 250,000 in the United States and as high as 1 in 40,000 in Japan. Studies have shown that XP patients often are born to parents who share a common ancestor.

This remote inbreeding is also referred to as consanguinity.

Variation in symptoms depends on the function of the specific NER system protein affected. Functions of specific NER system proteins implicated in XP fit within two NER subtypes, known as transcription-coupled repair (TCR), which works on damage in the genome undergoing transcription, and global genome repair (GGR), which works on damage in the entire genome and is slower than TCR. Recognition of damage in GGR occurs through XPC and XPE complexes. Recognition events in TCR and GGR activate unwinding of DNA through the XPB and XPD helicases. Subsequently, XPA binds and presents binding sites for the XPG nuclease and XPF-ERCC1 nuclease complex. Mutations in these specific genes generally lead to the corresponding form of XP, though mutations in other proteins that form complexes with XP proteins can lead to XP, such as the mutation in the DDB2 component of the XPE complex. The variant form of XP is the result of a mutation in DNA polymerase eta, also called hRAD30, which is not a part of the NER system, but functions after DNA replication. DNA polymerase eta is able to bypass many forms of DNA lesions that would stop the main DNA polymerase complex.

Of the eight forms of XP, mutations in genes found only in GGR, such as XPC, XPE, and DNA polymerase eta produce the fewest symptoms beyond an increased risk of cancer. However, mutations in one or more of the other five genes and other components of TCR known as CSA and CSB can produce more complicated arrays of symptoms, including neurodegenerative and developmental disorders.

DAMAGE TOLERANCE SYSTEMS

Not all DNA damage is or can be removed immediately; some of it may persist for a time. If a DNA replication complex encounters DNA damage such as a pyrimidine dimer, it will normally act as a block to further replication of that DNA molecule. In eukaryotes, however, DNA replication initiates at multiple sites and may be able to resume downstream of a damage site, leaving a "gap" of single-stranded, unreplicated DNA in one of the two daughter molecules. The daughter-strand gap is potentially just as dangerous as the original damage site, if not more so. The reason for this is that if the cell divides with

a gap in a DNA molecule, there will be no way accurately to repair that gap or the damage in one of its two daughter cells. To avoid this problem, cells have developed a way to repair daughter-strand gaps by recombination with an intact molecule of identical or similar sequence. The "recombinational" repair process, which requires a number of proteins, yields two intact daughter molecules, one of which still contains the original DNA damage. In addition to dealing with daughter-strand gaps, recombinational repair systems can also repair single- and double-strand breaks caused by the action of X rays and certain chemicals on DNA. Many of the proteins required for recombinational repair are also involved in the genetic recombination that occurs in meiosis, the process which produces sperm and egg cells in organisms which reproduce sexually. In 1997, it was shown that the products of the breast cancer susceptibility genes *BRCA1* and *BRCA2* participate in both recombinational repair and meiotic recombination.

An alternative choice for a DNA polymerase blocked at a DNA damage site is to change its specificity so that it can insert any nucleotide opposite the normally nonreadable damage and continue DNA replication. This type of "damage bypass" is very likely to cause a mutation, but if the cell cannot replicate its DNA, it will not be able to divide. In *Escherichia coli* bacteria, there is a set of genes that are turned on when the bacteria have received a large amount of DNA damage. Some of these gene products alter the DNA polymerase and allow damage bypass. This system has been termed the "SOS response" to indicate that it is a system of last resort. Other organisms, including humans, seem to have similar damage bypass mechanisms that allow a cell to continue growth despite DNA damage at the price of mutations. For this reason, damage bypass systems are sometimes referred to as "error-prone" or mutagenic repair systems.

IMPACT AND APPLICATIONS

DNA repair systems are an important component of the metabolism of cells. Studies in microorganisms have shown that as little as one unrepaired site of DNA damage per cell can be lethal or lead to permanent changes in the genetic material. The integrity of DNA is normally maintained by an elaborate series of interrelated checks and surveillance systems. The greatly increased risk of cancer suffered by humans with hereditary defects in DNA repair shows how important these systems are in avoiding genetic changes. As the relationship between mutations in DNA repair genes and cancer susceptibility becomes clearer, this information may be used in directing the course of cancer therapy and possibly in providing gene therapy to individuals with cancer.

Beth A. Montelone, Ph.D.;
updated by Andrew J. Reinhart, M.S.

FURTHER READING

Dizdaroglu, Miral, and Ali Esat Karakaya, eds. *Advances in DNA Damage and Repair: Oxygen Radical Effects, Cellular Protection, and Biological Consequences.* New York: Plenum Press, 1999. Covers advances in research and contains contributions from scientists working in the fields of biochemistry, molecular biology, enzymology, biomedical science, and radiation biology.

Gilchrest, Barbara A., and Vilhelm A. Bohr, eds. *The Role of DNA Damage and Repair in Cell Aging.* New York: Elsevier, 2001. Topics include aging in mitotic and post-mitotic cells, age-associated changes in DNA repair and mutation rates, human premature aging syndromes as model systems, and gene action at the Werner helicase locus. Illustrated.

Henderson, Daryl S., ed. *DNA Repair Protocols.* 2d ed. Methods in Molecular Biology. Totowa, N.J.: Humana Press, 2005. A collection of experimental protocols and techniques for detecting and studying DNA damage and repair.

Mills, Kevin D. *Silencing, Heterochromatin, and DNA Double Strand Break Repair.* Boston: Kluwer Academic, 2001. Presents new directions in research regarding the involvement of chromatin in the repair of broken DNA, concentrating on the study of the budding yeast *Saccharomyces cerevisiae* conducted in the laboratory of Leonard Guarente at the Massachusetts Institute of Technology.

Sancar, Aziz, Laura A. Lindsey-Boltz, Keziban Unsal-Kacmaz, and Stuart Linn. "Molecular Mechanisms of Mammalian DNA Repair and the DNA Damage Checkpoints." *Annual Review of Biochemistry* 73 (2004): 39-85. A thorough review of DNA damage recognition, DNA repair mechanisms, and DNA damage checkpoints.

Science, December 23, 1994. The magazine declared the DNA repair enzyme "Molecule of the Year" in 1994 and published three short reviews in this special issue that discuss three repair processes: "Mechanisms of DNA Excision Repair," by Aziz Sancar; "Transcription-Coupled Repair and Human Disease," by Philip C. Hanawalt; and "Mismatch Repair, Genetic Stability, and Cancer," by Paul Modrich.

Smith, Paul J., and Christopher J. Jones, eds. *DNA Recombination and Repair.* New York: Oxford University Press, 1999. Explores the cellular processes involved in DNA recombination and repair by highlighting current research, including strategies for dealing with DNA mismatches or lesions, the enzymology of excision repair, and the integration of DNA repair into cellular pathways.

Vaughan, Pat, ed. *DNA Repair Protocols: Prokaryotic Systems.* Totowa, N.J.: Humana Press, 2000. Divided into two sections, the book examines the classic identification, purification, and characterization of DNA repair enzymes and provides several protocols for the applied use of DNA repair proteins in the latest molecular biology techniques, including mutation detection, cloning, and genome diversification.

Weinberg, Robert A. "How Cancer Arises." *Scientific American* 197 (September, 1996). Discusses cancer and the roles of DNA repair genes.

WEB SITES OF INTEREST

Human DNA Repair Genes. Bioinformatics and Biostatistics, Cancer Research UK
http://www.cgal.icnet.uk/DNA_Repair_Genes
.html#DNA_glyco

Online Mendelian Inheritance in Man (OMIM). Xeroderma Pigmentosum, Complementation Group A; XPA
http://www.ncbi.nlm.nih.gov/entrez/dispomim
.cgi?id=278700

See also: Aging; Biochemical mutations; Breast cancer; Cancer; Chemical mutagens; DNA structure and function; Human genetics; Immunogenetics; Model organism: *Escherichia coli*; Mutation and mutagenesis; Oncogenes; Protein structure; Protein synthesis; RNA structure and function; RNA transcription and mRNA processing; Telomeres; Tumor-suppressor genes.

DNA replication

CATEGORY: Genetic engineering and biotechnology

SIGNIFICANCE: Cells and organisms pass hereditary information from generation to generation. To assure that offspring contain the same genetic information as their parents, the genetic material must be accurately reproduced. DNA replication is the molecular basis of heredity and is one of the most fundamental processes of all living cells.

KEY TERMS

replication: the process by which one DNA molecule is converted to two DNA molecules identical to the first

transcription: the process of forming an RNA according to instructions contained in DNA

translation: the process of forming proteins according to instructions contained in an RNA molecule

X-ray diffraction: a method for determining the structure of molecules which infers structure by the way crystals of molecules scatter X rays as they pass through

DNA STRUCTURE AND FUNCTION

The importance of chromosomes in heredity has been known since early in the twentieth century. Chromosomes consist of both DNA and protein, and in the early twentieth century there was considerable controversy concerning which component was the hereditary molecule. Early evidence favored the proteins. In 1944, however, a series of classic experiments by Oswald Avery, Maclyn McCarty, and Colin MacLeod lent strong support to the proponents favoring DNA as the genetic material. They showed that a genetic transforming agent of bacteria was DNA and not protein. In experiments reported in 1952, Alfred Hershey and Martha Chase provided evidence that DNA was the genetic material of bacteriophages (viruses that infect bacteria). Combined with additional circumstantial evidence from many sources, DNA became favored as the hereditary molecule, and a heated race began to determine its molecular structure.

In 1953, James Watson and Francis Crick published a model for the atomic structure of DNA. Their model was based on known chemical properties of DNA and X-ray diffraction data obtained from Rosalind Franklin and Maurice Wilkins. The structure itself made it clear that DNA was indeed the molecule of heredity and provided evidence for how it might be copied. The molecule resembles a ladder. The "rails" are composed of repeating units of sugar and phosphate, forming a backbone for the molecule. Each "rung" consists of a pair of nitrogenous bases, one attached to each of the two rails and held together in the middle through weak bonds called hydrogen bonds. Since there are thousands to hundreds of millions of units on a DNA molecule, the hydrogen bonds between each pair of bases add up to a strong force that holds the two strands together. DNA, then, consists of two strands, each consisting of a repeating sugar-phosphate backbone and nitrogenous bases with the two strands held together by base-pair interactions. The two strands are oriented in opposite directions. The ends of a linear DNA molecule can be distinguished by which part of the backbone sugar is exposed and are referred to as the 5′ (five prime) end and the 3′ end, named for a particular carbon atom on the ribose sugar. If one DNA strand is oriented 5′ to 3′, its complementary partner is oriented 3′ to 5′. This organization has important implications for the mechanism of DNA replication.

There are four different bases: adenine (A), guanine (G), cytosine (C), and thymine (T). They can be arranged in any order on a DNA strand, allowing the enormous diversity necessary to encode the blueprint of every organism. A key feature of the double-stranded DNA molecule is that bases have strict pairing restrictions: A can only pair with T; G can only pair with C. Thus if a particular base is known on one strand, the corresponding base is automatically known on the other. Each strand can serve as a template, or mold, dictating the precise sequence of bases on the other. This feature is fundamental to the process of DNA replication.

The genome (the complete DNA content of an organism) stores all the genetic information that determines the features of that organism. The features are expressed when the DNA is transcribed to a messenger molecule, mRNA, which is used to construct a protein. The proteins encoded by the organism's genes in its DNA carry out all of the activities of the cell.

THE CELL CYCLE

In eukaryotic organisms (most organisms other than bacteria), cells progress through a series of four stages between cell divisions. The stages begin with a period of growth (G_1 phase), followed by replication of the DNA (S phase). A second period of growth (G_2 phase) is followed by division of the cell (M phase). Each of the two cells resulting from the cell division goes through its own cell cycle or may enter a dormant stage (G_0 phase). The passage from one stage to the next is tightly regulated and directed by internal and external signals to the cell.

The transition from G_1 into S phase marks the beginning of DNA replication. In order to enter S phase, the cell must pass through a checkpoint or restriction point in which the cell determines the quality of its DNA: If there is any damage to the DNA, entry into S phase will be delayed. This prevents the potentially lethal process of beginning replication of a DNA molecule that has damage that would prevent completion of replication. If conditions are determined to be acceptable, a "molecular switch" is thrown, triggering the initiation of DNA replication. What is the nature of this molecular switch? There are many different proteins that participate in the process of DNA replication, and they can have their activity turned off and on by other proteins. Addition or removal of a chemical group called a phosphate is a common mechanism of chemical switching. This reaction is catalyzed by a class of enzymes called kinases. Certain key proteins are phosphorylated at the boundary of the G_1 and S phases of the cell cycle by kinases, switching on DNA replication.

ORIGINS AND INITIATION

If the human genome were replicated from one end to the other, it would take several years to complete the process. The DNA molecule is simply too large to be copied end to end. Instead, replication is initiated at many different sites called origins of replication, and DNA synthesis proceeds from each site in both directions until regions of copied DNA

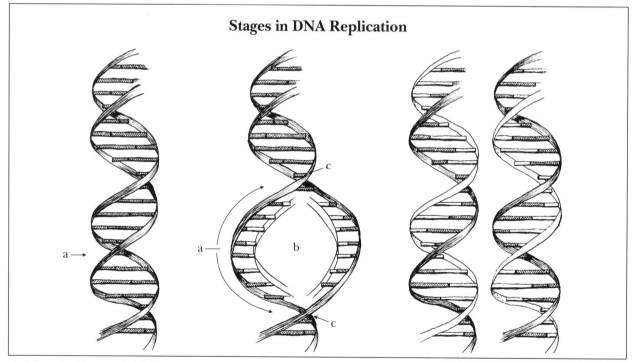

Stages in DNA Replication

At left, a double-stranded DNA molecule, with its sides formed by sugar-phosphate molecules and its "rungs" formed by base pairs. Replication begins at point (a), with the separation of a base pair as a result of the action of a special initiator protein (b). The molecule splits, or "unzips," in opposite directions (c) as each parental strand is used as a template for the daughter strand, which is formed when bases form hydrogen bonds with their appropriate "mate" bases to form new ladder "rungs." Finally (right), one parental strand and its newly synthesized daughter strand form a new double helix, while the other parental strand and its daughter strand form the second double helix. (Kimberly L. Dawson Kurnizki)

merge. The region of DNA copied from a particular origin is called a replicon. Using hundreds to tens of thousands of initiation sites and replicons, the genome can be copied in a matter of hours. The structure of replication origins has been difficult to identify in all but a few organisms, most notably yeast. Origins consist of several hundred base pairs of DNA comprising sequences that attract and bind a set of proteins called the origin recognition complex (ORC). The exact mechanism by which the origin is activated is still under investigation, but a favored model is supported by all of the available evidence.

The ORC proteins are believed to be bound to the origin DNA throughout the cell cycle but become activated at the G_1/S boundary through the action of kinases. Kinases add phosphate groups to one or more of the six ORC proteins, activating them to initiate DNA replication. Different replicons are initiated at different times throughout S phase. It is unclear how the proposed regulatory system distinguishes between replicons that have been replicated in a particular S phase and those that have not, since each must be used once and only once during each cell division cycle.

A number of different enzymatic activities are required for the initiation process. The two strands of DNA must be unwound or separated, exposing each of the parent strands so they can be used as templates for the synthesis of new, complementary strands. This unwinding is mediated by an enzyme called a helicase. Once unwound, the single strands are stabilized by the binding of proteins called single-strand binding proteins (SSBs). The resulting structure resembles a "bubble" or "eye" in the DNA strand. This structure is recognized by the DNA replication machinery that is recruited to the site, and DNA replication begins. As replication proceeds, the DNA continues to unwind through the action of helicase. The site at which unwinding and DNA synthesis are occurring is at either end of the expanding eye or bubble, called a replication fork.

DNA SYNTHESIS

The DNA synthesis machinery is not able to synthesize a strand of DNA from scratch; rather, a short stretch of RNA is used to begin the new strands. The synthesis of the RNA is catalyzed by an enzyme called primase. This short piece of RNA, or primer, is extended using DNA nucleotides by the enzymes

of DNA synthesis, called DNA polymerases. The RNA primer is later removed and replaced by DNA. Nucleotide monomers align with the exposed template DNA strand one at a time and are joined by the DNA polymerase. The joining of nucleotides into a growing DNA chain requires energy. This energy is supplied by the nucleotide monomers themselves. A high-energy phosphate bond in the nucleotide is split, and the breakage of this high-energy bond provides the energy to drive the polymerase reaction.

The two strands of DNA are not synthesized in the same way. The two strands are oriented opposite one another, but DNA synthesis only occurs in one direction: 5′ to 3′. Therefore, one strand, called the leading strand, is synthesized continuously in the same direction that the replication fork is moving, while the lagging strand is synthesized away from the direction of fork movement. Since the lagging-strand DNA synthesis and fork movement are in opposite directions, this strand of DNA must be made in short pieces that are later joined. Lagging-strand synthesis is therefore said to be discontinuous. These short intermediates are called Okazaki fragments, named for their discoverer, Reiji Okazaki. Overall, DNA replication is said to be semidiscontinuous.

The DNA synthesis machine operating at the replication fork is a complex assembly of proteins. Many different activities are necessary to carry out the process of DNA replication efficiently. Several proteins are necessary to recognize the unwound origin and assemble the rest of the complex. Primase must function to begin both new strands and is then required periodically throughout synthesis of the lagging strand. A doughnut-shaped clamp called PCNA functions as a "processivity factor" to keep the entire complex attached to the DNA until the job is completed. Helicase is continuously required to unwind the template DNA and move the fork along the parent molecule. As the DNA is unwound, strain is created on the DNA ahead of the replication fork. This strain is alleviated through the action of topoisomerase enzymes. Single-strand binding proteins are needed to stabilize the regions of unwound DNA that exist before the DNA is actually copied. Finally, an enzyme called ligase is necessary to join the regions replicated from different origins and to attach all of the Okazaki fragments of the lagging strand. All of

The Replication Process

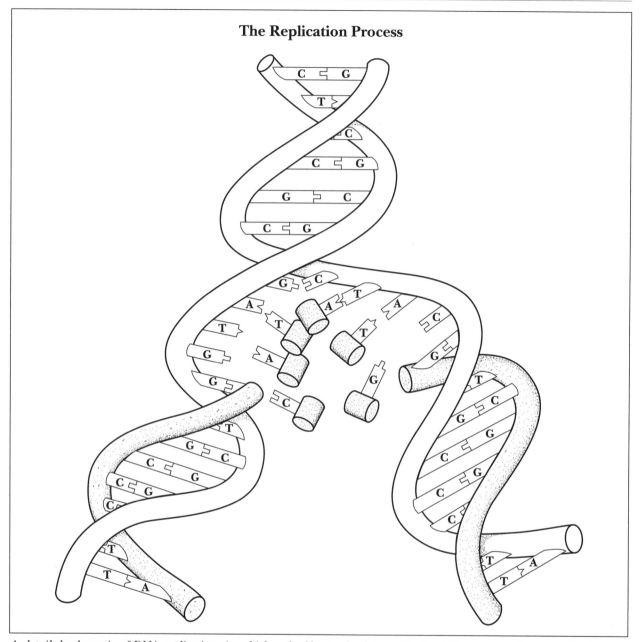

A detailed schematic of DNA replication, in which a double-standard parent helix splits apart and reassembles into two identical daughter helixes. The amino acid base pairs are reproduced exactly, because cytosine (C) pairs only with guanine (G), and adenine (A) pairs only with thymine (T). (Electronic Illustrators Group)

these proteins are part of a well-orchestrated, efficient machine ideally suited to its task of copying the genetic material.

DNA polymerases are not perfect. At a relatively low frequency, they can add an incorrect nucleotide to a growing chain, one that does not match the template strand as dictated by the base-pairing rules. However, because the DNA molecules are so extremely large, novel mechanisms for proofreading have evolved to ensure that the genetic material is copied accurately. DNA polymerases can detect the misincorporation of a nucleotide and use an additional enzymatic activity to correct the mistake. Specifically, the polymerase can "back up" and cut out the last nucleotide added, then try again. With this and other mechanisms to correct errors, the observed error rate for DNA synthesis is a remarkable one error in every billion nucleotides added.

IMPACT AND APPLICATIONS

DNA replication is a fundamental cellular process: Proper cell growth cannot occur without it. It must be carefully regulated and tightly controlled. Despite its basic importance, the details of the mechanisms that regulate DNA replication are poorly understood. Even with all of the checks and balances that have evolved to ensure a properly replicated genome, occasional mistakes do occur. Attempting to replicate a genome damaged by chemical or other means may simply lead to death of a single cell. Far more ominous are genetic errors that lead to loss of regulating mechanisms. Without regulation, cell growth and division can proceed without normal limits, resulting in cancer. Much of the focus for the study of cell growth and regulation is to set a foundation for the understanding of how cancer cells develop. This knowledge may lead to new techniques for selective inhibition or destruction of cancer cells.

Manipulation of DNA replication and cell cycle control are the newest tools for progress in genetic engineering. In early 1997, the first successful cloning of an adult mammal, Dolly the sheep, raised important new issues about the biology and ethics of manipulating mammalian genomes. The technology now exists to clone human beings, although such experiments are not likely to be carried out. More relevant is the potential impact on agriculture. It is now possible to select for animals that have the most desirable traits, such as lower fat content or disease resistance, and create herds of genetically identical animals. Of direct relevance to hu-mans is the potential impact on the understanding of fertility and possible new treatments for infertility.

A new class of genetic diseases was discovered in the 1980's called triplet repeat diseases. Regions of DNA consist of copies of three nucleotides (such as CGG) that are repeated up to fifty times. Through unknown mechanisms related to DNA replication, the number of repeats may increase from generation to generation, at some point reaching a threshold level at which disease symptoms appear. Diseases found to conform to this pattern include fragile X syndrome, Huntington's disease (Huntington's chorea), and Duchenne muscular dystrophy.

The process of aging is closely related to DNA replication. Unlike bacteria, eukaryotic organisms have linear chromosomes. This poses problems for the cell, both in maintaining intact chromosomes

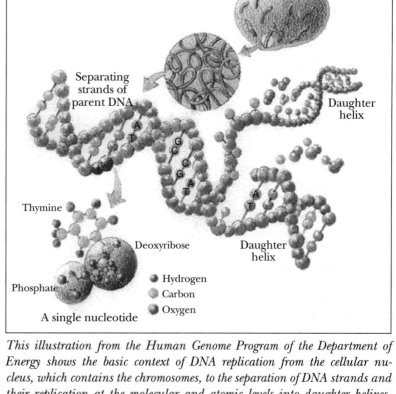

This illustration from the Human Genome Program of the Department of Energy shows the basic context of DNA replication from the cellular nucleus, which contains the chromosomes, to the separation of DNA strands and their replication at the molecular and atomic levels into daughter helixes. (U.S. Department of Energy Human Genome Program, http://www.ornl.gov/hgmis.)

(ends are unstable) and in replicating the DNA. The replication machinery cannot copy the extreme ends of a linear DNA molecule, so organisms have evolved alternate mechanisms. The ends of linear chromosomes consist of telomeres (short, repeated DNA sequences that are bound and stabilized by specific proteins), which are replicated by a separate mechanism using an enzyme called telomerase. Telomerase is inactivated in mature cells, and there may be a slow, progressive loss of the telomeres that ultimately leads to the loss of important genes, resulting in symptoms of aging. Cancer cells appear to have reactivated their telomerase, so potential anticancer therapies are being developed based on this information.

Michael R. Lentz, Ph.D.

FURTHER READING

Abstracts of Papers Presented at the 2007 Meeting on Eukaryotic DNA Replication and Genome Maintenance: September 5-September 9, 2007. Arranged by Stephen Bell and Joachim Li. Cold Spring Harbor, N.Y.: Cold Spring Harbor Laboratory, 2007. The laboratory regularly publishes abstracts of papers presented at its annual meeting on eukaryotic DNA replication.

Cann, Alan J. *DNA Virus Replication.* New York: Oxford University Press, 2000. Gives an analysis of protein-protein interactions in DNA virus replication, covering all major DNA virus groups: hepatitis B virus, papillomavirus, herpes simplex virus, Epstein-Barr virus, Kaposi's sarcoma herpesvirus (KSHV), human cytomegalovirus, and adenoviruses. Illustrated.

Cotterill, Sue, ed. *Eukaryotic DNA Replication: A Practical Approach.* New York: Oxford University Press, 1999. Serves as a comprehensive lab manual that describes key aspects of current techniques for investigating DNA replication in eukaryotes. Contains more than one hundred reliable protocols, including methods for studying the origin of replication, replication proteins, and the synthesis of telomeres.

DePamphilis, Melvin L., ed. *Concepts in Eukaryotic DNA Replication.* Cold Spring Harbor, N.Y.: Cold Spring Harbor Laboratory Press, 1999. A broad account of the basic principles of DNA replication and related functions such as DNA repair and protein phosphorylation. One chapter surveys advances in the field.

Drlica, Karl. *Understanding DNA and Gene Cloning: A Guide for the Curious.* 4th ed. Hoboken, N.J.: Wiley, 2004. An excellent introduction to the basic properties of DNA and its current applications. Consists of five sections: basic molecular genetics, manipulating DNA, molecular genetics, human genetics, and whole genomes.

Holmes, Frederic Lawrence. *Meselson, Stahl, and the Replication of DNA: A History of "the Most Beautiful Experiment in Biology."* New Haven, Conn.: Yale University Press, 2001. Traces the evolution of Matthew Meselson and Frank Stahl's 1957 landmark experiment, which confirmed that DNA replicates as predicted by the double helix structure Watson and Crick had proposed. Illustrations.

Kornberg, Arthur. *For the Love of Enzymes: The Odyssey of a Biochemist.* Reprint. Cambridge, Mass.: Harvard University Press, 1991. Kornberg discovered the enzymes that replicate DNA and was awarded the Nobel Prize for his work. This autobiography is a rich history of the process of science and discovery.

Kornberg, Arthur, and Tania A. Baker. *DNA Replication.* 2d ed. Sausalito, Calif.: University Science, 2005. The second edition of a classic text about DNA replication and related cellular processes.

Krebs, Jocelyn E., Elliott S. Goldstein, and Stephen T. Kilpatrick. "DNA Replication and Recombinations." In *Lewin's Essential Genes.* 2d ed. Sudbury, Mass.: Jones and Bartlett, 2010. This textbook includes a section on DNA replication.

Watson, James. *The Double Helix.* 1968. Reprint. New York: Simon & Schuster, 2001. Watson's account of the race to solve the structure of the DNA molecule.

WEB SITES OF INTEREST

National Center for Biotechnology Information: "What Is a Cell?"
http://www.ncbi.nlm.nih.gov/About/primer/genetics_cell.html

The site's science primer includes a page providing basic information about cells, including an explanation of DNA replication.

Scitable
http://www.nature.com/scitable

Scitable, a library of science-related articles compiled by the Nature Publishing Group, features several articles about DNA replication. Users can retrieve these articles by typing the words "DNA replication" into the site's search engine.

See also: Animal cloning; Cancer; Cell cycle; Cell division; Cloning; DNA sequencing technology; DNA structure and function; Genetic code; Genetic engineering; Molecular genetics; Mutation and mutagenesis; Protein structure; Protein synthesis; Restriction enzymes; RNA structure and function; RNA transcription and mRNA processing; Telomeres.

DNA sequencing technology

CATEGORY: Genetic engineering and biotechnology

SIGNIFICANCE: The genetic code is contained in the ordered, linear arrangement of the four nucleotides attached to the sugar-phosphate backbone of a strand of DNA: adenine, cytosine, guanine, and thymine. DNA sequencing is the determination of this ordered arrangement, and is used in basic biological research, as well as diagnostic applications, forensic investigations, and medical innovations.

KEY TERMS

automated fluorescent sequencing: a modification of chain-termination sequencing that uses fluorescent markers to identify the terminal nucleotides, allowing the automation of sequencing in which robots can carry out large-scale projects

base pair (bp): two nucleotides on opposite strands of DNA that are linked by a hydrogen bond; in DNA, adenine always pairs with thymine and guanine always pairs with cytosine; often used as a measure of the size of a DNA fragment or the distance along a DNA molecule between markers; both the singular and plural are abbreviated bp

Maxam-Gilbert sequencing: A method of base-specific chemical degradation to determine DNA sequence

primer: A short piece of single-stranded DNA that can hybridize to denatured DNA and provide a start point for extension of DNA by a DNA polymerase

Sanger sequencing: Also known as chain-terminator sequencing, a method using nucleotides that are missing the 3' hydroxyl group in order to terminate the polymerization of new DNA at a specific nucleotide

THE NEED FOR SEQUENCING

DNA was first discovered in 1869 as a viscous material in pus, and its basic chemical composition was well established by the 1930's. By 1950, the role of DNA as the hereditary material was clearly defined. In the 1950's, the classic papers by James Watson and Francis Crick and Matthew Meselson and Frank Stahl gave scientists a clear picture of the structure and function of DNA. In 1961, Crick demonstrated that the genetic code consisted of sets of three nucleotides in sequence (triplet codons) that identified specific amino acids. However, there was no system to read the sequence and uncover the actual words that spelled out the code of life.

The discovery of rapid sequencing methods in the 1970's created a flood of new discoveries in biology. The coding regions and control elements of DNA could be identified and compared. The sequence changes in different alleles of the same gene could be evaluated, homologous genes could be examined in divergent species, and evolutionary changes could be studied. Today, an entire genome can be sequenced, identifying every nucleotide in the correct order along every chromosome, in a matter of months. This ability to sequence the genomes of entire organisms has created a new field called genomics, the study and comparison of whole genomes of different organisms. Sequencing is now at the core of many of the new discoveries in biology. Modern DNA sequencing technology has allowed for the sequencing of the entire human genome, as well as many plant, animal, and microbial genomes.

The Human Genome Project, completed in 2003, isolated and identified the 3.2 billion base pairs (bp) comprising the entire set of genetic information contained in human DNA. The sequence has led to the discovery of genes associated with specific diseases, the isolation of DNA responsible for regulating cellular functions, and the development of gene-targeted drug therapies.

PRINCIPLES OF DNA SEQUENCING

Molecular biologists cannot observe DNA molecules directly, even through a microscope, so they must devise controlled chemical reactions whose outcomes are indicative of what occurs at the submicroscopic level. In DNA sequencing, the key is to use a chemical method that allows for the analysis of the base sequence one base at a time. Such a method needs to produce a collection of DNA fragments whose lengths can be used to detect the identity of the base located at the end of each different-sized fragment. For example, if fragments of the short DNA sequence ACGTCCGATCG can be predictably produced, then the size of each fragment could be used to determine the location of each base. If the fragment is cut to the right of each thymine base, fragments of 4 and 9 bp will be produced. Repeating the process for the other three nucleotides can identify their positions. The DNA sequence is obtained by reading from smallest to largest fragment and identifying which reaction generates each fragment. Although this is a very simple example, the principles apply to all current sequencing methods. Electrophoresis in denaturing polyacrylamide gels (to keep the DNA single-stranded) is used to separate fragments that are hundreds of base pairs in length but differ by only a single nucleotide. The DNA is labeled with either radioactive or fluorescent markers so that the bands of DNA fragments can be detected.

MAXAM-GILBERT SEQUENCING

Maxam-Gilbert sequencing, also known as chemical sequencing, was developed in the early 1970's and based on the chemical modification and cleavage of a strand of DNA at specific base pairs. To sequence DNA with this method, the DNA fragment to be sequenced is isolated and the 5′ end of one of the strands is labeled with a radioactive phosphorous-32 atom in the terminal phosphate group. This creates the endpoint for DNA elongation. In separate tubes, the DNA is reacted with chemicals that will cleave the backbone of the DNA strand at one of the four nucleotides. The method requires dangerous chemicals and does not easily lend itself to automation, so it is rarely used today.

SANGER SEQUENCING

Sanger sequencing, or chain-terminator sequencing, is named for its developer, Frederick Sanger. This method requires a short DNA segment of known sequence adjacent to the unknown region to be sequenced so that a short synthetic oligonucleotide can be made. The oligonucleotide acts as a primer for DNA synthesis in the direction of the DNA to be sequenced. The DNA to be sequenced is often cloned into a plasmid vector whose sequence is known, facilitating primer synthesis. The DNA is denatured and the primer is allowed to anneal to the DNA strand. A DNA polymerase is added to the reaction mixture, extending the DNA for a short distance in the presence of radioactive nucleotides, which labels the new DNA. The reaction is then divided into four equal parts and placed into four separate reaction tubes, each containing all four deoxynucleotides, the nucleotide precursors for DNA synthesis: deoxyadenosine triphosphate (dATP), deoxyguanosine triphosphate (dGTP), and deoxycytidine triphosphate (dCTP), and deoxythymidine triphosphate (dTTP). One modified dideoxynucleotide (ddATP, ddGTP, ddCTP, or ddTTP) is also added to each reaction. The dideoxynucleotides are missing the 3′ hydroxyl group; without the hydroxyl group, no more nucleotides can be added and DNA

Frederick Sanger developed one of the first methods for sequencing DNA and published the first genome sequence. (© The Nobel Foundation)

elongation terminates. Since the dideoxynucleotide constitutes only a small percentage of the available nucleotides, DNA elongation will proceed normally until the DNA polymerase inserts a dideoxynucleotide in place of the normal deoxynucleotide.

Since the terminated fragment is attached to the larger template strand, the DNA must be denatured by heat before electrophoresis on a polyacrylamide gel so that the size will correspond accurately to the position of the terminated base. Each of the four reactions is run in a separate lane on the gel and the DNA bands are visualized by autoradiography or UV light. The DNA sequence can then be read directly from the image, reading from top to bottom, or smallest fragment to largest fragment.

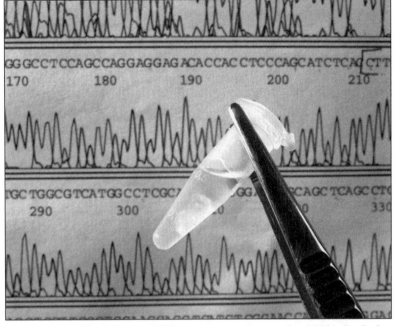

A chromatograph showing a DNA sequence and a sample of DNA for the human genome mapping project. (Getty Images)

AUTOMATED SEQUENCING

Automated sequencing methods are based on variations of chain-terminator methods. In dye-terminator sequencing, for example, each of the four dideoxynucleotides has a different fluorescent dye attached. When the DNA elongation is terminated, the fragment will be labeled with a specific color indicating which nucleotide is in the terminal position. As a result, only one reaction is needed instead of four separate reactions. Additionally, polymerase chain reaction (PCR), is often used in automated sequencing reactions, since it requires much smaller amounts of DNA than original sequencing methods and does not present a risk of sample contamination by cloning vectors. Modern automated sequencing also uses capillary electrophoresis, rather than gel electrophoresis. In this case, the reaction products are electrophoresed through a narrow capillary of polyacrylamide gel with a laser and fluorescence detector at the bottom. As the different-sized fragments reach the bottom, they pass the detector that registers the colors. The data are logged on a computer, which outputs the DNA sequence. This system can be automated so that robots move the samples into reaction tubes and load them into the capillaries. Computers compile and compare the sequence data. Automated sequencing methods can generate tens of thousands of bp of new sequence data per day, often with very little manpower.

IMPACT

New technologies have led to an increased volume of sequencing throughout the scientific community by simplifying sample preparation and increasing the accessibility to sequencing chemistries and equipment. However, current methods of DNA sequencing still sequence relatively short segments of DNA, often fewer than 1,000 bp. Future and emerging DNA sequencing strategies focus on larger-scale sequencing. Further, increased efficiency of high-throughput sequencing technologies will lower the cost of traditional DNA sequencing methods. New DNA sequencing technologies include in vitro cloning to amplify DNA molecules; parallel sequencing, in which DNA is bound to a solid surface and many samples are sequenced simultaneously; and sequencing by ligation, which uses the enzyme DNA ligase to identify nucleotides in a strand of DNA.

The goal of future DNA sequencing is to expand the scale of sequencing—possibly to entire chromosomes or large genomes all at once—and to enhance the precision and decrease the error rate of sequencing reactions. Together, new technologies will have far-reaching applications in the diagnosis and treatment of disease, the development of new biofuels, the protection against chemical and biological warfare agents, the study of anthropology and evolution, the determination of personalized genomes, and the optimization of agriculture, livestock breeding, and bioprocessing of food products.

J. Aaron Cassill, Ph.D.;
updated by Jennifer L. Gibson, Pharm.D.

FURTHER READING

Lister, R., B. D. Gregory, and J. R. Ecker. "Next Is Now: New Technologies for Sequencing of Genomes, Transcriptomes, and Beyond." *Current Opinion in Plant Biology* 12, no. 2 (April, 2009): 107-118. A discussion of the diverse applications of next-generation sequencing technologies.

Mardis, E. R. "Next-Generation DNA Sequencing Methods." *Annual Review of Genomics and Human Genetics* 9 (2008): 387-402. A review article describing recent scientific discoveries that resulted from next-generation sequencing technologies.

Maxam, Allan M., and Walter Gilbert. "A New Method for Sequencing DNA." *Proceedings of the National Academy of Sciences* 74 (1977): 560. The original description of sequencing by chemical cleavages.

Reilly, Philip R. *Abraham Lincoln's DNA and Other Adventures in Genetics.* Cold Spring Harbor, N.Y.: Cold Spring Harbor Laboratory Press, 2000. A series of brief articles about the social and moral implications of uncovering DNA information in humans.

Sanger, F., S. Nicklen, and A. R. Coulson. "DNA Sequencing with Chain-Terminating Inhibitors." *Proceedings of the National Academy of Sciences* 74 (1977): 5463. The original description of dideoxy termination sequencing.

Smith, Lloyd M., et al. "Fluorescence Detection in Automated DNA Sequence Analysis." *Nature* 321 (1986): 674. The original description of automated sequencing techniques.

WEB SITES OF INTEREST

How Do We Sequence DNA?
http://seqcore.brcf.med.umich.edu/doc/educ/dnapr/sequencing.html

Human Genome Project Information
http://www.ornl.gov/hgmis/home.shtml

NIH National Human Genome Research Institute
http://www.genome.gov

See also: Cloning; Cloning vectors; DNA replication; Genetic code; Genetic engineering; Genome libraries; Genome size; Genomics; Human Genome Project; Knockout genetics and knockout mice; Model organism: *Escherichia coli*; Molecular clock hypothesis; Polymerase chain reaction; Population genetics; Pseudogenes; Repetitive DNA; Restriction enzymes; Reverse transcriptase; RFLP analysis; Shotgun cloning; Synthetic genes; Transposable elements.

DNA structure and function

CATEGORY: Molecular genetics

SIGNIFICANCE: Structurally, DNA is a relatively simple molecule; functionally, however, it has wide-ranging roles in the cell. It functions primarily as a stable repository of genetic information in the cell and as a source of genetic information for the production of proteins. Greater knowledge of the characteristics of DNA has led to advances in the fields of genetic engineering, gene therapy, and molecular biology in general.

KEY TERMS

double helix: the molecular shape of DNA molecules, which resembles a ladder that twists, or spirals

gene expression: the processes (transcription and translation) by which the genetic information in DNA is converted into protein

transcription: the process by which genetic information in DNA is converted into messenger RNA (mRNA)

translation: the process by which the genetic information in an mRNA molecule is converted into protein

CHEMICAL AND PHYSICAL STRUCTURE OF DNA

Deoxyribonucleic acid (DNA) is the genetic material found in all cells. Chemically, it is classified as a nucleic acid, a relatively simple molecule composed of nucleotides. A nucleotide consists of a

The Four Nucleotides That Compose DNA

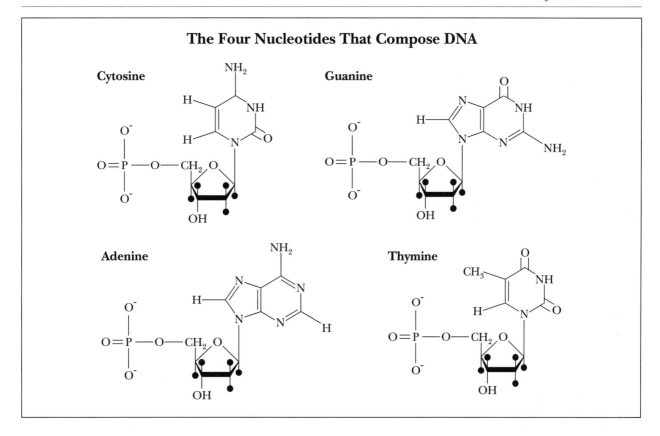

Cytosine

Guanine

Adenine

Thymine

sugar (deoxyribose), a phosphate group, and one of the nitrogenous bases: adenine (A), cytosine (C), guanine (G), or thymine (T). In fact, nucleotides differ only in the particular nitrogenous base that they contain. Ribonucleic acid (RNA) is the other type of nucleic acid found in the cell; however, it contains ribose as its sugar instead of deoxyribose and has the nitrogenous base uracil (U) instead of thymine. Nucleotides can be assembled into long chains of nucleic acid via connections between the sugar on one nucleotide and the phosphate group on the next, thereby creating a sugar-phosphate "backbone" in the molecule. The nitrogenous base on each nucleotide is positioned such that it is perpendicular to the backbone, as shown in the following diagram:

```
sugar — phosphate — sugar — phosphate — sugar — phosphate — sugar
   |                   |                   |                   |
 base                base                base                base
```

Any one of the four DNA nucleotides (A, C, G, or T) can be used at any position in the molecule; it is

therefore the specific sequence of nucleotides in a DNA molecule that makes it unique and able to carry genetic information. The genetic information is the sequence itself.

In the cell, DNA exists as a double-stranded molecule; this means that it consists of two chains of nucleotides side by side. The double-stranded form of DNA can most easily be visualized as a ladder, with the sugar-phosphate backbones being the sides of the ladder and the nitrogenous bases being the rungs of the ladder, as shown in the following diagram:

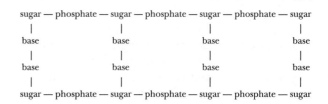

```
sugar — phosphate — sugar — phosphate — sugar — phosphate — sugar
   |                   |                   |                   |
 base                base                base                base
   |                   |                   |                   |
 base                base                base                base
   |                   |                   |                   |
sugar — phosphate — sugar — phosphate — sugar — phosphate — sugar
```

This ladder is then twisted into a spiral shape. Any spiral-shaped molecule is called a "helix," and since each strand of DNA is wound into a spiral, the complete DNA molecule is often called a "double helix." This molecule is extremely flexible and can be compacted to a great degree, thus allowing the cell to contain large amounts of genetic material.

THE DISCOVERY OF DNA AS THE GENETIC MATERIAL

Nucleic acids were discovered in 1869 by the physician Friedrich Miescher. He isolated these molecules, which he called nuclein, from the nuclei of white blood cells. This was the first association of nucleic acids with the nucleus of the cell. In the 1920's, experiments performed by other scientists showed that DNA could be located on the chromosomes within the nucleus. This was strong evidence for the role of DNA in heredity, since at that time there was already a link between the activities of chromosomes during cell division and the inheritance of particular traits, largely because of the work of the geneticist Thomas Hunt Morgan about ten years earlier.

However, it was not immediately apparent, based on this evidence alone, that DNA was the genetic material. In addition to DNA, proteins are present in the nucleus of the cell and are an integral part of chromosomes as well. Proteins are also much more complex molecules than nucleic acids, having a greater number of building blocks; there are twenty amino acids that can be used to build proteins, as opposed to only four nucleotides for DNA. Moreover, proteins tend to be much more complex than DNA in terms of their three-dimensional structure as well. Therefore, it was not at all clear in the minds of many scientists of the time that DNA had to be the genetic material, since proteins could not specifically be ruled out.

In 1928, the microbiologist Frederick Griffith supplied some of the first evidence that eventually led to the identification of DNA as the genetic material. Griffith's research involved the bacterium *Streptococcus pneumoniae*, a common cause of lung infections. He was working primarily with two different strains of this bacterium: a strain that was highly virulent (able to cause disease) and a strain that was nonvirulent (not able to cause disease). Griffith noticed that if he heat-killed the virulent strain and then mixed its cellular debris with the living, nonvirulent strain, the nonvirulent strain would be "transformed" into a virulent strain. He did not know what part of the heat-killed virulent cells was

responsible for the transformation, so he simply called it the "transforming factor" to denote its activity in his experiment. Unfortunately, Griffith never took the next step necessary to reveal the molecular identity of this transforming factor.

That critical step was taken by another microbiologist, Oswald Avery, and his colleagues in 1944. Avery essentially repeated Griffith's experiments with two important differences: Avery partially purified the heat-killed virulent strain preparation and selectively treated this preparation with a variety of enzymes to see if the transforming factor could be eliminated, thereby eliminating the transformation itself. Avery showed that transformation was prevented only when the preparation was treated with deoxyribonuclease, an enzyme that specifically attacks and destroys DNA. Other enzymes that specifically destroy RNA or proteins could not prevent transformation from occurring. This was extremely strong evidence that the genetic material was DNA.

Experiments performed in 1952 by molecular biologists Alfred Hershey and Martha Chase using the bacterial virus T2 finally demonstrated conclusively that DNA was indeed the genetic material. Hershey and Chase studied how T2 infects bacterial cells to determine what part of the virus, DNA or protein, was responsible for causing the infection, thinking that whatever molecule directed the infection would have to be the genetic material of the virus. They found that DNA did directly participate in infection of the cells by entering them, while the protein molecules of the viruses stayed outside the cells. Most strikingly, they found that the original DNA of the "parent" viruses showed up in the "offspring" viruses produced by the infection, directly demonstrating inheritance of DNA from one generation to another. This was an important element of the argument for DNA as the genetic material.

THE WATSON-CRICK DOUBLE-HELIX MODEL OF DNA

With DNA conclusively identified as the genetic material, the next step was to determine the structure of the molecule. This was finally accomplished when the double-helix model of DNA was proposed by molecular biologists James Watson and Francis Crick in 1953. This model has a number of well-defined and experimentally determined characteristics. For example, the diameter of the molecule, from one sugar-phosphate backbone to the other, is

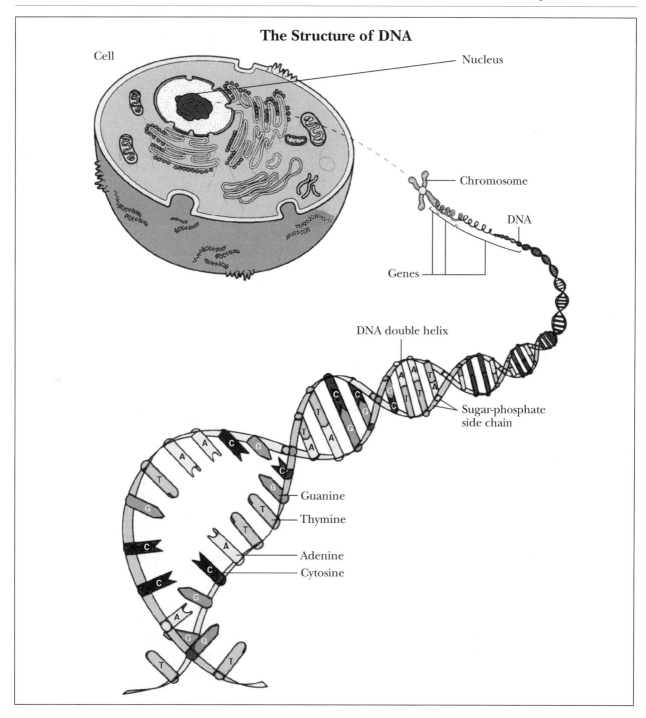

The Structure of DNA

Cell

Nucleus

Chromosome

DNA

Genes

DNA double helix

Sugar-phosphate
side chain

Guanine

Thymine

Adenine

Cytosine

20 angstroms. (There are 10 million angstroms in one millimeter, which is one-thousandth of a meter.) There are 3.4 angstroms from one nucleotide to the next, and the entire double helix makes one turn for every ten nucleotides, a distance of about 34 angstroms. These measurements were determined by the physicists Maurice Wilkins and Rosalind Franklin around 1951 using a process called X-ray diffrac-

tion, in which crystals of DNA are bombarded with X rays; the resulting patterns captured on film gave Wilkins and Franklin, and later Watson and Crick, important clues about the physical structure of DNA.

Another important aspect of Watson and Crick's double-helix model is the interaction between the nitrogenous bases in the interior of the molecule. Important information about the nature of this interaction was provided by molecular biologist Erwin Chargaff in 1950. Chargaff studied the amounts of each nitrogenous base present in double-stranded DNA from organisms as diverse as bacteria and humans. He found that no matter what the source of the DNA, the amount of adenine it contains is always roughly equal to the amount of thymine; there are also equal amounts of guanine and cytosine in DNA. This information led Watson and Crick to propose an interaction, or "base pairing," between these sets of bases such that A always base pairs with T (and vice versa) and G always base pairs with C. Another name for this phenomenon is "complementary base pairing": A is said to be the "complement" of T, and so on.

The force that holds complementary bases, and therefore the two strands of DNA, together is a weak chemical interaction called a hydrogen bond, which is created whenever a hydrogen atom in one molecule has an affinity for nitrogen or oxygen atoms in another molecule. The affinity of the atoms for each other draws the molecules together in the hydrogen bond. A-T pairs have two hydrogen bonds between them because of the chemical structure of the bases, whereas G-C pairs are connected by three hydrogen bonds, making them slightly stronger and more stable than A-T pairs. The entire DNA double helix, although it is founded upon the hydrogen bond, one of the weakest bonds in nature, is nonetheless an extraordinarily stable structure because of the combined force of the millions of hydrogen bonds holding most DNA molecules together. However, these hydrogen bonds can be broken under certain conditions in the cell. This usually occurs as part of the process of the replication of the double helix, in which the two strands of DNA must come apart in order to be duplicated. In the cell, the hydrogen bonds are broken with the help of enzymes. Under artificial conditions in the laboratory, hydrogen bonds in the double helix can easily be broken just by heating a solution of DNA to high temperatures (close to the boiling point).

OTHER FEATURES OF THE WATSON-CRICK MODEL

Watson and Crick were careful to point out that their double-helix model of DNA was the first model to immediately suggest a mechanism by which the molecule could be replicated. They knew that this replication, which must occur before the cell can divide, would be a necessary characteristic of the genetic material of the cell and that an adequate model of DNA must help explain how this duplication could occur. Watson and Crick realized that the mechanism of complementary base pairing that was an integral part of their model was a potential answer to this problem. If the double helix is separated into its component single-strand molecules, each strand will be able to direct the replacement of the opposite, or complementary, strand by base pairing properly with only the correct nucleotides. For example, if a single-strand DNA molecule has the sequence TTAGTCA, the opposite complementary strand will always be AATCAGT; it is as if the correct double-stranded structure is "built in" to each single strand. Additionally, as each of the single strands in a double-strand DNA molecule goes through this addition of complementary nucleotides, two new DNA double helices are produced where there was only one before. Further, these new DNA molecules are completely identical to each other, barring any mistakes that might have been made in the replication process.

A strand of DNA also has a certain direction built into it; the DNA double helix is often called "antiparallel" in reference to this aspect of its structure. "Antiparallel" means that although the two strands of the DNA molecule are essentially side by side, they are oriented in different directions relative to the position of the deoxyribose molecules on the backbone of the molecule. To help keep track of the orientation of the DNA molecule, scientists often refer to a 5′ to 3′ direction. This designation comes from numbering the carbon atoms on the deoxyribose molecule (from 1′ to 5′) and takes note of the fact that the deoxyribose molecules on the DNA strand are all oriented in the same direction in a head-to-tail fashion. If it were possible to stand on a DNA molecule and walk down one of the sugar-phosphate backbones, one would encounter a 5′ carbon atom on a sugar, then the 3′ carbon, and so on all the way down the backbone. If one were walking on the other strand, the 3′ carbon atom would always be encountered before the 5′ carbon. The concept of an antiparallel double helix has important implications for the ways that DNA is produced and used in the cell. Generally, the cellular enzymes that are involved in processes concerning DNA are restricted to recognizing just one direction. For example, DNA polymerase, the enzyme that is responsible for making DNA in the cell, can only make DNA in a 5′ to 3′ direction, never the reverse.

Watson and Crick postulated a right-handed helix as part of their double-helix model; this means that the strands of DNA turn to the right, or in a counterclockwise fashion. This is now regarded as the "biological" (B) form of DNA because it is the form present inside the nucleus of the cell and in solutions of DNA. However, it is not the only possible form of DNA. In 1979, an additional form of DNA was discovered by molecular biologist Alexander Rich that exhibited a zigzag, left-handed double helix; he called this form of DNA Z-DNA. Stretches of alternating G and C nucleotides most commonly give rise to this conformation of DNA, and scientists think that this alternative form of the double helix is important for certain processes in the cell in which various molecules bind to the double helix and affect its function.

The Function of DNA in the Cell

DNA plays two major roles in the cell. The first is to serve as a storehouse of the cell's genetic information. Normally, cells have only one complete copy of their DNA molecules, and this copy is, accordingly, highly protected. DNA is a chemically stable molecule; it resists damage or destruction under normal conditions, and, if it is damaged, the cell has a variety of mechanisms to ensure the molecule is rapidly repaired. Furthermore, when the DNA in the cell is duplicated in a process called DNA replication, this duplication occurs in a regulated and precise fashion so that a perfect copy of DNA is produced. Once the genetic material of the cell has been completely duplicated, the cell is ready to divide in two in a process called mitosis. After cell division, each new cell of the pair will have a perfect copy of the genetic material; thus these cells will be genetically identical to each other. DNA thus provides a mechanism by which genetic informa-

tion can be transferred easily from one generation of cells (or organisms) to another.

The second role of DNA is to serve as a blueprint for the ultimate production of proteins in the cell. This process occurs in two steps. The first step is the conversion of the genetic information in a small portion of the DNA molecule, called a gene, into messenger RNA (mRNA). This process is called

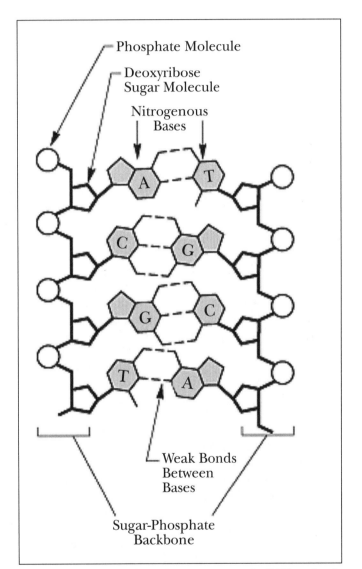

A schematic showing the major components of a DNA molecule, including the four bases that compose DNA—adenine (A) and thymine (T), cytosine (C) and guanine (G)—and how they form the "rungs" of the DNA "ladder" by forming hydrogen bonds. (U.S. Department of Energy Human Genome Program, http://www.ornl.gov/hgmis)

transcription, and here the primary role of the DNA molecule is to serve as a template for synthesis of the mRNA molecule. The second step, translation, does not involve DNA directly; rather, the mRNA produced during transcription is in turn used as genetic information to produce a molecule of protein. However, it is important to note that genetic information originally present in the DNA molecule indirectly guides the synthesis and final amino acid sequence of the finished protein. Both of these steps, transcription and translation, are often called gene expression. A single DNA molecule in the form of a chromosome may contain thousands of different genes, each providing the information necessary to produce a particular protein. Each one of these proteins will then fulfill a particular function inside or outside the cell.

IMPACT AND APPLICATIONS

Knowledge of the physical and chemical structure of DNA and its function in the cell has undoubtedly had far-reaching effects on the science of biology. However, one of the biggest effects has been the creation of a new scientific discipline: molecular biology. With the advent of Watson and Crick's double-helix model of DNA, it became clear to many scientists that, perhaps for the first time, many of the important molecules in the cell could be studied in detail and that the structure and function of these molecules could also be elucidated. Within fifteen years of Watson and Crick's discovery, a number of basic genetic processes in the cell had been either partially or completely detailed, including DNA replication, transcription, and translation. Certainly the seeds of this revolution in biology were being planted in the decades before Watson and Crick's 1953 model, but it was the double helix that allowed scientists to investigate the important issues of genetics on the cellular and molecular levels.

An increased understanding of the role DNA plays in the cell has also provided scientists with tools and techniques for changing some of the genetic characteristics of cells. This is demonstrated by the rapidly expanding field of genetic engineering, in which scientists can precisely manipulate DNA and cells on the molecular level to achieve a desired result. Additionally, more complete knowledge of how the cell uses DNA has opened windows of understanding into abnormal cellular processes such

as cancer, which is fundamentally a defect involving the cell's genetic information or the expression of that information.

Through the tools of molecular genetics, many scientists hope to be able to correct almost any genetic defect that a cell or an organism might have, including cancer or inherited genetic defects. The area of molecular biology that is concerned with using DNA as a way to correct cellular defects is called gene therapy. This is commonly done by inserting a normal copy of a gene into cells that have a defective copy of the same gene in the hope that the normal copy will take over and eliminate the effects of the defective gene. It is hoped that this sort of technology will eventually be used to overcome even complex problems such as Alzheimer's disease and acquired immunodeficiency syndrome (AIDS).

One of the most unusual and potentially rewarding applications of DNA structure was introduced by computer scientist Leonard Adleman in 1994. Adleman devised a way to use short pieces of single-stranded DNA in solution as a rudimentary "computer" to solve a relatively complicated mathematical problem. By devising a code in which each piece of DNA stood for a specific variable in his problem and then allowing these single-stranded DNA pieces to base pair with each other randomly in solution, Adleman obtained an answer to his problem in a short amount of time. Soon thereafter, other computer scientists and molecular biologists began to experiment with other applications of this fledgling technology, which represents an exciting synthesis of two formerly separate disciplines. It may be that this research will prove to be the seed of another biological revolution with DNA at its center.

Randall K. Harris, Ph.D.

FURTHER READING

Banaszak, Leonard. *Foundations of Structural Biology.* San Diego: Academic Press, 2000. Provides visualization skills with three-dimensional imaging to assist students in understanding the implications of the three-dimensional coordinates for a molecule with several thousand atoms.

Bates, Andrew D., and Anthony Maxwell. *DNA Topology.* 2d ed. New York: Oxford University Press, 2005. Begins with a basic account of DNA structure and then covers more complex concepts, including DNA supercoiling; the definitions and physical meanings of linking number, twist, and writhe; and the biological significance of the topological aspects of DNA structure.

Bradbury, E. Morton, and Sandor Pongor, eds. *Structural Biology and Functional Genomics.* Boston: Kluwer Academic, 1999. Topics include DNA repeats in the human genome, modeling DNA stretching for physics and biology, chromatin control of HIV-1 gene expression, and exploring structure space.

Calladine, Chris R., et al. *Understanding DNA: The Molecule and How It Works.* 3d ed. San Diego: Elsevier Academic Press, 2004. Provides an introduction to molecular biology for nonscientists.

Frank-Kamenetskii, Maxim D. *Unraveling DNA.* Rev. and updated paperback ed. Reading, Mass.: Addison-Wesley, 1997. Melds history, biographical details, and science to provide a general discussion of DNA and its basic structure and function.

McCarty, Maclyn. *The Transforming Principle: Discovering That Genes Are Made of DNA.* New York: W. W. Norton, 1985. Gives an insider's view of the circumstances surrounding Oswald Avery's pivotal experiments.

Maddox, Brenda. *Rosalind Franklin: The Dark Lady of DNA.* New York: HarperCollins, 2002. Tells the other side of the story in the discovery and structure of DNA, focusing on the often neglected role of Franklin. Illustrations, bibliography, index.

Rosenfield, Israel, Edward Ziff, and Borin van Loon. *DNA for Beginners.* New York: W. W. Norton, 1983. Provides an entertaining, yet factual, cartoon account of basic DNA structure and function, as well as more advanced topics in molecular biology.

Smith, Paul J., and Christopher J. Jones, eds. *DNA Recombination and Repair.* New York: Oxford University Press, 1999. Explores the cellular processes involved in DNA recombination and repair by highlighting current research, including strategies for dealing with DNA mismatches or lesions, the enzymology of excision repair, and the integration of DNA repair into other cellular pathways.

Watson, James. *The Double Helix.* 1968. Reprint. New York: Simon & Schuster, 2001. Watson's account of the race to solve the structure of the DNA molecule.

Watson, James, et al. *Recombinant DNA—Genes and Genomes: A Short Course.* 3d ed. New York: W. H. Freeman, 2007. Uses accessible language and exceptional diagrams to give a concise background on the methods, underlying concepts, and far-reaching applications of recombinant DNA technology.

WEB SITES OF INTEREST

Biology at University of Cincinnati Clermont College, DNA Structure and Function
http://biology.clc.uc.edu/courses/bio104/dna.htm
Describes the history of DNA research, including the Hershey-Chase experiment, and the structure and function of DNA. Provides links to related Web sites.

Left-Handed DNA Hall of Fame
http://www-lmmb.ncifcrf.gov/~toms/LeftHanded.DNA.html
Molecular information theorist Tom Schneider created this site to document media and book illustrations in which DNA is shown incorrectly twisting to the left.

University of Arizona Biology Learning Center, Introduction to DNA Structure
http://www.blc.arizona.edu/Molecular_Graphics/DNA_Structure/DNA_Tutorial.HTML
This online companion to introductory courses in biology and biochemistry includes an illustrated explanation of DNA structure.

University of Massachusetts, DNA Structure
http://www.umass.edu/molvis/tutorials/dna/
An interactive, animated, downloadable, tutorial on the molecular composition and structure of DNA for high school students and college freshmen. Available in Spanish, German, and Portuguese.

See also: Ancient DNA; Antisense RNA; Chromosome structure; DNA isolation; DNA repair; DNA replication; Genetic code; Genetic code, cracking of; Molecular genetics; Noncoding RNA molecules; One gene-one enzyme hypothesis; Protein structure; Protein synthesis; Repetitive DNA; RNA isolation; RNA structure and function; RNA transcription and mRNA processing.

Down syndrome

CATEGORY: Diseases and syndromes
ALSO KNOWN AS: Trisomy 21

DEFINITION

Down syndrome is one of the most common chromosomal defects in human beings. According to some studies, it occurs in one in seven hundred live births; other studies place the number at one in nine hundred. Further, it occurs in about one in every two hundred conceptions.

This syndrome (a pattern of characteristic abnormalities) was first described in 1866 by the English physician John Langdon Down. While in charge of an institution housing the profoundly mentally retarded, he noticed that almost one in ten of his patients had a flat face and slanted eyes causing Down to use the term "mongolism" to describe the syndrome; this term, however, is misleading. Males and females of every race and ethnicity can and do have this syndrome. To eliminate the unintentionally racist implications of the term "mongolism," Lionel Penrose and his colleagues changed the name to Down syndrome. Although Down syndrome was observed and reported in the 1860's, it was almost one hundred years before the cause was discovered.

RISK FACTORS

A woman's chance of having a child with Down syndrome increases with age because older egg cells are at greater risk of having improperly divided chromosomes. A pregnant woman who has given birth to one child with Down syndrome has about a 1 percent chance of having another child with the condition. In addition, both women and men who have the genetic translocation for Down syndrome can pass it on to their children.

ETIOLOGY AND GENETICS

In 1959, the French physician Jérôme Lejeune and his associates realized that the presence of an extra chromosome 21 was the apparent cause of Down syndrome. This fact places the syndrome in the broader category of aneuploid conditions. All human cells have forty-six chromosomes or strands made up of the chemical called deoxyribonucleic acid (DNA). The sections or subdivisions along these forty-six strands, called genes, are responsible for producing all the proteins that determine specific human characteristics. An aneuploid is a cell with forty-five or forty-seven or more chromosomes, with the missing or extra strands of DNA leaving the individual with too few or too many genes. This aneuploid condition then results in significant alterations in one's traits and a great number of potential abnormalities.

In a normal individual, the forty-six strands are actually twenty-three pairs of chromosomes that are referred to as homologous because each pair is the same size and contains the same genes. In most cases of Down syndrome, there are three copies of chromosome 21. An aneuploid with three of a particular chromosome is called trisomic; thus Down syndrome is often called trisomy 21. The extra chromosome is gained because either the egg or sperm that came together at fertilization contained an extra one. This error in gamete (egg or sperm) production is called nondisjunction and occurs during the process of meiosis. When meiosis proceeds normally, the homologous chromosome pairs are separated from each other, forming gametes with twenty-three chromosomes, one from each pair. If nondisjunction occurs, a pair fails to separate, producing a gamete type with twenty-two chromosomes and a second gamete type with twenty-four chromosomes. If the pair that has failed to separate is chromosome 21, then the potential exists for twenty-three chromosomes in a normal gamete to combine with a gamete containing twenty-four, creating a trisomic individual with forty-seven chromosomes.

Although this syndrome was recognized by Down in 1866, true understanding of it dates from the work that Lejeune began in 1953. The seemingly innocuous characteristic of abnormal palm prints and fingerprints fostered an important insight for him. Since those prints are laid down very early in the child's prenatal development, they suggest a profoundly altered embryological course of events. His intuition told him that not one or two altered genes but rather a whole chromosome's genes must be at fault. In 1957, he discovered, by the culturing of cells from children with Down syndrome in dishes in the laboratory, that those cells contained forty-seven chromosomes. This work eventually resulted in his 1959 publication, which was soon followed by the discovery that the extra chromosome present was a third copy of chromosome 21.

The subsequent development of more sophisticated methods of identifying individual parts of chromosomes has shed much light on the possible mechanisms by which the symptoms are caused. Some affected individuals do not have a whole extra chromosome 21; rather, they possess a third copy of some part of that chromosome. A very tiny strand of DNA, chromosome 21 contains only about fifteen hundred genes. Of these fifteen hundred, only a few hundred are consistently present in those who suffer from Down syndrome, namely the genes in the bottom one-third of the chromosome. Among those genes are several that could very likely cause certain symptoms associated with Down syndrome. A leukemia-causing gene and a gene for a protein in the lens of the eye that could trigger cataract formation have both been identified. A gene for the production of the chemicals called purines has been located. The overabundance of purines produced when three copies of this gene are present has been linked to the mental retardation usually seen.

Even the fact that individuals with Down syndrome have a greatly reduced life expectancy is validated by the presence of an extra gene for the enzyme superoxide dismutase, which seems involved in the normal aging process. Like Alzheimer's disease patients, Down syndrome patients who live past forty years of age have gummy tangles of protein strands called amyloid fibers in their brains. Since one form of inherited Alzheimer's is caused by a gene on chromosome 21, scientists continue to search for links between the impaired mental functioning characteristic of both diseases.

Other research has shed light on the long-recognized relationship between the age of the mother and an increased risk of having a child with Down syndrome. Using more and more elaborate methods of chromosome banding, geneticists can determine whether the extra chromosome 21 came from the mother or the father. In 94 percent of children, the egg brings the extra chromosome. Since the first steps of meiosis to produce her future eggs occur before the mother's own birth, the older the mother, the longer these egg cells have been exposed to potentially harmful chemicals or radiation. On the other hand, paternal age is not a factor because all the steps of meiosis in males occur in cells produced in the few weeks before conception. The continued study of the age factor, as well as new insights from genomics, are leading to a greater understanding for all those affected by Down syndrome.

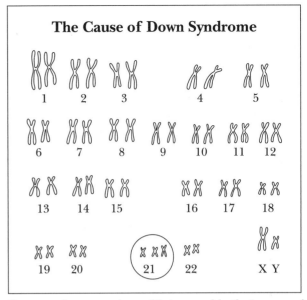

The Cause of Down Syndrome

Down syndrome, or trisomy 21, is caused by the presence of an abnormal third chromosome in pair 21. (Electronic Illustrators Group)

SYMPTOMS

The slanted appearance of the eyes first reported by Down is caused by a prominent fold of skin called an epicanthic fold (a fold in the upper eyelid near the corner of the eye). This fold of skin is accompanied by excess skin on the back of the neck and abnormal creases in the skin of the palm. In addition, the skull is wide, with a flat back and a flat face. The hair on the skull is sparse and straight.

The rather benign physical abnormalities are minor compared to the defects in internal organ systems. Almost 40 percent of Down syndrome patients suffer from serious heart defects. They are very prone to cancer of the white blood cells (acute leukemia), the formation of cataracts, and serious recurring respiratory infections. Short of stature with poorly formed joints, they often have poor reflexes, weak muscle tone, and an unstable gait. The furrowed, protruding tongue that often holds the mouth partially open is an external sign of the serious internal digestive blockages frequently present. These blockages must often be surgically repaired before the individual's first birthday. Many suffer from major kidney defects that are often irreparable. Furthermore, a suppressed immune system can easily lead to death from an infectious disease such as influenza or pneumonia.

With all these potential physical problems, it is not surprising that nearly 50 percent of Down syndrome patients die before the age of one. For those who live, there are enormous physical, behavioral, and mental challenges. The mental retardation that always accompanies Down syndrome ranges from quite mild to profound. This mental retardation makes all learning difficult and speech acquisition in particular very slow. Yet most individuals with Down syndrome have warm, loving personalities and enjoy art and music.

SCREENING AND DIAGNOSIS

The American College of Obstetricians and Gynecologists recommends that all pregnant women, regardless of their age, be screened for Down syndrome. Various prenatal screening tests can help detect whether the mother has a high risk of having a baby with this condition. A blood test usually is conducted around the sixteenth week of pregnancy to screen for Down syndrome and other disorders. Women who want an earlier assessment can receive a first-trimester combined test, which is conducted in two phases from week eleven to week thirteen of pregnancy. However, about 5 percent of women who undergo this test have a false-positive result, which means that they are incorrectly identified as being at risk for having a baby with Down syndrome. This first-trimester test consists of an ultrasound and blood tests.

Women who opt to wait for a later assessment can have full integrated testing, which is conducted in

Familial Down Syndrome

Down syndrome always involves either an extra portion of or a complete extra copy of chromosome 21. There are three mechanisms by which this can occur. Between 92 and 95 percent of cases result from nondisjunction during meiosis, in which two copies of chromosome 21 migrate to the same pole and end up in the same daughter cell. This most often happens in women, and if an egg with two copies of chromosome 21 is fertilized, the zygote will have three copies and all cells throughout the developing fetus will have an extra chromosome 21.

The second mechanism, mosaic trisomy 21, involves an error in cell division shortly after conception. This error produces two populations or lines of cells, some with 46 chromosomes and some with 47—the ones that have the additional chromosome 21. This mechanism occurs in 2 to 4 percent of Down syndrome live births. Covert mosaicism in parents used to be suspected as causing familial Down syndrome but is no longer indicated.

Between 1 and 4 percent of children with Down syndrome have translocation trisomy 21, which occurs when extra genetic material from chromosome 21 has been translocated to another chromosome. A family history of Down syndrome is an indication that this may be the cause of the defect. The occurrence of more than one case of Down syndrome in a family is relatively rare, but when it does occur, translocation trisomy is often suspected. Carrier parents usually do not display any genetic abnormalities. Not until there is miscarriage of a fetus with Down syndrome or birth of a child with Down syndrome do couples discover that one of them is a translocation carrier. Carriers can produce (1) noncarrier, chromosomally normal, children (which usually happens); (2) carrier children, just like the carrier parent, who are translocation heterozygotes; or (3) children with Down syndrome. Carrier mothers produce children with Down syndrome about 12 percent of the time. Carrier fathers produce children with Down syndrome about 3 percent of the time. Why greater risk exists for mothers is not clear.

Though maternal age is the most frequent predisposing factor for Down syndrome, it is uncorrelated with familial Down syndrome; translocation trisomy 21 occurs with equal frequency in younger and older women. Rarely, a carrier parent will have a translocation between both twenty-first chromosomes, a translocation carrier homozygote. This parent has a 100 percent chance for producing children with Down syndrome.

Ever since the genes on chromosome 21 were fully mapped, pedigree research (family recurrence studies) and epidemiological research (studies of chance occurrence among populations) have supported that these chromosomal abnormalities and uneven distributions of genetic material are inherited, and most often through mothers. Though cryptic parental mosaicism is no longer suspected and there is promising research investigating mitochondrial DNA in the form of cytoplasmic inheritance, the specific genetic mechanism of familial Down syndrome remains elusive.

Paul Moglia, Ph.D.

two phases during their first two trimesters of pregnancy. Only about 1 percent of women receive a false-positive result from this test. The first part of the integrated test, performed during the first trimester, includes an ultrasound and blood test; additional blood tests are performed during the second trimester, at fifteen to twenty weeks of pregnancy. If this test indicates a high risk of Down syndrome, a woman may receive a more invasive test in order to confirm whether her baby has the condition.

Women who do not want to have screening tests can choose to have diagnostic tests, such as amniocentesis, in which a sample of amniotic fluid is used to analyze the fetus's chromosomes; chorionic villus sampling (CVS), which uses cells from the mother's placenta to analyze fetal chromosomes; and percutaneous umbilical blood sampling (PUBS), in which blood is taken from a vein in the umbilical cord and examined for chromosomal defects.

After a child is born, an initial diagnosis of Down syndrome can be made at birth based upon the baby's appearance. The physician, using a stethoscope, may also hear a heart murmur in the baby's chest. Blood tests can then be conducted to locate the extra chromosome and confirm the diagnosis. Other diagnostic tests include an echocardiogram (the use of ultrasound to check for heart defects), an electrocardiogram (ECG) of the heart, and X rays of the chest and the gastrointestinal tract.

Individuals with Down syndrome must be closely screened for certain medical conditions. Children should have an eye exam every year during infancy and X rays of the upper or cervical spine between the ages of three and five. Patients with the syndrome also need to have a hearing test every six to twelve months, depending on age; dental exams every six months; and Pap smears and pelvic exams beginning in puberty or by the age of twenty-one.

TREATMENT AND THERAPY

There is no specific treatment for Down syndrome. Babies who are born with a gastrointestinal blockage may need major surgery immediately after birth, and some heart defects may also require surgery. The mother should be sure that the baby is well supported and fully awake during breast-feeding. Although poor tongue control could cause a baby to experience some milk leakage during feeding, many infants with Down syndrome can be successfully breast-fed.

Older children and adults with Down syndrome may have a problem with obesity. They should get plenty of exercise and avoid eating high-calorie foods. In addition, their necks and hips should be examined before they begin sports activities.

People with Down syndrome can benefit from behavioral training to help them cope with their anger, frustration, and compulsive behavior. Parents and caregivers should encourage a child with Down syndrome to be independent, and they should learn how to help the child deal with frustration.

Both males and females with Down syndrome are at risk of being abused sexually or in some other way. Adolescent girls and women with Down syndrome are usually able to become pregnant. For these reasons, it is important that individuals with Down syndrome learn about pregnancy and know how to take the proper precautions. They should also be taught how to take care of themselves in difficult situations.

A variety of social and educational services are available in many communities to help those with Down syndrome, including special education programs to address delayed mental development. Speech, physical, and occupational therapies can help with language and movement skills, feeding, and performing other tasks. In addition, children and their parents can receive mental health care in order to manage their moods or behaviors.

PREVENTION AND OUTCOMES

Down syndrome cannot be prevented. However, women at risk of giving birth to a child with this condition may seek genetic counseling before becoming pregnant.

Grace D. Matzen, M.A.; updated by Rebecca Kuzins

FURTHER READING

Beighton, Peter. *The Person Behind the Syndrome.* Rev. ed. New York: Springer-Verlag, 1997. Biographical details about John Langdon Down.

Cohen, William I., Lynn Nadel, and Myra E. Madnick, eds. *Down Syndrome: Visions for the Twenty-first Century.* New York: Wiley-Liss, 2002. Reviews findings from a 2000 conference, providing a comprehensive treatment of the current issues of self-determination, education, and advocacy, as well as new research developments.

Hogenboom, Marga. *Living with Genetic Syndromes Associated with Intellectual Disability.* Philadelphia: Jessica Kingsley, 2001. Addresses not only Down syndrome but also Williams, Angelman, and Prader-Willi syndromes from both a psychological and a practical standpoint. Valuable to the genetics student for its introductory material on the genetics of these syndromes.

Lubec, G. *The Molecular Biology of Down Syndrome.* New York: Springer, 1999. Twenty-five chapters examine different aspects of Down syndrome, including neuropathology, molecular pathology, mechanisms of neuronal death, oxidative stress, and apoptosis.

_____. *Protein Expression in Down Syndrome Brain.* New York: Springer, 2001. Both original research and current opinions on Down syndrome, with an emphasis on the molecular biology at the protein (rather than the nucleic acid) level, from studies using fetal brains with Down syndrome.

Newton, Richard. *The Down's Syndrome Handbook: A Practical Guide for Parents and Caregivers.* Rev. ed. London: Random House, 2004. Helpful advice about the capabilities of affected people. Combines medical knowledge with sympathetic common sense to provide help and advice to caregivers of young Down syndrome patients.

Rondal, Jean-Adolphe, and Alberto Rasore-Quartino, eds. *Therapies and Rehabilitation in Down Syndrome.* Hoboken, N.J.: J. Wiley and Sons, 2007. Essays analyze numerous forms of therapy and rehabilitation aimed at improving the lives of people with Down syndrome, including gene-based, pharmacological, medical, and hormonal therapies.

Rondal, Jean-Adolphe, Alberto Rasore-Quartino, and Salvatore Soresi, eds. *The Adult with Down Syndrome: A New Challenge for Society.* Philadelphia: Whurr, 2004. Discusses some of the questions raised by the increase in life expectancy for persons with Down syndrome, including issues of health, language, vocation, and aging.

Selikowitz, Mark. *Down Syndrome: The Facts.* 3d ed. New York: Oxford University Press, 2008. Covers the entire life span of Down syndrome patients, from infancy to adulthood, and deals with these individuals' developmental, educational, medical, and social-sexual needs.

Shannon, Joyce Brennfleck. *Mental Retardation Sourcebook: Basic Consumer Health Information About Mental Retardation and Its Causes, Including Down Syndrome, Fetal Alcohol Syndrome, Fragile X Syndrome, Genetic Conditions, Injury and Environmental Sources.* Detroit: Omnigraphics, 2000. Provides basic consumer health information about mental retardation, its causes, and prevention strategies. Topics include parenting issues, educational implications, health care needs, employment and economic matters, and legal issues. Glossary.

WEB SITES OF INTEREST

Dolan DNA Learning Center, Your Genes Your Health
http://www.ygyh.org
Sponsored by the Cold Spring Harbor Laboratory, this site, a component of the DNA Interactive Web site, offers information on more than a dozen inherited diseases and syndromes, including Down syndrome.

Genetics Home Reference, Down Syndrome
http://ghr.nlm.nih.gov/condition=downsyndrome
Fact sheet about Down syndrome, with information about the genetic changes related to the condition and the syndrome's inheritance patterns. Includes links to additional online resources.

Medline Plus, Down syndrome
http://www.nlm.nih.gov/medlineplus/ency/article/000997.htm
An online encyclopedia article describing the symptoms, diagnosis, treatment, and other aspects of Down syndrome.

National Down Syndrome Society
http://www.ndss.org
A comprehensive site that includes information on research into the genetics of the syndrome and links to related resources.

See also: Aging; Amniocentesis; Chorionic villus sampling; Congenital defects; Fragile X syndrome; Genetic counseling; Genetic testing; Genetics: Historical development; Intelligence; Mutation and mutagenesis; Nondisjunction and aneuploidy; Prenatal diagnosis; Proteomics.

DPC4 gene testing

CATEGORY: Techniques and methodologies

SIGNIFICANCE: The *DPC4* gene (also known as *SMAD4* and *MADH4*) encodes a protein, which in its normal state is involved in regulation of cell proliferation. However, defective *DPC4* is seen in juvenile polyposis syndrome (JPS) and several cancers. *DPC4* gene testing identifies patients with defective *DPC4* genes, enabling better disease management and clearer information to at-risk family members.

KEY TERMS

acquired mutation: mutation that arises in an individual after birth

denatured DNA: DNA that has been treated (usually by heat) to cause strand separation

germ line: genetic material passed from parents to offspring

hybridize: to form double-stranded DNA from single-stranded DNA molecules by mixing them in conditions where base pairs can form

polymerase chain reaction (PCR): a procedure using enzymes to amplify specific DNA sequences many fold to provide sufficient material, e.g., for sequencing

DPC4 BIOLOGY

Many genes that code for proteins that participate in signaling pathways that regulate functions such as cell proliferation promote cancer when altered. *DPC4* codes for a protein, SMAD4, that plays a central part in many transforming growth factor β (TGF-β) pathways. Normal function of these pathways is required for inhibition of the growth of many types of cells. When TGF-β pathway participants are defective, this inhibition is lost and malignancy can follow.

Defective *DPC4* genes are seen in various cancers, including pancreatic, colon, lung, breast, ovarian, and endometrial. Defective germ-line *DPC4* genes are also seen in about 29 percent of JPS patients and about 2 to 3 percent of hereditary hemorrhagic telangiectasia (HHT) patients.

JUVENILE POLYPOSIS SYNDROME

JPS is a rare autosomal dominant disorder characterized by gastrointestinal (GI) tract polyps most commonly in the colon and rectum. JPS usually, but not always, starts in childhood. It is diagnosed when there are more than five colorectal polyps (although there may be many more), any polyps in other GI tract locations, or a family history of JPS. JPS patients commonly suffer from GI bleeding, anemia, diarrhea, and abdominal pain. They have about a 68 percent risk of developing colorectal cancer by age sixty.

In one study, the average length of time between the first symptoms of JPS and JPS diagnosis was 5.5 years.

HHT is another rare autosomal dominant disorder with different symptoms. In HHT, blood vessel malformations lead to frequent episodes of bleeding from a variety of organs, including the GI tract. Sometimes JPS and HHT coexist in the same individual. All JP-HHT individuals whose DNA has been analyzed thus far carry an abnormal version of the *DPC4* gene. Germ-line mutations in *DPC4* are not seen in any other disorders.

TESTING FOR DEFECTIVE *DPC4*

Abnormalities of *DPC4* are either short deletions and insertions of a few base pairs and point mutations (about 75 percent) or large deletions (about 24 percent).

DNA sequencing is used to identify and map short abnormalities. *DPC4* DNA, purified from patient genomic DNA, is amplified using the polymerase chain reaction (PCR). The amplified DNA is purified and sequenced, usually with an automated sequencer. Sequences are compared to normal *DPC4* DNA sequences, and differences are noted.

Long deletions can be analyzed by comparative genomic hybridization (CGH). Denatured patient genomic DNA is hybridized to a large number of overlapping normal DNA probes of known map position in *DPC4*, which are each immobilized on glass. Probes that do not hybridize to patient DNA can locate deletions.

IMPACT

DPC4 gene testing aids in the diagnosis of JPS, a difficult-to-diagnose disorder. Diagnosis encourages increased surveillance for colorectal and gastric cancer.

JPS patients with defective *DPC4* are more likely to have noncolorectal GI polyps and gastric cancer, while gastric cancer is unlikely for patients with normal *DPC4*. Therefore, knowledge of *DPC4* status can emphasize or deemphasize surveillance at different GI sites.

DPC4 status of one family member can alert presymptomatic, first-degree relatives to get tested and take appropriate preventive measures or alternatively can eliminate unnecessary procedures. Knowledge of *DPC4* status can improve adherence to recommendations for endoscopy at three-year intervals and make it possible to tailor recommendations.

Patients with JPS-HHT are likely to be primarily under the care of cardiovascular specialists and may not know they have JPS and are at high risk for colorectal and gastric cancers. Knowledge of *DPC4* status advises them about the desirability of preventive endoscopy.

Although *DPC4* gene testing is not currently used in the diagnosis or prognosis of pancreatic or other cancers where *DPC4* mutations are acquired, it may be useful in the future, perhaps to help individualize chemotherapy.

Lorraine Lica, Ph.D.

FURTHER READING

Calva-Cerqueira, D., S. Chinnathambi, B. Pechman, et al. "The Rate of Germline Mutations and Large Deletions of *SMAD4* and *BMPR1A* in Juvenile Polyposis." *Clinical Genetics* 75 (2009): 79–85. Research paper that uses *DPC4* gene testing and summarizes the information about the genetic variations of JPS as of 2008.

Schouten, Jan P., Cathal J. McElgunn, Raymond Waaijer, et al. "Relative Quantification of Forty Nucleic Acid Sequences by Multiplex Ligation-Dependent Probe Amplification." *Nucleic Acids Research* 30 (2002): e57. A clear explanation of a technique to detect long genetic deletions.

Vogelstein, Bert, and Kenneth Kinzler. *The Genetic Basis of Human Cancer.* 2d ed. New York: McGraw-Hill, 2002. Large book with good introductory chapters on the genetics of cancer and a fourteen-page chapter on JPS.

WEB SITES OF INTEREST

Gene Tests: Juvenile Polyposis/Hereditary Hemorrhagic Telangiectasia Syndrome
http://www.ncbi.nlm.nih.gov/sites/GeneTests/lab/clinical_disease_id/293231?db=genetests
A list of labs offering *DPC4* testing.

Medscape's Clinical Management of Hereditary Colorectal Cancer Syndromes: Juvenile Polyposis Syndrome
http://www.medscape.com/viewarticle/466572_5
Good summary by prominent scientists.

UniProtKB/Swiss-Prot Q13485 (SMAD4_HUMAN)
http://www.uniprot.org/uniprot/Q13485#section_comments
Contains a lot of valuable information.

See also: Cancer; Genetic screening; Genetic testing; Genetic testing: Ethical and economic issues; Familial adenomatous polyposis; Hereditary diseases; Hereditary mixed polyposis; Hereditary nonpolyposis colorectal cancer; Mutagenesis and cancer; Mutation and mutagenesis; Oncogenes; Tumor-suppressor genes.

Duchenne muscular dystrophy

CATEGORY: Diseases and syndromes
ALSO KNOWN AS: DMD; pseudohypertrophic muscular dystrophy

DEFINITION

Duchenne muscular dystrophy (DMD) is a genetic disease that occurs mostly in boys. Symptoms typically appear between the ages of one and four. The main sign of DMD is muscle weakness that worsens over time. Before age five, the muscles in the legs, arms, and trunk begin to weaken. Later in the disease the heart and respiratory muscles weaken.

This is a progressive, serious condition that requires care from a child's doctor. Parents who suspect their child has this condition should contact the child's doctor promptly. There are many therapies used to treat the symptoms of this disease, and they should be started as soon as possible.

RISK FACTORS

Male children and children with a family history of DMD are at risk for the disease. Parents should tell their child's doctor if their child has any of these risk factors.

ETIOLOGY AND GENETICS

All individuals with Duchenne muscular dystrophy have a mutation in the *DMD* gene, which is located on the short arm of the X chromosome at position Xp21.2. This is one of the largest human genes known, spanning more than two million base pairs of deoxyribonucleic acid (DNA). Its protein product is called dystrophin, a large molecule found primarily in skeletal muscles and cardiac muscles, but small amounts of the protein are also present in nerve cells in the brain. In muscle cells, dystrophin is the major protein in a complex of proteins that serves to anchor the internal cytoskeleton to the cell membrane and ultimately to the extracellular matrix. It also appears to participate in aspects of cell signaling, since it interacts with other proteins that are known to transmit chemical signals within and between cells. In muscles that lack dystrophin altogether or have a nonfunctioning version, the cell membrane becomes permeable, allowing extracellular matrix elements to enter the cell. The damage to the membrane increases with repeated muscle contractions and relaxations until eventually the cell dies. The resulting immune response can exacerbate the damage.

The inheritance pattern of Duchenne muscular dystrophy is typical of all sex-linked recessive mutations (those found on the X chromosome). Mothers who carry the mutated gene on one of their two X chromosomes face a 50 percent chance of transmitting this disease to each of their male children. Female children have a 50 percent chance of inheriting the gene and becoming carriers like their mothers. Females usually do not exhibit this disorder, since in order to be affected they would have to inherit the defective gene from both parents. Affected males almost never live to reproduce.

SYMPTOMS

Parents whose child has any of these symptoms should not assume it is due to DMD. These symptoms may be caused by other conditions. Parents should tell their child's doctor if their child has any of the symptoms.

Children who are late in learning to walk, have larger than normal calf muscles, frequently fall, walk clumsily, have difficulty climbing stairs, have trouble running, walk on their toes or the balls of their feet, have trouble with balance, walk with their shoulders back and belly out, and do not run may have DMD. Additional symptoms include trouble keeping up with friends when playing and using their hands to climb up their bodies when getting up from floor (Gower's maneuver). Additional symptoms can occur later in the disease, including muscle contractures (a shortening of the muscle that traps a joint in a contracted position), scoliosis, weakness in respiratory muscles, cardiomyopathy (weakness of heart muscle), and learning disabilities.

SCREENING AND DIAGNOSIS

The doctor will ask about a child's symptoms and medical history. A physical exam will be done. The doctor will also ask if there is any family history of neuromuscular disease. The exam will focus on the child's muscles. The doctor will look for signs of weakness. Parents will likely be referred to a specialist.

Tests may include a muscle biopsy, a test that removes a small piece of muscle for examination and is used to detect abnormalities in the muscle; a creatine kinase (CK) level test, a blood test used to measure CK, an enzyme found in damaged muscle; genetic testing, a blood test that identifies the genetic mutation of DMD; and electromyography (EMG), a test that measures how well the nerves and muscles work and is used to detect muscle problems.

TREATMENT AND THERAPY

Parents should talk with their child's doctor about the best plan for their child. The disease worsens over time; a child may need different treatments as the disease progresses.

Options include physical therapy, which plays a large role in treating DMD. A child will work with a therapist to try to keep muscles strong. The disease causes contractures; this is when a muscle shortens, making it difficult to move. The therapist will focus on preventing this with range of motion exercises. Scoliosis is common in DMD. Exercises can help to keep the back as straight as possible.

Braces are used to keep the legs straight and prevent contractures. A walker and wheelchair may be needed later, when the leg muscles become too weak to walk.

The doctor may prescribe a steroid medication like prednisone. This can help to improve muscle strength and slow muscle weakening. Steroids can weaken bones; to keep bones healthy, a child will take vitamin D and calcium supplements. A child who experiences heart problems may be given certain medications to slow the damage.

As the disease progresses, the muscles that support breathing may weaken. A child may need a ventilator, which will deliver air through a mask, tube, or sometimes through a tracheotomy (a surgical hole in the windpipe).

Surgery is sometimes used to treat symptoms of DMD. For severe contractures, surgery may be performed to release specific tendons. Scoliosis can sometimes interfere with a child's breathing. In this case, back surgery may be done.

PREVENTION AND OUTCOMES

There are no known guidelines to prevent this progressive muscle disease.

Patricia Griffin Kellicker, B.S.N.;
reviewed by Marcin Chwistek, M.D.
"Etiology and Genetics" by Jeffrey A. Knight, Ph.D.

FURTHER READING

Brown, Robert H., Jr., Anthony A. Amato, and Jerry R. Mendell. "Muscular Dystrophies and Other Muscle Diseases." In *Harrison's Principles of Internal Medicine*, edited by Anthony S. Fauci et al. 17th ed. New York: McGraw-Hill Medical, 2008.

Chamberlain, Jeffrey S., and Thomas A. Rando, eds. *Duchenne Muscular Dystrophy: Advances in Therapeutics.* New York: Taylor & Francis, 2006

EBSCO Publishing. *DynaMed: Duchene Muscular Dystrophy.* Ipswich, Mass.: Author, 2009. Available through http://www.ebscohost.com/dynamed.

_____. *Health Library: Duchenne Muscular Dystrophy.* Ipswich, Mass.: Author, 2009. Available through http://www.ebscohost.com.

Emery, Alan E. H. *Muscular Dystrophy: The Facts.* 3d ed. New York: Oxford University Press, 2008.

Emery, Alan E. H., and Francesco Muntoni. *Duchenne Muscular Dystrophy.* 3d ed. New York: Oxford University Press, 2003.

WEB SITES OF INTEREST

Canadian Institutes of Health Research
http://www.cihr-irsc.gc.ca

Genetics Home Reference
http://ghr.nlm.nih.gov

Muscular Dystrophy Association
http://www.mda.org

Muscle Dystrophy Canada
http://www.muscle.ca

National Institute of Neurological Disorders and Stroke
http://www.ninds.nih.gov

See also: Congenital muscular dystrophy; Hereditary diseases.

Dwarfism

CATEGORY: Diseases and syndromes
ALSO KNOWN AS: Short stature; little person; achondroplasia; ACH; achondroplastic dwarfism

DEFINITION

"Dwarfism" in humans is a term used to describe adults who are less than 4 feet 10 inches in height. More than two hundred conditions are known to cause dwarfism, or short stature. Dwarfism is most often caused by genetic conditions, but it can also be related to endocrine malfunction, acquired conditions, or growth hormone deficiency. Individuals with dwarfism usually have normal intelligence and have an average life span. Little people can also give birth to children with normal stature. Dwarfism may result in multiple medical problems that can lead to death. The term "dwarf" is socially unacceptable by most people. Those with dwarfism prefer to be referred to as little people or as having short stature.

RISK FACTORS

Having a parent with a form of dwarfism, such as achondroplasia, a parent who carries a mutated *FGFR3* gene, or parents of advanced age can increase the risk of inheriting dwarfism. Other abnormalities such as damage or conditions of the pituitary gland, hormonal disorders, problems with absorption, malnutrition, kidney disease, or extreme emotional distress can increase the risk of developing dwarfism.

ETIOLOGY AND GENETICS

Dwarfism, of which there are several hundred forms, occurs in approximately one in every ten thousand births. Approximately 85 percent of little people are born to parents of average height. The most common type of dwarfism, achondroplasia, is an autosomal dominant trait, but in 80 percent of cases it appears in children born to normal parents as a result of mutations in the sperm or egg.

Dwarfisms in which body proportions are normal usually result from metabolic or hormonal disorders in infancy or childhood. Chromosomal abnormalities, pituitary gland disorders, problems with absorption, malnutrition, kidney disease, and extreme emotional distress can also interfere with normal growth. When body parts are disproportioned, the dwarfism is usually due to a genetic defect.

Participants at a convention of Little People of America, an organization for people with dwarfism, play bingo. (AP/Wide World Photos)

Skeletal dysplasias are the most common causes of dwarfism and are the major cause of disproportionate types of dwarfism. More than five hundred skeletal dysplasias have been identified. Chondrodystrophic dwarfism occurs when cartilage cells do not grow and divide as they should and cause defective cartilage cells. Most chondrodystrophic little people have abnormal body proportions. The defective cells occur only in the spine or only in the arms and legs. Short-limb dwarfism includes individuals with achondroplasia, diastropic dysplasia, and Hunter-Thompson chondrodysplasia.

Achondroplasia is the most common skeletal dysplasia and affects more than 70 percent of all dwarfs. It occurs in every 26,000 to 40,000 babies born of all races and ethnicities. Achondroplasia is caused by an autosomal dominant allele and is identified by a disproportionate short stature consisting

of a long trunk and short upper arms and legs. Eighty percent of all cases of achondroplasia result from a mutation on chromosome 4 in a gene that codes for a fibroblast growth factor receptor. Achondroplasia is seen in both males and females, occurs in all races, and affects approximately one in every twenty thousand births. If one parent has achondroplasia and the other does not, then a child born to them would have a 50 percent chance of inheriting achondroplasia. On the other hand, if both parents have achondroplasia, their offspring have a 50 percent chance of inheriting achondroplasia, a 25 percent chance of being normal, and a 25 percent chance of inheriting the abnormal allele from each parent and suffering often fatal skeletal abnormalities. Children who do not inherit the defective gene will never have achondroplasia and cannot pass it on to their offspring, unless a mutation occurs in the sperm or egg of the parents. Geneticists have observed that fathers who are forty years of age or older are more likely to have children with achondroplasia as a result of mutations in their sperm.

Diastrophic dysplasia is a relatively common form of short-limb dwarfism that occurs in approximately one in 100,000 births and is identified by the presence of short arms and calves, clubfeet, and short, broad fingers with a thumb that has a hitchhiker type appearance. Infant mortality can be high as a result of respiratory complications, but if they survive infancy, short-limbed dwarfs have a normal life span. Orthopedic dislocations of joints are common. Scoliosis is seen especially in the early teens, and progressive cervical kyphosis and partial dislocation of the cervical spine eventually cause compression of the spinal cord. Diastrophic dysplasia is an inherited autosomal recessive condition linked to chromosome 5. Parents have a 25 percent chance that each additional child will get diastrophic dysplasia.

Short-trunk dwarfism includes individuals with spondyloepiphyseal dysplasia, which results from abnormal growth in the spine and long bones that leads to a shortened trunk. It occurs in one of every 95,000 births. In spondyloepiphyseal dysplasia tarda, the lack of growth may not be recognized until five to ten years of age. Those affected have progressive joint and back pain and eventually develop osteoarthritis. Spondyloepiphyseal dysplasia congenita is caused by autosomal dominant gene mutations and is evidenced by a short neck and trunk, and barrel chest at birth. It is not uncommon for cleft palate, hearing loss, myopia, and retinal detachment to be present.

Morquio syndrome, which was first described in 1929, is classified as a mucopolysaccharidosis (MPS) disease caused by the body's inability to produce enzymes that help to break down and recycle dead cells. Consequently, wastes are stored in the body's cells.

Hunter-Thompson chondrodysplasia is a form of dwarfism caused by a mutation in growth factor genes. Affected individuals have shortened and misshapened bones in the lower arms, the legs, and the joints of the hands and feet. Fingers are shortened and toes are ball-shaped.

Growth hormone, a protein that is produced by the pituitary ("master") gland, is vital for normal growth. Hypopituitarism results in a deficiency of growth hormone and afflicts between ten thousand and fifteen thousand children in the United States. In panhypopituitarism, the gland does not produce any hormones. The pituitary gland shuts down and growth is stunted.

Turner syndrome affects one in every two thousand female infants and is characterized by the absence of or damage to one of the X chromosomes in most of the cells in the body. Short stature and the failure to develop sexually are hallmarks of Turner syndrome. Learning difficulties, skeletal abnormalities, heart and kidney problems, infertility, and thyroid dysfunction may also occur. Turner syndrome can be treated with human growth hormones and by replacing sex hormones.

SYMPTOMS

For inherited disorders at birth, a long trunk and shortened limbs will be noticeable. A child born with dwarfism may go on to exhibit delayed gross motor development and skills, breathing and neurologic problems, hydrocephalus (water on the brain), increased susceptibility to ear infections and hearing loss, weight problems, curvature of the spine (scoliosis), bowed legs, stiff arms, joint and back pain or numbness, and crowding of teeth. Portions of the face may be underdeveloped. Sleep apnea can develop as a result of compression of the spine. Adult height will be stunted (usually reaching 42-52 inches). Seeking proper medical care can help to relieve some of these symptoms and complications.

SCREENING AND DIAGNOSIS

Close monitoring by parents and doctors is necessary to record the constellation of symptoms for each unique case of dwarfism. Often an initial diagnosis can be made by observing physical characteristics. Magnetic resonance imaging (MRI) and computed tomography (CT) scans can illustrate spinal and other structural abnormalities before serious complications arise. Imaging techniques can also help to determine the type of dwarfism present. Molecular genetic testing can be done to detect a *FGFR3* mutation. Genetic testing is 99 percent sensitive and available in clinical laboratories. The Human Genome Project continues to investigate genetic links to dwarfism. Prenatal counseling and screening for traits of dwarfism, along with genetic counseling and support groups, are avenues to pursue for family and individual physical, psychological, and social well-being and to make informed choices.

TREATMENT AND THERAPY

Some forms of dwarfism can be treated through state-of-the-art surgical and medical interventions such as bone-lengthening procedures, reconstructive surgery, and growth and sex hormone replacement.

Short stature is the one quality all people with dwarfism have in common. After that, each of the many conditions that cause dwarfism has its own set of characteristics and possible complications. Fortunately, many of these complications are treatable, so that people of short stature can lead healthy, active lives. Continued follow-up with the physician team is essential.

For example, some babies with achondroplasia may experience hydrocephalus (excess fluid around the brain). They may also have a greater risk of developing apnea—a temporary stop in breathing during sleep—because of abnormally small or misshapen airways or, more likely, because of airway obstruction by the adenoids or the tonsils. Occasionally, a part of the brain or spinal cord is compressed. With close monitoring by doctors, however, these potentially serious problems can be detected early and surgically corrected.

PREVENTION AND OUTCOMES

Genetic counseling as well as family and public education regarding dwarfism and growth problems can bring greater awareness of dwarfism to communities and allow parents to make good choices. Inherited dwarfism is not preventable, but some cases caused by malnutrition, injury, absorption, or kidney conditions may be prevented.

The type, symptoms, and severity of complications vary from person to person, but most little people have an average life span. With a sense of support, self-esteem, and independence, a person with dwarfism can lead a very satisfying and productive life.

Sharon Wallace Stark, R.N., A.P.R.N., D.N.Sc.;
updated by Deanna M. Neff, M.P.H.

FURTHER READING

Ablon, J. *Living with Difference: Families with Dwarf Children.* Westport, Conn.: Greenwood, 1988. Exploration of developmental and medical problems, school experiences, social world of the dwarf children, and how dwarf children fit into family systems.

Apajasalo, M., et al. "Health-Related Quality of Life of Patients with Genetic Skeletal Dysplasias." *European Journal of Pediatrics* 157 (1998): 114-121. Presents tools for assessing the well-being of individuals with dwarfism and discusses results of a survey conducted by the authors.

EBSCO Publishing. *Patient Education Reference Center: Achondroplasia.* Ipswich, Mass.: Author, 2008. Available through http://www.ebscohost.com.

_____. *Patient Education Reference Center: Short Stature.* Ipswich, Mass.: Author, 2008. Available through http://www.ebscohost.com. Fact sheets that provide valuable information.

Krakow, D., et al. "Use of Three-Dimensional Ultrasound Imaging in the Diagnosis of Prenatal-Onset Skeletal Dysplasias." *Ultrasound in Obstetrics and Gynecology* 21 (2003): 4676-4678. Describes in detail the newest approach to prenatal detection of dysplasias.

Page, Nick. *Lord Minimus: The Extraordinary Life of Britain's Smallest Man.* New York: St. Martin's Press, 2002. Relates the exciting life led by Jeffrey Hudson, a dwarf, in the court of King Charles I and Queen Henrietta Maria.

Ranke, M., and G. Gilli. *Growth Standards, Bone Maturation, and Idiopathic Short Stature.* Farmington, Conn.: S. Karger, 1996. KABI International Growth Study to establish global guidelines and standards for diagnosis and treatment of growth disorders and definition of idiopathic short stature (ISS).

Richardson, John H. *In the Little World: A True Story of Dwarfs, Love, and Trouble.* San Francisco: HarperCollins, 2001. A wide-ranging look at the world of the "little people," as many of those with dwarfism prefer to be called.

Thorner, M., and R. Smith. *Human Growth Hormone: Research and Clinical Practice.* Vol. 19. Totowa, N.J.: Humana Press, 1999. Examines the use of human growth hormone therapies in the treatment of short stature and various diseases.

Ulijaszek, J. S., Francis E. Johnston, and Michael A. Preece. *Cambridge Encyclopedia of Human Growth and Development.* New York: Cambridge University Press, 1998. Broadly discusses genetic growth anomalies in relation to environmental, physiological, social, economic, and nutritional influences on human growth.

Vajo, Zoltan, Clair A. Francomano, and Douglas J. Wilkin. "The Molecular and Genetic Basis of Fibroblast Growth Factor Receptor 3 Disorders: The Achondroplasia Family of Skeletal Dysplasias, Muenke Craniosynostosis, and Crouzon Syndrome with Acanthosis Nigricans." *Endocrine Reviews* 21, no. 1 (2000): 23-39. Aimed at researchers.

Zelzer, Elazar, and Bjorn R. Olsen. "The Genetic Basis for Skeletal Diseases." *Nature* 423 (2003): 343-348. Aimed at researchers but understandable by a wider audience.

WEB SITES OF INTEREST

Centralized Dwarfism Resources
http://www.dwarfism.org
Offers information on types of dwarfism and links to other informative sites.

KidsHealth: "Dwarfism"
http://kidshealth.org/parent/medical/bones/dwarfism.html
Information for parents, children, and teenagers.

National Center for Biotechnology Information. Online Mendelian Inheritance in Man
http://www.ncbi.nlm.nih.gov/sites/entrez?db=omim
A catalog on genes and genetic disorders, including dwarfism, searchable by keyword.

See also: Congenital defects; Consanguinity and genetic disease; Hereditary diseases; Human growth hormone; Pedigree analysis.

Dyslexia

CATEGORY: Diseases and syndromes
ALSO KNOWN AS: Specific reading disability

DEFINITION

Dyslexia is a learning disability that can hinder a person's ability to read, write, spell, and sometimes speak. It is the most common learning disability in children and persists throughout life. The severity of dyslexia can vary from mild to severe.

The sooner dyslexia is treated, the more favorable the outcome. However, it is never too late for people with dyslexia to learn to improve their language skills.

RISK FACTORS

Because dyslexia runs in families, individuals should tell their doctors or pediatricians if they or other members of their families have the disability.

ETIOLOGY AND GENETICS

Dyslexia is a complex learning disability that results from a combination of genetic and environmental factors. The most recent estimates suggest that inherited factors account for up to 80 percent of the determinants involved in the development of dyslexia. The disability has long been known to run in families, and twin studies have confirmed that genetic factors are primarily responsible for the observed family clustering. While there is no single predictable pattern of inheritance, one report suggests that a child with one affected parent has a 40-60 percent risk of developing dyslexia. Another study found that 88 percent of dyslexics had a close relative who also had problems with spelling or reading.

Linkage analyses in families where two or more family members are dyslexic have helped to identify nine chromosomal regions that appear to contain a gene or genes for susceptibility to dyslexia. Four of these are particularly significant, since they are found in multiple large family samples: *DYX1*, at chromosomal position 15q21; *DYX2*, at position 6p21-p22; *DYX6*, at position 18p11; and *DYX8*, found on chromosome 1 at position 1p34-p36. Other *DYX* genes for susceptibility are found on chromosomes 2, 3, 11, and X. Molecular genetic studies have not yet revealed the exact locations or the protein products of these susceptibility genes, but this is an area of active current research, and researchers have high expectations for real progress in the coming decade.

SYMPTOMS

If a child experiences any of the following symptoms, his or her parents should not assume it is due to dyslexia. These symptoms may be caused by other health conditions. If a child experiences any one of them over time, his or her parents should see their physician or pediatrician. Symptoms include difficulty in the following areas: learning to speak, reading and writing at grade level, organizing written and spoken language, learning letters and their sounds, learning number facts, spelling, learning a foreign language, and correctly doing math problems.

SCREENING AND DIAGNOSIS

The doctor will ask about a parent or a child's symptoms and medical history and will perform a physical exam, including hearing and vision tests. The patient may then be referred to an expert in learning disabilities, such as a school psychologist or learning specialist, for additional testing to determine if he or she has dyslexia. Tests given by the specialist may include cognitive processing tests to measure thinking ability; intelligence quotient (IQ) tests to measure intellectual functioning; and tests to measure speaking, reading, spelling, and writing skills.

TREATMENT AND THERAPY

Most people with dyslexia need help from a teacher, tutor, or other trained professional. Parents should talk with their doctors or pediatricians and learning specialists about the best plans for them or their children.

Treatment options include remediation, a way of teaching that helps people with dyslexia to learn language skills. Concepts used in remediation include teaching small amounts of information at a time, teaching the same concepts many times ("overteaching"), and using all the senses—hearing, vision, voice, and touch—to enhance learning (multisensory reinforcement).

Compensatory strategies are ways to work around the effects of dyslexia. They include audiotaping classroom lessons, homework assignments, and texts; using flashcards; sitting in the front of the classroom; using a computer with spelling and grammar checks; and receiving more time to complete homework or tests.

PREVENTION AND OUTCOMES

There is little that can be done to prevent dyslexia, especially if it runs in a patient's family. However, early identification and treatment can reduce its effects. The sooner children with dyslexia get special education services, the fewer problems they will have learning to read and write at grade level. Under federal law, free testing and special education services are available for children in the public school system.

Alia Bucciarelli, M.S.;
reviewed by Theodor B. Rais, M.D.
"Etiology and Genetics" by Jeffrey A. Knight, Ph.D.

FURTHER READING

Brunswick, Nicky. *Dyslexia: A Beginner's Guide.* Oxford, England: Oneworld, 2009.

EBSCO Publishing. *Health Library: Dyslexia.* Ipswich, Mass.: Author, 2009. Available through http://www.ebscohost.com.

Hultquist, Alan M. *An Introduction to Dyslexia for Parents and Professionals.* Philadelphia: Jessica Kingsley, 2006.

Nicolson, Roderick I., and Angela J. Fawcett. *Dyslexia, Learning, and the Brain.* Cambridge, Mass.: MIT Press, 2008.

Reid, Gavin, ed. *The Routledge Companion to Dyslexia.* New York: Routledge, 2009.

WEB SITES OF INTEREST

Canadian Dyslexia Association
http://www.dyslexiaassociation.ca

International Dyslexia Association
http://www.interdys.org

KidsHealth from Nemours
http://kidshealth.org/kid

Learning Disabilities Association of Canada
http://www.ldac-taac.ca/index-e.asp

National Center for Learning Disabilities
http://www.ld.org

See also: Attention deficit hyperactivity disorder (ADHD); Autism.

E

Edwards syndrome

CATEGORY: Diseases and syndromes
ALSO KNOWN AS: Trisomy 18; complete trisomy 18

DEFINITION

Edwards syndrome is a congenital genetic anuploidic (abnormal number of chromosomes) abnormality arising when a fetus has an extra number 18 autosomal chromosome. The syndrome was initially described by John H. Edwards and associates in 1960. It occurs in 1 in 3,000-5,000 pregnancies and ends in spontaneous abortion, intrauterine death, or elective termination in approximately 95 percent of cases. In the 5 percent of affected fetuses surviving to term, multiple anomalies occur. The mortality rate of liveborn infants within the first year of life is 90 to 95 percent.

RISK FACTORS

Edwards syndrome can occur at any maternal age but occurs more frequently with advanced maternal age (more than thirty-five years). In addition, slightly more than 50 percent of infants with Edwards syndrome have a paternal age greater than forty years. There is no racial or ethnic predilection to the syndrome. The in utero female-to-male ratio is 1 to 1. However, more male fetuses die in utero or are spontaneously aborted and thus live births are approximately 80 percent female. Individuals with a translocation of chromosome 18 material are at a 50 percent risk of transmitting the complete syndrome (all cells affected) to their progeny.

ETIOLOGY AND GENETICS

An error in gamete (sperm or oocyte) division resulting in meiotoc nondisjunction (failure of a replicating chromosome to divide) is the typical etiology of Edwards syndrome. In 95 percent of cases, the somatic cells contain three copies of chromosome 18 rather than the normal two. This extra genetic material is responsible for the multiple anomalies and developmental and cognitive deficits present with this syndrome. The remaining 5 percent of cases exhibit mosaicism (trisomy in some but not all cells) or translocation (extra chromosome 18 genetic material is attached to a normal chromosome). In some infants with mosiacism or translocation, associated anomalies may be less and the individual may phenotypically appear unaffected. Developmental and cognitive function varies from severe to normal in these individuals. Complete trisomy (affecting all cells) and mosaicism are not inherited but result from a de novo (new) mutation. Translocation trisomic individuals have a 50 percent chance of each offspring inheriting complete trisomy 18.

SYMPTOMS

Prenatally, intrauterine growth deficiency accompanied by polyhydramnios (a large volume of amniotic fluid due to defective fetal swallowing) is common. Anomalies may be detected by fetal ultrasound.

Postnatally, the affected infant presents with classic signs and symptoms. They include central nervous system malformations (microcephaly with a prominent occiput, hydrocephaly, and neural tube defects); cardiac defects (ventricular and atrial septal defects, coarctation of the aorta); skeletal anomalies (growth retardation, clenched fist with index finger overlapping the middle finger and fifth finger overlapping the fourth, rocker bottom feet); gastrointestinal defects (omphalocele, malrotation); head and face issues (microphthalmia, micrognathia, microstomia, and low-set, malformed ears); and genitourinary obstruction. Feeding difficulties, developmental delay, and mental retardation are almost always present.

SCREENING AND DIAGNOSIS

Screening for anuploidy can be done during the first trimester of the pregnancy from weeks ten to fourteen. The evaluation includes maternal age, fetal nuchal translucency, fetal heart rate, maternal serum free beta human chorionic gonadotrophin (beta-hCG), and maternal serum pregnancy-associated plasma protein-A (PAPP-A). These factors are successful in predicting approximately 90 percent of affected fetuses with a 3 percent false positive rate. Positive screening is then followed by definitive prenatal testing, including the analysis of fetal cells obtained by either chorionic villus sampling or amniocentesis.

When prenatal screening is not performed, infants are diagnosed after delivery as a result of common prevalent features and anomalies and clinical instability. Cytogenetic testing confirms the diagnosis.

TREATMENT AND THERAPY

The diagnosis of Edwards syndrome requires thoughtful clinical decision-making. Because of the high mortality rate of this syndrome and inability to offer a cure, comfort measures only may be offered to the infant. In surviving infants, appropriate health care services are offered depending on the types of anomalies and degree of developmental delay and mental retardation present.

PREVENTION AND OUTCOMES

Currently, there are no known preventive strategies. The spontaneous prenatal death rate is high. With prenatal diagnosis, elective termination of the pregnancy is often performed.

For liveborn infants, the prognosis is grim, with the median survival of live born complete trisomic cases being less than one month. Between 90 and 95 percent die within the first year of life. Survival up to the third decade of life has been reported. Some affected individuals are institutionalized and others are cared for in the home.

Care providers should be alert to abnormal fetal growth patterns. Prenatal ultrasound and laboratory testing can be offered. After birth all infants require a thorough physical examination and follow-up of any abnormalities.

Wanda Todd Bradshaw, R.N., M.S.N.

FURTHER READING

Crider, Krista S., Richard S. Olney, and Janet D. Cragan. "Trisomies 13 and 18: Population Prevalences, Characteristics, and Prenatal Diagnosis, Metropolitan Atlanta, 1994-2003." *American Jour-* *nal of Medical Genetics, Part A* 146A (2008): 820-826.

Pont, Stephen J., et al. "Congenital Malformations Among Liveborn Infants with Trisomies 18 and 13." *American Journal of Medical Genetics, Part A* 140 (2006): 1749-1756.

Tucker, Megan E., Holly J. Garringer, and David D. Weaver. "Phenotypic Spectrum of Mosaic Trisomy 18: Two New Patients, a Literature Review, and Counseling Issues." *American Journal of Medical Genetics, Part A* 143A (2007): 505-517.

WEB SITES OF INTEREST

eMedicine: Trisomy 18
http://emedicine.medscape.com/article/943463-overview

Genetics Home Reference: Trisomy 18
http://ghr.nlm.nih.gov/condition=trisomy18

Prenatally and Postnatally Diagnosed Conditions Awareness Act
http://www.geneticsandsociety.org/downloads/InfoSheetBrownbackKennedyLegislation.pdf

Trisomy 18 Support Organization
http://www.trisomy18support.org

See also: Amniocentesis; Chorionic villus sampling; Congenital defects; Down syndrome; Fragile X syndrome; Genetic counseling; Genetic testing; Hereditary diseases; Intelligence; Mutation and mutagenesis; Nondisjunction and aneuploidy; Prenatal diagnosis; Proteomics.

Ellis-van Creveld syndrome

CATEGORY: Diseases and syndromes
ALSO KNOWN AS: EvC; chondroectodermal dysplasia; mesoectodermal dysplasia; six-fingered dwarfism

DEFINITION

Ellis-van Creveld syndrome (EvC) is a recessively inherited defect that affects development of several ectodermally and mesodermally derived structures in the body. It is commonly characterized by short stature, extra digits, tooth and nail defects, and heart malformations.

RISK FACTORS

There are no known risk factors for this disease. It is at an unusually high frequency in the Old Order Amish community of Lancaster County, Pennsylvania.

ETIOLOGY AND GENETICS

This syndrome was first formally described in the literature in 1940 by Richard Ellis of Edinburgh and Simon van Creveld of Amsterdam. The majority of mutations causing Ellis-van Creveld syndrome are located on the short arm of chromosome 4 in either of two adjacent genes, *EVC* or *EVC2*. These loci are proximal to the gene involved in achondroplasia, another form of inherited dwarfism. The normal functions of *EVC* and *EVC2* are not yet understood; however, the histopathology of fetuses with EvC syndrome showed chondrocyte disorganization in the growth zone of developing long bones and sometimes of vertebrae.

EvC syndrome is relatively uncommon. In the U.S. population, the frequency of homozygotes is approximately 1 in 60,000 (allele frequency 0.004 percent) which gives a frequency of 0.008 percent carriers (heterozygotes) in the population. In the Old Order Amish of Lancaster County, Pennsylvania, however, the frequency of homozygotes is approximately 1 in 229 (allele frequency 6.6 percent) which gives a carrier frequency of 12.3 percent in that genetically isolated population. The original mutant allele in the Old Order Amish can be traced to a single immigrant couple, Samuel King and his wife, who came to eastern Pennsylvania in 1744. The actual mutation in this population is in the fifth nucleotide of intron 13 of the *EVC* gene and causes an abnormal splicing of exons of this gene. Cases of the disorder outside the Old Order Amish subpopulation are at different positions in either the *EVC* or *EVC2* gene. Among those mutations are six mutations leading to a truncated protein and one mutation with a single amino acid deletion in *EVC* and one frame shift mutation, four truncating mutations, and one missence mutation in *EVC2*. It is interesting to note that a study by S. W. Thompson and colleagues in 2007 found that 31 percent of the EvC patients did not show a mutation in either the *EVC* or *EVC2* gene, indicating that other genes are most likely involved.

SYMPTOMS

A variably expressed clinical tetrad of symptoms defines this syndrome: chondrodystrophy (almost 100 percent of patients), polydactyly (almost 100 percent of patients), hidrotic ectodermal dysplasia (about 93 percent of patients), and congenital heart anomalies (about 60 percent of patients). Chondrodystrophy is usually expressed as short-limb dwarfism (approximate adult height 109 to 155 centimeters) and progressive distal limb shortening. Polydactyly (extra fingers and/or toes) is most commonly seen in the hands but can occasionally be seen in the feet. Hidrotic ectodermal dysplasia is seen in nails, teeth, and hair. Nails are small or absent and often malformed. Tooth abnormalities such as partial adontia, natal teeth, delayed tooth eruption, small teeth, and malformed teeth are common. Hair can occasionally be sparse. Cardiac abnormalities include common atrium, atrial, or ventricular septal defects and patent ductus arteriosus. The cardiac abnormalities are the leading cause of neonatal death. Approximately 50 percent of infants born with EvC syndrome die from these cardiac defects, but those who survive infancy have a relatively normal life expectancy. Other anomalies that may be present include musculoskeletal defects which may lead to misshapen bones and a narrow chest. The latter, when present, can lead to respiratory difficulties. Oral malformations, urogenital abnormalities, and (very rarely) mental retardation may also occur.

SCREENING AND DIAGNOSIS

EvC can be diagnosed prenatally by ultrasound at eighteen weeks. Fetal echocardiaography can also be used to detect cardiac abnormalities. DNA extracted by amniocentesis or chorionic vilus biopsy could be analyzed for mutations of the *EVC* and *EVC2* genes; however, this is not a generally available genetic test. Postnatally, skeletal surveys, chest X rays, ECG, MRI, echocardiography, and ultrasound are used to diagnose the clinical tetrad of EvC characteristics.

TREATMENT AND THERAPY

Orthopedic care needed to address polydactyly and bone malformations may include surgery, braces, and physical therapy. Cardiac surgery may be needed to repair heart anomalies. Dental care is often necessary and may include crowns to repair malformed or small teeth and partial dentures to replace missing teeth. For those patients with a smaller chest, respiratory care is often needed. Consultation with a psychologist/psychiatrist is sometimes recommended.

PREVENTION AND OUTCOMES

There is no way of preventing EvC syndrome. Genetic counseling of couples in high-risk groups is recommended. Prenatal testing for the skeletal and cardiac abnormalities is available. Many EvC pregnancies end in spontaneous abortion or stillbirth. Of those that make it to birth, about 50 percent of babies die in infancy from cardiac or respiratory problems. Those patients that survive infancy generally have a normal life span.

Richard W. Cheney, Jr., Ph.D.

FURTHER READING

McKusick, V. A. "Ellis-van Creveld Syndrome and the Amish." *Nature Genetics* 24 (2000): 203-204. An overview of EvC syndrome.

McKusick, V. A., J. A. Egeland, and R. Eldridge. "Dwarfism in the Amish." *Bulletin of the Johns Hopkins Hospital* 115 (1964): 306-336.

McKusick, V. A., R. Eldridge, J. A. Hostetler, U. Ruangwit, and J. A. Egeland. "Dwarfism in the Amish II: Cartilage-Hair Hypoplasia." *Bulletin of the Johns Hopkins Hospital* 116 (1965): 285-286. These two articles are early studies of the nature of EvC syndrome in the Old Order Amish.

Thompson, S. W., V. L. Ruiz-Perez, H. J. Blair, et al. "Sequencing *EVC* and *EVC2* Identifies Mutations in Two-thirds of Ellis-van Creveld Syndrome Patients." *Human Genetics* 120 (2007): 663-670. This article describes the genetic mutations in the *EVC* and *EVC2* genes seen in two thirds of EvC syndrome patients.

WEB SITES OF INTEREST

National Institutes of Health and the National Library of Medicine. MedlinePlus
http://www.nim.nih.gov/medlineplus/ency/article/001667.htm

Online Mendelian Inheritance in Man (OMIM)
http://www.ncbi.nlm.nih.gov/entrez/dispomim.cgi?id=225500

Orphanet Journal of Rare Diseases
http://www.pubmedcentral.nih.gov/articlerender.fcgi?artid+1891277

See also: Congenital defects; Dwarfism; Hereditary diseases; Polydactyly.

Emerging and reemerging infectious diseases

CATEGORY: Diseases and syndromes
ALSO KNOWN AS: EIDs; RIDs

INTRODUCTION

Emerging infectious diseases (EIDs) are diseases that are caused by newly discovered pathogens that are not known to have previously infected the human population. Five examples of EIDs that were first identified since the mid-1970's and the pathogens that cause the diseases are as follows: Ebola hemorrhagic fever caused by the Ebola virus; acquired immunodeficiency syndrome (AIDS) caused by the human immunodeficiency virus (HIV); Lyme disease caused by *Borrelia burgdorferi*; chronic gastritis and peptic ulcer disease caused by *Helicobacter pylori*; and severe acute respiratory syndrome (SARS) caused by the SARS-associated coronavirus (SARS-CoV).

Reemerging infectious diseases (RIDs) are diseases that were once controlled in the human population but have reemerged and caused the number of new cases, or incidence, of certain diseases to increase beyond their expected values. RIDs are caused by previously discovered pathogens that resurge and resume infecting susceptible human populations. Today, antibiotic resistance, which has a genetic basis, is one of the main causes of the reemergence of many infectious diseases. Five examples of RIDs and the pathogens that cause the diseases are as follows: skin and soft tissue infections caused by methicillin-resistant *Staphylococcus aureus* (MRSA); pseudomembranous colitis caused by *Clostridium difficile*; multidrug-resistant tuberculosis (MDR-TB) and extensively drug-resistant tuberculosis (XDR-TB) caused by *Mycobacterium tuberculosis*; invasive group A streptococcal (GAS) disease caused by group A streptococcus (*Streptococcus pyogenes*); and influenza caused by the influenza viruses.

GOALS OF PATHOGENS: MULTIPLICATION AND BALANCED PATHOGENICITY

The ability of a pathogen to infect a population is partially related to the pathogen's genetic makeup, or genotype. The pathogens that cause EIDs and RIDs are unique because they often display novel genetic features that amplify their virulence, or ability to cause disease in healthy hosts. Although many pathogens use their virulence to invade host cells, pathogens usually do not intend to kill their hosts immediately. The premature death of an infected host is evolutionarily unfavorable for the long-term survival of the pathogen. Because pathogens aim to prosper in their environments, they will usually strive for multiplication and balanced pathogenicity. Balanced pathogenicity is a type of host-pathogen interaction that promotes the transmission of the pathogen and long-term survival of the infected host. EIDs represent a new interaction between a new pathogen and host, whereas RIDs represent a continuation of a host-pathogen interactive tug-of-war. One way that pathogens can adapt to the demands of their environments is through genetic changes. Mutations, though not always beneficial for pathogens, are perhaps the most fundamental sources of genetic change.

MUTATIONS

A mutation is a structural alteration of the genetic sequence of a nucleic acid. Mutations can affect a single base pair in a strand of DNA or RNA, or they can involve much longer segments. The main classes of mutations are as follows: substitutions, deletions, and insertions. Substitutions occur when one base is replaced by another base. Substitutions usually result in one of the following outcomes: a silent mutation, missense mutation, or nonsense mutation. Substitutions are "silent" if the mutated nucleic acid sequence encodes the same amino acid as the original nucleic acid sequence. A missense mutation occurs when a mutated nucleic acid sequence encodes an amino acid different from the amino acid encoded by the original nucleic acid sequence. Lastly, a nonsense mutation happens when a mutated nucleic acid sequence encodes a stop codon instead of an amino acid. A deletion takes place when a portion of a gene is cut out and lost, and an insertion occurs when additional bases are inserted into the nucleic acid sequence of an existing gene.

Although mutations are sources of genetic variation, they can be very risky for pathogens. The wrong mutation, in the wrong place, at the wrong time can severely impair a pathogen's ability to infect a host. Although bacteria and viruses continually undergo mutations, they enhance their genetic variation using a variety of other methods as well.

HORIZONTAL GENE TRANSFER

Many bacteria are able to acquire new genes using a method called horizontal gene transfer, the movement of genetic material between bacteria of the same generation. The three main mechanisms of this transfer are as follows: transformation, transduction, and conjugation. It is important to remember that not all bacteria are able to use all three mechanisms. Nevertheless, horizontal gene transfer is still responsible for a tremendous amount of genetic variation within many bacterial species.

Transformation is the uptake of DNA from the external environment. Bacteria that are able to undergo transformation are called competent. Some bacteria are naturally competent, whereas other bacteria are not. In some cases, scientists can induce competence using a laboratory technique called electroporation. Electroporation can increase the permeability of the bacterial cell membrane and allow foreign DNA to enter the cell. Once the foreign DNA enters the cell, it can proceed to integrate into the bacterial chromosome by homologous recombination. *H. pylori* and *S. aureus* are two examples of naturally competent bacteria, and scientists can use electroporation to induce competence in *S. pyogenes*.

Transduction is the movement of DNA from a donor bacterial cell to a recipient bacterial cell via the transport of a bacteriophage, or bacterial virus. The process entails a phage binding to specific receptors on a donor bacterial cell and injecting its genetic material into the cell. Afterward, one of the following events usually takes place: the genetic material from the phage will either integrate into the bacterial chromosome (lysogenic cycle); or the genetic material will begin replicating inside the bacterial cell and produce a new generation of phages, which are released upon the lysis of the bacterial cell (lytic cycle).

Sometimes, environmental stimuli cause the phage

DNA segment of the bacterial chromosome, or prophage, to undergo an excision and enter the lytic cycle. When the prophage is excised from the chromosome, it can "tear off" adjacent genes on the bacterial chromosome and package the genes into phages. When the phages are released from the donor cell upon lysis, they can transport some of the donor genes into recipient cells. The integration of the new phage genome into recipient bacterial cells can lead to phenotypic changes, which may include antibiotic resistance and synthesis of exotoxins. For example, the gene for type A streptococcal exotoxin (*speA*) in *S. pyogenes* is located in a bacteriophage genome.

Conjugation is the transfer of DNA by direct cell-to-cell contact between donor and recipient cells. Plasmids and transposons are the two main types of conjugative units.

A plasmid is an extrachromosomal piece of DNA that is able to undergo autonomous replication. The fertility plasmid, or F-plasmid, is one example of a plasmid commonly transferred between bacteria. The purpose of the F-plasmid is to encode proteins for a sex pilus, which enables a bacterium to become a genetic donor.

Occasionally, the F-plasmid will integrate into the donor bacterial chromosome to form a high frequency of recombination (Hfr) strain. A portion of the Hfr strain is then transferred via a sex pilus to the recipient cell in the form of single-stranded DNA (ssDNA). The ssDNA, which may contain donor bacterial genes, later becomes circular, double-stranded DNA (dsDNA) and integrates by homologous recombination into the recipient chromosome. The recipient cell may now express its newly acquired bacterial genes. The plasmid composed of both transfer genes and chromosomal genes is called the F-prime, or F′, plasmid.

F′ plasmids function similarly but not identically to another group of plasmids called Resistance plasmids, or R-plasmids. R-plasmids are responsible for the transfer of multiple antibiotic resistance genes between bacterial cells. Antibiotic resistance genes can help promote the long-term survival and multiplication of pathogenic bacteria by encoding the following four defense mechanisms: restricted drug entry into the cell, drug efflux pumps, drug inactivating enzymes, and modified drug targets. MRSA is an example of a pathogen that uses both a drug inactivating enzyme and a modified drug target. MRSA can produce a -lactamase enzyme (for example, penicillinase), which can be used to inactivate -lactam antibiotics. Moreover, MRSA can express its *mecA* gene to encode a mutant -lactam binding protein, which causes -lactam antibiotics to bind to their target with low affinity.

R-plasmids comprise the following DNA segments: a resistance transfer factor (RTF) and r-determinant. The RTF segment carries genes used for DNA transfer and replication, whereas the r-determinant carries antibiotic resistance genes. R-plasmids are unique because they do not rely on the integration and excision of the RTF segment to acquire antibiotic resistance genes. Instead, the antibiotic resistance genes are usually part of transposons, or "jumping" genes, which specialize in transporting the antibiotic resistance genes between bacteria. The donor and recipient bacteria need not be members of the same species.

GENETIC CHANGES AND EIDs AND RIDs

The previously mentioned mechanisms of gene transfer help explain how bacteria acquire new abilities to enhance their pathogenicity. Viruses use other mechanisms to modify their genomes, some of which are demonstrated by the Influenza A virus (discussed later). We will now explore some of the pathogens that cause EIDs and RIDs.

PATHOGENS THAT CAUSE EIDs

Ebola hemorrhagic fever (EHF) is caused by the Ebola virus, which is a single-stranded RNA (ssRNA) virus. The first case of EHF was identified in Zaire (now called the Democratic Republic of the Congo) in 1976. The Ebola virus may have been transmitted to humans as they encroached upon the forests of central Africa. However, the natural reservoir of the Ebola virus remains unknown. The Ebola virus is currently one of the most deadly viruses in the world.

The first case of AIDS was diagnosed in 1981. AIDS is caused by HIV, which is an enveloped, ssRNA retrovirus. HIV binds to CD4+ T helper (Th) cells. Retroviruses are unique because they are able to convert RNA to DNA using an enzyme called reverse transcriptase. When HIV enters the host cell, it makes a dsDNA copy of ssRNA template. The dsDNA then transferred to the host cell nucleus, where it becomes circular and integrates into the chromosome. Next, the viral and host DNA are tran-

In the spring and early summer of 2003, a new pandemic, severe acute respiratory syndrome or SARS, emerged from China's Guangdong Province and was quickly spread across the globe by world travelers. After it was recognized as a new and highly infectious coronavirus, the World Health Organization issued guidelines for its containment. Here a woman wears a mask required for public transportation in Taipei, Taiwan, as she enters the subway. (AP/Wide World Photos)

scribed together into mRNA, and the mRNA is translated into proteins. Many of the proteins are then used to assemble new viruses. Eventually, the host's Th cell contraction decreases significantly, and the virus concentration increases. HIV patients are usually diagnosed with AIDS shortly afterward. According to the Centers for Disease Control and Prevention (CDC), the prevalence, or number of new and existing cases, of HIV in the United States in 2006 was 1.1 million. However, because of the asymptomatic stages of HIV infection, its actual prevalence is most likely greater.

B. burgdorferi, the pathogen that causes Lyme disease, was first identified in 1982. Symptoms of Lyme disease may include a bull's-eye-like rash and neurologic and cardiovascular abnormalities. *B. burgdorferi* is transmitted to humans by infected deer ticks.

H. pylori, the pathogen responsible for causing chronic gastritis and peptic ulcer disease, was first

cultured in 1982. *H. pylori* is able to neutralize the low pH environment of the gastric lumen by producing ammonia (a base). An *H. pylori* infection can be diagnosed using an interesting, noninvasive method called the urea breath test.

The first case of SARS was reported in Asia in 2003. SARS-CoV, the coronavirus that causes SARS, is transmitted from person-to-person by infected saliva. The CDC is currently working in collaboration with other public health agencies to develop methods to prevent and control future cases of the disease.

PATHOGENS THAT CAUSE RIDs

The two main types of MRSA are as follows: community-associated MRSA and healthcare-associated MRSA. MRSA is resistant to most -lactam antibiotics by the use of the β-lactamase enzyme and *mecA* gene.

C. difficile is a spore-forming bacterium that causes diarrhea and pseudomembranous colitis. When antibiotics (for example, ampicillin and clindamycin) disrupt the normal flora of the large intestine, they promote the overgrowth of *C. difficile* and its spores. The toxins produced by *C. difficile* cause fluid accumulation (toxin A) and cytotoxicity (toxin B). The following genes are responsible for encoding the A and B toxins: *tcdA* (toxin A) and *tcdB* (toxin B).

The multiple-drug-resistant strains of *M. tuberculosis* are MDR-TB and XDR-TB. Tuberculosis is transmitted by airborne respiratory droplets from infected patients. MDR-TB is resistant to at least isoniazid and rifampin (first-line drugs), whereas XDR-TB is resistant to isoniazid, rifampin, any fluoroquinolone, and at least one of three injectable second-line drugs. Patients who have HIV/AIDS are particularly susceptible to tuberculosis.

S. pyogenes is responsible for a variety of diseases including "strep" throat and invasive GAS disease. Necrotizing fasciitis is one example of invasive GAS disease.

There are three types of influenza viruses: A, B, and C. This discussion focuses exclusively on the influenza A virus, which is most frequently associated with pandemics. The influenza A virus, like the other orthomyxoviruses, is an enveloped virus with a segmented RNA genome. Segmented genomes do not come in one piece, like a chromosome; instead, they come in separate units. Therefore, if two or more strains of an influenza virus infect a host, they can exchange those units via genetic reassortment and produce a new virus strain that expresses characteristics of the parental strains. In 2009, a new influenza A (H1N1) virus appeared in the human population that later resulted in the World Health Organization declaring a phase 6 pandemic alert. Interestingly, the virus had undergone a unique reassortment process. The particular influenza A (H1N1) virus was composed of two gene segments from an influenza virus that normally infects European and Asian pigs in addition to gene segments from viruses that normally infect birds and humans; scientists called this particular influenza A (H1N1) virus a quadruple reassortant virus.

Brent M. Ardaugh; David M. Lawrence

FURTHER READING

DeSalle, Rob, ed. *Epidemic! The World of Infectious Disease.* New York: New Press, 1999. A discussion of infectious diseases caused by bacteria, viruses, and parasites. Additional topics include ecology and evolution, modes of transmission, infectious processes, outbreaks, and public health policies.

Drexler, Madeline. *Secret Agents: The Menace of Emerging Infections.* Washington, D.C.: Joseph Henry Press, 2002. Drexler discusses the ongoing war between humans and microbes; recounts the history of past microbial killers; and makes a case for increasing the ability of the public health system to respond to EIDs.

Garrett, Laurie. *The Coming Plague: Newly Emerging Diseases in a World Out of Balance.* New York: Penguin, 1995. The author, a reporter for *Newsday*, explains reasons infectious diseases remain a threat to humanity. The book provides a well-documented introduction to the topic of infectious diseases.

Heymann, David L., ed., and American Public Health Association. *Control of Communicable Diseases Manual.* Washington, D.C.: American Public Health Association, 2008. Provides an alphabetical listing of infectious diseases, and describes their pathogens, reservoirs, periods of communicability, modes of transmission, and prevention and control measures.

Kolata, Gina. *Flu: The Story of the Great Influenza Pandemic of 1918 and the Search for the Virus That Caused It.* New York: Simon & Schuster, 2001. Describes the history of Influenza and its effects on the world; the book also discusses the possibility of another flu pandemic.

Levy, Stuart B. *The Antibiotic Paradox: How Miracle Drugs Are Destroying the Miracle.* New York: Plenum Press, 1992. Levy presents a frightening and authoritative indictment of how misuse of antibiotics is leading to the emergence of drug-resistant microbes.

McNeill, William H. *Plagues and Peoples.* New York: Anchor Books, 1998. A historical perspective of the effects of infectious diseases on human populations.

Preston, Richard. *The Hot Zone.* New York: Anchor Books, 1995. A number-one *New York Times* best seller about the Ebola virus unexpectedly appearing in a U.S. animal laboratory.

Tierno, Phillip M. *The Secret Life of Germs: What They Are, Why We Need Them, and How We Can Protect Ourselves Against Them.* New York: Atria Books, 2001. Provides readers with entertaining information about their everyday encounters with pathogens.

Centers for Disease Control and Prevention (CDC)
http://www.cdc.gov
CDC aims to prevent and control diseases through research, health promotion, and emergency preparedness.

Emerging Infectious Diseases Journal
http://www.cdc.gov/ncidod/EID/index.htm
The *EID Journal* is published monthly by the CDC, and it features the latest scientific information on infectious diseases.

National Center for Health Statistics (NCHS)
http://www.cdc.gov/nchs
NCHS publishes morbidity and mortality data, which can be used to establish public health priorities.

The New England Journal of Medicine (NEJM)
http://content.nejm.org
NEJM is one of the world's most popular medical journals, and it provides readers with a clinical perspective of infectious diseases.

World Health Organization (WHO)
http://www.who.int/en
WHO monitors disease trends and collaborates with international government agencies to coordinate global health initiatives.

See also: Bacterial resistance and super bacteria; Biological weapons; Mutation and mutagenesis; Smallpox; Viral genetics; Viroids and virusoids.

Epidermolytic hyperkeratosis

CATEGORY: Diseases and syndromes
ALSO KNOWN AS: Bullous congenital ichthyosiform erythroderma; bullous ichthyosiform erythroderma; bullous congenital ichthyosiform erythroderma of Brocq; BCIE; BIE; EHK

DEFINITION

Epidermolytic hyperkeratosis is a rare congenital skin disorder which causes thick, scaly, red blistered skin in neonates. As the name suggests, epidermo-

lytic refers to fragile skin, while hyperkeratosis implies thickening of the outermost layer of the skin.

RISK FACTORS

There are no known risk factors for this disorder, but it may be inherited as an autosomal dominant trait or by a spontaneous mutation.

ETIOLOGY AND GENETICS

Epidermolytic hyperkeratosis was discovered in 1902 by Louis-Anne-Jean Brocq. Since then, genetic studies have determined that 50 percent of cases exhibit an autosomal dominant inheritance caused by spontaneous mutation. Mutations on the keratin-1 and/or keratin-10 gene clusters found on chromosomes 12q and 17q have been linked to this disorder. Keratin, which is produced by keratinocytes, provides integrity to skin, hair, and nail cells. Therefore, mutations to the keratin cells cause structural instability and weakness of the keratinocytes, leading to an unsupported epidermis. As a result, blistering, hyperkeratosis, and scaling of the epidermis occur after minimal trauma.

The diversity of epidermolytic hyperkeratosis is remarkable. Six clinical phenotypes have been characterized based on palmoplantar involvement (pertaining to the palms of the hand and soles of the feet). Three of these subtypes, which are classified as NPS-1, NPS-2, and NPS-3, have minimal palm or sole involvement. Patients have varying degrees of redness and scaling of the skin, balance abnormalities, and blistering. Spinelike rigid scales are often seen in this subgroup. Notably, mutations on keratin 1 have been associated with the NPS subtypes.

The remaining three subtypes, which are classified as PS-1, PS-2, and PS-3, have severe palm and sole involvement. Palmoplantar hyperkeratosis occurs in roughly 60 percent of patients. A typical smooth hyperkeratosis with a distinct border is often present. Tense hands and feet have been reported within the PS subtypes as well. Palmoplantar epidermolytic hyperkeratosis has been linked to mutations on the keratin 10 gene.

Lastly, a rare variant of epidermolytic hyperkeratosis is thought to occur by a spontaneous mutation during the early phase of embryogenesis and presents with a mosaic pattern of hyperkeratosis. In this particular subtype, normal skin is interchanged with streaks of hyperkeratosis. The distribution of the streaks can take on a tight and/or widespread pattern. It is interesting to note that the keratin mutations were not found on the portions of normal skin in individuals with a mosaic skin pattern. It is not surprising that carriers of this type of mutation can pass on the gene to subsequent generations.

SYMPTOMS

Mild to severe blistering, peeling, and erosive skin appears at birth. Over time, the redness and blistering of the skin subsides and hyperkeratosis persists. This results in thick, scaly, waxy, rigid skin predominantly at the joint flexures. Depending on the mutation, there can be scalp, palm, and/or sole involvement; the hair and nails are not affected. In rare instances, cardiac disturbances can occur in conjunction with skin disorders.

SCREENING AND DIAGNOSIS

The prevalence of epidermolytic hyperkeratosis is 1 in 200,000 to 300,000 persons in the United States. The distinct widespread thick erythroderma with blisters present at birth is typically the first indicator of epidermolytic hyperkeratosis. In addition, the skin usually gives off a pungent odor. Misdiagnosis often occurs with another genetic disorder known as nonbullous congenital ichthyosiform erythroderma, which lacks the presentation of blisters. Electron microscopy along with a skin biopsy is often performed to obtain a definitive diagnosis.

TREATMENT AND THERAPY

Treatment is limited to symptomatic relief with lactate lotion, topical/systemic retinoids (tretinoin), vitamin A derivatives, 10 percent glycerin and antibacterial soap. Excessive moisturization and soaking baths are important. During the neonatal period, prevention of sepsis and fluid/electrolyte imbalance is imperative to prevent mortality. Parents of newborns should be advised on proper wound and blister care. As the child grows, oral and topical antibiotics are used to control bacterial colonization due to open skin wounds. Fungal infections are also of concern. It has been hypothesized that gene therapy may be a potential cure for this disorder.

PREVENTION AND OUTCOMES

The probability of epidermolytic hyperkeratosis is unpredictable since half of the cases are spontaneous mutations. Prenatal detection can be useful via fetal skin biopsy or by examining DNA extracts with a confirmed family history. Similar to other autosomal dominant epidermolytic diseases, symptoms improve over time. In fact, by adolescence, blistering diminishes and mild redness and scaling of the skin follow. Many patients suffer from physical and psychological difficulties associated with the apparent scales, tender skin, as well as odor. In some cases, minor learning difficulties have been attributed to prenatal complications. Genetic counseling and support groups are also advised for parents with children who are affected.

Jigna Bhalla, Pharm.D.

FURTHER READING

Hall, John C. *Sauer's Manual of Skin Diseases.* 10th ed. Philadelphia: Wolter Kluwer/Lippincott Williams & Wilkins Health, 2010.

Sybert, Virginia P. *Genetic Skin Disorders.* New York: Oxford University Press, 1997.

Turkington, Carol A., and Jeffrey S. Dover. *Skin Deep: An A-Z of Skin Disorders, Treatments, and Health.* New York: Facts On File, 1998.

WEB SITES OF INTEREST

American Academy of Dermatology
www.aad.org

Foundation of Ichthyosis and Related Skin Types (FIRST)
www.scalyskin.org

National Organization for Rare Diseases
http://www.rarediseases.org

National Registry for Ichthyosis and Related Disorders
http://depts.washington.edu/ichreg/ichthyosis.registry

See also: Hereditary diseases.

Epilepsy

CATEGORY: Diseases and syndromes
ALSO KNOWN AS: Seizure disorder

DEFINITION

Epilepsy is a chronic brain disorder of the cerebrum whereby the brain produces sudden bursts of electrical energy abnormally interfering with consciousness and all types of sensations. It disrupts the nervous system, which can cause mental and physical disorders. Epilepsy with a known cause is called either secondary or symptomatic epilepsy. The most common type is the idiopathic or unknown type, with six out of ten persons suffering from it.

RISK FACTORS

Primary factors are head injuries, infections in the central nervous system, and tumors. Another factor is cerebrovascular disease, a condition characterized by its effects on the brain and its blood supply. A family history seems to influence the tendency toward epilepsy. Epilepsy and seizure disorders affect over three million Americans and many times that worldwide. It affects all age groups.

ETIOLOGY AND GENETICS

Symptomatic idiopathic (unknown) epilepsy has been discovered to start between the ages of two and fourteen. Seizures before the age of two usually indicate developmental defects.

Epilepsy is divided into two categories, partial and generalized, because of specific biological mechanisms at work. Partial, also called focal or localized, seizures occur more often than do generalized seizures. They can occur in one or more locations in the brain. Partial seizures can spread more widely in the brain, depending on the severity of the seizure.

Epilepsy is not a single disorder: It contains multiple gene factors that are influenced by the environment. Some types of epilepsy run in families, and children of have a risk range from 4 to 8 percent, some studies show. People with generalized seizures. in which both sides of the brain are involved, tend to have other family members affected by epilepsy more often than those with localized seizures.

Generalized seizures are more likely to be genetically based. Some types of epilepsy can be inherited. Epilepsies such as West syndrome or infantile spasms can cause delays in development of children between four and eight months of age. Juvenile myclonic epilepsy has another name, impulsive petit mal epilepsy, which is characterized as general seizures with spasmodic movements called myclonic jerks.

Several epilepsy syndromes start in the baby's first year of life. Studies in molecular genetics have identified problematic genes for some of them. A condition known as hypoxic-ischemic encephalopathy is the major cause of epilepsy in the first year of life. Other etiologies during infancy are chromosomal disorders and brain disorders.

Sometimes there is an absence of seizures in individuals between ages eight to twenty. Adult myclonic epilepsy is a distinct syndrome that involves the development of generalized epilepsy among people over forty.

Lennox-Gastaut syndrome is a very serious form of epilepsy in children, which causes multiple seizures and mental retardation. There are partial seizures and an absence of muscle control. A person with this condition has difficulty standing and sitting.

Progessive myclonic epilepsy is an inherited disorder that affects children from the age of six to fifteen. It is characterized by light sensitivity, and the seizures are tonic-clonic. Early studies suggested that the disorder would continue to worsen throughout their life, but better treatment has improved this outlook.

Autosomal dominant nocturnal frontal lobe epilepsy is an inherited syndrome that is rare, and typically occurs around the age of eleven. The onset varies among families, with twisting contractions or thrashing. These seizures are dystonic and occur at night for a short period of time. Landau-Kleffner syndrome is a epileptic condition that leads a person to have difficulty in writing and speaking (aphasia). Benign familial neonatal convulsions (BFNC) are very rare. This form of generalized seizures has a herediary factor to it.

Status epilepticus (SE) is a potentially life-threatening condition which can lead to chronic epilepsy. It affects 100,000 to about 150,000 people, 50 percent of them being children, and can cause death or permanent brain damage if not not treated correctly and quickly. This condition is associated with recurrent convulsions that can last for more than twenty minutes, interrupted only briefly for partial relief. Generalized or tonic-clonic type is the most serious form of SE. The trigger is unknown.

Dozens of genetic syndromes covering a variety of seizure patterns may cause different types of epilepsy. Some genetic causes have been identified for a few cases of juvenile myclonic epilepsy, which constitute 10 percent of all epilepsy cases. Some research has suggested that the GABA signaling system is an important element in many cases of epilepsy.

Some epilepsy syndromes have a genetic inclination, which has created the possibility of genetic testing. Genetic testing has drawbacks, however, as only monogenetic (single-gene) epilepsies can be confirmed. Dravet syndrome (severe myoclonic epilepsy in infancy) and benign familial neonatal seizures (BFNS) are easy to recognize, and early genetic testing may benefit these syndromes. Other epileptic syndromes have complex patterns that would not be aided by genetic testing.

SYMPTOMS

The hallmark of epilepsy is seizures. Some temporary symptoms can be loss of awareness and movement, which includes vision, hearing, taste, mood, and mental function. Seizure problems often create physical problems (such as bruising easily or breaking bones) as well as psychosocial or mental problems.

SCREENING AND DIAGNOSIS

A diagnosis of epilepsy comes in multiple steps. A blood workup and neurological exams are needed. A complete medical history is important and a family history to gather as much information as possible. The health care provider should find out when the seizure began and what it looked like. The next step is to identify the type of seizure and whether it falls under a recognizable syndrome. A clinical evaluation is important to determine the source of the epilepsy.

Doctors will use an electroencephalograph (EEG), which measures brain waves, to determine any abnormal patterns in the brain. A computed tomography (CT) scan uses more sensitive imaging equipment than a single X ray. It provides clear images of organs such as the brain and heart and is another tool to help identify seizure activity.

TREATMENT AND THERAPY

Antiepilepsy drugs (AEDs) are the main treatment regimen for epilepsy, but only about 65 percent of patients' seizures are controlled by these medications. Stimulation of the vagus nerve (a large nerve in the back of the neck) was approved late in the twentieth century for adults and children; this procedure is still being used in the twenty-first century when medications to control seizures fail. It is designed to send light electrical pulses regularly to the back of the brain via the vagus nerve to prevent further seizures. One of the complications with vagus nerve stimulation is that it does not eliminate seizures in all patients. In fact, it can cause shortness of breath, sore throat, vomiting, nausea, and ear and throat problems. Epilepsy surgery is used when the area of the brain in which the seizures originally started can be isolated or removed.

PREVENTION AND OUTCOMES

For the most part, what causes epilepsy is still unknown. Idiopathic epilepsy cannot be prevented. However, it is known that head injuries, a common cause of epilepsy, can be prevented, such as by wearing a helmet when riding a bicycle, motorcycle, or horse or taking part in any activity for which head protection would be beneficial. Scientists from all over the world constantly seek out the best antiepilepsy drugs and study the way in which neurotransmitters react to brain cells to control nerve firing. Scientists are contiuing to improve MRI and other diagnostic tools. Some studies suggest that children can have fewer seizures if they maintain a ketogenic diet, which consists of a high intake of fats and a low one of carbohydrates. Scientists are also working with stem cells to further improve the treatment of epilepsy.

Marvin L. Morris, LAc, M.P.A.

FURTHER READING

Foldvary-Schaefer, N., and E. Wyllie. "Epilepsy." In *Textbook of Clinical Neurology*, edited by C. Goetz. 3d ed. Philadelphia: Saunders Elsevier, 2007.

Krebs, P. P. "Psychogenic Nonepileptic Seizures." *American Journal of Electroneurodiagnostic Technology* 47, no. 1 (March, 2007): 20-28.

Kwan, P., and M. J. Brodie. "Emerging Drugs for Epilepsy." *Expert Opinions on Emerging Drugs* 12, no. 3 (September, 2007): 407-422.

Scheffer, I. E., et al. "Temporal Lobe Epilepsy and GEFS+ Phenotypes Associated with *SCN1B* Mutations." *Brain* 130 (2007): 100-109.

Steinlein, O. K., C. Corad, and B. Weidner. "Benign Familial Neonatal Convulsions: Always Benign?" *Epilepsy Research* 73 (2007): 245-249.

WEB SITES OF INTEREST

American Academy of Neurology
www.aan.com

American Epilepsy Society
www.aesnet.org

Epilepsy Foundation
www.EpilepsyFoundation.org

National Library of Medicine
www.nlm.nih.gov/hinfo

See also: Amyotrophic lateral sclerosis; Ataxia telangiectasia; Batten disease; Essential tremor; Friedreich ataxia; Hereditary diseases; Huntington's disease; Parkinson disease.

Epistasis

CATEGORY: Classical transmission genetics

SIGNIFICANCE: Epistasis is the interaction of genes such that the alleles at one locus can modify or mask the expression of alleles at another locus. Dihybrid crosses involving epistasis produce progeny ratios that are non-Mendelian, that is, different from the kinds of ratios discovered by Gregor Mendel.

KEY TERMS

allele: an alternate form of a gene at a particular locus; a single locus can possess two alleles

dihybrid cross: a cross between parents that involve two specified genes, or loci

F_1: first filial generation, or the progeny resulting from the first cross in a series

F_2: second filial generation, or the progeny resulting from the cross of the F_1 generation

locus (pl. *loci*): a more precise word for gene; in diploid organisms, each locus has two alleles

DEFINITION AND HISTORY

The term "epistasis" is of Greek and Latin origin, meaning "to stand upon" or "stoppage." The term was originally used by geneticist William Bateson at the beginning of the twentieth century to define genes that mask the expression of other genes. The gene at the initial location (locus) is termed the epistatic gene. The genes at the other loci are "hypostatic" to the initial gene. In its strictest sense, it describes a nonreciprocal interaction between two or more genes, such that one gene modifies,

suppresses, or otherwise influences the expression of another gene affecting the same phenotypic (physical) character or process. By this definition, simple additive effects of genes affecting a single phenotypic character or process would not be considered an epistatic interaction. Similarly, interactions between alternative forms (alleles) of a single gene are governed by dominance effects and are not epistatic. Epistatic effects are interlocus interactions. Therefore, in terms of the total genetic contribution to phenotype, three factors are involved: dominance effects, additive effects, and epistatic effects. The analysis of epistatic effects can suggest ways in which the action of genes can control a phenotype and thus supply a more complete understanding of the influence of genotype on phenotype.

A gene can influence the expression of other genes in many different ways. One result of multiple genes is that more phenotypic classes can result than can be explained by the action of a single pair of alleles. The initial evidence for this phenomenon came out of the work of Bateson and British geneticist Reginald C. Punnett during their investigations on the inheritance of comb shape in domesticated chickens. The leghorn breed has a "single" comb, brahmas have "pea" combs, and wyandottes have "rose" combs. Crosses between brahmas and wyandottes have "walnut" combs. Intercrosses among walnut types show four different types of F_2 (second-generation) progeny, in the ratio 9 walnut: 3 rose: 3 pea: 1 single. This ratio of phenotypes is consistent with the classical F_2 ratio for dihybrid inheritance. The corresponding ratio of genotypes, therefore, would be 9 $A\_ B\_$:3 $A\_ bb$:3 $aa B\_$:1 $aa bb$, respectively. (The underscore is used to indicate that the second gene can be either dominant or recessive; for example, $A\_$ means that both AA and Aa will result in the same phenotype.) In this example, one can recognize that two independently assorting genes can affect a single trait. If two gene pairs are acting epistatically, however, the expected 9:3:3:1 ratio of phenotypes is altered in some fashion. Thus, although the preceding example involves interactions between two loci, it is not considered a case of epistasis, because the phenotype ratio is a classic Mendelian ratio for a dihybrid cross. Five basic examples of two-gene epistatic interactions can be described: complementary, modifying, inhibiting, masking, and duplicate gene action.

A Punnett Square Showing Flower Pigmentation

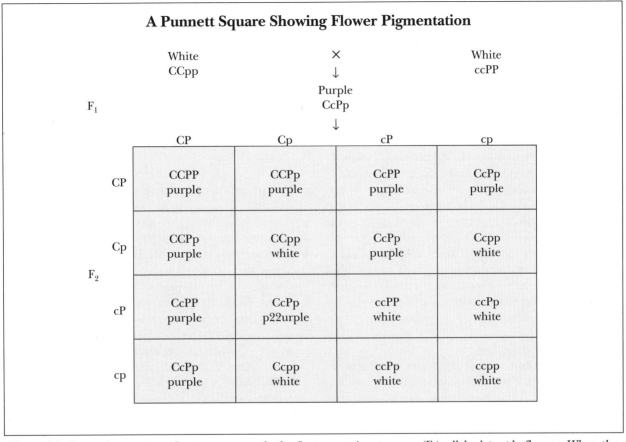

When white-flowered sweet pea plants were crossed, the first-generation progeny (F₁) all had purple flowers. When these plants were self-fertilized, the second-generation progeny (F₂) revealed a ratio of nine purple to seven white. This result can be explained by the presence of two genes for flower pigmentation, P (dominant) and p (recessive) and C or c. Both dominant forms, P and C, must be present in order to produce purple flowers.

COMPLEMENTARY GENE ACTION

For complementary gene action, a dominant allele of two genes is required to produce a single effect. An example of this form of epistasis again comes from the observations of Bateson and Punnett of flower color in crosses between two white-flowered varieties of sweet peas. In their investigation, crosses between these two varieties produced an unexpected result: All the F₁ (first-generation) progeny had purple flowers. When the F₁ individuals were allowed to self-fertilize and produce the F₂ generation, a phenotypic ratio of nine purple-flowered to seven white-flowered individuals resulted. Their hypothesis for this ratio was that a homozygous recessive genotype for either gene (or both) resulted in the lack of flower pigmentation. A simple model to explain the biochemical basis for this type of flower pigmentation is a two-step process, each step controlled by a separate gene and each gene having a recessive allele that eliminates pigment formation. Given this explanation, each parent must have had complementary genotypes (AA bb and aa BB), and thus both had white flowers. Crosses between these two parents would produce double heterozygotes (Aa Bb) with purple flowers. In the F₂ generation, $\frac{9}{16}$ would have the genotype A_ B_ and would have purple flowers. The remaining $\frac{7}{16}$ would be homozygous recessive for at least one of the two genes and, therefore, would have white flowers. In summary, the phenotypic ratio of the F₂ generation would be 9:7.

MODIFYING GENE ACTION

The term "modifying gene action" is used to describe a situation whereby one gene produces an effect only in the presence of a dominant allele of a second gene at another locus. An example of this type of epistasis is aleurone color in corn. The aleurone is the outer cell layer of the endosperm (food storage tissue) of the grain. In this system, a dominant gene ($P\_$) produces a purple aleurone layer only in the presence of a gene for a red aleurone ($R\_$) but expresses no effect in the absence of the second gene in its dominant form. Thus, the corresponding F_2 phenotypic ratio is 9 purple:3 red:4 colorless. The individuals without aleurone pigmentation would, therefore, be of the genotype $P\_ rr$ ($\frac{3}{16}$) or *pp rr* ($\frac{1}{16}$). Again, a two-step biochemical pathway for pigmentation can be used to explain this ratio; however, in this example, the product of the second gene (R) acts first in the biochemical pathway and allows for the production of red pigmentation and any further modifications to that pigmentation. Thus, the phenotypic ratio of the F_2 generation would be 9:3:4.

INHIBITING GENE ACTION

Inhibiting action occurs when one gene acts as an inhibitor of the expression of another gene. In this example, the first gene allows the phenotypic expression of a gene, while the other gene inhibits it. Using a previous example (the gene R for red aleurone color in corn seeds), the dominant form of the first gene R does not produce its effect in the presence of the dominant form of the inhibitor gene I. In other words, the genotype $R\_ i\_$ results in a phenotype of red aleurone ($\frac{3}{16}$), while all other genotypes result in the colorless phenotype ($\frac{13}{16}$). Thus gene R is inhibited in its expression by the expression of gene I. The F_2 phenotypic ratio would be 13:3. This ratio, unlike the previous two examples, includes only two phenotypic classes and highlights a complicating factor in determining whether one or two genes may be influencing a given trait. A 13:3 ratio is close to a 3:1 ratio (the ratio expected for the F_2 generation of a monohybrid cross). Thus it emphasizes the need to look at an F_2 population of sufficient size to discount the possibility of a single gene phenomenon over an inhibiting epistatic gene interaction.

MASKING GENE ACTION

Masking gene action, a form of modifying gene action, results when one gene is the primary determinant of the phenotype of the offspring. An example of this phenomenon is fruit color in summer squash. In this example, the F_2 ratio is 12:3:1, indicating that the first gene in its dominant form results in the first phenotype (white fruit); thus this gene is the primary determinant of the phenotype. If the first gene is in its recessive form and the second gene is in its dominant form, the fruit will be yellow. The fruit will be green at maturity only when both genes are in their recessive form ($\frac{1}{16}$ of the F_2 population).

DUPLICATE GENE INTERACTION

Duplicate gene interaction occurs when two different genes have the same final result in terms of their observable influence on phenotype. This situation is different from additive gene action in that either gene may substitute for the other in the expression of the final phenotype of the individual. It may be argued that duplicate gene action is not a form of epistasis, since there may be no interaction between genes (if the two genes code for the same protein product), but this situation may be an example of gene interaction when two genes code for similar protein products involved in the same biochemical pathway and their combined interaction determines the final phenotype of the individual. An example of this type of epistasis is illustrated by seed capsule shape in the herb shepherd's purse. In this example, either gene in its dominant form will contribute to the final phenotype of the individual (triangular shape). If both genes are in their recessive form, the seed capsule has an ovoid shape. Thus, the phenotypic ratio of the F_2 generation is 15:1.

IMPACT AND APPLICATIONS

Nonallelic gene interactions have considerable influence on the overall functioning of an individual. In other words, the genome (the entire genetic makeup of an organism) determines the final fitness of an individual, not only as a sum total of individual genes (additive effects) or by the interaction between different forms of a gene (dominance effects) but also by the interaction between different genes (intragenomic or epistatic effects). This situation is something akin to a chorus: Great choruses not only have singularly fine voices, but they also perform magnificently as finely tuned and coordinated units. Knowledge of what contributes to a superior genome would, therefore, lead to a fuller understanding of the inheritance of quantitative characters and more directed approaches to genetic improvement. For example, most economically important characteristics of agricultural species (such as yield, pest and disease resistance, and stress tolerance) are quantitatively inherited, the net result of many genes and their interactions. Thus an understanding of the combining ability of genes and their influence on the final appearance of domesticated breeds and crop varieties should lead to more efficient genetic improvement schemes. In addition, it is thought that many important human diseases are inherited as a complex interplay among many genes. Similarly, an understanding of genomic functioning should lead to improved screening or therapies.

Henry R. Owen, Ph.D.

FURTHER READING

Frankel, Wayne N., and Nicholas J. Schork. "Who's Afraid of Epistasis?" *Nature Genetics* 14 (December, 1996): 371-373. A reexamination of the entire concept of epistasis, with statistical implications.

Russell, Peter J. *Fundamentals of Genetics.* 2d ed. San Francisco: Benjamin Cummings, 2000. Introduces the three main areas of genetics: transmission genetics, molecular genetics, and population and quantitative genetics.

Snustad, D. Peter, and Michael J. Simmons. "Epistasis." In *Principles of Genetics.* 5th ed. Hoboken, N.J.: John Wiley and Sons, 2009. This textbook provides an explanation of epistasis within the context of gene action from genotype to phenotype.

Wolf, Jason B., Edmund D. Brodie III, and Michael J. Wade. *Epistasis and the Evolutionary Process.* New York: Oxford University Press, 2000. Primary focus is on the role of gene interactions (epistasis) in evolution. Leading researchers examine how epistasis impacts the evolutionary processes in overview, theoretical, and empirical chapters. Illustrations, index.

Wood, Stacey J., and Alison A. Motsinger-Reif. "Epistasis: Understanding and Evaluating the Phenomenon in Human Genetics Disease Mapping." In *Genetic Predisposition to Disease,* edited by Sara L. Torres and Marta S. Marin. New York: Nova Science, 2008. Focuses on the impact of epistasis on human genetic disease.

WEB SITE OF INTEREST

Scitable

http://www.nature.com/scitable

Scitable, a library of science-related articles compiled by the Nature Publishing Group, features several articles about epistasis. Users can retrieve these articles by typing the word "epistasis" into the site's search engine

See also: Chromosome structure; Chromosome theory of heredity; Classical transmission genetics; Complete dominance; Dihybrid inheritance; Extra-chromosomal inheritance; Hybridization and introgression; Incomplete dominance; Lamarckianism; Mendelian genetics; Monohybrid inheritance; Multiple alleles; Nondisjunction and aneuploidy; Parthenogenesis; Penetrance; Polygenic inheritance; Quantitative inheritance.

Essential tremor

CATEGORY: Diseases and syndromes
ALSO KNOWN AS: Benign essential tremor; familial tremor

DEFINITION

Benign essential tremor is a movement disorder most commonly characterized by shaking in the hands. It occurs in as many as 10 percent of people over the age of sixty. It may also cause shaking of the head, voice, arms, and trunk, and, less often, of the legs and feet.

Two types of tremor are common with essential tremor. Postural tremor is shaking only in certain positions, such as with arms outstretched. Kinetic or action tremor is shaking that gets worse during activities, such as eating or shaving.

In some cases, essential tremor can be socially isolating. It may interfere with normal daily activities, such as writing or speaking. If so, a patient should contact his or her doctor for an evaluation.

RISK FACTORS

A family history of tremors is the only known risk factor for essential tremor. Although the condition may occur at any age, it is more likely to occur in people older than forty years old.

ETIOLOGY AND GENETICS

Familial essential tremor is a condition in which multiple environmental and genetic factors play a contributing part. Approximately 50 percent of affected individuals report one or more family members who are similarly affected. The inheritance pattern may vary, but in most families an autosomal dominant mode of transmission is observed, meaning that a single copy of the mutation is sufficient to cause expression of the trait. An affected individual has a 50 percent chance of transmitting the mutation to each of his or her children. Many cases of essential tremor, however, result from a spontaneous new mutation, so in these instances affected individuals will have unaffected parents. The age of onset is variable, but virtually all individuals who carry the mutation will show some expression by age seventy.

Two genes have been identified with a direct association with essential tremor, and other genes are expected to play minor roles as well. The first gene to be discovered is known variously as either *DRD3* or *FET1*, and it is located on the long arm of chromosome 3 at position 3q13.3. This gene encodes the dopamine receptor protein D3, which is expressed in nerve cells in the brain. It responds to the neurotransmitter dopamine and triggers a signal to produce physical movement. A mutation in the *FET1* gene may cause the receptor protein to react more strongly to dopamine, causing the involuntary shaking characteristic of the condition.

The second gene shown to be associated with essential tremor is *HS1BP3*, found on the short arm of chromosome 2 at position 2p24.1. Its protein product is the hematopoietic-specific protein 1 binding protein 3. Localized primarily in the cerebellum region of the brain, this protein helps regulate the chemical signaling involved in coordinating movements of muscles by motor neurons. A third gene on the short arm of chromosome 4, at position 4p14, is involved in only those individuals who have both Parkinson disease and essential tremor.

SYMPTOMS

Essential tremor is generally not serious, but its severity may vary and worsen over time. Symptoms may include a tremor that occurs when standing or moving the limbs, but not usually at rest; uncontrollable, rhythmic, up-and-down movement; shaking in hands, arms, head, voice, trunk, legs, or feet on both sides; shaking only in certain positions or during activity; and trouble with fine motor skills, such as drawing, sewing, or playing an instrument. Other symptoms may include shaking that gets worse from caffeine, stress, fatigue, or heat; hearing loss (some cases are associated with hearing loss); and problems with social, functional, or occupational abilities (more severe cases interfere with these abilities). To be considered as having essential tremor, an individual's tremors must not be related to other health conditions.

SCREENING AND DIAGNOSIS

The doctor will ask about a patient's symptoms and his or her medical and family history. The doctor will also do a physical exam, paying particular attention to the patient's central nervous system. At this time, there are no special tests to diagnose essential tremor. However, patients may have blood, urine, or other tests, such as a magnetic resonance imaging (MRI) scan, to rule out other causes, like Parkinson disease, elevated thyroid hormone, low blood sugar, stroke, and medications.

TREATMENT AND THERAPY

Most people with essential tremor do not require treatment. Mild tremors may be relieved or even eliminated by simple measures, including staying well rested, avoiding caffeine, avoiding stimulants often found in over-the-counter medications such as cold remedies, and avoiding temperature extremes.

Individuals should talk to their doctors about any medications that may be contributing to their symptoms. If a patient's symptoms are troubling, treatment options that may be helpful include beta blockers, such as propranolol (a blood pressure medication); antiseizure medications, such as primidone (Mysoline), gabapentin (Neurontin), or topiramate (Topamax); and sedatives (benzodiazepines).

Botulinum injections may be used in rare situations. In rare cases where tremors are very disabling and medications do not help, surgery may be an option. Two approaches are possible: deep brain stimulation (DBS) and thalamotomy. DBS transmits painless electrical pulses to the brain, interrupting faulty signals. Thalamotomy destroys a tiny part of the brain generating the tremors; it is less commonly performed than DBS.

PREVENTION AND OUTCOMES

There is no known way to prevent essential tremor.

Annie Stuart;
reviewed by J. Thomas Megerian, M.D., Ph.D., F.A.A.P.
"Etiology and Genetics" by Jeffrey A. Knight, Ph.D.

FURTHER READING

EBSCO Publishing. *Health Library: Essential Tremor.* Ipswich, Mass.: Author, 2009. Available through http://www.ebscohost.com.

Jankovic, J., and K. M. Shannon. "Movement Disorders." In *Neurology in Clinical Practice*, edited by Walter G. Bradley et al. 5th ed. 2 vols. Philadelphia: Butterworth-Heinemann/Elsevier, 2008.

Lorenz, D., and G. Deuschl. "Update on Pathogenesis and Treatment of Essential Tremor." *Current Opinion in Neurology* 20, no. 4 (August, 2007): 447-452.

Lyons, Kelly E., and Rajesh Pahwa, eds. *Handbook of Essential Tremor and Other Tremor Disorders.* Boca Raton, Fla.: Taylor & Francis, 2005.

Plumb, Mark, and Peter Bain. *Essential Tremor: The Facts.* New York: Oxford University Press, 2007.

WEB SITES OF INTEREST

"Essential Tremor." FamilyDoctor.org
http://familydoctor.org/online/famdocen/home/common/brain/disorders/807.html

"Essential Tremor." International RadioSurgery Association
http://www.irsa.org/essential_tremor.html

"Essential Tremor." MayoClinic.com
http://www.mayoclinic.com/health/essential-tremor/DS00367

Genetics Home Reference
http://ghr.nlm.nih.gov

Health Canada
http://www.hc-sc.gc.ca/index-eng.php

International Essential Tremor Foundation
http://essentialtremor.org

National Institute of Neurological Disorders and Stroke
http://www.ninds.nih.gov

Parkinson Society Canada
http://www.parkinson.ca

WE MOVE (Worldwide Education and Awareness for Movement Disorders)
http://www.wemove.org

See also: Hereditary diseases; Parkinson disease.

Eugenics

CATEGORY: History of genetics; Human genetics and social issues

SIGNIFICANCE: The eugenics movement sought to speed up the process of natural selection through the use of selective breeding and led to the enactment of numerous laws requiring the sterilization of "genetically inferior" individuals and limiting the immigration of supposedly defective groups. Such flawed policies were based on an inadequate understanding of the complexity of human genetics, an underestimation of the role of the environment in gene expression, and the desire of certain groups to claim genetic superiority and the right to control the reproduction of others.

KEY TERMS

biometry: the measurement of biological and psychological variables

negative eugenics: improving human stocks through the restriction of reproduction

positive eugenics: improving human stocks by encouraging the "naturally superior" to breed extensively with other superior humans

THE FOUNDING OF THE EUGENICS MOVEMENT

With the publication of Charles Darwin's *On the Origin of Species by Means of Natural Selection* (1859), the concept of evolution began to revolutionize the way people thought about the human condition. Herbert Spencer and other proponents of what came to be known as social Darwinism adhered to the belief that social class structure arose through natural selection, seeing class stratification in industrial societies, including the existence of a permanently poor underclass, as a reflection of the underlying, innate differences between classes.

During this era there was also a rush to legitimize all sciences by using careful measurement and quantification. There was a blind belief that attaching numbers to a study would ensure its objectivity.

Francis Galton, an aristocratic inventor, statistician, and cousin of Darwin, became one of the primary promoters of such quantification. Obsessed with mathematical analysis, Galton measured everything from physiology and reaction times to boredom, the efficacy of prayer, and the beauty of women. He was particularly interested in the differences between human races. Galton eventually founded the field of biometry by applying statistics to biological problems.

A hereditarian, Galton assumed that talent in humans was subject to the laws of heredity. Although Galton did not coin the term "eugenics" until 1883, he published the first discussion of his ideas in 1865, in which he recognized the apparent evolutionary paradox that those of talent often have few, if any, children and that civilization itself diminishes the effects of natural selection on human populations. Fearing that medicine and social aid would lead to the propagation of weak individuals, Galton advocated increased breeding by "better elements" in the population (positive eugenics), while at the same time discouraging breeding of the "poorer elements" (negative eugenics).

Like most in his time, Galton believed in "blending inheritance," whereby hereditary material would mix together like different colors of paint. Trying to reconcile how superior traits would avoid being swamped by such blending, he came up with the statistical concept of the correlation coefficient, and in the process connected Darwinian evolution to the "probability revolution." His work focused on the bell-shaped curve or "normal distribution" demonstrated by many traits and the possibility of shifting the mean by selection pressure at either extreme. His statistical framework deepened the theory of natural selection. Unfortunately, the mathematical predictability he studied has often been misinterpreted as inevitability. In 1907, Galton founded the Eugenics Education Society of London. He also carefully cataloged eminent families in his *Hereditary Genius* (1869), wherein the Victorian world was assumed to be the ultimate level that society could attain and the cultural transmission of status, knowledge, and social connections were discounted.

EARLY EUGENICS IN BRITAIN

Statistician and social theorist Karl Pearson was Galton's disciple and first Galton Professor of Eugenics at the Galton Laboratory at the University of London. His *Grammar of Science* (1892) outlined his belief that eugenic management of society could prevent genetic deterioration and ensure the existence of intelligent rulers, in part by transferring resources from inferior races back into the society. According to philosopher David J. Depew and biochemist Bruce H. Weber, even attorney Thomas Henry Huxley, champion of Darwinism, balked at this "pruning" of the human garden by the administrators of eugenics. For the most part, though, British eugenicists focused on improving the superior rather than eliminating the inferior.

Another of Galton's followers, comparative anatomist Walter Frank Weldon, like Galton before him, set out to measure all manner of things, showing that the distribution of many human traits formed a bell-shaped curve. In a study on crabs, he showed that natural selection can cause the mean of such a curve to shift, adding fuel to the eugenicists' conviction that they could better the human race through artificial selection.

Population geneticist Ronald A. Fisher was Pearson's successor as the Galton Professor of Eugenics. Fisher cofounded the Cambridge Eugenics Society and became close to Charles Darwin's sons, Leonard and Horace Darwin. In a speech made to the Eugenics Education Society, Fisher called eugenicists the "agents of a new phase of evolution" and the "new natural nobility," with the view that humans were becoming responsible for their own evolution. The second half of his book *The Genetical Theory of Natural Selection* (1930) deals expressly with eugenics and the power of "good-making traits" to shape society. Like Galton, he believed that those in the higher social strata should be provided with financial subsidies to counteract the "resultant sterility" caused when upper class individuals opt to have fewer children for their own social advantage.

British embryologist William Bateson, who coined the terms "genes" and "genetics," championed the Mendelian genetics that finally unseated the popularity of Galton's ideas in England. In a debate that lasted thirty years, those that believed in Austrian monk Gregor Mendel's particulate inheritance argued against the selection touted by the biometricians, and vice versa. Bateson, who had a deep distrust of eugenics, successfully replicated Mendel's experiments. Not recognizing that the two arguments were not mutually exclusive, Pearson and Weldon rejected genetics, thus setting up the standoff between the two camps.

Fisher, on the other hand, tried to model the trajectory of genes in a population as if they were gas molecules governed by the laws of thermodynamics, with the aim of converting natural selection into a universal law. He used such "genetic atomism" to propose that continuous variation, natural selection, and Mendelian genetics could all coexist. Fisher also mathematically derived Galton's bell-shaped curves based on Mendelian principles. Unfortunately, by emulating physics, Fisher underestimated the degree to which environment dictates which traits are adaptive.

EARLY EUGENICS IN THE UNITED STATES

While Mendelians and statisticians were debating in Britain, in the United States, Harvard embryologist Charles Davenport and others embarked on a mission of meshing early genetics with the eugenics movement. In his effort, Davenport created the Laboratory for Experimental Evolution at Cold Springs Harbor, New York. The laboratory was closely linked to his Eugenics Record Office (ERO), which he established in 1910. Davenport raised much of the money for these facilities by appealing to wealthy American families who feared unrestricted immigration and race degeneration. Though their wealth depended on the availability of cheap labor guaranteed by immigration, these American aristocrats feared the cultural impact of a flood of "inferior immigrants."

Unlike the British, U.S. eugenicists thought of selection as a purifying force and thus focused on how to stop the defective from reproducing. Davenport wrongly felt that Mendelian genetics supported eugenics by reinforcing the effects of inheritance over the environment. He launched a hunt to identify human defects and link specific genes (as yet poorly understood entities)

to specific traits. His primary tool was the family pedigree chart. Unfortunately, these charts were usually based on highly subjective data, such as questionnaires given to schoolchildren to determine the comparative social traits of various races.

The Eugenics Research Association was founded in 1913 to report the latest findings. In 1918, the Galton Society began meeting regularly at the American Museum of Natural History in New York, and in 1923 the American Eugenics Society was formed. These efforts paid off. By the late 1920's and early 1930's, eugenics was a topic in high school biology texts and college courses across the United States.

Among eugenics supporters was psychologist Lewis M. Terman, developer of the Stanford-Binet intelligence quotient (IQ) test, and Harvard psychologist Robert M. Yerkes, developer of the Army IQ test, who both believed that IQ test performance (and hence intelligence) was hereditary. The administration of such tests to immigrants by eugenicist Henry Goddard represented a supposedly "objective and quantitative tool" for screening immigrants for entry into the United States. Biologist Garland Allen reports that Goddard, in fact, determined that

In the first half of the twentieth century, thousands of people in the United States, many of them teenagers thought to be weak or abnormal, were sterilized to prevent their genes from passing on to the next generation. Here Sarah Jane Wiley revisits the Virginia Colony for the Epileptic and Feebleminded in Lynchburg, Virginia, where she and her brother were both sterilized in 1959. (AP/Wide World Photos)

more than 80 percent of the Jewish, Hungarian, Polish, Italian, and Russian immigrants were mentally defective.

Fear that immigrants would take jobs away from hardworking Americans, supported by testimony from ERO's superintendent, Harry Laughlin, and the findings of Goddard's IQ tests, resulted in the Johnson Act of 1924, which severely restricted immigration. In the end, legal sterilization and immigration restrictions became more widespread in the United States than in any country other than Nazi Germany. By 1940, more than thirty states in the United States had enacted compulsory sterilization laws. Most were not repealed until after the 1960's.

EUGENICS AND THE PROGRESSIVE ERA

During the Progressive Era, the eugenics movement became a common ground for such diverse groups as biologists, sociologists, psychologists, militarists, pacifists, socialists, communists, liberals, and conservatives. The progressive ideology, exemplified by Theodore Roosevelt's Progressive Party, sought the scientific management of all parts of society. Eugenics attracted the same crowd as preventive medicine, since both were seen as methods of harnessing science to reduce suffering and misfortune. For example, cereal entrepreneur John Harvey Kellogg founded the Race Betterment Foundation, mixing eugenics with hygiene, diet, and exercise. During this period, intellectuals of all stripes were attracted by the promise of "the improvement of the human race by better breeding." The genetics research of this time focused on improving agriculture, and eugenics was seen as the logical counterpart to plant and animal husbandry.

Davenport did not hesitate to play on their sympathies by making wild claims about the inheritance of "nomadism," "shiftlessness," "love of the sea," and other "traits" as if they were single Mendelian characteristics. Alcoholism, pauperism, prostitution, rebelliousness, criminality, feeblemindedness, chess expertise, and industrial sabotage were all claimed to be determined by one or two pairs of Mendelian genes. In particular, the progressives were lured by the idea of sterilizing the "weak minded," especially after the publication of articles about families in Appalachia and New Jersey that supposedly documented genetic lines cursed by a preponderance of habitual criminal behavior and mental weakness.

Having the allure of a "social vaccination," the enthusiasm to sterilize the "defective" spread rapidly among intellectuals, without regard to political or ideological lines. Sweden's Social Democrats forcibly sterilized some sixty thousand Swedes under a program that lasted from 1935 to 1976 organized by the state-financed Institute for Racial Biology. Grounds for sterilization included not only "feeblemindedness" but also "gypsy features," criminality, and "poor racial quality." The low class or mentally slow were institutionalized in the Institutes for Misled and Morally Neglected Children and released only if they would agree to be sterilized. Involuntary sterilization policies were also adopted in countries ranging from Switzerland and Austria to Belgium and Canada, not to be repealed until the 1970's.

Hermann Müller, a eugenicist who emigrated to the Soviet Union (and later returned to the United States), attacked Davenport's style of eugenics at the International Eugenics Congress in 1932. Müller, a geneticist who won the 1946 Nobel Prize in Physiology or Medicine for his discovery of the mutagenic power of X rays, instead favored the style of eugenics envisioned by English novelist Aldous Huxley's *Brave New World* (1932), with state nurseries, artificial insemination, and the use of other scientific techniques to produce a genetically engineered socialist society.

According to journalist Jonathan Freedland, the British left, including a large number of socialist intellectuals such as playwright George Bernard Shaw and philosopher Bertrand Russell, was convinced that it knew what was best for society. Concerned with the preservation of their higher intellectual capacities, they joined the fashionable and elitist Eugenics Society in the 1930's, where they advocated the control of reproduction, particularly favoring the idea of impregnating working-class women with sperm of men with high IQs.

THE AMERICAN MOVEMENT SPREADS TO
NAZI GERMANY

The eugenics movement eventually led to grave consequences in Nazi Germany. Negative eugenics reached its peak there, with forced sterilization, euthanasia or "mercy killing," experimentation, and ultimately genocide being used in the name of "racial hygiene." Eugenicists in the United States and Germany formed close and direct alliances, especially after the Nazis came to power in 1933. The

ERO's Laughlin gave permission for his article "Eugenical Sterilization" to be reprinted in German in 1928. It soon became the basis of Nazi sterilization policy. Davenport even arranged for a group of German eugenicists to participate in the three hundredth anniversary of Harvard's founding in 1936.

Inspired by the U.S. eugenics movement and spurred by economic hardship that followed World War I, the Nazi Physician's League took a stand that those suffering from incurable disease caused useless waste of medications and, along with the crippled, the feebleminded, the elderly, and the chronic poor, posed an economic drain on society. Hereditary defects were considered to be the cause of such maladies, and these people were dubbed "lives not worth living." In 1933, the German Law for the Prevention of Genetically Diseased Offspring made involuntary sterilization of such people, including the blind, deaf, epileptic, and poor, legal. The Nazis set up "eugenics courts" to decide cases of involuntary sterilization. Frederick Osborn, secretary of the American Eugenics Society, wrote a 1937 report summarizing the German sterilization programs, indicative of the fascination American eugenicists had for the Nazi agenda and the Nazi's ability to move this experiment to a scale never possible in the United States.

THE DEMISE OF EUGENICS

With the Great Depression in 1929, the U.S. eugenics movement lost much of its momentum. Geneticist and evolutionary biologist Sewall Wright, although himself a member of the American Eugenics Society, found fault with the genetics and the ideology of the movement: "Positive eugenics seems to require . . . the setting up of an ideal of society to aim at, and this is just what people do not agree on." He also wrote several articles in the 1930's challenging the assumptions of Fisher's genetic atomism model. In a speech to the Eugenics Society in New York in 1932, Müller pointed out the economic disincentive for middle and upper classes to reproduce, epitomized by the failure of many eugenicists to have children. Galton himself died childless. This inverse relationship between fertility and social status, coupled with the apparent predatory nature of the upper class, seemed to doom eugenics to failure.

Evolutionary biologist Stephen Jay Gould claimed that the demise of the eugenics movement in the United States was more a matter of Adolf Hitler's use of eugenic arguments for sterilization and racial purification than it was of advances in genetic knowledge. Once the Holocaust and other Nazi atrocities became known, eugenicists distanced themselves from the movement. Depew and Weber have written that Catholic conservatives opposed to human intervention in reproduction and progressives, who began to abandon eugenics in favor of behaviorism (nurture rather than nature), were political forces that began to close down the eugenics movement, while Allen points out that the movement had outlived its political usefulness. Russian geneticist Theodosius Dobzhansky had by this time recognized the prime importance of context in genetics and consequently rejected the premise of eugenics, helping to push it into the realm of phony genetics.

IMPLICATIONS

The term "euphenics" is used to describe human genetic research that is aimed at improving the human condition, replacing the tainted term eugenics. Euphenics deals primarily with medical or genetic intervention that is designed to reduce the impact of defective genotypes on individuals (such as gene therapy for those with cystic fibrosis). However, in this age of increasing information about human genetics, it is necessary to keep in mind the important role played by environment and the malleability of human traits.

Allen argues that the eugenics movement may reappear (although probably under a different name) if economic problems again make it attractive to eliminate "unproductive" people. His hope is that a better understanding of genetics, combined with the lessons of Nazi Germany, will deter humans from ever again going down that path that journalist Jonathan Freedland calls "the foulest idea of the 20th century."

Lee Anne Martínez, Ph.D.

FURTHER READING

Allen, Garland E. "Science Misapplied: The Eugenics Age Revisited." *Technology Review* 99, no. 6 (August/September, 1996): 22. Discusses the connection between the eugenics movement and periods of economic or social hardship.

Depew, David, and Bruce Weber. *Darwinism Evolving: Systems Dynamics and the Genealogy of Natural Selection.* Boston: MIT Press, 1995. Discusses the relationship between eugenics and Darwinian evolution and the role played by statistics in the origin of this movement.

Duster, Troy. *Backdoor to Eugenics.* 2d ed. New York: Routledge, 2003. Updated edition of a book originally published in 1990. Considers the social and political implications—and the hazards—of contemporary genetic technologies.

Gillham, Nicholas Wright. *A Life of Sir Francis Galton: From African Exploration to the Birth of Eugenics.* New York: Oxford University Press, 2001. A biography of the founder of the eugenics movement.

Kevles, Daniel J. *In the Name of Eugenics: Genetics and the Uses of Human Heredity.* Cambridge, Mass.: Harvard University Press, 1995. Traces the history of eugenics, mainly in the United States and Britain, from the nineteenth century to the late twentieth century. Individuals such as Karl Pearson, C. B. Davenport, R. A. Fisher, and J. B. S. Haldane, who have been associated with eugenics in various ways, are discussed.

Kühl, Stefan. *The Nazi Connection: Eugenics, American Racism, and German National Socialism.* New York: Oxford University Press, 1994. Exposes the ties between the American eugenics movement and the Nazi program of "racial hygiene."

Lynn, Richard. *Eugenics: A Reassessment.* Westport, Conn.: Praeger, 2001. Provides a historical overview of the eugenics movement. Argues that denunciation of the movement was too extreme and that eugenics needs to be reconsidered.

Mazumdar, Pauline Margaret. *Eugenics, Human Genetics, and Human Failings.* London: Routledge, 1991. A thorough historical approach that examines the eugenics movement from its origin to its heyday as the source of a science of human genetics.

Pernick, Martin S. "Eugenics and Public Health in American History." *American Journal of Public Health* 87, no. 11 (November, 1997): 1767-1772. A fascinating exploration of the overlap between the goals, values, and concepts of public health and the eugenics movements in the early twentieth century.

Stern, Alexandra Minna. *Eugenic Nation: Faults and Frontiers of Better Breeding in Modern America.* Berkeley: University of California Press, 2005. A history of the eugenics movement in California in the twentieth century, describing the influence of the "better breeding" concept in sterilization policies, school segregation, the environmental movement, and other phenomena.

Witkowski, Jan A., and John R. Inglis, eds. *Davenport's Dream: Twenty-first Century Reflections on Heredity and Eugenics.* Cold Spring Harbor, N.Y.: Cold Spring Harbor Laboratory Press, 2008. Collection of essays maintaining that many of the issues raised in Davenport's book *Heredity in Relation to Eugenics* are being raised almost one hundred years after the book's publication. The essays discusss human genetic variation, mental illness, nature versus nurture, and human evolution within the context of the twenty-first century.

WEB SITES OF INTEREST

Cold Spring Harbor Laboratory, Image Archive on the American Eugenics Movement
http://www.eugenicsarchive.org
Comprehensive and extensively illustrated site that covers the eugenics movement in the United States, including its scientific history and origins, research methods and flaws, and sterilization laws.

Future Generations
http://www.eugenics.net
Future Generations describes itself as a site "about humanitarian eugenics," a movement which "strives to leave a genuine legacy of love to future generations," and adds that most of its ideas are "politically incorrect." The site provides articles on the case for eugenics, the "mismeasures" of Stephen Jay Gould, reviews of books, and other information in support of current eugenic theory.

National Reference Center for Bioethics Literature
http://bioethics.georgetown.edu/nrc
Users can search the center's database and other online resources to retrieve bibliographies listing books, articles, and other sources of information about eugenics.

Race and Membership: The Eugenics Movement
http://www.facinghistorycampus.org/campus/rm.nsf
Facing History and Ourselves, an organization offering support to teachers and students in the areas of history and social studies, created this site that traces the history of the eugenics movement in the United States and Germany.

See also: Artificial selection; Bioethics; Bioinformatics; Biological determinism; Cloning: Ethical issues; Eugenics: Nazi Germany; Evolutionary biology; Gene therapy: Ethical and economic issues; Genetic counseling; Genetic engineering: Social and ethical issues; Genetic screening; Genetic testing: Ethical and economic issues; Heredity and environment; Human genetics; Insurance; Intelligence; Miscegenation and antimiscegenation laws; Natural selection; Patents on life-forms; Paternity tests; Race; Sociobiology; Stem cells; Sterilization laws.

Eugenics
Nazi Germany

CATEGORY: History of genetics; Human genetics and social issues

SIGNIFICANCE: Fueled by economic hardship and racial prejudice, the largest-scale application of eugenics occurred in Nazi Germany, where numerous atrocities, including genocide, were committed in the name of genetic improvement of the human species. The German example raised worldwide awareness of the dangers of eugenics and did much to discredit eugenic theory.

KEY TERMS

Aryan: a "race" believed by Nazis to have established the civilizations of Europe and India

euthanasia: the killing of suffering people, sometimes referred to as "mercy" killing

Nordic: the northernmost of the Aryan groups of Europe, believed by the Nazis to be the highest and purest racial group

ORIGINS OF NAZI EUGENIC THOUGHT

Nazi eugenic theory and practice grew out of two traditions: the eugenics movement, founded by British scientist Francis Galton, and racial theories of human nature. Most historians trace the origin of modern racial theories to French diplomat and writer Joseph-Arthur de Gobineau, who maintained that all great civilizations had been products of the Aryan, or Indo-Germanic, race. Through the late nineteenth and early twentieth centuries, German thinkers applied Galton's ideas to the problem of German national progress. The progress of the nation, argued scientists and social thinkers, could be best promoted by improving the German people through government-directed control of human reproduction. This type of eugenic thinking became known as "racial hygiene"; in 1904, eugenicists and biologists formed the Racial Hygiene Society in Berlin.

The Aryan mythology of Gobineau also grew in popularity. In 1899, an English admirer of Germany, Houston Stewart Chamberlain, published a widely read book entitled *The Foundations of the Nineteenth Century.* Chamberlain, heavily influenced by Gobineau, maintained that Europe's accomplishments had been the work of ethnic Germans, members of a healthy and imaginative race. Opposed to the Germans were the Jews, who were, according to Chamberlain, impure products of crossbreeding among the peoples of the Middle East.

BASICS OF NAZI EUGENICS

The Law for the Prevention of Genetically Disease Offspring, requiring sterilization of people with hereditary diseases and disabilities, was drafted and decreed in Germany in 1933. Before the Nazis came to power, many segments of German society had supported sterilization as a way to improve future generations, and Adolf Hitler's emergence as a national leader provided the pressure to ensure the passage of the law. Between 1934 and 1945, an estimated 360,000 people (about 1 percent of the German population) who were believed to have hereditary ailments were sterilized. Despite this law, the Nazis did not see eugenics primarily as a matter of discouraging the reproduction of unhealthy individuals and encouraging the reproduction of healthy individuals. Following the theories of Chamberlain, Adolf Hitler and his followers saw race, not individual health or abilities, as the distinguishing characteristic of human beings.

The Schutzstaffel (SS) organization was a key part of Nazi eugenic activities. In January, 1929, Heinrich Himmler was put in charge of the SS, a police force aimed at establishing order among the street fighters who formed a large part of the early Nazi Party. In addition to disciplining rowdy Nazis, the SS quickly emerged as a racial elite, the spearhead of an intended German eugenic movement. Himmler recruited physicians and biologists to help ensure that only those of the purest Nordic heritage could serve in his organization. In 1931, the agricul-

On Wehrmacht Day, 1935, in Nazi Germany (from left): German chancellor Adolf Hitler, head of the air force Hermann Göring, army commander Wernher von Fritsch, minister of war Werner Eduard Fritz von Blomberg, commander of the navy Erich Raeder, and other Nazi officials. During the late 1930's and early 1940's, the Nazi government conducted the extreme and brutal form of eugenics that culminated in the Holocaust and the murder of millions of innocent Jews and other "undesirables." (Library of Congress)

turalist R. Walther Darré helped Himmler draw up a marriage code for SS men, and Himmler appointed Darre head of an SS Racial Office. Himmler hoped to create the seeds of a German super race by directing the marriages and reproduction of the "racially pure" members of the SS.

Since the Nazis saw Germans as a "master race," a race of inherently superior people, they attempted to improve the human stock by encouraging the birth of as many Germans as possible and by encouraging those seen as racially pure to reproduce. The Nazis declared that women should devote themselves to bearing and caring for children. Hitler's mother's birthday was declared the Day of the German Mother. On this day, public ceremonies awarded medals to women with large numbers of children. The SS set up and maintained an organization of maternity homes for unmarried mothers of acceptable racial background and orphanages for their children; these institutions were known as the

Lebensborn ("fountain of life"). There is some evidence that young women with desired racial characteristics who were not pregnant were brought to the Lebensborn to have children by the SS men to create "superior" Nordic children.

IMPACT

In addition to encouraging the reproduction of those seen as racially pure, the Nazis sought to eliminate the unhealthy and the racially undesirable. In August, 1939, a committee of physicians and government officials, operating under Hitler's authority, issued a secret decree under which all doctors and midwives would have to register births of malformed or handicapped children. By October of that year, Hitler had issued orders for the "mercy killing" of these children and all those with incurable diseases. This euthanasia movement expanded from sick and handicapped children to those believed to belong to "sick" races. The T4 euthanasia organization, designed for efficient and secret killing, experimented with lethal injections and killing by injection and became a pilot program for the mass murder of the Jews during the Holocaust.

German racial hygienists had long advocated controlling marriages of non-Jewish Germans with Jews in order to avoid "contaminating" the German race. In July, 1941, Nazi leader Hermann Göring appointed SS officer Reinhard Heydrich to carry out the "final solution" of the perceived Jewish problem. At the Wannsee Conference in January, 1942, Hitler and his close associates agreed on a program of extermination. According to conservative estimates, between four million and five million European Jews died in Nazi extermination camps. When the murderous activities of the Nazis were revealed to the world after the war, eugenics theory and practice fell into disrepute.

Carl L. Bankston III, Ph.D.

FURTHER READING

Gasman, Daniel. *The Scientific Origins of National Socialism.* New York: American Elsevier, 1971. Reprint. New Brunswick, N.J.: Transaction, 2004. Describes how the ideas of Ernst Haeckel, a German scientist and supporter of Charles Darwin's theory of evolution, and other German scientists led to the Nazis' eugenic policies.

Goldhagen, Daniel J. *Hitler's Willing Executioners: Ordinary Germans and the Holocaust.* New York: Random House, 1996. Argues that the German people participated in the mass murder of Jews because Germans had come to see Jews as a racial disease.

Henry, Clarissa, and Marc Hillel. *Of Pure Blood: An Investigation into the Creation of a Super Race.* Video, produced by Maryse Addison and Peter Bate. Maljack Productions, 1976. Oak Forest, Ill.: MPI Home Video, 1985. Investigates the Lebensborn organization, a Nazi plan to breed and distill the German children into a pure Aryan race.

Kühl, Stefan. *The Nazi Connection: Eugenics, American Racism, and German National Socialism.* New York: Oxford University Press, 2002. Exposes the ties between the American eugenics movement and the Nazi program of racial hygiene.

Kuntz, Dieter. *Deadly Medicine: Creating the Master Race.* Washington, D.C.: United States Holocaust Memorial Museum, 2004. A companion to an exhibit at the U.S. Holocaust Memorial Museum.

Contains essays tracing the progression of the eugenics movement in Nazi Germany from its initial reformist impulse to its eventual justification of genocide.

Laffin, John. *Hitler Warned Us: The Nazis' Master Plan for a Master Race.* Totowa, N.J.: Barnes and Noble Books, 1998. Using photographs and propagandist ephemera, Laffin, a military historian, questions why Hitler was allowed by other leaders and nations to engage in his destructive drive for power and domination.

Weikart, Richard. *From Darwin to Hitler: Evolutionary Ethics, Eugenics, and Racism in Germany.* New York: Palgrave Macmillan, 2004. Describes how Charles Darwin's theory of evolution was adapted by German biologists, social thinkers, and Adolf Hitler to justify the Nazis' eugenic policies.

Weindling, Paul. *Health, Race, and German Politics Between National Unification and Nazism, 1870-1945.* 1989. Reprint. Cambridge, Mass.: Cambridge University Press, 1993. Offers a definitive history of the origins, social composition, and impact of eugenics in the context of the social and political tension of the rapidly industrializing Nazi empire.

WEB SITES OF INTEREST

Deadly Medicine: Creating the Master Race
http://www.ushmm.org/museum/exhibit/online/deadlymedicine
An online, multimedia version of an exhibition at the U.S. Holocaust Memorial Museum about the Nazis' "racial hygiene" policies. Describes the scientific ideas prevalent in Germany from 1919-1945, contains profiles of German physicians and scientists, provides video testimony from individuals who describe various aspects of the Nazis' eugenic policies, and offers a bibliography and Web links for further information.

National Reference Center for Bioethics Literature
http://bioethics.georgetown.edu/nrc
Users can search the center's database and other online resources for bibliographies of sources about eugenics and about Nazi Germany.

Race and Membership: The Eugenics Movement
http://www.facinghistorycampus.org/campus/rm.nsf
Facing History and Ourselves, an organization offering support to teachers and students in the areas

of history and social studies, created this site that traces the history of the eugenics movement in the United States and Germany.

See also: Bioethics; Bioinformatics; Biological determinism; Eugenics; Evolutionary biology; Gene therapy: Ethical and economic issues; Genetic counseling; Genetic engineering: Social and ethical issues; Genetic screening; Genetic testing: Ethical and economic issues; Heredity and environment; Human genetics; Insurance; Intelligence; Miscegenation and antimiscegenation laws; Patents on life-forms; Paternity tests; Race; Sociobiology; Stem cells; Sterilization laws.

Evolutionary biology

CATEGORY: Evolutionary biology; Population genetics

SIGNIFICANCE: While the existence of evolutionary change is firmly established, many questions remain about how evolution proceeded in the past and how it operates in the present in particular groups of organisms. The science of evolutionary biology focuses on reconstructing the actual history of life and on understanding how evolutionary mechanisms operate in nature.

KEY TERMS

adaptation: a genetically based characteristic that confers on an organism the ability to survive and reproduce under prevailing environmental conditions

evolution: the process of change in the genetic structure of a population over time; descent with modification

fitness: the relative reproductive contribution of one individual to the next generation as compared to that of others in the population

genetic drift: chance fluctuations in allele frequencies within a population, resulting from random processes in gamete formation and sampling, and variation in the number and genotypes of offspring produced by different individuals

genotype: the genetic constitution of an individual or group

natural selection: the phenomenon of differing sur-

vival and reproduction rates among various genotypes in response to external factors; the frequencies of alleles carried by favored genotypes, and the phenotypes conferred by those alleles, increase in succeeding generations

phylogeny: the history of descent of a group of species from a common ancestor

AN EVOLUTIONARY CONTEXT

Life is self-perpetuating, with each generation connected to previous ones by the thread of DNA passed from ancestors to descendants. Life on Earth thus has a single history much like the genealogy of an extended family, the shape and characteristics of which have been determined by internal and external forces. The effort to uncover that history and describe the forces that shape it constitutes the field of evolutionary biology.

As an example of the need for this perspective, consider three vertebrates of different species, two aquatic (a whale and a fish) and one terrestrial (a deer). The two aquatic species share a torpedolike shape and oarlike appendages. These two species differ, however, in that one lays eggs and obtains oxygen from the water using gills, while the other produces live young and must breathe air at the surface. The terrestrial species has a different, less streamlined, shape and appendages for walking, but it too breathes air using lungs and produces live young. All three species are the same in having a bony skeleton typical of vertebrates. In order to understand why the various organisms display the features they do, it is necessary to consider what forces or historical constraints influence their genotypes and subsequent phenotypes.

It is logical to hypothesize that a streamlined shape is beneficial to swimming creatures, as is the structure of their appendages. This statement is itself an evolutionary hypothesis; it implies that streamlined individuals will be more successful than less streamlined ones and so will become prevalent in an aquatic environment. It may initially be difficult to reconcile the differences between the two aquatic forms swimming side-by-side with the similarities between one of them and the terrestrial species walking around on dry land. However, if it is understood that the whale is more closely related to the terrestrial deer than it is to the fish, much of the confusion disappears. Using this comparative approach, it is unnecessary, and scientifically unjustified, to construct an elaborate scenario whereby breathing air at the surface is more advantageous to a whale than gills would be; the simpler explanation is that the whale breathes air because it (like the deer) is a mammal, and both species inherited this trait from a common ancestor sometime in the past.

Organisms are thus a mixture of two kinds of traits. Ecological traits are those the particular form of which reflects long-term adaptation to the species' habitat. Two species living in the same habitat might then be expected to be similar in such features and different from species in other habitats. Evolutionary characteristics, on the other hand, indicate common ancestry rather than common ecology. Here, similarity between two species indicates that they are related to each other through common ancestry, just as familial similarity can be used to identify siblings in a crowd of people. In reality, all traits are somewhere along a continuum between these two extremes, but this distinction highlights the importance of understanding the evolutionary history of organisms and traits. The value of an evolutionary perspective comes from its comparative and historical basis, which allows biologists to place their snapshot-in-time observations within the broader context of the continuous history of life.

Charles Darwin is credited as the father of the theory of natural selection, on which modern evolutionary biology is based. (Library of Congress)

EARLY EVOLUTIONARY THOUGHT

Underlying evolutionary theory is Mendelian genetics, which provides a mechanism whereby genes conferring advantageous traits can be passed on to offspring. Both Mendelian genetics and the theory of evolution are, at first glance (and in retrospect), remarkably simple. The theory of evolution, however, is paradoxical in that it leads to extremely complex predictions and thus is often misunderstood, misinterpreted, and misapplied.

It is important to distinguish between the phenomenon of evolution and the various processes or mechanisms that may lead to evolution. The idea that species might be mutable, or subject to change over generations, dates back to at least the mid-eighteenth century, when the French naturalist Georges-Louis Leclerc, comte de Buffon, the Swiss naturalist Charles Bonnet, and even the Swedish botanist Carolus Linnaeus suggested that species (or at least "varieties") might be modified over time by intrinsic biological or extrinsic environmental factors. Other biologists after that time also promoted the idea that populations and species could

evolve. Nevertheless, with the publication of *On the Origin of Species by Means of Natural Selection* in 1859, Charles Darwin became the most prominent of those who proposed that all species had descended from a common ancestor and that there was a single "tree of life." These claims regarding the history of evolution, however, are distinct from the problem of how, or through what mechanisms, evolution occurs.

In the first decade of the nineteenth century, Jean-Baptiste Lamarck promoted a hypothesis of inheritance of acquired characteristics to explain how species could adapt over time to their environments. His famous giraffe example illustrates the Lamarckian view: Individual giraffes acquire longer necks as a result of reaching for leaves high on trees, then pass that modified characteristic to their offspring. According to Lamarck's theories, as a result of such adaptation, the species—and, in fact, each individual member of the species—is modified over time. While completely in line with early nineteenth century views of inheritance, this view of the mechanism of evolution has since been shown to be incorrect.

DARWINIAN EVOLUTION: NATURAL SELECTION

In the mid-nineteenth century, Darwin and Alfred Russel Wallace independently developed the theory of evolution via natural selection, a theory that is consistent with the principles of inheritance as described by Gregor Mendel. Both Darwin's and Wallace's arguments center on four observations of nature and a logical conclusion derived from those observations (presented here in standard genetics terminology, although Darwin and Wallace used different terms).

First, variation exists in the phenotypes of different individuals in a population. Second, some portion of that variation is heritable, or capable of being passed from parents to offspring. Third, more individuals are produced in a population than will survive and reproduce. Fourth, some individuals are, because of their particular phenotypes, better able to survive and reproduce than others. From

Alfred Russel Wallace is now considered the coauthor of modern evolutionary theory along with Darwin. (National Library of Medicine)

this, Darwin and Wallace deduced that because certain individuals have inherited variations that confer on them a greater ability to survive and reproduce than others, these better-adapted individuals are more likely to transmit their genetically inherited traits to the next generation. Therefore, the frequency of individuals with the favored inherited traits would increase in the next generation, though each individual's genetic constitution would remain unchanged throughout its lifetime. This process would continue as long as new genetic variants continued to arise and selection favored some over others. The theory of natural selection provided a workable and independently testable natural mechanism by which evolution of complex and sometimes very different adaptations could occur within and among species.

EVOLUTIONARY BIOLOGY AFTER DARWIN

Despite their theoretical insight, Darwin and Wallace had an incomplete and partially incorrect understanding of the genetic basis of inheritance. Mendel published his work describing the fundamental principles of inheritance in 1866 (he had reported the results before the Brünn Natural History Society earlier, in February and March of 1865), but Darwin and Wallace were unaware throughout their lives that the correct mechanism of inheritance had been discovered. In fact, Mendel's work went almost entirely unnoticed by the scientific community for thirty-four years; it was rediscovered, and its significance appreciated, in the first decade of the twentieth century. Over the next three decades of the twentieth century, theoreticians integrated Darwin's theory of natural selection with the principles of genetics discovered by Mendel and others. Simultaneously, Ernst Mayr, G. Ledyard Stebbins, George Gaylord Simpson, and Julian Huxley demonstrated that the evolution of species and the patterns in the fossil record were consistent with each other and could be readily explained by Darwinian principles. This effort culminated in the 1930's and 1940's in the "modern synthesis," a fusion of thought that resulted in the development of the field of population genetics, a discipline in which biologists seek to describe and predict, quantitatively, evolutionary changes in populations of sexually reproducing organisms.

Since the modern synthesis (also called the neo-Darwinian synthesis), biologists have concentrated their efforts on applying the theories of population genetics to understand the evolutionary dynamics of particular groups of organisms. More recently, techniques of phylogenetic systematics have been developed to provide a means of reconstructing phylogenetic relationships among species. This effort has emphasized the need for a comparative and evolutionary approach to general biology, which is essential to correct interpretation of biological classification.

In the 1960's, Motoo Kimura proposed the neutral theory of evolution, which challenged the "selectionist" view that patterns of genetic and phenotypic variation in most traits are determined by natural selection. The "neutralist" view maintains that much genetic variation, especially that seen in the numerous alleles of enzyme-coding genes, has little effect on fitness and therefore must be controlled by mechanisms other than selection. Advances in molecular biology, particularly those from genomics projects, have allowed testing of the selectionist and neutralist views and have provided evidence that natural selection has a powerful effect on certain variations in DNA, whereas other variations in DNA are subject to neutral evolution. An ongoing effort for a unified model of evolution is integration of evolutionary theory with the understanding of the processes of development (dubbed "evo-devo"), a field that also has benefited greatly from genome projects.

Evolutionary Mechanisms

Natural selection as described by Darwin and Wallace leads to the evolution of adaptations. However, many traits (perhaps the majority) are not adaptations; that is, differences in the particular form of those traits from one member of the species to the next do not lead to differences in fitness among those individuals. Such traits are mostly uninfluenced by natural selection, yet they can and do evolve. Thus, there must be mechanisms beyond natural selection that lead to changes in the genetic structure of biological systems over time.

Evolutionary mechanisms are usually envisioned as acting on individual organisms within a population. For example, natural selection may eliminate some individuals while others survive and produce a large number of offspring genetically similar to themselves. As a result, evolution occurs within those populations. A key tenet of Darwinian evolution (which distinguishes it from Lamarckian evolution) is that populations evolve, but the individual organisms that constitute that population do not, in the sense that their genetic constitution remains essentially constant even though their environments may change. Although evolution of populations is certainly the most familiar scenario, this is not the only level at which evolution occurs.

Richard Dawkins energized the scientific discussion of evolution with his book *The Selfish Gene*, first published in 1976. Dawkins argued that natural selection could operate on any type of "replicator," or unit of biological organization that displayed a faithful but imperfect mechanism of copying itself and that had differing rates of survival and reproduction among the variant copies. Under this definition, it is possible to view individual genes or strands of DNA as focal points for evolutionary mechanisms such as selection. Dawkins used this framework to consider how the existence of DNA selected to maximize its chances of replication (or "selfish DNA") would influence the evolution of social behavior, communication, and even multicellularity.

Recognizing that biological systems are arranged in a hierarchical fashion from genes to genomes (or cells) to individuals through populations, species, and communities, Elisabeth Vrba and Niles Eldredge in 1984 proposed that evolutionary changes could occur in any collection of entities (such as populations) as a result of mechanisms acting on the entities (individuals) that make up that collection. Because each level in the biological hierarchy (at least above that of genes) has as its building blocks the elements of the preceding one, evolution may occur within any of them. Vrba and Eldredge further argued that evolution could be viewed as resulting from two general kinds of mechanisms: those that introduce genetic variation and those that sort whatever variation is available. At each level, there are processes that introduce and sort variation, though they may have different names depending on the level being discussed.

Natural selection is a sorting process. Other mechanisms that sort genetic variation include sexual selection, whereby certain variants are favored based on their ability to enhance reproductive success (though not necessarily survival), and genetic drift, which is especially important in small populations. Although these forces are potentially strong engines for driving changes in genetic structure, their action—and therefore the direction and magnitude of evolutionary changes that they can cause—is constrained by the types of variation available and the extent to which that variation is genetically controlled.

Processes such as mutation, recombination, development, migration, and hybridization introduce variation at one or more levels in the biological hierarchy. Of these, mutation is ultimately the most important, as changes in DNA sequences constitute the raw material for evolution at all levels. Without mutation, there would be no variation and thus no evolution. Nevertheless, mutation alone is a relatively weak evolutionary force, only really significant in driving evolutionary changes when coupled with processes of selection or genetic drift that can quickly change allele frequencies. Recombination, development, migration, and hybridization introduce new patterns of genetic variation (initially derived from the mutation of individual genes) at the genome, multicellular-organism, population, and species levels, respectively.

THE REALITY OF EVOLUTION

It is impossible to absolutely prove that descent with modification from a common ancestor is responsible for the diversity of life on earth. In fact, this dilemma of absolute proof exists for all scientific theories; as a result, science proceeds by constructing and testing potential explanations, gradually accepting those best supported by the accumulation of observation and evidence, and their logical interpretations, until theories are either clearly refuted or replaced by modified theories more consistent with the data.

Darwin's concept of a single tree of life is supported by vast amounts of scientific evidence. In fact, the theory of evolution is among the most thoroughly tested and best-supported theories in all of science. The view that evolution has and continues to occur is not debated by biologists; there is simply too much evidence to support its existence across every biological discipline.

On a small scale, it is possible to demonstrate evolutionary changes experimentally or through direct observation. Spontaneous mutations that introduce genetic variation are well documented; the origination and spread of drug-resistant forms of viruses, bacteria, and other pathogens is clear evidence of this potential. Agricultural breeding

programs and other types of artificial selection demonstrate that the genetic structure of lineages containing heritable variation can be changed over time through agents of selection. For example, work by John Doebley begun in the late 1980's suggested that the evolution of corn from a wild ancestor resembling modern teosinte may have involved changes in as few as five major genes and that this transition likely occurred as a result of domestication processes established in Mexico between seven thousand and ten thousand years ago. The effects of natural selection can likewise be observed in operation: Peter Grant and his colleagues discovered that during drought periods, when seed is limited, deep-billed individuals of the Galápagos Island finch *Geospiza fortis* increase in proportion to the general population of the species, as only the deep-billed birds can crack the large seeds remaining after the supply of smaller seeds is exhausted. These and similar examples demonstrate that the evolutionary mechanisms put forward by Darwin and others do occur and lead to microevolution, or evolutionary change within single species.

Attempts to account for larger-scale macroevolutionary patterns, such as speciation and the origin of major groups of organisms, rely to some extent on direct observation but for the most part are based on indirect tests using morphological and genetic comparisons among different species, observed geographic distributions of species, and the fossil record. Such comparative studies rely on the concept of homology, the presence of corresponding and similarly constructed features among species, as well as similar DNA sequences and chromosomal rearrangements, which are a consequence of inheritance from common ancestry.

At the most basic level, organization of the genetic code is remarkably similar across species; only minor variations exist among organisms as diverse as archaea (bacteria found in extreme environments such as hot springs, salt lakes, and habitats lacking in oxygen), bacteria, and eukaryotes (organisms whose cells contain a true nucleus, including plants, animals, fungi, and their unicellular counterparts). This genetic homology extends as well to the presence of shared and similarly functioning gene sequences across biological taxa, such as homeotic genes, common within major groups of eukaryotes. The near-universal nature of the genetic code can be best explained if it arose once during the early evolution of the first forms of life and has been transmitted through inheritance and preserved through natural selection to the present in all organisms.

Morphological homologies are also widespread; the limbs of mammals, birds, amphibians, and reptiles, for example, are all built out of the same fundamental arrangement of bones. The particular shapes, and even number, of these bones can vary among groups, often as adaptations to the widely varying functions of these bones. For example, if the bones in the pectoral fins of dolphins are compared to the bones in the human arm and hand, the same arrangement of bones is immediately evident but the bones differ in their relative sizes in accordance with the different functions of these forelimbs.

Genetic, cytological, and molecular studies have greatly enhanced the understanding of evolution. In general, these studies support previously reconstructed evolutionary histories derived from anatomical comparisons, geographic distributions, and the fossil record, while refining many of the details and clarifying the molecular mechanisms of evolution. As methods for chemical staining and microscopic examination of chromosomes were developed, cytologists noticed that the chromosomes of related species are highly similar and, in many cases, can be aligned with one another. The aligned chromosomes of related species, however, frequently differ by noticeable rearrangements, such as inversions, translocations, fusions, and fissions. For instance, human and chimpanzee chromosomes differ by nine inversions and one chromosome fusion. Molecular evidence has revealed that the fusion and two of the inversions happened in the human ancestral lineage, whereas seven of the inversions happened in the chimpanzee ancestral lineage since the two lineages diverged from common ancestry. Comparative chromosomal analyses have allowed scientists to reconstruct the chromosomal constitutions of several now-extinct common ancestral species.

Genome projects have generated massive amounts of DNA sequence data that reveal in exquisite detail the molecular evolutionary history of genomes. As an example, three primate genomes (human, chimpanzee, and rhesus macaque) have been sequenced and annotated. They show that gene duplication followed by mutational divergence is a principal mechanism for the evolution of new genes. Pseudogenes (nonfunctional, mutated copies of genes) are as numerous as functional genes in these genomes, and millions of transposable elements constitute approximately 43 percent of their DNA. Nearly all genes, pseudogenes, and transposable elements are in the same chromosomal locations in all three of these genomes, indicating that they arose in a common ancestor. Those that differ are highly similar to functional genes or currently active transposable elements, evidence that they arose recently, since the divergences of these species' lineages from common ancestry.

The conclusion that emerges from this weight of independent evidence is that structural, chromosomal, and genomic homologies reflect an underlying evolutionary homology, or descent from common ancestry.

PUNCTUATED EQUILIBRIUM

Although the order of appearance of organisms in the fossil record is consistent with evolutionary theory in general, evolution does not always proceed in a gradual, predictable way. Paleontologists have long emphasized that gradualism—that is, evolution by gradual changes proceeding at more or less a constant rate, eventually producing major changes—is often not supported by the fossil record. The fossil record more often shows a pattern of relatively minor change over long periods of time, punctuated by much shorter periods of rapid change. Stephen Jay Gould and Niles Eldredge, both paleontologists, offered a hypothesis called punctuated equilibrium to explain this discrepancy.

Gould and Eldredge's hypothesis recognizes the fact that the fossil record shows long periods of relative stasis (little change) punctuated by periods of rapid change, and consider this the principal mode for evolution. Instead of the strict neo-Darwinian view of gradual changes leading to large changes over time, Gould and Eldredge suggest that large changes are the result of a series of larger steps over a much shorter period of time. When first proposed, the punctuated equilibrium theory was subject to considerable skepticism, but it has gained more acceptance over time.

THE PRACTICE OF EVOLUTIONARY BIOLOGY

Contemporary evolutionary biology builds upon the theoretical foundations of Darwinian evolution by natural selection, the modern synthesis of Darwinian evolution with Mendelian inheritance, augmentation of evolutionary theory with research on its mechanisms and processes, such as punctuated equilibrium and biological development, and integration of an enormous body of data from molecular studies and genome projects. Although the reality of evolution is no longer in doubt, considerable research is underway on the relative importance of various evolutionary mechanisms in the history of particular groups of organisms. Much effort continues to be directed at reconstructing the particular historical path that life on earth has taken and that has led to the enormous diversity of species in the past and present. Likewise, scientists seek a fuller understanding of how new species arise, as the process of speciation represents a watershed event separating microevolution and macroevolution.

Unlike many other fields of biology, evolutionary biology is not always amenable to tests of simple cause-and-effect hypotheses. Much of what evolutionary biologists are interested in understanding occurred in the past and over vast periods of time. In addition, the evolutionary outcomes observed in nature depend on such a large number of environmental, biological, and random factors that re-creating and studying the circumstances that could have led to a particular outcome is virtually impossible. Finally, organisms are complex creatures exposed to conflicting evolutionary pressures, such as the need to attract mates while simultaneously attempting to remain hidden from predators; such compromise-type situations are hard to simulate under experimental conditions.

Many evolutionary studies rely on making predictions about the patterns one would expect to observe in nature if evolution in one form or another were to have occurred, and such studies often involve synthesis of data derived from fieldwork, theoretical modeling, and laboratory analysis. While such indirect tests of evolutionary hypotheses are not based on the sort of controlled data that are generated in direct experiments, if employed appropriately the indirect tests can be equally valid and powerful. Their strength comes from the ability to formulate predictions based on one species or type of data that may then be supported or refuted by examining additional species or data from another area of biology. In this way, evolutionary biologists are able to use the history of life on earth as a natural experiment, and, like forensic scientists, to piece together clues to solve the greatest biological mystery of all.

Doug McElroy, Ph.D., and Bryan Ness, Ph.D.;
updated by Daniel J. Fairbanks, Ph.D.

FURTHER READING

Carroll, Sean B. *Endless Forms Most Beautiful: The New Science of Evo Devo.* New York: W. W. Norton, 2006. Explores the complementary sciences of evolution and biological development and how their integration tells a fascinating story of how species have evolved.

_____. *The Making of the Fittest: DNA and the Ultimate Forensic Record of Evolution.* New York: W. W. Norton, 2007. Reviews how molecular studies in a wide variety of animal species confirm the reality of evolution and reveal its mechanisms.

Coyne, Jerry A. *Why Evolution Is True.* New York: Viking, 2009. An up-to-date summary of current evidence from various fields of biology supporting evolution, written by one of the world's leading evolutionary biologists.

Darwin, Charles. *On the Origin of Species by Means of Natural Selection.* 1859. Reprint. New York: Modern Library, 1998. While difficult (partly as a result of its nineteenth century language and style), Darwin's seminal work is an enormously thorough and visionary treatise on evolution and natural selection.

Dawkins, Richard. *The Blind Watchmaker: Why the Evidence of Evolution Reveals a Universe Without Design.* New York: W. W. Norton, 1996. Argues the case for Darwinian evolution, criticizing the prominent punctuationist school and taking issue with the views of creationists and others who believe that life arose by design of a deity.

_____. *Climbing Mount Improbable.* New York: W. W. Norton, 1997. Using "Mount Improbable" as a metaphor, discusses genetics, natural selection, and embryology for hundreds of species spanning millions of years in a fascinating, instructive way.

_____. *The Selfish Gene.* 2d ed. New York: Oxford University Press, 1990. This pathbreaking book reformulated the notion of natural selection by positing the existence of true altruism in a genetically "selfish" world. This edition contains two new chapters.

Eldredge, Niles, and Stephen Jay Gould. "Punctuated Equilibria: An Alternative to Phyletic Gradualism." In *Models in Paleobiology,* edited by Thomas J. M. Schopf. San Francisco: Freeman, Cooper, 1972. The 1972 paper that introduced the theory of punctuated equilibrium to the scientific community. Illustrations, bibliography.

Fairbanks, Daniel J. *Relics of Eden: The Powerful Evidence of Evolution in Human DNA.* Amherst, N.Y.: Prometheus Books, 2007. Focuses on evidence from human DNA and comparisons of the human and chimpanzee genomes to show how the human genome evolved.

Freeman, Scott, and Jon C. Herron. *Evolutionary Analysis.* 4th ed. San Francisco: Benjamin Cummings, 2007. An excellent textbook that presents evolutionary biology as a dynamic field of scientific inquiry.

Gould, Stephen Jay. *Eight Little Piggies.* New York: W. W. Norton, 1994. In this collection of essays originally published by Gould in *Natural History,* the author of the theory of punctuated equilibrium considers the potential for mass extinctions of species in the face of ongoing degradation of the environment.

Quammen, David. *Song of the Dodo.* New York: Simon & Schuster, 1997. Chronicles the rich experiences of the unsung theorist of evolution Alfred Russel Wallace, whose research paralleled that of Charles Darwin.

Singh, Rama S., and Costas B. Krimbas, eds. *Evolutionary Genetics: From Molecules to Morphology.* New York: Cambridge University Press, 2000. Focuses on the necessary role of evolutionary genetics in evolutionary biology. Published in recognition of Richard Lewontin's work in evolutionary biology. Illustrations, bibliography, tables, diagrams, and index.

Weiner, Jonathan. *The Beak of the Finch: A Story of Evolution in Our Time.* New York: Random House, 1995. Describes the work of Peter and Rosemary Grant on the evolution of Charles Darwin's finches in the Galápagos Islands.

WEB SITES OF INTEREST

Literature.org. The Origin of Species (Charles Darwin)
http://www.literature.org/authors/darwin-charles/the-origin-of-species
A full-text, free, online version of Darwin's *On the Origin of Species.*

PBS. Evolution: A Journey into Where We're from and Where We're Going
http://www.pbs.org/wgbh/evolution
Site associated with the PBS series *Evolution.* Has links to numerous resources on evolutionary topics.

Talk.Origins Archive: Exploring the Creationism/ Evolution Controversy
http://www.talkorigins.org
A site dealing with the evolution/creation controversy from a scientific perspective, offering excellent summaries of the scientific evidence of evolution.

Understanding Evolution
http://evolution.berkeley.edu
A comprehensive, one-stop site on modern evolutionary biology maintained by the University of California, Berkeley.

See also: Ancient DNA; Artificial selection; Classical transmission genetics; Genetic code; Genetic code, cracking of; Genetics: Historical development; Hardy-Weinberg law; Human genetics; Lamarckianism; Mendelian genetics; Molecular clock hypothesis; Mutation and mutagenesis; Natural selection; Population genetics; Punctuated equilibrium; Repetitive DNA; RNA world; Sociobiology; Speciation; Transposable elements.

Extrachromosomal inheritance

CATEGORY: Cellular biology

SIGNIFICANCE: Extrachromosomal inheritance refers to the transmission of traits that are controlled by genes located in nonnuclear organelles such as chloroplasts and mitochondria, or in genes contained within extrachromosomal elements such as plasmids or viruses. In animals nuclear or chromosomal traits are determined equally by both parents, but the site of nonnuclear DNA, the cytoplasm, is almost always contributed by the female parent. The understanding of this extrachromosomal inheritance is crucial, since many important traits in plants and animals—as well as mutations implicated in disease and aging—display this type of transmission. Nonnuclear traits do not demonstrate Mendelian inheritance.

KEY TERMS

genome: hereditary material in the nucleus or organelle of a cell

mitochondria: small structures enclosed by double membranes found in the cytoplasm of all higher cells, which produce chemical power for the cells and harbor their own DNA. Since mitochondria are contributed only by the egg, inheritance is exclusively maternal.

plasmagene: a self-replicating gene in a cytoplasmic organelle

plasmon: the entire complement of genetic factors in the cytoplasm of a cell (plasmagenes or cytogenes); a plastid plasmon is referred to as a "plastome"

plastid: organelles, including chloroplasts, that are located in the cytoplasm of plant cells and that form the site for metabolic processes such as photosynthesis

DISCOVERY OF EXTRACHROMOSOMAL INHERITANCE

Carl Correns, one of the three geneticists who rediscovered Austrian botanist Gregor Mendel's laws of inheritance in 1900, and Erwin Baur first described, independently, extrachromosomal inheritance of plastid color in 1909. However, they did not know then that they were observing the transmission patterns of organelle genes. Correns studied the inheritance of plastid color in the albomaculata strain of four-o'clock plants (*Mirabilis jalapa*), whereas Baur investigated garden geraniums (*Pelargonium zonate*). Correns observed that seedlings resembled the maternal parent regardless of the color of the male parent (uniparental-maternal inheritance). Seeds obtained from plants with three types of branches—with green leaves, white leaves, and variegated (a mixture of green and white) leaves—provided interesting results. Seeds from green-leaved branches produced only green-leaved seedlings, and seeds from white-leaved branches produced only white-leaved seedlings. However, seeds from branches with variegated leaves resulted in varying ratios of green-leaved, white-leaved, and variegated-leaved offspring. The explanation is that plastids in egg cells of the green-leaved branches and white-leaved branches were only of one type (homoplasmic or homoplastidic)—that is, normal chloroplasts in the green-leaved cells and white plastids (leukoplasts) in the white-leaved cells. The cells of the variegated branches, on the other hand, contained both chloroplasts and leukoplasts (heteroplasmic or heteroplastidic) in varying proportions. Some descendants of the heteroplastidic cells received only chloroplasts, some received only leukoplasts, and some received a mixture of the two types of plastids in varying proportions in the next generation, hence variegation.

Baur observed similar progeny from reciprocal crosses between normal green and white *Pelargonium* plants. Progeny in both cases were of three types: green, white, and variegated, in varying ratios. This indicated that cytoplasm was inherited from the male as well as the female parent; however, the transmission of plastids was cytoplasmic. Male transmission of plastids has also been observed in oenothera, snapdragons, beans (*Phaseolus*), potatoes, and rye. Rye is the only member of the grass family that exhibits both maternal and paternal inheritance of plastids.

The investigations on plastid inheritance also clearly established that in plants exhibiting uniparental-maternal inheritance, a variegated maternal parent always produces green, white, and variegated progeny in varying proportions because of its heteroplastidic nature. Crosses between green and white plants always yield green or white progeny, depending upon the maternal parent, when the parental plants are homoplasmic for plastids.

EXTRACHROMOSOMAL INHERITANCE VS. NUCLEAR INHERITANCE

Extrachromosomal inheritance has been found in many plants, including barley, maize, and rice. Traits are inherited through chloroplasts, mitochondria, or plasmids (small, self-replicating structures). Inheritance of traits that are controlled by organelle genomes (plasmons) can be called nonnuclear or cytoplasmic. The cytoplasm contains, among other organelles, mitochondria in all higher organisms and mitochondria and chloroplasts in plants. Because cytoplasm is almost always totally contributed by the female parent, this type of transmission may also be called maternal or uniparental inheritance.

Most chromosomally inherited traits obey Mendel's law of segregation, which states that a pair of alleles or different forms of a gene separate from each other during meiosis (the process that halves the chromosome number in gamete formation). They also follow the law of independent assortment, in which two alleles of a gene assort and combine independently with two alleles of another gene. Such traits may be called Mendelian traits. Extrachromosomal inheritance is one of the exceptions to Mendelian inheritance. Thus, it can be called non-Mendelian inheritance. (Mendel only studied and reported on traits controlled by nuclear genes.) Mendelian heredity is characterized by regular ratios in segregating generations for qualitative trait differences and identical results from reciprocal crosses. On the contrary, non-Mendelian inheritance is characterized by a lack of regular segregation ratio and nonidentical results from reciprocal crosses.

The mitochondria are the sites of aerobic respiration (the breaking down of organic substances to release energy in the presence of oxygen) in both plants and animals. They are, like plastids, self-replicating entities and exhibit genetic continuity. The mitochondrial genes do not exhibit the Mendelian segregation pattern either. Mitochondrial genetics began around 1950 with the discovery of "petite" mutations in baker's yeast (*Saccharomyces cerevisiae*). Researchers observed that one or two out of every one thousand colonies grown on culture medium were smaller than normal colonies. The petite colonies bred true (produced only petite colonies). The petite mutants were respiration deficient under aerobic conditions. The slow growth of the petite colonies was related to the loss of a number of respiratory (cytochrome) enzymes that occur in mitochondria. These mitochondrial mutants, termed "vegetative petites," can be induced with acriflavine and related dyes. Another type of mutation, called a "suppressive petite," was found to be caused by defective, rapidly replicating mitochondrial DNA (mtDNA). Petite mutants that are strictly under nuclear gene control have also been reported and are called segregational petite mutants. Most respiratory enzymes are under both nuclear and mitochondrial control, which is indicative of collaboration between the two genetic systems.

Mitochondria also play a role in the programmed cell death—known as apoptosis—of eukaryotic cells, most notably in the death of cells that have accumu-lated potentially lethal genetic mutations. Apoptosis is also observed in ontological processes such as metamorphosis during the maturation of tadpoles into frogs, or in the disappearance of webbing between fingers and toes during embryonic development. In each of these examples, mitochondria undergo degeneration in the early stages.

In the fungus *Neurospora*, mitochondrial inheritance has been demonstrated for mutants referred to as "poky" (a slow-growth characteristic). The mutation resulted from an impaired mitochondrial function related to cytochromes involved in electron transport. The mating between poky female and normal male yields only poky progeny, but when the cross is reversed, the progeny are all normal, confirming maternal inheritance for this mutation.

According to a 1970 study, cytoplasmic male sterility is found in about eighty plant species. The molecular basis of cytoplasmic male sterility in maize through electrophoretic separation of restriction-endonuclease-created fragments of DNA was traced to mitochondrial DNA. Cytoplasmic male sterility can be overcome by nuclear genes. The plasmids that reside in mitochondria are also important ex-

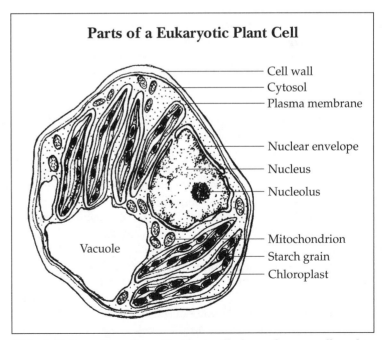

Parts of a Eukaryotic Plant Cell

Cell wall
Cytosol
Plasma membrane

Nuclear envelope
Nucleus
Nucleolus

Mitochondrion
Starch grain
Chloroplast

Vacuole

This depiction of a eukaryotic plant cell shows the organelles where extrachromosomal DNA is found: mitochondria, and in plants chloroplasts and other plastids. (Kimberly L. Dawson Kurnizki)

trachromosomal DNA molecules that are especially important in antibiotic resistance. Plasmids have been found to be extremely useful in genetic engineering.

MUTATOR GENES

Plastome mutations can be induced by nuclear genes. A gene that increases the mutation rate of another gene is called a "mutator." One such gene is the recessive, nuclear iojap (*ij*) mutation in maize. In the homozygous (*ij ij*) condition, it induces a plastid mutation. The name "iojap" has been derived from "Iowa" (the maize strain in which the mutation is found) and "japonica" (a type of striped variety that the mutation resembles). Once the plastid gene mutation caused by the *ij* gene has been initiated, the inheritance is non-Mendelian, and it no longer depends on the nuclear *ij* gene. As long as the iojap plants are used as female parents, the inheritance of the trait is similar to that for plastids in the albomaculata variety of four o'clock plants.

The *chm* mutator gene causes plastid mutations in the plant *Arabidopsis*, and mutator "striata" in barley causes mutations in both plastids and mitochondria. Cases of mutator-induced mutations in the plastome have also been reported in rice and catnip.

CHLOROPLAST AND MITOCHONDRIAL DNA

Plastids contain DNA, have their own DNA polymerase (the enzyme responsible for DNA replication), and undergo mutation. The chloroplast DNA (cpDNA) is a circular, self-replicating system approximately 140-200 kilobase pairs (kb) in size that carries genetic information which is transcribed (from DNA to RNA) and translated (from RNA to protein) in the plastid. It replicates in a semiconservative manner—that is, an original strand of DNA is conserved and serves as the template for a new strand in a manner similar to replication.

The soluble enzyme ribulose biphosphate carboxylase/oxygenase (Rubisco) is involved in photosynthetic carbon dioxide fixation. In land plants and green algae, its large subunit is a cpDNA product, while its small subunit is controlled by a nuclear gene family. Thus, the Rubisco protein is, as are chloroplast ribosomes, a product of the cooperation between the nuclear and chloroplast genes. In all other algae, both the large and small subunits of Rubisco are encoded in cpDNA.

Mitochondrial DNA (mtDNA) molecules are also circular and self-replicating. Human, yeast, and higher plant mtDNAs are the major systems that have been studied. The size of mitochondrial DNA and the number of RNA or protein-encoding genes varies significantly both within, and between species. Genome size among fungi ranges from 19-100 kb. In plants, the size of mitochondrial DNA ranges from 186-366 kb, and in animals from 16-17 kb. The human mtDNA has a total of 16,569 base pairs (16.6 kb, while yeast mtDNA is five times larger than that (84 kilobases), and maize mtDNA is much larger than the yeast mtDNA. Every base pair of human mtDNA may be involved in coding for a mitochondrial messenger RNA (mRNA) for a protein, a mitochondrial ribosomal RNA (rRNA), or a mitochondrial transfer RNA (tRNA). It is compact, showing no intervening, noncoding base sequences between genes. It has only one major promoter (a DNA region to which an RNA polymerase binds and initiates transcription) on each strand. Most codons—triplets of nucleotides (bases) in messenger RNA carrying specific instructions from DNA—have the same meaning as in the universal genetic code, except the following differences: UGA represents a "stop" signal (universal), but represents tryptophan in yeast and human mtDNA; AUA represents isoleucine (universal), but methionine in human mtDNA; CUA represents leucine (universal), but threonine in yeast mtDNA; and CGG represents arginine (universal), but tryptophan in plant mtDNA.

The mtDNA carries the genetic code (plasmagene names in parentheses) for proteins, such as cytochrome oxidase subunits I (coxl), II (cox2), and III (cox3); cytochrome B (cytb); and ATPase subunits 6 (atp6), 8 (atp8), and 9 (atp9). It also contains the genetic codes for several ribosomal RNAs, such as mtrRNA 16s and 12s in the mouse; mtrRNA 9s, 15s, and 21s in yeast; and mtrRNA 5s, 18s, and 26s in maize. In addition, twenty-two transfer RNAs in mice, twenty-four in yeast, and three in maize are encoded in mtDNA. Most mitochondria contain between two and ten copies of the genome, all of which are identical (homoplasmy).

CHLAMYDOMONAS REINHARDTII

Chlamydomonas reinhardtii is a unicellular green alga in which chloroplast and mitochondrial genes show uniparental transmission. In 1954, Ruth Sager discovered the chloroplast genetic system. Resistance to high levels of streptomycin (a trait controlled by chloroplast genes) has been shown to be transmitted uniparentally by the mt⁺ mating type parent. The mt⁻ mating type transmits the mitochondrial genes uniparentally. Mutants in chloroplasts have been identified for antibiotic and herbicide resistance. Genetic recombination is common in *C. reinhardtii*, which occurs in zygotes (the fused gametes of opposite sexes) when biparental cytogenes are in a heterozygous (union of unlike genes) state. This is an ideal system among plants for recombination studies, since there is only one large plastid per cell. In higher plants, study of genetic recombination is difficult because of a large number of plastids in cells and a lack of genetic markers.

Mutations in mitochondria of *C. reinhardtii* can be induced with acriflavine or ethidium bromide dyes. Point mutations for myxothiazol resistance mapping in the *cytb* gene have been isolated. The mitochondrial genome of this species of algae has been completely sequenced. It encodes five of more than twenty-five subunits of the reduced nicotinamide-adenine dinucleotide (NADH) dehydrogenase of complex I (nad1, nad2, nad4, nad5, and nad6), the COX I subunit of cytochrome oxidase (cox1), and the apocytochrome b (cob) subunit of complex III. All of these proteins have a respiratory function.

ORIGIN OF PLASTID AND MITOCHONDRIAL DNA

According to the endosymbiont theory, plastids and mitochondria in eukaryotes are the descendants of prokaryotic organisms that invaded primitive eukaryotes. Subsequently, they developed a symbiotic relationship and became dependent upon each other. There is much support for this theory. Researchers in 1972 showed homology (genetic similarity) between ribosomal RNA from photosynthetic blue-green bacteria (cyanobacteria) and DNA from the chloroplasts of *Euglena gracilis*. The organization of rRNA genes is also similar to their counterpart in cyanobacteria. This provided support for chloroplasts as the descendants of cyanobacteria. Mitochondria are believed to have evolved from a variety of primitive bacteria and plastids from cyanobacteria. Molecular evidence strongly supports the endosymbiotic origin of mitochondria both from alpha purple bacteria as well as intracellular bacteria such as *Rickettsia*, the genus which includes the etiological agents for typhus and spotted fevers.

In 1981, Lynn Margulis summarized evidence for this theory. There are many similarities between prokaryotes and organelles: Both have circular DNA and the same size ribosomes, both lack histones and a nuclear membrane, and both show similar response to antibiotics that inhibit protein synthesis. Both also show a primitive mode of translation that begins with formulated methionine. The discovery of promiscuous DNA (DNA segments that have been transferred between organelles or from a mitochondrial genome to the nuclear genome) in eukaryotic cells also lends support to this theory.

IMPACT AND APPLICATIONS

Genetic investigations have helped tremendously in constructing a genetic map of maize cpDNA. Important features of the map, including two large, inverted, repeat segments containing several rRNA and tRNA genes, are now known. Detection and quantification of mutant mtDNA are essential for the diagnosis of diseases and for providing insights into the molecular basis of pathogenesis, etiology, and ultimately the treatment of diseases. This should help enhance the knowledge of mitochondrial biogenesis. Mitochondrial dysfunction, resulting partly from mutations in mtDNA, may play a central role in organismal aging.

A number of human diseases associated with defects in mitochondrial function have been identified since their first description in 1988. Mitochondria appear to be particularly sensitive to genetic mutations—more so than nuclear DNA—perhaps resulting from the absence of efficient repair mechanisms for mtDNA, the lack of histones, or due to the accumulation of free radicals as a by-product of cell respiration by the organelle. More than 150 different types of mutations have now been identified. Large-scale deletions and tRNA point mutations (base changes) in mtDNA are associated with clinical mitochondrial encephalomyopathies. Heteroplasmy (the coexistence of more than two types of mtDNA) has provided experimental systems in which the transmission of mtDNA in animals can be studied. Numerous deleterious point mutations of mtDNA are associated with various types of human disorders involving deficiencies in the mitochondrial oxidative phosphorylation (respiration) apparatus. Leigh disease is caused by a point mutation in mtDNA. Deletions of mtDNA have been associated with diseases such as isolated ocular myopathy, chronic progressive external ophthalmoplegia, Kearns-Sayre syndrome, and Pearson syndrome. Mitochondrial defects have also been reported in Alzheimer's and Huntington's diseases.

Treatment of mitochondrial diseases is primarily palliative, allowing the relief of symptoms but not eliminating the underlying defect. Genetic diagnosis of fertilized eggs provides a mechanism, albeit highly inefficient on a large scale, of allowing only the development of "healthy" embryos. Medical experiments using nonhuman primates (rhesus monkeys) have shown it is theoretically possible to replace defective mitochondria in oocytes with normal mtDNA. The process is not yet feasible in humans as the volume of cytoplasmic material sufficient to eliminate the problem of heteroplasmy is unrealistic.

The influence of the mitochondrial genome and mitochondrial function on nuclear gene expression is poorly understood, but progress is being made toward understanding why a few genes are still sequestered in the mitochondria and toward developing new tools to manipulate mitochondrial genes. The most obvious difference between mitochondrial DNA, as well as chloroplast DNA, and the DNA in free-living bacteria is in their relative size: The smallest free-living bacteria contain approximately eight hundred genes. Molecular evidence suggests mitochondrial genes began transferring to the cell nucleus early in the evolution of endosymbiosis. The process seems to still be taking place in some plants; hundreds of genes in the *Arabidopsis* nucleus appear similar to those in the chloroplasts.

Manjit S. Kang, Ph.D.; updated by Richard Adler, Ph.D.

FURTHER READING

Attardi, Giuseppe M., and Anne Chomyn, eds. *Methods in Enzymology: Mitochondrial Biogenesis and Genetics.* Vols. 260, 264. San Diego: Academic Press, 1995. One hundred authors contribute to thirty-six chapters, presenting a wealth of new methods and data and covering the significant developments that have expanded the scope of enzyme chemistry.

Cummings, Michael J. *Human Heredity: Principles and Issues.* 5th ed. Pacific Grove, Calif.: Brooks/Cole, 2000. College text that surveys topics such as genetics as a human endeavor; cells, chromosomes, and cell division; transmission of genes from generation to generation; cytogenics; the source of genetic variation; cloning and recombinant DNA; genes and cancer; genetics of behavior; and genes in populations.

Krebs, Jocelyn, Elliott Goldstein, and Stephen Kilpatrick. *Lewin's Essential Genes.* 2d ed. Sudbury, Mass.: Jones and Bartlett, 2010. Contains an updated description of mitochondrial and chloroplast genetic information: sizes, essential genes.

Scheffler, Immo. *Mitochondria.* 2d ed. Hoboken, N.J.: John Wiley & Sons, 2008. An updated description of mitochondrial genetics and the role played by the organelle in the cell.

WEB SITES OF INTEREST

BioPortfolio. "*Extrachromosomal Inheritance*"
http://www.bioportfolio.com/indepth/Extrachromosomal_Inheritance.html
Provides links to recent publications on extrachromosomal inheritance.

Physician's Guide to the Spectrum of Mitochondrial Diseases
http://biochemgen.ucsd.edu/mmdc/ep-3-10.pdf
Regularly updated source of links to research into extrachromosomal inheritance.

See also: Ancient DNA; Chloroplast genes; Genetic code; Human genetics; Mitochondrial diseases; Mitochondrial genes; Model organism: *Chlamydomonas reinhardtii*; RNA world.

F

Fabry disease

CATEGORY: Diseases and syndromes

ALSO KNOWN AS: Alpha-galactosidase A deficiency; Anderson-Fabry disease; angiokeratoma corporis diffusum; angiokeratoma diffuse; ceramide trihexosidase deficiency; gla deficiency; glycolipid lipidosis; hereditary dystopic lipidosis

DEFINITION

Fabry disease is an inherited genetic disorder caused by a defective gene. The disease causes fatty deposits in several organs of the body.

Males who inherit the defective gene will express the disease. Females who have a single copy of the gene are called carriers and most are asymptomatic (do not have symptoms). However, some women do have symptoms, and the severity of these symptoms can vary widely. On occasion, women may be as severely affected as men.

RISK FACTORS

The primary risk factor for Fabry disease is having family members who have the disease or are carriers of the disease.

ETIOLOGY AND GENETICS

Fabry disease results from a mutation in the *GLA* gene, located on the long arm of the X chromosome at position Xq22. This gene encodes a protein called alpha-galactosidase A, which normally breaks down a fatty metabolic intermediate known as globotriaosylceramide. When the gene is missing or mutated, the enzyme function is absent or severely reduced, and as a result globotriaosylceramide will build up in cells over time. The endothelial cells lining the blood vessels in the heart, kidney, and nervous system are particularly prone to this fatty accumulation, and the consequent damage to these cells reduces blood flow to the organs.

The inheritance pattern of Fabry disease is typical of all sex-linked recessive mutations (those found on the X chromosome). Mothers who carry the mutated gene on one of their two X chromosomes face a 50 percent chance of transmitting this disease to each of their male children. Female children have a 50 percent chance of inheriting the gene and becoming carriers like their mothers. Affected males will pass the mutation on to all of their daughters but none of their sons. Female carriers with one mutant and one normal copy of the gene often exhibit a mild form of the disease, although there is considerable variability in the degree of expression, and some carriers remain totally asymptomatic. The severe classic form of the disease is found almost exclusively in males.

SYMPTOMS

Symptoms of Fabry disease may begin in childhood or early adulthood. Common symptoms include pain and burning sensations in the hands and feet, often provoked by exercise, fatigue, or fever; spotted, dark red skin lesions (angiokeratomas) that generally are found in the area between the belly button and the knees (they may also be found elsewhere); inability to sweat; and changes in the eyes, such as corneal opacities and cataracts.

As adults, males may experience symptoms due to blood vessel blockage, including kidney problems, often requiring dialysis or transplant; risk of early stroke or heart attack; chest pain; hypertension; heart failure, left ventricular hypertrophy; mitral valve prolapse or insufficiency; frequent bowel movements after eating; and diarrhea. Additional symptoms in adult males due to heart vessel blockage may include joint or back pain, ringing in the ears (tinnitus) or dizziness (vertigo), chronic bronchitis or shortness of breath, osteoporosis, delayed puberty or retarded growth, and stroke.

SCREENING AND DIAGNOSIS

The doctor will ask about a patient's symptoms and medical history and will perform a physical exam. Diagnosis is usually made on the basis of the symptoms listed above. A test to measure the enzyme GALA or a DNA analysis can confirm Fabry disease.

TREATMENT AND THERAPY

There is no cure for Fabry disease. However, in 2003, the U.S. Food and Drug Administration (FDA) approved the use of Fabrazyme (recombinant alpha-galactosidase), an enzyme replacement therapy, as treatment for Fabry disease. While the long-term effects and risks of this treatment are not yet known, treatment is currently recommended for all adults with Fabry disease and for all adult women who are known carriers. Preliminary pediatric data is somewhat encouraging, but enzyme replacement in children is still an experimental procedure. The National Institutes of Health (NIH) is conducting ongoing research into the use of Fabrazyme in children.

Currently, medications or procedures are used to treat symptoms of Fabry disease, including carbamazepine (Tegretol), which is used to treat pain. According to the FDA, patients of Asian ancestry who have a certain gene, called *HLA-B*1502*, and take carbamazepine are at risk for dangerous or even fatal skin reactions. The FDA recommends that patients of Asian descent get tested for this gene before taking carbamazepine. Patients who have been taking this medication for a few months with no skin reaction are at low risk of developing these reactions. Patients should talk to their doctors before stopping this medication.

Other medications used to treat the pain of Fabry disease are Dilantin (phenytoin) and Neurontin (gabapentin). Lipisorb, the brand name for a nutritional supplement with medium chain triglyceride (MCT); Reglan (metoclopramide); and Cotazym (pancrelipase) treat stomach hyperactivity. Anticoagulants can be used to treat certain heart disorders, and hemodialysis and kidney transplantation can treat kidney disease.

PREVENTION AND OUTCOMES

There is no known way to prevent Fabry disease. Individuals who have Fabry disease or have a family history of the disorder can talk to a genetic counselor when deciding to have children.

Michelle Badash, M.S.;
reviewed by Rosalyn Carson-DeWitt, M.D.
"Etiology and Genetics" by Jeffrey A. Knight, Ph.D.

FURTHER READING

Desnick, Robert J. "Fabry Disease: Alpha-Galactosidase A Deficiency." In *The Molecular and Genetic Basis of Neurologic and Psychiatric Disease*, edited by Roger N. Rosenberg et al. 4th ed. Philadelphia: Wolters Kluwer Health/Lippincott Williams & Wilkins, 2008.

EBSCO Publishing. *Health Library: Fabry Disease.* Ipswich, Mass.: Author, 2009. Available through http://www.ebscohost.com.

Khan, M. Gabriel. "Anderson-Fabry Disease." In *Encyclopedia of Heart Diseases*. Burlington, Mass.: Elsevier Academic, 2006.

Kleigman, Robert M., et al., eds. *Nelson Textbook of Pediatrics*. 18th ed. Philadelphia: Saunders Elsevier, 2007.

WEB SITES OF INTEREST

Canadian Fabry Association
http://www.fabrycanada.com

Fabry Society of Canada
http://www.fabrysociety.org

Fabry Support and Information Group
http://www.fabry.org

Genetics Home Reference
http://ghr.nlm.nih.gov

National Institute of Neurological Disorders and Stroke
http://www.ninds.nih.gov

National Tay-Sachs and Allied Diseases Association
http://www.ntsad.org

U.S. Food and Drug Administration: "Information on Carbamazepine (Marketed as Carbatrol, Equetro, Tegretol, and Generics) with FDA Alerts."
http://www.fda.gov/Drugs/DrugSafety/PostmarketDrugSafetyInformationforPatientsandProviders/ucm107834.htm

See also: Hereditary diseases; Inborn errors of metabolism.

Familial adenomatous polyposis

CATEGORY: Diseases and syndromes
ALSO KNOWN AS: FAP

DEFINITION

Familial adenomatous polyposis (FAP) is a rare, inherited type of colorectal cancer. FAP results in the development of hundreds of polyps inside the large intestine.

RISK FACTORS

The primary risk factor for FAP is having family members with this condition.

ETIOLOGY AND GENETICS

Mutations in either of two different genes are known to cause familial adenomatous polyposis. In the majority of cases, the mutation is localized in the *APC* gene, which is found on the long arm of chromosome 5 at position 5q21-q22. This gene encodes a very large protein (2,843 amino acids) that is partitioned into several domains, each with its own distinct function. It serves as a tumor suppressor in cells by antagonizing one signaling pathway while being an essential component of a second signaling pathway for the production of beta-catenin. Beta-catenin is a protein necessary for the development and continuity of the epithelial tissue that lines organ surfaces. One of its functions is to regulate normal cell growth and behavior, and altered or missing copies of this protein can lead to polyp formation and ultimately colorectal cancer.

A second gene associated with FAP is the *MUTYH* gene, located on the short arm of chromosome 1 at position 1p34.3-p32.1. This gene specifies an important enzyme in a deoxyribonucleic acid (DNA) repair pathway that is active during DNA replication. Individuals with a mutation in this gene that causes a nonfunctional enzyme to be produced will be unable to repair some replication errors, and the accumulated mistakes increase the likelihood of overgrowth in the intestinal epithelia, leading to polyp formation.

The inheritance pattern for FAP critically depends on the gene in which the mutation (or mutations) occurs. Mutations in the *APC* gene are inherited in an autosomal dominant fashion, meaning that a single copy of the mutation is sufficient to cause full expression of the disease. An affected individual has a 50 percent chance of transmitting the mutation to each of his or her children. Many cases of *APC*-associated FAP, however, result from a spontaneous new mutation, so in these instances affected individuals will have unaffected parents.

Mutations in the *MUTYH* gene are inherited in an autosomal recessive pattern. In this case, both copies of the gene must be deficient in order for the individual to be afflicted. Typically, an affected child is born to two unaffected parents, both of whom are carriers of the recessive mutant allele. The probable outcomes for children whose parents are both carriers are 75 percent unaffected and 25 percent affected. If one parent has *MUTYH*-associated FAP and the other is a carrier, there is a 50 percent probability that each child will be affected.

SYMPTOMS

In the early stages, there may be no symptoms of FAP. When symptoms do occur, they may include bright red blood in the stool, diarrhea, constipation, cramping pain in the stomach, consistent decrease in the size of stool, weight loss, bloating, and fatigue.

SCREENING AND DIAGNOSIS

The doctor will ask about a patient's symptoms and medical history and will perform a physical exam. Tests may include a DNA analysis, in which blood samples are taken from members of the patient's family to determine if the patient has the defective gene; and an endoscopy, in which a thin, lighted, telescope-like tube is inserted into the colon to check for polyps. The endoscopy may be a limited exam of the rectum with a proctoscope, a more extensive exam of the rectum and sigmoid colon with a sigmoidoscope, or a complete exam of the entire large intestine with a colonoscope. If a polyp is found during endoscopy, a small sample will be removed and sent to a lab for testing; this test is called a biopsy.

TREATMENT AND THERAPY

FAP is treated with surgery. Since FAP causes so many polyps, they cannot be removed individually. Therefore, the goal of surgery is to remove the portion of the intestine that contains the cancerous or precancerous polyps. The surgical procedure used depends on the length of intestine involved.

For reasons not entirely understood, rectal polyps will often regress or disappear after a more limited surgery that does not require the removal of the rectum. Therefore, the surgeon often will leave the rectum in place and remove the rest of the large bowel. If the polyps in the rectum do not disappear, then the rectum will likely need to be removed as well.

There are three main surgical treatments for FAP. A colectomy with ileorectal anastomosis (IRA) is the most common procedure for patients with few polyps in the rectum. The colon is removed, but five inches of the rectum remain. The small intestine is surgically joined to the upper rectum. This procedure preserves sphincter tone and allows for relatively normal sensation of the need to have a bowel movement.

In a restorative proctocolectomy (pouch) the colon and rectum are removed, leaving the anal canal and anal sphincter muscles. An artificial rectum (pouch) is created from the lower end of the small intestine. The pouch is attached to the anus in order to control bowel actions. This operation is usually done in two stages.

In a total proctocolectomy with permanent ileostomy the colon, rectum, and anus are removed. After that, a permanent ileostomy (opening in the abdomen) is created. A bag is attached to collect waste through the ileostomy. This type of surgery is not used very often, unless the rectum contains many polyps and they do not regress after a lesser surgery.

If only a portion of the bowel is removed at surgery, the remaining bowel will need to be inspected by endoscopy every three to six months for the rest of a patient's life. Because the risk of developing other polyps that could grow to become cancer is so high, it is crucial for a patient's doctor to keep a very close watch on the remaining bowel. If more polyps arise, further surgery may be required.

PREVENTION AND OUTCOMES

There are no guidelines for the prevention of FAP. There are some preliminary studies evaluating the use of cyclooxygenase antagonists, such as Vioxx or Celebrex, to prevent the development of colorectal polyps. However, it is too early to tell if these drugs have any effect on the development of cancerous polyps in FAP.

Michelle Badash, M.S.; reviewed by Daus Mahnke, M.D.
"Etiology and Genetics" by Jeffrey A. Knight, Ph.D.

FURTHER READING

Delaini, Gian Gaetano, ed. *Inflammatory Bowel Disease and Familial Adenomatous Polyposis: Clinical Management and Patients' Quality of Life*. New York: Springer, 2006.

EBSCO Publishing. *Health Library: Familial Adenomatous Polyposis*. Ipswich, Mass.: Author, 2009. Available through http://www.ebscohost.com.

Finn, Robert. "Treatment Is Key to Cancer Prevention in FAP Patients." *Family Practice News* 34, no. 8 (April 15, 2004): 23.

Lefevre, J. H., et al. "*APC, MYH*, and the Correlation Genotype-Phenotype in Colorectal Polyposis." *Annals of Surgical Oncology* 16, no. 4 (April, 2009): 871-877.

WEB SITES OF INTEREST

Canadian Cancer Society
http://www.cancer.ca

Colon Cancer Canada
http://www.coloncancercanada.ca

"Familial Adenomatous Polyposis." Cleveland Clinic
http://www.clevelandclinic.org/registries/inherited/fap.htm

Genetics Home Reference
http://ghr.nlm.nih.gov

National Cancer Institute
http://www.cancer.gov

Polyposis Registry
http://www.polyposisregistry.org.uk

See also: *APC* gene testing; Cancer; Colon cancer; Hereditary diseases.

Familial Mediterranean fever

CATEGORY: Diseases and syndromes
ALSO KNOWN AS: Familial paroxysmal polyserositis; benign paroxysmal peritonitis; FMF

DEFINITION

Familial Mediterranean fever (FMF) is a rare autosomal recessive disorder characterized by recurrent fevers and painful inflammation of the abdominal peritoneum. Serosal, synovial, and cutaneuous inflammation may also cause pleuritis, skin lesions, arthritis, and rarely pericarditis. Long-term complications include renal amyloidosis, which may lead to renal failure. People of Mediterranean origins are more commonly affected.

RISK FACTORS

The greatest risk factor for FMF is having genetic origins from the Mediterranean basin. Sephardic Jews, North African Arabs, Armenians, Turks, Greeks, and Italians are at higher risk. With the advent of genetic testing, FMF has also been increasingly identified in Ashkenazi Jews. Occasional cases are confirmed in people without known Mediterranean ancestry. Up to 50 percent of people with FMF have a family history of the disorder.

ETIOLOGY AND GENETICS

FMF is caused by mutations in the *MEFV* gene on the short arm of chromosome 16. More than 50 mutations in the gene have been identified. The *MEFV* gene is responsible for encoding a 781-amino acid protein called pyrin or marenostrin. Pyrin is expressed in circulating neutrophils and it is believed to be responsible for blunting the inflammatory response. The FMF gene mutation may alter the pyrin molecule in such a way that it fails to suppress unknown triggers in the inflammatory response, resulting in spontaneous episodes of neutrophil-predominant inflammation.

FMF is almost always inherited in an autosomal recessive pattern. DNA testing reveals that carrier frequencies may be as high as 1 in 3 in affected populations. In addition to the *MEFV* gene mutation, normal variations in the *SAA1* gene may influence the course of FMF. A version of SAA1 called the alpha variant may increase the risk of developing amylioidosis in some people with FMF.

SYMPTOMS

The first signs and symptoms of FMF usually appear between the ages of five and fifteen, although initial attacks have been documented in infancy. Ninety percent of people with FMF have their first attack before the age of twenty. The attacks commonly last from twenty-four to seventy-two hours. The frequency of attacks is unpredictable and can be as frequent as two attacks per week or as infrequent as one attack per year. Spontaneous remis-

sions of several years may occur. Physical exertion, emotional stress, and menses have been noted as contributing factors; pregnancy has been cited as a remitting factor.

Fever and abdominal pain are the most common manifestations. Fever is nearly always present and may range from 100 degrees Fahrenheit (37.8 degrees Celsius) to 104 degrees Fahrenheit (40 degrees Celsius). Acute abdominal pain typically starts in one quadrant and then spreads to the whole abdomen. Rebound tenderness, guarding, and decreased bowel sounds may lead to exploratory laparotomy which shows only a small amount of sterile peritoneal exudate.

Pleurisy occurs in about 30 percent of cases and may cause unilateral, stabbing chest pain. Arthritis, which may be seen in 25 percent of attacks, is manifested by single joint pain in the knee, ankle, or hip. The skin rash of FMF is a raised, erythematous rash that appears on the foot, ankle, or lower leg. Less common manifestations include muscle aches, scrotal inflammation, pericarditis, and vasculitis. Although the acute manifestations can be severe, most patients recover completely and are asymptomatic between attacks.

SCREENING AND DIAGNOSIS

FMF is usually diagnosed by its clinical presentation. Genetic testing is available and is useful for atypical cases. Genetic testing may also have some prognostic value. Certain types of *MEFV* mutations have been correlated with earlier age of onset, higher frequency of rash, and greater likelihood of developing amyloidosis. Because of the possibility of nonpenetrance and the potential impact on insurability, genetic screening of unaffected individuals is not recommended.

Nonspecific findings that may help lead to a diagnosis of FMF include elevation of white blood cells with neutrophil predominance, elevated ESR, C-reactive protein, and fibrinogen. Renal amyloidosis may be indicated by proteinuria. The incidence of FMF in Mediterranean populations ranges between 1 in 250 to 1 in 1,000 people, and the disorder is slightly more common in men than in women.

TREATMENT AND THERAPY

Although there is no cure for FMF, the drug colchicine decreases the frequency of attacks and the intensity of attacks in about 85 percent of people diagnosed with FMF. Colchicine taken as a daily oral medication provides complete remission from attacks in most people. For those who have infrequent attacks, cochicine may be used to abort an attack if there are prodromal symptoms, but once an attack is well underway, colchicine is no longer effective. Colchicine also prevents the development of amyloidosis and it can safely be taken during pregnancy.

PREVENTION AND OUTCOMES

There is no way to prevent FMF. Widespread use of prophylactic colchicine has led to a dramatic reduction in attacks and in complications. In most cases FMF patients can lead normal and productive lives. The most common complication from untreated FMF is amyloidosis, which can lead to nephrotic syndrome and renal failure. Inflammation caused by FMF can also lead to infertility due to involvement of reproductive organs.

Chris Iliades, M.D.

FURTHER READING

Fauci, Anthony S. *Harrison's Principles of Internal Medicine.* 17th ed. New York: McGraw-Hill, 2008. The definitive textbook for internal medicine.
Ochs, Hans D., C. I. E. Smith, and J. M. Puck. *Primary Immunodeficiency Diseases: A Molecular and Genetic Approach.* 2d ed. New York: Oxford University Press, 2007. Gene identification, mutation detection, and clinical research for more than one hundred human genetic disorders.
Warrel, David A., T. M. Cox, and J. D. Firth. *Oxford Textbook of Medicine.* Vol. 2. 4th ed. New York: Oxford University Press, 2005. A comprehensive guide to all aspects of internal medicine.

WEB SITES OF INTEREST

Genetics Home Reference
http://ghr.nlm.nih.gov

National Human Genome Research Institute
http://www.genome.gov

National Organization of Rare Disorders
http://www.rarediseases.org

See also: Autoimmune disorders; Hereditary diseases.

Fanconi anemia

CATEGORY: Diseases and syndromes
ALSO KNOWN AS: Fanconi's anemia; Fanconi pancytopenia; Fanconi panmyelopathy; aplastic anemia with congenital anomalies; congenital aplastic anemia; congenital pancytopenia; hypoplastic congenital anemia; Diamond-Blackfan anemia

DEFINITION

Fanconi anemia is a rare recessive genetic disorder that affects the blood and bone marrow, often leading to total bone marrow failure. In addition to pancytopenia (decreased levels of all types of blood cells), the disease may result in a variety of congenital anomalies and increased susceptibility to leukemia and other cancers. At the cellular level, Fanconi anemia is characterized by chromosomal instability caused by defects in the maintenance and repair of DNA.

RISK FACTORS

Children are at risk if both parents are carriers; for each child conceived by two carriers, the risk of being affected will be one in four. The severity of congenital malformations, the age at which bone marrow failure begins, and life expectancy are uncertain. In different populations, carrier frequency may range from 1 in 600 to 1 in 100. The disease occurs in about 1 in 360,000 live births but may be as high as 1 in 40,000 births in certain groups.

ETIOLOGY AND GENETICS

Fanconi anemia occurs in males and females in all ethnic groups throughout the world. The inheritance pattern is usually autosomal recessive, but at least one Fanconi anemia gene is located on the X chromosome. Since the 1990's, researchers have discovered a series of Fanconi anemia genes, their chromosomal location, and their gene products. At least thirteen Fanconi anemia genes have been identified; twelve of them have been cloned. Mutations in three of the most common of these genes account for about 85 percent of Fanconi anemia patients.

The proteins encoded by Fanconi anemia genes form a complex involved in fundamental cellular activities, including the maintenance of genomic integrity. Mutations in Fanconi anemia genes lead to increased susceptibility to chemicals that damage DNA. Cells that carry mutated Fanconi anemia genes exhibit a high level of chromosomal aberrations. In addition to aplastic anemia, patients are at very high risk of developing acute myeloid leukemia and squamous cell cancers. The discovery of Fanconi anemia-like genes in common experimental animals, such as mice and chickens, should expedite basic research.

SYMPTOMS

Clinical histories indicate that there is no typical Fanconi anemia patient, but certain signs and symptoms call for specific diagnostic tests. Birth defects may include malformed thumbs, skeletal abnormalities, microcephalus (small head), heart defects, kidney problems, patchy discolorations of the skin, defects of the eyes and ears, underdevelopment of the bone marrow, and abnormal red blood cells. Bone marrow failure usually appears between five and ten years of age. The lack of platelets, red blood cells, and white blood cells impairs the body's ability to form blood clots, oxygenate tissues, and fight infection. Bone marrow failure is the main cause of death.

SCREENING AND DIAGNOSIS

Because Fanconi anemia is so rare, pediatricians and family doctors may be unfamiliar with the disease. Early diagnosis and referral to appropriate experts is important because of the risk of bone marrow failure. Some cases are recognized at birth, because of characteristic physical anomalies, but some patients are not diagnosed until adulthood.

Physical examinations and blood tests may detect symptoms that suggest Fanconi anemia, but specific tests are needed to visualize the disease at the cellular level. A definitive diagnosis requires studies of chromosome breakage and hypersensitivity to DNA damaging agents. A test for gene mutations is also possible. Genetic tests can be performed on embryos in affected families.

TREATMENT AND THERAPY

Improvements in therapy have increased life expectancy and improved the quality of life for most patients, but there is no definitive cure for the full spectrum of problems associated with the disease. Blood, blood products, androgens, corticosteroids, and growth factors have been used to treat early

bone marrow failure. Stem cell transplants using umbilical cord blood or bone marrow have significantly increased life expectancy. Transplants from a healthy sibling of the same tissue type are most successful, but recent developments are making it possible to use less perfectly matched donors. Successful transplants cure the aplastic anemia, but other anomalies and the high risk of cancers remain. Researchers hope that gene therapy will eventually be used to correct the genetic defect.

PREVENTION AND OUTCOMES

Genetic tests make it possible to identify carriers and affected embryos. If both parents are known carriers, then a fetus with Fanconi anemia can be identified during pregnancy by testing fetal cells obtained by amniocentesis or chorionic villi sampling for sensitivity to chromosome breakage. Parents can use in vitro fertilization and preimplantation genetic diagnosis to select embryos that are free of the disease. Genetic screening with embryo diagnosis can be used to select healthy embryos that match the tissue type of an affected child in order to provide umbilical cord blood cells or bone marrow.

Studies of this rare genetic disease are providing insights into broader questions about genomic maintenance, the genesis of cancers, and the mechanism of aging.

Lois N. Magner, Ph.D.

FURTHER READING

Ahmad, Shamin I., and Sandra H. Kirk, eds. *Molecular Mechanisms of Fanconi Anemia.* New York: Springer, 2006. Studies of the genetic basis of the disease, clinical patterns, and therapies.

Schindler, Detlev, and Holger Hoehn, eds. *Fanconi Anemia: A Paradigmatic Disease for the Understanding of Cancer and Aging.* New York: Karger, 2007. A comprehensive survey of the relationship between genetic discoveries and treatment options.

WEB SITES OF INTEREST

Fanconi Anemia Research Fund, Inc.
http://www.fanconi.org/family/Centers.htm

International Fanconi Anemia Registry
http://www.rockefeller.edu/labheads/auerbach/auerbach.html

Online Mendelian Inheritance of Man
http://www.ncbi.nlm.nih.gov/sites/entrez?db
=omim&TabCmd=Limits

See also: ABO blood group system; Hereditary diseases; Sickle-cell disease.

Farber disease

CATEGORY: Diseases and syndromes

ALSO KNOWN AS: Farber's disease; Farber's lipogranulomatosis; ceramidase deficiency; acid ceramidase deficiency

DEFINITION

Farber disease is a severe, progressive inherited disorder that affects the throat, lungs, joints, liver, skin, and brain. The symptoms of Farber disease are caused by the harmful buildup of fatty substances in the body's cells.

RISK FACTORS

Farber disease is a very rare genetic disease caused by the inheritance of a nonworking *ASAH1* gene from both parents. Although the exact number of individuals affected by Farber disease is unknown, approximately fifty cases have been reported in the literature worldwide. The disease is present in males and females at equal rates. This condition is not caused by infections.

ETIOLOGY AND GENETICS

Farber disease is caused by the lack of an enzyme known as N-acylsphingosine amidohydrolase (ceramidase) in a small cellular organelle called the lysosome. The lysosome is the recycling center of the cell. When the enzyme is missing, a fatty substance called chondoritin sulphate B builds up in the cells of the body.

Farber disease is an autosomal recessive genetic condition. It occurs when a child inherits two copies of the nonworking *ASAH1* gene. Individuals with only one copy of a nonworking *ASAH1* gene are known as carriers and have no problems related to the condition. In fact, all people carry between five and ten nonworking genes for harmful, recessive conditions. When two people with the same nonworking recessive *ASAH1* gene meet, however, there is a chance, with each pregnancy, for the child to inherit two copies, one from each parent. That child then has no working copies of the *ASAH1* gene and therefore has the signs and symptoms associated with Farber disease.

SYMPTOMS

Typically, affected children begin having symptoms of Farber disease in the second or third week after birth. Early symptoms of the disease often include a hoarse cry, problems feeding, irritability, impaired mental and physical abilities, swollen lymph nodes, cherry-red macular spots in the eye, and skin bumps (called erythematous periarticular swellings). These symptoms progressively become worse, with painful joint swelling, shortened muscles around joints, frequent infections, breathing difficulties, and then heart and kidney failure.

Although the most common form of Farber disease is the severe, infant-onset form, Farber disease symptoms can vary from person to person. Accordingly, affected individuals are grouped into at least seven different types of Farber disease. Patients with type 1 disease are very severely affected with symptoms starting in the weeks after birth. Most type 1 patients die during the first years of life. Type 2 and 3 patients do not have liver or lung symptoms, are of normal intelligence, and have a longer life expectancy than do type 1 patients. Type 4 patients have very severe disease with symptoms including an enlarged spleen and liver. These patients rarely live past six months of age. Type 5 patients have classic symptoms beginning at one to two years of age. Types 6 and 7 are severe varieties of Farber disease that result from the lack of ceramidase and one or more other lysosomal enzymes.

SCREENING AND DIAGNOSIS

Screening for Farber disease is not part of routine testing in the prenatal or newborn periods of life. Since the symptoms of Farber disease are present in the first few weeks of life, diagnosis is most often made on the basis of disease signs and symptoms. Biochemical testing is available to confirm the diagnosis through identification of the low or missing enzymes. Molecular genetic testing can help identify the changes in the *ASAH1* gene in carriers and affected individuals. This disease can be misdiagnosed as colic, juvenile rheumatoid arthritis, or sarcoidosis.

TREATMENT AND THERAPY

At this time, there is no cure or disease specific treatment for Farber disease. Therapy for Farber disease focuses on the treatment of each symptom individually. For example, corticosteroids may be given to reduce painful swelling of the joints and nodes could be surgically removed. In individuals affected by a mild form of Farber disease without brain involvement, bone marrow transplants may help relieve physical symptoms of the disease. Researchers are working on gene therapy to provide individuals affected by Farber disease with a working copy of the altered *ASAH1* gene, but this has not yet been proved to be effective in affected humans.

PREVENTION AND OUTCOMES

Carrier testing can be conducted for individuals who are interested in learning if they carry an altered *ASAH1* gene. Genetic counseling is available for parents who have an affected child or are concerned about being a carrier for the *ASAH1* gene. Although the severity and symptoms of Farber disease varies from individual to individual, death from Farber disease symptoms most often occurs by two years of age as a result of lung disease. Rare individuals affected by a much milder form of the disease may live past their twenties.

Dawn A. Laney, M.S.

FURTHER READING

Gonick, Larry, and Mark Wheelis. *The Cartoon Guide to Genetics.* New York: HarperPerennial, 1991.

Moser, Hugo W., et al. "Acid Ceramidase Deficiency: Farber Lipogranulomatosis." In *The Metabolic and Molecular Bases of Inherited Disease*, edited by Charles Scriver et al. New York: McGraw-Hill, 2001.

Willett, Edward. *Genetics Demystified.* New York: McGraw-Hill, 2005.

WEB SITES OF INTEREST

Hide and Seek Foundation for Lysosomal Disease Research
http://www.hideandseek.org

National Institute of Neurological Disorders and Stroke (NINDS). Metachromatic Leukodystrophy Information Page
http://www.ninds.nih.gov/disorders/
metachromatic_leukodystrophy

United Leukodystrophy Foundation
http://www.ulf.org

See also: Chronic granulomatous disease; Hereditary diseases; Infantile agranulocytosis.

Fibrodysplasia ossificans progressiva

CATEGORY: Diseases and syndromes
ALSO KNOWN AS: FOP; myositis ossificans progressiva

DEFINITION

Fibrodysplasia ossificans progressiva (FOP) is a very rare genetic condition in which muscle, tendons, and ligaments are transformed into bone. This process of heterotopic ossification, meaning the development of normal bone in abnormal places, leads to the formation of a second ectopic skeleton that immobilizes the joints and severely restricts movement.

RISK FACTORS

Most cases of FOP are sporadic and result from a new gene mutation. FOP is inherited as an autosomal dominant trait. Only a few multigenerational families exist, due to the low reproductive fitness. There are no ethnic patterns, sexual predilection, or predisposing risk factors.

ETIOLOGY AND GENETICS

FOP is characterized by a progressive transformation of skeletal muscle and connective tissue into ectopic bone. This process of one tissue type being transformed into another is a clinical feature unique to FOP. It is similar to the formation of bone in the developing fetus and during the healing of a fracture. This normal process, called endochondral ossification, involves the formation of bone from a cartilage model. In FOP, this same process occurs in the wrong places and at the wrong time. Progenitor cells in connective tissue and skeletal muscle are transformed into endochondral bone to form a second skeleton. This process of heterotopic ossification occurs in similar anatomic and temporal patterns to that seen in the fetus, beginning in the head, neck, and shoulders and progressing caudally to the hips and distally through the limbs. It characteristically spares the face, eyes, heart, and tongue. Heterotopic ossification is also induced by tissue injury and inflammation.

The transformation process in FOP has been well characterized. A flareup begins with the appearance of a painful connective tissue swelling. T lymphocytes migrate into the skeletal muscle and cause cell destruction. B lymphocytes then proliferate around blood vessels and stimulate the formation of new vessels. The final stages include cartilage and endochondral bone formation. These stages are similar to embryonic skeletal development and early fracture healing, except for the involvement of inflammatory cells.

Bone morphogenetic proteins (BMPs) are a family of extracellular signaling proteins that regulate cell differentiation in a variety of tissues. BMPs act by binding specific receptors in the cell membrane of target cells, resulting in the activation of an intracellular signaling pathway. Activin receptor type I (ACVR1) is one type of BMP receptor found in many tissues of the body, including skeletal muscle and cartilage. Binding of BMPs to the ACVR1 receptor results in activation of the BMP signaling pathway and transcription of genes required for cartilage and bone cell differentiation.

FOP is caused by a mutation of the gene that encodes the ACVR1 receptor. The *ACVR1* gene is located at chromosome band 2q23-q24. All patients with the classic features of FOP have the same R206H mutation in one copy of the gene. This mutation causes the substitution of histidine for arginine in a glycine/serine-rich domain of the receptor. Studies predict that this amino acid substitution leads to a change in the shape of the receptor that alters its sensitivity and function. FOP cells demonstrate constant activation of the ACVR1 receptor and dysregulated BMP signaling pathways, resulting in excessive cartilage and bone cell differentiation.

SYMPTOMS

All patients with FOP are born with a characteristic malformation of the great toe. Heterotopic ossification begins in early childhood with an episode of painful soft tissue swelling, followed by metamorphosis into ectopic bone. These episodes, called flare-ups, occur intermittently throughout life, resulting in progressive fusion of the joints and spine in fixed positions and associated immobility.

SCREENING AND DIAGNOSIS

The clinical diagnosis of FOP is based on the presence of heterotopic ossification in characteristic anatomic patterns and the presence of the great toe malformation. Plain radiographs may also demonstrate the presence of extraskeletal bony lesions, and other minor bone malformations, including cervical spine abnormalities and tibial osteochondromas. The soft tissue swellings of FOP are often misdiagnosed as cancer; however, a biopsy is not necessary for diagnosis and will trigger a flare-up. Definitive testing for FOP is based on DNA testing of the *ACVR1* gene for the specific R206H mutation.

TREATMENT AND THERAPY

At the present time, there is no medication or therapy that can stop the progressive formation of ectopic bone in FOP patients. Medical management is limited to symptomatic relief of painful flare-ups. Supportive care is important for the progressive disability resulting from spinal deformity and joint immobilization, restrictive cardiopulmonary function, recurrent pulmonary infections, hearing loss, and poor nutrition. Future therapy may involve medications that decrease ACVR1 receptor activation, in addition to the identification of affected children before the onset of heterotopic ossification.

PREVENTION AND OUTCOMES

Unnecessary surgical procedures, injections, and dental procedures are contraindicated; falls and injuries should be prevented. Patients are wheelchair bound before age thirty. Progressive immobility of the chest wall leads to thoracic insufficiency syndrome and life-threatening pulmonary complications. The average life span is forty-five years.

Lynne A. Ierardi-Curto, M.D., Ph.D.

FURTHER READING

Gorlin, R. J., M. M. Cohen, and R. C. M. Hennekam. *Syndromes of the Head and Neck.* 4th ed. New York: Oxford University Press, 2001.

Kaplan, F. S., E. M. Shore, and J. M. Connor. "Fibrodysplasia Ossificans Progressiva." In *Connective Tissue and Its Heritable Disorders: Molecular, Genetic, and Medical Aspects,* edited by P. M. Royce and B. U. Steinmann. 2d ed. New York: John Wiley and Sons, 2002.

Kaplan, F. S., M. Xu, D. L. Glaser, et al. "Early Diagnosis of Fibrodysplasia Ossificans Progressiva." *Pediatrics* 121, no. 5 (2008): e1295-1300.

Web Sites of Interest

International Fibrodysplasia Ossificans Progressiva Association (IFOPA)
http://www.ifopa.org

Weldon FOP Research Fund
http://www.weldonfop.org

See also: Crouzon syndrome; Diastrophic dysplasia; Hereditary diseases.

Fluorescence in situ hybridization (FISH)

CATEGORY: Molecular genetics; Techniques and methodologies

SIGNIFICANCE: Fluorescence in situ hybridization (FISH) is a technique used to visualize and map the location of a specific gene on the chromosome in a cell. FISH is a powerful tool used to diagnose various genetic disorders and different forms of cancer.

Key terms

bacterial artificial chromosome: cloning vector that contains large genome fragments grown in bacteria

chromatin: complex structure consisting of DNA, RNA, and protein

chromosome: coiled structure made of DNA and protein

denature: separation of double-stranded DNA

fluorophore: molecule that emits fluorescence after excitation by light

hybridization: binding of two complementary DNA strands

in situ: in the natural position

interphase: cell cycle phase in which the DNA is duplicated

metaphase: stage where condensed chromosomes are aligned in the middle of the cell

probe: short single-stranded DNA used to identify complementary DNA sequence

Probes and Hybridization

The two complementary deoxyribonucleic acid (DNA) strands are bound by hydrogen bonds. Heat and chemicals break the hydrogen bonds but they re-form when the conditions are favorable; this is the basis of nucleic acid hybridization. The probe is either tagged with biotin or digoxigenin, and they are detected by fluorophore conjugated streptavidin or antidigoxigenin antibody, respectively. Fluorophores are tagged directly to the probe, thus enabling rapid visualization of the target DNA. Fluorescent-labeled probes are safe, simple to use, and provide low background and high resolution. There are mainly three types of probes: The locus specific probe is used to locate the position of a particular gene on the chromosome, the centromeric repeat probe binds to the repetitive sequences found in the centromere of the chromosome, and the whole chromosome probe maps different regions along the length of any given chromosome. Thousands of bacterial artificial chromosome (BAC) clones obtained from the Human Genome Project are used as probes to map chromosomes. Probes are also available commercially.

Target Chromosome Preparation

FISH can be performed on cells, tissues, and solid tumors. The different types of target chromosome preparations are metaphase preparation, interphase preparation, and fibre FISH. In the metaphase preparation, the cells are captured in mitosis; the probes are large fragments that cover up to 5 megabases (Mb) and are used to map the entire chromosome. Interphase preparation is useful to study nondividing cells like those found in solid tumors. Hybridization occurs in the nucleus, thus enabling scientists to study the genome organization and location in its "natural" environment. The DNA is significantly less condensed in the interphase, which allows the probes to bind to their target DNA with greater resolution. The probes usually cover 50 kilobases (kb) to 2 megabases (Mb) of the chromosome. In fibre FISH, the interphase DNA is stripped of all proteins by either chemicals or mechanical shear. The released chromatin fibre can thus unfold and stretch into a straight line on a glass slide. This provides the highest resolution, from 5 kb to 500 kb. Fibre FISH is useful to study small rearrangements within the chromosome.

TECHNIQUE

The target chromosome preparations are usually attached to a glass slide. The fluorescent-labeled probe and target chromosome DNA are denatured. The denatured probe is then applied to the target DNA and incubated for approximately twelve hours; this allows the probe to hybridize with its complementary sequence on the target chromosome DNA. The glass slide is washed several times to remove all unhybridized probes. The fluorescence in situ hybridization is then visualized by fluorescence microscopy. Advanced FISH techniques include multifluor (M) FISH, comparative genome hybridization (CGH), and microarray FISH.

APPLICATIONS

In molecular biology, FISH is used to count the number of chromosomes in the cell. FISH visualizes chromosomal rearrangements such as translocation, inversion, and truncation. FISH is used to map genes and study the genome organization and structure in the cell. In the field of medicine FISH is used for prenatal and postnatal diagnosis of genetic disorders, cancer cytogenetics, and determination of infectious diseases. It plays a major role in understanding the chromosomal rearrangements that occurred during evolution and in developmental biology. FISH also plays a role in the field of microbial ecology. It is widely used for microorganism identification in drinking water and biofilms.

IMPACT

FISH played a major role in mapping genes on human chromosomes; this information was used during the annotation phase of the Human Genome Project. FISH is routinely used to diagnose and evaluate prognosis of cancers such as chronic myeloid leukemia, acute lymphoblastic leukemia, chronic lymphocytic leukemia, bladder cancer, breast cancer, and ovarian cancer. It is useful in diagnosing genetic disorders such as Down syndrome. FISH is also used to diagnose diseases such as the Charcot-Marie-Tooth disease, Angelman syndrome, and Prader-Willi syndrome. It is used to screen donated blood for the presence of HIV-infected cells as well as in the clinical diagnosis of the infection.

Anuradha Pradhan, Ph.D.

FURTHER READING

Andreeff, Michael, and Daniel Pinkel, eds. *Introduction to Fluorescence In Situ Hybridization: Principles and Clinical Applications.* New York: Wiley-Liss, 1999. This book covers the basic principles and techniques of FISH and describes in detail the applications of this technology to human cancer.

Liehr, Thomas, ed. *Fluorescence In Situ Hybridization (FISH): Application Guide.* New York: Springer, 2009. This book provides an overview about the principles and the basic techniques of FISH.

Speicher, Michael R., and Nigel P. Carter. "The New Cytogenetics: Blurring the Boundaries with Molecular Biology." *Nature Reviews: Genetics* 6 (2005): 782-792. This review discusses the history of cytogenetics and the exciting advances in FISH.

WEB SITES OF INTEREST

National Institutes of Health. National Human Genome Research Institute
http://www.genome.gov/10000206

Scitable by Nature Education. Genetics: Fluorescence In Situ Hybridization (FISH)
http://www.nature.com/scitable/topicpage/Fluorescence-In-Situ-Hybridization-FISH-327

See also: DNA sequencing technology; Genetic testing.

Forbes disease

CATEGORY: Diseases and syndromes
ALSO KNOWN AS: Cori disease; glycogen storage disease Type III; glycogen debrancher deficiency; amylo-1,6-glucosidase deficiency; limit dextrinosis

DEFINITION

Forbes disease, one of a dozen glycogen storage diseases, is a rare genetic defect that prevents the normal breakdown of glycogen in liver, muscles, and heart. Glycogen is largely unavailable for use in the body and builds up in these tissues, leading to their enlargement and impairing their function.

RISK FACTORS

The disease exhibits a familial association and is due to a deleterious mutation in the gene for glycogen debranching enzyme, also called amylo-1,6-glucosidase (AGL). The condition is rare (1 in 400,000 live births), but it is frequent among non-Ashkenazi Jews in North Africa (1 in 5,400) and among inhabitants of the Faroe Islands (1 in 3,600). It is widely distributed geographically and ethnically; it affects boys and girls equally.

ETIOLOGY AND GENETICS

It is named for Gilbert B. Forbes, who first characterized it in 1953. Forbes disease is an autosomal recessive condition involving a mutation in the gene for AGL, which is located on chromosome 1 in the region 1p21. Over a dozen separate mutations that lead to an inactive or unstable enzyme have been identified.

Glycogen, the storage form of carbohydrate in the body, is a highly branched polymer of glucose molecules. Glycogen phosphorylase, the enzyme that breaks down glycogen, removes glucose molecules one at a time from the end of a glycogen strand but is unable to do so at a branch point. AGL removes a strand of glucose molecules at a branch point, reattaching it to the end of a strand, and permits glycogen phosphorylase to continue working. If the debranching enzyme is lacking, then only the glucose molecules from the outermost ends can be released, leaving a glycogen structure referred to as limit dextrin. Most of the glycogen remains unavailable to the body. In addition, when more dietary glucose is available, more glucose is attached to the end of the strands and more branching is added, leading to larger and larger amounts of glycogen in the tissues, which cannot be effectively used when needed. Because glycogen is normally stored in the liver, muscles, and heart, these tissues are affected in this disease, leading to their enlargement and impairing their function.

Glycogen in the liver is used primarily between meals as a source of glucose for the body. At these times, patients with Forbes disease are not able to make full use of such glycogen and are dependent on gluconeogenesis, making glucose from noncarbohydrate sources. Over time, the buildup of glycogen may lead to cirrhosis and liver failure. Glycogen in muscles is used as a source of energy when needed for heavy exercise, and patients with Forbes disease have difficulty under that condition; they can have muscle weakness that may worsen with age. Glycogen in the heart is not an important source of energy, but excessive heart glycogen impairs its function.

Forbes disease exhibits considerable variability. While Type IIIa affects the liver, muscles, and heart, Type IIIb involves only the liver. Some patients have no measurable AGL, whereas others may have 15 percent of normal. Some are affected shortly after birth, while others manifest the condition later in life. Some improve around puberty, but others get worse with age.

SYMPTOMS

The first symptom is usually an enlarged liver, which may be so severe as to distend the belly. Low blood glucose after an overnight fast is sometimes seen, but it is less severe than in von Gierke disease. Growth may be delayed during childhood, but adult height is usually reached. Muscle weakness is often seen and can get progressively worse. Involvement of the heart will result in an abnormal electrocardiogram.

SCREENING AND DIAGNOSIS

Definitive diagnosis requires a biopsy of the liver and/or muscle and the demonstration of abnormal glycogen (limit dextrin, with short outer branches) and a deficiency of AGL. The latter can also be measured in skin cells or white blood cells. DNA tests are now available for many known mutations in the *AGL* gene; these tests are particularly effective with the Type IIIb disorder.

TREATMENT AND THERAPY

Treatment of Forbes disease is less demanding than for von Gierke disease. Any low blood glucose can be rectified by frequent high protein meals and overnight infusion of protein supplements, as these provide substrates for glucose synthesis via gluconeogenesis, while minimizing more glycogen deposition. No current treatment is available to treat the muscle and heart problems associated with this disease. Liver transplants have been performed in patients with highly compromised liver function.

PREVENTION AND OUTCOMES

Prenatal diagnosis of the disease is possible, especially when a familial association has been shown. Early neonatal diagnosis is desirable to minimize preventable side effects. Type IIIb, which only involves the liver, generally has a benign prognosis.

When muscles are involved, it is considered a muscular dystrophy and patients may qualify for services provided by the Muscular Dystrophy Association.

James L. Robinson, Ph.D.

FURTHER READING

Devlin, Thomas M. *Textbook of Biochemistry with Clinical Correlations.* 5th ed. New York: Wiley-Liss, 2005. Textbook for medical students clearly explains the basis for glycogen storage diseases.

Fernandes, John, Jean-Marie Saudubray, George van den Berghe, and John H. Walker. *Inborn Metabolic Diseases: Diagnosis and Treatment.* 4th ed. Berlin: Springer, 2006. Written for the physician, understandable by the nonprofessional, and describes glycogen storage diseases.

Shannon, Joyce B. *Endocrine and Metabolic Disorders Sourcebook.* 2d ed. Detroit: Omnigraphics, 2007. Basic consumer health information about metabolic disorders, including a section on glycogen storage diseases.

WEB SITES OF INTEREST

Association for Glycogen Storage Disease
http://www.agsdus.org/html/typeiiicori.htm

Muscular Dystrophy Association
http://www.mda.org/disease/dbd.html

See also: Andersen's disease; Glycogen storage diseases; Hereditary diseases; Hers disease; Inborn errors of metabolism; McArdle's disease; Pompe disease; Tarui's disease; Von Gierke disease.

Forensic genetics

CATEGORY: Human genetics and social issues

SIGNIFICANCE: Forensic genetics uses DNA or the inherited traits derived from DNA to identify individuals involved in criminal or civil legal cases. Blood tests and DNA testing are used to determine the source of evidence, such as blood stains or semen, left at a crime scene. Forensic DNA analysis is also used to determine paternity or other kinship.

KEY TERMS

alleles: alternative versions of genes at a genetic locus that determine an individual's traits

DNA fingerprinting: a DNA test used by forensic scientists to aid in the identification of criminals or to resolve paternity disputes

forensic science: the application of scientific knowledge to analyze evidence used in civil and criminal law, especially in court proceedings

kinship: genetic relatedness between persons

paternity testing: determination of a child's biological father

FORENSIC SCIENCE AND DNA ANALYSIS

Forensic scientists use genetics for two primary legal applications: identifying the source of a sample of blood, semen, or other tissue, and establishing the biological relationship between two people in paternity or other kinship lawsuits. Forensic scientists are frequently called upon to testify as expert witnesses in criminal trials. One of the most useful sources of inherited traits for forensic science purposes is blood. Such traits include blood type, proteins found in the plasma, and enzymes found in blood cells. The genes in people that determine such inherited traits have many different forms (alleles), and the specific combination of alleles for many of the inherited blood traits can be used to identify an individual. The number of useful blood group systems is small, however, which means that a number of individuals might have blood groups identical to those of the subject being tested.

The ultimate source of genetic information for identification of individuals is the DNA found in the chromosomes. Using a class of enzymes known as restriction enzymes, technicians can cut strands of DNA into segments, forming bands similar to a supermarket bar code that vary with individuals' family lines. The pattern, termed a DNA "fingerprint" or profile, is inherited as are the alleles for blood traits. DNA fingerprinting can be used to establish biological relationships (including paternity) with great reliability, because a child cannot have a variation that is not present in one of the parents. Because DNA is a relatively stable biological material and can be reliably tested in dried blood or semen even years after a crime has been committed, DNA fingerprinting has revolutionized the solution of criminal cases in which biological materials are the primary evidence. The likelihood of false matches

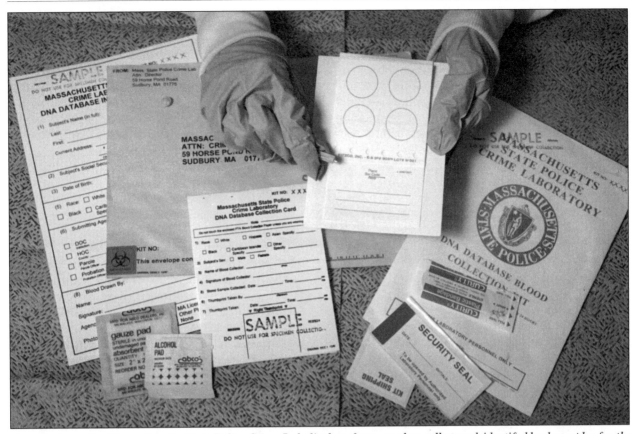

A serologist at the Massachusetts State Police Crime Lab displays forms used to collect and identify blood samples for the state's DNA database of people convicted of certain crimes. In 1998, a group of prisoners brought a suit against the state to overturn a law requiring blood samples from anyone convicted of any of thirty-three different crimes. (AP/Wide World Photos)

ranges from one per million to one per billion. These numbers, however, do not include the possibility of mishandling of evidence, laboratory errors, or planting of evidence.

CRIMINAL CASES INVOLVING DNA EVIDENCE

On November 6, 1987, serial rapist Tommy Lee Andrews became the first American ever convicted in a case involving DNA evidence. Samples of semen left at the crime scene by the rapist and blood taken from Andrews were sent to a New York laboratory for testing. Using the techniques of DNA fingerprinting, the laboratory isolated DNA from each sample, compared the patterns, and found a DNA match between the semen and the blood. Andrews was sentenced to twenty-two years in prison for rape, aggravated battery, and burglary.

The 1990-1991 *United States v. Yee* homicide trial

in Cincinnati, Ohio, was the first major case that challenged the soundness of DNA testing methods. DNA analysis by the Federal Bureau of Investigation (FBI) showed a match between blood from the victim's van and from Steven Yee's car. The defense claimed that the matching DNA data were ambiguous or inconsistent, citing what they claimed to be errors, omissions, lack of controls, and faulty analysis. However, after a fifteen-week hearing, the judge accepted the DNA testing as valid.

In 1994, former football star O. J. Simpson was arrested and charged with the murders of his ex-wife Nicole Brown and her friend, Ronald Goldman. Blood with DNA that matched Simpson's was found at Brown's home and blood spots in Simpson's car contained DNA matching Brown's, Goldman's, and Simpson's. Furthermore, blood at Simpson's home contained DNA that matched Brown's and Goldman's. For the most part, the defense admitted the accuracy of the DNA tests and did not scientifically challenge the results of the DNA fingerprinting. Instead, they argued that the biological evidence had been contaminated by shoddy laboratory work and that some evidence had even been planted; the jury found Simpson not guilty of the charges against him. In 1997, a jury in a civil trial unanimously found Simpson liable for wrongful death and battery, in part because the burden of proof was less onerous according to the "preponderance of evidence" test.

NOTABLE PATERNITY TESTING

In February, 2007, former Playboy playmate Anna Nicole Smith died, leaving behind a five-month-old daughter and two men claiming to be the child's father. Before her death, Smith had been ordered by the Los Angeles Superior Court to submit the child Dannielynn for paternity testing in response to a lawsuit by Larry Birkhead, who claimed to be the biological father, although Smith's lawyer Howard K. Stern was listed as the father on the birth certificate. The legal wranglings moved from California to Florida to the Bahamas, where the child was born and residing; finally, a Bahamian judge ap-

pointed Dr. Michael Baird, laboratory director of the DNA Diagnostics Center (DDC) in Fairfield, Ohio, as the court's DNA expert and ordered the paternity testing. In April, 2007, the DDC results confirmed that Birkhead was the biological father, and he was subsequently awarded custody of the girl.

OTHER APPLICATIONS

Forensic genetics professionals have also been called on in recent years to identify victims in situations with mass fatalities, most notably the 2004 Indian Ocean tsunami and the 2005 Atlantic Ocean hurricane Katrina. In addition to natural disasters, mass casualties may result from transportation accidents and terrorist attacks. Forensic genetics professionals are brought in to collect and process remains for DNA identity-testing; bone and teeth fragments are the most reliable sources of DNA, but soft tissue may be used as well. Laboratories then establish separate information management systems specifically for this type of forensic DNA analysis.

IMPACT

DNA evidence is used in thousands of criminal investigations and tens of thousands of paternity tests annually in the United States. In addition, forensic DNA testing has been used to free previously convicted and incarcerated individuals, with an average sentence served of twelve years. The Innocence Project, a nonprofit organization founded in 1992 by Barry Scheck and Peter Neufeld, claims that as of June 2009, 240 wrongfully convicted people in thirty-three states and Washington, D.C., have been exonerated through DNA testing. This includes seventeen people who were sentenced to death. However, in June, 2009, the U.S. Supreme Court ruled that an inmate has no automatic right to receive access to the DNA evidence used in his or her conviction for additional analysis at personal expense.

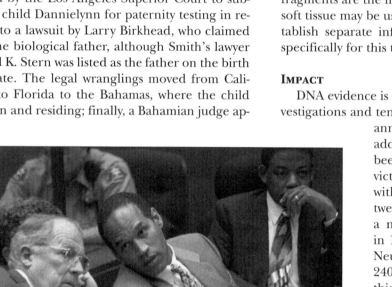

O. J. Simpson and attorneys discuss strategy for cross-examining a forensic scientist during Simpson's 1995 murder trial. Despite DNA evidence that blood found near Simpson's home and in his car matched that of the murder victims, Simpson was acquitted by a jury upon testimony that the evidence might have been contaminated. (AP/Wide World Photos)

Databases containing DNA profiles of people already convicted of particular crimes are available to local, state, and national law enforcement officials; when investigating a crime, they are now able to test DNA collected at the scene to see if it matches that of anyone in the database. The Combined DNA Index System (CODIS) is one such database. It contains DNA profiles from convicted individuals, evidence collected in unsolved crimes, and missing persons. Such information may link serial crimes to each other as well as past unsolved cases to present ones. In addition, investigators may reopen cold cases using methods for testing DNA evidence that were not available at the time of the crime.

Alvin K. Benson, Ph.D.;
updated by Bethany Thivierge, M.P.H.

FURTHER READING

Burke, Terry, R. Wolf, G. Dolf, and A. Jeffreys, eds. *DNA Fingerprinting: Approaches and Applications.* Boston: Birkhauser, 2001. Describes repetitive DNA and the broad variety of practical applications to law, medicine, politics, policy, and more. Aimed at the layperson.

Butler, John M. *Forensic DNA Typing.* 2d ed. New York: Academic Press, 2005. A comprehensive reference book that covers the history, biology, and technology of forensic DNA typing.

Coleman, Howard, and Eric Swenson. *DNA in the Courtroom: A Trial Watcher's Guide.* Seattle: Gene-Lex Press, 1994. Gives a good overview of DNA fingerprinting, expert evidence in court, and applications of forensic genetics.

Connors, Edward, et al. *Convicted by Juries, Exonerated by Science: Case Studies in the Use of DNA Evidence to Establish Innocence After Trial.* Washington, D.C.: U.S. Department of Justice, Office of Justice Programs, National Institute of Justice, 1996. Provides case studies in the use of DNA evidence to establish innocence after conviction in a trial.

Fridell, Ron. *DNA Fingerprinting: The Ultimate Identity.* New York: Scholastic, 2001. The history of the technique, from its discovery to early uses. Aimed at younger readers and nonspecialists.

Goodwin, William, Adrian Linacre, and Sibte Hadi. *An Introduction to Forensic Genetics.* Hoboken, N.J.: John W. Wiley & Sons, 2007. Intended to introduce undergraduate students to the subject.

Hummel, Susanne. *Fingerprinting the Past: Research on Highly Degraded DNA and Its Applications.* New York: Springer-Verlag, 2002. Manual about typing ancient DNA.

Jarman, Keith, and Norah Rudin. *An Introduction to* *Forensic DNA Analysis.* 2d ed. Boca Raton, Fla.: CRC Press, 2001. Emphasizes the advantages and limitations of various DNA techniques used in the analysis of forensic evidence.

Semikhodskii, Andrei. *Dealing with DNA Evidence: A Legal Guide.* London: Routledge Cavendish, 2007. Presents how DNA evidence is collected and analyzed and describes its strengths and shortcomings as evidence in criminal cases.

U.S. National Research Council. *The Evaluation of Forensic DNA Evidence.* Rev. ed. Washington, D.C.: National Academy Press, 1996. Evaluates how DNA is interpreted in the courts, includes developments in population genetics and statistics, and comments on statements made in the original volume that proved controversial or that have been misapplied in the courts.

WEB SITES OF INTEREST

Department of Justice. Federal Bureau of Investigation.
DNA Analysis Unit
http://www.fbi.gov/hq/lab/html/dnau1.htm

Forensics.ca: The Forensics Science Portal
http://forensics.ca/phpcode/web/index.php

The Innocence Project
http://www.innocenceproject.org

International Society of Forensic Genetics
http://www.isfg.org

See also: Biological determinism; Criminality; DNA fingerprinting; Eugenics; Eugenics: Nazi Germany; Human genetics; Insurance; Paternity tests; Sociobiology; Sterilization laws.

Fragile X syndrome

CATEGORY: Diseases and syndromes

DEFINITION

Sex chromosomes, the chromosomes X and Y, determine sex; the presence of two X chromosomes codes for females and an X chromosome paired with a Y chromosome codes for males. These chromosomes are received from an individual's parents, each of whom contributes one sex chromosome to their offspring. In 1969, geneticists studied a family of four mentally retarded brothers who had X chromosomes whose tips appeared to be detached from the rest of the chromosome. It is now recognized that this fragile site occurs in the vicinity of the *FMR1* gene. There are more than fifty mental retardation disorders associated with the X chromosome, but their frequencies are rare. Fragile X syndrome is the most common inherited form of mental retardation, affecting an estimated 1 in 1,500 males and 1 in 2,500 females.

RISK FACTORS

Individuals who have family members (especially male relatives) with fragile X syndrome are at risk for the condition.

ETIOLOGY AND GENETICS

In males, any abnormal gene on the X chromosome is expressed because males have only one X chromosome. In females, two copies of the fragile X chromosome must be present for them to be affected. This is the classic pattern for X-linked, or sex-linked, traits (traits whose genes are located on the X chromosome.)

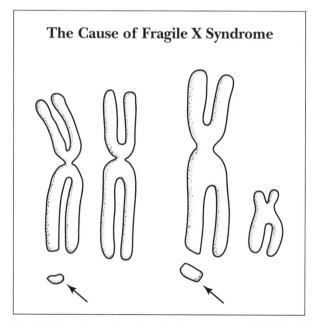

The Cause of Fragile X Syndrome

Fragile X syndrome in a female chromosome pair (left) and in a male pair (right). Note the apparently detached tips of the X chromosomes, the defect that gives the syndrome its name. (Electronic Illustrators Group)

The pattern of inheritance for fragile X is unusual. Fragile X syndrome increases in severity through successive generations. This is explained by a worsening of the defect in the *FMR1* gene as it is passed from mothers to sons. Since males contribute the Y chromosome to their sons, fathers do not pass the fragile X gene to their sons. They will, however, contribute their X chromosome to their daughters. Because these daughters also receive an X chromosome from their mothers, they generally appear normal or only mildly affected. It is only when these daughters have a son that the condition is expressed.

An explanation for this increasing severity through generations was discovered by analyzing the DNA sequence of the *FMR1* gene. The molecules composing DNA are adenine (A), thymine (T), cytosine (C), and guanine (G) and are referred to collectively as "bases." In fragile X syndrome, a sequence in which the three bases CGG are repeated over and over was found. The repetitive sequence is found in normal copies of the *FMR1* gene, but in individuals with fragile X syndrome there are many times more copies of the CGG triplet. The longer repetitive sequence in the *FMR1* gene prevents it from being expressed. Individuals not having the fragile X syndrome have a working *FMR1* gene.

SYMPTOMS

Males affected with fragile X syndrome have moderate to severe mental retardation and show distinctive facial features, including a long and narrow face, large and protruding ears, and a prominent jaw. Additional features include velvet-like skin, hyperextensible finger joints, and double-jointed thumbs. These features are generally not observed until maturity. Prior to puberty, the only symptoms a child may have are delayed developmental milestones, such as sitting, walking, and talking. Children with fragile X syndrome may also display an abnormal temperament marked by tantrums, hyperactivity, or autism.

A striking feature of most adult males with fragile X syndrome is an enlarged testicular volume (macroorchidism). This enlargement is not a result of testosterone levels, which are normal. Men with fragile X syndrome are fertile, and offspring have been documented, but those with significant mental retardation rarely reproduce.

The intelligence quotient (IQ) of the majority of affected males is in the moderate to severely retarded range. Only a few affected males have IQs above seventy-five. Males with fragile X syndrome frequently show delayed speech development and language difficulties. Repetitive speech patterns may also be present.

SCREENING AND DIAGNOSIS

Fragile X syndrome is often evident from an individual's appearance, intelligence, and behavior. External signs may include a large head circumference in babies, oversized testes in males during puberty, mental retardation, and subtle differences in facial characteristics. For females, the only external sign of the condition may be excessive shyness. Fragile X can be diagnosed with a polymerase chain reaction (PCR), a test that looks for the triplet repeat mutation in the *FMR1* gene.

High school football player Jake Porter, who was born with fragile X syndrome, receives an honorary ESPY award from actor Dennis Haysbert in 2003. (AP/Wide World Photos)

TREATMENT AND THERAPY

There is no specific treatment for people with fragile X syndrome, but there are a variety of ways to minimize the symptoms. Affected individuals can receive special education and training, as well as speech, physical, occupational, and behavioral therapies to address the educational, physical, social, emotional, language, and sensory problems associated with the condition.

PREVENTION AND OUTCOMES

There currently is no cure for fragile X syndrome, but various educational and treatment programs are available to alleviate or eliminate its symptoms. Individuals with a family history of the syndrome may wish to seek genetic counseling before deciding to have a child.

Linda R. Adkison, Ph.D.; Bryan Ness, Ph.D.;
updated by Rebecca Kuzins

FURTHER READING

Dykens, Elisabeth M., Robert M. Hodapp, and Brenda M. Finucane. *Genetics and Mental Retardation Syndromes: A New Look at Behavior and Interventions.* Baltimore: Paul H. Brookes, 2000. Reviews the genetic and behavioral characteristics of fragile X and eight other mental retardation syndromes, giving in-depth information on genetic causes, prevalence, and physical and medical features of these syndromes.

Hagerman, Randi Jenssen, and Paul J. Hagerman. *Fragile X Syndrome: Diagnosis, Treatment, and Research.* 3d ed. Baltimore: Johns Hopkins University Press, 2002. Discusses the clinical approach to diagnosing fragile X. Presents research in epidemiology, molecular biology, and genetics, and provides information on genetic counseling, pharmacotherapy, intervention, and gene therapy.

Hirsch, David. "Fragile X Syndrome." *The Exceptional Parent* 25 (June, 1995). Answers parental concerns about fragile X syndrome genetic testing.

_____. "Fragile X Syndrome: Medications for Aggressive Behavior?" *The Exceptional Parent* 26 (October, 1996). Addresses parental concerns about medications for males displaying aggressive behavior.

Hogenboom, Marga. *Living with Genetic Syndromes Associated with Intellectual Disability.* Philadelphia: Jessica Kingsley, 2001. Gives an accessible introduction to genetics before detailing the ways in which young people are affected by genetic conditions. Discusses the extent to which these individuals' behavior is determined, the difficulties they face, and the ways in which they can achieve independence and fulfillment.

Parker, James N., and Philip M. Parker, eds. *The 2002 Official Parent's Sourcebook on Fragile X Syndrome.* San Diego: Icon Health, 2002. Draws from public, academic, government, and peer-reviewed research to guide parents about where and how to look for information covering virtually all topics related to fragile X syndrome.

Schmidt, Michael A. "Fragile X Syndrome: Diagnosis, Treatment, and Research." *Journal of the American Medical Association* 277, no. 14 (April 9, 1997): 1169. Provides a detailed discussion of fragile X syndrome.

Shannon, Joyce Brennfleck, ed. *Mental Retardation Sourcebook: Basic Consumer Health Information About Mental Retardation and Its Causes, Including Down Syndrome, Fetal Alcohol Syndrome, Fragile X Syndrome, Genetic Conditions, Injury, and Environmental Sources.* Detroit: Omnigraphics, 2000. Reviews causes, prevention, family life, education, specific health care issues, and legal and economic concerns for health care consumers.

Warren, Stephen T. "Trinucleotide Repetition and Fragile X Syndrome." *Hospital Practice* 32, no. 4 (April 15, 1997): 73. Provides detail about CGG repeats in fragile X syndrome.

Wells, Robert D., and Tetsuo Ashizawa, eds. "Fragile X Syndrome." In *Genetic Instabilities and Neurological Diseases.* 2d ed. Burlington, Mass.: Academic Press, 2006. Contains three articles discussing the syndrome in humans and in mice and the mechanisms of chromosomal fragile sites.

WEB SITES OF INTEREST

Dolan DNA Learning Center, Your Genes Your Health
http://www.ygyh.org
Sponsored by the Cold Spring Harbor Laboratory, this site, a component of the DNA Interactive Web site, offers information on more than a dozen inherited diseases and syndromes, including fragile X syndrome.

FRAXA Research Foundation
http://www.fraxa.org
The foundation supports research aimed at the treatment of fragile X syndrome, and its Web site offers information on managing the disorder.

Medline Plus, Fragile X Syndrome
http://www.nlm.nih.gov/medlineplus/fragilexsyndrome.html
Provides links to a wide range of online resources about the syndrome.

National Fragile X Foundation: Xtraordinary Accomplishments
http://www.nfxf.org
General information about the disorder, with advice for caregivers on testing, medical treatment, education, and life planning.

See also: Attention deficit hyperactivity disorder (ADHD); Autism; Behavior; Chromatin packaging; Classical transmission genetics; Congenital defects; DNA replication; Down syndrome; Intelligence; Repetitive DNA.

Friedreich ataxia

CATEGORY: Diseases and syndromes
ALSO KNOWN AS: Friedreich's ataxia

DEFINITION

Friedreich ataxia is a very rare inherited disease that causes progressive damage to the nervous system. Ataxia refers to coordination problems and unsteadiness. Friedreich ataxia causes degeneration of neurons in the spinal cord that control movement, as well as the sensory nerves that assist coordination. In later stages, additional cell injury can develop in the heart and pancreas.

RISK FACTORS

There are no environmental risk factors for Friedreich ataxia. The disease is inherited.

ETIOLOGY AND GENETICS

Friedreich ataxia results from a mutation in the *FXN* gene, which is located on the long arm of chromosome 9 at position 9q13. The protein product of this gene is known as frataxin, and it is normally produced in the mitochondria of some cells, particularly nerve and muscle cells. In the absence of normal levels of frataxin, these cells cannot efficiently produce energy in the form of ATP by oxidative phosphorylation, and there is an accumulation of toxic metabolites that leads to what physicians call oxidative stress. In 98 percent of cases, the mutational event is known as a GAA triplet repeat.

In patients with Friedreich ataxia, a sequence of three deoxyribonucleic acid (DNA) bases (GAA) near the beginning of the gene is repeated from seventy to more than one thousand times. Normal *FXN* genes have this triplet repeated anywhere from five to thirty times. Apparently the expanded triplet repeat region interferes with the normal process of frataxin synthesis in cells, and a greatly reduced amount of the protein is made. In a very small percentage of cases (2 percent), the gene defect appears to be a point mutation that also drastically reduces the amount of frataxin produced.

The inheritance pattern for Friedreich ataxia is typical of other autosomal recessive conditions, in that both copies of the *FXN* gene must be deficient in order for the individual to be afflicted. Typically, an affected child is born to two unaffected parents, both of whom are carriers of the recessive mutant allele. The probable outcomes for children whose parents are both carriers are 75 percent unaffected and 25 percent affected. If one parent has Friedreich ataxia and the other is a carrier, there is a 50 percent probability that each child will be affected. A simple blood test to check for carrier status is available. It is interesting to note that although triplet repeat expansions have been implicated in several diseases that show dominant inheritance, Friedreich ataxia is the only known recessive disease to have this particular molecular etiology.

SYMPTOMS

Symptoms may vary. The onset of the disease generally occurs in individuals under the age of twenty-five, usually in their early teenage years. Early symptoms include progressive leg weakness (difficulty walking); ataxia—incoordination and imbalance affecting limbs and gait; impaired sensation, especially "position sense" in the feet; and loss of tendon reflexes in the legs. Heart failure and diabetes develop as the disease progresses.

Later symptoms include difficulty speaking and swallowing (decreased coordination of the tongue); loss of tendon reflexes in all limbs; atrophy of muscles; scoliosis, or curving of the spine, which affects 85 percent of people with this condition; foot deformities; and foot ulcers. By age forty-five, 95 percent of those affected are confined to a wheelchair. Other symptoms include hearing loss and/or vision loss (for more than 10 percent of those affected); eye movement abnormalities; and movement disorders, such as tremor, dystonia, and chorea.

SCREENING AND DIAGNOSIS

The doctor will ask about a patient's symptoms, including medical history, family history, and medications. The doctor will also perform a physical exam. If this disorder is suspected, a patient may also see a neurologist, a doctor who specializes in the nervous system.

Tests may include electromyography (EMG) and nerve conduction studies to assess the function of the muscles and nerves; a computed tomography (CT) scan, a test that uses a computer to make cross-sectional images of the head; a magnetic resonance imaging (MRI) scan, a test that uses magnetic waves to make pictures of structures inside the brain and spinal cord; an electrocardiogram (ECG) and twenty-four-hour holter monitoring to assess the electrical activity of the heart; and an echocardiogram, a test that uses high-frequency sound waves (ultrasound) to examine the size, shape, and motion of the heart. Other tests include genetic testing for the frataxin gene, blood (diabetic testing) and urine tests, and a sural nerve biopsy.

TREATMENT AND THERAPY

There is no known cure for this condition. Long-term management is aimed at maximizing function and controlling symptoms. Management may include physical therapy and rehabilitation to cope with muscle weakness, the use of orthotics (devices that go in a patient's shoes) to provide stability and to help with weakness, surgery for correcting foot abnormalities and scoliosis, and periodic testing for associated conditions of diabetes and cardiomyopathy.

As the exact role of the protein frataxin is clarified, treatments may emerge. Studies are ongoing to assess the role of antioxidants, coenzyme Q10, and vitamin E.

PREVENTION AND OUTCOMES

There is no known way to prevent this condition.
Dianne Scheinberg, M.S., RD, LDN;
reviewed by J. Thomas Megerian, M.D., Ph.D., F.A.A.P.
"Etiology and Genetics" by Jeffrey A. Knight, Ph.D.

FURTHER READING

Bradley, Walter G., et al., eds. *Neurology in Clinical Practice.* 5th ed. 2 vols. Philadelphia: Butterworth-Heinemann/Elsevier, 2008.

Cooper, J. M., and J. L. Bradley. "Friedreich's Ataxia." In *Mitochondrial Function and Dysfunction,* edited by Anthony H. V. Schapira. London: Academic Press, 2002.

EBSCO Publishing. *DynaMed: Friedreich's Ataxia.* Ipswich, Mass.: Author, 2009. Available through http://www.ebscohost.com/dynamed.

_____. *Health Library: Friedreich Ataxia.* Ipswich, Mass.: Author, 2009. Available through http://www.ebscohost.com.

Koenig, Michel, and Alexandra Dürr. "Friedreich's Ataxia." In *Handbook of Ataxia Disorders,* edited by Thomas Klockgether. New York: Marcel Dekker, 2000.

WEB SITES OF INTEREST

Friedreich's Ataxia Research Alliance
http://www.curefa.org

Genetic Alliance
http://www.geneticalliance.org

Genetics Home Reference
http://ghr.nlm.nih.gov

International Network of Ataxia Friends
http://internaf.org/groups/canada.html

MedLine Plus: "Friedreich's Ataxia"
http://www.nlm.nih.gov/medlineplus/ency/article/001411.htm

Muscular Dystrophy Canada
http://www.muscle.ca

National Ataxia Foundation
http://www.ataxia.org

National Institute of Neurological Disorders and Stroke
http://www.ninds.nih.gov

See also: Amyotrophic lateral sclerosis; Ataxia telangiectasia; Batten disease; Epilepsy; Essential tremor; Hereditary diseases; Parkinson disease.

G

Galactokinase deficiency

CATEGORY: Diseases and syndromes
ALSO KNOWN AS: GALK deficiency; galactosemia II

DEFINITION

Galactokinase deficiency is a rare genetic disorder in which galactose obtained from the diet cannot be phosphorylated in the cell due to a defect in the enzyme galactokinase. The disorder is linked to mutations in the gene encoding galactokinase and manifests in the newborn with cataracts and galactosuria.

RISK FACTORS

Galactokinase deficiency is an autosomal recessive disease. It manifests if the patient's parents each carry one copy of the mutated *GALK1* gene, even if they do not show signs and symptoms; both sexes are equally affected. The familial risk factor is increased in consanguineous marriages. The incidence of galactokinase deficiency has been found to be higher (approximately 1 per 10,000) in the Roma (Gypsy) population of Eastern Europe with endogamous traditions. In the United States, the estimated incidence is 1 per 50,000-100,000 live births.

ETIOLOGY AND GENETICS

Galactokinase deficiency is associated with mutations in the *GALK1* gene, located on the long (q) arm of chromosome 17, from base pair 71,265,612 to 71,272,874. The gene contains eight exons and spans approximately 7.3 kilobases (kb) of genomic DNA. It encodes the enzyme galactokinase (EC 2.7.1.6), which phosphorylates galactose to galactose-1-phosphate. Phosphorylation is the first of three enzymatic steps in galactose metabolism. Galactose-1-phosphate uridyltransferase (EC 2.7.7.10; GALT) and UDP-galactose-4-epimerase (EC 5.1.3.2; GALE) are responsible in this order for the next two metabolic steps. Inborn metabolic errors linked to impaired activity of each enzyme lead to galactosemia. Galactosemia linked to galactokinase deficiency results in the least severe symptoms compared to classic galactosemia linked to GALT deficiency (with life-threatening signs and symptoms) and galactosemia linked to GALE deficiency (the rarest kind, which causes damages to tissues and organs).

More than twenty-three different mutations within the *GALK1* gene have been identified in galactokinase deficiency. Most of these are missense (a codon for one amino acid is substituted by the codon for a different amino acid) or deletion type (part of a chromosome or sequence of DNA is missing) and cause changes in the stability and activity of the enzyme. Private mutations, found only in the kindred of patients, are not uncommon in galactokinase deficiency. A founder mutation was identified in Roma patients as the *P28T* mutation (proline at position 28 substituted by threonine).

SYMPTOMS

The most consistent symptoms of galactokinase deficiency include congenital cataracts in infants and presenile cataracts in adults. The disease also results in galactosemia and galactosuria. Pseudotumor cerebri (idiopathic intracranial hypertension) is rare but consistently reported in galactokinase-deficient patients. Both cataract and pseudotumor cerebri can be ascribed to accumulation of galactitol, a product of an alternative route of galactose utilization, which results in osmotic swelling; both resolve with therapy. A variety of clinical abnormalities has also been reported, but a causal relationship with GALK deficiency could not be determined.

SCREENING AND DIAGNOSIS

Galactokinase deficiency is rare, and the diagnosis is not immediately apparent. Unlike classic galactosemia, galactokinase deficiency does not present with severe manifestations; thus, most cases are diagnosed after the development of lens opacity in the infant. Because the disease is rare, genetic screening is not usually done. The diagnosis is established by demonstrating deficient activity of the galactokinase enzyme in erythrocytes.

TREATMENT AND THERAPY

The only treatment for GALK deficiency is specifically directed to the restriction of galactose (and sugars containing galactose units, such as lactose) in the diet. This is usually effective in reversing symptoms.

PREVENTION AND OUTCOMES

The development of early cataracts in homozygous affected infants is fully preventable through early diagnosis and treatment with a galactose-restricted diet. Ideally, screening programs in genetically at-risk populations would allow prevention of galactokinase deficiency. In reality, such programs are not usually available, and the disease is not identified until cataract and blindness develop.

According to some studies, depending on milk consumption later in life, heterozygous carriers of galactokinase deficiency may be prone to presenile cataracts at twenty to fifty years of age. The general outcome for patients with galactokinase deficiency is positive, and with dietary precautions the patients lead a normal life.

Donatella M. Casirola, Ph.D.

FURTHER READING

Bosch, A. M., et al. "Clinical Features of Galactokinase Deficiency: A Review of the Literature." *Journal of Inherited Metabolic Diseases* 25 (2002): 629-634. A review article in a scientific journal for biomedical researchers, mostly understandable to nonprofessionals.

Kalaydjieva, Luba, et al. "A Founder Mutation in the *GK1* Gene Is Responsible for Galactokinase Deficiency in Roma (Gypsies)." *American Journal of Human Genetics* 65 (1999): 1299-1307. An article in a scientific journal for biomedical researchers, mostly understandable to nonprofessionals.

Novelli, Giuseppe, and Juergen K. V. Reichardt. "Molecular Basis of Disorders of Human Galactose Metabolism: Past, Present, and Future." *Molecular Genetics and Metabolism* 71 (2000): 62-65. A mini-review in a scientific journal for biomedical researchers, mostly understandable to nonprofessionals.

Segal, Stanton, and Gerard T. Berry. "Disorders of Galactose Metabolism." In *The Metabolic and Molecular Basis of Inherited Disease*, edited by C. R. Scriver, A. L. Beaudet, W. S. Sly, and D. Valle. 7th ed. New York: McGraw-Hill, 1995. A chapter in a book for biomedical professionals, mostly understandable to nonprofessionals with basic knowledge of genetics.

WEB SITES OF INTEREST

Genetics Home Reference: Galactosemia
http://www.ghr.nlm.nih.gov/condition=galactosemia

Genetics Home Reference: GALK1 Gene
http://www.ghr.nlm.nih.gov/gene=galk1

National Institutes of Health, Genetic and Rare Diseases Information Center (GARD)
http://rarediseases.info.nih.gov/GARD

See also: Alkaptonuria; Andersen's disease; Diabetes; Diabetes insipidus; Fabry disease; Forbes disease; Galactosemia; Gaucher disease; Glucose galactose malabsorption; Glucose-6-phosphate dehydrogenase deficiency; Glycogen storage diseases; Hemochromatosis; Hereditary diseases; Hereditary xanthinuria; Hers disease; Homocystinuria; Inborn errors of metabolism; Kearns-Sayre syndrome; Krabbé disease; Lactose intolerance; Lesch-Nyhan syndrome; McArdle's disease; Maple syrup urine disease; Menkes syndrome; Metachromatic leukodystrophy; Niemann-Pick disease; Phenylketonuria (PKU); Pompe disease; Tarui's disease; Tay-Sachs disease.

Galactosemia

CATEGORY: Diseases and syndromes

DEFINITION

Classic galactosemia is an inherited disease. Due to a defective gene, there is a deficiency of the enzyme galactose-1-phosphate uridyltransferase. This enzyme is necessary for the conversion of galactose to glucose. Galactose is a simple sugar found in milk products. (The main sugar in milk is called lactose.

It is made up of two simple sugars: galactose and glucose.) Glucose is the usable form of sugar in the human body.

Normally, the body converts galactose-1-phosphate into glucose, which it then uses for energy. In galactosemia, galactose builds up in the blood. A buildup of galactose-1-phosphate can cause severe damage to the liver, kidneys, central nervous system, and other body systems. If undetected, galactosemia is fatal.

A less severe form of this disease is due to galactokinase deficiency. This type may be managed with a few dietary restrictions; it does not carry the risk of neurologic or liver damage. However, the eye lens may be damaged, which can lead to cataracts.

RISK FACTORS

The primary risk factor for galactosemia is having parents who carry the genes for this condition.

ETIOLOGY AND GENETICS

Classic galactosemia (type I) results from a mutation in the *GALT* gene, which is located on the short arm of chromosome 9 at position 9p13. This gene encodes the enzyme galactose-1-phosphate uridyltransferase, which normally catalyzes an essential step in the conversion of galactose to glucose. The most common mutations in the *GALT* gene result in the production of a completely nonfunctional enzyme. It is the accumulation of galactose and related metabolites in tissues that causes the severe symptoms associated with the disease.

Two other genes specify enzymes in the galactose breakdown pathway, and other less common forms of galactosemia (types II and III) can result if there are mutations in either of these genes. *GALK1*, on chromosome 17 at position 17q24, encodes the enzyme galactokinase-1, and *GALE*, on the short arm of chromosome 1 at position 1p36-p35, codes for the enzyme UDP-galactose-4-epimerase.

For all types of galactosemia, the inheritance pattern is characteristic of an autosomal recessive mutation. Both copies of the gene must be deficient in order for the individual to be afflicted. Typically, an affected child is born to two unaffected parents, both of whom are carriers of the recessive mutant allele. The probable outcomes for children whose parents are both carriers are 75 percent unaffected and 25 percent affected. If one parent has galactosemia and the other is a carrier, there is a 50 percent probability that each child will be affected

SYMPTOMS

An infant with classic galactosemia usually appears normal at birth. If galactosemia is not detected at birth with testing, symptoms usually occur within the first few days or weeks of life after the baby drinks breast milk or a lactose-containing formula. Early symptoms may include jaundice (yellowing) of the skin and whites of the eyes, vomiting, poor weight gain, low blood sugar (hypoglycemia), feeding difficulties, irritability, lethargy, and convulsions.

If left untreated, later signs and symptoms may include opaque lenses of the eyes (cataracts), enlarged liver, enlarged spleen, mental retardation, sepsis caused by a specific bacteria (Escherichia coli), cirrhosis (scarring of the liver), liver failure, kidney problems, and swelling of the extremities or stomach.

If diet restrictions are started immediately, it may be possible to prevent acute toxicity. However, long-term complications may still occur. These may include poor growth, learning disabilities, speech and language problems, fine and gross motor skill delays, ovarian failure (in girls), cataracts (usually regress with dietary treatment, leaving no residual visual impairment), and decreased bone mineral density.

SCREENING AND DIAGNOSIS

Today, most American infants are screened for galactosemia at birth with a simple blood test. A small sample of blood is taken with a heel prick. It is also possible to diagnose galactosemia during pregnancy with an amniocentesis.

The diagnosis may also be suggested if a urine test shows the presence of a reducing substance. The diagnosis can be confirmed with a blood test or with a biopsy of the liver or other tissues.

TREATMENT AND THERAPY

Galactosemia cannot be cured. However, patients can take steps to prevent or minimize galactosemia symptoms and complications. The treatment is the strict avoidance of all sources of galactose. The most common source is lactose, which is the milk sugar that breaks down to galactose and glucose.

To avoid all sources of galactose, patients should closely monitor their diets. They should avoid all products that contain or produce galactose. This includes milk or milk by-products, such as milk, casein, lactose (milk sugar), dry milk solids, curds, and whey. This also includes some nonmilk products, such as fermented soy products, legumes, organ meats, and hydrolyzed protein.

Lactose or galactose may be used as an additive in some food products. Therefore, patients should always read food labels carefully in order to avoid these foods. Because galactose is so commonly found in foods, parents of a child who has galactosemia will nearly always require the services of a dietician skilled in advising about the management of this rare disorder.

Patients should check with their pharmacists to avoid medications that have fillers that contain galactose or lactose. They should also avoid supplements, unless prescribed by their doctors, because fillers and inactive ingredients are not required to be listed in supplements. If patients do take a supplement, they should check with their pharmacists to ensure that there are no hidden sources of galactose in the product.

PREVENTION AND OUTCOMES

There is no known way to prevent galactosemia. Individuals who have galactosemia or have a family history of the disorder can talk to a genetic counselor when deciding to have children.

Michelle Badash, M.S.;
reviewed by Rosalyn Carson-DeWitt, M.D.
"Etiology and Genetics" by Jeffrey A. Knight, Ph.D.

FURTHER READING

Beers, Mark H., et al. *The Merck Manual of Diagnosis and Therapy.* 18th ed. Whitehouse Station, N.J.: Merck Research Laboratories, 2006.

Calcar, Sandra van, and John Wolf. "Galactosemia." In *Pediatric Nutrition in Chronic Diseases and Developmental Disorders: Prevention, Assessment, and Treatment,* edited by Shirley W. Ekvall and Valli K. Ekvall. 2d ed. New York: Oxford University Press, 2005.

EBSCO Publishing. *Health Library: Galactosemia.* Ipswich, Mass.: Author, 2009. Available through http://www.ebscohost.com.

Kleigman, Robert M., et al., eds. *Nelson Textbook of Pediatrics.* 18th ed. Philadelphia: Saunders Elsevier, 2007.

WEB SITES OF INTEREST

American Liver Foundation
http://www.liverfoundation.org

Genetics Home Reference
http://ghr.nlm.nih.gov

Parents of Galactosemic Children, Inc.
http://www.galactosemia.org

Save Babies Through Screening Foundation of Canada
http://www.savebabiescanada.org/ehome.htm

Sick Kids
http://www.sickkids.ca

See also: Alkaptonuria; Andersen's disease; Diabetes; Diabetes insipidus; Fabry disease; Forbes disease; Galactokinase deficiency; Gaucher disease; Glucose galactose malabsorption; Glucose-6-phosphate dehydrogenase deficiency; Glycogen storage diseases; Hemochromatosis; Hereditary diseases; Hereditary xanthinuria; Hers disease; Homocystinuria; Inborn errors of metabolism; Kearns-Sayre syndrome; Krabbé disease; Lactose intolerance; Lesch-Nyhan syndrome; McArdle's disease; Maple syrup urine disease; Menkes syndrome; Metachromatic leukodystrophy; Niemann-Pick disease; Phenylketonuria (PKU); Pompe disease; Tarui's disease; Tay-Sachs disease.

SALEM HEALTH
GENETICS
& INHERITED CONDITIONS

CATEGORY INDEX

BACTERIAL GENETICS

Anthrax, 65
Bacterial genetics and cell structure, 105
Bacterial resistance and super bacteria, 111
Gene regulation: Bacteria, 467
Gene regulation: *Lac* operon, 475
MLH1 gene, 813
Model organism: *Escherichia coli*, 833
Transposable elements, 1193

BIOETHICS

Bioethics, 133
Chorionic villus sampling, 230
Cloning: Ethical issues, 272
Gene therapy: Ethical and economic issues, 487
Genetic engineering: Risks, 528
Genetic engineering: Social and ethical issues, 532
Genetic testing: Ethical and economic issues, 547
Insurance, 718
Miscegenation and antimiscegenation laws, 795
Patents on life-forms, 954

BIOINFORMATICS

Bioinformatics, 140
cDNA libraries, 197
Genomic libraries, 565
Icelandic Genetic Database, 684
Microarray analysis, 794

CELLULAR BIOLOGY

Archaea, 79
Bacterial genetics and cell structure, 105
Cell culture: Animal cells, 201
Cell culture: Plant cells, 204
Cell cycle, 207
Cell division, 210
Chromosome mutation, 238
Chromosome structure, 241
Cytokinesis, 323
Extrachromosomal inheritance, 427
Gene regulation: Bacteria, 467
Gene regulation: Eukaryotes, 471
Gene regulation: *Lac* operon, 475
Gene regulation: Viruses, 478
Harvey *ras* oncogene, 595
Mitosis and meiosis, 809
Nondisjunction and aneuploidy, 909
RNA interference, 1086
Stem cells, 1150
Telomeres, 1176
Totipotency, 1184

CLASSICAL TRANSMISSION GENETICS

ABO blood group system, 2
BRCA1 and *BRCA2* genes, 173
Chromosome structure, 241
Chromosome theory of heredity, 247
Classical transmission genetics, 259
Complete dominance, 290
Dihybrid inheritance, 347

Epistasis, 406
Incomplete dominance, 708
Mendelian genetics, 783
Monohybrid inheritance, 860
Multiple alleles, 866
Polygenic inheritance, 986
SRY gene, 1147

DEVELOPMENTAL GENETICS

Developmental genetics, 332
Hermaphrodites, 631
Homeotic genes, 641
Model organism: *Danio rerio*, 825
Steroid hormones, 1158
Von Gierke disease, 1219
X chromosome inactivation, 1247

DISEASES AND SYNDROMES

Aarskog syndrome, 1
Achondroplasia, 5
Adrenoleukodystrophy, 7
Adrenomyelopathy, 9
Agammaglobulinemia, 11
Alagille syndrome, 21
Albinism, 23
Alcoholism, 25
Alexander disease, 28
Alkaptonuria, 30
Allergies, 31
Alpha-1-antitrypsin deficiency, 34
Alport syndrome, 36
Alzheimer's disease, 40
Amyotrophic lateral sclerosis, 47
Andersen's disease, 54
Androgen insensitivity syndrome, 55

EVOLUTIONARY BIOLOGY